Handbook
of
Molecular Sieves

"It is a rare event in the mining world when a little known mineral leaps to the forefront of technological interest"

F.A. Mumpton

Proc. of the Sixth International Zeolite Conference
1984

Handbook
of
Molecular Sieves

R. Szostak

VAN NOSTRAND REINHOLD
————————— New York

04930974

CHEMISTRY

Copyright © 1992 by Van Nostrand Reinhold

Library of Congress Catalog Card Number 91-45180
ISBN 0-442-31899-5

Manufactured in the United State of America

Published by Van Nostrand Reinhold
115 Fifth Avenue
New York, New York 10003

Chapman and Hall
2-6 Boundary Row
London, SE 1 8HN, England

Thomas Nelson Australia
102 Dodds Street
South Melbourne 3205
Victoria, Australia

Nelson Canada
1120 Birchmount Road
Scarborough, Ontario M1K 5G4, Canada

16 15 14 13 12 11 10 9 8 7 6 5 4 3 2 1

Library of Congress Cataloging-in-Publication Data
Szostak, Rosemarie, 1952–
    Handbook of molecular sieves / Rosemarie Szostak.
      p. cm.
    Includes bibliographical references and index.
    ISBN 0-442-31899-5
    1. Molecular sieves. I. Title.
TP159.M6S96   1992
660′.2842--dc20   91-45180
                                 CIP

To my sister Carlene and her family,
Cliff, Michael, and Rachael Lavin,
for their love and encouragement

Contents

Preface

This book evolved out of personal necessity. In trying to keep abreast of the literature, I found it increasingly difficult to keep track of the expanding list of molecular sieve materials. Is CoAPO-5 in the MeAPO or the XAPO patent? Is TPZ-3 related to ZSM-5 or ZSM-50? Is FAPO-41 known? If so, what organic is used in the crystallization of the structure? A continually growing list taped to the wall of my office was the only reference I had. Most of these new materials have been reported in the patent literature, which is a problem when access to the patents is difficult or expensive. Alternatively, many structure and characterization reports are found only in conference proceedings, some of limited availability. I rapidly tired of searching for my copy of a patent or conference proceeding to look up an X-ray diffraction pattern, an infrared spectrum, or adsorption data for a material of interest. I wanted information the easy way, everything under one cover. Thus I began the data compilation for this book. I hope that the other users of this book will also find it a convenient and quick source of information on zeolites and other crystalline microporous materials.

Compiling this information was not a simple task. Between 1967 and 1990 over 42,000 papers and patents had "zeolite" or "molecular sieve" as a key term. A number of decisions had to be made. I chose a historical approach and tended to work with the earliest reference claiming the composition of matter for each material. An exhaustive review of the data reported about each material was not possible. I have tried to include information of general interest for the scientist working in this field in an encyclopedic format.

R. Szostak
Zeolite Research Program
Georgia Institute of Technology
Atlanta, Georgia

Acknowledgments

Without the Georgia Tech Research Institute and Dr. T. L. Thomas's encouragement, this book could not have been written. Many thanks go to Joanna Kennedy and Ann Campbell of the Georgia Tech Library for their help in searching out the many patents and references needed to compile this handbook and to zeolite program students Bryan Duncan, Earl Seugling, David Gwinup, Kristin Vinje, Flaviano Testa, Michelle Shelton, and Madhu Chaudhary for providing the X-ray, TGA/DTA, and infrared data for many of the materials listed. A very special thanks goes to Alicia Long with her bucket of commas for reviewing and editing the original manuscript and to Valerie L. Sisom for her help with the artwork. Without the assistance of many members of the zeolite community around the world who helped me track down obscure references and patents this work could not have been completed. I am especially grateful to my mother, who insisted that I learn to type in high school.

Introduction

What's in a Name?

Nature has provided less than 40 aluminosilicate zeolite molecular sieve topologies, with a large number of compositional variants. Over the last 30 years scientists working in both academic and industrial laboratories have added several hundred microporous crystalline materials to the list. The names given to the natural zeolites have generally reflected the location where they first were found, such as Bikitaite from Bikita, Zimbabwe. Names also are given to honor people. For example, barrerite was named in honor of the eminent zeolite scientist Professor R. M. Barrer. Of course, some names have a more obscure meaning. Scolecite comes from the Greek word for worm because it was observed to curl like a worm when placed in a blowpipe.

Some of the synthetic molecular sieves have been named on a rational basis, such as synthetic ferrierite or synthetic bikiatite, indicating the man-made form of the respective natural minerals. Unfortunately, no systematic naming system has been followed for synthetic materials, especially materials with no natural counterpart. Each research group has taken on the task of defining its own naming system. Greek and arabic letters of the alphabet had been a popular means of naming molecular sieve materials. Barrer and his coworkers identified products of their synthesis as A, B, C, and so on. The inorganic cation that was used in the synthesis of a structure was applied as a prefix. Thus, in a sodium cation–containing system, Na-D was the fourth material identified. That material, which happened to be microporous, had an X-ray powder diffraction pattern similar to that of the natural zeolite mordenite. In the strontium cation–containing system, the fourth material identified, Sr-D, had the ferrierite structure. Thus Na-D and Sr-D identified two different products. In this case, both have natural counterparts.

Union Carbide scientists also used an alphabetical nomenclature. The zeolite that was designated as D had the chabazite structure and was not related to Barrer's Na-D or Sr-D. The company used "Linde," "Linde type," or "LZ" to identify a product; hence Linde D indicates it is a material from Union Carbide. Prefixes also were applied, such as P-A and N-A, that identified a material with the Linde type A topology but, in the former, a material prepared in the presence of phosphorous, and in the latter, a zeolite prepared using a cationic organic amine. A compositional designation is used by Union Carbide for its aluminophosphate and metal sulfide–based molecular sieves: $AlPO_4$ for the aluminophosphates, SAPO for the silicoaluminophosphates, and GS for the gallium sulfides.

Workers at Mobil Oil Corporation initially utilized the Greek letters system, producing the zeolites alpha and beta. Dr. George Kerr synthesized a number of zeolites that bear the name "ZK" for "zeolite Kerr." The familiar designation "ZSM" stands for "Zeolite Socony Mobil." ZSM designates many of Mobil's silicate and aluminosilicate materials. A more recent designation is that of "MCM," which represents "Mobil Composition of Matter," covering other materials.

Workers at Chevron used "SSZ" for "Socal Silica Zeolites," whereas Imperial Chemical Industries used the Greek letters "sigma" and "NU." "AG" and "FU" also were employed to designate ICI materials. "EU" is a shorthand notation for materials prepared at Edinburgh University. Exxon has used "ECR" for "Exxon Corporate Research," and Amoco used "AMS" for "Amoco molecular sieve."

Several academic institutes have applied school initials to identify molecular sieves: "GTRI" for Georgia Tech Research Institute ($AlPO_4$-H1(GTRI)), "TAMU" for Texas A and M University ($AlPO_4$-TAMU-12), and "VPI" for Virginia Polytechnic Institute (VPI-5).

There are over 500 materials listed in this book. The materials identified by a name need not each represent a unique framework topology, but they may represent a unique material. SSZ-13, AlPO$_4$-17, and willhendersonite are all of the chabasite (CHA) framework topology; however, they differ in composition. The first is a silica-rich composition, the second an aluminophosphate, and the third a potassium-rich form of the mineral.

This summary of molecular sieve nomenclature is by no means exhaustive. It is meant merely to provide a sense of the dilemma facing workers in the molecular sieve field today when naming a new material or trying to locate related materials.

Using This Handbook

One of the fundamental difficulties in the zeolite and molecular sieve field is the lack of ready access to information on compositional, structural, and physical properties of these materials. Most of the information is found only in the patent literature. Only a small fraction of the structures reported have been solved by X-ray diffraction techniques. Searching the molecular sieve literature for information on a framework topology turns the scientist into Sherlock Holmes searching for clues, as many of the materials are not cross-referenced by structure type, especially in the patent literature. Those well versed in the molecular sieve literature may be able to ferret out the desired information with a combination of searching standard texts and flipping through a few conference proceedings and abstract clippings. A good memory is critical. A neophyte, even with excellent library facilities available, would find the task insurmountable. Thus, it is the objective of this book to bring together in a simple reference form pertinent information on composition, structure, synthesis, and physical properties of the many crystalline microporous materials cited in the literature.

From A to Z

The book is organized in alphabetical order by the names given to the molecular sieves. Using the example stated above, "Na-D", "Sr-D," and "Linde D" are listed as such and not under "D." "D" is listed but only to refer the reader to "Linde D," as the term zeolite "D" has also been used to describe the "Linde D" material in the literature. Though it would be reasonable to list the aluminophosphate materials separately from the aluminosilicates and silicates, this was not done. Such a separation based on composition, such as the silicates versus aluminophosphate, could make it difficult for users of this book who are unfamiliar with the field to locate a material by name without prior knowledge of its composition. The chosen method of organization has led to an apparent redundancy in information, as data for GaAPO-5 and AsAPO-5 are similar to data for AlPO$_4$-5. Each, however, is considered a unique material based on the combination of composition plus diffraction pattern; so each has been treated separately in this text.

Following the name identifying the material is the patent number and assignee if the material is a patented composition. Unless otherwise noted, the information that follows is taken from the citation listed after the name of the material. Attempts were made to provide the earliest patent cited. In the case of the natural zeolites, the word "natural" appears. A brief history of the natural mineral follows. Listed also is the name of the structure type under which the material also can be cataloged as the three-letter code designated by the Structure Commission of the International Zeolite Association. Thus by the heading "EU-1" is the assigned three-letter code "EUO." Related materials include TPZ-3 and ZSM-50. Under ZSM-50, the three-letter code "EUO" is given to designate the framework topology, and the reference to related materials is EU-1, where the listing of all the other materials with that topology can be found.

Obsolete and improper names are provided for reference because they occasionally appear during searches of older literature. A number of Union Carbide "LZ" materials are listed and briefly described, as references to them have been made in the literature, and quickly identifying which topology they represent, can be helpful.

Structure

As both composition and X-ray powder diffraction data are used to identify a material, this information is provided in the section called "Structure." The X-ray powder diffraction data are given as d-spacing and intensity. The X-ray data listed generally are taken from the first example for the patented materials. The reader is cautioned that materials claimed in patents can (and do) sometimes include peaks due to impurities. Thus the diffraction pattern listed may not represent a single, pure phase; however, it does represent the patented phase. Many references use only s, m, and w for strong, medium, and weak intensities (or f, FF, etc. for materials described in the French literature). As such designations have no universal standard, it is suggested that the user of this information refer to the original reference for the accurate meaning of the terms. When possible, the published X-ray powder diffraction pattern is provided. The use of line representation and reporting of the theoretical diffraction pattern was avoided. If the structure is known, a description of it follows. Attempts were made to provide recently published structural data, with a preference for single crystal studies. Recent publications would cite earlier structural studies of the material.

Synthesis

Reports of molecular sieve synthesis are voluminous and virtually impossible to represent in concise form. Thus the syntheses reported in this book cover the methods of synthesis given in the original paper and occasionally include additional synthesis methods. The method reported may not represent the ideal method of synthesizing the material, but does represent the early and successful attempts. The organic additives that have been employed also are listed. For patented materials the ranges, either broad or preferred also are provided. Formulas or ratios are used, depending on how they were originally represented. No attempt was made to standardize the method of reporting a synthesis. When possible, common impurity phases that can arise in the course of the synthesis are noted.

Thermal Properties

Reported thermal properties and stabilities are provided when possible. Transformations to other phases are noted.

Spectroscopy

Infrared spectroscopy has been applied to the identification of structures, and such information has been included when available. The NMR data for these materials are not included, as such data are strongly dependent on a large number of factors, in both the materials studied and the instrumentation used.

Adsorption and Ion Exchange

Two fundamental features of the molecular sieve materials are their abilities to selectively adsorb molecules and to exchange their nonframework cations. These properties are included in this book. Generally, the data are from the original patent or early report.

References

For the sake of brevity, the references do not list authors and are abbreviated as much as possible. In general, the abbreviations follow standard practice, with a few exceptions. These are shown below.

Patents

Country of origin, patent number, and year patent was granted are included. The following abbreviations are used:

BP—British Patent Application
EP—European Patent Application
Fr—French Patent
Ger—German Patent
IP—Italian Patent
US—United States Patent

VP—Venezuelan Patent Application
WP—World Patent Application

Books

Title and year are given. The following abbreviations are used:

AZST—Atlas of Zeolite Structure Types, W. M. Meier, D. H. Olson, Butterworths, London 1987.

Hydro. Chem. Zeo.—Hydrothermal Chemistry of Zeolites, R. M. Barrer, Academic Press, London, 1982.

Nat. Zeo. (1978)—*Natural Zeolites: Occurrence, Properties, Use*, L. B. Sand, F. A. Mumpton, eds., Pergamon Press, Oxford, 1978.

Nat. Zeo. (1985)—*Natural Zeolites*, G. Gottardi, E. Galli, Springer-Verlag, Berlin, 1985.

Zeo. Clay Min.—Zeolites and Clay Minerals as Sorbents and Molecular Sieves, R. M. Barrer, Academic Press, London, 1978.

Zeo. Mol. Sieve.—Zeolite Molecular Sieves, D. W. Breck, John Wiley and Sons, New York, 1974 (original edition); R. E. Krieger, Malabar, Florida 1984 (new edition).

Nonstandard Reference Designations

ACS Symp. Ser.—American Chemical Society Symposium Series, American Chemical Society, New York.

BZA Meeting Report—Abstracts from a British Zeolite Association Meeting.

JACS—Journal of the American Chemical Society.

JCS—Journal of the Chemical Society.

JCS Chem. Commun.—Journal of the Chemical Society, Chemical Communications.

JCS Dalton—Journal of the Chemical Society, Dalton Transactions.

JCS Faraday—Journal of the Chemical Society, Faraday Transactions.

PP—poster paper from the specified International Zeolite Conference.

Proc. 5th IZC—Proceedings of the Fifth International Conference on Zeolites, L. V. Rees, ed., Heyden, London, 1980.

Proc. 6th IZC—Proceedings of the Sixth International Zeolite Conference, D. Olson, A. Bisio, eds., Butterworths, Guildford, Surrey, England, 1984.

RR—research reports from the specified International Zeolite Conference.

SSSC—Studies in Surface Science and Catalysis, Elsevier, Amsterdam.

A

A(BW) (ABW)
JCS1267(1951)

Related Materials: CsAlSiO$_4$
RbAlSiO$_4$

This structure first was reported by the name "A" or "Li-A" by Barrer and White in 1951 and was found to be structurally unique. To avoid confusion with zeolite A patented by Union Carbide, this material was given the designation A(BW) or Li-A(BW). Several groups have prepared Tl-A(BW). It was referred to in one report as Tl-B (*JCS* 1949:1253(1949)) and another as Tl-C (*JCS Dalton* 934(1974)).

Structure

(*Z. Kristallogr.* 176:671(1986))

CHEMICAL COMPOSITION: Li$_4$[Al$_4$Si$_4$O$_{16}$]·4H$_2$O
SYMMETRY: Orthorhombic
SPACE GROUP: Pna2$_1$
UNIT CELL CONSTANTS: a = 10.311(1) A, b = 8.196(1) A, c = 4.995(1) A from single crystal data (*SSSC*28:443(1986))
PORE VOLUME: 0.28 cm^3 H$_2$O/cm^3
D$_c$: 2.27 g/cm^3
PORE STRUCTURE: eight-member rings along [001]; 3.4 × 3.8 A

A(BW) Fig. 1S: Framework topology.

Powder Diffraction Data: (*JCS* 1951: 1267(1951)) (d(Å) (I/I_o)) 6.42(vs), 5.21(mw), 4.29(vs), 4.06(vw), 3.27(vw), 3.15(vs), 3.03(vs), 2.490(s), 2.392(w), 2.326(m), 2.243(vw), 2.173(mw), 2.042(w), 1.952(mw), 1,868(vw), 1.754(m), 1.725(mw), 1.556(w), 1.524(mw), 1.474(m), 1.445(vw), 1.405(mw), 1.371(vw), 1.348(vw), 1.348(vw), 1.327(vw), 1.300(vw), 1.270(w), 1.244(vw).

The structure of Li-A(BW) was solved in 1974 by powder methods (*Z. Kristallogr.* 139:186(1974)) and later from single crystal data (*Z. Kristallogr.* 176:67(1986)). The A(BW) framework is made up of four-rings, which define the eight-ring channel system.

Lattice constants in Å, unit cell volumes in Å^3, and framework densities FD (number of tetrahedral atoms per 1000 Å^3) in materials with the ABW structure (number in parentheses are standard deviations) (*Zeo.* 11:149(1991)):

	a	b	c	V	FD
LiA(BW)	10.314(1)	8.194(1)	4.993(1)	421.93	19.0
TlABW(1)	8.297(1)	9.417(1)	5.413(1)	420.80	19.0
TlABW(2)	8,287(1)	9.396(1)	5.404(1)	420.78	19.0
RbABW	8.741	9.226	5.337	430.40	18.6
CsABW	8.907(2)	9.435(1)	5.435(1)	457.00	17.5

Atomic coordinates ($\times 10^4$) and equivalent isotropic temperature factors (Å^2) with standard deviations in parentheses $B_{eq} = 4/3\Sigma_{(i,j)}b_{ij}(a_ia_j)$ (*S. Kristallogr.* 176:67(1986)):

	x	y	z	B_{eq}
Li	1862(11)	6849(15)	2520(4)	1.57(22)
Al	1593(2)	810(2)	2500	0.40(11)
Si	3544(1)	3757(2)	2492(7)	0.44(09)
01	65(4)	1584(6)	1970(12)	0.79(30)
02	2736(5)	2198(6)	1391(13)	0.97(30)
03	1912(5)	399(6)	5907(12)	0.76(29)
04	1804(5)	−1008(6)	689(12)	0.68(28)
05	5891(6)	903(8)	−2395(28)	2.93(53)

The lithium ions lie in the center of a nearly regular tetrahedron of oxygen atoms, three from the aluminosilicate framework and one from the water molecule. No deviation from complete ordering of the Si and Al can be found in the framework. The only framework atom not co-ordinated to the lithium ion (O(1)) is an acceptor in a weak hydrogen bond from the water molecule adsorbed in the structure. Neutron diffraction was also used to identify proton positions in the ABW framework with the gallosilicate composition [LiGaSiO$_4$] · D$_2$O (*JCS Chem. Commun.* 1295(1986)). The structure of this material is well ordered at both 19 and 298°K.

The Rb and Cs cations in RbABW and CsABW are situated in only one position in the structure, whereas three positions partially occupied by Tl$^+$ were found in Tl-ABW (*Zeo.* 11:149(1991)).

Synthesis

ORGANIC ADDITIVES
generally none
tetramethylammonium$^+$ (TMA)

Crystals of this structure have been prepared from Li, (Li,K), (Li,Na), and (Li,Cs,TMA)–containing reaction mixtures; however, this structure shows a distinct preference for lithium for its synthesis (*Zeo.* 1:130 (1981)). Using metakaolin as a source of aluminosilicate, this zeolite has been prepared from the composition range 1 (metakaolin) : 0 to 8 SiO$_2$: 2.5 to 45 LiOH : 275 H$_2$O in the temperature range between 80 and 170°C (*JCS Dalton*, 2534(1972)), with the best crystalline products appearing at 110°C. With the composition of the gel Li$_2$O : Al$_2$O$_3$:

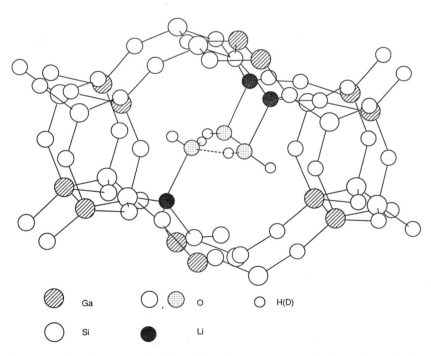

| Ga | O | H(D) |
| Si | Li | |

A(BW) Fig. 2:   View along the channel of the ABW framework gallosilicate at 298K showing the lithium cation position, the position of the sorbed water molecule, and the weak hydrogen bonding (*JCS Chem. Commun.* 1295 (1986)). (Reproduced with permission of the Chemical Society)

1 to 4 SiO_2 : x H_2O, the best materials were prepared at 250°C with a gel SiO_2/Al_2O_3 of 2 (*JCS* 1267(1951)). Species that also formed from these gels include alpha-eucryptite and alpha-petalite, and, at temperatures above 450°C, beta-

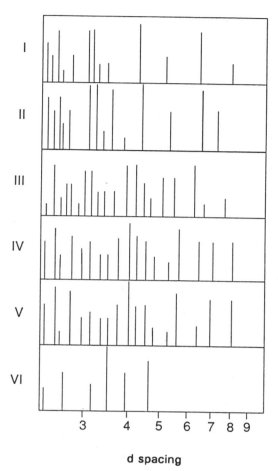

I, Species A.
II,   "   heated at 150°.
III,   "   "   200°.
IV,   "   "   250°.
V,   "   "   360°.
VI,   "   "   ~ 800°.

A(BW) Fig. 3: Changes in the principal lattice spacing on thermal treatment of A(BW). I, A(BW); II, heated at 150°C; III, heated at 200°C; IV, heated 250°C; V, heated at 360°C; VI, heated at ca. 800°C (*JCS*, 1267 (1951)). (Reproduced with permission of the Chemical Society)

spodumene becomes the primary phase. Hydrothermal treatment of the isolated crystals at 150°C in the presence of NaCl results in a recrystallization to analcime. Large crystals of this zeolite were prepared by hydrothermal treatment of sodium bromosodalite with lithium bromide at 260°C over a 72-hour period (*Z. Kristallogr.* 176:67 (1986)).

Thermal Properties

The X-ray diffraction pattern of this zeolite remains constant with slight changes observed at 200°C. At 800°C coversion to beta-eucryptite is observed (*JCS* 1267 (1951)). See A(BW) Fig. 3.

Ion Exchange

Good ion exchange properties have been observed for zeolite ABW. Silver readily replaces

Adsorption

Sorption behavior of zeolite ABW in various ion-exchanged forms (*JCS* 1267 (1951)):

Exchange form	Outgassing temp. (°C)	Gas	Temp. of isotherm (°C)	Max. sorption, cc at STP/g
Parent material	220	NH_3	0	50
		NH_3	150	30
		N_2	−180	6.5
		C_3H_8	0	2.0
		H_2	−186	0.3
	300	NH_3	21	3.3
Regenerated	230	NH_3	0	84
LiA (BW)		NH_3	25	53
		NH_3	50	25
Ba^{+2}	230	NH_3	0	ca.5.0
AG^+	230	NH_3	0	ca.15
		NH_3	0	
		NH_3	50	
		NH_3	100	
Gel**	230	NH_3	0	21[#]
		NH_3	50	18.5[#]
		N_2	−186	14.2
		C_3H_8	0	3.1

*Not true equilibrium
** $Li_2O,Al_2O_3,3SiO_2$.
[#] Saturation not reached.

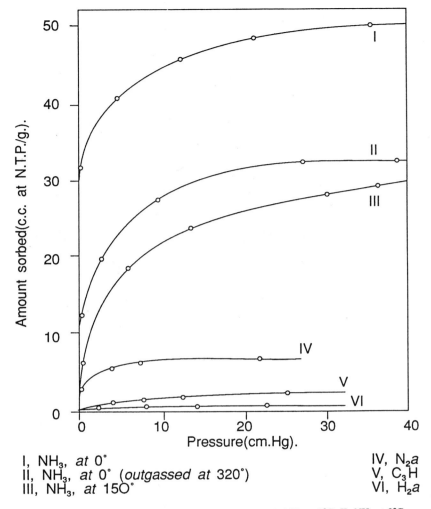

I, NH₃, at 0°
II, NH₃, at 0° (outgassed at 320°)
III, NH₃, at 150°

IV, N₂a
V, C₃H
VI, H₂a

A(BW) Fig. 4:   Typical sorption isotherms in A(BW): I, NH₃ at 0°C; II, NH₃ at 0°C; VI,H₂ at −186°C (*JCS* 1267 (1951)). (Reproduced with permission of the Chemical Society)

lithium when treated at 110°C for 2 days. Some decomposition is observed when barium nitrate or silver nitrate is fused with zeolite ABW. The Silver exchanged form has been treated with metallic chlorides at 110°C to produce other ion exchange forms. The cationic forms that can be obtained by this method include Tl^+, Ca^{+2}, Ba^{+2}, Na^+, K^+, and Rb^+ (*JCS* 1267(1951)). Limited exchange was observed for NH_4^+.

Ion exhange data for ABW (*JCS* 1267 (1951)):

Exchange cations	% Exchange
Li*(original)	—
Li*(regenerated from Ag*A)	100*
Na^+	33.8
$Ca^+{}_2$	12.9
Ag*	88.1
K^+	28.1

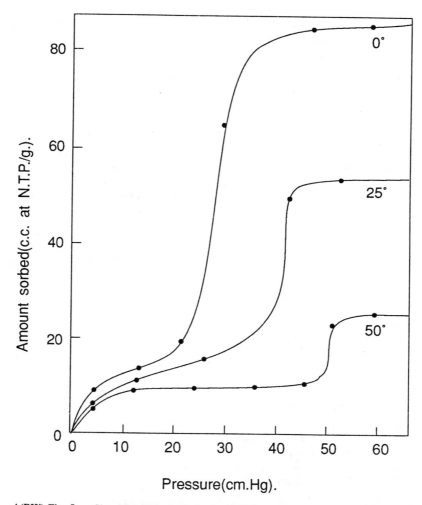

A(BW) Fig. 5:   Sigmoid isotherms obtained with NH₃ on species A(BW) converted into Ag-A and then reconverted into A(BW) by ion-exchange Li⁺ = Ag⁺ (*JCS* 1267 (1951)). (Reproduced with permission of the Chemical Society).

Exchange cations	% Exchange
Ba^{+2}	85.6
Tl^{+}	79.5
Rb^{+}	—
NH$_4^{+}$	11.1

*On the basis of complete absence of Ag⁺.

Acadialite (CHA)
(natural)

Obsolete name for chabazite.

Afghanite (AFG)
(natural)

Afghanite is a natural aluminosilicate discovered at the lapis lazuli mine in Afghanistan (*Bull. Soc. Fr. Mineral. Cristallogr.* 91:23 (1968)).

Structure

(*Bull. Soc. Fr. Mineral. Cristallogr.* 91:34 (1968))

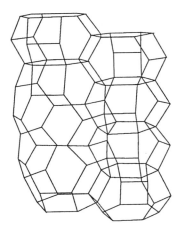

Afg Fig. 1S: Framework topology.

CHEMICAL COMPOSITION: $(Na,Ca,K)_{12}(Si,Al)_{16}O_{34}$
  $(Cl,SO_4,CO_3)_4 \cdot 0.6H_2O$
SYMMETRY: hexagonal
SPACE GROUP: $P6_3/mmc$ (or $P6_3/mc$, P62C)
UNIT CELL CONSTANTS: (Å) $a = 12.77$,
  $c = 21.35$.
UNIT CELL VOLUME: 3015 Å³

X-Ray Powder Diffraction Data: (*Bull. Soc. Fr. Mineral. Cristallogr.* 91:34(1968) (*d*(Å) (*I/I*ₒ)) 11.10(fff), 10.70(fff), 9.90(fff), 6.00 (mF), 5.36(f), 4.82(FF), 4.18(ff), 4.105(f), 3.997(F), 3.850(fff), 3.688(FFF), 3.609(ff), 3.388(fff), 3.298(FFF), 3.202(fff), 3.125(fff), 3.039(mf), 2.951(f), 2.821(f), 2.799(ff), 2.769(mf), 2.744 (mf), 2.685(F), 2.663(ff), 2.606(ff), 2.580(f), 2.495(f), 2.473(fff), 2.460(mF), 2.418(f), 2.325(f), 2.294(fff), 2.170(f), 2.130(f), 2.130(F), 2.094(fff), 2.057(mF), 1.950(fff), 1.925(mf), 1.881(fff), 1.864(f), 1.845(fff), 1.812(fff), 1.803(fff), 1.792(F), 1.771(fff).

The afghanite structure contains only four- and six-member rings. Access to the cages is through six-rings only.

AG-2 (OFF)
(Ger 2,248,626 (1973))
Imperial Chemical Industries

Related Material: offretite

Structure

CHEMICAL COMPOSITION: $1.1 \pm 0.4M_2O$:
  Al_2O_3 : 4–12 SiO_2. (M is alkali cations K & Na)

X-Ray Powder Diffraction Data: (*d*(Å) (*I/I*ₒ))
11.40(76), 8.26(2.8), 7.53(14.8), 6.84(4.2), 6.607(47.9), 6.281(8.3), 5.71(16.7), 4.56(22.6), 3.31(32.4), 3.27(18.3), 3.17(23.9), 2.99(22.5), 2.866(71.8), 2.848(74.6), 2.68(19.7), 2.52(3.5), 2.50(12.7).

AG-2 exhibits properties of a pure offretite without the presence of erionite intergrowths.

Synthesis

ORGANIC ADDITIVES
tetramethylammonium⁺
tetraethylammonium⁺

AG-2 crystallizes from a reactive gel with molar oxide ratios: $SiO_2/Al_2O_3 = 8$ to 24, $R_2O/M_2O = 0.02$ to 0.5, $M_2O/SiO_2 = 0.2$ to 1.0, $H_2O/M_2O = 14$ to 160, and $K_2O/K_2O + Na_2O = 0.5$ to 1.0. Crystallization occurs after 72 hours at 95°C (R = organic; M = Na + K).

AlAsO-1 (APD)
(*RR 8th IZC* 48(1989))

Related Materials: $AlPO_4$-D

Structure

CHEMICAL COMPOSITION: $AlAsO_4 \cdot 0.5$ EAN
  (EAN = ethanolamine)
SYMMETRY: orthorhombic
SPACE GROUP: Pcab
UNIT CELL PARAMETERS (Å): $a = 8.7812$
  $b = 10.2615$
  $c = 20.4332$
PORE STRUCTURE: eight-member rings

X-ray Powder Diffraction Data: (*d*(Å) (*I/I*ₒ))
10.1 (36), 7.24(100, 6.32(8), 5.07(40), 4.36(8),

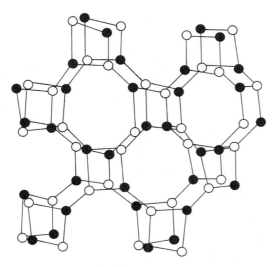

AlAsO$_4$-1 **Fig. 1S:** Framework topology.

4.0(28), 3.72(10), 3.60(10), 3.38(20), 3.23(31), 2.15(10), 3.02(14), 2.83(6), 2.69(8), 2.55(34), 2.45(8), 2.41(8), 2.04(6), 1.98(4), 1.85(4), 1.80(9), 1.77(6), 1.75(5), 1.70(9).

The structure of AlAsO$_4$-1 consists of As atoms in tetrahedral coordination and Al atoms in both tetrahedral and octahedral coordination. Two of the coordination sites on the aluminum are occupied by EAN molecules. The As–O bond lengths lie in the range 1.6490 to 1.698 A, and the O–As–O angles vary from 101.5° to 116.4°. The Al–O lengths range from 1.718 to 1.751 A for tetrahedral Al and 1.844 to 1.949 Å for the octahedral. The channels are one-dimensional eight-member ring channels, and the EAN molecules occupy sites within the channels. AlAsO$_4$-1 has the same framework topology as AlPO$_4$-D (*JCS Chem. Commun.* B19 (1989)).

Synthesis

ORGANIC ADDITIVES
ethanolamine (EAN)

The aluminoarsinate structure was prepared by using pyroarsenic acid, aluminum-isopro-

pylate, and water. Ethanolamine (EAN) was used as the organic additive. The composition of the reactive gel consisted of:

$$0.8 \text{ EAN} : Al_2O_3 : As_2O_5 : 40 \text{ H}_2O$$

Crystallization occurred after 5 days at 200°C.

Alpha (LTA)
(US3,375,205(1968))
Mobil Oil Corporation

Related Materials: Linde type A

Structure

Chemical Composition: 0.2 to 0.5 $R_{2/m}O$: 0.8 $M_{2/n}O$: Al_2O_3 : > 4.0 to 7.0 SiO_2 : Y H_2O (R = TMA or other cations, M = Na or other cations, m & n = 1 or 2).

X-Ray Powder Diffraction Data: ($d(\text{Å})$ (I/I_0)) 12.03(68), 8.51(69), 6.94(70), 6.00(8), 5.38(26), 4.91(5), 4.26(38), 4.02(86), 3.80(9), 3.63(100), 3.34(63), 3.21(64), 3.01(2), 2.91(87), 2.83(25), 2.68(10), 2.62(10), 2.56(35), 2.45(8), 2.40(2), 2.36(4), 2.31(7), 2.19(5), 2.12(22), 2.09(3), 2.06(4), 2.030(3), 2.003(19), 1.953(3), 1.877(9), 1.854(9), 1.812(2), 1.791(22), 1.711(2), 1.699(21), 1.650(16), 1.635(4), 1.592(12), 1.565(9), 1.539(1), 1.491(2), 1.479(3), 1.470(1), 1.463(3), 1.453(3).

Alpha exhibits properties of a small-pore silica-rich type A material. See also N-A.

Synthesis

ORGANIC ADDITIVES
tetramothylammonium$^+$ (TMA)

This material is claimed to crystallize from a batch composition of SiO_2/Al_2O_3 from 15 to 60, Na_2O/Na_2O + TMA_2O from 0.01 to 0.3, H_2O/Na_2O + TMA_2O from 30 to 60, and Na_2O + TMA_2O/SiO_2 from 0.5 to about 1. Crystallization occurs at pH between 9 and 12 and at temperatures around 90°C. Six days of crystallization at that temperature have been employed.

Adsorption

Comparison of the adsorption properties of zeolite NaA, ZK-4, and alpha:

	Molar Composition			Sorption Capacity in Wt.%		
Zeolite	Na$_2$O	Al$_2$O$_3$	SiO$_2$	cyclo-hexane	n-hexane	H$_2$O
NaA	1.0	1.0	2.0	<1	<1	24
NaZK-4	0.8	1.0	3.5	<1	13	26
Na-alpha	0.6	1.0	6.0	2	17	30
Ca-alpha	0.1	1.0	6.0	2	15	26

AlPO$_4$-5 (AFI)
(US4,310,440(1982))
Union Carbide Corporation

Related Materials: SSZ-24
SAPO-5
and related metal containing aluminophosphates

Structure

(*ACS* Symp. Ser. 218:109(1983))

CHEMICAL COMPOSITION: TPAOH*12AlPO$_4$
(TPAOH) = tetrapropylammonium hydroxide)
SYMMETRY: hexagonal
SPACE GROUP: P6cc
UNIT CELL PARAMETERS: a = 13.726 Å,
c = 8.484 Å, γ = 120°
PORE STRUCTURE: unidimensional 12-ring system

X-Ray Powder Diffraction Data: (d(Å) (I/I_0))
11.8(100),   6.84(11),   5.93(28),   4.50(66),
4.24(63), 3.97(94), 3.43(37), 3.08(21), 2.97(22),
2.60(19), 2.40(13).

AlPO$_4$-5 contains alternating Al and P atoms forming four- and six-member rings. The pore system consists of nonconnecting parallel channels of 12-member rings. In the as-synthesized form, this molecular sieve contains tetrapro-

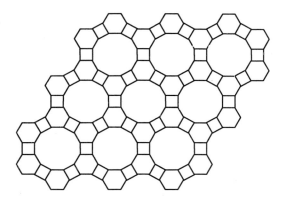

AlPO$_4$-5 Fig. 1S: Framework topology.

pylammonium hydroxide species when prepared using the TPAOH as the organic additive. The AlPO$_4$-5 structure is polar, as is the tripod-shaped TPAOH. The tripod orientation of the TPAOH in the channel system of this aluminophosphate differs significantly from that observed in silicalite, where the propyl groups radiate from the channel intersection into the four channel openings (*ACS* Symp. Ser. 218:109 (1983)).

Atomic positions of TPAOH-AlPO$_4$-5 (*ACS* Symp. Ser. 218:109 (1983)):

Atom	x	y	z
P	0.4524(3)	0.3283(2)	0.078(3)
Al	0.4577(4)	0.3377(3)	0.450(3)
O1	0.4210(5)	0.2137(7)	0.028(4)
O2	0.4555(8)	0.3312(10)	0.25*
O3	0.3670(7)	0.3584(8)	0.026(4)
O4	0.5684(7)	0.4126(8)	0.014(4)
N	0	0	0.350
C1	0	0	0.106
C2	0.000	0.106	0.106
C3	0.999	0.105	0.923
C4	0.122	0.079	0.409
C5	0.126	0.177	0.504
C6	0.150	0.166	0.679
O	0	0	0.620

*fixed to define origin

Synthesis

ORGANIC ADDITIVES   (*Proc. 6th IZC* 97(1984))
(C$_3$H$_7$)$_4$N$^+$
(C$_6$H$_{11}$)$_2$NH
(C$_2$H$_5$)$_2$NCH$_2$CH$_2$N(C$_2$H$_5$)$_2$

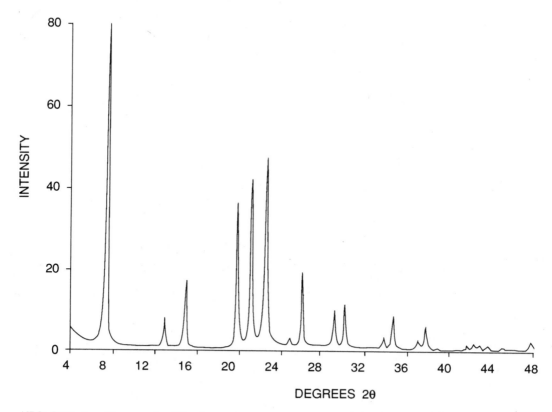

AlPO₄-5 Fig. 2:   X-ray powder diffraction pattern of AlPO₄-5 (*Zeo.* 6:349 (1986)). (Reproduced with permission of Butterworth Publishing Co.)

(C₃H₇)₃N
(C₂H₅)₃N (*JACS* 104:1146(1982))
(CH₃)₂NCH₂C₆H₅
(HOCH₂CH₂)₃N
(CH₃CH₂)₄N⁺
N,N′-dimethylpiperazine
C₆H₁₁NHCH₃
1,4-diazabicyclo[2,2,2]octane
(C₂H₅)₂NCH₂CH₂OH
CH₃N(CH₂CH₂OH)₂
Cyclohexylamine
(CH₃)₃NCH₂CH₂OH
2-methylpyridine
3-methylpyridine
4-methylpyridine
N-methylpiperidine
3-methylpiperidine
piperidine
(CH₃)₂NCH₂CH₂OH
CH₃NHCH₂CH₂OH
and others

The synthesis of the AlPO₄-5 structure occurs readily using aluminum hydroxide or boehmite as the source of aluminum and phosphoric acid as the source of phosphorus. Over 23 different organic additives have been successfully employed to produce this structure. Crystallization occurs rapidly at temperatures between 150 and 200°C, usually complete within 4 to 8 hours. The pH range of the initial starting gel can be acidic, with crystallization occurring from gels with pH as low as 3.0. Final pH ranges are generally slightly basic (ca. pH = 8). At the more elevated temperatures (190–200°C), crystallization to the aluminophosphate analogs of cristobalite and tridymite do occur with time. Crystallization at temperatures below 125°C results in the crystallization of other nonzeolite-type phases. These include H3, metavariscite, and variscite. Using cyclohexylamine as the organic will result in the crystallization of AlPO₄-5 at lower temperatures (150°C) and AlPO₄-17 at higher temperatures. Conversely, in the presence of tetraethylammonium hydrox-

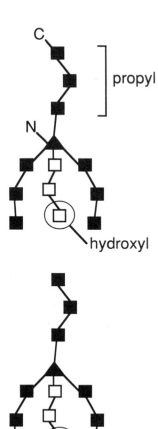

AlPO₄-5 Fig. 3:   In the cylindrical channel in AlPO₄-5, the encapsulated tetrapropylammonium hydroxide takes on a tripod shape (*Zeo.* 6:349 (1986)). (Reproduced with permission of Butterworth Publishing Company)

ide, AlPO₄-18 crystallizes at the lower temperature and AlPO₄-5 at more elevated temperatures. Generally, gel compositions that effectively produce the AlPO₄-5 structure are: 1.0 R : 1.0 Al₂O₃ : 1.0 P₂O₅ : 40 H₂O (*Proc. 6th IZC* 97(1984)). (R = organic)

Infrared Spectra

Mid-infrared Spectral Data for AlPO₄-5 Crystallized in the Presence of HF (from author) (*cm⁻¹*):   1230sh, 1050–1150br, 730w, 700mw, 630w, 580sh, 560m, 510m.

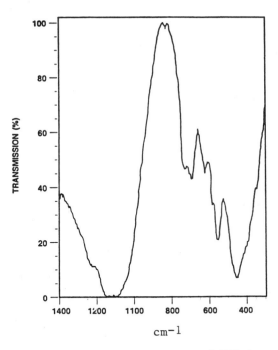

AlPO₄-5 Fig. 4:   Mid-infrared spectrum of AlPO₄-5 crystallized in the presence of HF. (Reproduced with permission of K. Vinje, U. of Oslo)

Adsorption

Adsorption capacities of AlPO₄-5 (US 4,310,440 (1982)):

	Kinetic diameter, Å	Pressure, Torr	Temp., °C	Wt. % adsorbed
O₂	3.46	97	−183	14.6
O₂		750	−183	21.3
Neopentane	6.2	102	24	6.5
(C₄F₉)₃N (after 4 hr)	10	0.073	24	1.2
H₂O	2.65	4.6	24	6.5
H₂O		18.5	23	32.6

AlPO₄-8 (AET) (US4,310,440(1982)) Union Carbide Corporation

Structure

(*Zeo.* 10:522(1990))

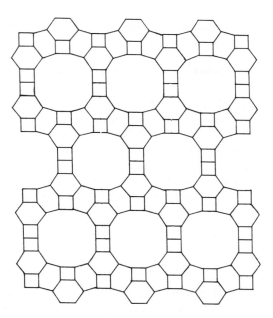

AlPO₄-8 Fig. 1S: Framework topology.

CHEMICAL COMPOSITION: $Al_2O_3 : 1.0 \pm 0.2\ P_2O_5$
SYMMETRY: orthorhombic
SPACE GROUP: $Cmc2_1$
UNIT CELL CONSTANTS: $a = 33.29(2)$ Å
                                          $b = 14.76(2)$
                                          $c = 8.257(4)$
FRAMEWORK DENSITY: 1.8g/cm³
PORE STRUCTURE: unidimensional 14-member rings;
  7.9 × 8.7 Å

X-Ray Powder Diffraction Data: (d(Å) (I/I_0))
16.7(80), 13.6(100), 8.84(17), 8.19(2), 6.07(4),
5.56(16), 4.72(2), 4.48(8), 4.40(12), 4.19(82),
4.06(18), 3.97(39), 3.92(sh), 3.77(3), 3.68(11),
3.58(11), 3.29(2), 3.16(5), 2.853(4), 2.722(3),
2.622(1), 2.522(3), 2.368(9), 2.344(3sh),
2.094(2), 1.937(1), 1.841(3).

The framework topology of AlPO₄-8 has been
determined from model building/DLS refinement (*Zeo.* 10:522(1990)). The structure contains a unidimensional 14-member ring channel

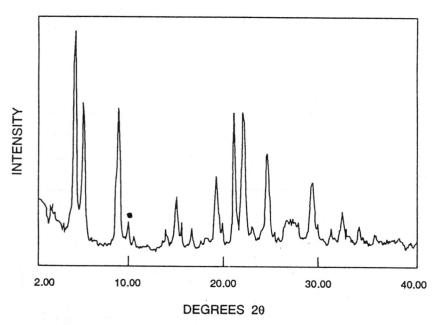

AlPO₄-8 Fig. 2: X-ray powder diffraction pattern of AlPO₄-8 with small AlPO4-H1
impurity (•). (Reproduced with permission of author)

system. This framework may be generated by a simple transformation of the VPI-5 network. Stacking faults are observed in a number of samples of this material, resulting in significantly reduced adsorption properties (*Cat. Lett.* 6:209 (1990)). When hydrated, the aluminum ions in the double 4-member ring are octahedral with two associated water molecules (*Appl. Catal.* 69:L7(1991)).

DLS atomic positions for AlPO$_4$-8 (*Zeo.* 10:522 (1990)):

Atom	x	y	z
Al1	0.00000	0.06261	0.43990
P1	0.08258	0.07711	0.60903
Al2	0.15415	0.20434	0.48142
P2	0.22196	0.08355	0.61285
P13	0.31054	0.07860	0.48222
P3	0.00000	−0.13164	0.57511
Al4	0.08381	0.12380	0.98454
P4	0.15503	0.24276	0.10731
Al5	0.27698	0.38458	0.48721
P5	0.18821	0.38456	0.60545
O1	0.00000	−0.05472	0.45101
O2	0.04302	0.10462	0.52881
O3	0.00000	0.09276	0.24085
O4	0.08990	−0.02441	0.59196
O5	0.11639	0.13563	0.55050
O6	0.07595	0.09538	0.78650
O7	0.14426	0.23518	0.28374
O8	0.19939	0.14507	0.49443
O9	0.15920	0.30452	0.59271
O10	0.21048	−0.01464	0.58186
O11	0.26670	0.09545	0.59065
O12	0.21049	0.10989	0.78459
O13	0.32877	−0.02999	0.52379
O14	0.30203	0.09376	0.27931
O15	−0.03932	−0.18081	0.54576
O16	0.12454	0.19355	0.00370
O17	0.19653	0.20347	0.07755
O18	0.34517	0.15771	0.55363
O19	0.27365	0.13859	0.01161

Synthesis

ORGANIC ADDITIVES   (US4,310,440(1982))
(n-C$_4$H$_9$)$_4$N$^+$
(n-C$_5$H$_{11}$)$_4$N$^+$
(n-C$_4$H$_9$)$_2$NH
(n-C$_5$H$_{11}$)$_2$NH
(n-C$_3$H$_7$)$_2$NH

This aluminophosphate has been synthesized under conditions:

$$0.5(R)_2O: Al_2O_3 : P_2O_5 : 52\ H_2O$$

from several hours to six days at 150°C. After isolation from the crystallization vessel, the crystals are dried at 100°C. Only trace amounts of the organic can be found in the solid crystalline material. Dried at 100°C, AlPO$_4$-8 is formed via a solid state thermal transformation of the aluminophosphates VPI-5 and AlPO$_4$-H1.

Thermal Properties

Thermal treatment of AlPO$_4$-8 to 600°C in air shows little loss in structure, based on examination of the x-ray powder diffraction pattern after calcination (US4,310,440(1982)).

Infrared Spectrum

Infrared Spectral Data (cm^{-1}): 1263w, 1240vw, 1205sh, 1150vs, 1088s, 726w, 540wsh, 456m (from author).

Adsorption

Adsorption capacities of AlPO$_4$-8 (US 4,310,440 (1982)):

	Kinetic diameter, Å	Pressure, Torr	Temp., °C	Wt. % adsorbed
O$_2$	3.46	101	−183	8.9
O$_2$		755	−183	18.6
n-Butane	4.3	768	24	5.0
Neopentane	6.2	501	24	4.5
(C$_4$F$_9$)$_3$N (after 4 hr)	10	0.073	25	8.2
H$_2$O	2.65	4.6	24	18.5
H$_2$O	2.65	20.0	24	31.9

*See also AlPO$_4$-H1 and VPI-5, thermally treated.

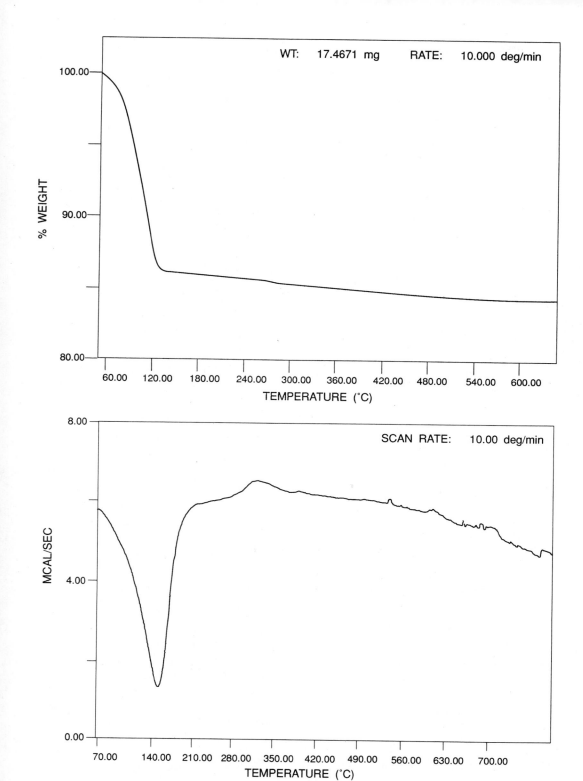

AlPO₄-8 Figs. 3a and 3b: TGA. (Fig. 3a) /DTA. (Fig. 3b) Trace of AlPO₄-8. (Reproduced with permission of K. Vinje, U. of Oslo)

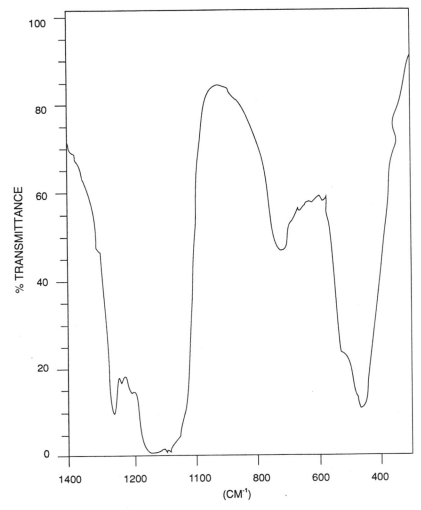

AlPO$_4$-8 Fig. 4:   Infrared Spectrum of AlPO$_4$-8. (Reproduced with permission of K. Vinje, U. of Oslo, Norway)

AlPO$_4$-9
(US4,310,440(1982))
Union Carbide Corporation

Structure

Chemical Composition: Al$_2$O$_3$ 1 ± 0.2 P$_2$O$_5$.

X-Ray Powder Diffraction Data: (d(Å) (I/I_0))
10.5(12), 7.97(43), 7.08(18), 6.33(6), 5.87(5), 5.28(11), 4.85(29), 4.60(11), 4.23(100), 4.06(30), 3.99(18), 3.90(30), 3.51(38), 3.47(79), 3.36(13), 3.22(21), 3.13(29), 2.92(5), 2.843(10), 2.814(27), 2.763(<1), 2.702(12), 2.660(18), 2.614(5), 2.578(3), 2.529(4), 2.468(1), 2.380(4), 2.347(10), 2.341(sh), 2.315(10), 2.186(2), 2.144(6), 2.122(4), 2.034–2.029(3), 1.953(2), 1.895(3), 1.881(1), 1.833(7), 1.,817(9), 1.791(1), 1.768–1.762(sh), 1.708–1.704(4), 1.660(7).

Synthesis

ORGANIC ADDITIVES
diaminobicyclooctane (DABCO)

AlPO$_4$-9 is prepared from a batch composition with a molar oxide ratio of: 1.0 DABCO : Al$_2$O$_3$: P$_2$O$_5$: 40 H$_2$O. Crystallization occurs at 200°C within 14 days using DABCO as the organic. The DABCO remains associated with the aluminosilicate after crystallization (US4, 310,440(1982)).

AlPO$_4$-11 (AEL)
(US4,310,440(1982))
Union Carbide Corporation

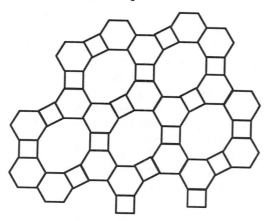

AlPO$_4$-11 Fig. 1S: Framework topology.

Related Materials: SAPO-11
and related metal-substituted aluminophosphates

Structure

(*Zeo.* 7:160(1987))

CHEMICAL COMPOSITION: Al$_{20}$P$_{20}$O$_{80}$
SYMMETRY: orthorhombic
SPACE GROUP: Icmm
UNIT CELL CONSTANTS: a = 13.5 Å
b = 18.5 Å
c = 8.4 Å

VOLUME = 2094 A^3
CHANNEL SYSTEM: unidimensional 10-ring system
PORE STRUCTURE: 6.7 × 4.4 Å

X-Ray Powder Diffraction Data: (d(Å) (I/I_0)
10.85(34), 9.31(49), 6.66(16), 5.64(30), 5.42(5), 4.67(6), 4.32(50), 4.23(100), 4.00(58), 3.93(75), 3.83(67), 3.62(10), 3.59(11), 3.38(13), 3.34(17), 3.13–3.11(15), 3.06(6), 3.02(9), 2.84(10), 2.71(15), 2.61(11), 2.51(3), 2.46(6), 2.39–2.37(14), 2.28(4), 2.23(2), 2.11(5), 2.02(2), 2.01(4), 1.99(6), 1.89(2), 1.86(4), 1.80(3),

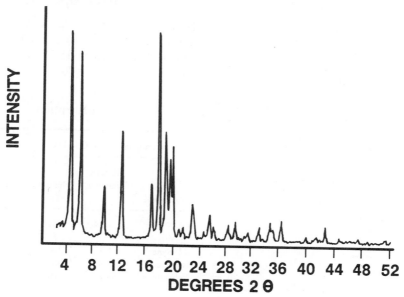

AlPO$_4$-11 Fig. 2:   X-ray powder diffraction pattern of AlPO$_4$-11 as synthesized in the presence of di-*n*-propylamine (*ACS Sym. Ser.* 218:79 (1983)). (Reproduced with permission of the American Chemical Society)

1.68(4); (after thermal treatment to 600°C) 11.1(14), 9.03(40), 7.56(3), 6.94(20), 6.49(6), 6.03(5), 5.50(54), 5.05(2), 4.55(17), 4.47(23), 4.29(8), 4.06(100), 4.02(44), 3.95(48), 3.79(54), 3.71(16), 3.68(10), 3.47(20), 3.44(9), 3.34(13), 3.27(14), 3.23(20), 3.13(8), 3.02(29), 2.95(16), 2.82(8), 2.75(22), 2.64(7), 2.61(5), 2.53(9), 2.42(8), 2.36(6), 2.32(14), 2.29(2), 2.27(2), 2.20(8), 2.18(5), 2.08(3), 2.03(5), 2.00(6), 1.851.84(8), 1.81(3), 1.75(1), 1.71(5), 1.68(2).

The aluminophosphate AlPO$_4$-11 consists of an open AlPO$_4$ framework containing unidimensional 10-ring channels. There are four unique T atom sites in the framework structure. The topology consists of sheets of six-ring–six-ring–four-ring units. This structure is related to AlPO$_4$-5 by removing one set of the four rings in the framework (a reverse sigma transformation) (*Zeo.* 7:160(1987)). The structure of AlPO$_4$-11 was solved for the calcined, organic-free material.

Atomic coordinates for AlPO$_4$-11 (*Zeo.* 7:160 (1987)):

Atom	x	y	z	B
T1	0.1489(10)	0.0281(8)	0.1931(14)	1.30
T2	0.9501(9)	0.1060(11)	0.3077(15)	2.60
T3	0.8590(9)	0.2500	0.1846(13)	−2.61
O1	0.1419(16)	0.0396(12)	0.0000	7.16
O2	0.9431(10)	0.1010(12)	0.5000	2.88
O3	0.8454(18)	0.2500	0.0000	2.90
O4	0.2500	0.0717(7)	0.2500	1.97
O5	0.0635(7)	0.0946(8)	0.2563(16)	3.70
O6	0.1207(7)	0.9556(7)	0.2526(22)	4.68
O7	0.9174(10)	0.1851(7)	0.2502(25)	5.95
O8	0.7500	0.2500	0.2500	2.18

Synthesis

ORGANIC ADDITIVES

i-pr$_2$-NH
n-pr$_2$-NH
n-pr(et)-NH
n-bu$_2$NH
n-pent$_2$NH

To date, five different secondary amines have been found to induce the crystallization of the AEL structure. Successful crystallization in the presence of cationic amines has not been reported. Sources of aluminum that have been effective in producing this structure include pseudoboehmite (Al(OOH)) and bayerite (Al(OH)$_3$), the latter producing better-quality crystalline products. Other phases that have been observed with AlPO$_4$-11 include VPI-5, AlPO$_4$-31, AlPO$_4$-41, and AlPO$_4$-C. Batch compositions that produce quality AlPO$_4$-11 include:

1.0 Pr$_2$NH : 1.0 Al$_2$O$_3$: 1.0 P$_2$O$_5$: 40 H$_2$O

with an aging of the gel at 90°C for 24 hours prior to crystallization at 200°C (*Zeo.* 8:183(1988)). AlPO$_4$-11 also readily crystallizes at 150°C over one to two days. Large needle-shaped crystals greater than 500 microns in length have been prepared from clear aluminophosphate solutions in the presence of HF (Al:F ca. 1:1) and excess phosphoric acid, using dipropylamine as the organic additive.

Thermal Properties

Infrared Spectra

Infrared Spectral Data for AlPO$_4$-11 (from author): (as synthesized) cm^{-1}: 730, 705, 620, 585, 545, 462; (calcined 550°C) cm^{-1}: 712, 695, 660, 620, 590, 498, 470.

Adsorption

Adsorption data for AlPO$_4$-11:

Adsorbate	Kinetic diameter, Å	Pressure, Torr	Temp., °C	Wt. % adsorbed
O$_2$	3.46	101	−183	9.22
O$_2$		755	−183	10.7
n-Butane	4.3	304	24	4.35
Isobutane	5.0	502	24	4.71
Neopentane	6.2	301	24	1.22
Cyclohexane	6.0	30	24	5.30
H$_2$O	2.65	4.6	24	11.8
H$_2$O		20.0	24	16.4

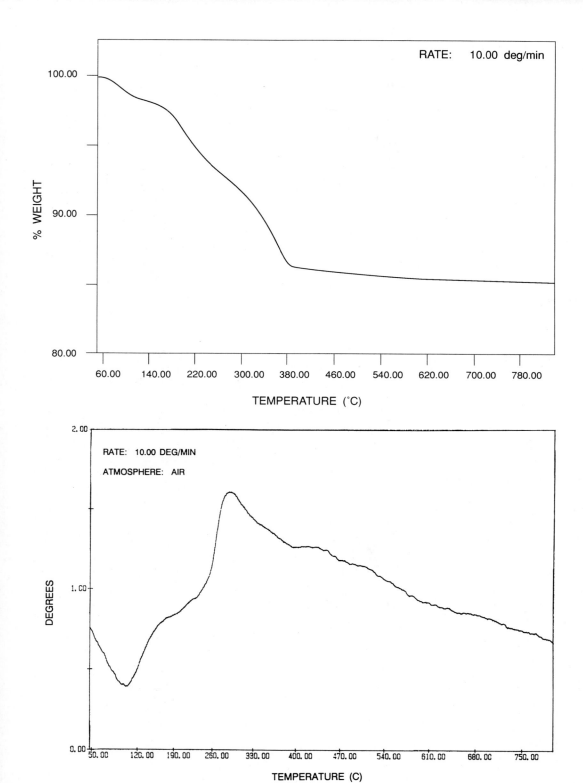

AlPO₄-11 Figs. 3a and 3b: TGA. (Fig. 3a) /DTA. (Fig. 3b) AlPO₄-11 from di-*n*-propylamine system. (Reproduced with permission of K. Vinje, U. of Oslo)

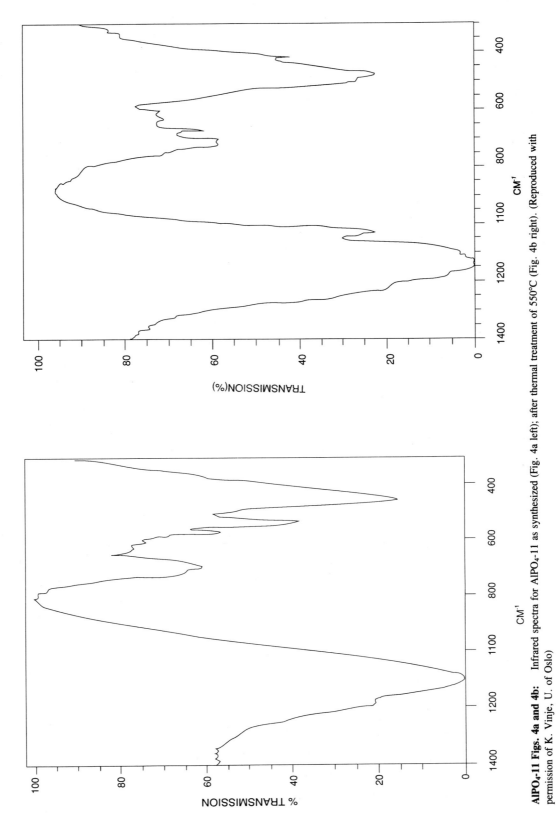

AlPO₄-11 Figs. 4a and 4b: Infrared spectra for AlPO₄-11 as synthesized (Fig. 4a left); after thermal treatment of 550°C (Fig. 4b right). (Reproduced with permission of K. Vinje, U. of Oslo)

AlPO$_4$-12
(US4,310,440(1982))
Union Carbide Corporation

Structure

(*JCS Chem. Commun.* 1449(1984))

CHEMICAL COMPOSITION: Al$_3$P$_3$O$_{11}$(OH)$_2$N$_2$C$_2$H$_8$
SYMMETRY: monoclinic
SPACE GROUP: P2$_1$/d
UNIT CELL CONSTANTS (Å):   (*JCS Chem. Commun,*
1449(1984))
$a = 14.54$
$b = 9.43$
$c = 9.63$
$\beta(°) = 98.21$

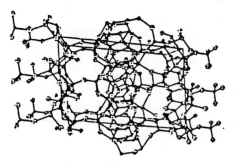

AlPO$_4$-12 Fig. 1S: Framework topology.

2.16(7), 2.12sh, 2.08(5), 2.04(5), 2.04(5),
1.99(4), 1.95(7), 1.89(9), 1.87(5), 1.82(5),
1.76(7), 1.74(4), 1.68(9), 1.66(2).

X-Ray Powder Diffraction Data: (d(Å) (I/I_0)
14.37(25), 7.90(7), 7.20(22), 6.66(56), 6.28(5),
5.83(4), 5.16(20), 4.80(27), 4.70(13), 4.46sh,
4.25(100),   3.99(44),   3.86(31),   3.73(55),
3.55(35), 3.36(31), 3.34sh, 3.19sh, 3.16(20),
3.08(11), 3.00(25), 2.94sh, 2.90(44), 2.87sh,
2.75(7), 2.71(5), 2.70sh, 2.64(18), 2.58(18),
2.38(4), 2.33(11), 2.29(9), 2.25(7), 2.22(9),

The structure of AlPO$_4$-12 is open, consisting of sheets of four-coordinate phosphorus alternating with aluminum, [AlPO$_4$(OH)]$^-$ alternating along [100] with a layer of four-coordinate phosphorus linked with five-coordinate aluminum [Al$_2$P$_2$O$_7$(OH)]$^+$. This configuration results in channels in the [010] and [001] direction. The layers are linked via Al–O–P, and the organic, ethylenediamine, is positioned between

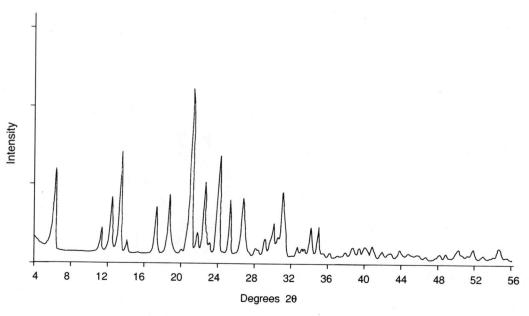

Degrees 2θ

AlPO$_4$-12 Fig. 2:   X-ray powder diffraction pattern of AlPO$_4$-12 (*Zeo.* 6:349 (1986)). (Reproduced with permission of Butterworth Publishing Company)

them. It is hydrogen-bonded to the O(1) hydroxyl group via N–H . . . O(1) [H . . . O(1) 2.0Å] and N–H . . . O(1) [H . . . O(1) 2.17 Å] bridges (*JCS Chem. Commn.* 1449(1984)).

Synthesis

ORGANIC ADDITIVES
ethylenediamine
2-imidazolidone

This material is prepared from orthophosphoric acid (H_3PO_4) and the pseudoboehmite phase of hydrated aluminum oxide. Ethylenediamine or 2-imidazolidone has been used to induce the crystallization of this structure. The reaction mixture has a molar oxide ratio of:

$$0.5 \ C_2H_8N_2 : Al_2O_3 : P_2O_5 : 40 \ H_2O$$

Crystallization occurs within 24 hours at 200°C.

AlPO₄-12-TAMU (ATT)
(*J. Phys. Chem.* **90:6122(1986)**)

Related Materials: AlPO₄-33

Structure

CHEMICAL COMPOSITION: $(AlPO_4)_3(CH_3)NOH$
SYMMETRY: orthorhombic
SPACE GROUP: $P2_12_12$
UNIT CELL CONSTANTS (Å): $a = 10.3325(1)$
$b = 14.6405(2)$
$c = 9.5112(1)$
DENSITY (g/cm³):   2.019
PORE STRUCTURE: eight-member rings

See AlPO₄-33 for X-ray Powder Diffraction Data.

AlPO₄-12-TAMU consists of a structure based on four-, six-, and eight-member rings. In the *xz* direction there is alternation of four- and eight-rings, while in the *yz* plane rows of four- and six-rings alternate. In the *xz* direction, four- and eight-rings alternate. The large cavities in AlPO₄-12-TAMU are the 12-hedron type. The organic amine sits in the center of each of the eight-ring cavities. There are two independent tetramethyl-

ammonium hydroxide molecules per asymmetric unit.

Final atomic and thermal parameters for $(AlPO_4)_3(CH_3)_4NOH$ (*J. Phys. Chem.* 90:6122 (1986)):

	x	y	z
P1	0.339(6)	0.770(3)	0.157(4)
P2	0.354(5)	0.112(3)	0.368(5)
P3	0.343(5)	0.415(3)	0.369(5)
Al1	0.354(6)	0.608(3)	0.387(6)
Al2	0.366(6)	0.896(3)	0.383(5)
Al3	0.349(7)	0.249(3)	0.139(5)
O1	0.338(9)	0.696(4)	0.272(7)
O2	0.311(9)	0.864(4)	0.218(7)
O3	0.482(7)	0.758(6)	0.086(5)
O4	0.249(9)	0.764(7)	0.018(8)
O5	0.318(9)	0.137(5)	0.210(7)
O6	0.370(6)	0.012(4)	0.381(8)
O7	0.485(7)	0.168(4)	0.412(7)
O8	0.244(8)	0.143(5)	0.466(8)
O9	0.300(7)	0.506(4)	0.317(7)
O10	0.337(10)	0.349(5)	0.242(7)
O11	0.482(6)	0.412(4)	0.425(8)
O12	0.254(8)	0.381(5)	0.488(8)
N1	0.0	0.0	0.100(12)
C1	−0.024(32)	0.089(4)	0.196(12)
C2	−0.131(5)	−0.017(2)	0.004(12)
O13	0.0	0.0	0.284(15)
N2	0.0	0.5	0.158(12)
C3	−0.018(26)	0.587(4)	0.249(12)
C4	−0.129(6)	0.482(18)	0.065(12)
O14	0.0	0.5	−0.024(15)

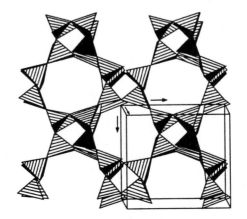

AlPO₄-12-TAMU Fig. 1S: Framework topology.

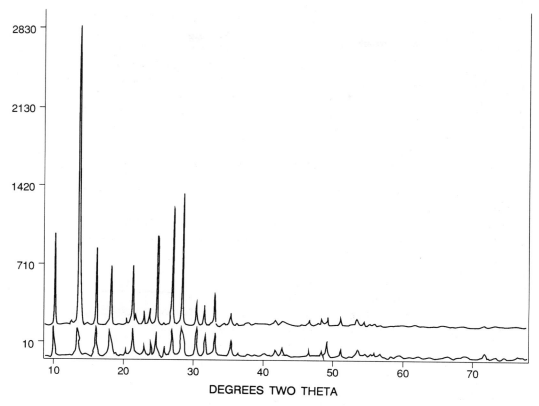

AlPO₄-12-TAMU Fig. 2: Rietveld refinement plot of AlPO₄-12-TAMU (*J. Phys. Chem.* 90: 6122 (1986)). (Reproduced with permission of the American Chemical Society)

Synthesis

ORGANIC ADDITIVE
tetramethylammonium⁺ (TMA)

AlPO₄-12-TAMU was prepared from aluminum chloride/ammonia/phosphoric acid solutions containing $(CH_3)_4NOH$ with the following batch composition:

$$2TMAOH : Al_2O_3 : P_2O_5 : 252 H_2O$$

Direct crystallization occurred after 7 days at 200°C.

AlPO₄-14
(US4,310,440(1982))
Union Carbide Corporation

Related Materials: GaPO₄-14

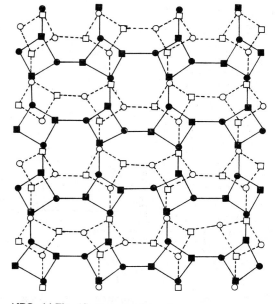

AlPO₄-14 Fig. 1S: Framework topology.

Structure

Chemical Composition: 0.49 *i*-PrNH$_2$: 1.00 Al$_2$O$_3$: 1.00 P$_2$O$_5$: 1.03 H$_2$O.

X-Ray Powder Diffraction Data: (as synthesized with *t*-BuNH$_2$) (d(Å) (I/I_0) 9.61(100), 9.41(sh), 7.90(18), 6.76(17), 6.61(sh), 5.98(3), 5.61(23), 4.93(12), 4.72(1), 4.62(1), 4.25(5), 4.11(10), 4.00(22), 3.92(36), 3.80(1), 3.75(1), 3.54(2), 3.41(20), 3.29(9), 3.22(2), 3.13(5), 3.03(12), 2.96(8), 2.93(sh), 2.89(4), 2.87(sh), 2.80(1), 2.76(1), 2.70(sh), 2.67(6), 2.63(1), 2.56(1), 2.53(2), 2.47(3), 2.36(3), 2.32(5), 2.28(1), 2.23(5), 2.21(1), 2.15(2), 2.13(2), 2.08(1), 2.06(3), 2.02(2), 1.98(2), 1.96(2), 1.90(2), 1.88(2), 1.83(1), 1.77(2), 1.74(1), 1.71(4), 1.65(2); (after thermal treatment at 550°C) d(Å) (I/I_0) 9.83(100), 9.21sh, 7.69(29), 7.38(9), 6.71(45), 6.66sh, 6.51(32), 6.19(10), 5.99(5), 5.54(10), 5.47sh, 5.28(2), 4.87(35), 4.77(38), 4.70sh, 4.46(5), 4.29(9), 4.25(9), 4.11sh, 4.04(23), 3.95(29), 3.90sh, 3.83(7), 3.79(6), 3.68(3), 3.59sh, 3.55(6), 3.44(3), 3.35(12), 3.30(20), 3.25(16), 3.21(11), 3.19(11), 3.12sh, 3.06(11), 3.01(20), 2.94(14), 2.88(9), 2.77(4), 2.71(6), 2.64(4), 2.58(4), 2.52(2), 2.44(5), 2.39(3), 2.34(2), 2.30(3), 2.25(3), 2.22(3), 2.16(1), 2.12(3), 2.10(2), 2.05(6), 2.00(1), 1.95(1), 1.92(2), 1.87(3), 1.82(3), 1.76(3), 1.70(4).

See GaPO$_4$-14 for a description of the framework topology.

Synthesis

ORGANIC ADDITIVES
t-C$_4$H$_9$NH$_2$
i-C$_3$H$_7$NH$_2$

From aluminophosphate gels containing *t*-BuNH$_3$ in a ratio:

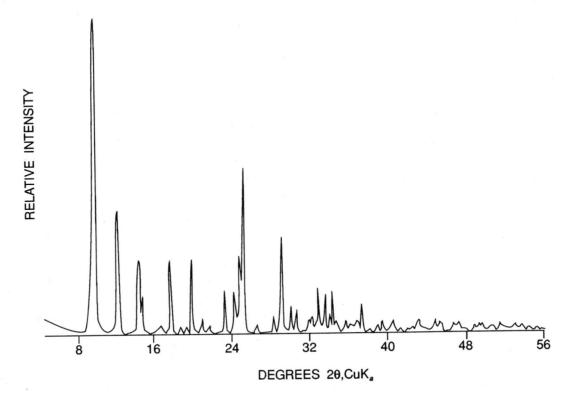

AlPO$_4$-14 Fig. 2: X-ray powder diffraction pattern of AlPO$_4$-14 (*Zeo.* 6: 349 (1986)). (Reproduced with permission of Butterworth Publishing Company)

1.0 *t*-BuNH₂ : Al₂O₃ : P₂O₅ : 40 H₂O

the AlPO₄-14 structure crystallizes after 96 hours at 150°C.

Adsorption

Adsorption properties of AlPO₄-14:

Adsorbate	Kinetic diameter, Å	Pressure, Torr	Temp., °C	Wt. % adsorbed
O₂	3.46	102	−183	15.53
O₂		763	−183	21.56
n-Hexane	4.3	45	26	0.25
Neopentane	6.2	499	24	0.37
H₂O	2.65	4.6	24	21.46
H₂O		21.0	24	28.66
N₂	3.64	100	−196	11.28
N₂		747	−196	14.99

AlPO₄-14A
(*Acta Crystallogr.* C43:866(1987))

Structure

Chemical Composition: Al₇P₇O₂₈(OH)₂*2C₃H₇NH₂

X-Ray Powder Diffraction Data: (*Zeo.* 11:477(1991)) (*d*(Å) (*I/I₀*)) 12.36(100), 11.94(48), 6.98(94), 6.18(41), 6.00(6), 5.56(1), 5.12(21), 4.71(2), 4.61(6), 4.56(5), 4.35(19), 4.16(20), 4.12(53), 4.00(67), 3.98(68), 3.92(5), 3.81(1),[a] 3.72(11), 3.66(9), 3.62(16), 3.50(13), 3.40(6), 3.23(62), 3.21(60), 3.17(12), 3.07(12), 3.04(24), 2.95(4), 2.84(7), 2.80(2), 2.73(4), 2.71(6), 2.67(20), 2.64(5), 2.27(9), 2.23(5), 2.20(1), 2.16(3) (a—peak that could be common with KBr).

Crystallographic experimental conditions and results (*Zeo.* 11:477 (1991); *Acta Crystallogr.* C43:866 (1987)):

space group	*a* (Å)	*b* (Å)	*c* (Å)	β (°)	V (Å³)
C2/c	24.085(1)	14.393(1)	8.7122(4)	94.260(4)	3012
monoclinic	24.085(2)	14.409(2)	8.721(1)	94.318(8)	3018

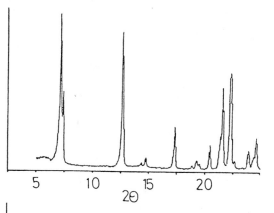

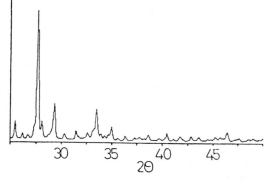

AlPO₄-14A Fig. 1:   X-ray powder diffraction pattern of as-synthesized AlPO4-14A (*Zeo.* 11:477 (1991)). (Reproduced with permission of Butterworth Publishing Co.).

Synthesis

ORGANIC ADDITIVE
i-C₃H₇NH₂

AlPO₄-14A crystallizes from a reactive gel with the batch composition:

Al₂O₃ : P₂O₅ : 2 *i*-C₃H₇NH₂ : 0.5HF : 100 H₂O

Crystallization occurs after 1 hour of aging at 20°C followed by 200°C for 24 hours. Pseudoboehmite was used as the source of aluminum in this crystallization (*Zeo.* 11:477(1991)).

Thermal Properties

The thermal decomposition occurs in two stages of weight loss: between 25 and 300°C and be-

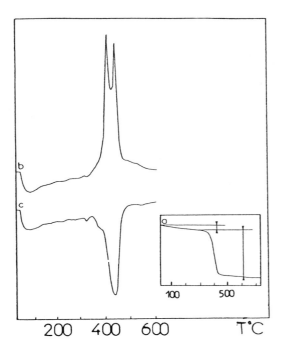

AlPO₄-14A Fig. 2: Thermal analysis of AlPO₄-14A. (a) TG in argon atmosphere; (b) DTA in air; (c) DTA in argon (*Zeo.* 11:477 (1991)). (Reproduced with permission of Butterworth Publishing Co.)

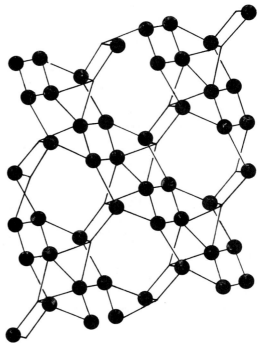

AlPO₄-15 Fig. 1S: Framework topology.

to water loss and the second to decomposition of the organic species. The structure is lost at temperatures around 400°C (*Zeo.* 11:477(1991)).

AlPO₄-15
(*Acta Crystallogr.* C40:2008(1984))

Related Materials: Leucophosphite
GaPO₄*2H₂O

Structure

(*Acta Crystallogr.* C40:1641(1984))

CHEMICAL COMPOSITION: Al₂[NH₄](OH) (PO₄)₂*2H₂O
SYMMETRY: monoclinic
SPACE GROUP: P2₁/n
UNIT CELL CONSTANTS (Å): $a = 9.55$
$b = 9.58$
$c = 9.61$
VOLUME: 855.1 Å³
DENSITY (D$_x$): 2.45 g/cm³
PORE SYSTEM: very small pores

X-Ray Diffraction Lines: (from author) (d(Å) (I/I_0)) 7.57(42), 6.70(100), 5.95(35), 5.15(3), 4.66(7), 4.25(5), 4.19(6), 4.06(2), 3.23(35), 3.52(2), 3.34(3), 3.08(3), 3.03(9), 2.97(7), 2.84(6), 2.80(10), 2.66(6), 2.63(4), 2.52(26).

This structure consists of columns of Al-centered corner- and edge-shared octahedra linked via PO₄ tetrahedra to provide channels approximately parallel to the *b* axis. The ammonium cations in the composition occupy these channels and are hydrogen-bonded to framework oxygen atoms and water molecules (*Acta Crystallogr.* C40:1641(1984)).

The phosphate tetrahedra are only slightly distorted, and the average P–O distances (1.53(2) and 1.53(1) Å for P(1) and P(2)) and angles (109(1)°) are similar to those reported for other dihydrated aluminophophates such as variscite and metavariscite. The Al-centered octahedra are significantly distorted. These distortions are more pronounced than those found for GaPO₄*2H₂O (GaPO₄-14).

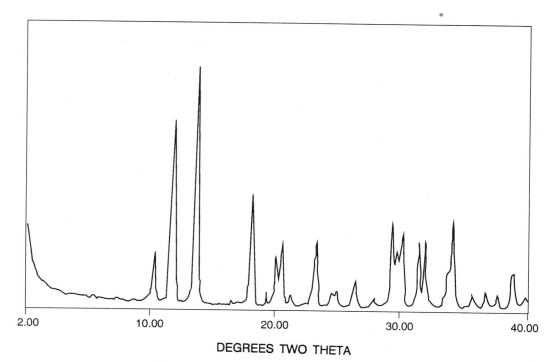

X-ray diffraction pattern of AlPO₄-15. (Reproduced with permission of author)

Atomic positions of AlPO₄-15 excluding hydrogen atom positions (*Acta Crystallogr*. C40:2008 (1984)):

	x	y	z
P(1)	0.34830(2)	0.53183(2)	0.69853(2)
P(2)	0.13673(2)	0.30797(2)	0.29463(2)
Al(1)	0.37209(3)	0.53719(3)	0.38618(3)
Al(2)	0.31503(3)	0.22986(3)	0.60731(3)
O(1)	0.29674(6)	0.58593(6)	0.54355(6)
O(2)	0.19438(7)	0.12742(7)	0.69302(7)
O(3)	0.48683(6)	0.48033(6)	0.26314(6)
O(4)	0.28440(7)	0.38569(6)	0.70988(7)
O(5)	0.16583(7)	0.26792(7)	0.45379(6)
O(6)	0.20999(6)	0.44859(6)	0.27945(6)
O(7)	0.30331(7)	0.69867(6)	0.29189(7)
O(8)	0.47629(6)	0.17560(7)	0.73909(7)
O(9)	0.44566(6)	0.36468(6)	0.50306(6)
O(10)	0.35324(8)	0.06688(8)	0.49997(8)
O(11)	0.53720(9)	0.14339(9)	0.31701(10)
N	0.10388(10)	0.80953(11)	0.48671(10)

Synthesis

ORGANIC ADDITIVES
NH₄⁺

This aluminophosphate structure does not require organic additives to promote crystallization. Synthesis has been successful in the t-C₄H₉NH₂ and i-C₃H₇NH₂ and diaminobutane–containing gel systems; however, the ammonia decomposition product of these organic amines appeared as the stronger structure promoter in this system. Rapid crystallization is observed in the absence of organic additives when only NH₃ (NH₄⁺) is added. Crystallization occurs between 150 and 200°C within several hours to several days. Large well-formed crystals result (from author).

Thermal Properties

A two-step weight loss is observed upon heating AlPO₄-15. Between 60 and 200°C, 5% of the volatile material is lost, and 14% additional decrease in weight is observed between 200 and 350°C.

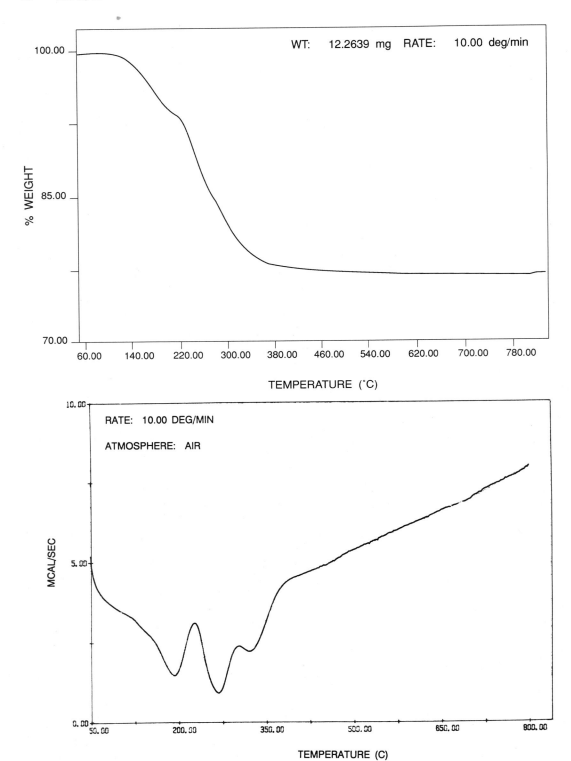

AlPO$_4$-15 Figs. 3a and 3b: TGA, (Fig. 3a top) /DTA. (Fig. 3b bottom) Of AlPO$_4$-15. (Reproduced with permission of A. Long, Georgia Tech)

Infrared Spectra

Mid-infrared Vibrations (cm^{-1}): 1200w, 1115s, 1070m, 1020s, 890m, 665m, 610m, 585w, 542wsh, 512m, 462w, 440w (from author).

Adsorption

Adsorption properties of AlPO$_4$-15 (from author):

Adsorption properties of AlPO$_4$-15 (from author):

		Kinetic diameter (Å)	Pressure, Torr	Temp., °C	Wt% adsorbed
AlPO$_4$-15	O$_2$	3.46	755	− 183	3.0

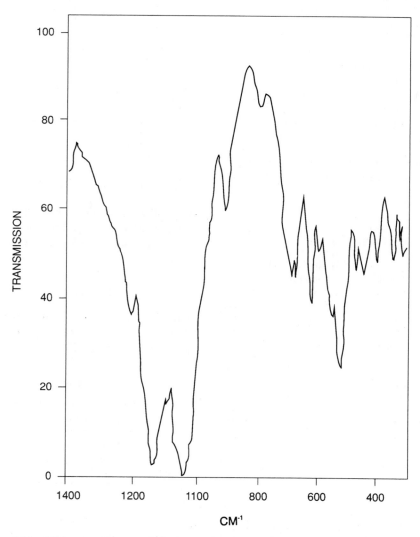

AlPO$_4$-15 Fig. 4:   Infrared spectrum of as-synthesized ALPO$_4$-15. (Reproduced with permission of K. Vinje, U. of Oslo)

AlPO₄-16 (AST)
(US4,310,440(1982))
Union Carbide Corporation

Structure

(*Zeo.* 11:503(1991))

CHEMICAL COMPOSITION: $[Al_{20}P_{20}O_{80}]*4R*16H_2O$
  (R = 1,4-ethylene piperidine)
SYMMETRY: cubic
SPACE GROUP: F23
UNIT CELL CONSTANTS (Å): $a = 13.4$
PORE STRUCTURE: entrance through six-member
  rings only

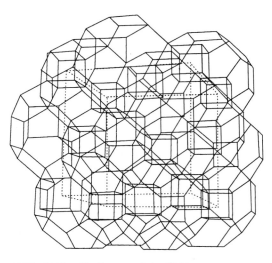

AlPO₄-16 Fig. 1S: Framework topology.

X-Ray Powder Diffraction Pattern: (d(Å))

(I/I_0)) 7.83(62), 5.72(2), 5.13(2), 4.75(50), 4.06(100), 3.875(9sh), 3.357(27), 3.23(2), 3.19(2), 3.08(12), 3.00(28), 2.735(4), 2.585(5), 2.374(8), 2.276(2), 2.049(2), 1.877(6), 1.746(3), 1.675(3).

ORGANIC ADDITIVE
quinuclidine

AlPO₄-16 is prepared from a gel composition with a molar oxide ratio of:

1.0 quinuclidine : Al_2O_3 : P_2O_5 : 40 H_2O

Crystallization occurs after 24 to 48 hours at temperatures between 150 and 200°C. At the elevated temperature a small amount of AlPO₄-17 also is formed.

Final positional parameters of AlPO₄-16 (*Zeo.* 11:503 (1991)):

Atom	x	y	z
Al1	0.1186(17)	0.1186	0.1186
P1	−0.1116(13)	0.1116	0.1116
P2	0.25	0.25	0.25
Al2	−0.25	0.25	0.25
O1	−0.0078(27)	0.1285(13)	0.1495(13)
O2	0.1915(18)	0.1915	0.1915
O3	−0.1704(21)	0.1704	0.1704
N1	0.0560	0.4329	0.0424
C1	0.0056	0.3916	−0.0462
C2	−0.0601	0.4680	−0.0973
C3	0.1159	0.5193	0.0103
C4	0.0529	0.5988	−0.0393
C5	−0.0214	0.4589	0.1118
C6	−0.0878	0.5472	0.0646
C7	−0.0552	0.5662	−0.0418

Parameters without estimated standard deviations in the least significant figures (given in parentheses) were not refined. The population parameter is 1.0 for each framework atom (Al, P, or O) and 0.0833 for each quinuclidine atom.

Thermal Properties

After thermal treatment for several hours at 500°C, no change in the X-ray diffraction pattern is observed.

Adsorption

Adsorption properties of AlPO₄-16:

Adsorbate	Kinetic diameter, Å	Pressure, Torr	Temp., °C	Wt. % adsorbed
O₂	3.46	101	−183	1.2
O₂		755	−183	11.6
n-Butane	4.3	768	24	2.0
Neopentane	6.2	301	25	1.4
H₂O	2.65	4.6	24	19.0
H₂O	2.65	20	24	36.3

The framework topology of AlPO₄-16 can be described as an idealized zunyite framework. It is built from double four-rings. The largest pore opening is a six-member ring.

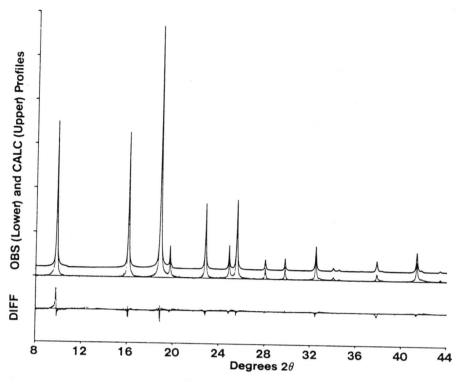

AlPO₄-16 Fig. 2:   Results of the Rietveld refinement for as-synthesized AlPO4-16 using synchrotron data (wavelength 1.32365 Å). The upper lines are the calculated profile (smooth line) and the experimental powder diffraction profile. Below these, the difference between the experimental and the calculated profile is shown (*Zeo.* 11:503 (1991)). (Reproduced with permission of Butterworth Publishing Co.)

AlPO₄-17 (ERI)
(US4,310,440(1982))
Union Carbide Corporation

Related Materials: erionite

Structure

(*Acta Crystallogr.* C42:283(1986))

CHEMICAL COMPOSITION: $Al_{18}P_{18}O_{72}$*
   $4(C_5H_{11}H*H_2O)$
SYMMETRY: hexagonal
SPACE GROUP P6₃/m
UNIT CELL CONSTANTS (Å) a = 13.24
                          c = 14.77
VOLUME: 2241.4 Å³
DENSITY: 1.93 g/cm³
PORE STRUCTURE: eight-member-ring pore openings

X-Ray Powder Diffraction Data: (as synthesized) (d(Å) (I/I_0)) 11.55(100), 9.12(41), 7.79(2),

7.44(3), 6.63(39), 6.24(13), 6.03(2), 5.75(62), 5.34(33), 4.93(20), 4.72(2), 4.53(67), 4.33(93), 4.15(50), 4.02(2), 3.95(15), 3.82(34), 3.74(39), 3.68(3), 3.52(55), 3.38(35), 3.26(20), 3.18(5), 3.11(20), 2.92(17), 2.87(29), 2.81(68), 2.76(2), 2.67(18), 2.64(2), 2.59(2), 2.55(2), 2.50(7), 2.47(6), 2.44(2), 2.41(2), 2.37(3), 2.29(2), 2.26(7), 2.23(4), 2.19(3), 2.14(3), 2.11(1), 2.08(9), 2.04(2), 1.98(5), 1.96(1), 1.95(4),

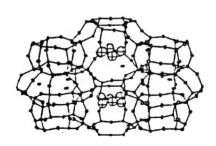

AlPO₄-17 Fig. 1S: Framework topology.

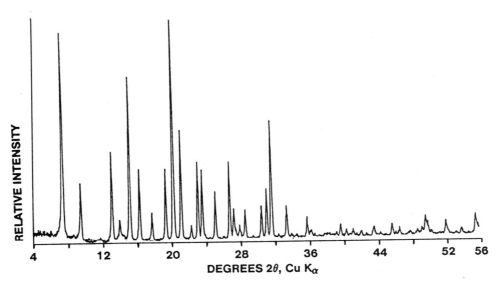

AlPO$_4$-17 Fig. 2:   X-ray powder diffraction pattern of ALPO$_4$-17 (as-synthesized with quinuclidine) (*ACS Sym. Ser.* 218:79 (1983)). (Reproduced with permission of the American Chemical Society)

1.90(3), 1.87(2), 1.85(8), 1.84(11), 1.82(2), 1.78(1),, 1.76(11), 1.72(1), 1.70(5); (after 550°C calcination) 11.5(77), 9.1(46), 7.63–7.47(13), 6.58(100), 6.33(21), 6.24(28), 5.99(2), 5.68(16), 5.28(6), 5.22(3), 4.62(5), 4.52(12), 4.33–4.27(42), 4.11(22), 4.00(2), 3.93(2), 3.76–3.72(29), 3.65–3.63(22), 3.56(7), 3.52(12), 3.40(7), 3.26(20), 3.16(15), 3.10(10), 3.038–3.018(7), 2.969(7), 2.903(7), 2.849(24), 2.805–2.788(15), 2.660(11), 2.550(1), 2.488–2.471(7), 2.281(1), 2.186(3), 2.151(2), 2.118(4), 2.067(1), 2.019(1), 1.973(1), 1.953(2), 1.972(3), 1.895(2), 1.852(2), 1.817(7), 1.781(7), 1.734(2), 1.698(1), 1.661(2).

The X-ray powder diffraction pattern of this molecular sieve is very similar to that of natural erionite. Differences in intensity are observed between the two compositions, attributable to Al-P ordering and to scattering contributions by the cations in the zeolite structure.

The structure of calcined AlPO$_4$-17 has the erionite framework topology with alternating Al and P in the T atom positions. The as-synthesized AlPO$_4$-17 contains a modified erionite framework with the organic, piperidine, trapped within the pores. Extra nonframework oxygen also is located in the structure and is attributed to the presence of hydroxide anions of a piperidinium cation. Such a complex appears rea-sonable because of the basicity of the piperidine molecule. The hydroxide also interacts with one of the framework aluminum cations, producing a distorted tetrahedral arrangement around that aluminum. Disorder is observed in the structurally trapped piperidine. Elemental analysis shows that there are four organic molecules per unit cell.

Atomic populations positions, and displacements of AlPO$_4$-17 (*Acta Crystallogr.* C42:283 (1986)) (fractional coordinates are multiplied by 10^4):

	x	y	z
P(1)*	9989(2)	2369(2)	938(2)
P(2)	5675(3)	9079(3)	2500
Al(1)	7708(2)	9983(2)	1180(2)
Al(2)	999(3)	4214(3)	2500
O(1)	365(6)	3435(5)	1529(4)
O(1')	6373(5)	9669(5)	1649(4)
O(2)	826(5)	1910(5)	1053(5)
O(3)	1432(5)	2640(6)	6244(4)
O(4)	2773(6)	30(6)	9970(4)
O(5)	2468(7)	4693(7)	2500
O(6)	4584(7)	9171(4)	2500
O(7)$^\$$	1721(12)	1876(12)	2500
C(1)$^{\$\$}$	3770	7930	9599
C(2)$^{\$\$}$	4590	7490	9920

*In the erionite structure P(1), Al(1); P(2), Al(2); O(1), O(1') are equivalent.
$^\$$Only 4.14(14) of the six sites are occupied.
$^{\$\$}$C(1) and C(2) are disordered (fixed in refinement).

Synthesis

ORGANIC ADDITIVES
1-azabicyclo[2,2,2]octane (*Proc. 6th IZC* 97(1983))
2,2-dimethylpropylamine (*Proc. 6th IZC* 97(1983))
cyclohexylamine (*Proc. 6th IZC* 97(1983))
piperidine (*Proc. 6th IZC* 97(1983))

AlPO$_4$-17 synthesis is sensitive to both crystallization time and temperature. AlPO$_4$-16 is a common impurity when quinuclidine is used, and AlPO$_4$-5 will crystallize in the presence of cyclohexylamine. Changing the organic requires a modification of the temperature needed to induce crystallization of this structure. In the presence of neopentylamine, a 150°C cystallization temperature is required, whereas 200°C is needed when cyclohexylamine and piperidine are used. Temperatures up to 250°C have been used to generate this structure. Crystallization occurs after 168 hours (US4,310,440(1982)).

Thermal Properties

AlPO$_4$-17 is both thermally and hydrothermally stable. Weight loss with increasing temperature is due to burn-off of the organic additive used in the synthesis.

Adsorption

Adsorption properties of AlPO$_4$-17:

	Kinetic diameter, Å	Pressure, Torr	Temp., °C	Wt. % adsorbed
O$_2$	3.46	101	−183	22.2
O$_2$		724	−183	23.1
n-Hexane	4.3	45	23	7.7
Isobutane	5.0	101	23	0.2
Neopentane	6.2	308	23	9.3
H$_2$O	2.65	4.6	22	24.9
H$_2$O		18	22	27.8

AlPO$_4$-18 (AEI)
(US4,310,440(1982))
Union Carbide Corporation

Related Materials: metal-substituted aluminophosphates

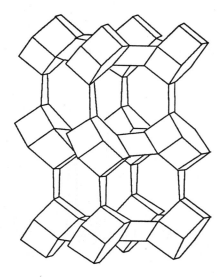

AlPO$_4$-18 Fig. 1S: Framework topology.

Structure

(*BZA Meeting Abstract,* 1990)

CHEMICAL COMPOSITION:
  (US4,310,440(1982)): 0.14 (TEA)20 : 1.00 Al$_2$O$_3$: 0.8 P$_2$O$_5$: 0.9 H$_2$O (TEA-Tetraethylammonium)
SYMMETRY: monoclinic
SPACE GROUP: C2/c
UNIT CELL CONSTANTS: (Å) a = 13.711
                          b = 12.731
                          c = 18.570
                          β = 90.01°
PORE STRUCTURE: three-dimensional eight-member rings

X-Ray Powder Diffraction Data: (d(Å) (I/I_0))
9.21(100), 8.47(8), 8.04(9), 6.76(6), 6.33(8), 5.99(10), 5.72(27), 5.25–5.22(61), 4.96(20), 4.60–4.55(17), 4.41(35), 4.24(45), 4.02–3.99(17), 3.82(5), 3.73(6), 3.65(14), 3.58(9), 3.51(6), 3.41(13), 3.37–3.33(12), 3.19(16), 3.08(7), 2.98(20), 2.91(14), 2.86(14), 2.81–2.76(24), 2.68(5), 2.60(3), 2.51–2.48(3), 2.36(2), 2.24(1), 2.17(3), 2.11(5), 1.90(3), 1.87(2), 1.84(4), 1.79(4), 1.76(2), 1.69(4), 1.67(2); (after thermal treatment to 600°C) (d(Å) (I/I_0)) 9.31(100), 8.85sh, 8.51sh, 8.35(14), 7.83(4), 6.83(9), 6.56(8), 6.11(4), 5.50(11), 5.22(18), 5.16(17), 4.65(5), 4.51(7), 4.44(6), 4.29(12), 4.17(15), 4.06(5), 3.95(8), 3.88(9), 3.72(12), 3.66sh, 3.58(3), 3.53(3), 3.47sh,

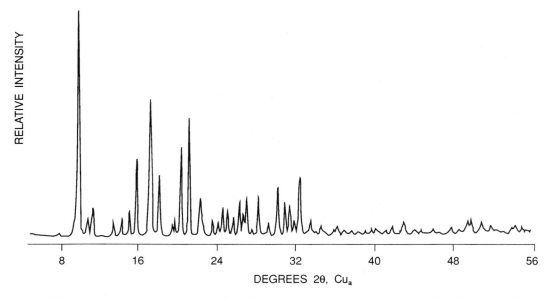

AlPO₄-18 Fig. 2:   X-ray powder diffraction pattern for AlPO₄-18, as-synthesized (*Proc. Sixth Int. Zeo. Conf.* 97 (1984)). (Reproduced with permission of Butterworth Publishing Company)

3.43(8), 3.39(7), 3.29(4), 3.21(7), 3.07(7), 2.98(8), 2.94sh, 2.88(13), 2.83(5), 2.78(7), 2.57sh, 2.71sh, 2.66(4), 2.59(3), 2.43(2), 2.32(2), 2.09(2), 1.86(4), 1.66(3), 1.65(2).

AlPO₄-18 exhibits properties characteristic of a small-pore molecular sieve.

Synthesis

ORGANIC ADDITIVES
$(C_2H_5)_4N^+$ (TEA)

From a batch composition of Al_2O_3 : 1.5–2.5 P_2O_5 : 2.0–4.5 TEAOH : 60–1200 H_2O,

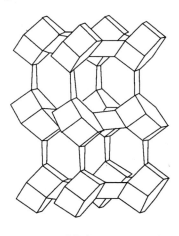

AlPO₄-18

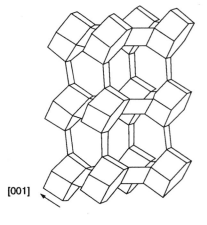

[001]

CHA

AlPO₄-18 Fig. 3:   A comparison between the structure of AlPO₄-18 and chabazite (CHA). (*Proc. Sixth Int. Zeo. Conf.* 97 (1984)). (Reproduced with permission of Butterworth Publishing Company)

this structure will form within the temperature range of 130°C to 200°C (*PP 8th IZC* 21(1989)). The product has an Al/P ratio of 1.35. Addition of HCl does not appear to hinder crystallization (US4,310,440(1982)).

Thermal Properties

Upon thermal treatment to 600°C, the X-ray diffraction pattern changes. Reversible changes occur with dehydration and rehydration of this molecular sieve (*PP 8th IZC* 21(1989)).

Adsorption

Adsorption properties of AlPO₄-18 (US 4,310,440 (1982)):

	Kinetic diameter, Å	Pressure, Torr	Temp., °C	Wt. % adsorbed
O₂	3.46	130	−183	23.0
O₂		697	−183	27.9
n-Butane	4.3	718	24	16.2
Isobutane	5.0	101	24	0.1
H₂O	2.65	4.6	24	30.3
H₂O		21.0	24	36.9

AlPO₄-20 (SOD)
(US4,310,440(1982))
Union Carbide Corporation

Related Materials: sodalite

Structure

(*JACS* 104:1146(1982))

CHEMICAL COMPOSITION: 0.20TMA₂O*Al₂O₃* 0.98P₂O₅*2.17H₂O (TMA = tetramethylammonium)
SPACE GROUP: PP43n
UNIT CELL CONSTANT (Å): *a* = 8.9
PORE STRUCTURE: very small, six-ring openings only

X-Ray Powder Diffraction Data: (as-synthesized) (*d*(Å) (*I/I₀*)) 6.326(31), 4.462(44), 3.986(16), 3.633(100), 3.164(25), 2.831(18), 2.585(18), 2.238(4), 2.099(5), 1.903(4), 1.752 (10); (after 600°C calcination) 6.19(100),

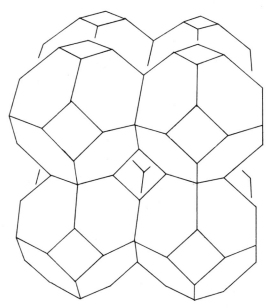

AlPO₄-20 Fig. 1S: Framework topology.

4.37(27), 3.92(9), 3.59(57), 3.14(20), 2.79(19), 2.56(13), 2.21(4), 1.74(6).

AlPO₄-20 is structurally analogous to the aluminosilicate sodalite. The four- and six-rings in this structure contain alternating tetrahedrally coordinated AlO₂ and PO₂ units. A discussion of the framework topology is found in the section "Sodalite." The TMA cations used in the synthesis of this structure are trapped within the cavities. Access to the cavities is through six-member rings only, thus limiting the adsorptive ability of this material (*JACS* 103:1146(1982)).

Synthesis

ORGANIC ADDITIVES
tetramethylammonium⁺ (TMA)
tetrapropylammonium + tetramethylammonium (*JACS* 110:2127(1988))

AlPO₄-20 crystallizes from a batch composition: 1.0 Al₂O₃ : 1.5–2.5 P₂O₅: 2.0–4.5 ROH: 60–1200 H₂O (*PP 8th IZC* 21(1989)). Pseudoboehmite (hydrated aluminum oxide) is used as a source of aluminum. The cationic amine, TMA⁺, aids the crystallization of this structure. Cystallization occurs within 71 hours at

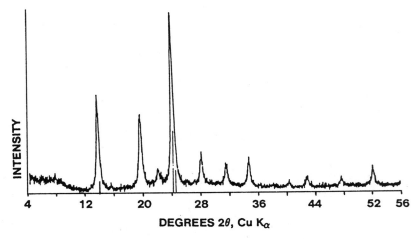

AlPO₄-20 Fig. 2:   X-ray powder diffraction pattern of AlPO₄-20 (*ACS Sym. Ser.* 218:79 (1983). (Reproduced with permission of the American Chemical Society)

150°C. A minor unidentified crystalline impurity is reported under these conditions (US4,310,440(1982)).

Thermal Properties

Crystallinity is not lost upon heating to 600°C, though a change in the X-ray diffraction pattern peak positions and intensities is noted (*PP 8th IZC* 21(1989)).

Infrared Spectrum

See AlPO₄-20 Figure 4.

Adsorption

Adsorption properties of AlPO₄-20:

	Kinetic diameter, Å	Pressure, Torr	Temp., °C	Wt. % adsorbed
O₂	3.46	97	−184	2.7
O₂		750	−183	11.5
n-Hexane	4.3	45	24	1.7
Neopentane	6.2	30.3	24	1.5
Cyclohexane	6.0	11	24	1.3
H₂O	2.65	4.6	24	22.6
H₂O		18.5	24	37.2

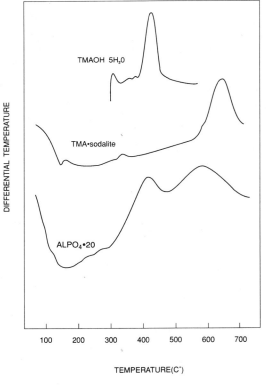

AlPO₄-20 Fig. 3a: DTA of AlPO₄-20 (see Fig. 3b).

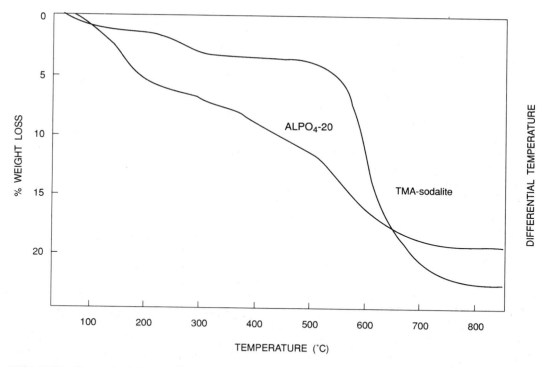

AlPO$_4$-20 Fig. 3b: TGA. Of AlPO$_4$-20. A comparison is made to TMA-sodalite (*JACS* 110:2127 (1988)). (Reproduced with permission of the American Chemical Society)

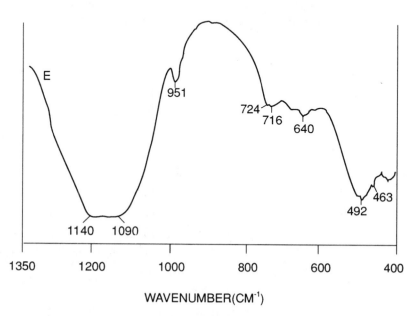

AlPO$_4$-20 Fig. 4:   Infrared spectrum of AlPO$_4$-20 (*JACS* 110:2127 (1988)). (Reproduced with permission of the American Chemical Society)

AlPO₄-21
(US4,310,440(1982))
Union Carbide Corporation

Structure

(*Inorg. Chem.* 24:188(1985); *Acta Crystallogr.* C41:515(1985))

CHEMICAL COMPOSITION: $Al_3P_3O_{32}OH*1.33N_2C_7H_{21}$
SYMMETRY: monoclinic
SPACE GROUP: $P2_1/n$

UNIT CELL CONSTANTS (Å):	TMPD	en	py
$a =$	10.33	8.47	8.67
$b =$	17.52	17.75	17.56
$c =$	8.67	9.06	9.19
$beta =$	123.4	106.7	107.75
DENSITY (D_x):	2.60	2.26	2.27 g/cm³

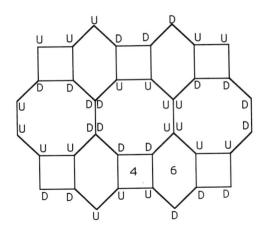

AlPO₄-21 Fig. 1S: Framework topology.

2.797(35), 2.747(23), 2.722(38), 2.637(15), 2.550(2), 2.501(6), 2.417(12), 2.380(3), 2.321(5), 2.243(7), 2.040(7), 1.910(8), 1.797(7), 1.759(9), 1.740(11), 1.722(8), 1.698(7).

X-Ray Powder Diffraction Data: $(d(Å))$ (I/I_0)
8.845(51), 8.425(7), 7.830(14), 7.500(13), 7.255(36), 6.707(6), 6.237(36), 6.067(27), 5.574(31), 5.277(6), 4.796(4), 4.529(11), 4.353(25), 4.210(25), 4.171(30), 3.934(97), 3.900(88), 3.754(23), 3.548(100), 3.453(6), 3.401(19), 3.351(53), 2.232(47), 3.164(21), 3.079(11), 3.038(3), 2.921(5), 2.885(8),

The proposed AlPO₄-21 structure contains tetrahedral phosphorus and both tetrahedral and trigonal-bipyramidal aluminum. Some unique features of this structure include a three-ring composed of two 5-coordinated Al atoms and one tetrahedrally coordinated phosphorus atom, two types of five-rings containing two and three Al atoms, four types of four-rings (two of which

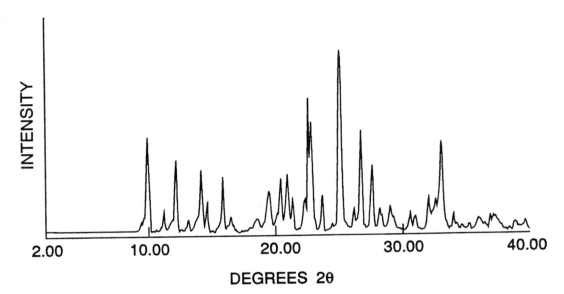

AlPO₄-21 Fig. 2:   X-ray powder diffraction pattern of AlPO₄-21. (Reproduced with permission of author)

combine into an extended double crankshaft chain), a chain of edge-sharing three- and five-rings, and a tesellation composed of three-, four-, five-, and eight-rings (*Inorg. Chem.* 24:188(1985)). The aluminum and phosphorus oxides form ribbons of edge-shared three- and five-member rings along [101], which are joined along the [010] direction via four-member rings, thus forming corrugated sheets of $Al_2P_2O_7 \cdot H_2O$ or $Al_2P_2O_7OH$. These sheets are crosslinked by crankshaft-shaped single chains of strictly alternating aluminum and phosphorus, giving an open network of channels in the (010) bounded by eight-membered ring apertures.

The organic TMPD used in the synthesis of this structure decomposes during synthesis, resulting in three triatomic species which are trapped in the structure (*Inorg. Chem.* 24:1288 (1985)). When AlPO₄-21 is synthesized using en or py, the organic remains intact and the en and py are located at the eight-ring cavities (*Acta Crystallogr.* C41:515(1985)).

Synthesis

ORGANIC ADDITIVES
pyrrolidine
trimethylamine
1,3-dimethyl piperazine
3-(di-*n*-butylamino)propylamine
N,N,N′,N′-tetramethyl-1,3-propanediamine
N,N-dimethyl-ethanolamine
n-propylamine
N,N,N′,N′-tetramethylethylenediamine
N-methyl-ethanolamine

Mixtures of pseudoboehmite and phosphoric acid and pyrrolidine with the batch composition:

$$1.0(CH_2)_4NH : Al_2O_3 : P_2O_5 : 40\ H_2O$$

are mixed and heated at 150°C for 150 hours. Isolation of the crystals via filtration and drying in air at 110°C results in the X-ray diffraction pattern shown above. Triemthylamine addition to the AlPO₄ gel results in AlPO₄-21 at 150°C. Higher temperatures (ca. 200°C) were used to prepare this material when the other amine based

Atomic positions for AlPO₄-21 (*Inorg. Chem.* 24:188 (1985)):

	x	y	z
P(1)	0.1335(1)	0.0708(1)	0.9738(2)
P(2)	0.0269(1)	0.2112(1)	0.3235(2)
P(3)	0.4967(1)	0.1646(1)	0.7462(2)
Al(1)	0.3082(1)	0.1714(1)	0.3192(2)
Al(2)	0.3372(1)	0.3982(1)	0.9739(2)
Al(3)	0.2057(1)	0.2037(1)	0.7707(2)
O(1)	0.3375(3)	0.4885(2)	0.0519(4)
O(2)	0.2331(4)	0.0912(2)	0.1807(4)
O(3)	0.4210(4)	0.2188(2)	0.2580(4)
O(4)	0.0904(4)	0.1947(2)	0.5237(4)
O(5)	0.0484(3)	0.2523(2)	0.7775(4)
O(6)	0.4254(4)	0.1411(2)	0.5439(4)
O(7)	0.4636(3)	0.4153(2)	0.9012(4)
O(8)	0.1819(3)	0.1203(2)	0.8696(4)
O(9)	0.3221(4)	0.2905(2)	0.8647(5)
O(10)	0.3759(3)	0.1525(2)	0.7924(4)
O(11)	0.1316(3)	0.3898(2)	0.8634(4)
O(12)	0.1589(4)	0.2283(2)	0.2955(5)
O(13)	0.4353(4)	0.3575(2)	0.2081(4)
N(1)	0.2155(9)	0.4998(4)	0.2852(11)
C(1)	0.3305(13)	0.4857(5)	0.4786(16)
C(2)	0.0868(9)	0.4358(4)	0.2039(11)
C(3)	0.177(2)	0.4828(9)	0.362(3)
C(4)	0.125(3)	0.4639(12)	0.477(3)
H(1)	0.381(6)	0.286(3)	0.854(8)

organics (listed above) were used. Crystallization times between 94 and 336 hours were reported.

Thermal Properties

Upon heating, AlPO₄-21 transforms into AlPO₄-25 (*Inorg. Chem.* 24,188(1985); US 4,310,440(1982)). A two step weight loss is observed with 4.9% loss when the sample is heated to 120°C and a further 14.1% weight loss when the sample is heated above 400°C. The temperature at which this transformation takes place is 400°C. The organic also is lost during the conversion of AlPO₄-21 to AlPO₄-25 (from author).

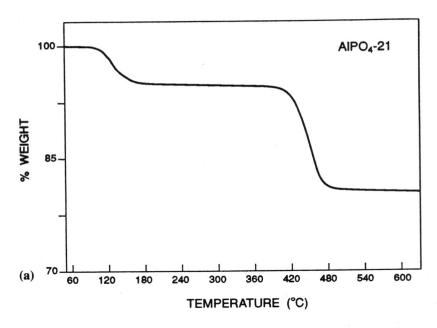

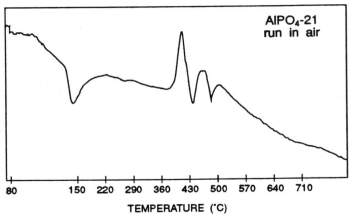

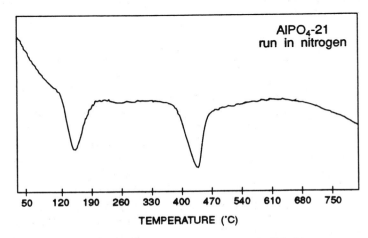

(b)

AlPO₄-21 Figs. 3a and 3b: TGA. (Fig. 3a) /DTA. (Fig. 3b) Of AlPO₄-21.
(Reproduced with permission of B. Duncan, Georgia Tech)

38

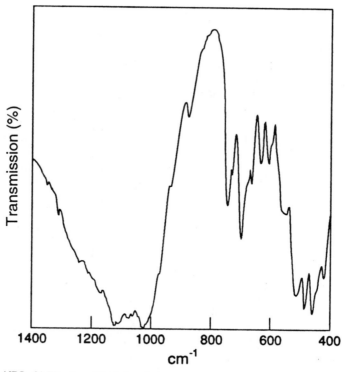

AlPO₄-21 Fig. 4: Mid-infrared spectrum of AlPO₄-21. (Reproduced with permission of B. Duncan, Georgia Tech)

Infrared Spectrum

Adsorption

Adsorption of AlPO₄-21 (from author):*

Adsorbate	Kinetic diameter, Å	Pressure, Torr	Temp., °C	Wt. % adsorbed
O₂	3.46	75	−183	0.112
O₂		75	−183	0.425
H₂O	2.65	17.5	22	1.067
H₂O		17.5	22	1.122

*The AlPO₄-21 samples were activated at 200°C for 4 hours under vacuum.

AlPO₄-22 (AWW)
(US4,310,440(1982))
Union Carbide Corporation

Structure

Chemical Composition: 0.31 DDO : 1.0 Al₂O₃ : 1.03 P₂O₅ : 0.31 H₂O.

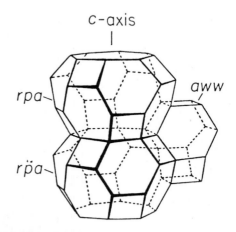

AlPO₄-22 Fig. 1S: Framework topology of AlPO4-22 indicating the aww and rpa polyhedral units.

X-Ray Powder Diffraction Data: (as-synthesized) (d(Å) (I/I_0)) 10.34(25), 9.83(2sh), 9.72(57), 7.76(1), 6.97sh, 6.81(9), 6.07(4), 5.68(6), 5.13(26), 4.80(100), 4.30(43), 4.15(1sh), 4.08(19), 3.93(6), 3.75(22), 3.7(23), 3.68(8), 3.59(24), 3.40(36), 3.30sh, 3.27(20),

3.21(8), 3.12(10), 3.05(30), 2.969(8), 2.835(16), 2.780(2), 2.714(2), 2.698(2), 2.625(1), 2.571(11), 2.525(5), 2.404sh, 2.392(9), 2.304(1), 2.295(1), 2.268(2), 2.235(4), 2.191(2), 2.146(2), 2.071(1), 2.045(4), 2.027(2), 2.006(4), 1.957(1), 1.933(2), 1.881(4), 1.868(13), 1.834(5), 1.817(2), 1.778(<1), 1.722(3), 1.698(5), 1.687(2); (after thermal treatment in air at 600°C) 10.28sh, 9.65(100), 7.03–6.86(42), 6.15–6.02(21), 5.13(50), 4.80(92), 4.35(64), 4.12(23), 3.95(11), 3.72(34), 3.59(18), 3.44–

3.41(sh), 3.27(24), 3.24sh, 3.14(12), 3.07(31), 2.867–2.849(19), 2.722(4), 2.596(15), 2.571(14), 2.529(5), 2.411(3), 2.275(1), 2.212(3), 2.176(1), 2.040(4), 2.019(2), 1.945(3), 1.926(2), 1.903(3), 1.888(3), 1.859(2), 1.737(2), 1.716(4).

The structure of AlPO$_4$-22 contains two new polyhedral units identified as the rpa and ww units. The aww cages have two topologically distinct four-rings and four six-rings. The aww

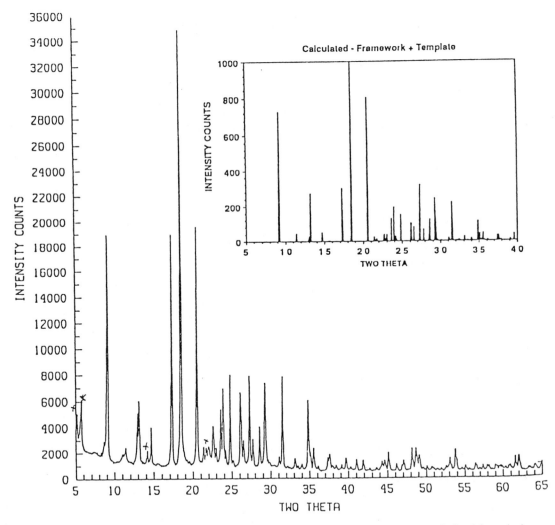

AlPO$_4$-22 Fig. 2:   X-ray powder diffraction pattern of AlPO$_4$-22 (Cu radiation), and pattern calculated from single-crystal structural model (insert). Weak lines at low angles from an unidentified impurity are shown by crosses (*Naturwissenschaften* 76:467 (1989)). (Reproduced with permission of Springer-Verlag Inc.)

cage encapsulate a phosphate tetrahedron. The presence of the occluded phosphate results in a P/Al ratio greater than unity. The rpa cage is composed of eight four-rings, eight six-rings, and two eight-rings and contains the quinuclidine molecule (*Naturwissenschaften* 76:467 (1989)).

Synthesis

ORGANIC ADDITIVES
N,N′-dimethyl-1,4-diazabicyclo[2,2,2]octane dihydroxide (DDO)

This structure is prepared from the batch composition:

$$1.7 \text{DDO} : \text{Al}_2\text{O}_3 : \text{P}_2\text{O}_5 : 40 \text{ H}_2\text{O}$$

The synthesis employs pseudoboehmite (hydrated aluminum oxide) as the source of aluminum. Crystallization occurs at 200°C after 72 hours. Some of the DDO remains trapped within the crystals after synthesis.

Thermal Properties

Crystallinity of this material remains after thermal treatment to 600°C to remove the trapped organic species. The X-ray powder diffraction pattern changes slightly upon thermal treatment.

AlPO₄-23
(US4,310,440(1982))
Union Carbide Corporation

Structure

Chemical Composition: 0.64 $(CH_2)_4NH$: 1.00 Al_2O_3 : 1.04 P_2O_5 : 0.79 H_2O.

X-Ray Powder Diffraction Data: $(d(\text{Å}) \ (I/I_0))$
11.946(47), 8.588(24), 8.268(64), 7.628(54), 7.437(43sh), 6.607(4), 6.026(4), 5.906(6), 5.277(6), 4.671(21), 4.353(28), 4.171(100), 3.834(65), 3.708(9), 3.633(6sh), 3.401(16), 3.278(11), 3.209(12), 3.121(36), 3.018(30), 2.903(11), 2.780(74), 2.652(2), 2.571(7), 2.481(8), 2.404(4), 2.298(2), 2.217(7), 2.090(4), 1.888(8), 1.831(4), 1.765(2), 1.734(5).

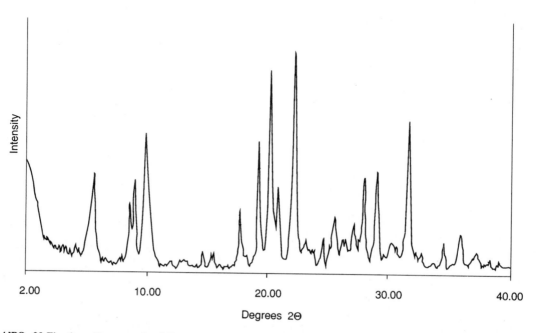

AlPO₄-23 Fig. 1:   X-ray powder diffraction pattern for AlPO₄-23. (Reproduced with permission of author)

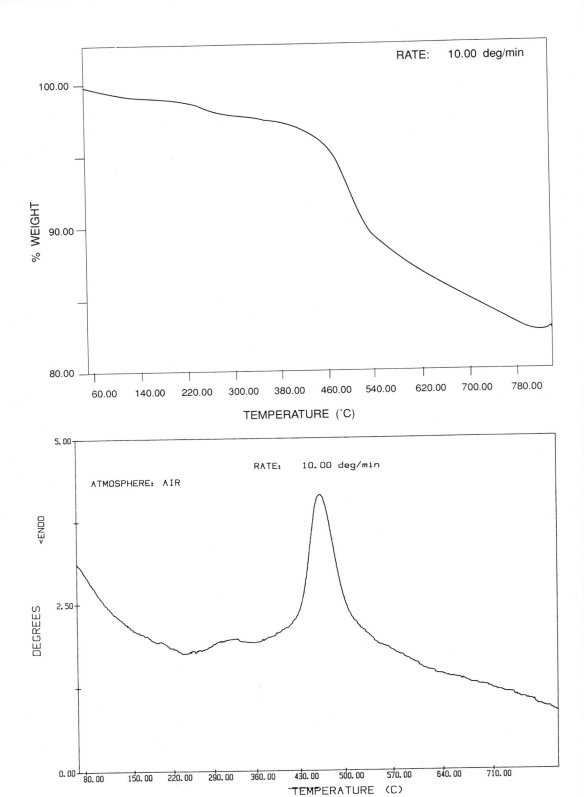

AlPO₄-23 Figs. 2a and 2b: TGA. (Fig.2a) and DTA (Fig. 2b) for AlPO₄-23. (Reproduced with permission of B. Duncan, Georgia Tech)

Synthesis

ORGANIC ADDITIVE
pyrrolidine (CH$_2$)$_4$NH

AlPO$_4$-23 is prepared from a reactive gel with the composition:

$$(CH_2)_4NH : Al_2O_3 : P_2O_5 : 50\ H_2O$$

Pseudoboehmite is used as the source of aluminum, and phosphoric acid is the phosphorus source. Crystallization occurs at 200°C after 91 hours. After recovery of the solid, the crystals are dried at 110°C (US4,310,440(1982)).

Thermal Properties

Calcination of AlPO$_4$-23 to 600°C to remove the entrapped organic from the material results in a substantial change in the X-ray diffraction pattern and the generation of the AlPO$_4$-28 structure (US4,310,440(1982)). An 18% reduction in sample weight is observed (from author).

Infrared Spectrum

Mid-infrared Vibrations (cm^{-1}): 1110(s), 1075(s), 1040(m), 730(w), 670(w), 630(w), 570(w), 520(w), 460(w).

AlPO$_4$-24 (ANA)
(JACS 104:1146(1982))
Union Carbide Corporation

AlPO$_4$-24 is topologically related to analcime.

AlPO$_4$-25 (ATV)
(US4,310,440(1982))
Union Carbide Corporation

Structure

(Acta Crystallogr. C41:515(1985))

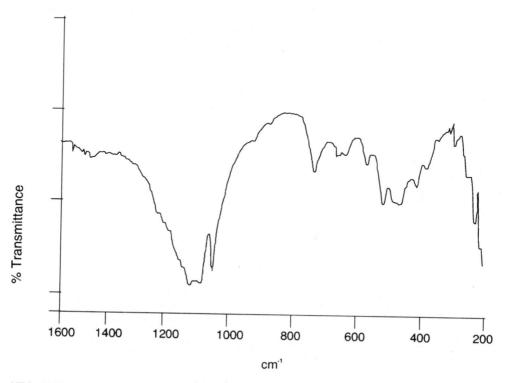

AlPO$_4$-23 Fig. 3:   Mid-infrared spectrum of AlPO$_4$-23. (Reproduced with permission of B. Duncan, Georgia Tech)

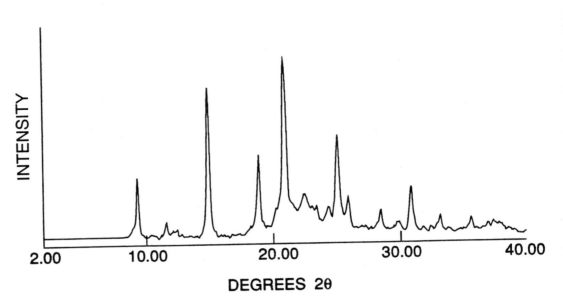

AlPO₄-25 Fig. 1S: Framework topology (proposed).

CHEMICAL COMPOSITION: (US4,310,440(1982)):

$Al_2O_3 : 1.0 \pm 0.2\ P_2O_5$

SPACE GROUP: B2/a

UNIT CELL PARAMETERS (Å): $a = 15.24$
$b = 18.91$
$c = 8.45$

PORE STRUCTURE: eight-member-ring pores proposed

X-Ray Powder Diffraction Data: $(d(\text{Å})\ (I/I_0))$

11.632(3), 9.213(83), 8.346(1), 7.500(14), 7.025(3), 5.829(100), 4.647(71), 4.171(84), 3.917(17), 3.770(8), 3.619(6), 3.490(43), 3.414(14), 3.121(10), 2.988(3), 2.885(23),

2.698(8), 2.515(7), 2.404(8), 2.374(8), 2.094(3), 1.719(3).

The AlPO₄-25 structure contains the same 2-D net as AlPO₄-21, but it has an up-down-up-down chain instead of the up-up-down-down chain of AlPO₄-21 (*J. Phys. Chem.* 94: 3365(1990)).

Atomic coordinates of AlPO₄-25 at 593°K[a]:

atom	site	sym	x	y	z
T(1)	16o	1	0.3535(13)	0.0982(6)	0.1987(12)
T(2)	8m	m	0.1566(15)	0.2500	0.3142(12)
O(1)	8l	2	0.3017(8)	0.0000	0.2500
O(2)	8k	2	0.5000	0.1275(5)	0.2500
O(3)	16o	1	0.2317(8)	0.1625(5)	0.2497(13)
O(4)	8m	m	0.3308(10)	0.0971(6)	0.0000
O(5)	4e	2/m	0.0000	0.2500	0.2500
O(6)	4g	mm	0.1076(24)	0.2500	0.5000

a. $T = Al_{0.5}P_{0.5}$; $a = 9.4489(4)$ Å, $b = 15.2028(5)$ Å; $c = 8.4084(3)$ Å; Acmm(67) (*J. Phys Chem.* 94:3365 (1990)).

See AlPO₄-21 for hydrothermal synthesis. Conversion of AlPO₄-21 to AlPO₄-25 occurs upon calcination to 500–600°C (US4, 310,440(1982)).

AlPO₄-25 Fig. 2: X-ray powder diffraction pattern for AlPO₄-25. (Reproduced with permission of author)

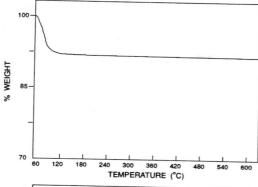

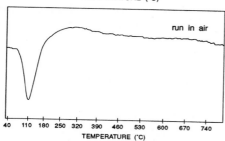

AlPO$_4$-25 Fig. 3: TGA trace (top) with 3 wt% loss before 140°C; DTA trace (bottom) of AlPO$_4$-25, 10°C/min in air. (Reproduced with permission of B. Duncan, Georgia Tech)

Thermal Properties

AlPO$_4$-25 loses 8% of its mass between room temperature and 120°C. (See AlPO$_4$-25 Fig. 3.)

Infrared Spectrum

Mid-infrared Vibrations (cm^{-1}): 1100(s), 700(w), 625(vw), 450(w). (See AlPO$_4$-25 Fig. 4 below.)

Adsorption

Adsorption properties of AlPO$_4$-25:

Adsorbate	Kinetic diameter, Å	Pressure, Torr	Temp., °C	Wt. % adsorbed
O$_2$	3.46	103	−183	4.9
O$_2$		761	−183	5.9
n-Hexane	4.3	28	25	0.3
Neopentane	6.2	310	25	0.4
H$_2$O	2.65	4.6	25	4.4
H$_2$O		20.0	25	16.6

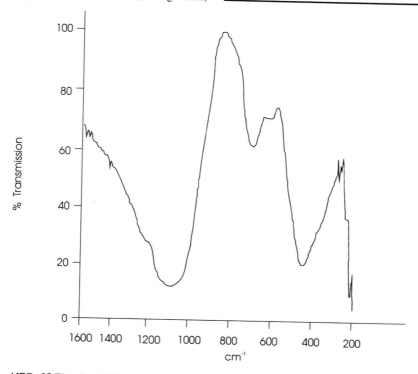

AlPO$_4$-25 Fig. 4:   Mid-infrared spectrum of AlPO$_4$-25. (Reproduced with permission of B. Duncan, Georgia Tech)

AlPO$_4$-26
(US4,310,440(1982))
Union Carbide Corporation

Structure

Chemical Composition: Al$_2$O$_3$: 1 ± 0.2 P$_2$O$_5$ (organic-containing composition not reported).

X-Ray Powder Diffraction Data: (d(Å) (I/I_0)) 10.59(100), 8.98(14), 8.67sh, 8.38(68), 7.50(1), 6.56(8), 6.11(6), 5.73(3), 5.28(10), 5.02(15), 4.91(18), 4.67(1), 4.46(11), 4.37(1), 4.10(4), 3.99(31), 3.86–3.88(74), 3.60(6), 3.527(10), 3.446(2), 3.278(10), 3.249(6), 3.192(6), 3.058(4), 3.008(1), 3.036(9), 2.8938–2.875(12), 2.83(2), 2.776(8), 2.739(1), 2.652(1), 2.614(6), 2.564(1), 2.536(1), 2.485(<1), 2.468(<1), 2.401(2), 2.332(<1), 2.309(<1), 2.287(1), 2.254(<1), 2.227(2), 2.186(2), 2.132(2), 2.099(4), 2.049(<1), 1.985(<1), 1.969(1), 1.903(1), 1.890(2), 1.855(2), 1.834(<1), 1.801(<1), 1.778(2), 1.743(1), 1.731(<1), 1.707(2), 1.687(3), 1.664(1).

Synthesis

ORGANIC ADDITIVE
polymeric quaternary ammonium salt [(C$_{14}$H$_{32}$N$_2$)(OH)$_2$]$_x$

AlPO$_4$-26 crystallizes from reactive gels containing the polymeric quaternary ammonium salt. Pseudoboehmite and phosphoric acid are used to prepare the aluminophosphate gel with the following composition:

1.0(C$_{14}$H$_{32}$N$_2$)(OH)$_2$: Al$_2$O$_3$: P$_2$O$_5$: 105 H$_2$O

Crystallization occurs at 200°C after 24 hours. The solid is recovered and dried at 110°C.

AlPO$_4$-28
(US4,310,440(1982))
Union Carbide Corporation

Structure

Chemical Composition: Al$_2$O$_3$: 1.0 ± 0.2 P$_2$O$_5$.

X-Ray Powder Diffraction Data: (d(Å) (I/I_0)) 11.191(41), 8.934(7), 7.255(100), 6.657(22), 5.644(7), 4.770(47), 4.623(25), 4.230(9), 4.058(30), 3.786(7), 3.507(20), 3.414(13), 3.302(16), 3.008(13), 2.894(13), 2.788(11), 2.536(6), 2.468(2).

AlPO$_4$-28 exhibits properties characteristic of a very small-pore material.

Synthesis

This material is prepared by calcination of AlPO$_4$-23 at 600°C (see AlPO$_4$-23 for synthesis details).

Thermal Properties

This material represents the high-temperature phase of AlPO$_4$-23 as described above. Examination of stability beyond 600°C has not been reported.

Adsorption

Adsorption capacities of AlPO$_4$-28 (US 4,310,440 (1982)):

	Kinetic diameter, Å	Pressure, Torr	Temp., °C	Wt. % adsorbed
O$_2$	3.46	103	−183	1.0
O$_2$		761	−183	2.5
n-Hexane	4.3	28	24	0.5
Neopentane	6.2	310	24	0.5
H$_2$O	2.65	4.6	24	11.1
H$_2$O	2.65	20.0	24	21.4

AlPO$_4$-31 (ATO)
(US4,310,440(1982))
Union Carbide Corporation

Related Materials: SAPO-31 and metal-substituted aluminophosphates

Structure

(*BZA Meeting Abstract,* 1990)

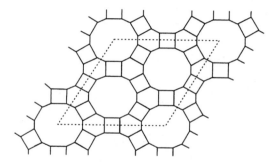

AlPO$_4$-31 Fig. 1S: Framework topology.

CHEMICAL COMPOSITION: (US4,310,440(1982)):
  0.18 Pr$_2$NH : 1.00 Al$_2$O$_3$: 0.99 P$_2$O$_5$: 0.56 H$_2$O
SPACE GROUP: $R3$
UNIT CELL CONSTANTS: (Å) $a = b = 20.83$
                                          $c = 5.00$
                                          $\gamma = 109°$
PORE STRUCTURE: 12-member ring,
  unidimensional channels

The AlPO4-31 structure consists of non-planar layers of 4^2.6^2.12-rings linked with staggered four-rings (M. Bennett, *BZA Meeting Abstract*, 1990). The pore size determined from adsorption pore gauging is predicted to be 6.5 Å. Entry into the structure is through 12-member rings.

X-Ray Powder Diffraction Pattern: (d(Å) (I/I_0))
10.40(85), 9.31(12), 6.63(6), 6.42(2), 5.99(2), 5.64(6), 5.19(7), 4.85(7), 4.44(—), 4.37(48), 4.21(28), 4.08(26), 3.98 (100 3.59(7), 3.48(5), 3.99(11), 3.75(7), 3.02(9), 2.99(9), 2.85, 2.83(14), 2.55(10), 2.51(6), 2.37(8), 2.25(4).

Synthesis

ORGANIC ADDITIVE
Pr$_2$NH

AlPO$_4$-31 crystallizes from batch composition:

$$1.0 \ n\text{-Pr}_2\text{NH} : \text{Al}_2\text{O}_3 : \text{P}_2\text{O}_5 : 40\text{H}_2\text{O}$$

Pseudoboehmite is used as the source of aluminum, and phosphoric acid is the source of phosphorus in this crystallization mixture. Crystallization occurs at 200°C after 2 days (US4,310,440(1982)). Lower temperatures and/or lower amounts of n-Pr$_2$NH result in the crystallization of AlPO$_4$-11 *SSSC* 49: 169 (1989)).

Thermal Properties

This molecular sieve is stable to temperatures as high as 1000°C.

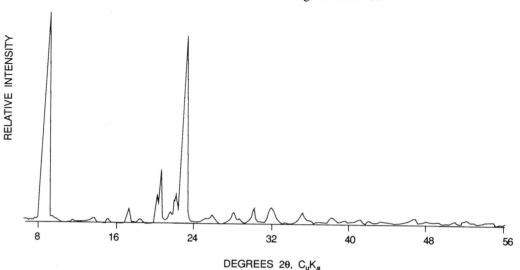

DEGREES 2θ, C$_u$K$_a$

AlPO$_4$-31 Fig. 2:   X-ray powder diffraction pattern of AlPO$_4$-31 (*6IZC*, 97 (1983)). (Reproduced with permission of Butterworth Publishing Company)

Adsorption

Adsorption capacities of AlPO₄-31 (US 4,310,440 (1982)):

	Kinetic diameter, Å	Pressure, Torr	Temp., °C	Wt. % adsorbed
O_2	3.46	99	−183	8.1
O_2		711	−183	11.5
n-Butane	4.3	107	24	4.0
Cyclohexane	6.0	54	24	5.3
Neopentane	6.2	109	24	3.1
H_2O	2.65	4.6	24	5.3
H_2O	2.65	14.0	24	13.9

AlPO₄-33 (ATT)
(US4,473,663(1984))
Union Carbide Corporation

Related Materials: AlPO₄-12-TAMU

Structure

AlPO₄-33 exhibits properties of a small-pore molecular sieve. AlPO₄-33 consists of four-, six-, and eight-member rings with large gismondine-type cavities separated by smaller 14-member polyhedra. The structure is related to the theoretical lattice number, 102, by Smith (*Am. Mineral.* 64:551(1979)). A more detailed description of the structure is found under AlPO₄-12-TAMU.

X-Ray Powder Diffraction Data: $(d(\text{Å})\ (I/I_0))$ (as synthesized) 12.10(.9*), 9.42(21), 8.65(0.7*), 6.98(100), 6.30(0.7*), 5.77(9.6), 5.18(13.9), 5.05(11.6), 4.89(2.4), 4.55(2.7), 4.31(23.3), 4.23(2.8), 4.13(1), 3.99(2.1), 3.86(4.5), 3.71(21.5), 3.65(0.7*), 3.55(2.5). 3.41(18.3), 3.32(1.3sh), 3.26(86), 3.02(11.2), 2.90(6.2), 2.84(0.7sh), 2.79(5.9), 2.75(1.2), 2.61(7.3), 2.57(1*), 2.52(2) [* peak may contain an impurity]; (after thermal treatment to 600°C) 9.3(20), 6.66(100), 6.23(2), 6.03(2), 5.64(12), 4.90(39), 4.82(32), 4.43(16), 4.27(11*), 4.20(1sh), 4.00(3br*), 3.90(2),

AlPO₄-33 Figs. 1Sa and 1Sb: Framework topology (two figures).

3.68(12), 3.62(13), 3.51(4br), 3.35(39br), 3.27(2sh), 3.08(21), 3.02(11), 2.96(21), 2.84(6sh), 2.80(43sh), 2.72(4), 2.64(5).

Synthesis

ORGANIC ADDITIVE
$(CH_3)_4N^+$ (TMA)

AlPO₄-33 was prepared using pseudoboehmite as the source of aluminum and orthophosphoric acid as the source of phosphorus. The batch composition that produces this structure is:

$$1.0(TMA)_2O : Al_2O_3 : P_2O_5 : 50\ H_2O$$

Crystallization occurs at 200°C after 72 hours. After centrifuging and washing, this molecular sieve is dried in air at 100°C before being subjected to X-ray analysis.

Thermal Properties

Thermal treatment of AlPO$_4$-33 results in a change in the X-ray diffraction pattern. A minor component found after calcination is AlPO$_4$-25.

Adsorption

Adsorption capacities of AlPO$_4$-33:

	Kinetic diameter, Å	Pressure, Torr	Temp., °C	Wt. % adsorbed
O$_2$	3.46	100	−183	17.5
Xenon	3.96	753	25	22.7
n-Butane	4.3	100	25	0.4
H$_2$O	2.65	4.6	25	18.6

AlPO$_4$-34 (CHA)
(EP 293,939(1988))
Union Carbide Corporation

Related Materials: chabazite

Structure

Chemical Composition: (EP292,939(1988)) 0.092 TEAOH $(Al_{0.46}P_{0.54})O_2$ (TEA = tetraethylammonium)

X-Ray Powder Diffraction Data: $(d(Å)(I/I_0))$ (as-synthesized) 9.24(100), 6.80(12), 6.27(11), 5.69(7), 5.48(23), 5.27(3), 4.94(14), 4.64(3), 4.25(41), 4.02(4), 3.94(3), 3.81(3), 3.54(15), 3.39(9), 3.16(3), 2.959(4), 2.896(22), 2.753(3), 2.647(3), 2.570(3), 2.465(2), 2.258(2), 2.090(2), 1.842(2), 1.782(3); (calcined 600°C) 9.10(100), 8.65(34), 7.40(3), 6.88(19), 5.73(4), 5.15(4), 4.55(29), 4.48(26), 4.30(23), 4.13(5), 3.91(7), 3.65(14), 3.57(7), 3.44(5), 3.30(5), 3.25(3), 3.17(8), 3.05(17), 2.883(15), 2.799(10), 2.714(7), 2.347(3), 2.250(3), 2.097(2), 1.838(2), 1.726(3).

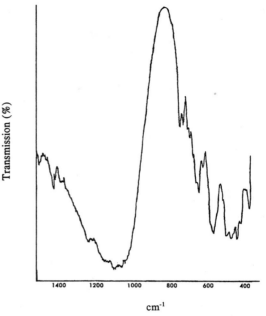

AlPO$_4$-34.

AlPO$_4$-34 exhibits properties characteristic of a small-pore chabazite molecular sieve.

Synthesis

ORGANIC ADDITIVE
tetraethylammonium$^+$
morpholine (*SSSC* 37 (1988))
l-propylamine (author)

AlPO$_4$-34 crystallizes from a reactive mixture with a batch composition:

10.0 TEAOH : Al$_2$O$_3$: 5.0 P$_2$O$_5$: 179 H$_2$O

AlPO$_4$-5 or AlPO$_4$-34 seed crystals were added to this batch composition to encourage crystallization of the AlPO$_4$-34 phase. Crystallization occurred at 100°C after 10 days. Small amounts of AlPO$_4$-5 were observed to form. When the seeds were not added to the original mixture, AlPO$_4$-18 crystallized. Monoaluminum phosphate was used as the source of both aluminum and phosphorus. Phosphoric acid also was used as a further source of phosphorus. In the presence of i-propylamine and HF rapid crystallization of this structure occurs at 190°C (author).

Adsorption

Adsorption properties of AlPO$_4$-34:

Adsorbate	Kinetic diameter, Å	Pressure, Torr	Temp., °C	Wt. % adsorbed
O$_2$	3.46	106	− 183	21.8
O$_2$		705	− 183	31.7
Isobutane	5.0	704	23	0.9
n-Hexane	4.3	44	22	9.7
H$_2$O	2.65	4.6	22	24.9
H$_2$O	2.65	19	23	37.0

AlPO$_4$-39 (ATN)
(US4,663,139(1987))
Union Carbide Corporation

Related Material: MAPO-39

Structure

X-Ray Powder Diffraction Data: (d(Å) (I/I$_0$))
(as synthesized) 9.29(81), 6.58(49), 4.79(42), 4.17(100), 3.89(94), 3.26(5), 3.18(5), 3.16(7), 3.11(13), 2.95(34), 2.60(6), 2.57(11), 2.35(10), 2.33(7); (calcined at 600°C) 9.28(30), 6.53(100), 4.76(33), 4.64(25), 4.13(41), 3.93(22), 3.86(21), 3.82(22), 3.05(14), 3.02(10).

AlPO$_4$-39 exhibits properties characteristic of a small-pore molecular sieve with eight-ring pore openings. See MAPO-39 for a description of the framework topology.

Synthesis

ORGANIC ADDITIVES
di-*n*-propylamine

AlPO$_4$-39 is prepared from gels of di-*n*-propylamine and acetic acid with the following composition:

$$2.0\,\text{di-}n\text{Pr}_2\text{NH}:\text{Al}_2\text{O}_3:\text{P}_2\text{O}_5:\text{CH}_3\text{COOH}:40\,\text{H}_2\text{O}$$

Pseudobohemite is the source of aluminum, and crystallization occurs between 24 and 48 hours

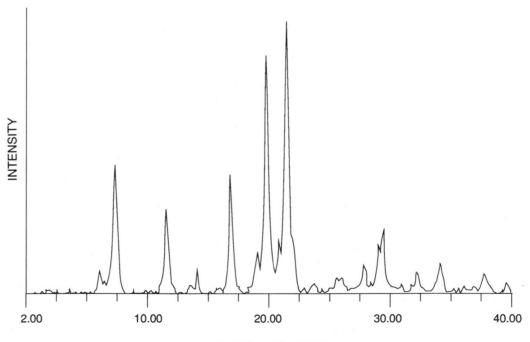

AlPO$_4$-39 Fig. 1:   X-ray powder diffraction pattern of AlPO$_4$-39. (Reproduced with permission of K. Vinje, U. of Oslo)

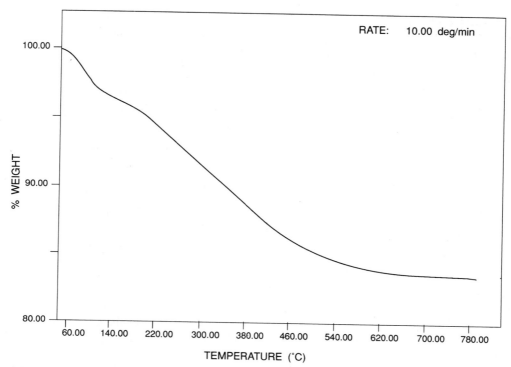

AlPO$_4$-39 Fig. 2:   TGA trace of AlPO$_4$-39 with 17% weight loss to 700°C. (Reproduced with permission of K. Vinje; U. of Oslo).

at 150°C. The resulting solid is centrifuged, washed, and dried at room temperature.

Thermal Properties

See AlPO$_4$-39 Fig. 2 above.

Infrared Spectrum

Mid-infrared   Vibration   Bands   (cm^{-1}): 1100sbr, 700wbr, 640 mw, 555wsh, 515 mbr.

Adsorption

Adsorption capacities of AlPO$_4$-39:

	Kinetic diameter, Å	Pressure, Torr	Temperature, °C	Wt. % adsorbed
Oxygen	3.46	100	−183	10.2
n-Butane	4.3	100	25	1.7
i-Butane	5.0	100	25	0.8
H$_2$O	2.65	4.6	25	18.1

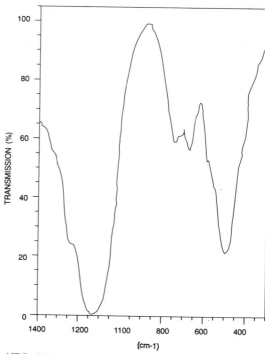

AlPO$_4$-39 Fig. 3:   Mid-infrared spectrum of AlPO$_4$-39. (Reproduced with permission of K. Vinje, U. of Oslo)

AlPO$_4$-41 (AFO)
(EP 254,075(1988))
Union Carbide Corporation

Related Materials: SAPO-41
related metal-substituted
materials

Structure

(*BZA Meeting Abstract*, 1990)

CHEMICAL COMPOSITION: (EP254,075(1988)): 0.31
  Pr$_2$NH : Al$_2$O$_3$: 1.09
P$_2$O$_5$:   1.49 H$_2$O
SPACE GROUP: Cmc^{2}1
UNIT CELL CONSTANTS: (Å) a = 9.72
                                    b = 25.83
          alpha = beta = gamma = 90°
PORE STRUCTURE: 10-member rings,
  unidimensional

X-Ray Powder Diffraction Data: (d(Å) (I/I_0))
(as-synthesized) 12.89(49), 9.11(38), 6.41(40),
4.84(25),   4.27(25),   4.20(43),   3.99(100),
3.86(60), 3.81(26), 3.51(19), 3.42(47), 3.20(10),
3.17(5), 3.15(5), 3.02(32), 2.99(8), 2.84(8),
2.56(6), 2.45(5), 2.42(7), 2.38(11), 1.90(6);
(calcined 600°C) 12.99(13), 9.30(35), 8.42(48),
6.48(40), 5.90(16), 4.70(19), 4.32(27), 4.20(23),
4.09(100), 3.91(7), 3.82(36), 3.74(23), 3.68(12),
3.65(7), 3.55(25), 3.46(31), 3.26(12), 3.09(12),
2.96(14), 2.89(29), 2.72(7), 2.48(9), 2.38(9),
2.35(5), 2.05(6), 1.85(6).

The framework topology of AlPO$_4$-41 consists of 4.6^2.5.10-layers connected by UDUD

chains (*BZA Meeting Abstract*, 1990). AlPO$_4$-
41 exhibits properties characteristic of a medium-pore molecular sieve.

Synthesis

ORGANIC ADDITIVE
di-*n*-propylamine

AlPO$_4$-41 crystallizes from a reactive mixture with a batch composition:

$$2.0\ Pr_2NH : Al_2O_3 : P_2O_5 : 50H_2O$$

AlPO$_4$-31 or AlPO$_4$-41 seed crystals were added
to this batch composition to encourage crystallization of the AlPO$_4$-41 phase. Crystallization
occurred at 200°C after 18 hours. Small amounts
of AlPO$_4$-11 were observed to form. Pseudoboehmite was used as the source of aluminum
and phosphoric acid as the source of phosphorus.

Adsorption

Adsorption properties of AlPO$_4$-41:

Adsorbate	Kinetic diameter, Å	Pressure, Torr	Temp., °C	Wt. % adsorbed
O$_2$	3.46	102	−183	9.1
Cyclohexane	6.0	45	22	9.0
n-Hexane	4.3	45	23	6.0
Neopentane	6.2	302	22	3.3
H$_2$O	2.65	4.6	22	3.3

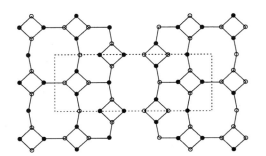

AlPO$_4$-41 Fig. 1S: Framework topology.

AlPO$_4$-52
(US4,851,204(1989))
Union Carbide Corporation

Structure

(*SSSC* 49:731(1989))

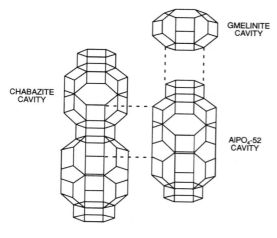

GMELINITE CAVITY

CHABAZITE CAVITY

AlPO₄-52 CAVITY

AlPO₄-52 Fig. 1S: Framework topology.

Idealized parameters for the AABBCCAACCBB sequence used to simulate the powder pattern (*SSSC* 49:731 (1989)):

atom	x	y	z
Al1	0.23336	0.00520	0.19772
P1	0.23444	0.22575	0.19531
Al2	0.32971	0.43494	0.46961
P2	0.32993	0.43476	0.36196
Al3	0.43234	0.32231	0.63538
P3	0.43160	0.32244	0.52781
O11	0.27681	0.03818	0.25489
O111	0.21384	0.11290	0.17491
O112	0.10570	−0.12330	0.19480
O12	0.33640	−0.00627	0.16623
O22	0.29912	0.39486	0.41202
O23	0.33935	0.32945	0.50000
O231	0.45841	−0.43782	0.47337
O232	0.22185	0.45316	0.49314
O31	0.32471	0.32361	0.16629
O321	0.45737	0.21842	0.65844
O322	0.55493	−0.54506	0.63924
O33	0.39210	0.29090	0.57790

52 cage. The framework density (15.2 TO2/10000 Å³) is intermediate between erionite (16.1) and chabazite (14.7).

CHEMICAL COMPOSITION: 0.2(TEA)(H₂PO₄)* [Al₀.₅P₀.₅]O₂ (TEA = tetraethylammonium)
SYMMETRY: hexagonal
SPACE GROUP: (idealized) P31c
UNIT CELL CONSTANTS   (Å): $a = 13.73$
$c = 28.95$
$\gamma = 120°$
PORE STRUCTURE: eight-member ring pores

X-Ray Powder Diffraction Data: (d(Å) (I/I_0)) (as-synthesized) 11.80(13), 11.00(91), 9.21(54), 6.86(20), 6.21(24), 5.50(35), 5.07(26), 4.85(50), 4.44(34), 4.29(100), 4.08(47), 3.96(25), 3.93(20), 3.48(32), 3.43(24), 3.12(22), 2.91(32), 2.83(23), 2.62(22); (at 600°C) 10.80(62), 9.14(100), 7.51(44), 6.79(54), 5.46(13), 5.05(10), 4.91(26), 4.41(17), 4.27(32), 4.06(21), 3.41(18), 3.16(11), 2.90(16), 2.84(11), 2.83(11).

This discussion is based on an idealized topology for AlPO₄-52. The (calcined) AlPO₄-52 topology is a member of the ABC six-ring family of structures. It contains the longest known repeat sequence (12 layers). These repeat units consist of AABBCCAACCBB.

There are three different cages contained within the structure, all governed by eight-ring windows. These cages include the gmelinite cage, the chabazite cage, and the 8*3(2.4.6) or AlPO₄-

Synthesis

ORGANIC ADDITIVE (*SSSC* 49:731(1989))
TEA⁺/tripropylamine

AlPO₄-52 was synthesized from a mixed organic-containing gel with the composition:

1.0 TEAOH : 2.0 Pr₃N : Al₂O₃ : 1.25 P₂O₅ : 40 H₂O

This structure crystallizes after 120 hours at 150°C. AlPO₄-18 has been used as a seed to encourage crystallization of this material. Very little of the tripropyl amine was found to occlude in this structure during synthesis. The organic located within the channels of this material is the phosphate salt TEA*H₂PO₄. There are 6.9 to 7.6 TEA per unit cell.

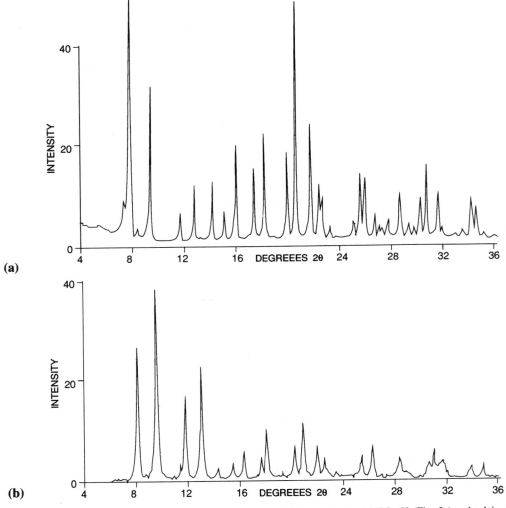

AlPO$_4$-52 Figs. 2a and 2b: X-ray powder diffraction pattern of as-synthesized AlPO$_4$-52 (Fig. 2a) and calcined (Fig. 2b) (*SSSC* 49:731 (1989)). (Reproduced with permission of Elsevier Publishing Company)

Adsorption

Adsorption capacities of AlPO$_4$-52:

	Kinetic diameter, Å	Pressure, Torr	Temp., °C	Wt. % adsorbed
O$_2$	3.46	700	−183	30.1
n-Butane	4.2	700	23	12.9*
Isobutane	5.0	700	23	0.5
H$_2$O	2.65	4.6	23	30.8

*Sample failed to reach equilibrium.

AlPO$_4$-54 (VFI)
US5,013,535(1991)
UOP

Related Material: VPI-5

Structure

(*RR 8th IZC*, 116(1989))

SYMMETRY: hexagonal
SPACE GROUP: P6$_3$/mcm

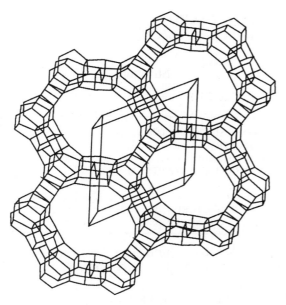

AlPO₄-54 Fig. 1S: Framework topology.

UNIT CELL CONSTANTS (Å):
  (as-synthesized): $a = 19.009(2)$
                     $c = 8.122(1)$
  (dehydrated): $a = 18.549(1)$
                     $c = 8.404(1)$
PORE STRUCTURE: unidimensional 18-member
  rings, very large-pore

The AlPO₄-54 structure has been examined using neutron diffraction on the dehydrated material. Bond distances in this material range from 1.43 to 1.70, with an average value of 1.60 Å. This structure is consistent with the net 81(1) as shown for VPI-5 to contain an 18-member ring channel system. See VPI-5 for a further description of the framework topology.

Atomic coordinates of AlPO₄-54 (*RR 8th IZC*, 116 (1989)):

	x	y	z
T(1)	0.5815(7)	0.000	0.0533(12)
T(2)	0.1739(9)	0.6603(6)	−0.0486(14)
O(1)	0.5000	0.0000	0.0000
O(2)	0.5858(10)	0.0000	0.2500
O(3)	0.0843(7)	0.6536(4)	−0.0227(22)
O(4)	0.5930(3)	0.1860(7)	0.0000
O(5)	0.8372(12)	0.3428(8)	0.2500
O(6)	0.2465(3)	0.4930(6)	0.0000

AlPO₄-A
(*Bull. Chem. Soc. Fr.* **1762(1961)**)

Structure

Chemical Composition: AlPO₄

X-Ray Powder Diffraction Data: (d(Å)(I))
5.77(f), 5.64(fff), 4.47(F), 4.36(FFF), 4.23(m), 4.07(FF), 3.94(ff), 3.76(mf), 3.64(FF), 3.20(f), 3.05(mF), 2.945(mf), 2.895(m), 2.57(F), 2.46(m), 2.41(ff), 2.39(fff), 2.355(ff), 2.325(f), 2.31(f), 2.225(f), 2.235(ff), 2.18(f), 2.12(mf), 2.08(f), 2.045(ff), 2.035(f), 1.98(fff), 1.965(ff), 1.925(ff), 1.88(m), 1.84(ff), 1.82(m), 1.805(ff).

Synthesis

AlPO₄-A is prepared by dehydration of metavariscite between 70 and 120°C. Rehydration may return the metavariscite structure. Heating to 450°C results in the formation of tridymite.

AlPO₄-B
(*Bull. Chem. Soc. Fr.* **1762(1961)**)

Structure

Chemical Composition: AlPO₄.

X-Ray Powder Diffraction Data: (d(Å)(I))
5.19, 4.97, 4.30, 3.95, 3.85, 3.58, 3.26, 3.04, 2.912, 2.627, 2.595, 2.514, 2.481, 2.386, 2.319, 2.237, 2.150, 2.100, 2.088, 2.038, 1.971, 1.922, 1.886.

Synthesis

AlPO₄-B is prepared by dehydration of variscite between 100 and 200°C. Rehydration returns the variscite structure. Heating to 400°C results in the formation of tridymite and a small quantity of cristobalite.

AlPO₄-C (APC)
(*Bull. Chem. Soc. Fr.* **1762(1961)**)

Structure

(*Zeo.* 6:349(1986))

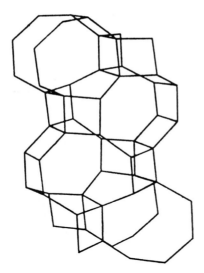

AlPO$_4$-C Fig. 1S: Framework topology.

CHEMICAL COMPOSITION: [Al$_{16}$P$_{16}$O$_{64}$]
SYMMETRY: orthorhombic
SPACE GROUP: Pbca
UNIT CELL CONSTANTS (Å): $a = 19.8$
$b = 10.0$
$c = 8.9$
PORE STRUCTURE: two eight-ring channel systems
(3.4 × 3.7 Å) and (2.9 × 5.7 Å)

X-Ray Powder Diffraction Data: (d(Å)(I))
7.04(FFF), 6.28(FF), 5.50(F), 4.98(FF), 4.69(ff),
4.43(FF), 4.33(m), 4.23(FF), 4.10(mf), 3.97(F),
3.75(m),   3.45(fff),   3.41(mf),   3.31(mF),
3.26(mf),   3.15(FF),   3.07(f),   2.964(m),
2.935(mf), 2.843(f), 2.804(mf), 2.726(f),
2.615(f), 2.563(f), 2.550(f), 2.519(f), 2.484(f),
2.448(fff), 2.380(f), 2.346(f), 2.331(ff).

The entire AlPO$_4$-C structure is composed of single four- and eight-member rings (*Zeo.* 6:349(1986)).

Synthesis

AlPO$_4$-C is prepared by dehydration of AlPO$_4$-H3 between 100 and 180°C. Increasing the temperature to 200°C results in the formation of AlPO$_4$-D. The transformation between AlPO$_4$-C and AlPO$_4$-H3 is reversible.

AlPO$_4$-D (APD)
(*Bull. Chem. Soc. Fr.* 1762(1961))

Structure

(*AZST* (1987))

CHEMICAL COMPOSITION: [Al$_{16}$P$_{16}$O$_{64}$]
SYMMETRY: orthorhombic
SPACE GROUP: Pca2$_1$
UNIT CELL CONSTANTS (Å): $a = 19.2$
$b = 8.6$
$c = 9.8$
PORE STRUCTURE: two eight-ring channel systems
(2.1 × 6.3) and (1.3 × 5.8)

X-Ray Powder Diffraction Data: (d(Å)(I))
6.84(FFF), 6.12(mf), 5.35(f), 5.11(mf), 4.89(m),
4.78(mF),   4.54(m),   4.35(mF),   4.28(FFF),
4.15(mF), 3.88(mf), 3.84(mf), 3.64(m), 3.54(f),
3.49(f), 3.42(mF), 3.29(mf), 3.22(f), 3.19(f),
3.08(mF),   3.05(f),   3.03(mF),   2.90(?fff),
2.86(fff), 2.75(ff), 2.69(mf), 2.67(m), 2.52(mf),
2.51(f), 2.473(f), 2.447(m), 2.387(f), 2.331(fff),
2.303(f),   2.278(f),   2.,233(f),   2.208(fff),
2.173(ff), 2.146(m), 2.126(f), 2.076(f), 2.040(f),
2.013(f).

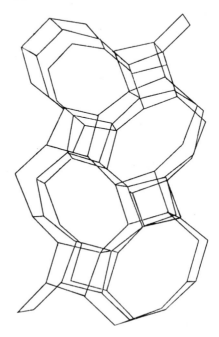

AlPO$_4$-D Fig. 1S: Framework topology.

The entire AlPO$_4$-D structure is composed of single four- and eight-member rings.

Synthesis

AlPO$_4$-D is prepared by heating either AlPO$_4$-H3 or AlPO$_4$-C to 200°C. With further heating to temperatures between 600 and 850°C a mixture of tridymite and cristobalite results.

AlPO$_4$-E
(Bull. Chem. Soc. Fr. 1762(1961))

Structure

Chemical Composition: AlPO$_4$.

X-Ray Powder Diffraction Data: $(d(Å)(I))$ 7.72(FF), 7.02(ff), 6.29(fff), 5.76(m), 5.68(f), 5.18(f), 4.97(fff), 4.76(mf), 4.52(mf), 4.29(F), 4.17(FFF), 3.96(mf), 3.85(F), 3.67(m), 3.58(mf), 3.51(mF), 3.46(mf), 3.37(f), 3.19(mf), 3.09(mf), 2.99 (mF), 2.87(m), 2.82(m), 2.78(mf).

Synthesis

AlPO$_4$-E is prepared by conversion of AlPO$_4$-H2 through dehydration at room temperature. Heating to 500°C results in the tridymite structure.

AlPO$_4$-EN3
(SSSC 24:271(1985))

Structure

(SSSC 24:271(1985))

CHEMICAL COMPOSITION: Al$_6$(PO$_4$)$_6$*4H$_2$O*en
SPACE GROUP: P2$_1$2$_1$2$_1$
UNIT CELL PARAMETERS (Å): $a = 10.292(2)$
$b = 13.636(2)$
$c = 17.344(3)$
PORE STRUCTURE: intersecting eight-ring pores

The AlPO$_4$-EN3 structure is composed of sheets of three-, four-, five-, and eight-member rings linked together via chains of Al and P

tetrahedra. These chains are in a zigzag conformation. This differs from AlPO$_4$-21, which has a crankshaft conformation in these chains. The ethylenediamine organic is located at the

Atomic positional parameters for AlPO$_4$-EN3(en) ($\times 10^4$) (SSSC 24:271 (1985)):

	x	y	z
P1	8796(4)	60(3)	476(2)
P2	1315(4)	3183(3)	1724(2)
P3	1041(4)	8166(3)	−1724(2)
P4	8779(4)	4999(3)	−418(2)
P5	1139(4)	7153(3)	841(2)
P6	1367(4)	2143(3)	−811(2)
Al1	1160(5)	−72(4)	−572(3)
Al2	8632(4)	−1892(3)	1424(3)
Al3	1247(4)	4861(3)	529(3)
Al4	8912(4)	3103(3)	−1428(3)
Al5	9094(4)	7307(4)	−494(3)
Al6	9371(4)	2329(3)	522(3)
O1	8625(10)	2537(8)	−433(7)
O2	8301(10)	−2337(8)	397(6)
O3	9344(10)	1045(6)	728(6)
O4	9217(12)	6025(6)	−641(7)
O5	8772(10)	4356(7)	−1132(5)
O6	8778(12)	−650(8)	1149(6)
O7	1233(13)	6027(6)	929(8)
O8	1114(11)	1049(6)	−1000(7)
O9	990(11)	2372(9)	14(5)
O10	9671(11)	4566(9)	202(7)
O11	10055(10)	2764(8)	1403(6)
O12	9640(10)	−375(8)	−167(5)
O13	154(9)	7485(8)	1443(8)
O14	2402(9)	2440(7)	1695(7)
O15	1704(12)	4089(8)	1244(7)
O16	9766(10)	7858(8)	−1363(6)
O17	716(10)	7416(8)	2(6)
O18	1102(10)	3498(8)	2552(6)
O19	7388(8)	213(7)	118(6)
O20	2062(9)	7331(7)	−1651(6)
O21	9151(10)	3511(8)	−2451(6)
O22	2467(10)	−2386(8)	978(6)
O23	7780(11)	2638(9)	940(7)
O24	579(10)	2733(8)	−1406(6)
O25	8456(10)	4073(8)	−3719(6)
O26	7594(12)	4989(9)	−5109(7)
N1	10179(16)	5556(12)	2605(10)
N2	6897(27)	4236(21)	2663(21)
N2*a	8164(87)	3442(41)	2513(50)
OW1	7262(18)	1300(13)	−7474(17)
OW2	4316(26)	4419(20)	2533(15)
C1	8825(22)	5274(18)	2333(14)
C2	8343(24)	4456(20)	2772(16)

[a]Occupation factor for N$_2$ is 0.25(3).

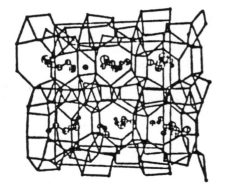

AlPO₄-EN3 Fig. 1S: Framework topology.

intersection of the eight-rings in this structure and arranged in a *trans* configuration so that Nl-C-C-N2 is extended along the straight eight-ring channel.

Synthesis

AlPO₄-EN3 is prepared from a starting gel composition of:

0.5 en : Al_2O_3 : P_2O_5 : 40 H_2O

and is heated for 200°C (*SSSC* 24:271(1985)).

AlPO₄-H1 (VFI)
(*Bull. Chem. Soc. Fr.* 1762(1961))

Related Materials: VPI-5

Structure

Chemical Composition: AlPO₄* $\geq$ 2 H_2O.

X-Ray Powder Diffraction Data: (d(Å)(I/I_0))
16.53 (FFF), 8.23(f), 6.16(f), 3.97(f), 3.93(m), 3.28(m), 2.95(f), 2.74(f)

Synthesis

See VPI-5 for description of the framework topology.

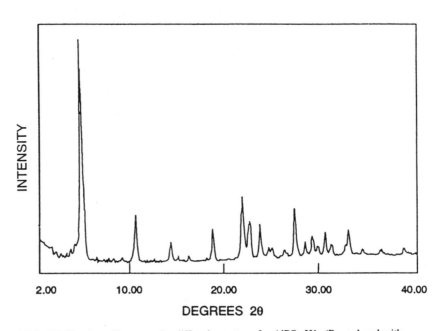

AlPO₄-H1 Fig. 1:   X-ray powder diffraction pattern for AlPO₄-H1. (Reproduced with permission of B. Duncan, Georgia Tech)

ORGANIC ADDITIVES
none

AlPO₄-H1 is prepared from reactive aqueous aluminophosphate gels at pH of 1.6 and P_2O_5/Al_2O_3 ratio of 2.73. Crystallization begins at temperatures between 80 and 100°C after 2 hours. This molecular sieve appears as a minor product with AlPO₄-H2 as the major product under these conditions (*Bull. Chem. Soc. Fr.* 1762 (1961)). AlPO₄-H1(GTRI) has been prepared in pure form from the batch composition: $Al_2O_3 : 0.8\ P_2O_5 : 1\ HCl : 50\ H_2O$. Crystallization occurring readily in less than 1 day at 140 to 150°C (*Cat. Lett.* 7:367(1990)).

Thermal Properties

The mixture of AlPO₄-H1 and AlPO₄-H2 converts to AlPO₄-trydimite upon heating (Bull. Chem. Soc. Fr. 1762(1961)). The pure AlPO₄-H1(GTRI) converts to AlPO₄-8 upon treatment to 100°C (*Cat. Lett.* 7:367(1990)). See AlPO₄-H1 Fig. 2.

Infrared Spectra

Mid-infrared Vibration Bands (cm⁻¹): 1265(m), 1160(s), 760(w), 610(w), 520(m), 470(m), 420(m). See AlPO₄-H1 Fig. 3 on following page.

Adsorption

Water adsorption properties of (a) AlPO₄-H1 (b) AlPO₄-H1 calcined to AlPO₄-8, (c) AlPO₄-H1 calcined to AlPO₄-8 and steamed (kinetic diameter of H_2O = 2.65 angstroms):

Pressure, Torr	% Weight adsorbed (20°C)	% Weight adsorbed (35°C)	% Weight adsorbed (50°C)
(a) H₂O adsorption data for AlPO₄-H1			
0.5	4.95	1.67	0.28
1.0	12.12	4.33	0.99
2.0	17.03	12.11	3.98
5.0	19.22	15.88	6.55
10.0	20.80	17.58	8.17
17.0	22.48	18.51	9.36
(b) H₂O adsorption data for AlPO₄-H1 calcined (AlPO₄-8)			
0.5	0.95	0.58	0.07
1.0	1.70	1.28	0.12
2.0	3.05	2.09	0.16
5.0	9.85	4.88	0.29
10.0	13.65	9.11	0.42
17.0	14.40	13.76	0.84
(c) H₂O adsorption data for AlPO₄-H1 calcined (AlPO₄-8) and steamed			
0.5	0.66	0.17	0.17
1.0	1.41	0.41	0.25
2.0	5.63	1.08	0.66
5.0	11.84	5.62	1.24
10.0	12.41	11.41	1.99
17.0	13.25	12.32	8.53

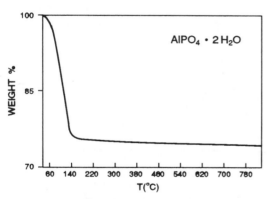

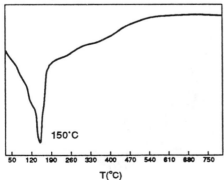

AlPO₄-H1 Figs. 2a and 2b: TGA. (Fig. 2a) /DTA. (Fig. 2b) Of AlPO₄-H1. Weight loss of ca. 23% observed. (Reproduced with permission of B. Duncan, Georgia Tech)

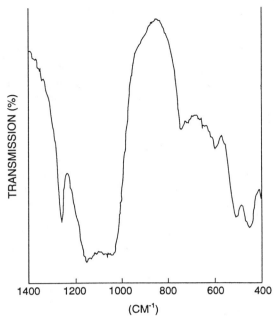

AlPO$_4$-H1 Fig. 3: Mid-infrared spectrum of AlPO$_4$-H1. (Reproduced with permission of B. Duncan, Georgia Tech)

AlPO$_4$-H1(GTRI) (VHI)
(Cat. Lett. 7:367(1990))

Related Material: VPI-5

A designation used for the pure single-phase

AlPO$_4$-H1 prepared in the presence of HCl. See AlPO$_4$-H1.

X-Ray Powder Diffraction Data: (d(Å) (I/I_0))
16.54(100), 9.52(2), 8.23(23), 6.20(5), 5.49(3), 4.75(11), 4.12(18), 4.06(9), 3.96(9), 3.87(9), 3.77(17), 3.60(6), 3.42(1), 3.28(18), 3.16(7), 3.08(6), 2.95(13), 2.90(5), 2.74(12).

AlPO$_4$-H2
(Bull. Chem. Soc. Fr. 1762(1961))

Structure

Chemical Composition: AlPO$_4$* $\geq$ 2H$_2$O.

X-Ray Powder Diffraction Data: (d(Å) (I_0))
8.48(FFF), 5.87(m), 4.96(mf), 4.74(f), 4.55(fff), 4.25(?f), 4.09(m), 4.06(FF), 4.04(f), 3.75(F), 3.66(F), 3.64(f), 3.24(m), 3.14(f), 3.08(F), 3.01(f), 2.933(ff), 2.875(f), 2.819(mF), 2.697(mf), 2.666(ff), 2.581(fff), 2.533(fff), 2.482(ff), 2.453(f), 2.373(ff), 2.352(f).

Synthesis

ORGANIC ADDITIVES
none

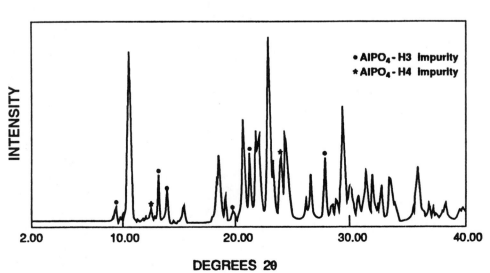

AlPO$_4$-H2 Fig. 1:   X-ray powder diffraction pattern for AlPO$_4$-H2. AlPO$_4$-H3 and -H4 impurities are noted. (Reproduced with permission of B. Duncan, Georgia Tech)

AlPO₄-H2 is prepared from reactive aqueous aluminophosphate gels at pH of 1.6 and P_2O_5/Al_2O_3 ratio of 2.73. Crystallization begins at temperatures between 80 and 100°C after 2 hours. This molecular sieve appears as the major product with impurities of AlPO₄-H1 and AlPO₄-H3. With longer crystallization times, variscite and metavariscite are observed (*Bull. Chem. Soc. Fr.* 1762(1961)).

Thermal Properties

Upon dehydration at room temperature, AlPO₄-H2 converts to AlPO₄-E. This transformation is reversible. Heating to above 500°C results in the formation of AlPO₄-tridymite.

Infrared Spectrum

See AlPO₄-H2 Figure 3.

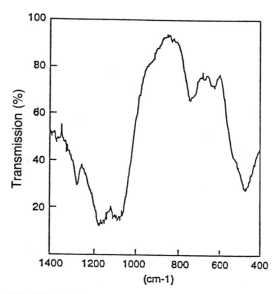

AlPO₄-H2 Fig. 3: Mid-infrared spectrum for AlPo₄-H2 containing trace impurties of AlPO₄-H1. (Reproduced with permission of B. Duncan, Georgia Tech)

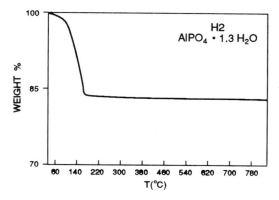

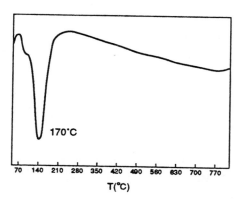

AlPO₄-H2 Figs. 2a and 2b: TGA trace. (Fig. 2a) /DTA. (Fig. 2b) For AlPO₄-H2. (Reproduced with permission of B. Duncan, Gerogia Tech)

AlPO₄-H3
(*Bull. Soc. Chem. Fr.* **372:1762(1961)**)

Related Structures: MCM-1

Structure

(*Nature* 318:165(1985))

CHEMICAL COMPOSITION: AlPO₄*1.52H₂O
SYMMETRY: orthorhombic
SPACE GROUP: Pbca
UNIT CELL CONSTANTS (Å): $a = 19.352(1)$
                            $b = 0.721(1)$
                            $c = 9.762(1)$

X-Ray Powder Diffraction Data: (from author) (d(Å) (I/I_0)) 9.71(37), 6.88(96), 6.5(87), 4.88(93), 4.26(100), 3.96(14), 3.39(49), 3.06(91), 2.68(52).

The AlPO₄ framework of H3 consists of PO₄ alternating with AlO₄ tetrahedra and AlO₄(H₂O)₂ octahedra. The AlO₄(H₂O)₂ units in this structure are similar to those found in metavariscite. Two-dimensional sheets composed of six-mem-

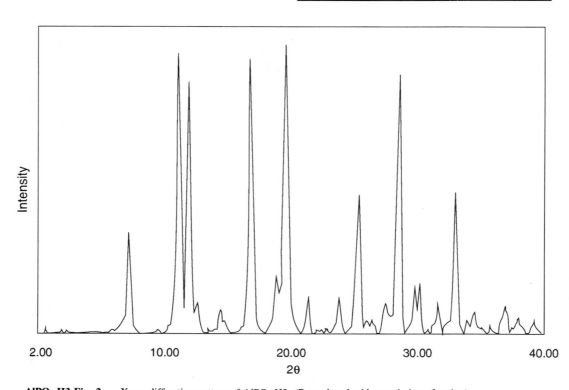

AlPO₄-H3 Fig. 1S: Framework topology.

ber rings and two-dimensional sheets of four- and eight-member rings alternate in this structure (*Nature* 318:165(1985)).

Atomic positions ($\times$ 10⁵) of AlPO₄-H3 (*Nature* 318:165 (1985)):

	x/a	y/b	z/c
P(1)	44,769(4)	34,072(8)	13,498(9)
P(2)	29,050(4)	6,565(8)	33,879(9)
Al(1)	44,808(4)	8,470(9)	32,419(10)
Al(2)	28,207(5)	33,224(9)	14,654(10)
O(1)	37,433(10)	37,652(22)	17,006(24)
O(2)	46,047(12)	16,616(22)	48,037(24)
O(3)	49,520(10)	44,967(21)	19,709(24)
O(4)	46,889(11)	20,038(20)	19,416(23)
O(5)	27,397(10)	5,819(20)	49,128(22)
O(7)	28,329(11)	21,582(20)	29,532(23)
O(8)	25,855(10)	47,583(20)	25,919(22)
O(9)	18,446(12)	28,311(27)	11,931(31)
O(10)	30,512(12)	17,298(23)	3,398(26)
O(11)	11,929(14)	14,398(33)	36,268(37)
H(1)	15,752(12)	29,879(27)	5,232(31)
H(2)	16,096(12)	25,744(27)	18,827(31)
H(3)	28,123(12)	10,248(23)	5,406(26)
H(4)	30,108(12)	19,016(23)	−5,101(26)

AlPO₄-H3 Fig. 2:   X-ray diffraction pattern of AlPO₄-H3. (Reproduced with permission of author).

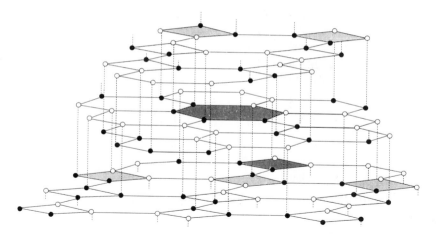

AlPO₄-H3 Fig. 3:   Alternating sheets of four-member and eight-member rings with sheets of six-member rings comprising the AlPO₄-H3 structure. (*Nature* 318:165 (1985)).

Synthesis

ORGANIC ADDITIVE
none

AlPO₄-H3 is prepared from reactive aqueous gels at Ph of 1.60 and P_2O_5/Al_2O_3 ratio of 2.73. Crystallization begins at temperatures between 80 and 100°C after 1 hour, with maximum crystallization occurring after 3 hours. Longer reaction times result in conversion to variscite and metavariscite. Impurities of AlPO₄-H1 and AlPO₄-H2 are observed in the system containing no organic additives (*Bull. Chem. Soc. Fr.* 1762(1961)). VPI-5 and AlPO₄-11 are observed when dipropylamine is present (*SSSC* 52:193(1989)). AlPO₄-H3 is a common impurity phase in many AlPO₄ molecular sieve syntheses.

Thermal Properties

When AlPO₄-H3 is heated between 100 and 180°C, the X-ray diffraction pattern changes, giving a new form of AlPO₄ identified as AlPO₄-C. This form is very hygroscopic, and rehydration returns the structure to AlPO₄-H3. At temperatures above 200°C, AlPO₄-C transforms to AlPO₄-D, which is stable to 550°C. Rehydration

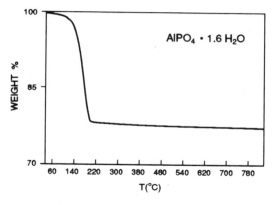

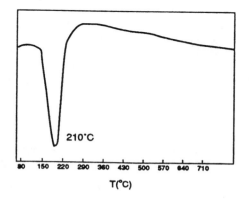

AlPO₄-H3 Figs. 4a and 4b: TGA. (Fig. 4a) /DTA. (Fig. 4b) For AlPO₄-H3. (Reproduced with permission of B. Duncan, Georgia Tech)

of this species results in the formation of AlPO₄-H6, which has the composition AlPO4*1.4 to 1.5 H_2O. This latter step is reversible. Temperatures above 800 to 900°C result in the formation of AlPO₄-tridymite and AlPO₄-crystobalite.

Infrared Spectrum

Mid-infrared Bands (cm⁻¹): 1150s, 1070s, 890wbr, 725w, 655w, 630w, 540sh, 520sh, 490m, 465m.

Adsorption

Water adsorption properties of AlPO₄-H3 (from author):

Pressure, Torr	% Weight adsorbed (20°C)	% Weight adsorbed (35°C)	% Weight adsorbed (50°C)
0.5	8.08	0.44	0.0
1.0	14.38	6.44	0.0
2.0	15.63	13.92	3.89
5.0	16.28	15.61	12.40
10.0	17.04	16.16	15.30
17.0	17.96	16.68	15.91

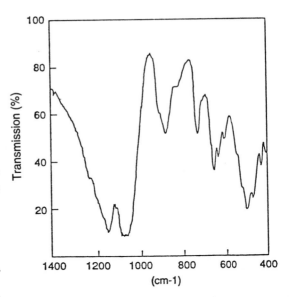

AlPO₄-H3 Fig. 5: Mid-infrared spectrum of AlPO₄-H3. (Reproduced with permission of K. Vinje, U. of Oslo)

AlPO₄-H4
(Bull. Chem. Soc. Fr. 1762(1961))

Structure

Chemical Composition: AlPO4*1.2 H_2O.

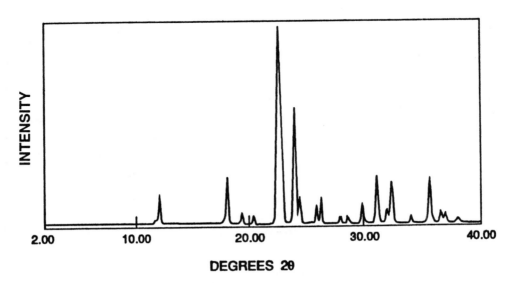

AlPO₄-H4 Fig. 1: X-ray powder diffraction pattern for AlPO₄-H4. (Reproduced with permission of K. Vinje, U. of Oslo, Norway)

X-Ray Powder Diffraction Data: $(d(\text{Å}) (I))$
7.28(m), 4.90(F), 4.58(mf), 4.36(f), 3.92(FFF),
3.69(FF), 3.45(m), 3.39(m), 3.19(mf), 3.11(f),
2.995(m), 2.860(F), 2.794(ff), 2.758(mF),
2.708(fff), 2.648(ff), 2.511(F), 2.452(mf),
2.423(ff), 2.251(mf), 2.229(mf), 2.177(mf).

Synthesis

ORGANIC ADDITIVES
none

AlPO$_4$-H4 is prepared from reactive aqueous
aluminophosphate gels at pH of 3.09, P$_2$O$_5$/Al$_2$O$_3$
ratio of 2.73, and P/Al ratio of 1.5. Aluminum
hydroxide was used as the source of aluminum,
and 0.4% Na$_2$O also was present in the aluminum source. Crystallization begins at temperatures between 80 and 100°C after 24 hours. Under these conditions, AlPO$_4$-H4 appears in trace
quantities together with traces of AlPO$_4$-H2.
AlPO$_4$-H4 also can be prepared under conditions
used to prepare VPI-5 (dipropylamine as the
organic additive) with the addition of HCl to
the reaction mixture or in the presence of acetic
acid with extended crystallization times (from
author).

Thermal Properties

See AlPO$_4$-H4 Figures 2a and 2b.

Infrared Spectrum

Mid-infrared Bands (cm^{-1}):   1236s, 1201s,
1160s, 1095s, 1041s, 780mw, 741mw, 714mw,
663mw, 617w, 515m, 492m, 453m, 410m,
383m.

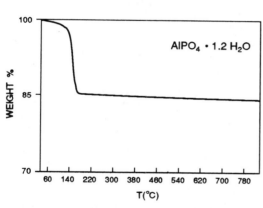

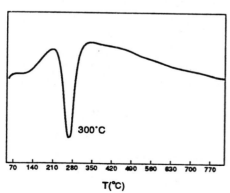

AlPO$_4$-H4 Figs. 2a and 2b: TGA. (Fig. 2a) /DTA.
(Fig. 2b) Of AlPO$_4$-H4. (Reproduced with permission of
K. Vinje, U. of Oslo, Norway)

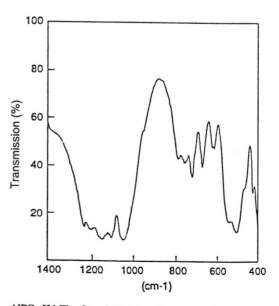

AlPO$_4$-H4 Fig. 3:   Mid-infrared spectrum of AlPO$_4$-
H4. (Reproduced with permission of K. Vinje, U. of
Oslo, Norway)

Adsorption

H₂O adsorption data for AlPO₄-H4 (from author):

Pressure, Torr	% Weight adsorbed (20°C)	% Weight adsorbed (35°C)	% Weight adsorbed (50°C)
0.5	0.12	0.0	0.0
1.0	0.12	0.0	0.0
2.0	0.18	0.0	0.0
5.0	0.30	0.0	0.0
10.0	0.41	0.0	0.0
17.0	0.77	0.0	0.0

AlPO₄-H5
(Bull. Chem. Soc. Fr. 1762(1961))

Structure

Chemical Composition: AlPO₄*5/3H₂O.

X-Ray Powder Diffraction Data: (d(Å) (I))
6.65(f), 5.70(fff), 5.16(mF), 4.66(FF,FFF), 4.08(FFF,FF), 3.85(m), 3.54(F), 3.47(FF), 3.17(m), 3.08(mf), 2.874(mf), 2.843(F,FF), 2.778(m), 2.681(fff), 2.626(mF), 2.570(ff), 2.514(m), 2.444(m), 2.340(mf).

Synthesis

ALPO₄-H5 represents a partial hydration product of AlPO₄-B that was prepared from variscite.

AlPO₄-H6
(Bull. Chem. Soc. Fr. 1762(1961))

Structure

Chemical Composition: AlPO₄*3/2H₂O.

X-Ray Powder Diffraction Data: (d(Å) (I))
7.07(FF), 6.11(fff), 5.44(fff), 5.12(f), 4.90(f), 4.67(f), 4.53(fff), 4.43(fff), 4.23(FFF), 4.14(mf), 4.10(ff), 3.91(ff), 3.62(m), 3.44(f), 3.37(ff), 3.26(f), 3.21(m), 3.12(m), 2.82(ff), 2.79(ff), 2.75(ff), 2.61(fff), 2.56(ff), 2.52(f), 2.44(ff), 2.11(f).

Synthesis

ALPO₄-H6 represents a partial hydration product of AlPO₄-D that was prepared originally from AlPO₄-H3.

AlPO₄ (Other Phases)
(Soil Science Society Proceedings 15:76(1951))

Structure

Empirical formula and classification of aluminum phosphate minerals:

Product	Empirical formula*	Group$
A	2K₂O • 3Al₂O₃ • 5P₂O₅ • 26H₂O	1
B	2K₂O • 3Al₂O₃ • 5P₂O₅ • 20H₂O	1
C	K₂O • Al₂O₃ • P₂O₅ • 4H₂O	2
G	(NH₄)₂O • Al₂O₃ • 2P₂O₅ • 3H₂O	3
H	K₂O • 2Al₂O₃ • 2P₂O₅ • 5H₂O	4
HH	K₂O • 2Al₂O₃ • 2P₂O₅ • 5H₂O	4
J	(NH₄)₂O • 2Al₂O₃ • 2P₂O₅ • 5H₂O	4
MM	Al₂O₃ • P₂O₅ • 4H₂O	6
Q	Al₂O₃ • P₂O₅	7
R	K₂O • Al₂O₃ • P₂O₅ • 4H₂O	8
S	K₂O • Al₂O₃ • P₂O₅ • 2H₂O	8
T	K₂O • 2Al₂O₃ • 2P₂O₅ • 1.6HF • 6.2H₂O	9

*Empirical formula according to Z. Kristallogr. 103:228(1941)).
$The only mineral found in the literature that corresponded to these compounds were: Group 1—Taranakite; group 6—variscite, barrandite, and strengite; group 7—berlinite; and group 9—minyulite.

X-ray Powder Diffraction Data (d(Å) (I/I):

(Group 1)
A 15.7(100), 7.88(35), 7.35(68), 5.81(29), 5.03(13), 4.61(8), 4.30(37), 4.14(11), 3.99(9), 3.79(71), 3.55(41), 3.34(22), 3.27(25), 3.13(65), 2.93(16), 2.81(48), 2.72(24), 2.61(40), 2.56(10), 2.53(8), 2.46(6), 2.42(5), 2.39(22), 2.34(8), 2.14(6), 2.08(14), 2.05(18), 2.03(8), 1.97(10).

B 13.8(100), 7.35(68), 6.79(51), 6.05(15), 5.46(68), 4.26(40), 4.14(51), 3.89(4), 3.64(24), 3.50(16), 3.39(100), 3.13(49), 3.01(40), 2.90(64), 2.84(11), 2.79(38), 2.76(40), 2.74(25), 2.68(16), 2.64(24), 2.59(9), 2.47(9), 2.45(11), 2.35(11), 2.29(20), 2.22(22), 2.15(7), 2.11(9), 2.08(15), 2.04(11), 2.01(20), 1.98(11), 1.95(15), 1.83(15), 1.76(17).

Chemical composition and density of aluminum phosphate minerals:

Product	Chemical Composition, %					
	P$_2$O$_5$	Al$_2$O$_3$	K$_2$O	(NH$_4$)$_2$O	Ignition loss§	Density, g/cc
A	42.0	17.5	11.5		29.0	2.09
B*	45.7	19.0	12.5		23.0	2.24
C	51.2	16.7	16.2		13.4	2.52
G	59.7	20.8		8.6	19.9	2.47
H	42.5	29.8	15.0		13.8	2.55
HH	42.3	28.9	14.7		13.3	2.55
J	45.0	30.2		7.8	24.7	
MM	44.5	30.1			24.6	
R	35.0	24.6	23.7		17.2	2.29
S	40.1	28.8	20.9		10.6	
T	38.0	28.4	12.7	(F,4.1)	21.0	2.44

*Produced by drying product A to constant weight at 95°C.
§At 1000°C.

(Group 2)
C 9.80(3), 7.88(100), 6.79(17), 6.40(8), 5.46(9), 4.86(5), 4.56(3), 4.34(7), 4.07(3), 3.89(17), 3.64(26), 3.52(13), 3.42(13), 3.37(15), 3.25(13), 3.09(15), 2.99(19), 2.81(18), 2.74(3), 2.68(12), 2.60(3), 2.53(16), 2.52(5), 2.24(7), 2.22(3), 2.20(5), 2.16(7), 2.11(3).

(Group 3)
G 8.02(8), 7.61(100), 6.40(9), 6.13(15), 5.59(3), 4.81(2), 4.34(2), 4.14(3), 4.07(7), 3.89(3), 3.79(4), 3.70(3), 3.50(3), 3.25(3), 3.20(3), 3.13(3), 3.03(24), 2.90(27), 2.84(6), 2.68(16), 2.63(20), 2.39(4), 2.28(3), 2.20(6), 2.17(3), 2.09(4).

(Group 4)
H 6.79(59), 6.59(72), 5.46(45), 5.08(6), 4.81(33), 4.56(22), 4.26(37), 4.14(58), 3.92(13), 3.79(11), 3.32(45), 3.20(56), 3.13(47), 3.07(13), 3.03(20), 2.90(100), 2.77(72), 2.68(6), 2.63(19), 2.59(26), 2.54(45), 2.49(15), 2.41(9), 2.34(8), 2.31(7), 2.28(17), 2.22(38), 2.16(8), 2.08(56), 2.02(7), 1.95(6), 1.94(7), 1.83(15), 1.76(17).

J 7.48(11), 6.69(100), 5.89(37), 4.66(19), 4.22(8), 4.14(16), 4.03(3), 3.70(11), 3.50(3), 3.32(4), 3.27(3), 3.01(17), 2.97(16), 2.95(12), 2.93(16), 2.82(7), 2.79(7), 2.61(16), 2.50(3), 2.44(3), 2.39(4), 2.31(7), 2.26(3), 2.14(6), 2.09(6), 1.96(3), 1.89(7), 1.76(3), 1.64(6), 1.57(6).

(Group 6)
MM 5.66(47), 4.81(17), 4.26(100), 3.89(26), 3.61(10), 3.20(5), 3.18(6), 3.03(87), 2.90(44), 2.86(20), 2.77(12), 2.56(14), 2.47(11), 2.45(17), 2.39(6), 2.33(6), 2.28(7), 2.11(8), 2.08(11), 2.04(6), 2.01(5), 1.94(10), 1.91(6), 1.85(7), 1.84(8), 1.75(13), 1.71(6), 1.67(6), 1.60(19), 1.59(7).

(Group 7)
Q 2.46(56), 3.96(5), 3.32(3), 2.93(7), 2.74(4), 2.46(15), 2.29(5), 2.24(7), 2.19(2), 2.14(17), 1.98(7), 1.91(2), 1.83(10), 1.72(2), 1.68(4), 1.55(11).

(Group 8)
R 11.6(100), 7.4(21), 5.5(21), 4.9(10), 4.26(4), 3.76(10), 3.42(26), 3.16(64), 2.90(26), 2.76(26), 2.49(8), 2.39(21), 2.34(21), 2.26(10), 2.07(15), 1.85(6).

S 6.59(100), 5.74(48), 4.97(9), 4.61(25), 4.22(18), 4.10), 3.70(29), 3.55(29), 3.39(14), 3.27(39), 2.97(75), 2.93(63), 2.91(47), 2.84(32), 2.76(7), 2.72(14), 2.64(12), 2.59(66), 2.44(30), 2.39(18), 2.29(45), 2.25(9), 2.22(7), 2.11(20), 2.09(9), 2.03(5), 1.92(9), 1.88(14), 1.83(7), 1.80(4), 1.78(7), 1.76(11), 1.73(7), 1.64(18).

T 6.69(100), 5.52(70), 4.86(4), 4.66(11), 4.18(15), 3.64(17), 3.55(4), 3.39(29), 3.34(42), 3.05(33), 2.95(31), 2.86(9), 2.76(6), 2.66(20), 2.60(18), 2.54(3), 2.44(7), 2.39(6), 2.35(5), 2.31(5), 2.26(28), 2.13(17), 2.10(16), 2.00(5), 1.90(5), 1.89(6), 1.80(5), 1.77(4), 1.74(4), 1.68(15), 1.66(5), 1.61(7), 1.58(4), 1,56(5).

Synthesis

Group 1 materials were prepared from partial dehydration of taranakite at 95°C. Taranakite crystallizes at temperatures below 95°C within a pH range of 1.7 to 5.5. Phosphorus concentrations in the solution range between 0.001 and 1.0 M. Crystallinity is lost completely at 125°C.

Group 2 materials were prepared between 95 and 145°C and at pH ranging from 2 to 3. Phosphate concentrations are higher (from 1.5 to 2.5 M). Group 3 minerals are crystallized at 27, 95, and 145°C in a pH range between 2.0 and 3.5.

Phosphate concentrations were set at 3.0 M. Minerals of Group 4 were crystallized between 75 and 145°C, with a pH range between 2.5 and 6.0. Phosphate concentrations were between 1.0 and 3.5 M. Group 5 materials were crystallized at temperatures from 27 to 95°C and pH from 0.5 to 2.5. Phosphate concentrations ranged from 0.5 to 3.0 M. Group 6 minerals were crystallized in the presence of lithium, magnesium, and calcium phosphate at temperatures from 27 to 145°C, and in the presence of cesium phosphate at temperatures to 95°C. Crystal properties of MM agree with the literature data for variscite. Berlinite, the AlPO$_4$ analog of quartz, was a product in Group 7 crystallization. It was precipitated with calcium phosphate solution (0.5 M) at pH of 1.5 at 145°C. Group 8 materials were prepared at higher pH, between 7 and 8.5, at temperatures between 27 and 145°C. These minerals appeared to hydrolyze readily in water. In the presence of fluoride ion (KF), the Group 9 compound was prepared. The x-ray diffraction spacing corresponds to those of the mineral minyulite. Crystallization occurs at 95°C and pH around 3.0. In all preparations, crystallization occurred between 1 and 54 days.

AlPO$_4$-P1

Synthetic pollucite analog containing Cs. Cs$_{12}$ [Al$_{24}$P$_{24}$O$_{96}$]·(OH)$_{12}$

Amicite (GIS)
(natural)

Related Materials: gismondine

Amicite, first identified in 1979, was named for G. B. Amici, the inventor of the Amici-lens, by Alberti and Vezzalini (*Acta Crystallogr.* B335:2886(1979)).

Structure

(*Acta Crystallogr.* B335:2886(1979))

CHEMICAL COMPOSITION: Na$_4$K$_4$(Al$_8$Si$_8$O$_{32}$)* 10H$_2$O
SPACE GROUP: I2

UNIT CELL CONSTANTS: (Å) $a = 10.23$
$b = 10.42$
$c = 9.88$
$\beta = 88°19'$

X-Ray Powder Diffraction Data: (*Nat. Zeo.* 1985)) (d(Å) (I/I_0)) 7.30(55), 7.20(10), 5.11(40), 4.94(28), 4.22(90), 4.18(40), 4.12(8), 4.05(4), 3.647(7), 3.585(5), 3.289(30), 3.238(45), 3.141(80), 2.965(10), 2.769), 2.722(100), 2.704(50), 2.674(20), 2.605(40), 2.470(5), 2.424(8), 2.390(7), 2.355(3), 2.324(7), 2.305(7), 2.249(5), 2.243(7), 2.183(4), 2.165(3), 2.112(10), 2.090(10), 2.025(5), 2.006(3), 1.997(3).

Amicite has perfect (Si,Al) order within the structure. The four sodium ions occupy two of the sites where calcium is located in the gismondine framework, whereas the four potassium ions occupy sites that are void of cations and filled only by water in gismondine (*Acta Cryst.* B335:2886(1979)).

Ammonioleucite (ANA)

See analcime.

AMS-1B (MFI)
(US4,269,813(1981))
AMOCO Chemical Company

Related Materials: ZSM-5

Structure

CHEMICAL COMPOSITION: 0.9 M$_{2/n}$ O : B$_2$O$_3$: Y SiO$_2$:Z H$_2$O (m = H or alkali cations)
UNIT CELL CONSTANTS: (Å) variable with B content

X-Ray Powder Diffration Data: (d(Å) (I/i_0)) 11.32(71), 10.11(46), 6.01(21), 5.60(14), 5.01(11), 4.27(15), 4.01(16), 3.84(100), 3.73(53), 3.65(40), 3.45(18), 3.32(14), 3.05(15), 2.98(19), 1.99(19).

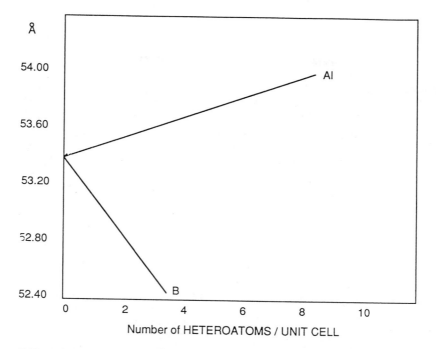

AMS-1B Fig. 1:   Unit cell volume as a function of the number of heteroatoms per unit cell. A comparison between AMS-1B (B) and ZSM-5 (Al) (*J. Catal.* 91, 352 (1985)). (Reprinted with permission of the American Chemical Society)

The borosilicate molecular sieve AMS-1B exhibits a contraction in the unit cell with increasing boron content. See ZSM-5 and silicalite for topological description of the oxide framework.

Synthesis

ORGANIC ADDITIVES (US4,285,919(1981))
tetrapropyl ammonium$^+$
tetrapropyl ammonium$^+$ tetramethylammonium$^+$
lower alkyl primary or secondary amines
  (US4,514,416(1985)) diaminoenthane (EP 74,900(1983))

AMS-1B crystallizes between 8 hours and 3 weeks from reaction mixtures with a range of compositions:

(1) SiO_2/B_2O_3 1–400
(2) $R_2O/[R_2O + (NH_4)_2O]$ 0.01–1
(3) $NH_4^+/[NH_4^+ + Metal\ cation]$ 0.7–1
(4) NH_3/SiO_2 0.02–20
(5) H_2O/NH_3 2–2000

Crystallization in this composition range occurs between 95 and 225°C, at a pH ranging from 8.4 to 11.2. After isolation of the solid, drying is done at 110°C for 16 hours. Crystals with diameters of 2 microns are obtained under such conditions.

Thermal Properties

Samples of AMS-1B are calcined in forced air at 538°C with a heating rate of 111°C per hour to produce the organic-free materials. Ammonia desorbs upon thermal treatment of the NH_4^+ form around 200°C.

Analcime (ANA)
(natural)

Related Structures: AlPO₄-24
AlPO₄-pollucite
Na-B
Ca-D
beryllosilicate pollucite
iron silicate pollucite
hsianghualite
kehoeite
leucite
pollucite
SAPO-analcime
viseite
wairakite

Analcime has been known from ancient times, as it is a common and easily recognizable mineral. It was first reported in 1784.

Structure

CHEMICAL COMPOSITION: $Na_{16}[Al_{16}Si_{32}O_{96}]*16$ H_2O
SYMMETRY: cubic (*Z. Kristallogr.* 74:1(1930)); orthorhombic (*Zeo.* 8:247 (1988).
SPACE GROUP: Ia3d ; Ibca
UNIT CELL CONSTANTS: a = 13.7 Å ; a = 13.720 Å, b = 13.715 Å, c = 13.709 Å

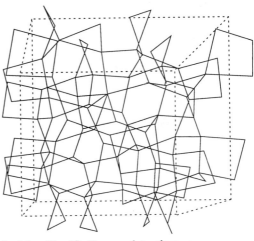

Analcime Fig. 1S: Framework topology.

FRAMEWORK DENSITY: 1.85 g/cc
VOID VOLUME: 0.18 cc H_2O/g
PORE STRUCTURE: irregular one-dimensional, highly distorted eight-rings. Free opening of 2.6 Å.

X-Ray Powder Diffraction Data: (*Nat. Zeol.* 321 (1985) (d(Å)(I/I_0)) 6.87(<10), 5.61(80), 4.86(40), 3.67(20), 3.43(100), 2.925(80), 2.801(20), 2.693(50), 2.505(50), 2.426(30), 2.226(40), 2.168(<10), 2.115(<10), 2.022(10), 1.940(<10), 1.904(50), 1.867(40), 1.833(<10), 1.743(60), 1.716(30), 1.689(40), 1.644(10),

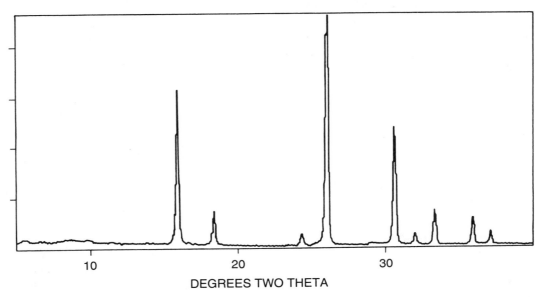

Analcime Fig. 2:   X-ray powder diffraction pattern of analcime. (Reproduced with permission of author).

1.618(20), 1.596(30), 1.498(20), 1.480(20), 1,463(10), 1.447(10), 1.415(40), 1.386(<10), 1.372(10), 1.359(40), 1.308(10), 1.285(20), 1.263(20), 1.220(30).

Analcime forms a dense structure relative to the other zeolites and thus is a border between the zeolites and the feldsphathoids. The cubic (or orthorhombic) structure consists of four-, six-, and eight-member rings, giving 16 large interconnected cavities that form nonintersecting channels (*Atti. Acad. Sci. Torino Cl. Sci. Fis. Mat. Nat.* 98, 424 (1964)). There are also 24 smaller cavities in this structure. Sixteen of the small cages are occupied by sodium cations, with water located in the larger cages. The mean value of Si(Al)–O bonds = 1.6467 A, Na–O = 2.4754 A, O–Si(Al)–O = 109.49°, and O–Na–O = 84.19°. Some variability in the SiO_2/Al_2O_3 does occur in this mineral, and the water content varies linearly with silica content. Leucite is the potassium form of this structure, pollucite is the cesium form, and the calcium form is known as wairakite. The calcium-rich form of this aluminosilicate, when dehydrated, can occlude molecules such as methane and ethane (*Z. Kristallogr.* 128:352(1969)). Lucite and pollucite are usually anhydrous.

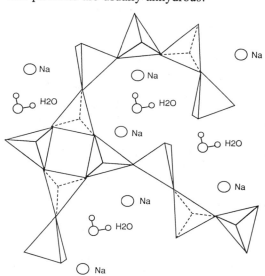

Analcime Fig. 3: Cation and water locations in the crystal structure of orthorhombic analcime (*Zeo.* 8:247 (1988)). (Reprinted with permission of Butterworth Publishing Company)

Fractional atomic coordinates for natural orthorhombic analcime (*Zeo.* 8:247 (1988)):

	x	y	z
Si(Al)1	0.1265(1)	0.1624(1)	0.4122(1)
Si(Al)2	0.0876(1)	0.3757(1)	0.3377(1)
Si(Al)3	0.1625(2)	0.0877(2)	0.6252(2)
O1	0.1048(2)	0.3676(2)	0.2189(2)
O2	0.1176(2)	0.1452(2)	0.5314(2)
O3	0.2207(2)	0.1040(2)	0.3642(2)
O4	0.1458(3)	0.4704(3)	0.3851(3)
O5	0.1350(3)	0.2811(3)	0.3957(3)
O6	0.0309(3)	0.1147(3)	0.3541(3)
Na1	0.1247(3)	0.0	0.25
Na2	0.25	0.1252(3)	0.0
Na3	0.1236(4)	0.1268(4)	0.1252(4)

Synthesis

ORGANIC ADDITIVES
none required

Though occurring readily in nature, analcime can be synthesized from a variety of cation-containing mixtures. These include Na, K, Rb, Cs, Tl, ammonia, Ca, Sr, and the mixed (Na,K), (Na,Rb), (Na,Cs), (Na,Tl), (K,Rb), (Rb,Tl), and (Li,Cs). Little preference is seen for any one cation system (*Zeo.* 1:130(1981)). In gels with SiO_2/Al_2O_3 around 4, this structure is dominant (*JCS* 4035(1953); *SSSC* 28:263 (1986)). Crystallization can occur at 100°C under atmospheric pressure over a time span of 120 hours or less (*Am. Mineral.* 64:172(1979)).

Crystallization occurring from clear solutions has been shown for this zeolite (*Am. Mineral.* 64:172(1979)). Crystals form at 100°C after 120 hours, with the composition governed by the initial ratios of hydroxide, alumina, and silica in the gel and appearing independent of the depletion of the solution species with time. Compositions that produced analcime from this system include: $10\ Na_2O : 0.1\ Al_2O_3 : 7\ SiO_2 : 370\ H_2O$.

Thermal Properties

Analcime exhibits a continual weight loss due to water (8.7%) by 400°C. An endotherm is observed in the DTA between 200 and 400°C.

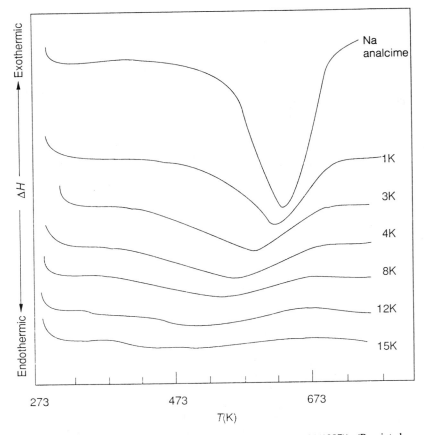

Analcime Fig. 4:   DSC curves for Na and K analcime (*Zeo* 7:191(1987)). (Reprinted with permission of Butterworth Publishing Company)

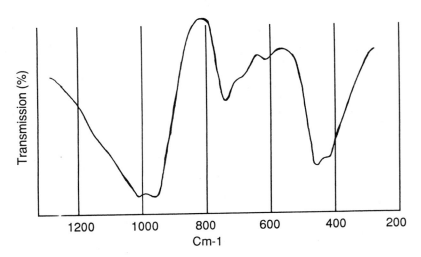

Analcime Fig. 5:   Infrared spectrum of zeolite C (analcime) (*ACS* 101:201(1971)). (Reproduced with permission of the American Chemical Society)

Structural collapse occurs around 700°C (*J. Phys. Chem.* 41:1029(1937); *Proc. Roy. Soc. A* 167:393(1938); *Mineral. J.* 1:36(1953)).

Infrared Spectra

Infrared Spectral Data for Synthetic Analcime (C) (cm⁻¹): (*ACS* 101:201(1971)) (SiO_2/Al_2O_3 = 4) 1162vwsh, 1021s, 952s, 740m, 686wb, 615vw, 442ms, 410msh.

Adsorption

Because of the size of the pore openings in this zeolite, molecules larger than 3 Å cannot be adsorbed.

Sorption properties of natural analcime (*Nat. Zeo.* 353 (1978)):

Molecule adsorbed	Wt%
H_2O	8
N_2	0
n-C_4H_{10}	0
C_6H_6	0

Ion Exchange

Room-temperature exchange does not occur readily in this material. At elevated temperatures Na^+ can be completely replaced by K^+, Ag^+, Tl^+, NH_4^+, and Rb^+ cations. Very little exchange is observed with Li^+, Cs^+, Mg^{+2}, Ca^{+2}, and Ba^{+2} cations. A very distinct ion sieving effect between Rb and Cs is observed (*JCS* 2342(1950)). Analcime (Ag^+ exchanged form) has been used to purify CsCl solutions contaminated with NaCl, KCl, or RbCl (US3,010,785 (1961)).

Analcite (ANA)

Improper common name for analcime.

Arduinite (MOR)
(natural)

Obsolete synonym for mordenite.

AsAPO
(US4,913,888(1990))
UOP

AsAPO compositionally represents a family of arsenic-aluminophosphates with microporous structures. See AsAPO Fig. 1 on following page for composition.

AsAPO-5 (AFI)
(US4,913,888(1990))
UOP

Related Materials: AlPO₄-5

Structure

Chemical Composition: See AsAPO.

X-Ray Powder Diffraction Data: ($d(Å)(I/I_0)$) 12.1–11.56 (m–vs), 4.55–4.46 (m–2), 4.25–4.17 (m–vs), 4.00–3.93 (w–ws), 3.47–3.40 (w–m).

Synthesis

AsAPO-5 crystallizes from a batch composition of: 1.0–2.0 TPA : 0.05–0.2 As_2O_q : 0.5–1.0 Al_2O_3 : 0.5–1.0 P_2O_5 : 40–100 H_2O (*q* denotes the oxidation state of arsenic). (TPA = tetrapropylammonium⁺)

Adsorption properties of AsAPO-5:

Adsorbate	Kinetic diameter, Å	Pressure, Torr	Temp., °C	Wt. % adsorbed*
O_2	3.46	100	−183	7
O_2		750		10
Neopentane	6.2	700	24	4
H_2O	2.65	4.3	24	4
H_2O		20	24	12

*Typical amount adsorbed.

AsAPO-11 (AEL)
(US4,913,888(1990))
UOP

Related Material: AlPO₄-11

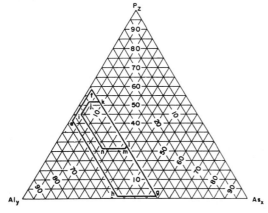

AsAPO Fig. 1:   Ternary diagram relating to the preferred composition in mole fractions of the arsenic-aluminophosphate molecular sieves (US4,913,888 (1990)).

Structure

Chemical Composition: See AsAPO.

X-Ray Powder Diffraction Data: $(d(\text{Å})\,(I/I_0))$
10.81(51), 9.29(86), 6.67(24), 5.63(53), 4.66(9), 4.33(68), 4.21(90), 4.00(59), 3.92(56), 3.90(62), 2.83(100), 3.59(14), 3.37(13sh), 3.34(24), 3.190(5), 3.142(18), 3.109(21), 3.061(6), 3.021(6), 2.833(8), 2.718(22), 2.615(12), 2.457(8), 2.436(5), 2.385(12), 2.374(15), 2.344(6), 2.321(8), 2.283(5), 2.009(6), 1.867(7), 1.797(5).

Synthesis

AsAPO-11 crystallizes from a batch composition of: 1.0 $(C_6H_{15}N)$: 0.5 As_2O_5: 1.0 Al_2O_3 : 0.5 P_2O_5: 50 H_2O. Crystallization occurs at 200°C after 108 hours.

Adsorption

Adsorption properties of AsAPO-11:

Adsorbate	Kinetic diameter, Å	Pressure, Torr	Temp., °C	Wt. % adsorbed*
O_2	3.46	100	−183	5.8
O_2		750		8.9
Cyclohexane	6.0	50	22.8	5.1

*Typical amount adsorbed.

AsAPO-16 (AST) (US4,913,888(1990)) UOP

Related Material: AlPO$_4$-16

Structure

Chemical Composition: See AsAPO.

X-Ray Powder Diffraction Data: $(d(\text{Å})(i/I_0))$
7.76(75),   4.74(50),   4.05(100),   3.87(17), 3.353(27),   3.077(10),   3.000(26),   2.733(7), 2.578(6),   2.369(11),   2.263(3),   2.041(3), 1.876(7),   1.745(2),   1.674(2).

Synthesis

AsAPO-16 crystallizes from a batch composition typically of: 1.0 quinuclidine : 0.5 As_2O_q : 1.00 Al_2O_3 : 0.5 P_2O_5 : 50 H_2O. Crystallization occurs after 23 hours at 200 °C.

Adsorption

Adsorption properties of AsAPO-16:

Adsorbate	Kinetic diameter, Å	Pressure, Torr	Temp., °C	Wt. % adsorbed
O_2	3.46	100	−183	1.0
O_2		750		3.8
H_2O	2.65	20	24	29.6
H_2O		4.6	24	17.6

*Typical amount adsorbed.

AsAPO-17 (ERI) (US4,913,888(1990)) UOP

Related Materials: AlPO$_4$-17
                                      erionite

Structure

Chemical Composition: See AsAPO.

X-Ray Powder Diffraction Data: $(d(Å)(I/I_o))$
11.5–11.4(vs), 6.61(s–vs), 5.72–5.70(s), 4.52–
4.51(w–s), 4.33–4.31(vs), 2.812–2.797(w–s).

Synthesis

AsAPO-17 crystallizes from a batch composition of: 1.0–2.0 quinuclidine : 0.05–0.2 As_2O_q : 0.5–1.0 Al_2O_3 : 0.5–1.0 P_2O_5 : 40–100 H_2O.

Adsorption

Adsorption properties of AsAPO-17:

Adsorbate	Kinetic diameter, Å	Pressure, Torr	Temp., °C	Wt. % adsorbed
O_2	3.46	100	−183	10
O_2		750		12
n-Butane	4.3	100	24	4
H_2O	2.65	4.3	24	13
H_2O		20	24	14

*Typical amount adsorbed.

AsAPO-18 (AEI)
(US4,913,888(1990))
UOP

Related Material: $AlPO_4$-18

Structure

Chemical Composition: See AsAPO.

X-Ray Powder Diffraction Data: $(d(Å)(I/I_o))$
9.27(100), 8.61(5), 8.15(9), 6.82(6), 6.37(7),
6.02(7), 5.76(8), 5.26(24), 5.01(9), 4.44(22),
4.27(25), 4.04(9), 3.94(5), 3.68(8), 3.600(6),
3.445(10), 3.363(8), 3.205(7), 3.006(8), 2.933(8),
2.883(10), 2.791(12).

Synthesis

AsAPO-18 crystallization from a reactive gel with a composition of: Et_4NOH : 0.75 As_2O_5 : 1.00 Al_2O_3 : 0.25 P_2O_5 : 45 H_2O : 6.0 i-C_3H_7OH. Aluminum isopropoxide is used as the source of aluminum. Crystallization occurs after 65 hours at 200°C. The product is not pure.

Adsorption

Adsorption properties of AsAPO-18:

Adsorbate	Kinetic diameter, Å	Pressure, Torr	Temp., °C	Wt. % adsorbed
O_2	3.46	100	−183	18.3
O_2		750		24.2
n-hexane	4.3	45	21.5	9.3
Isobutane	5.0	101	21.3	0.4
H_2O	2.65	4.3	24	25.9
H_2O		20	24	34.8

*Typical amount adsorbed.

AsAPO-31 (ATO)
(US4,913,888(1990))
UOP

Related Materials: $AlPO_4$-31

Structure

Chemical Composition: See AsAPO.

X-Ray Powder Diffraction Data: $(d(Å)(I/I_o))$
10.40–10.28(m–s), 4.40–4.37(m), 4.06–4.02(w–m), 3.93–3.92(vs), 2.823–2.814(w–m).

Synthesis

AsAPO-31 crystallizes from a reactive mixture of: 1.0–2.0 DPA : 0.05–0.2 As_2O_q : 0.4–1.0 Al_2O_3 : 0.5–1.0 P_2O_5 : 40–100 H_2O where q denotes the oxidation state of the arsenic, and DPA represents di-n-propylamine.

Adsorption

Adsorption properties of AsAPO-31:

Adsorbate	Kinetic diameter, Å	Pressure, Torr	Temp., °C	Wt. % adsorbed*
O_2	3.46	100	−183	4
O_2		750		6
Cyclohexane	6.0	90	24	3
Neopentane	6.2	700	24	3
H_2O	2.65	4.3	24	3
H_2O		20	24	10

*Typical amount adsorbed.

AsAPO-34 (CHA)
(US4,913,888(1990))
UOP

Related Material: AlPO₄-34
chabazite

Structure

Chemical Composition: See AsAPO.

X-Ray Powder Diffraction Data: $(d(\text{Å})\ (I/I_0))$,
9.41–9.17(s–vs),   5.57–5.47(vw–m),   4.97–
4.82(w–s), 4.37–4.25(m–vs), 3.57–3.51(vw–
s), 2.95–2.90(w–s).

Synthesis

AsAPO-34 crystallizes from a reactive gel with
a composition of: 1.0–2.0 TEAOH : 0.5–0.2
As_2O_q : 0.5–1.0 Al_2O_3 : 0.5–1.0 P_2O_5 : 40–100
H_2O (TEA = tetraethylammonium$^+$).

Adsorption

Adsorption properties of AsAPO-34:

Adsorbate	Kinetic diameter, Å	Pressure, Torr	Temp., °C	Wt. % adsorbed
O_2	3.46	100	−183	13
O_2		750		18
n-Butane	4.3	100	24	6
H_2O	2.65	4.3	24	15
H_2O		20	24	21

*Typical amount adsorbed.

AsAPO-44 (CHA)
(US4,913,888(1990)
UOP

Related Material: chabazite
SAPO-44

Structure

Chemical Composition: See AsAPO.

X-Ray Powder Diffraction Data: $(d(\text{Å})\ (I/I_0))$
9.41–9.26(vs), 6.81–6.76(w–m), 5.54–5.47(w–
m), 4.31–4.26(s–vs), 3.66–3.65(w–vs), 2.912–
2.889(w–s).

Synthesis

AsAPO-44 crystallizes from a batch composition of: 1.0–2.0 CHA : 0.5–0.2 As_2O_q : 0.5–
1.0 Al_2O_3 : 0.5–1.0 P_2O_5 : 40–100 H_2O (CHA
= cyclohexylamine.

Adsorption

Adsorption properties of AsAPO-44:

Adsorbate	Kinetic diameter, Å	Pressure, Torr	Temp., °C	Wt. % adsorbed
O_2	3.46	100	−183	13
O_2		750		16
n-Hexane	4.3	100	24	2
H_2O	2.65	4.3	24	15
H_2O		20	24	17

*Typical amount adsorbed.

AsBeAPO-31 (ATO)
(EP 158,976(1985))
Union Carbide Corporation

Related Materials: AlPO₄-31

Structure

Chemical Composition: See FCAPO.

X-Ray Powder Diffraction Data: $(d(\text{Å})\ (I/i_0))$
10.40(85), 9.31(12), 6.63(6), 6.42(2), 5.99(2),
5.64(6), 5.19(7), 4.85(7), 4.44(—), 4.37(48),
4.21(28), 4.08(26), 3.98(100), 3.59(7), 3.48(5),
3.99(11), 3.75(7), 3.02(9), 2.99(9),   2.85,
2.83(14), 2.55(10), 2.51(6), 2.37(8), 2.25(4).

The   arsenic-beryllium-aluminophosphate,
AsBeAPO-31, exhibits properties of a medium-
pore molecular sieve.

Synthesis

ORGANIC ADDITIVE
di-n-propylamine

AsBePO-31 crystallizes from a reactive gel
with a batch composition of: 1.0–2.0 R : 0.05–

0.2 M_2O_q : 0.5–1.0 Al_2O_3 : 0.5–1.0 P_2O_5 : 40–100 H_2O (R represents the organic additive, and q denotes the oxidation state of the arsenic and the beryllium).

Adsorption

Adsorption properties of AsBeAPO-31:

Adsorbate	Kinetic diameter, Å	Pressure, Torr	Temp., °C	Wt. % adsorbed
O_2	3.46	100	−183	4
O_2		750	−183	6
Cyclohexane	6.0	90	24	3
Neopentane	6.2	700	24	3
H_2O	2.65	4.3	24	3
H_2O		20	24	10

*Typical amount adsorbed.

AsGaAPO-5 (AFI)
(EP 158,976(1985))
Union Carbide Corporation

Related Materials: $AlPO_4$-5

Structure

Chemical Composition: See FCAPO.

X-Ray Powder Diffraction Data: (d(Å) (I/I_0)) 12.1–11.56(m–vs), 4.55–4.46(m–s), 4.25–4.17(m–vs), 4.00–3.93(w–vs), 3.47–3.40(w–m).

For a more detailed description of the framework topology see $AlPO_4$-5.

Synthesis

ORGANIC ADDITIVE
tripropylamine

The arsenic-gallium aluminophosphate, AsGaAPO-5, crystallizes from a reactive gel with a batch composition of: 1.0–2.0 R : 0.05–0.2 M_2O_q : 0.5–1.0 Al_2O_3 : 0.5–1.0 P_2O_5 : 40–

100 H_2O (R represents the organic additive, and q denotes the oxidation state of the arsenic and the gallium).

Adsorption

Adsorption properties of AsGaAPO-5:

Adsorbate	Kinetic diameter, Å	Pressure, Torr	Temp., °C	Wt. % adsorbed
O_2	3.46	100	−183	7
O_2		750	−183	10
Neopentane	6.2	700	24	4
H_2O	2.65	4.3	24	4
H_2O		20	24	12

*Typical amount adsorbed.

AsVBeAPO-34 (CHA)
(EP 158,976(1985))
Union Carbide Corporation

Related Materials: chabasite
$AlPO_4$-34

Structure

Chemical Composition: See FCAPO

X-Ray Powder Diffraction Data: (d(Å) (I/I_0)) 9.41–9.17(s–vs), 5.57–5.47(vw–m), 4.97–4.82(w–s), 4.37–4.25(m–vs), 3.57–3.51(vw–s), 2.95–2.90(w–s).

See $AlPO_4$-34 for a description of the framework topology.

Synthesis

ORGANIC ADDITIVE
tetraethylammonium$^+$

The arsenic-vanadium-beryllium aluminophosphate, AsVBeAPO-34, crystallizes from a reactive gel with a batch composition of: 1.0–2.0 R : 0.05–0.2 M_2O_q : 0.5–1.0 Al_2O_3 : 0.5–1.0 P_2O_5 : 40–100 H_2O (R represents the organic additive, and q the oxidation states of the arsenic, vanadium, and beryllium).

Adsorption

Adsorption properties of AsVBeAPO-34:

Adsorbate	Kinetic diameter, Å	Pressure, Torr	Temp., °C	Wt. % adsorbed*
O_2	3.46	100	−183	13
O_2		750	−183	18
n-Hexane	4.2	100	24	6
H_2O	2.65	4.3	24	15
H_2O		20	24	21

*Typical amount adsorbed

AZ-1 (MFI)
(EP 113,116(1983))
Asahi Kasei Kogyo Kabushiki Kaisha

Related Materials: ZSM-5

Structure

Chemical Composition: $M_{2/n}O$: Al_2O_3 : at least 10 SiO_2.

X-Ray Powder Diffraction Data: (d(Å) (2 theta)) 7.8(5–30), 8.7(90–100), 8.9(90–100), 17.5(5–30), 17.5(5–30), 23.1(30–80), 23.6(20–50).

Synthesis

ORGANIC ADDITIVE
1,8-diamino-4-aminomethyloctane

AZ-1 crystallizes from a reactive gel with a batch composition in the range: $SiO_2Al_2O_3$ of 10 to 1000; Na/SiO_2 between 0.5 and 1.0; H_2O/SiO_2 of 5 to 200 and R/SiO_2 between 0.1 and 10. Crystallization occurs between 5 and 200 hours when the temperature is between 120 and 200°C.

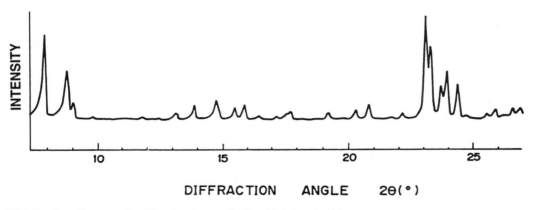

AZ-1 Fig. 1:   X-ray powder diffraction pattern of AZ-1 (EP 113,116(1983)).

B

B (GIS)
(US3,008,803(1961))
Union Carbide Corporation

Structure

Chemical Composition: $M_{2/n}[1–2.5\ SiO_2 : 0.51Al_2O_3]$. [M = alkali or alkaline earth cation]

See Na-P1 for X-ray power diffraction pattern.

Ba-G (LTL)
(*JCS 2296 (1964)*)

Related Materials: Linde type L

Structure

CHEMICAL COMPOSITION: Si/Al = 1.5
SYMMETRY: tetragonal
UNIT CELL CONSTANTS: a = 18.89 Å
c = 15.16 Å

X-Ray Powder Diffraction Data: $d(Å)$ (I/I_0)
16.15(s), 9.35(vvw), 8.10(vvw), 7.53(m), 6.83(mw), 5.85(vw), 5.38(w), 4.85(w), 4.72(w), 3.95(vs), 3.84(w), 3.75(vvw), 3.49(m), 3.32(m), 3.23(m), 3.19(m), 3.08(s), 2.92(s), 2.87(vw), 2.817(vw), 2.753(vw), 2.690(s), 2.635(mw), 2.502(w), 2.447(mw), 2.390(vw), 2.294(vvw), 2.224(m), 2.203(m), 2.187(w), 1.874(m).

Ba-G is the aluminum-rich analog of Linde type L. See Linde type L for a description of the framework topology.

Synthesis

ORGANIC ADDITIVES
none

Ba-G crystallizes from barium aluminosilicate gels with the composition: $BaO : Al_2O_3 : 2–3\ SiO_2$ at temperatures between 150 and 200°C (*JCS 2296(1964)*). It is not obtained in pure phase but does occur as the major product after crystallization for between 3 and 4 weeks. The reproducibility of this synthesis is reported to be moderate to poor. Harmotome is observed as a co-crystallizing phase in addition to unreacted gel and barium carbonate. Crystallization is improved when KOH is added to the batch composition.

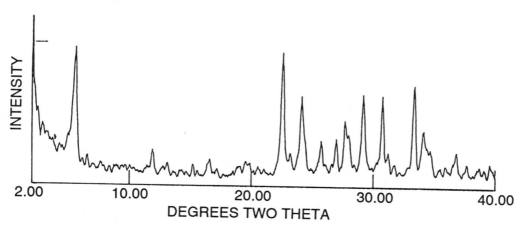

Ba-G Fig. 1:   X-ray powder diffraction pattern for zeolite Ba-G. (Reproduced with permission of author)

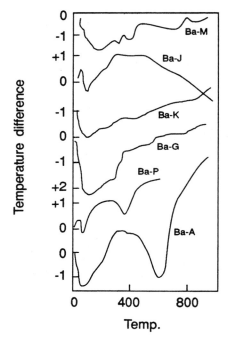

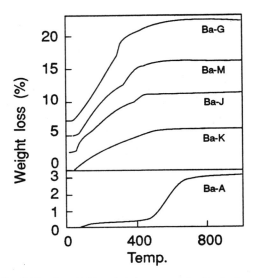

Ba-G Fig. 2:   Differential thermal analysis curves for
Ba-G, -M, -J, and -K at 10°C per min. Ba-A is not a
zeolitic phase (*JCS* 2296(1964)). (Reproduced with
permission of the Chemical Society)

Ba-G Fig. 3:   TGA weight loss curves for Ba-G, -J,
-K, and -M at 10°C/min. Ba-P and -A are not zeolite
phases (*JCS* 2296(1964)). (Reproduced with permission
of the Chemical Society)

Thermal Properties

Total weight loss is 15%. This material has a
pore volume of 0.15 cc/g and is thermally stable.

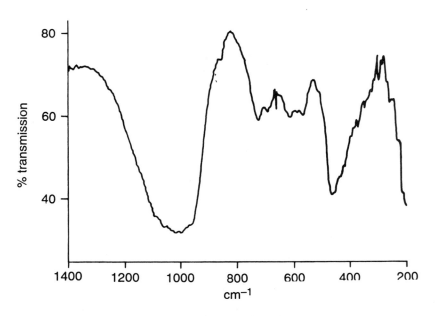

Ba-G Fig. 4:   Mid-infrared spectrum of Ba-G. (Reproduced with permission of E.
Seugling, Georgia Tech)

Ion Exchange

It has been reported that the barium ions in this zeolite structure can be readily replaced with sodium, exhibiting little change in the X-ray diffraction pattern.

Adsorption

For a sample containing only 60% Ba-G, 50 cc of oxygen (at STP and 90K) was adsorbed (*JCS* 2296(1964)).

Infrared Spectrum

See Ba-G Fig. 4.

Ba-J
(*JCS* 2296(1964))

Structure

Chemical Composition: Si/Al = 4.

X-Ray Powder Diffraction Data: (d(Å) (I/I_0))
11.39(s),   10.53(vw),   10.22(w),   7.39(m), 7.04(w), 6.23(m), 6.08(vw), 5.84(m), 5.73(mw), 5.63(vw), 4.79(m), 4.58(s), 4.52(s), 4.30(vw), 4.19(w), 4.08(w), 4.02(ms), 3.97(m), 3.91(ms), 3.81(m), 3.70(w), 3.56(ms), 3.51(s), 3.45(s), 3.28(w), 3.20(w), 3.16(w), 3.12(vs), 2.98(m), 2.82(w), 2.88(m), 2.82(w), 2.797(w), 2.709(m), 2.664(m), 2.636(wv), 2.605(w), 2.455(ms), 2.419(ms), 2.370(ms), 2.334(w), 2.307(vw), 2.287(vw), 2.262(m), 2.213(vw), 2.189(vw), 2.169(vw), 2.125(vvw), 2.107(vvw), 2.028(vw), 1.987(vw), 1.953(vw), 1.847(ms), 1.819(w), 1.794(w), 1.775(vw), 1.708(m), 1.670(mw), 1.632(w), 1.596(w).

Synthesis

ORGANIC ADDITIVES
none

Ba-J crystallizes in a range of SiO_2/Al_2O_3 between 7:1 and 9:1 at temperatures between 250 and 300°C. The best results were obtained from gels with a composition of: BaO : Al_2O_3 : $8SiO_2$. Impurity phases include harmotome and Ba-K. Crystallization occurs be-

tween 6 and 20 days. Reproducibility of this synthesis is not high, as Ba-K or mixtures of Ba-J and Ba-K were found. Crystals of Ba-J have a fibrous habit.

Thermal Properties

Thermogravimetric analysis of Ba-J shows a weight loss of 8.9% between room temperature and 450°C. A further 1% loss in weight is observed between 450 and 1000°C. This zeolite becomes amorphous above 600°C. It has a pore volume of 0.10 cc/g.

Adsorption

At 90K, outgassed Ba-J adsorbs 59 cc of O_2 (at STP) per gram of hydrated sample.

Ba-K
(*JCS* 2296(1964))

Structure

Chemical Composition: Si-Al not stated.

X-Ray Powder Diffraction Data: (d(Å) (I/I_0))
10.59(ms),   7.39(mw),   6.87(w),   6.07(ms), 5.58(ms), 5.29(w), 5.08(w), 4.72(m), 4.59(ms), 4.30(ms),   4.21(ms),   3.89(w),   3.67(vw), 3.53(ms),   3.49(m),   3.45(w),   3.34(vw), 3.29(mw), 3.16(s), 3.06(w), 3.03(ms), 2.94(m), 2.88(vw),   2.79(mw),   2.72(m),   2.648(w), 2.613(mw),   2.553(w),   2.493(m),   2.469(w), 2.433(w),   2.336(w),   2.301(m),   2.234(vw), 2.206(w),   2.117(w),   1.965(w),   1.919(w), 1.846(m),   1.801(w),   1.760(w),   1.730(w), 1.694(vw), 1.650(vw), 1.559(vw), 1.539(w), 1.525(vw), 1.509(vw), 1.286(w), 1.271(vw).

Synthesis

ORGANIC ADDITIVE
none

Ba-K has been prepared under composition ranges similar to those of Ba-J. Crystallization occurs at 300°C with SiO_2/Al_2O_3 of 7–8:1 with a batch composition of: BaO : Al_2O_3 : 8 SiO_2. Crystallization occurs over 6 to 20 days. Lower temperatures result in co-crystallization with Ba-J or Ba-P. Crystals are typically spherulitic.

Thermal Properties

Ba-K loses 7% of its weight upon treatment to 950°C, with the major portion of the loss taking place below 500°C. It has a pore volume of 0.08 cc/g.

Adsorption

Ba-K adsorbs 46 cc of oxygen per gram of sample (STP) at 90K.

(Ba,Li)-ABW

ABW topology prepared from Ba/Li containing gels.

(Ba,Li)-G,L

LTL topology prepared from A Ba/Li containing gel.

(Ba,Li)-M

Phillipsite topology prepared from Ba/Li containing gels.

(Ba,Li)-Q

Yugawaralite topology prepared from Ba/Li containing gels.

Ba-M (PHI)
(*JCS* 2296(1964))

Related Materials: phillipsite

Structure

Chemical Composition: Si-Al not stated.

X-Ray Powder Diffraction Data: (d(Å) (I/I_0))
8.24(vs), 7.56(vw), 7.18(s), 7.03(w), 6.39(vs), 6.11(vw), 5.02(ms), 4.29(ms), 4.10(s), 4.06(s), 4.03(s), 3.89(m), 3.66(w), 3.57(vvw), 3.52(w), 3.46(mw), 3.23(s), 3.19(mw), 3.16(ms), 3.12(vs), 3.07(m), 2.91(ms), 2.838(w), 2.745(w), 2.724(s), 2.674(s), 2.668(s), 2.624(w), 2.557(w), 2.525(w), 2.058(w), 2.464(m), 2.368(m), 2.315(ms), 2.295(m), 2.256(w), 2.236(m), 2.213(w), 2.147(m), 2.119(w), 2.084(vw), 2.062(m), 2.052(m), 2.021(mw), 2.008(vw), 1.996(vw), 1.964(w), 1.947(m), 1.921(w), 1.894(mw), 1.840(vw), 1.818(mw), 1.802(w), 1.76(m)b, 1.71(ms)b, 1.670(m), 1.639(w), 1.530(m), 1.473(w).

Synthesis

ORGANIC ADDITIVES
none

Ba-M crystallizes in a range of SiO_2/Al_2O_3 of 7–8:1 at temperatures ranging between 200 and 250°C. At lower temperatures and ratios, mixtures with Ba-G are observed. Crystallization requires 3 weeks. Crystals display a cruciform twinning much like natural harmotome.

Thermal Properties

Total weight loss from Ba-M is 11.5%, with most of the loss occurring below 450°C. There are three endothermic peaks in the DTA below this temperature, which are similar to those observed in synthetic calcium harmotome. This material has a pore volume of 0.13 cc/g.

Adsorption

No significant adsorption is observed for this zeolite.

Ba-N
(*JCS Dalton Trans.* 1259(1972))

Structure

Chemical Composition: Si/Al = 1.

X-Ray Powder Diffraction Data: (d(Å))(I/I_0))
7.1(m), 5.6(ms), 5.2(ms), 4.5(w), 4.35(vw), 3.50(m), 3.40(s), 3.15(mw), 2.92(ms), 2.84(w), 2.79(w), 2.69(w), 2.59(w), 2.57(ms), 2.44(m), 2.40(w), 2.33(w), 2.24(m),, 2.230(w), 2.165(w), 2.078(mw), 2.051(mw), 2.020(mw), 1.917(w), 1.893(m), 1.845(mw), 1.823(w), 1.762(mw).

Synthesis

ORGANIC ADDITIVES
none

Ba-N crystallizes from aluminosilicate gels with a SiO_2/Al_2O_3 ratio of 2. $Ba(OH)_2$, metakolin, and silica gel are used as the sources of reagents, and crystallization occurs at 80°C after 7 days.

Thermal Properties

This material exhibits limited thermal stability.

BAPO-5 (AFI)
(EP 158,976(1985))
Union Carbide Corporation

Related Materials: AlPO$_4$-5

Structure

Chemical Composition: See FCAPO

X-Ray Powder Diffraction Data: (d(Å) (I/I_0))
12.1–11.56(m–vs),   4.55–4.46(m–s),   4.25–4.17(m–vs), 4.00–3.93(w–vs), 3.40–3.47(w–m).

BAPO-5 exhibits properties similar to those of large-pore molecular sieves. For a description of the framework topology see AlPO$_4$-5.

Synthesis

ORGANIC ADDITIVE
tripropylamine

BAPO-5 crystallizes from a reactive gel with a batch composition of: 1.0–2.0 R : 0.05–0.2 B$_2$O$_3$: 0.5–1.0 Al$_2$O$_3$: 0.5–1.0 P$_2$O$_5$: 40–100 H$_2$O (R refers to the organic additive).

Adsorption

Adsorption properties of BAPO-5:

Adsorbate	Kinetic diameter, Å	Pressure, Torr	Temp., °C	Wt. % adsorbed*
O$_2$	3.46	100	– 183	7
O$_2$		750	– 183	10
Neopentane	6.2	700	24	4
H$_2$O	2.65	4.3	24	4
H$_2$O		20	24	12

*Typical amount adsorbed.

BAPO-11 (AEL)
(EP 158,976(1985))
Union Carbide Corporation

Related Materials: AlPO$_4$-11

Structure

Chemical Composition: See FCAPO

X-Ray Powder Diffraction Data: (d(Å) (I/I_0))
9.51–9.17(m–s), 4.40–4.31(m–s), 4.25–4.17(s–vs), 4.04–3.95(m–s), 3.95–3.92(m–s), 3.87–3.80(m–vs).

BAPO-11 exhibits properties characteristic of a medium-pore molecular sieve. See AlPO$_4$-11 for a description of the topology.

Synthesis

ORGANIC ADDITIVE
di-*n*-propylamine

BAPO-11 crystallizes from a reactive gel with a batch composition of: 1.0–2.0 R : 0.05–0.2 B$_2$O$_3$: 0.5–1.0 Al$_2$O$_3$: 0.5–1.0 P$_2$O$_5$: 40–100 H$_2$O (R refers to the organic additive).

Adsorption

Adsorption properties of BAPO-11:

Adsorbate	Kinetic diameter, Å	Pressure, Torr	Temp., °C	Wt. % adsorbed*
O$_2$	3.46	100	– 183	5
O$_2$		750	– 183	6
Cyclohexane	6.0	90	24	4
H$_2$O	2.65	4.3	24	6
H$_2$O		20	24	8

*Typical amount adsorbed.

Barrerite (STI)
(natural)

Related Materials: stilbite

Barrerite, the orthorhombic sodium-rich stilbite from Sardinia, was first named by Passaglia and

Pongiluppi in 1975, in honor of R. M. Barrer (*Miner. Mag.* 40:208(1975)).

Structure

(*Bull. Soc. Fr. Miner. Crist.* **98**, 331(1975))

CHEMICAL COMPOSITION: $Na_8(Al_8Si_{28}O_{72})*26H_2O$
SYMMETRY: orthorhombic
SPACE GROUP: Amma
UNIT CELL CONSTANTS: (Å) $a = 13.64$
$b = 18.20$
$c = 17.84$

X-Ray Powder Diffraction Data: (*Nat. Zeo.* 344(1985)) (d(Å) (I/I_0)) 9.10(>100), 6.83(4), 5.30(6), 5.23(5), 4.66(21), 4.55(6), 4.46(3), 4.29(12), 4.05(100), 4.01(13), 3.732(12), 3.566(1), 3.483(12), 3.408(11), 3.393(11), 3.189(16), 3.103(7), 3.028(78), 3.004(25), 2.974(4), 2.885(2), 2.871(3), 2.824(3), 2.805(3), 2.773(22), 2.727(3), 2.709(3), 2.609(4), 2.563(6), 2.524(1), 2.507(5), 2.487(3), 2.354(3), 2.347(3), 2.274(1), 2.267(1), 2.225(3), 2.204(2), 2.164(1br), 2.122(4), 2.097(3), 2.084(<1),

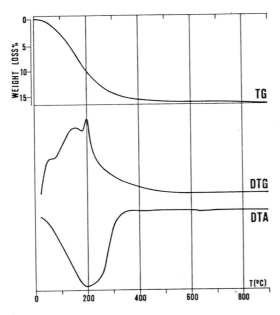

Barrerite Fig. 1: Thermal curves for barrerite from Capo Pula, Sardinia, Italy; in air, heating rate 20°C/min (*Nat. Zeo.* 296(1985)). (Reproduced with permission of Springer-Verlag Inc.)

2.075(1br), 2.062(3), 2.040(2), 2.034(3), 2.025(3), 2.018(2), 2.012(<1), 1.893(3).

Thermal Properties

There are three main water losses in barrerite, at 50, 150, and 200°C. Dehydroxylation occurs between 500 and 900°C. A lattice shrinkage is observed with the water loss.

Ba-T
(*JCS Dalton Trans.* **1259(1972)**)

Structure

Chemical Composition: Si/Al = 1.

X-Ray Powder Diffraction Data: (d(Å))(I/I_0)) 7.4(w), 7.2(w), 5.2(ms), 5.0(w), 4.3(m), 4.24(mw), 4.20(mw), 4.16(m), 4.00(vw), 3.79(s), 3.70(w), 3.68(s), 3.52(m), 3.37(ms), 3.30(mw), 3.22(s), 3.18(s), 3.15(w), 2.45(m), 2.43(ms), 2.38(mw), 2.36(mw), 2.35(mw), 2.33(m), 2.32(vw), 2.31(vw), 2.27(w), 2.25(mw), 2.24(w), 2.22(m), 2.18(m), 2.15(w), 2.09(w), 2.07(w), 2.04(w), 2.03(m), 2.02(m), 2.00(m), 1.99(mw), 1.97(mw), 1.96(mw), 1.95(w), 1.92(w), 1.89(w), 1.83(m), 1.78(m).

Synthesis

ORGANIC ADDITIVES
none

Ba-T crystallizes from aluminosilicate gels with a SiO_2/Al_2O_3 ratio of 2. $Ba(OH)_2$, metakolin, and silica gel are used as the sources of reagents, and crystallization occurs at 80°C.

Thermal Properties

The pore volume (cc/g) is 0.06. The material has low thermal stability.

B-C_n
(*SSSC* **49A:143(1989)**)

The B-C_n series consists of some materials that are aluminoborates, some with microporous properties.

Structure

Chemical Composition: Compositionally all the materials are boron-rich, with B_2O_3/Al_2O_3 ranging between 5.85 and 11.70.

X-Ray Power Diffraction Data:

B-C$_7$: (d(Å) (I/I_0)) 11.6(100), 6.76(17), 6.54(16.5), 5.81(17), 5.16(11), 4.30(11), 4.01(15.5), 3.77(20), 3.50(8.5), 3.43(25).

B-C$_8$: (d(Å) (I/I_0)) 9.82(100), 7.21(12), 5.99(12), 5.36(10), 4.72(15), 4.44(43), 4.33(24), 3.98(20), 3.89(15), 3.68(9).

B-C$_9$: (d(Å) (I/I_0)) 16.83(78), 8.08(100), 5.59(56), 5.54(58), 4.42(5), 3.63(6), 3.43(15), 3.38(17), 3.08(11.5), 2.84(5).

B-C$_{10}$: (d(Å) (I/I_0)) 12.24(48), 9.42(100), 7.14(11), 6.13(20), 5.69(67), 4.95(13), 4.48(17), 4.34(18), 3.90(65), 3.65(19).

In general, these materials exhibit properties characteristic of medium-pore molecular sieves.

Synthesis

ORGANIC ADDITIVES
tetraethylammonium$^+$
tripropylamine
triethylamine

B-Cn crystallizes from hydro-gels with the molar oxide ratios shown in the table.

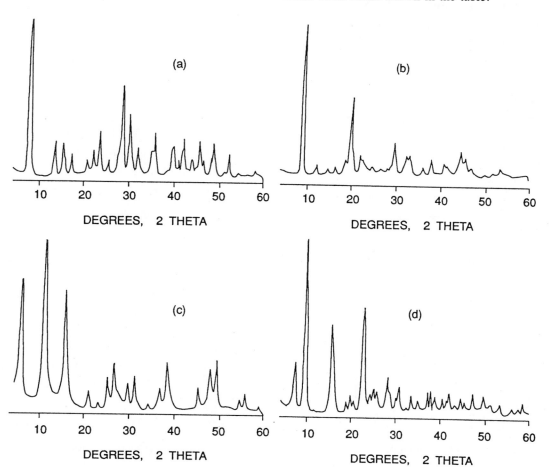

B-Cn Fig. 1: X-ray powder diffraction patterns for B-C7 (a), B-C8 (b), B-C9 (c), and B-C10 (d) (*SSSC*. 49A:143(1989)). (Reproduced with permission of Elsevier Publishing Company)

Composition limit of initial mixture for B-C$_n$:

B-Cn	B$_2$O$_3$	Al$_2$O$_3$	CaO	Na$_2$O	H$_2$O	R
			Composition Limit of Reactant (mol.)			
B-C7	5–6	0.5–0.7	1.2		200–250	0.5–1.0
B-C8	5–7	0.6–0.8	0.7–1.5	0.1–0.4	180–250	0.5–1.0
B-C9	6–8	0.5–1.0	0.6–0.9	0.7–1.1	150–200	0.5–1.0
B-C10	3–4	0.6–1.2	1.4–1.8	0.2–0.4	170–200	0.5–1.0

Crystallization occurs between 120 and 480 hours at temperatures between 160 and 210°C, depending on the structure prepared. The pH of the initial mixture is between 2.7 and 4.5. The B-C7 structure is favored in a Ca^{+2} system, whereas B-C8 and B-C10 crystallize with Ca^{+2} and Na^{+} as well as Mg^{+2} and Na^{+}. B-C9 also favors Ca^{+2} and Na^{+}, as well as K^{+}.

Thermal Properties

DTA/TG analysis indicates that structure collapse occurs in these materials at 733, 465, 626, and 425°C for B-C7, B-C8, B-C9, and B-C10, respectively.

Infrared Spectra

Trigonal BO$_3$ in borates has very strong stretching vibrational bands in the range of 1450 to 1200 cm^{-1}, and tetrahedral BO$_4$ has medium absorption bands in the range of 1180 to 900 cm^{-1}. The infrared bands around 668 to 655 cm^{-1} shift to higher frequency with increasing B/Al ratio. These bands are assigned to the bending vibration of the B–O–Al linkages.

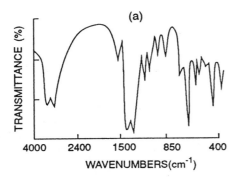

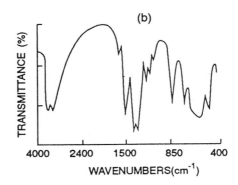

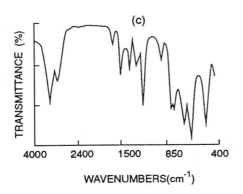

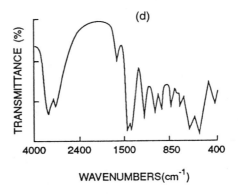

B-Cn Fig. 2:   Infrared spectra of B-C7 (a), B-C8 (b), B-C9 (c), and B-C10 (d) (*SSSC*. 49A:143(1989)). (Reproduced with permission of Elsevier Publishing Company)

Adsorption

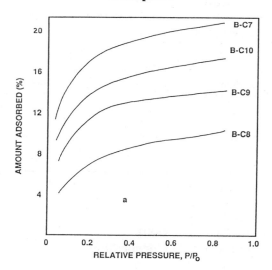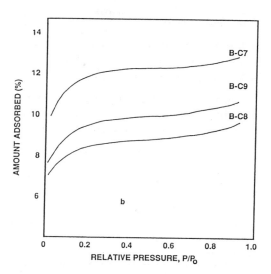

B-Cn Fig. 3:   Adsorption isotherms for B-Cn. (a) H_2O isotherms at 20°C. (b) N_2 isotherms at − 196°C (*SSSC*. 49A:143(1989)). (Reproduced with permission of Elsevier Publishing Company)

BeAPO-5 (AFI)
(EP 158,976(1985))
Union Carbide Corporation

Related Materials: AlPO$_4$-5

Structure

Chemical Composition: See FCAPO

X-Ray Powder Diffraction Data: $(d(A)(I/I_0))$
12.1–11.56(m–vs),   4.55–4.46(m–s),   4.25–4.17(m–vs), 4.00–3.93(w–vs), 3.47–3.40(w–m).

BeAPO-5 exhibits properties characteristic of a large-pore molecular sieve. See AlPO$_4$-5 for a description of the framework topology.

Synthesis

ORGANIC ADDITIVE
tripropylamine

BeAPO-5 crystallizes from a reactive gel with a batch composition of: 1.0–2.0 R : 0.1–0.4

BeO : 0.5–1.0 Al$_2$O$_3$: 0.5–1.0 P$_2$O$_5$: 40–100 H$_2$O (R refers to the organic additive).

Adsorption

Adsorption properties of BeAPO-5:

Adsorbate	Kinetic diameter, Å	Pressure, Torr	Temp., °C	Wt. % adsorbed*
O$_2$	3.46	100	− 183	7
O$_2$		750	− 183	10
Neopentane	6.2	700	24	4
H$_2$O	2.65	4.3	24	4
H$_2$O		20	24	12

*Typical amount adsorbed.

BeAPO-17 (ERI)
(EP 158,976(1985))
Union Carbide Corporation

Related Materials: erionite
                        AlPO$_4$-17

Structure

Chemical Composition: See FCAPO

X-Ray Powder Diffraction Data: $(d(Å) (I/I_0))$
11.5–11.4(vs), 6.61(s–vs), 5.72–5.70(s), 4.52–
4.51(w–s), 4.33–4.31(vs), 2.812–2.797(w–s).

BeAPO-17 exhibits properties characteristic
of a small-pore molecular sieve. See SAPO-17
for a description of the framework topology.

Synthesis

ORGANIC ADDITIVE
quinuclidine

BeAPO-17 crystallizes from a reactive gel
with a batch composition of: 1.0–2.0 R : 0.1–
0.4 BeO : 0.5–1.0 Al_2O_3 : 0.5–1.0 P_2O_5 : 40–
100 H_2O (R refers to the organic additive).

Adsorption

Adsorption properties of BeAPO-17:

Adsorbate	Kinetic diameter, Å	Pressure, Torr	Temp., °C	Wt. % adsorbed*
O_2	3.46	100	−183	10
O_2		750	−183	12
n-Butane	4.3	100	24	4
H_2O	2.65	4.3	24	13
H_2O		20	24	14

*Typical amount adsorbed.

BeAPSO
(US4,737,353(1988))
Union Carbide Corporation

Structure

BeAPSO represents structures with beryllium
silicoaluminophosphate compositions. See
BeAPSO Fig. 1.

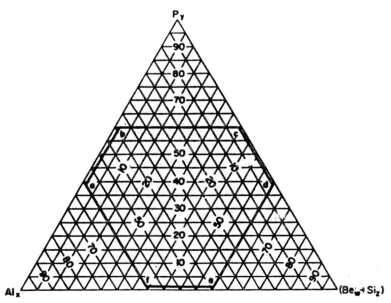

BeAPSO Fig. 1:   Ternary diagram of the preferred compositions for BeAPSO
molecular sieves (US4,737,353(1988)).

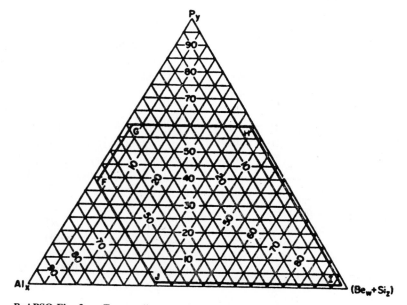

BeAPSO Fig. 2:   Ternary diagram relating to parameters used in synthesis mixture employed in the preparation of BeAPSO molecular sieves (US4,737,353(1988)).

Synthesis

ORGANIC ADDITIVES
BeAPSO-5:   tripropylamine
BeAPSO-11: di-*n*-propylamine
BeAPSO-17: quinuclidine
BeAPSO-20: tetramethylammonium hydroxide
BeAPSO-31: di-*n*-propylamine
BeAPSO-34: tetraethylammonium hydroxide (mixture with BeAPSO-5 produced)
BeAPSO-44: cyclohexylamine

BeAPSO molecular sieves crystallize from reaction mixtures containing beryllium sulfate as the source of beryllium. See BeAPSO Fig. 2.

BeGeAPO-11 (AEL)
(EP 158,976(1985))
Union Carbide Corporation

Related Materials: AlPO$_4$-11

Structure

Chemical Composition: See FCAPO

X-Ray Powder Diffraction Data: $(d(\text{Å})\,(I/I_0))$
9.51–9.17(m–s), 4.40–4.31(m–s), 4.25–4.17(s–vs), 4.04–3.95(m–s), 3.95–3.92(m–s), 3.87–3.80(m–vs).

See AlPO$_4$-11 for further information concerning the BeGeAPO-11 framework topology.

Synthesis

ORGANIC ADDITIVE
di-*n*-propylamine

BeGeAPO-11 crystallizes from a reactive gel with a batch composition of: 1.0–2.0 R : 0.05–0.2 M$_2$O$_q$: 0.5–1.0 Al$_2$O$_3$: 0.5–1.0 P$_2$O$_5$: 40–100 H$_2$O (R refers to the organic additive, and q denotes the oxidation state of beryllium and germanium).

Adsorption

Adsorption properties of BeGeAPO-11:

Adsorbate	Kinetic diameter, Å	Pressure, Torr	Temp., °C	Wt. % adsorbed*
O$_2$	3.46	100	−183	5
O$_2$		750	−183	6
Cyclohexane	6.0	90	24	4
H$_2$O	2.65	4.3	24	6
H$_2$O		20	24	8

*Typical amount adsorbed.

Beryllophosphate-E (EDI)
(*SSSC* 49A:411(1989))

Related Materials: edingtonite

Structure

CHEMICAL COMPOSITION: $K_{10}Be_{10}P_{10}O_{40}*10H_2O$
UNIT CELL CONSTANTS: (Å) $a = 9.17$
$c = 12.30$

X-Ray Powder Diffraction Data: $(d(Å)\ (I/I_0))$
6.48(100),   6.13(21),   3.89(37),   3.74(40),
3.410(8),   3.244(36),   3.047(7),   2.903(40),
2.869(42),   2.826(7),   2.780(99),   2.625(41),
2.378(7),   2.227(11),   2.188(6),   1.168(11),
2.111(25),   1.958(9),   1.726(6),   1.621(6).

Beryllophosphate-E is structurally analogous to zeolite edingtonite.

Synthesis

ORGANIC ADDITIVE
tetraethylammonium$^+$

Beryllophosphate-E crystallizes from a reactive gel with a Be/P ratio of 1.0. Upon mixing of beryllium sulfate and phosphoric acid, a clear solution of pH 1 forms. Addition of potassium chloride follows. The clear solution becomes gelatinous upon addition of tetraethylammonium hydroxide to adjust the pH to 6. Crystallization of this phase was observed after 7 days at 100°C. The pH and the source of alkali are the important factors in this synthesis.

Ion Exchange

Beryllophosphate-E does not exhibit ion exchange capacity.

Beryllophosphate-G (GIS)
(*SSSC* 49A:411(1989))

Related Materials: gismondine

Structure

CHEMICAL COMPOSITION: $Na_8Be_8P_8O_{32}*10H_2O$
UNIT CELL CONSTANTS: (Å) $a = 9.32$
$b = 8.48$
$c = 9.13$
$\beta = 90.8°$

X-Ray Powder Diffraction Data: $(d(Å)\ (I/I_0))$
6.27(100), 6.21(31), 4.56(24), 4.23(8), 3.839(6),
3.745(34),   3.713(67),   3.563(32),   3.282(6),
3.237(19), 3.106(17), 2.917(53), 2.862(45),
2.705(60),   2.596(6),   2.573(17),   2.453(9),
2.443(6),   2.424(2),   2.395(2),   2.332(27),
2.322(4), 2.190(7), 2.187(4), 2.175(6), 2.137(8),
2.120(4), 2.072(2), 2.063(9), 2.043(4), 2.018(3),
1.945(4), 1.923(3), 1.886(4), 1.853(3), 1.783(5),
1.767(5), 1.771(7), 1.568(3), 1.555(2), 1.553(2).

Beryllophosphate-G is structurally analogous to zeolite gismondine.

Synthesis

ORGANIC ADDITIVE
tetraethylammonium$^+$

Beryllophosphate-G crystallizes from a reactive gel with a Be/P ratio of 1.0. Upon mixing of beryllium sulfate and phosphoric acid, a clear solution of pH 1 forms. After addition of the alkali metal cation (chloride), the clear solution becomes gelatinous upon addition of tetraethylammonium hydroxide to adjust the pH to 6. Crystallization of this phase was observed after 7 days at 100°C. The pH and the source of alkali are the important factors in this synthesis.

Ion Exchange

Beryllophosphate-G exchanges Li, K, Rb, and NH_4^+ cations; Cs^+, Ca^{+2}, Ba^{+2}, and Ag^+ do not exchange into this structure.

Beryllophosphate-H (BPH)
(SSSC 49A:411(1989))

Structure

CHEMICAL COMPOSITION: $Na_7K_7Be_{14}P_{14}O_{56}$*
2OH$_2$O
UNIT CELL CONSTANTS: (Å)a = 12.59
c = 12.46
γ = 120°
SPACE GROUP: P321

X-Ray Powder Diffraction Data: $(d(Å) (I/I_0))$
12.4(60), 10.90(100), 8.20(1), 6.28(2), 6.23(2),
5.58(3), 5.41(4), 4.429(1), 4.151(2), 4.120(5),
4.101(2), 3.911(25), 3.880(10), 3.488(5),
3.465(4),   3.436(7),   3.144(2),   3.137(8),
3.050(11),   3.022(2),   2.937(8),   2.923(8),
2.807(17),   2.790(30),   2.723(8),   2.500(4),
2.483(6), 2.268(1), 2.180(4), 2.170(4), 2.143(3),
2.098(4), 2.065(4), 1.957(5), 1.923(4), 1.721(2),
1.605(2), 1.557(2).

Beryllophosphate-H is structurally unrelated
to any previously known zeolite structure. This
structure can be considered to be composed of
an eight-member ring and a cage. There are four
cation sites contained within this structure. The
first site is in the middle of the small cage that
forms the basis of the structure. The sodium

cations are found in the center of this cage, and
the Na–O bond distance is 2.4 A, the ideal coordination distance for Na$^+$. The K–O bond
distance is 2.8 A, similar to its ideal value.

Synthesis

ORGANIC ADDITIVE
tetraethylammonium$^+$

Beryllophosphate-H crystallizes from a reactive gel with a Be/P ratio of 1.0. Upon mixing
of beryllium sulfate and phosphoric acid, a clear
solution of pH 1 forms. Addition of a mixture
of sodium and potassium chloride follows. The
clear solution becomes gelatinous upon addition
of tetraethylammonium hydroxide to adjust the
pH to 6. Crystallization of this phase was observed after 21 days at 100°C. The pH and the
source of alkali are the important factors in this
synthesis.

Ion Exchange

Beryllophosphate-H exchanges only sodium and
potassium; NH$_4$$^+$, calcium, and silver ions do
not exchange.

Beryllophosphate-P (ANA)
(SSSC 49A:411(1989))

Related Materials: analcime

Structure

CHEMICAL COMPOSITION: $Cs_{16}Be_{24}P_{24}O_{96}$
UNIT CELL CONSTANT: (Å) a = 13.11

X-Ray Powder Diffraction Data: $(d(Å) (I/I_0))$
5.34(29), 4.64(1), 4.14(5), 3.782(7), 3.502(69),
3.276(100), 3.084(2), 2.931(2), 2.794(41),
2.571(3), 2.393(16), 2.317(24), 2.126(24),
1.932(6), 1.891(3), 1.784(14), 1.665(12),
1.637(10), 1.569(3).

Beryllophosphate-G is structurally analogous
to zeolite pollucite.

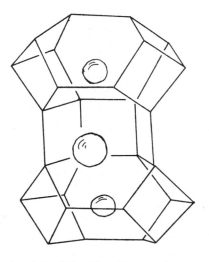

Beryllophosphate-H Fig. 1S: Framework topology.

Synthesis

ORGANIC ADDITIVE
tetraethylammonium$^+$

Beryllophosphate-P crystallizes from a reactive gel with a Be/P ratio of 1.0. Upon mixing of beryllium sulfate and phosphoric acid, a clear solution of pH 1 forms. Addition of cesium chloride follows. The clear solution becomes gelatinous upon addition of tetraethylammonium hydroxide to adjust the pH to 6. Crystallization of this phase was observed after 7 days at 200°C. The pH and the source of alkali are the important factors in this synthesis.

Ion Exchange

Beryllophosphate-P does not exhibit ion exchange capacity.

Beryllophosphate-R (RHO)
(*SSSC* 49A:411(1989))

Related Materials: Rho

Structure

CHEMICAL COMPOSITION: $Li_{24}Be_{24}P_{24}O_{96}$*40 H_2O
UNIT CELL CONSTANT: (Å) $a = 13.78$

X-Ray Powder Diffraction Data: (d(Å) (I/I_0))
9.61(97), 5%5(40), 4.81(7), 4.30(28), 3.929(23), 3.637(76), 3.405(3), 3.208(70), 3.044(66), 2.902(100), 2.777(30), 2.669(31), 2.485(32), 2.405(5), 2.335(5), 2.269(19), 2.208(17), 2.154(9), 2.101(7), 2.052(8), 2.006(6), 1.964(1), 2.925(7), 1.887(5), 1.852(4), 1.819(5), 1.728(4), 1.702(4), 1.676(11), 1.652(5), 1.626(2), 1.605(4), 1.583(12), 1.542(3).

Beryllophosphate-R is structurally analogous to zeolite rho.

Synthesis

ORGANIC ADDITIVE
tetraethylammonium$^+$

Beryllophosphate-R crystallizes from a reactive gel with a Be/P ratio of 1.0. Upon mixing of beryllium sulfate and phosphoric acid, a clear solution of pH 1 forms. After addition of the alkali metal cation (chloride), the clear solution becomes gelatinous upon addition of tetraethylammonium hydroxide to adjust the pH to 6. Crystallization of this phase was observed after 7 days at 100°C. The pH and the source of alkali are the important factors in this synthesis.

Ion Exchange

Beryllophosphate-R exchanges Na, K, and NH_4^+ cations; neither Ca^{+2} nor Ag^+ exchanges into this structure.

Beta (BEA*)
(US3,308,069(1967))
Mobile Oil Corporation
*disordered framework

Related Material: tschernichite
                      Nu-2

Structure

CHEMICAL COMPOSITION: (US3,308,069 (1967))
   (TEA, Na)$_2$O • Al_2O_3 • 5–100 SiO_2 • $4H_2O$
   (TEA-tetraetylammonum)
UNIT CELL PARAMETERS: (*Zeo.* 8:446 (1988)) (Å)

Polytype A	Polytype B	Polytype C
P4$_1$22	C2/c	P2/c
$a = 12.469$	$a = 17.634$	$a = 12.470$
$b = 12.469$	$b = 17.635$	$b = 12.470$
$c = 26.330$	$c = 14.416$	$c = 27.609$
	beta $= 107.52°$	beta $= 114.04°$

PORE STRUCTURE: 12-member rings, intersecting 6.5 × 5.6 and 7.5 × 5.7 Å

X-Ray Powder Diffraction Data: (d(Å) (I/I_0))
14.3(mw)   11.5(s),   7.44(vw),   6.86(nw), 6.165(vw), 564(m), 5.51(w), 509(mw), 4.75(w), 4.29(m), 4.17(m), 3.98(vs), 3.54(w), 3.42(w), 3.33(mw), 3.12(w), 3.03(m), 2.90(m), 2.74(vw), 2.69(w), 2.58(w), 2.49(w), 2.41(fw), 2.34(w), 2.25(vw), 2.08(w), 2.03(vw).

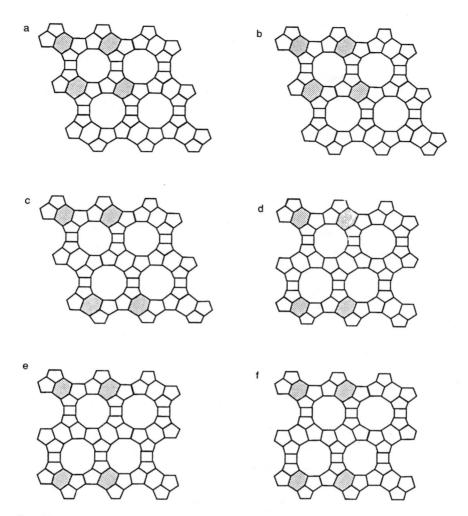

Beta Fig. 1S: 100 and 010 projections of the A (P4₁22) and C (P2) polytypes and 110 and
−110 projections of the B (C2/c) polytype. The sharded six-rings outline primitive unit cell
projections. Note that the three ordered polytypes exhibit only two different projections: one
inclined and one orthogonal (a) Polytype B 100, (b) polytype B 010, (c) polytype C 110, (d)
polytype C −110, (e) polytype A 100, (f) polytype A 010 (*Zeo.* 8:446(1988)). (Reproduced with
permission of Butterworth Publishing Company)

The tetrahedral framework structure of zeolite beta is disordered along [001] (*Zeo.* 8.446(1988); *Nature* 332:249(1988)). The disordered structure and the three simple ordered polytypes are related through layer displacements on 001 planes. These polytypes have mutually perpendicular 12-ring channel systems, and beta exhibits properties characteristic of the presence of open 12-member-ring channels. The smaller building units are double six-ring units connected by two four-rings and four five-rings. These are connected to form chains along the [001] direction. It is interesting to note that the double six-ring building unit does not correspond to any of the proposed SBU's of Gramlich-Meier and Meier (*J. Solid State Chem.* 44:41(1982)). Polytypes A and B contain 9 unique T sites, whereas polytype C contains 32.

DLS atomic coordinates ($\times 10^4$) of polytype A (*Zeo.* 8:446 (1988)):

	x	y	z
T1	4535	2956	0544
T2	4533	0431	0542
T3	2099	2915	0563
T4	2103	0453	0555
T5	8310	2894	0605
T6	8326	0419	0608
T7	0257	3637	1256
T7	0286	−0286	1250
T9	6369	3631	1250
O1	5000	3304	0000
O2	4576	1692	0619
O3	3308	3368	0575
O4	5197	3521	0998
O5	5000	0086	0000
O6	3313	0004	0565
O7	5195	−0162	0986
O8	2116	1680	0737
O9	1616	3007	0002
O10	1395	3575	0968
O11	1604	0334	0000
O12	1420	−0228	0960
O13	8324	1657	0774
O14	9324	3504	0844
O15	7242	3469	0810
O16	7262	−0169	0811
O17	9344	−0197	0840

DLS atomic coordinates ($\times 10^4$) of polytype A (*Zeo.* 8:446 (1988)):

	x	y	z
O15	3833	2468	3929
O16	4659	1653	3149

DLS atomic coordinates ($\times 10^4$) of polytype C (*Zeo.* 8:446 (1988)):

	x	y	z
T1	8725	−0428	3119
T2	8725	7094	3109
T3	2419	−0446	3041
T4	2435	7090	3054
T5	4857	0418	3033
T6	4877	7047	3038
T7	7154	0251	3732
T8	7188	6369	3737
T9	1121	0270	3763
T10	1103	6336	3765
T11	1621	2081	0573
T12	0861	2082	4460
T13	8390	−1688	4374
T14	0885	−1696	4404
T15	8366	4533	4436
T16	0920	4524	4478
O1	8843	−1669	3283
O2	1537	−0344	2486
O3	7722	0135	3283
O4	9864	0212	3389
O5	8438	6993	2503
O6	7737	6536	3285
O7	9870	6459	3374
O8	2542	−1672	3228
O9	3637	−0013	3046
O10	1995	0254	3437
O11	3658	6657	3063
O12	2015	6421	3461
O13	5000	−0043	2500
O14	4970	−1682	3097
O15	5807	0157	3490
O16	5000	6677	2500
O17	5843	6496	3499
O18	2555	1387	0982
O19	7600	−0687	4136
O20	7688	7233	4179
O21	7493	5192	3983
O22	1321	1382	4075
O23	1367	−0706	4158
O24	1322	7219	4208
O25	1234	5167	4030
O26	0489	2070	0738
O27	2053	3303	0592
O28	1369	1607	0011
O29	1299	3293	4455
O30	1303	1593	5019
O31	9533	−1670	4218
O34	9599	4597	4381
O36	1627	4997	5020

DLS atomic coordinates ($\times 10^4$) of polytype B (*Zeo.* 8:446 (1988)):

	x	y	z
T1	1735	8370	8880
T2	2032	5412	1127
T3	1981	4133	2552
T4	3917	0203	6202
T5	5178	1438	6254
T6	5000	2795	7500
T7	3031	2888	3960
T8	4220	1642	3932
T9	5000	1114	2400
O1	2136	9186	8840
O2	5819	3319	7967
O3	2306	7706	8774
O4	6674	3305	9960
O5	2293	4736	1948
O6	1515	6027	1457
O7	1475	5103	0011
O8	1801	4534	3444
O9	2708	3513	3063
O10	1155	3729	1776
O11	4658	0815	6563
O12	0736	5600	3268
O13	5023	2270	6600
O14	4888	1426	5042

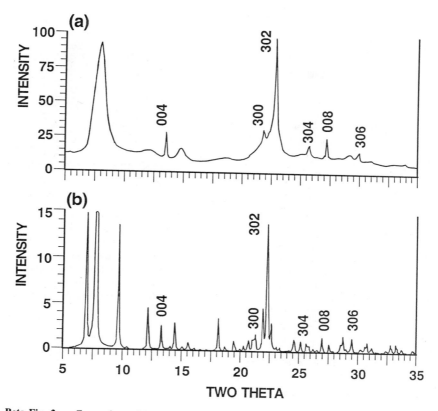

Beta Fig. 2:   Comparison of (a) the X-ray powder diffraction pattern of zeolite beta (CuK$_{alpha}$) with (b) the calculated pattern for the ordered tetragonal A polytype. Six sharp reflections on the beta pattern are indexed on the tetragonal unit cell (*Zeo.* 8:446(1988)). (Reproduced with permission of Butterworth Publishing Company)

Synthesis

ORGANIC ADDITIVES
tetraethylammonium$^+$

The synthesis of zeolite beta occurs successfully in the system: $TEA_2O : SiO_2 : Al_2O_3 : Na_2O : H_2O$. The broad range over which zeolite beta is claimed to crystallize is a SiO_2/Al_2O_3 from 10 to 200, $Na_2O/TEAOH$ from 0 to 1.0, $TEAOH/SiO_2$ from 0.1 to 1.0, and $H_2O/TEAOH$ from 20 to 75. ZSM-20 has been observed as an impurity phase in this system. Crystallization occurs between 100 and 150°C.

The SiO_2/Al_2O_3 range of 30 to 50 for crystals of zeolite beta lies between TEA-mordenite (10 to 30) and ZSM-12 (ca. 60). Both ZSM-12 and TEA-mordenite are crystallized from a Na^+/TEA^+ system. The efficiency of the synthesis and the chemical composition of the crystal are dependent on the concentration of aluminum in the liquid phase and the SiO_2/Al_2O_3 ratio of the gel. Ethanol encourages crystallization (*Zeo.* 8:46(1988)).

Thermal Properties

The weight and energy changes during the thermal decomposition of as-synthesized samples are shown in Beta Fig. 4. Four distinct zones of exothermic weight loss are seen in the DTA/DTG curves, with maxima at 318, 377, 450, and 627°C (*Zeo.* 9:231(1989), *Appl. Catal.* 31:35(1987)), attributed to the decomposition and removal of the occluded TEA species.

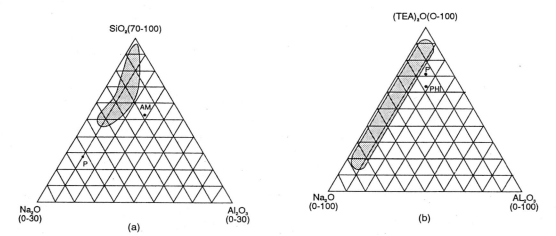

Beta Fig. 3:   (a) Molar proportiosns of SiO_2, Al_2O_3, and Na_2O in reaction mixtures for the synthesis of zeolite beta (($(TEA)_2O/SiO_2$ = 0.25, K_2O/SiO_2 = 0.018; synthesis time: 4 weeks; AM = amorphous, P = zeolite P, shaded area = zeolite beta). (b) Molar proportions of $(TEA)_2O$, Na_2O, and Al_2O_3 in reaction mixture for the synthesis of zeolite beta ($K_2O/(TEA)_2O$) = 0.071, SiO_2/Al_2O_3 = 100–10; PHI = zeolite PHI, P = zeolite P, shaded area = zeolite beta (*Appl. Catal.* 31:35 (1987)). (Reproduced with permission of Academic Press)

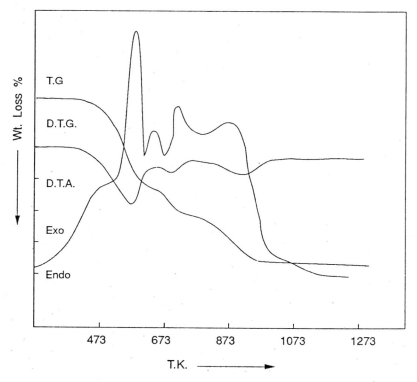

Beta Fig. 4:   TG, DTG, and DTA curves of as-synthesized zeolite beta, heating rate 10 K/min (*Zeo.* 9:231 (1989)). (Reproduced with permission of Butterworth Publishing Company)

Adsorption

Adsorption in beta ($T = 25°C$, $P/P_0 = 0.5$):

Adsorbate	Kinetic diameter, Å	% Adsorption	Pv^a (ml/g)
Water	2.65	21.3	0.21
n-Hexane	4.3	19.0	0.29
Cyclohexane	6.0	20.5	0.26
o-Xylene	6.8	21.8	0.25
m-Xylene	7.0	25.0	0.29
1,2,4-TMB[b]	7.5	24.0	0.28

[a]Pv = pore volume.
[b]TMB = trimethylbenzene.

Bicchulite (SOD)
(natural)

Related Materials: sodalite

Structure

(*Z. Kristallogr*. 152:13(1980))

CHEMICAL COMPOSITION: $Ca_2[Al_2Si_6](OH)_2$
SYMMETRY: cubic
SPACE GROUP: $I\overline{4}3m$
UNIT CELL CONSTANTS (A): $a = 8.825$

Bicchulite has the sodalite framework structure, with Al and Si distributed statistically over the tetrahedral sites. Because the Al:Si ratio is 2:1, Loewenstein's rule cannot be obeyed in this structure. The calcium ions and empty $(OH)_4$ tetrahedra occupy the cavities of the framework (*Z. Kristallogr*. 152:13(1980)).

Positional parameters for bicchulite (*Z. Kristallogr*. 152:13 (1980)):

	x	y	z
Ca	0.1434(1)	0.1434(1)	0.1434(1)
(Si,Al)	1/4	1/2	0
O1	0.1407(1)	0.1407(1)	0.4220(2)
O2	0.3845(2)	0.3845(2)	0.3845(2)
H	0.328(3)	0.328(3)	0.328(3)

Bikitaite (BIK)
(natural)

Related Materials: synthetic bikitaite

Bikitaite is a natural zeolite with a cation composition that is essentially all lithium. Deposits of this mineral have been found in Bikita, Zimbabwe as well as in North Carolina, USA. The mineral was first identified in 1957 (Am. Mineral. 43:768(1958)). This structure has been synthesized in the cesium form (see bikitaite (syn)).

Structure

CHEMICAL COMPOSITION: $Li_2Al_2Si_4O_{12} \cdot 2H_2O$
   (Nat. Zeo. 252 (1985))
SPACE GROUP: P1
UNIT CELL CONSTANTS (Å): at 295K

$a = 8.6071$
$b = 4.9540$
$c = 7.5972$
$\alpha = 89.900$
$\beta = 114.437$
$\gamma = 89.988$

at 13K

$a = 8.5971$
$b = 4.9395$
$c = 7.6121$
$\alpha = 89.850$
$\beta = 114.520$
$\gamma = 90.004$
(*Zeo.* 9:303 (1989))

DENSITY: 2.28g/cm³
PORE STRUCTURE: eight-member rings 2.8 × 3.7 Å

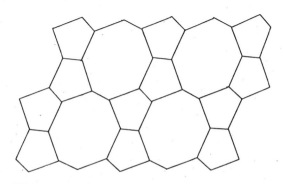

Bikitaite Fig. 1S: Framework topology

X-Ray Powder Diffraction Data: (*Nat. Zeo.* 339(1985)) (*d*(A) (*I/I*ₒ)) 7.865(80), 6.930(50), 6.732(30), 4.374(40), 4.265(10), 4.195(90), 4.022(20), 3.991(10), 3.926(10), 3.806(30), 3.462(100), 3.371(100), 3.284(40), 3.215(40), 3.076(40), 3.023(10), 2.930(10), 2.870(20), 2.749(10), 2.739(10), 2.629(10), 2.523(20), 2.479(90), 2.423(10), 2.364(20), 2.337(10), 2.323(10), 2.316(10), 2.240(10), 2.167(10), 2.141(10), 2.097(10), 2.094(10), 2.077(20), 2.012(10), 2.005(10).

Both X-ray diffraction and neutron diffraction have been applied to determining the structure of natural bikitaite (*Acta Crystallogr.* 13:1002(1960); *J. Am. Mineral.* 59:71(1974)); *Zeo.* 9:303(1989)). There are six unique T atom sites within the structure, with some ordering of the Si/Al observed in the natural material. Both five- and eight-member rings comprise this framework. The two independent water molecules are linked to each other via hydrogen bonding and also linked to the Li^+ cations. Only one hydrogen of each water molecule is hydrogen-bonded. The water molecules are positioned in the channels of this structure so as to form a continuous chain connected via hydrogen bonds.

Atomic coordinates ($\times 10^4$) for bikitaite (*Zeo.* 9:303(1989)):

	x	y	z	g
at 295K				
Si11	1057(3)	8645(5)	975(3)	0.957(12)
Si12	1038	8006	5049	0.981(13)
Al13	3808(3)	8749(5)	9374(3)	1.061(14)
Al21	8985(3)	3639(5)	9065(3)	1.066(15)
Si22	8920(2)	2984(5)	4877(3)	0.981(13)
Si23	6187(3)	3748(5)	630(3)	0.975(12)
O11	2641(2)	7404(4)	569(3)	
O12	830(2)	1853(4)	413(3)	
O13	1584(3)	8281(5)	3279(3)	
O14	557(3)	4868(5)	5173(3)	
O15	2612(3)	8920(5)	6947(3)	
O16	4496(2)	1940(4)	282(3)	
O21	7311(2)	2433(4)	9552(3)	
O22	9283(2)	7021(4)	9708(3)	
O23	8414(3)	3281(5)	6655(3)	
O24	9378(3)	9872(5)	4657(3)	

Atomic coordinates ($\times 10^4$) for bikitaite (*Zeo.* 9:303(1989)):

	x	y	z	g
at 295K				
O25	7318(3)	3929(5)	2966(3)	
O26	5593(2)	6797(4)	9807(3)	
Li1	3095(5)	3657(9)	1469(7)	0.910(28)
Li2	6995(6)	8665(10)	8722(7)	0.947(29)
O17	4083(3)	3224(6)	4347(3)	
H11	3230(9)	2849(17)	4786(10)	
H12	4870(8)	1784(12)	4863(8)	
O27	5991(3)	8225(6)	5871(4)	
H21	6853(8)	7872(19)	5420(10)	
H22	5214(8)	6763(12)	5343(8)	
at 13K				
Si11	1062(3)	8651(4)	979(3)	0.974(14)
Si12	1038	8006	5049	1.004(14)
Al13	3801(3)	8778(5)	9377(3)	1.050(15)
Al21	8991(3)	3650(5)	9073(3)	1.046(16)
Si22	8917(2)	2991(4)	4878(3)	0.983(13)
Si23	6195(3)	3773(4)	637(3)	0.985(13)
O11	2646(2)	7397(4)	569(3)	
O12	847(2)	1883(4)	441(3)	
O13	1591(2)	8250(4)	3277(3)	
O14	591(2)	4842(4)	5208(3)	
O15	2594(2)	9007(4)	6946(3)	
O16	4508(2)	1960(4)	324(3)	
O21	7314(2)	2426(4)	9562(3)	
O22	9275(2)	7050(4)	9605(3)	
O23	8410(2)	3246(4)	6658(3)	
O24	9344(2)	9843(4)	4618(3)	
O25	7339(2)	4016(4)	2970(3)	
O25	5589(2)	6823(4)	9778(3)	
Li1	3117(5)	3667(8)	1506(6)	0.932(29)
Li2	6977(5)	8692(9)	8691(6)	0.926(29)
O17	4084(3)	3260(5)	4392(3)	
H11	3194(5)	2991(9)	4825(6)	
H12	4837(5)	1717(7)	4895(5)	
O27	5988(2)	8253(4)	5830(3)	
H21	6881(5)	8004(9)	5397(6)	
H22	5252(5)	6691(7)	5330(5)	

Thermal Properties

Continuous weight loss is observed between 150 and 400°C. The bikitaite structure is decomposed at 400°C in the presence of steam to produce eucryptite and petalite. This mineral can be reversibly hydrated and dehydrated to temperatures up to 240°C; above this temperature only partial rehydration is possible (*Am. Miner.* 59:71(1974)).

Infrared Spectrum

Mid-infrared Vibrations (cm⁻¹): 1105(sh), 968(s), 782(w), 680(m), 460(s) (*Zeo.* 4:369(1984)).

Adsorption

Adsorption measurements have shown that after dehydration at 230°C, bikitaite exhibits no adsorptive properties.

Ion Exchange

The Li^+ ions in bikitaite do not exchange with NH_4^+ ions.

Bikitaite (Syn) (BIK)
(US4,820,502(1989))
Mobile Oil Corporation

Related Materials: bikitaite (natural)

Structure

CHEMICAL COMPOSITION: (0.8–1.0) Cs_2O : (0–0.2) Na_2O : Al_2O_3 : 17–100 SiO_2.

X-Ray Powder Diffraction Data: $(d(Å)$ (I/I_o)
7.42(w), 6.72(w), 4.28(s–vs), 4.02(w), 3.68(w), 3.35(s–vs), 3.29(m–vs), 3.14(w–m), 2.946(w), 2.791(w), 2.716(w), 2.532(w–s), 2.381(w), 2.357(w), 2.182(w), 2.013(w), 1.946(w), 1.911(w), 1.843(w), 1.815(w), 1.688(w), 1.672(w), 1.640(w), 1.587(w), 1.565(w).

The structure of a synthetic cesium-aluminosilicate with the bikitaite framework and a composition of $Cs_{0.35}Al_{0.35}Si_{2.65}O_6$ also has been identified (*Z. Kristallogr.* 116:301(1984)). The Cs cations are located in the large eight-mem-

ber-ring channels (*Z. Kristallogr.* 116:301 (1984)).

CHEMICAL COMPOSITION: $Cs_{0.35}Al_{0.35}Si_{2.65}O_6$
SPACE GROUP: $B2_1$
D_c = 2.55g/cm
D_m = 2.56 g/cm
UNIT CELL CONSTANTS   (A): a = 7.3585(4)
b = 5.0334(3)
c = 90.778(9)°
PORE STRUCTURE: eight-member ring openings

Positional parameters with estimated standard deviations (*Z. Kristallogr.* 116:301 (1984)):

	x	y	z
Cs	0.4971(3)	0.7585(10)	0.4642(1)
Si(1)	0.4998(4)	0.7500(0)	0.0577(2)
Si(2)	0.2094(4)	0.8418(9)	0.1964(2)
Si(3)	0.7934(4)	0.7567(10)	0.1975(2)
O(1)	0.5544(10)	0.0029(22)	0.0012(6)
O(2)	0.2859(11)	0.6478(19)	0.2688(5)
O(3)	0.7283(12)	0.0487(19)	0.2279(5)
O(4)	0.3271(10)	0.8333(20)	0.1132(4)
O(5)	0.6728(10)	0.6754(19)	0.1158(4)
O(6)	0.0032(9)	0.7718(21)	0.1693(4)

Synthesis

ORGANIC ADDITIVES
none

Synthetic bikitaite has been claimed to crystallize from reactive gels in a preferred range containing SiO_2/Al_2O_3 between 30 and 100, H_2O/SiO_2 between 24 and 40, OH/SiO_2 of 0.1 to 0.4, and $Cs/(Cs + Na)$ between 0.2 and 1. Crystallization requires between 24 hours and 60 days for completion. Higher temperatures (ca. 225°C) result in more rapid crystallization, whereas lower temperatures (ca. 80°C) result in crystallization only after an extended period of time. Quartz is a major impurity phase, observed at the higher crystallization temperatures.

Adsorption

Adsorption properties of synthetic bikitaite (wt %):

	Cyclohexane	*n*-Hexane	Water
Sample 1	1.3	1.7	1.3
Sample 2	1.6	1.2	1.0

Boggsite (BOG)
(natural)

Boggsite, a name suggested to the IMA Commission on New Minerals and Mineral Names, is a rare zeolite mineral found near Goble, Oregon in small cavities in the Columbia basalt (*PP 8 IZC* 45(1989)). Boggsite occurs in basalt percolated by rain water. It is associated with another zeolitic phase, tschernichite.

Structure

(*PP 8th IZC* 45(1989))

CHEMICAL COMPOSITION:
 $Ca_{7.4}Na_{3.7}Al_{18.5}Si_{77.5}O_{192} * 74H_2O$
SYMMETRY: orthorhombic
SPACE GROUP: Imma
UNIT CELL CONSTANTS (Å): $a = 20.236$
                           $b = 23.798$
                           $c = 12.798$
PORE STRUCTURE: 12- and 10-member rings

X-Ray Powder Diffraction Data: (calculated data provided, experimental data not reported, *Am. Mineral.* 75:501(1990)) (d(Å) (I/I_0)) 11.899(33), 11.271(100), 10.816(27), 10.118(24), 8.004(6), 7.708(3), 6.742(15),
5.967(10),   5.910(3),   5.636(7),   5.408(2),
5.059(5),   4.924(5),   4.556(6),   4.461(34),
4357(31),   4.213(4),   4.174(2),   4.007(11),
4.002(2),   3.966(3),   3.878(2),   3.859(86),
3.757(13),   3.753(2),   3.693(6),   3.672(13),
3.605(31),   3.522(7),   3.417(5),   3.386(31),
3.373(2),   3.371(63),   3.323(7),   3.303(8),
3.200(3), 3.198(9), 3.177(4), 3.141(2), 3.083(5),
3.051(3),   2.975(9),   2.936(6),   2.875(6),
2.8870(3),   2.818(5),   2.778(5),   2.778(5),
2.668(4),   2.633(15),   2.590(11),   2.548(2),
2530(2), 2.500(2), 2.471(2), 2.436(5), 2.393(2),
2.368(2), 2.360(4), 2.298(2), 2.247(3), 2.133(3),
2.049(3),   2.024(7),   1.9843(7),   1.9832(2),
19788(2), 1.9511 (7), 1.8702(2), 1.8204(4),
1.7940(2), 1.6429(4), 1.6306(2), 1.6044(2),
1.6007(2), 1.5844(2), 1.5745(2).

Boggsite contains 4-, 5-, 6-, 10-, and 12-member rings. The 12- and 10-member rings are similar to those found in gmelinite and ferrierite. The bifurcated four-rings reduce the free diameter of the 10-rings to approximately that of a circular nine-ring (*Am. Mineral.* 75:501 (1990)).

Positional parameters for boggsite (*PP 8 IZC* 45(1989)):

	x	*y*	*z*
Si1	0.18897(10)	0.18520(8)	0.6722(2)
Si2	0.18984(11)	0.02409(9)	0.3297(2)
Si3	0.07709(10)	0.18525(8)	0.8354(2)
Si4	0.07750(10)	0.02213(9)	0.1644(2)
Si5	0.22112(9)	0.08322(9)	0.5379(2)
Si6	0.12275(10)	0.08370(9)	0.9657(2)
O1	0.1893(5)	0.25	0.6289(6)
O2	0.1197(3)	0.1709(2)	0.7307(5)
O3	0.1956(3)	0.1454(2)	0.5688(4)
O4	0.1905(3)	0.0708(3)	0.4234(4)
O5	0.1185(3)	0.0321(3)	0.2718(5)
O6	0.0898(4)	0.25	0.8735(7)
O7	0.0	0.1744(4)	0.8044(8)
O8	0.0	0.0262(4)	0.1971(7)
O9	0.1941(4)	0.0385(3)	0.6213(5)
O10	0.1000(3)	0.1466(2)	0.9333(5)
O11	0.0955(3)	0.0726(3)	0.0828(5)
O12	0.2009(3)	0.0805(4)	0.9678(7)
O13	0.0948(4)	0.0385(3)	0.8851(5)
O14	0.25	0.1746(3)	0.75
O15	0.25	−0.0382(5)	0.75

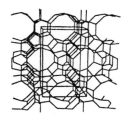 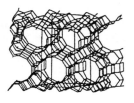

Boggsite Fig. 1S: Framework topology

Boralites
(It. Pat. App. #24884 A/78)
ENI

Boralites are topologically related to the zeolites with a borosilicate composition.

Boralites and aluminosilicate analogs (*Proc. 5th IZC* 40 (1980)):

Phase	Organic base used in synthesis	Aluminosilicate analog
BOR-A	[N(CH$_3$)$_4$]OH [N(CH$_3$)$_3$(CH$_2$-C$_6$H$_5$)]OH	Nu-1
BOR-B	[N(C$_2$H$_5$)$_4$]OH	Beta
BOR-C	[N(C$_2$H$_5$)$_4$]OH [N(C$_3$H$_7$)$_4$]OH [N(C$_5$H$_{11}$)$_4$]OH H$_2$N(CH$_2$)$_2$NH$_2$	Silicalite ZSM-5
BOR-D	[N(C$_4$H$_9$)$_4$]OH	ZSM-11

Boron Beta (BEA*)
(WO 91/00777(1991))
Chevron

Related Material: beta

Structure

Chemical Composition: (1.0–5.0)R$_2$O : (0.1–2.0) M$_2$O : W$_2$O$_3$: (greater than 10) YO$_2$; M = alkali metal cation, W is boron, Y is silicon, germanium, or mixture and R is diquaternary ammonium ion.

X-Ray Powder Diffraction Data: (d(Å) (I/I_0)) 11.5(85), 6.52(9), 5.96(12), 4.80(3), 4.07(15), 3.89(100), 3.26(10), 3.05(6), 2.97(8).

See beta for a description of the framework topology.

Synthesis

ORGANIC ADDITIVE
bis(1-azonia bicyclo[2.2.2]octane) (DABCO)

Boron beta crystallizes from a reactive gel system at at least 140°C after 6 to 11 days. Seeds of (B) Beta have been used to encourage nucleation. ZSM-12 can appear as an impurity phase in this system.

Breck Structure Six (EMT)
(Zeo. Mol. Sieves 1974)

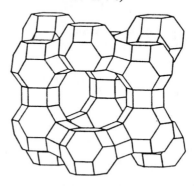

Breck Structure Six Fig. 1S: Framework topology of structure six.

Related Material: EMC-2

The theoretical structure six proposed by D. W. Breck is based upon double six-rings and beta cage units. The hexagonal structure is produced by taking alternate layers of the truncated octahedral beta cages as they occur in the (111) planes of the faujasite structure, rotating by 60° and relinking them. This structure has also been referred to as hexagonal faujasite. The synthetic material is called EMC-2.

Brewsterite (BRE)
(natural)

The naturally occurring brewsterite was named in honor of Sir David Brewster in 1822. Deposits were found in Strontian, Argyllshire, Scotland. Brewsterite occurs with other zeolites such as stilbite, heulandite,and natrolite in the USSR, with thomsonite, mesolite, and analcime in Canada, and with epistilbite in the United States (*Nat. Zeo.* 300(1985)).

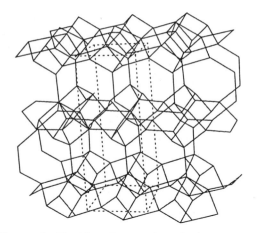

Brewsterite Fig. 1S: Framework topology.

ages gives rise to two sets of intersecting channels defined by the eight-member rings. Random occupancy is observed for two of the silicons and one aluminum, and regular occupancy is observed for one of the silicon tetrahedra. The strontium cations occupy sites near the intersections and are coordinated with five H_2O molecules and four framework oxygen atoms. Such an incomplete hydration sphere is typical for zeolites with small cavities.

Atomic coordinates for brewsterite (*Acta Crystallogr.* 17:857(1964)):

	x	y	z
Sr	0.2507	0.250	0.1778
TD	0.9106	0.0527	0.6412
TC	0.5566	0.1584	0.5345
TB	0.4053	0.0568	0.2105
TA	0.3187	0.0814	0.8220
O1	0.3461	0.1057	0.0286
O2	0.4244	0.1227	0.3600
O3	0.7909	0.1214	0.5466
O4	0.4519	0.1406	0.7141
O5	0.0731	0.0915	0.7604
O6	0.2212	0.9970	0.2379
O7	0.3835	0.9920	0.7959
O8	0.0000	0.000	0.500
O9	0.5767	0.250	0.4986
H2O1	0.0545	0.250	0.4703
H2O2	0.9319	0.1469	0.1537
H2O3	0.5896	0.250	0.0256
H2O4	0.0544	0.250	0.8642

Structure

(*Acta Crystallogr* 17:857(1964))

CHEMICAL COMPOSITION:
  $(Sr,Ba,Ca)_2[(AlO_2)_4(SiO_2)_{12}]*10H_2O$
SYMMETRY: monoclinic
SPACE GROUP: $P2_1/m$
FRAMEWORK DENSITY (g/cc): 1.77
VOID FRACTION: 0.26
UNIT CELL CONSTANTS: (Å) $a = 6.77$
                            $b = 17.51$
                            $c = 7.74$
                            beta $= 94°$
PORE STRUCTURE: two-dimensional 2.7 × 4.1 ; 2.3 × 5.0 Å

X-Ray Powder Diffraction Data: (*Nat. Zeo.* 345(1985)) (d(Å) (I/I_o)) 6.81(30), 6.15(90), 4.98(40), 4.53(100), 3.87(70), 3.71(10), 3.48(10), 3.35(10), 3.21(80), 3.02(10), 2.885(90), 2.667(30), 2.549(30), 2.442(30), 2.309(30), 2.243(30), 2.191(20), 2.103(30), 1.989(40), 1.933(20), 1.866(10), 1.824(10), 1.771(10), 1.728(10), 1.642(10), 1.595(20), 1.542(20), 1.514(10), 1.471(10), 1.435(20), 1.386(20), 1.359(20), 1.324(20), 1.302(20), 1.272(20).

The brewsterite framework consists of layers of tetrahedra linked through four-, six-, and eight-member rings. The crosslinking of these layers is through tilted five-member rings. Such link-

Thermal Properties

The DTG curve shows a shoulder at 50°C and three peaks appearing at 190, 260, and 380°C. Desorption of water is nearly complete at 500°C, with some tailing in the TGA observed to 800°C (*Nat. Zeo.* 300(1985)).

Adsorption

Only H_2O is adsorbed in this material, with the quantity of water desorbed being greater than that adsorbed (12 wt% adsorbed, 15 wt% desorbed) (*Zeo. Mol. Sieves.* 211(1974)).

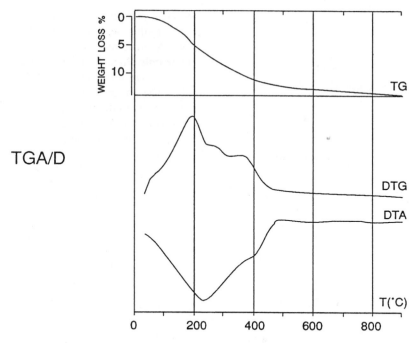

Brewsterite Fig. 2:   Thermal curves of brewsterite from Strontian, Scotland, in air, heating rate 20°C/min (*Nat. Zeo.* 304 (1985)). (Reproduced with permission of Springer Verlag Inc.)

BSD (SOD)
(Ger 35,37,998(1987))
BASF

Related Materials: borosilicate sodalite

Structure

CHEMICAL COMPOSITION: $M_{2/n}O * B_2O_3 * 2–15$ $SiO_2 * 0–80H_2O$
SYMMETRY: Cubic
UNIT CELL CONSTANTS: (A) $a = 8.77$

X-Ray Powder Diffraction Data: ($d(\text{Å}) (I/I_o)$)
6.21(100)   4.38(15),   3.58(67),   3.10(12), 2.77(13), 2.53(10), 2.35(3).

Synthesis

ORGANIC ADDITIVE
trioxane ($C_3H_6O_3$)

BSD crystallizes from reactive gels with SiO_2/B_2O_3 ratios between 2 and 15, Na_2O/B_2O_3 of 2.5 to 3, and $C_3H_6O_3/B_2O_3$ ratios between 18 and 300. Crystallization occurs between 110 and 130°C after 7 days.

B-SGT (SGT)
(*RR 8th IZC 20 (1989)*)

Related Materials: sigma-2

Structure

SYMMETRY: tetragonal
SPACE GROUP: $14_1/amd$
   UNIT CELL CONSTANTS (Å): $a = 10.2431(4)$
                                           $c = 34.3184(24)$

X-Ray Powder Diffraction Data: $d(\text{Å}) (I/I_o)$)
9.825(70), 8.590(10), 7.632(9), 4.541(100), 4.490(92),   4.399(35),   4.292(9),   4.253(9), 3.812(15), 3.398(17), 3.337(38), 3289(19), 3.057(11), 2.986), 2.757(9), 2.625(11).

B-SGT is the borosilicate analog of the zeolite sigma-2

Synthesis

ORGANIC ADDITIVES
1-adamantane amine
2-adamantane amine

Synthesis of B-SGT used silica sol as the source of silica and boric acid as the source of boron. Adamantane amines were used to encourage crystallization of this structure. The pH of the reactive mixture was in the range of 9.0 to 13.5. Crystallization at 150°C required 10 to 30 days. If aluminum or gallium is substituted for boron in the synthesis mixture, the sigma-1 or DDR structure is obtained instead (*RR 8th IZC*, 20 (1989)).

C

C (ANA)

Early name identifying synthetic analcime from Union Carbide (*Zeo. Mol. Sieve* 1974). See analcime.

Cacoxenite
(natural)

The natural iron aluminophosphate mineral, cacoxenite was first reported in 1825 (*Encyclopedia of Minerals* 98(1974)). Cacoxenite occurs lining the cavities in limonite. It crystallizes as radiating globular masses with bundles of fibers reaching 6 mm in diameter. The needles are golden yellow because of the presence of octahedral iron contained within the structure. Crystals have been found in Arkansas, Alabama, Tennessee, and Pennsylvania. Cacoxenite crystals also have been associated with hematite from deposits in New York and Nevada. Other countries with reported deposits in-

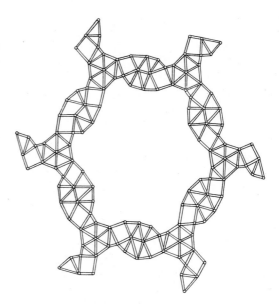

Cacoxenite Fig. 1S: Framework topology.

clude France, Germany, Czechoslovakia, and Sweden. The synthesis of cacoxenite has not yet been reported.

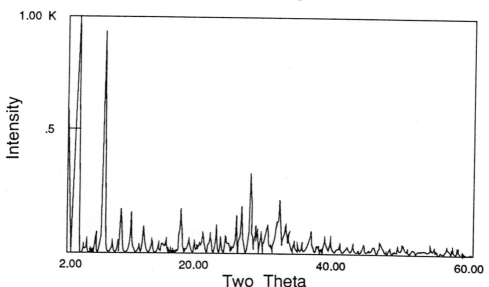

Cacoxenite Fig. 2:   X-ray powder diffraction pattern of cacoxenite. (Reproduced with permission of author)

Structure

(Nature 306:356(1983))

CHEMICAL COMPOSITION: [Al(Al,Fe)₃Fe₂₁O₆(OH)₁₂(PO₄)₁₇(H₂O)₂₄]

CHEMICAL COMPOSITION: $[Al(Al,Fe)_3Fe_{21}O_6(OH)_{12}(PO_4)_{17}(H_2O)_{24}]$
SYMMETRY: hexagonal
SPACE GROUP: $P6_3/m$
UNIT CELL CONSTANTS (Å): $a = 27.7$
$c = 10.66$
DENSITY: 2.26 g/cc
PORE STRUCTURE: unidimensional channels with 14 Å pore opening

X-Ray Powder Diffraction Data: (*Am. Mineral.* 51:1811 (1966)) (calculated). $(d(\text{Å})\ (I/I_0))$

23.962, 13.830, 11.981, 9.736, 9.057, 7.987, 6.901, 4.863, 4.592, 4.221, 4.245, 4.157, 3.9897, 3.8344, 3.7293, 3.3456, 3.1933, 3.1281, 3.0169.

The most distinctive feature of cacoxenite is the large free 14 Å diameter of the pore system, oriented parallel to the *c*-axis. The structure of this material can be thought of in terms of two fundamental building blocks. The first is related to the central segment in the Keggin structure.

The second resembles the octahedral tetramer in leucophosphite. This tetramer shares edges defined by inversion to form infinite chains or bundles parallel to the *c*-axis. This highly hydrated basic ferric oxyphosphate mineral contains associated water in several different environments within the structure.

Atomic coordinates for cacoxenite (*Nature* 306:356 (1983)):

	x	y	z
Fe1	0.4759(2)	0.1140(2)	1/4
Fe2	0.6542(2)	0.9040(2)	3/4
Fe3	0.5717(2)	0.2330(2)	3/4
Fe4	0.5400(1)	0.0525(1)	0.0525(1)
Fe5	0.6758(1)	0.2229(1)	0.6066(4)
Al1	2/3	1/3	1/4
Al2	0.3831(4)	0.0107(4)	3/4
P1	2/3	1/3	0.562(1)
O1	2/3	1/3	0.422(3)
O2	0.6060(6)	0.2979(6)	0.614(2)
P2	0.6555(4)	0.1129(4)	1/4
O3	0.6731(10)	0.0672(10)	1/4
PO4	0.7063(10)	0.1709(10)	1/4
O5	0.6203(7)	0.1022(7)	0.128(2)
P3	0.4061(2)	0.0107(2)	0.049(1)
O6	0.3839(6)	0.0171(6)	−0.077(2)

(continued)

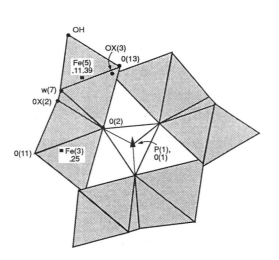 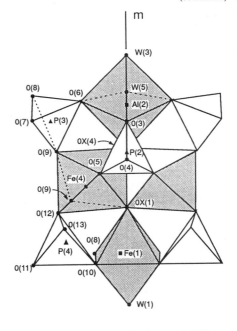

Cacoxenite Fig. 3:   Building units for cacoxenite: (left) Keggin-like structure; (right) leucophosphite-type building unit (*Nature* 306:356(1983)). (Reproduced with permission of McMillan Journals Ltd.)

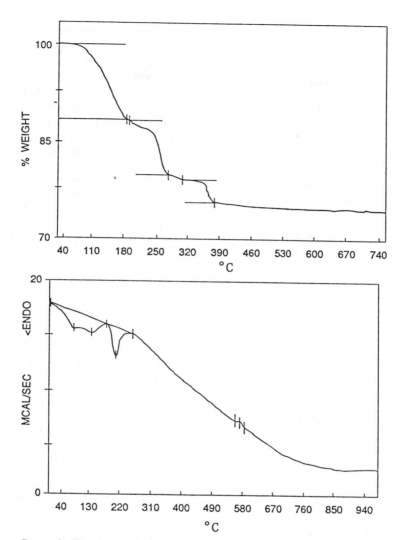

Cacoxenite Fig. 4:   TGA/DTA for cacoxenite. (Reproduced with permission of author)

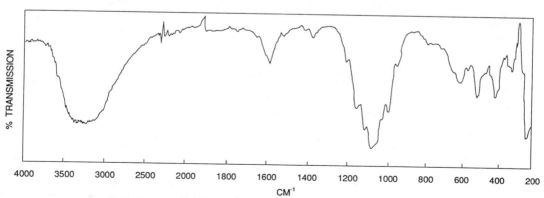

Cacoxenite Fig. 5:   Infrared spectrum of natural cacoxenite. (Reproduced with permission of author)

Atomic coordinates for cacoxenite (*Nature* 306:356 (1983)):

	x	y	z
O7	0.3591(7)	-0.0398(7)	0.119(2)
O8	0.4256(7)	0.0650(7)	0.122(2)
O9	0.4572(6)	-0.0001(6)	0.031(1)
P4	0.5522(2)	0.1657(3)	0.000(1)
O10	0.5239(7)	0.1706(7)	0.122(2)
O11	0.5284(6)	0.1817(6)	-0.113(2)
O12	0.5403(6)	0.1053(7)	-0.021(2)
O13	0.6155(6)	0.2062(6)	0.015(2)
OH	0.6680(7)	0.1496(7)	0.615(2)
OX1	0.5144(10)	0.0740(10)	1/4
OX2	0.5172(9)	0.2646(9)	3/4
OX3	0.6294(9)	0.2150(9)	3/4
OX4	0.5520(11)	0.0025(11)	3/4
W1	0.427(1)	0.156(1)	1/4
W2	0.570(1)	0.079(1)	3/4
W3	0.325(1)	0.034(1)	3/4
W4	0.376(1)	0.262(1)	3/4
W5	0.443(1)	0.094(1)	3/4
W6	0.620(3)	0.256(3)	1/4
W6d	0.656(4)	0.269(1)	1/4
W7	0.510(1)	0.269(1)	0.047(2)

Thermal Properties

The X-ray diffraction pattern of natural cacoxenite changes with thermal treatment. Recrystallization to a condensed phase material begins by 750°C. Water loss occurs in four steps. Between room temperature and 60°C, 7.13% of the weight is lost from this mineral. Between 60 and 175°C, a further weight reduction of 10.6% is observed. Between 177 and 269°C and between 306 to 381°C, 7.63% and 3.31 wt% weight losses occur, respectively. See Cacoxenite Fig. 4 on previous page.

Change in low-angle peak positions of cacoxenite after selected thermal treatment (SSSC 49A 439 (1989)):

Sample	$d(\text{Å})$
Parent mineral	24.2
200°C	22.3, 20.5
350°C	19.7
750°C	—

Infrared Spectrum

Mid-infrared Vibrations of Cacoxenite (cm⁻¹):
Mid-infrared Vibrations of Cacoxenite (cm^{-1}):
1212, 1155, 1110, 1070, 1040, 1010, 980, 933, 580, 538, 482, 385 (from author).

Adsorption

Results obtained from the adsorption studies on the crystalline microporous mineral, cacoxenite. (SSSC 49A: 439 (1989)):

Adsorbent*	color	act. temp.	act. time	act. press.	ads. time	ads. press.	%wt adsorbed
sample 1:							
n-hexane	brown	198°C	1hr	5 mtorr	2hr	70torr	5.17
cyclohexane	brown	198°C	3hr	—	2hr	50torr	0.64
n-hexane	brown	198°C	2hr	—	2hr	70torr	0.55
sample 2:							
n-hexane	yellow	50°C	overnight	90 mtorr	2hr	70torr	2.88
cyclohexane	yellow	50°C	2 days	—	2hr	50torr	0.86
O₂	yellow	50°C	2 days	50 mtorr	2hr	75torr	9.76
n-hexane	yellow	50°C	2 days	20 mtorr	2hr	70torr	0.34
O₂	yellow	50°C	2 days	15 mtorr	2hr	75torr	10.63
H₂O	yellow	50°C	overnight	15 mtorr	2hr	17.5torr	14.82
O₂	brown	200°C	1 hr	20 mtorr	2hr	75torr	17.15
sample 3:							
O₂	brown	200°C	1hr	—	2hr	75torr	13.3

*Adsorption was run at room temperature for n-hexane, cyclohexane and water, at liquid nitrogen temperatures for O₂ adsorption.

Ca-D (ANA)
(*JCS* 983 (1961))

Related Materials: (tetragonal) analcime

Structure

CHEMICAL COMPOSITION: Si/Al = 2
UNIT CELL CONSTANT: (Å) a = 13.62

X-Ray Powder Diffraction Data: $(d(\text{Å})\ (I/I_0))$
6.82(m), 5.56(vs), 4.82(m), 3.63(m), 3.40(vs),
3.37(s), 2.90(s), 2.89(m), 2.66(w), 2.47(w),
2.40(w),   2.20(vw),   1.88(vw),   1.85(vw),
1.72(vw).

Synthesis

ORGANIC ADDITIVE
none

Zeolite Ca-D was prepared from reactive gels
in the CaO : Al$_2$O$_3$: SiO$_2$: H$_2$O system. The
SiO$_2$/Al$_2$O$_3$ of the reactive gel was between 3
and 4. Crystallization did not readily occur below 225°C, with this phase observed between
220 and 310°C. Both silica sol and powdered
silica glass were used in this crystallization.

Ca-E (ANA)
(JCS 983 (1961))

Related Materials: (cubic) analcime

Structure

Chemical Composition: aluminosilicate composition not specified.

X-Ray Powder Diffraction Data: $(d(\text{Å})\ I/I_0))$
6.80(m), 5.54(vs), 4.80(m), 3.63(w), 3.40(vs),
2.90(vs), 2.67(m), 2.48(m), 2.40(w), 2.21(w),
1.88(w), 1.85(w), 1.68(w), 1.58(w).

Synthesis

ORGANIC ADDITIVE
none

Zeolite Ca-E was crystallized from CaO :
Al$_2$O$_3$: SiO$_2$: H$_2$O systems requiring high temperature because of the difficulty of using calcium aluminosilicate gels in crystallizing zeolitic material. Temperatures between 250 and
450°C were employed, and SiO$_2$/Al$_2$O$_3$ ratios
between 4 and 8 resulted in the crystallization
of this phase. Materials with intracrystalline pore
volumes of this 0.18 cm^3/cm^3 were obtained.

Ca-I (THO)
(*JCS* 983(1961))

Related Materials: thompsonite

Structure

Chemical Composition: aluminosilicate composition not specified.

X-Ray Powder Diffraction Data: $(d(\text{Å})\ (I/I_0))$
4.63(m), 3.50(m), 3.19(w), 2.94(m), 2.85(m),
2.67(m), 2.25(w), 2.17(w), 1.76(vw), 1.62(vw).

Synthesis

ORGANIC ADDITIVES
none

This impure form of thompsonite was prepared from a calcium aluminosilicate reactive
gel system in a SiO$_2$/Al$_2$O$_3$ range of 1 to 2. Silica
sol and hydrous alumina were used. Crystallization occurred after a month at 245°C.

Ca-J (EPI)
(*JCS* 983(1961))

Related Materials: epistilbite

Structure

CHEMICAL COMPOSITION: aluminosilicate
  composition not specified
SYMMETRY: near orthorhombic
UNIT CELL CONSTANTS: (Å) a = 15.0
                         b = 17.0
                         c = 10.25
                         β = 90°

X-Ray Powder Diffraction Data: $(d(Å)\ (I/I_0))$
8.89(s), 6.88(s), 5.54(mw), 4.89(s), 4.48(w),
4.31(mw), 3.87(s), 3.44(vs), 3.33(w), 3.25(w),
3.20(ms), 2.91(mw), 2.85(m), 2.78(w), 2.68(m),
2.50(m),   2.41(w),   2.21(vw),   2.09(vw),
1.86(vw), 1.77(w).

Synthesis

ORGANIC ADDITIVES
none

Zeolite Ca-J crystallizes from a reactive aluminosilicate gel in the presence of calcium oxide in a SiO_2/Al_2O_3 ratio between 6 and 7. Silica glass and hydrous alumina are used as reactants. Crystallization takes place after one month at 250°C.

Ca-L (PHI)
(*JCS* 983(1961))

Related Materials: phillipsite

Structure

CHEMICAL COMPOSITION: aluminosilicate
  composition not specified
SYMMETRY: tetragonal
UNIT CELL CONSTANTS: (Å) $a = 10.01$
                       $c = 9.89$
PORE VOLUME: 0.36 cm³/cm³

X-Ray Powder Diffraction Data: $(d(Å)\ (I/I_0))$
7.14(vs), 4.96(ms), 4.14(s), 4.07(w), 3.24(m),
3.14(vs),   2.67(ms),   2.56(vw),   2.33(vw),
2.12(vw),   2.07(vw),   1.98(vw),   1.94(vw),
1.81(vw),   1.78(mw),   1.74(vw),   1.71(w),
1.67(vw), 1.65(vw), 1.26(w).

Synthesis

ORGANIC ADDITIVES
none

Ca-L crystallizes from a reactive gel prepared from calcium oxide, silica glass, and hydrous alumina, with SiO_2/Al_2O_3 between 3 and 4. At 250°C, it took one month to produce a crystalline product.

Cancrinite (CAN)
(natural)

Related Materials: ECR-5
                       tiptopite

Cancrinite is a member of the aluminosilicate family of minerals, which are intermediate between zeolites and felspathoids.

Structure

(*Z. Kristallogr.* 122:407(1965))

CHEMICAL COMPOSITION: $Na_{6.3}Si_6Al_6O_{24}$ *
  $Ca_{0.91}Fe_{0.106}(CO_3)_{1.47}$ * $(CO_3)_{1.47}$
SYMMETRY: hexagonal
SPACE GROUP: $P6_3$
UNIT CELL CONSTANTS: (Å) $a = 12.75$
                           $c = 5.14$

X-Ray Powder Diffraction Data: $(d(Å)\ (I/I_0))$
11.00(30),   6.32(60),   5.47(10),   4.64(80),
4.13(30),   3.75(10),   3.65(50),   3.22(100),
3.03(30), 2.974(30), 2.732(60), 2.610(50),
2.564(60), 2.500(30), 2.410(50).

The cancrinite structure consists of AB stacking with channels around the threefold axis and

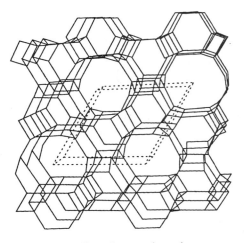

Cancrinite Fig. 1S: Framework topology.

the 6_3 axis. The cations and occluded anions such as H_2O and CO_3 are enclosed within these channels. Si/Al ordering is observed (*Z. Kristallogr.* 122:407(1965)). The presence of these salt moieties results in a closed-pore material. Cancrinite commonly has been found as an intergrowth with other zeolite structures such as erionite, offretite, and sodalite (*Proc. R. Soc. Lond.* A:399:57(1985)).

Atomic parameters for cancrinite (*Z. Kristallogr.* 122:407 (1965)):

	x	y	z
Si	0.0750(2)	0.4124(2)	0.7500(8)
Al	0.3272(2)	0.4101(2)	0.7500(8)
Na2(Ca)	0.1232(4)	0.2486(4)	0.2920(14)
Na1(Ca)	0.666$_7$	0.333$_3$	0.1440(28)
O1	0.2039(7)	0.4045(7)	0.6589(22)
O2	0.1141(8)	0.5623(8)	0.7229(23)
O3	0.0269(7)	0.3485(7)	0.0619(20)
O4	0.3150(8)	0.3587(8)	0.0407(22)
C	0.0	0.0	0.6835(10)
O5	0.0584(16)	0.1200(16)	0.6857(42)
H_2O	0.6150(38)	0.3130(50)	0.6492(14)

Synthesis

ORGANIC ADDITIVES
tetramethylammonium$^+$ (TMA)

The cancrinite structure has been prepared in the presence of Na, (Na,Li), (Na,TMA), Sr, (Li,Cs,TMA), (Cs,Li), and (Rb,Li), with sodium being the preferred cation (*Zeo.* 1:130(1981)). Generally, ions can become trapped within the cages during synthesis. For example, in the presence of sulfide or sulfate, sulfur ions are found to occupy the cavities (Ger 131,979(1983)). When ions such as hydroxide are trapped in the cavities, the materials generally are referred to as basic cancrinites or hydroxy cancrinites. Cancrinite generally crystallizes in less silica-rich systems at temperatures between 140 and 200°C. Batch compositions of $1.4\ Li_2O : 2.11\ M_2O : 2\ SiO_2 : 13\ H_2O$, where M is Cs or Rb, have been used. The crystalli-

zation rate in the presence of Cs ion is six times higher than when Rb is used. Cancrinite crystals with SiO_2/Al_2O_3 ratios of around 3.5 have been observed to form.

Crystallization with salt pairs, each 0.1 M (2 g kaolinite, 200 ml 4M NaOH), 80°C, agitated in plastic bottles[a] (*JCS* A:1516 (1970)):

Salt anions	Products	Salt anions	Products
Cl^-/CrO_4^{-2}	$S^0 + C$	WO_4^{-2}/NO_3^-	C
Cl^-/NO_3^-	$C^0 + S$	WO_4^{-2}/MoO_4^{-2}	S
Cl^-/MO_4^{-2}	$S^0 + C$	CO_3^{-2}/CrO_4^{-2}	S
Br^-/CrO_4^{-2}	$S^0 + C$	CO_3^{-2}/NO_3^-	S
Br^-/NO_3^-	$S^0 + C$	CO_3^{-2}/MoO_4^{-2}	$S^0 + C$
Br^-/MoO_4^{-2}	$S^0 + C$	ClO_4^-/CrO_4^{-2}	S
ClO_4^-/CrO_4^{-2}	S	ClO_3^-/NO_3^-	$S^0 + C$
ClO_4^-/NO_3^-	$S^0 + C$	ClO_3^-/MoO_4^{-2}	S
ClO_4^-/MoO_4^{-2}	$S^0 + C$	NO_3^-/CrO_4^{-2}	C
WO_4^{-2}/CrO_4^{-2}	C	NO_4^-/MoO_4^{-2}	C

[a]S = sodalite; C = cancrinite. ^{0} indicates major yields.

Thermal Properties

Salt-filled cancrinites and their thermal stabilities (*JCS* A:1516 (1970)):

Stability in DTA	Salt
Stable above 850°C	Na_2SO_4
Stable at 850°C	Na_2SeO_4
Lattice breakdown at 850°	NaOH
	Na_2CrO_4
	$NaNO_3$
	$NaMoO_4$
	Na_2FeO_4
	$NaMnO_4$
	Na_2VO_4
	Na_2TeO_4
	NaN_3

Infrared Spectrum

Mid-infrared Vibrations (cm^{-1}): hydroxycancrinite (SiO_2/Al_2O_3 = 2) 1095mw, 1035msh, 1000s, 965msh, 755w, 680m, 624m, 567m,

498mw, 458ms, 429ms, 390mw, 353w♭ (*ACS* 101:201(1971)).

Ca,N-G (GME)
JCS 983(1961)

Name identifying a gmelinite structure prepared in the presence of an organic amine and calcium cations.

CAPO
(US4,759,919(1988))
Union Carbide Corporation

Structure

The CAPO molecular sieves represent the chromium aluminophosphate compositions. See CAPO Fig. 1.

Synthesis

ORGANIC ADDITIVES
CAPO-5:   TEAOH/n-Pr$_2$NH
          TEAOH
          quinuclidine
          tripropylamine
CAPO-11: di-*n*-propylamine
CAPO-17: quinuclidine

CAPO-18: TEAOH
CAPO-31: di-*n*-propylamine
CAPO-34: TEAOH
CAPO-41: di-*n*-propylamine
CAPO-44: cyclohexylamine

The CAPO molecular sieves are prepared by using Cr$_3$(OH)$_2$(CH$_3$COO)$_7$ as the source of chromium. See CAPO Fig. 2.

CAPO-5 (AFI)
(EP 158,976(1985))
Union Carbide Corporation

Related Materials: AlPO$_4$-5

Structure

Chemical Composition: See FCAPO and CAPO.

X-Ray Powder Diffraction Data: (d(Å) (I/I_0))
12.1–11.56 (m–vs), 4.55–4.46(m–s), 4.25–4.17(m–vs),   4.00–3.93(w–vs),   3.47–3.40 (w–m).

The chromium-aluminophosphate, CAPO-5, exhibits properties characteristic of a large-pore molecular sieve. See AlPO$_4$-5 for a description of the framework topology.

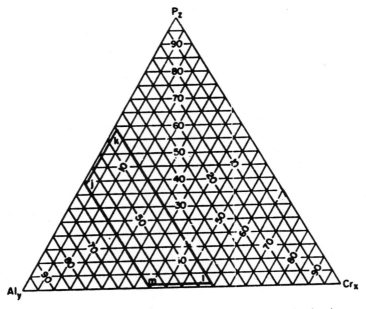

CAPO Fig. 1:   The preferred composition range for CAPO molecular sieves (US4,759,919(1988)).

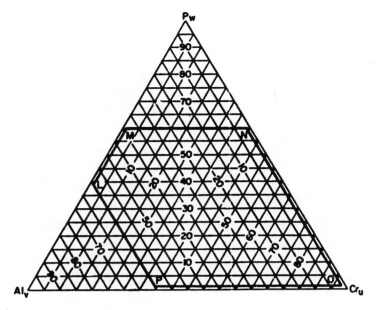

CAPO Fig. 2:   Ternary diagram with the parameters with the parameters relating to the reaction mixtures employed in the preparation of the CAPO compositions (US4,759,919(1988)).

Synthesis

ORGANIC ADDITIVE
tripropylamine

CAPO-5 crystallizes from a reactive gel with a batch composition of: (1.0–2.0) R : (0.05–0.2) Cr_2O_q : (0.5–1.0) Al_2O_3 : (0.5–1.0) P_2O_5 : (40–100) H_2O (R represents the organic additive; q denotes the oxidation state of the chromium).

Adsorption

Adsorption properties of CAPO-5:

Adsorbate	Kinetic diameter, Å	Pressure, Torr	Temp., °C	Wt. % adsorbed*
O_2	3.46	100	− 183	7
O_2		750	− 183	10
Neopentane	6.2	700	24	4
H_2O	2.65	4.3	24	4
H_2O		20	24	12

*Typical amount adsorbed.

CAPO-31 (ATO)
(EP 158,976(1985))
Union Carbide Corporation

Related Materials: $AlPO_4$-31

Structure

Chemical Composition: See FCAPO and CAPO.

X-Ray Powder Diffraction Data: (d(Å) (I/I_0))
10.40(85), 9.31(12), 6.63(6), 6.42(2), 5.99(2), 5.64(6), 5.19(7), 4.85(7), 4.44(—), 4.37(48), 4.21(28), 4.08(26), 3.98(100), 3.59(7), 3.48(5), 3.99(11), 3.75(7), 3.02(9), 2.99(9), 2.85, 2.83(14), 2.55(10), 2.51(6), 2.37(8), 2.25(4).

The chromium-aluminophosphate CAPO-31 exhibits properties characteristic of a restricted large-pore molecular sieve. See $AlPO_4$-31 for a description of the framework topology.

Synthesis

ORGANIC ADDITIVE
di-*n*-propylamine

CAPO-31 crystallizes from a reactive gel with a batch composition of: (1.0–2.0) R : (0.05–0.2) Cr_2O_q : (0.5–1.0) Al_2O_3 : (0.5–1.0) P_2O_5 : (40–100) H_2O (R represents the organic additive; q denotes the oxidation state of the chromium).

Adsorption

Adsorption properties of CAPO-31:

Adsorbate	Kinetic diameter, Å	Pressure, Torr	Temp., °C	Wt. % adsorbed*
O_2	3.46	100	−183	4
O_2		750	−183	6
Cyclohexane	6.0	90	24	3
Neopentane	6.2	700	24	3
H_2O	2.65	4.3	24	3
H_2O		20	24	10

*Typical amount adsorbed.

Caporcianite (LAU)

Obsolete name for laumontite.

CAPSO
(US4,738,837(1988))
Union Carbide Corporation

Structure

CAPSO designates the chromium silicoaluminophosphate molecular sieve compositions. See CAPSO Fig. 1.

Synthesis

ORGANIC ADDITIVES
CAPSO-5:   tripropylamine
CAPSO-11: di-*n*-propylamine
CAPSO-16: quinuclidine
CAPSO-17: quinuclidine
CAPSO-31: di-*n*-propylamine
CAPSO-34: tetraethylammonium hydroxide
CAPSO-44: cyclohexylamine

See CAPSO Fig. 2.

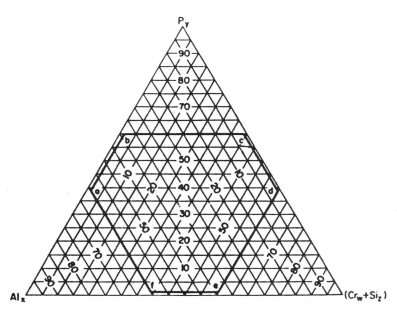

CAPSO Fig. 1: Ternary diagram of preferred compositions in mole fractions of the CAPSO molecular sieves (US4,738,837(1988)).

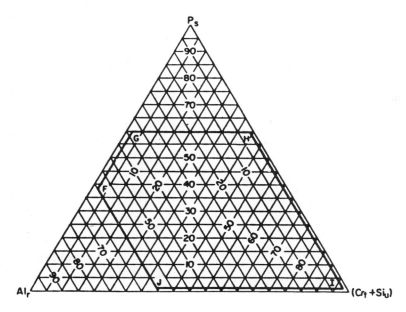

CAPSO Fig. 2:   Ternary diagram of parameters relating to the reaction mixtures employed in preparation of the chromium silicoaluminophosphate molecular sieves (US4,738,837(1988)).

Ca-Q (MOR)
(*JCS* 983(1961))

Related Materials: mordenite

Structure

Chemical Composition: CaO : Al$_2$O$_3$: 10 SiO$_2$: (6–7) H$_2$O.

X-Ray Powder Diffraction Data: none listed.

Synthesis

ORGANIC ADDITIVES
none

Ca-Q first was synthesized from reactive gels containing calcium oxide, silica, sol, and hydrous alumina, with SiO$_2$/Al$_2$O$_3$ of 7. Crystallization occurred at 390°C. This material was not prepared as a major phase under these conditions.

CF-3 (MTN)
(*J. Incl. Phenom.* 4:121(1986))

See ZSM-39.

CFAP-7
(*J. Incl. Phenom.* 5:363(1987))

Structure

Chemical Composition: Al$_2$O$_3$: 0.95 P$_2$O$_5$.

X-Ray Powder Diffraction Data: (d(Å) (I/I_0))

CFAP-7(A)   6.91(78),   6.70(100),   5.94(11), 5.53(72), 4.92(36), 4.87(5), 4.67(36), 4.35(52), 4.31(25), 4.19(52), 3.97(16), 3.85(9), 3.74(3), 3.35(16), 3.24(72), 3.18(47), 3.11(28), 3.08(22), 3.06(24), 3.03(7), 2.95(37), 2.92(47), 2.81(49), 2.65(12), 2.61(28), 2.58(24), 2.57(17).

CFAP-7(B)   7.56(12),   6.70(100),   6.32(61), 5.57(6), 5.27(33), 5.03(3), 4.67(21), 4.29(55), 4.15(52), 3.80(45), 3.53(38), 3.34(9), 3.13(12), 2.98(45), 2.85(36), 2.78(11), 2.65(39).

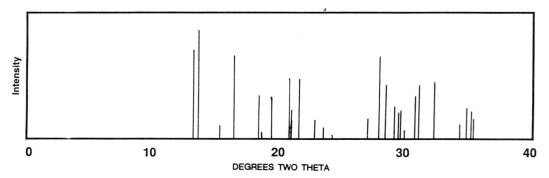

CFAP-7 Fig. 1:   X-ray powder diffraction pattern for CFAP-7(A) (*J. Incl. Phenom.* 5:363(1987)). (Reproduced with permission of Reidel Publishing Company)

CFAP-7(C) 6.75(19), 6.28(100), 5.24(78), 4.79(17), 4.29(94), 4.15(94), 4.11(72), 3.80(94), 3.53(83), 3.12(21), 2.99(17), 2.97(17), 2.85(61), 2.66(29).

Two other unknown phases also were reported:
    Unknown 1: 5.02, 3.50, 3.02, 2.99.
    Unknown 2: 10.27, 5.21, 5.01, 4.72, 4.60, 4.33, 3.50, 3.02, 3.00.

Synthesis

ORGANIC ADDITIVE
di-*n*-butylamine

CFAP-7(A) is prepared from a batch composition of: DBA : 3 $(NH_4)_2O$: Al_2O_3 : P_2O_5 : 67 H_2O between 120 and 140°C for 24 hours.

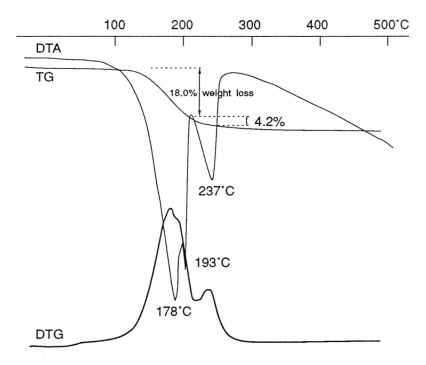

CFAP-7 Fig. 2:   DTA/TGA of CFAP-7 (*J. Incl. Phenom.* 5:363(1987)). (Reproduced with permission of Reidel Publishing Company)

Crystals consist of block aggregates up to 10 to 15 microns.

Thermal Properties

DTA, TG, and DTG curves indicate three endotherms at 178, 193, and 274°C, with a weight loss of 22.2%. The X-ray powder diffraction pattern after calcination to 170°C (CFAP-7(B))

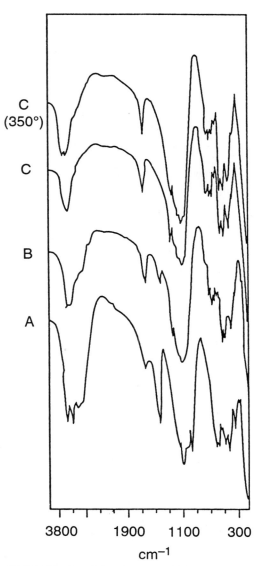

CFAP-7 Fig. 3:   Infrared spectra of CFAP-7(A–C) (*J. Incl. Phenom.* 5:363(1987)). (Reproduced with permission of Reidel Publishing Company)

differs from that of the parent material. With further treatment to 240°C for three hours, a third phase is identified (CFAP-7(C)). CFAP-7(C) is thermally stable to 500°C.

Infrared Spectra

Two vibrations at 3507 and 3344 cm^{-1} are assigned to the N–H stretch and the vibration at 1421 cm^{-1} to the N–H deformation of DBA, with the vibration at 1624 cm^{-1} thought to be due to water.

Adsorption

Little change in the adsorption properties is observed despite the change in the X-ray diffraction pattern upon thermal treatment. See CFAP-7 Fig. 4 on following page.

Chabazite (CHA)
(natural)

Related Materials: MAPO-44
MAPO-47
LZ-218
Linde D
Linde R
SAPO-34
willhendersonite
ZK-14
ZYT-6
acadialite
herschelite
haydenite
seebachite

Chabazite first was described by von Born in 1772. The name "chabasie" was proposed in 1792 and later "chabasit," which is still being used in Germany. References to chabasite also can be found in a spelling used by the French to the nineteenth century. Phakolite (from the Greek, meaning lentil) was used to describe a chabazite from Leipa, Bohemia, which exists in a particular pseudohexagonal habit. The term "phakolitic habit" still can be found to describe some chabazite minerals (*Nat. Zeo.*175 (1985)).

Natural chabazite is a very common zeolite. Other zeolites found associated with it include

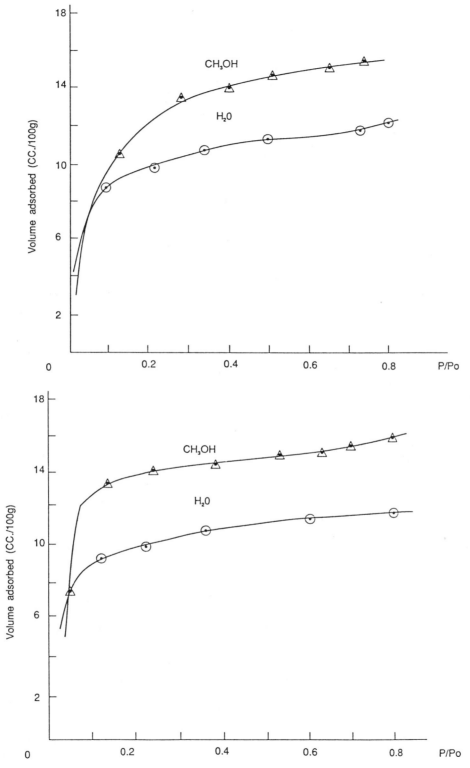

CFAP-7 Fig. 4: Isotherms of adsorption on CFAP-7(B) at 23°C (top); isotherms of adsorption on CFAP-7(C) at 23°C (bottom) (*J. Incl. Phenom.* 5:363(1987)). (Reproduced with permission of Reidel Publishing Company)

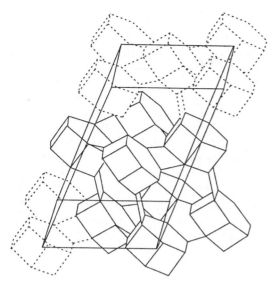

Chabazite Fig. 1S: Framework topology.

phillipsite and analcime, and it has been observed with clinoptilolite, erionite, and mordenite, depending on the location and the environment.

Structure

CHEMICAL COMPOSITION: $Ca_2[(AlO_2)_4(SiO_2)_8]$ * $13H_2O$
SPACE GROUP: $R\bar{3}m$
UNIT CELL PARAMETERS (Å):   (*Zeo.* 2:303 (1982))

	natural	Na	K	Ca	Sr
$a =$	9.428	9.425	9.442	9.401	9.406
$\alpha =$	94.14	96.06	94.29	94.39	93.87

PORE STRUCTURE: eight-member rings 3.8 A × 3.8 Å

X-Ray Powder Diffraction Data: (*Nat. Zeo.* 333(1985)) (*d*(Å) (*I/I₀*)) 9.351(50), 6.894(10), 6.384(5),   5.555(9),   5.021(30),   4.677(6), 4.324(76),   4.044(1),   3.976(2),   3.870(28), 3.590(23),   3.448(13),   3.235(6),   3.190(5), 3.033(2),   2.925(100),   2.890(30),   2.842(3), 2.776(4),   2.690(7),   2.605(10),   2.574(2), 2.507(11),   2.361(2),   2.358(2),   2.310(3), 2.300(4), 2.277(1), 2.233(1), 2.123(2), 2.119(2), 2.090(6), 2.016(1), 1.941(1), 1.911(3), 1.871(3).

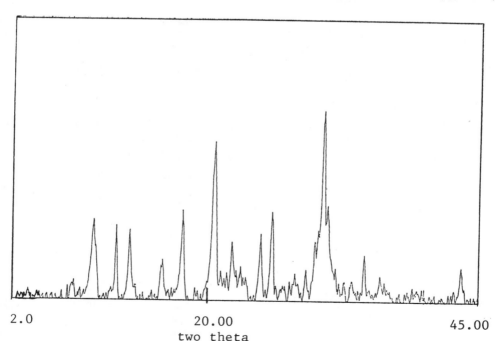

2.0   20.00   45.00
                   two theta

Chabazite Fig. 2:   X-ray powder diffraction pattern of natural chabazite. (From Bowie)

The chabasite framework consists of D6R units arranged in layers in the sequence ABCABC. The D6R units are linked by tilted four-rings. This produces a framework with large ellipsoidal cavities; entrance to these cavities is via six eight-ring pores. The ordering of the Si and Al atoms in the tetrahedral positions controls the Si/Al ratio, and the state of hydration determines the location of the cations contained within this structure.

There are four cation sites in the structure: one at the center of the hexagonal prism, two in the large cage along the 111 diagonal, and a fourth near the eight-ring window. This site is split into two sites close to each other in the monovalent ion exchange forms. In the Na and K exchanged chabasite, the site at the D6R is empty, as is one of the two sites along the 111 direction. Water is spread throughout the crystal, with 100% occupancy at the eight-ring windows but with low occupancy within the large cage. Cation movement upon dehydration results in a deformation of the framework (*Zeo.* 4:323(1984)). For dehydrated Na-CHA, a change in space group is observed from the rhombohedral R$\bar{3}$m to the monoclinic C2/m (*Mat. Res. Bull.* 12:1001(1977)). Occupancies in both cation and water molecule sites vary greatly in the different cationic forms. Structural data have been obtained for Ag-CHA (*Zeo.* 4:323(1984);

Positional parameters ($\times 10^4$) occupancies (%) for chabazite in various ion-exchanged forms (*Zeo.* 2:303 (1982)):

Site		Nat	Na	K	Ca	Sr
T	x/a	1035(2)	1032(1)	1040(1)	1045(1)	1030(1)
12	y/b	3338(2)	3311(1)	3332(1)	3336(1)	3338(1)
1	z/c	8741(2)	8763(1)	8755(1)	8763(1)	8726(1)
O1	x/a	2651(4)	2640(2)	2642(2)	2611(4)	2662(5)
6	y/b	−2651	−2640	−2642	−2611	−2662
2	z/c	0	0	0	0	0
O2	x/a	1536(4)	1443(2)	1494(4)	1533(3)	1541(4)
6	y/b	1536	1443	−1494	−1533	−1541
2	z/c	1/2	1/2	1/2	1/2	1/2
O3	x/a	2494(5)	2564(2)	2497(4)	2545(4)	2465(4)
6	y/b	2494	2546	2497	2545	2465
m	z/c	8947(7)	8873(3)	8919(6)	8927(6)	8933(7)
O4	x/a	0214(5)	0250(2)	0226(4)	0268(4)	0196(4)
6	y/b	0214	0250	0226	0268	0196
m	z/c	3234(7)	3078(3)	3200(6)	3262(6)	3253(6)
C1	occ.	16(1)			10(1)	11(1)
1	x/a	0			0	0
3m	y/b	0			0	0
	z/c	0			0	0
C2	occ.	47(1)	62(1)	96(1)	15(1)	31(1)
2	x/a	1965(6)	2013(4)	2249(2)	193(1)	1918(3)
3m	y/b	1965	2013	2249	193	1918
	z/c	1965	2013	2249	193	1918

Positional parameters ($\times 10^4$) occupancies (%) for chabazite in various ion-exchanged forms (*Zeo.* 2:303 (1982)):

Site		Nat	Na	K	Ca	Sr
C3	occ.	30(1)	21(1)	8(1)	17(1)	35(1)
2	x/a	406(1)	431(1)	388(6)	408(1)	4133(3)
3m	y/b	406	431	388	408	4133
	z/c	406	431	388	388	4133
C4	occ.	14(1)	17(1)	11(1)	17(1)	10(1)
6	x/a	579(2)	572(1)	561(3)	579(1)	571(1)
m	y/b	579	572	561	579	571
	z/c	233(2)	233(3)	239(4)	231(1)	270(1)
C4a	occ.		4(1)			
12	x/a		602(5)			
1	y/b		504(6)			
	z/c		185(5)			
C4b	occ.			12(1)		
6	x/a			529(2)		
m	y/b			529		
	z/c			107(3)		
W1	occ.	100	90(2)	82(4)	97(2)	60(2)
3	x/a	1/2	1/2	1/2	1/2	1/2
2/m	y/b	1/2	1/2	1/2	1/2	1/2
	z/c	0	0	0	0	0
W2	occ.	54(1)	17(1)	33(2)	28(3)	50(2)
6	x/a	256(2)	277(2)	264(2)	237(4)	238(2)
m	y/b	256	277	264	237	238
	z/c	509(3)	539(3)	526(3)	519(5)	496(3)
W3	occ.	33(1)	22(1)	20(1)	23(2)	26(1)
12	x/a	199(2)	197(2)	196(4)	246(3)	148(4)
1	y/b	357(2)	305(2)	341(3)	342(3)	382(4)
	z/c	507(2)	469(2)	518(4)	543(3)	502(4)
W4**	occ.	29(1)	23(1)	19(1)	9(1)	23(2)
6	x/a	3367(2)	354(3)	352(5)	365(6)	354(5)
2	y/b	−376	−354	−379(5)	−365	−354
	z/c	1/2	1/2	432(4)	1/2	1/2
W5	occ.	26(1)	15(1)		13(1)	
6	x/a	585(4)	621(2)		584(4)	
m	y/b	585	621		584	
	z/c	316(6)	386(3)		334(6)	
W6	occ.	77(2)	25(1)		60(2)	55(2)
2	x/a	248(1)	253(2)		251(1)	237(1)
3m	y/b	248	253		251	237
	z/c	248	253		251	237
W7	occ.		9(1)	15(2)		
6	x/a		368(4)	370(7)		
m	y/b		368	370		
	z/c		193(6)	086(9)		

**Number of positions and point symmetry for the K-chab are 12 and 1, respectively. Esd's on the last significant digit in parentheses.

Zeo. 3:205(1983)), cobalt and copper exchanged chabazite (both hydrated and dehydrated) (*Zeo.* 4:251(1984)), Cd and Ba exchanged chabazite (*Zeo.* 2:200(1982)), hydrated and dehydrated Mn-CHA (*Zeo.* 5:317(1985)), and Cs exchanged CHA (*Zeo.* 6:137(1986)). In the hydrated Ag chabasite, only 61% of the ions are located in sites 1 and 4; the remainder appear to be spread out in the zeolite cages. In Cs exchanged chabasite, the Ca ions occupy sites along the [111] direction in the hydrated material and site III at the center of the D6R after dehydration. In the Co exchanged material, the Co ions move during dehydration to positions on the threefold axis near the six-ring and to the center of the D6R cage. The silver ions occupy eight independent positions, near the center of the six-ring, at the center of the D6R cage, and near the eight-ring.

Synthesis

ORGANIC ADDITIVES
tetramethylammonium$^+$

Synthetically, this structure has been found to crystallize from K-, Sr-, (K,Na)-, (K,Ba)-, and (K,Na,TMA)-containing reaction mixtures, but a preference seems to be for systems containing potassium (*Zeo.* 1:130(1981)). Good crystals have been formed when the ratio of K/(K + Na) was 0.8 or higher (*Rend. Soc. Ital. Miner. Petrol.* 31:641(1975)). For ratios between 0.5 and 0.8, mixtures of chabazite and gmelinite appear. The pH of the crystallizing mixture was reported to be around 10.5, and stilbite can be found as a common impurity phase. Other phases that have appeared after longer crystallization times include faujasite and Na-P (*Nat. Zeo.*[175] (1985)). Mixtures of chabazite and gmelinite were crystallized from $Na_2O : Al_2O_3 : (7-9) SiO_2$ at 80°C and from $CaO : Al_2O_3 : 2 SiO_2 : 3 N(CH_3)_4OH$ gels at 200°C (*JCS* London 195(1959); *JCS* London 983(1961)).

Thermal Properties

The thermal curves are influenced by the major cation present in chabazite. In the sodium-rich chabazite, a very broad area in the DTA is observed, with two peaks, at 170 and 250°C; calcium forms give four peaks, at 100, 150, 220, and 280°C. Dehydration is nearly complete at 400°C for all exchange forms of chabazite (*Nat. Zeo.*[175] (1985)). Quantitative studies of the dehydroxylation of the NH_4^+ chabazite were based on TG/MS investigations. The maximum temperature of the DTA peak for dehydroxylation was found to be around 647°C (*JCS Faraday Trans.* I,77:2877(1981)). See Chabazite Fig. 3.

Infrared Spectrum

Evaluation of all the studied vibrational spectra of chabazite (*Zeo.* 3:329 (1983)):

Type of vibration	Type of Spectrum			
	ir	FIR	REF	R
1	3610–3490s	—	—	3520w 3480w 3420w
2	2400vw	—	—	—
3	1610m	—	—	1680vw
4	1080–990vs	—	980–820	1080w 1040w 1010m
5	780vw 740w	— —	— —	790vw 740vw 710s
6	1110w	—	—	1210vw 1180w
7	420w 405w	420m	420vw	400w
8	505w	—	510w	520w 505m
9	620w	—	—	—
10	—	260w	—	290vw
11	—	205w 180w	—	210w
12	—	98vw 82vw	—	100w 80m
13	—	155vw 140m	—	—
14	—	110vw	—	110vw

(*continued*)

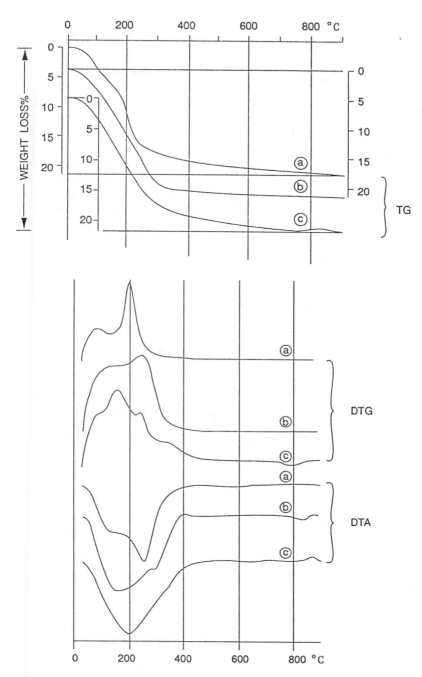

Chabazite Fig. 3:   Thermal curves of (a) Sr- and Al-rich chabazite from Vallerano, Rome, Italy, (b) herschelite" (Na-rich) from Poggio Rufo, Palagonia, Sicily, Italy, and (c) typical Ca-rich chabazite from Keller, Nidda, Germany; in air, heating rate 20°C/min (*Nat. Zeo*. (1985)). (Reproduced with permission of Springer-Verlag Inc.)

Evaluation of all the studied vibrational spectra of chabazite (*Zeo.* 3:329 (1983)):

Type of vibration	Type of Spectrum			
	ir	FIR	REF	R
15	380w 220m	370w 360m 330w 220w	380w	230w

ir = infrared absorption spectra, FIR = far-infrared absorption spectra; REF = infrared reflectance spectra, R = Raman scattering spectra; vs = very strong, s = strong, m = medium, w = weak, vs = very weak band.
 1. Antisymmetric and symmetric stretching vibration O–H.
 2. Combination band H_2O.
 3. Bending H–O–H.
 4. Antisymmetric stretching Si(Al)–O.
 5. Symmetric stretching Si(Al)–O.
 6. External TO_4.
 7. Bending O–Si(Al)–O.
 8. "Pore opening" oxygen rings.
 9. Vibration H_2O.
10. Ca–O.
11. Na–O.
12. Ca–H_2O complex.
13. Na–H_2O complex.
14. K–O.
15. Optical translational mode of lattice.

Ion Exchange

For the natural as well as the synthetic chabazites, ion exchange follows the series:

(monovalent ions) Tl > K > Ag > Rb > NH_4 > Pb > Na
(divalent ions) Ba > Sr > Ca

The Na^+ form is prepared from a 1.0 M aqueous solution of NaCl contacted with the crystals for 15 days at 100°C. Cobalt exchange from the sodium form requires heating to 110°C for 5 days in a 1.0 M aqueous solution of $Co(CH_3COO)_2$ (*Zeo.* 4:251(1984)). This zeolite has been proposed to selectively remove trace amounts of cobalt ions from water (JP 86,202,195(1986)).

Adsorption

The natural mineral chabazite cannot adsorb molecules with a kinetic diameter greater than 4.3 A, as the apertures are 3.7 × 4.1 Å in size. When larger cations are exchanged, sorption is further suppressed.

Adsorption properties of chabazite (*Zeo. Mol. Sieve* (1974)):

Adsorbate	t(K)	P(Torr)	x/m	P(Torr)	x/m	P(Torr)	x/m
Argon	77	1	0.23	10	0.27		
Krypton	423	7600	26	38,000	50*		
Xenon	423	7600	41	23,000	50*		
Oxygen	90	0.1	0.116	10	0.202	100	0.217
		700	0.22				
Nitrogen	77	10	0.134	100	0.206	700	0.206
H_2O	298	0.05	0.156	1	0.22	10	0.266
		20	0.271				
CO_2	298	1	0.037	20	0.14	100	0.218
		700	0.24				
	195	10	0.26	100	0.29	700	0.31
C_2H_4	298	50	0.07	200	0.08	700	0.085
C_3H_8	not adsorbed						
C_3H_6	298	2	0.010	50	0.050	200	0.073
		700	0.095				
n-Butene	298	700	<0.005				
n-C_4H_{10}	373	adsorbed above 373K					
I_2	393	90	0.59				

x/m in g/g
*High-pressure adsorption isotherms to 50 atm and temperature range of 150 to 450°C. Absolute adsorption.

Chiavennite (-CHI)
(natural)

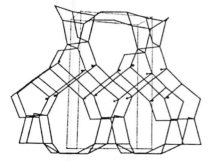

Chiavennite Fig. 1S: Framework topology.

Structure

(*AZST* (1987))

CHEMICAL COMPOSITION: $Ca_4Mn_4[Be_8Si_{20}O_{52}$
(OH)$_8$]*8H$_2$O
SYMMETRY: orthorhombic
SPACE GROUP: P2$_1$ab
UNIT CELL CONSTANTS (Å): a = 8.7
                             b = 31.3
                             c = 4.9

PORE STRUCTURE: unidimensional nine-member
rings, 3.9 × 4.3 Å

Christianite (PHI)
(natural)

This was the name given to a group of phillipsite minerals. The name is in honor of the Danish King Christian VII (*Nat. Zeo.* (1985)).

CJS-1
(*JCS Chem. Commun.* 884(1990))

Structure

See CJS-1 Fig. 1 below.

Synthesis

ORGANIC ADDITIVE
piperazine (pip)

CJS-1 crystallizes from a reactive aluminosilicate gel in the presence of piperazine and HF. The molar composition of the reaction mixture is: 0.5 pip : 0.04 Al$_2$O$_3$: SiO$_2$ HF : H$_2$O.

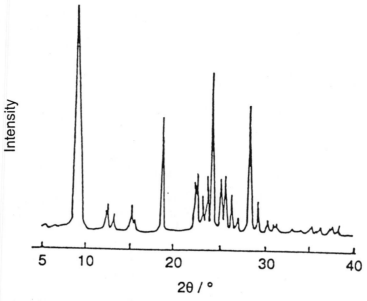

CJS-1 Fig. 1:   X-ray powder diffraction pattern for CJS-1 (*JCS Chem. Commun.* 884(1990). (Reproduced with permission of the Chemical Society)

Crystallization occurs at 150°C after 52 days. The organic can be removed from the structure through calcination at 510°C. This material is thermally stable to 900°C for 2 hours, collapsing at 1122°C. CJS-1 exhibits adsorption properties toward nitrogen with uptake values of 10 wt% (*JCS Chem. Commun.* 884(1990)).

Clathrasils
(general)

Clathrasils, or clathrate compounds, consist of three-dimensional 4-connected frameworks of silica and represent a class of porous tectosilicates that exhibit no exchange properties for guest species, either ionic or neutral, so that they are a distinct class from the zeolites and other molecular sieves (*Zeo.* 3:191(1983)). The presence of cagelike voids, which are generally occupied with guest species present during crystallization of the structure, is a common feature of these materials. There are several known types of clathrasils: the melanophlogites, dodecasils 3C, dodecasil 1H, nonasils, deca-dodecasils 3R, and silica-sodalite.

Designations of the known clathrasil families (*Zeo.* 6:373 (1986)):

Nonasils (non)   — [[5^46^4][4^15^8]][[5^86^{12}]
88[SiO_2] ∗ $8M^8$ ∗ $8M^9$
∗ $4M^{20}$

Deca-dodecasils-R (DDR)   — [[4^35^661]][$5^{12}[4^35^{12}6^18^3$]
120[SiO_2] ∗ $6M^{10}$ ∗ $9M^{12}$
∗ $6M^{19}$

Melanophlogites (MEP)   — [[5^{12}][$5^{12}6^2$]]
46[SiO_2] ∗ $2M^{12}$ ∗ $6M^{14}$

Dodecasils 3C (MTN)   — [[5^{12}]][$5^{12}6^4$]
136[SiO_2] ∗ $16M^{12}$ ∗ $8M^{16}$

Dodecasils 1H (DOH)   — [[5^{12}]][$4^35^66^3$][$4^{12}6^8$]
34[siO_2] ∗ $3M^{12}$ ∗ $2M^{12}$
∗ $1M^{20}$

Silica-sodalites (SOD)   — [[4^66^8]]
12[SiO_2] ∗ $2M^{14}$

Sigma-2 (SGT)
[seeSigma-2]

Linde type N (LTN)
[see Linde type N]

Cage characteristics of Clathrasils (*Ind. J. Chem.*) 27A:380 (1988)):

Cage type	Shape	Free diameter of 'windows' in Å (approx)	Effective inner volume in Å^3 (approx)
[5^46^4]-8-hedron	Ellipsoid	2.5	25
[4^15^8]-9-hedron	Sphere	1.5	30
[$4^35^66^1$]-10-hedron	Sphere	2.5	35
[$4^35^66^3$]-12-hedron	Sphere	2.5	50
[5^{12}]-12-hedron	Sphere	1.5	80
[$5^{12}6^2$]-14-hedron	Ellipsoid	2.5	160
[$5^{12}6^4$]-16-hedron	Sphere	2.5	250
[$4^35^{12}6^18^3$]-19-hedron	Ellipsoid	4.0	350
[$5^{12}6^8$]-20-hedron	Ellipsoid	2.5	430
[5^86^{12}]-20-hedron	Ellipsoid	2.5	290

Clinoptilolite (HEU)
(natural)

Related Materials: heulandite

Clinoptilolite is a silica-rich member of the heulandite family of minerals. It was discovered first in the Hoodoo mountains in Wyoming in the 1930s (*Nat. Zoo.* 256(1985)).

Structure

(*Nat. Zeo.* 256(1985))

CHEMICAL COMPOSITION: $(Na,K)_6(Al_6Si_{30}O_{72})$·
20H_2O
SYMMETRY: monoclinic
SPACE GROUP: C2/m
UNIT CELL CONSTANTS: (Å) $a = 17.668$
$b = 17.986$
$c = 7.417$
$\beta = 116°32'$
FRAMEWORK DENSITY: 0.34
DENSITY: 2.16 g/cm^3
PORE STRUCTURE: 10-member rings 4.4 × 7.2 Å

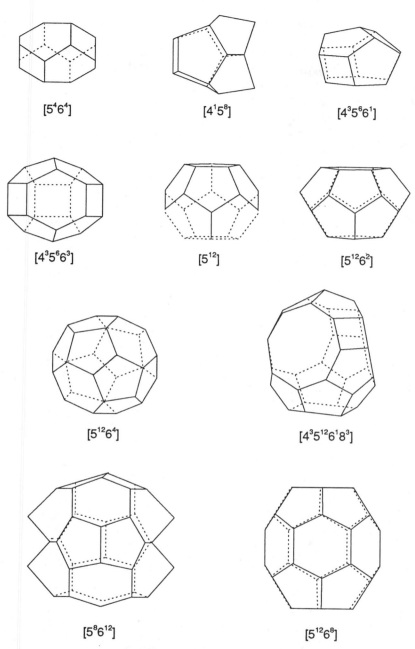

$[5^46^4]$ $[4^15^8]$ $[4^35^66^1]$

$[4^35^66^3]$ $[5^{12}]$ $[5^{12}6^2]$

$[5^{12}6^4]$ $[4^35^{12}6^18^3]$

$[5^86^{12}]$ $[5^{12}6^8]$

Clathrasils Fig. 1:   Cages present in different clathrasils (silicon atoms are located at the corners of the polyhedra and the oxygen atoms ear the center of the lines connecting the silicon atoms e.g., $[5^{12}6^8]$ means that a 20-hedron consisting of 12 five-membered rings and 8 six-membered rings of [SiO$_4$] units) (*Ind. J. Chem.* 27A:380 (1988)).

X-Ray Powder Diffraction Data: (*Nat. Zeo.* 341(1985)) (d(Å) (I/I_0)) 8.95(100), 7.93(13), 6.78(9), 5.94(3), 5.59(5), 5.24(10), 5.12(12), 4.65(19), 4.35(5), 3.976(61), 3.955(63), 3.905(48), 3.835(7), 3.738(6), 3.707(5), 3.554(9), 3.513(4), 3.424(18), 3.392(12), 3.316(6), 3.170(16), 3.120(15), 3.074(9), 2.998(18), 2.971(47), 2.795(16), 2.730(16), 2.667(4), 2.527(6), 2.485(3), 2.458(3), 2.437(8), 2.422(5), 2.319(2), 2.089(3), 2.056(2), 2.016(2), 1.974(4).

Clinoptilolite and heulandite differ only in framework and exchange ion composition, not in framework structure. Clinoptilolite represents the higher silica form of this structure (*Am. Mineral.* 57:1463(1972)). For a discussion of the framework see heulandite.

Atomic coordinates for natural clinoptilolite (*SSSC* 28:449 (1986)):

	x	y	z
T(1)	0.17835(8)	0.17073(8)	0.0956(2)
T(2)	0.21482(8)	0.41052(8)	0.5067(2)
T(3)	0.20787(8)	0.19069(8)	0.7151(2)
T(4)	0.06668(8)	0.29933(8)	0.4163(2)
T(5)	0.0	0.2187(1)	0.0
O(1)	0.1991(4)	0.5	0.4640(8)
O(2)	0.2325(2)	0.1205(2)	0.6145(6)
O(3)	0.1845(3)	0.1559(2)	0.8878(5)
O(4)	0.2315(2)	0.1057(2)	0.2503(5)
O(5)	0.0	0.3246(3)	0.5
O(6)	0.0803(2)	0.1650(2)	0.0539(6)
O(7)	0.1258(3)	0.2331(2)	0.5488(6)
O(8)	0.01.3(2)	0.2710(2)	0.1863(6)
O(9)	0.2128(2)	0.2519(2)	0.1856(6)
O(10)	0.1205(2)	0.3721(2)	0.4187(6)
M(1)	0.1467(6)	0.0	0.670(1)
M(2)	0.0415(4)	0.5	0.229(1)
M(3)	0.2476(7)	0.5	0.053(2)
M(4)	0.0	0.0	0.5
W(1)	0.289(1)	0.0	0.023(3)
W(2)	0.079(4)	0.0	0.84(1)
W(3)	0.0793(6)	0.4219(7)	0.966(2)
W(4)	0.5	0.0	0.5
W(5)	0.0	0.101(2)	0.5
W(6)	0.076(2)	0.0	0.290(7)
W(7)	0.118(4)	0.0	0.70(1)
W(8)	0.052(3)	0.0	0.06(1)

Clinoptilolite Fig. 1S: Framework topology.

Contents in the extra-framework atomic sites (*SSSC* 28:449 (1986)):

M(1)	0.49(1) Na
	0.01(1) Ca(Sr)
M(2)	0.23(1) Na
	0.16(1) Ca(Sr)
M(3)	0.0925 K
	0.0375 Ba
M(4)	0.40 Mg

Occupancies of Al atoms in the tetrahedral sites; the Al contents, per 18 oxygen atoms, in the structure are listed together (*SSSC* 28:449 (1986); *Z. Kristallogr.* 145:216 (1977)):

Tetrahedral site	Kamloops CTL	Kuruma CTL	Agoura CTL
Al contents	1.7	1.68	1.55
T(1)	13%	20%	17%
T(2)	32	42	31
T(3)	21	9	13
T(4)	15	6	11
T(5)	9	14	10

Synthesis

ORGANIC ADDITIVES
none

Clinoptilolite crystallizes from aluminosilicate gels with SiO_2/Al_2O_3 ranging between 7 and 18. Crystallization has been observed in the presence of lithium hydroxide with gel compositions of: $0.5-3\ Li_2O : Al_2O_3 : 8-18\ SiO_2 : x\ H_2O$. Crystallization occurs after one month at temperatures between 300 and 350°C. (*Am. Mineral.* 48:1374(1963); *USSR* 1:353,729 (1987)). In the presence of Ca, Sr, and Ba, clinoptilolite can form along with paracelsian and sanbornite (*Mat. Res. Bull.* 2:951 (1967)). Under milder conditions (170–200°C), this structure will form in Na/K-containing gel systems (*Am. Mineral.* 62:330(1971)). A low temperature synthesis (100°C) also has been reported, using blast furnace slag as a source of aluminosilicate (JP 83,167,421 (1983)).

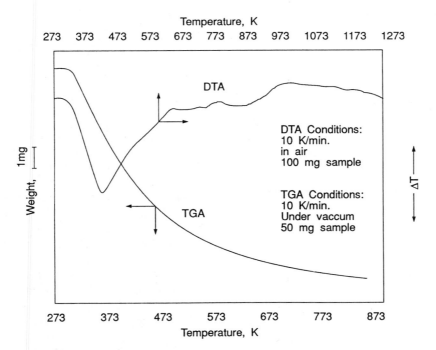

Clinoptilolite Fig. 2:   DTA and TGA curves of clinoptilolite (*Nat Zeo.* 99(1988). (Reproduced with permission of Springer-Verlag Inc.)

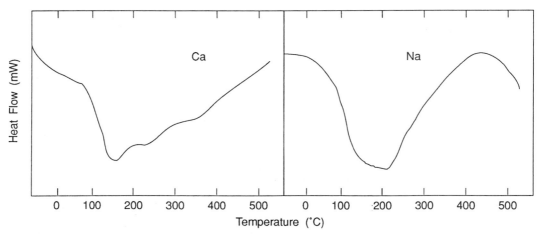

Clinoptilolite Fig. 3:   DSC curves for Ca- and Na-exchanged clinoptilolite (*Nat. Zeo.* 565(1988)). (Reproduced with permission of Springer-Verlag Inc.)

Thermal Properties

Because of the higher silica/alumina ratio found in clinoptilolite, it exhibits thermal stability to 700°C in air. The amount of water lost below 800°C varies from 7.8 wt% for cesium containing clinoptilolite to about 15.5% for Mn, Li, and Sr clinoptilolite. The water association is dependent on the exchangeable cations (*Nat. Zeo.* 565(1988)).

Adsorption

The adsorption capacities for CO_2, SO_2, N_3, and NO_x were examined for natural clinoptilolite (*Zeo.* 3:259(1983)). natural clinoptilolite also has been shown to be effective in removing hydrogen from hydrogen–nitrogen mixtures at high pressures (*Zeo.* 5:211(1985)).

Adsorption properties for two natural clinoptilolites (*Zeo.* 3:259 (1983)):

Zeolite	Gas velocity (m/s)	CO_2	SO_2 (g/100 g)	NH_3 (g/100 g)	NO_x
Clino-1	0.067	3.47	5.67	4.86	0.39
	0.133	2.88	5.94	3.98	0.45
	0.234	1.82	2.98	—	—
Clino-2	0.067	4.55	8.43	4.90	1.44
	0.133	3.62	8.15	4.70	1.33
	0.234	—	4.41	—	—

Adsorption equilibrium data for clinoptilolite from Hector, California (*Zeo. Mol Sieve.* 625)

Adsorbate	Activation temperature (°C)	T (°)	P (Torr)	Wt %
H_2O	350	25	1	10
			20	16
CO_2	350	25	100	1.63
			700	8.2
O_2	350	−185	700	1.9
N_2	350	−196	700	1.4

Ion Exchange

Clinoptilolite is highly selective for the ammonium cation. The natural clinoptilolite that has been converted into the ammonium ion form has a high exchange capacity for lead and cadmium, near equivalency for lead and somewhat lower for cadmium. This exchange is irreversible (*Zeo.* 7:153(1987)). Clinoptilolite is highly selective for cesium and strontium. These properties make it desirable for application in radioactive waste treatment (*Zeo.* 4:215(1984)). In the sodium form, complete replacement of all of the Na^+ is accomplished by methylamine, dimethylamine, ethylamine, and propylamine, but not by trimethylamine, isopropylamine, and butylamine. EtA > M_2A,Ma,NH_4 > *n*-PA > *n*-BA (*J. Inorg. Nucl. Chem.* 29:2047(1967)). The selectivity effects are attributed mainly to

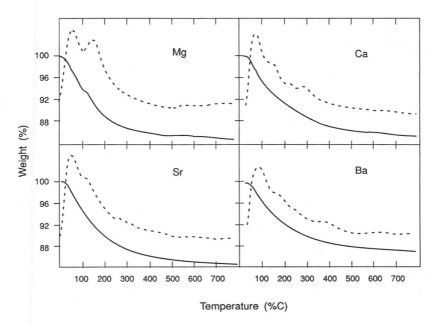

Clinoptilolite Fig. 4:   TGA curves for divalent cation clinoptilolites. Solid lines are weight-loss curves, and dashed lines are derivative curves (%/min) (*Nat. Zeo.* 565(1988)). (Reproduced with permission of Springer-Verlag Inc.)

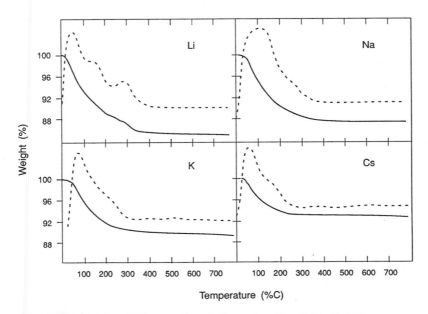

Clinoptilolite Fig. 5:   TGA curves for univalent cation clinoptilolite. Solid lines are weight-loss curves, and dashed lines are derivative curves (%/min) (*Nat. Zeo.* 565(1988)). (Reproduced with permission of Springer-Verlag Inc.)

the size of these ions. The selectivity of this material for ammonia and iron has been applied to the purification of drinking water (*Zeo.* 3: 188(1983)).

Exchange of silver(I) ions (*Nat. Zeo.* 463 (1988))

	Static ion-exchange capacity [meq/g]
"Potassium" clinoptilolite	
natural form	0.79(03)
lithium form	0.81(03)
sodium form	0.83(04)
calcium form	0.69(02)
ammonium form	0.72(03)
"Calcium" clinoptilolite	
natural form	1.01(04)
lithium form	1.25(05)
sodium form	1.28(05)
calcium form	1.08(04)
ammonium form	1.04(03)

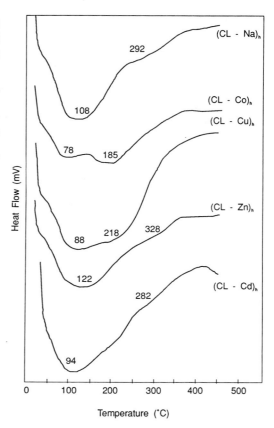

Clinoptilolite Fig. 6:   DSC curves for various transition metal ion-exchanged forms of clinoptilolite (*J. Serb. Chem. Soc.* 54:547(1989)).

Cloverite
(Fr Appl. 91-03378(1991))
Ecole Nationale Superieure de Chimie

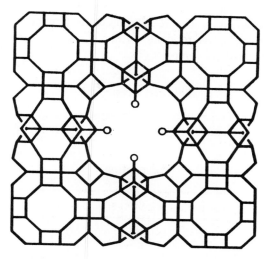

Cloverite Fig. 1.

Atomic coordinates for cloverite (*Nature* 352:320 (1991)):

	x	y	z
P(1)	0.0888	0.3656	0.2818
Ga(1)	0.1341	0.3275	0.2780
P(2)	0	0.3226	0.2769
Ga(2)	0.0370	0.3507	0.3134
P(3)	0	0.3876	0.3414
Ga(3)	0.0958	0.3654	0.2236
P(4)	0.1336	0.3210	0.2180
Ga(4)	0	0.3242	0.2140
P(5)	0.0446	0.3097	0.3556
Ga(5)	0	0.3415	0.3801
O(1)	0.0800	0.3669	0.2542
O(2)	0.1059	0.3885	0.2891
O(3)	0.1036	0.3404	0.2866
O(4)	0.0657	0.3679	0.2999
O(5)	0.1360	0.3124	0.2458
O(6)	0.1339	0.2970	0.2951
O(7)	0.1631	0.3439	0.2914
O(8)	0.0237	0.3376	0.2830
O(9)	0	0.3158	0.2475
O(10)	0	0.2979	0.2916
O(11)	0.0521	0.3286	0.3360
O(12)	0.0242	0.3795	0.3277
O(13)	0	0.3738	0.3670
O(14)	0.1091	0.3344	0.2142
O(15)	0.0675	0.3692	0.2010
O(16)	−0.0289	0.3436	0.2115
O(17)	0.0283	0.3231	0.3764
O(18)	0	0.4161	0.3461
O(19)	0	0.3461	0.4150
F(1)*	0.1335	0.3665	1/4
F(2)	0	0.3325	0.3297

Structure

CHEMICAL COMPOSITION: $[Ga_{768} P_{768} O_{2,976}$ $(OH)_{192}] \cdot 192$ RF (RF = quinuclidinium fluoride)

SYMMETRY: cubic

SPACE GROUP: $Fm\overline{3}c$

UNIT CELL CONSTANT: (Å) a = 52.712 (Nature 352: 320(1991))

PORE STRUCTURE: 20 member rings

The cloverite structure is extremely open and consists of a cubic arrangement of alpha cages connected by two rpa cages. The entire structure can be generated from double four-member rings.

Half of these D4R's are not fully connected giving rise to an interrupted structure.

Synthesis

ORGANIC ADDITIVE
quinuclidine

Cloverite crystallizes from a batch composition:

$$1 Ga_2O_3 : 1 P_2O_5 : 6 \text{ quinuclidine} : 0.75 \text{ HF} : 64 H_2O$$

Crystallization occurs after 24 hrs at 150°C. Gallium sulfate is the source of gallium.

CoAPO-5 (AFI)
(US4,567,029(1986))
Union Carbide Corporation

Related Materials: AlPO$_4$-5

Structure

Chemical Composition: See MeAPO.

X-Ray Powder Diffraction Data: ($d(\text{Å})$ (I/I_0))
11.6(100), 6.86(14), 5.95(21), 4.51(51), 4.24(36), 3.97(82), 3.93(4), 3.44(30), 3.09(15), 2.98(21), 2.675(4), 2.607(3), 2.416(3), 2.392(10), 2.146(1), 1.910(4).

The cobalt-aluminophosphate CoAPO-5 exhibits properties characteristic of a large-pore molecular sieve. For a description of the aluminophosphate framework topology see AlPO$_4$-5.

Synthesis

ORGANIC ADDITIVES
N,N-diethylethanolamine
tetraethylammonium$^+$
tripropylamine

CoAPO-5 crystallizes from a reactive gel composition with a molar oxide ratio of: R : 0.167 CoO : 0.917 Al$_2$O$_3$: P$_2$O$_4$: 0.33 CH$_3$COOH

: 5.5 C$_3$H$_7$OH : 45 H$_2$O. Crystallization occurred after 48 hours at 200°C. Cobalt acetate tetrahydrate and cobalt sulfate both have been successfully employed in the crystallization of this molecular sieve. Aluminum isopropoxide was used as a source of aluminum. Crystalline impurities were identified in some of the crystallized gel compositions.

Thermal Properties

CoAPO-5 maintains its framework structure to temperatures of 600°C.

Adsorption

Adsorption capacities of CoAPO-5:

	Kinetic diameter, Å	Pressure, Torr	Temp., °C	Wt. % adsorbed
O$_2$	3.46	12	−183	10.8
O$_2$		710	−183	13.1
Neopentane	6.2	12	24	4.9
		710	24	5.2
H$_2$O	2.65	4.6	23	5.2
H$_2$O		20.6	23	20.2

CoAPO-11 (AEL)
(US4,567,029(1986))
Union Carbide Corporation

Related Materials: AlPO$_4$-11

Structure

Chemical Composition: See MeAPO.

X-Ray Powder Diffraction Data: ($d(\text{Å})$ (I/I_0))
10.8(20), 9.34(37), 6.66(10), 5.61(25), 5.44(3), 4.65(3), 4.19(100), 3.99(57), 3.95(50), 3.90(50), 3.83(67), 3.59(10), 3.36(27), 3.11(13), 2.83(7), 2.73(13).

The cobalt-aluminophosphate CoAPO-11 exhibits properties characteristic of a medium-pore molecular sieve. For a description of the framework topology of CoAPO-11, see AlPO$_4$-11.

Synthesis

ORGANIC ADDITIVES
diisopropylamine
di-n-propylamine

CoAPO-11 crystallizes from a reactive gel with a molar oxide ratio of: R : 0.167 CoO : 0.917 Al$_2$O$_3$: P$_2$O$_5$: 0.33 CH$_3$ COOH : 5.5 C$_3$H$_7$OH : 50 H$_2$O. Crystallization occurs after 24 hours at 150 to 200°C. Impurity products were observed under these conditions in this system.

CoAPO-16 (AST)
(US4,456,029(1986))
Union Carbide Corporation

Related Materials: AlPO$_4$-16

Structure

Chemical Composition: See MeAPO.

X-Ray Powder Diffraction Data: (d(Å) (I/I_0))
7.76(64),   4.77(49),   4.04(100),   3.88(11), 3.35(25), 3.08(11), 3.00(26), 2.73(4), 2.571(4), 2.368(9), 2.254(4), 1.873(8).

The cobalt-aluminophosphate CoAPO-16 exhibits properties characteristic of a very small-pore molecular sieve. For a description of the CoAPO-16 topology, see AlPO$_4$-16.

Synthesis

ORGANIC ADDITIVES
quinuclidine

CoAPO-16 crystallizes from a reactive gel with a molar composition of: R : 0.167 CoO : 0.917 Al$_2$O$_3$: P$_2$O$_5$: 0.167 H$_2$SO4 : 40 H$_2$O. Crystallization occurs readily after 24 hours at 150°C. Cobalt sulfate has been used success-

fully as the source of cobalt. Hydrated aluminum oxide is used as a source of aluminum.

CoAPO-34 (CHA)
(US4,567,029(1986))
Union Carbide Corporation

Related Materials: AlPO$_4$-34
chabazite

Structure

X-Ray Powder Diffraction Data: (d(Å) (I/I_0))
9.31(79), 6.92(19), 6.28(16), 5.54(43), 4.93(29), 4.31(100), 4.00(5), 3.87(3), 3.53(30), 3.45(21), 3.24(5), 3.15(5), 3.03(5), 2.93(38), 2.87(25), 2.65(3), 2.614(10), 2.481(5), 2.276(3), 2.103(3), 2.090(3), 1.918(4), 1.859(6), 1.794(5), 1.728(3); (after thermal treatment to 500°C) (d(Å) (I/I_0)) 9.12(100), 6.76(35), 6.19(3), 5.44(19), 4.87(19), 4.58(3), 4.23(43), 3.95(1), 3.79(1), 3.49(14), 3.38(16), 3.10(5), 2.87(24), 2.82(11).

The cobalt-aluminophosphate CoAPO-34 exhibits properties characteristic of a small-pore molecular sieve.

Synthesis

ORGANIC ADDITIVES
tetraethylammonium$^+$
diisopropylamine

CoAPO-34 crystallizes readily from a reactive gel with a molar oxide ratio of: R : 0.167 CoO : 0.917 Al$_2$O$_3$: P$_2$O$_5$: 0.167 H$_2$SO$_4$: 40 H$_2$O. Crystallization occurs at 100 to 150°C after 24 hours. CoAPO-34 is the major phase under these conditions. Cobalt sulfate or cobalt acetate is used as a source of cobalt, and aluminum isopropoxide or hydrated aluminum oxide as a source of aluminum.

Thermal Properties

CoAPO-34 remains crystalline with a shift in X-ray diffraction peaks upon thermal treatment to 500°C.

Adsorption

Adsorption capacities of CoAPO-34*:

	Kinetic diameter, Å	Pressure, Torr	Temp., °C	Wt. % adsorbed
O_2	3.46	11	−183	21.9
O_2		684	−183	25.4
Butane	4.3	746	23	9.2
Xenon	4.0	759	23	21.7
H_2O	2.65	4.6	24	28.6
H_2O		21	23	32.8

*Contains impurity.

CoAPO-35 (LEV)
(US4,567,029(1986))
Union Carbide Corporation

Related Materials: SAPO-35
                        levyne

Structure

Chemical Composition: See MeAPO.

X-Ray Powder Diffraction Data: $(d(Å)\ (I/I_0))$
10.3(21), 8.04(48), 7.50(3), 6.6627), 5.57(9), 5.43(79), 4.98(12), 4.19(58), 4.06(100), 3.93(21), 3.75(9), 3.55(6), 3.34(24), 3.12(27), 3.10(sh), 2.78(48), 2.585(9), 2.508(6), 1.873(9), 1.841(6), 1.774(6), 1.656(6).

The cobalt-aluminophosphate CoAPO-35 exhibits properties characteristic of a small-pore molecular sieve. For a description of the CoAPO-35 framework topology, see levyne.

Synthesis

ORGANIC ADDITIVES
quinuclidine

CoAPO-35 crystallizes from a reactive gel with a molar oxide composition of: R : 0.4 CoO : 0.8 Al_2O_3 : P_2O_5 : 0.8 CH_3COOH : 4.8 C_3H_7OH : 50 H_2O. Crystallization takes place at 100°C after 72 hours. Cobalt acetate or cobalt sulfate is used as a source of cobalt, and aluminum isopropoxide is an effective source of aluminum.

Thermal Properties

When heated to 600°C, CoAPO-35 maintains its crystallinity with little change observed in the X-ray powder diffraction pattern.

Adsorption

Adsorption capacities of CoAPO-35:

	Kinetic diameter, Å	Pressure, Torr	Temp., °C	Wt. % adsorbed
O_2	3.46	13	−183	8.2
O_2		710	−183	11.4
H_2O	2.65	4.6	24	14.3
H_2O		20.3	23	21.4
Butane	4.3	751	24	0.1

CoAPO-36 (ATS)
(US4,567,029(1986))
Union Carbide Corporation

Related Materials: other metal-substituted molecular sieve aluminophosphates

Structure

Chemical Composition: See MeAPO.

X-Ray Powder Diffraction Data: $(d(Å)\ (I/I_0))$
11.2(100), 10.9(sh), 6.56(5), 5.57(12), 5.37(33), 4.65(57), 4.27(40), 4.10(sh), 4.04(45), 3.95(45), 3.88(sh), 3.82(12), 3.28(19), 3.23(10), 3.16(10), 3.08(10), 2.96(7), 2.805(10), 2.578(21), 2.508(2).

The cobalt aluminophosphate CoAPO-36 exhibits properties characteristic of a large-pore molecular sieve.

Synthesis

ORGANIC ADDITIVES
tripropylamine

CoAPO-36 crystallizes from a reactive gel with the molar composition: 1.5 R : 0.167

CoO : 0.917 Al$_2$O$_3$: P$_2$O$_5$: 0.33 CH$_3$COOH : 40 H$_2$O. Crystallization occurs readily after 24 hours at 150°C.

Thermal Properties

Crystal stability is maintained to at least 500°C. Little change in the X-ray powder diffraction pattern is observed.

Adsorption

Adsorption capacities of CoAPO-36:

	Kinetic diameter, Å	Pressure, Torr	Temp., °C	Wt. % adsorbed
O$_2$	3.46	12	−183	15.0
O$_2$		710	−183	23.7
Neopentane	6.2	12	24	5.4
Neopentane		710	24	7.6
H$_2$O	2.65	4.6	24	18.6
H$_2$O		20.6	23	29.5

CoAPO-39 (ATN)
(US4,567,029(1986))
Union Carbide Corporation

Related Materials: AlPO$_4$-39

Structure

Chemical Composition: See MeAPO.

X-Ray Powder Diffraction Data: (d(Å) (I/I_0))
9.51(64),   6.71(32),   4.85(36),   4.23(92), 3.92(100), 3.33(6), 3.12(8), 3.04(20), 2.98(24), 2.75(8), 2.61(8), 2.38(8).

The cobalt aluminophosphate CoAPO-39 exhibits properties of a small-pore molecular sieve. See MAPO-39 for a description of the framework topology.

Synthesis

ORGANIC ADDITIVES
di-*n*-propylamine

CoAPO-39 crystallizes from reactive gels with a molar oxide ratio: R : 0.2 CoO : 0.9 Al$_2$O$_3$: P$_2$O$_5$: 0.4 CH$_3$COOH : 5.4 C$_3$H$_7$OH : 40 H$_2$O. Crystallization occurs readily at 150°C after 72 hours. The resulting solid was a mixture of phases, with CoAPO-39 present as the major phase.

CoAPO-43
(US4,853,197(1989))
UOP

Structure

Chemical Composition: See MeAPO.

X-Ray Powder Diffraction Data: (d(Å) (I/I_0))
7.15(100),   5.11(10),   4.12(41),   3.23(40), 3.17(20), 2.73(9), 2.70(7), 2.003(3), 1.79(4), 1.72(5).

Synthesis

ORGANIC ADDITIVES
di-*n*-propylamine

CoAPO-43 crystallizes from a reactive mixture with a molar oxide ratio of: 2.5 R : 0.3 CoO : 0.8 Al$_2$O$_3$: 1.0 P$_2$O$_5$: 0.6 CH$_3$COOH : 50 H$_2$O. Crystallization occurs at 150°C after 168 hours. Cobalt acetate is used as the source of cobalt in this crystallization.

CoAPO-44 (CHA)
(US4,567,029(1986))
Union Carbide Corporation

Related Materials: SAPO-44
chabazite

Structure

Chemical Composition: See MeAPO.

X-Ray Powder Diffraction Data: (d(Å) (I/I_0))
9.41(100), 6.81(18), 6.42(1), 5.50(41), 5.14(2), 4.67(5), 4.29(77), 4.10(77), 3.97(6), 3.87(8), 3.65(59), 3.41(18), 3.24(10), 3.02(5), 2.98(18), 2.91(59), 2.76(5), 2.72(8), 2.53(10), 1.895(8), 1.873(5), 1.821(10), 1.707(5).

The cobalt aluminophosphate CoAPO-44 exhibits properties characteristic of a small-pore molecular sieve. CoAPO-44 crystallizes with the chabazite framework topology. The partial occupancy of tetrahedral sites by cobalt is expected to distort the framework thus lowering the symmetry. Symmetry reduction also can be caused by a distortion of the four-member rings. In a triclinic unit cell there are three chabazite cavities. Between one and two (protonated) cyclohexylamine molecules occupy either of two positions in each cavity. Positional disorder of the amino group is observed. Both cyclohexylamine and potassium are observed in the CoAPO-44 channel system (communication from J. M. Bennett).

Synthesis

ORGANIC ADDITIVE
cyclohexylamine

CoAPO-44 crystallizes from a reactive gel composition of: R : 0.40 CoO : 0.80 Al$_2$O$_3$: P$_2$O$_5$: 0.8 CH$_3$COOH : 50 H$_2$O. Crystallization occurs readily at 150°C after 24 hours. CoAPO-44 is present as a major phase in the preparation.

CoAPO-47 (CHA)
(US4,567,029(1986))
Union Carbide Corporation

Related Materials: chabazite
other metal-substituted
aluminophosphates

Structure

Chemical Composition: See MeAPO.

X-Ray Powder Diffraction Data: (d(Å) (I/I_0))
9.41(100), 6.86(7), 6.33(4), 5.50(11), 5.04(3), 4.67(12), 4.33(45), 4.06(4), 3.97(1), 3.87(3), 3.60(14), 3.44(10), 3.24(3), 3.19(2), 3.03(3), 2.92(21), 2.89(sh), 2.69(2), 2.600(2), 2.508(3), 2.338(8), 2.270(2), 2.122(1), 1.910(1), 1.873(24), 1.814(5), 1.722(2); (after thermal treatment to 600°C) (d(Å) (I/I_0)) 9.41(100), 6.92(9), 6.33(2), 5.54(5), 4.96(5), 4.65(4), 4.29(26), 4.00(2), 3.85(2), 3.53(5), 3.43(5), 2.91(10), 2.85(6).

The cobalt aluminophosphate CoAPO-47 exhibits properties of a small-pore molecular sieve.

Synthesis

ORGANIC ADDITIVES
N,N-diethylenethanolamine

CoAPO-47 crystallizes from a reactive gel with a molar oxide ratio of: 2 R : 0.167 CoO : 0.917 Al$_2$O$_3$: P$_2$O$_5$: 0.33 CH$_3$COOH : 4.4 C$_3$H$_7$OH : 45 H$_2$O. Crystallization occurs readily after 120 hours at 150°C. Substantial quantities of cobalt have been shown to be incorporated into this material, producing a pure crystalline phase.

Adsorption

Adsorption capacities of CoAPO-47:

	Kinetic diameter, Å	Pressure, Torr	Temp., °C	Wt. % adsorbed
O$_2$	3.46	12	−183	17.1
O$_2$		704	−183	19.7
Butane	4.3	692	24	1.5
Xenon	4.0	754	23	17.4
H$_2$O	2.65	20	23	17.4

CoAPO-50 (AFY)
(US4,853,197(1989))
Union Carbide Corporation

Structure

X-Ray Powder Diffraction Data: (d(Å) (I/I_0))
10.90(100), 8.97(38), 6.33(4), 3.77(6), 3.67(7), 3.40(2), 3.00(2), 2.40(3), 1.83(2).

The cobalt aluminophosphate CoAPO-50 contains a 12-member-ring unidimensional pore system with interconnecting eight-member rings. This material is somewhat lower in stability than other aluminosilicate-based molecular sieves.

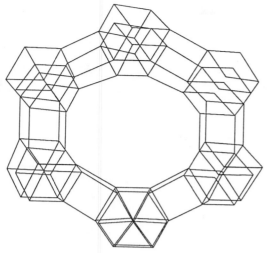

CoAPO-50 Fig. 1S: Framework topology.

This difference is thought to be due to the presence of tetrahedral T atoms capping the six-member rings present in the sheets that contain the 12-ring pores. The number of vertical bonds holding these sheets together is small. Half of these bonds are cobalt–oxygen–phosphorus

bonds, while the rest are aluminum–oxygen–phosphorus bonds (communication from J. M. Bennett).

CoMnGS-2 (-STO)
(*SSSC* 49A:375(1989))

Structure

CHEMICAL COMPOSITION: $R_{0.39}Co_{0.03}Mn_{0.08}Ge_{0.89}S_2$
SYMMETRY: rhombohedral
SPACE GROUP: R3
UNIT CELL CONSTANTS (Å): $a = b = c = 17.878$
                          alpha = beta =
                          gamma = $109.47°$
FRAMEWORK DENSITY: 7.27 T/1000 Å³

See CuGS-2 for a description of the GS-2 X-ray diffraction pattern. CoMnGS-2 crystallizes as large, single, tris-tetrahedral crystals. The basic secondary building unit in this structure is an adamantine Me_4S_6 unit connected by MeS_3SH moieties in a "pinwheel" arrangement to give trimers. The trimers are stacked in alternating right-hand/left-hand configuration,

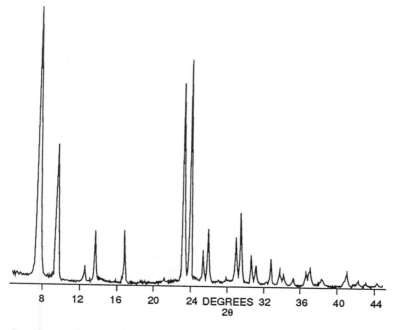

CoAPO-50 Fig. 2:   X-ray powder diffraction pattern of CoAPO-50 (*BZA Meeting abstracts* (1990)). (Reproduced with permission of M. J. Bennett)

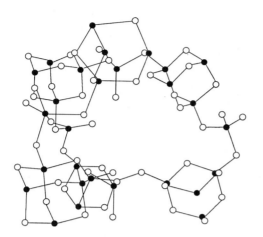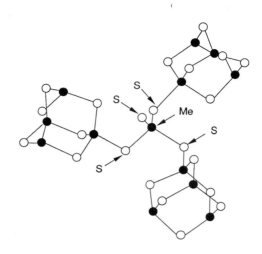

CoMnGs-2 Fig. 1S: Framework topology.

forming chains along the principal axis. The unit cell contains six adamantine units and eight 3-connected tetrahedral units, as well as 12 TMA cations.

Comptonite (THO)
(natural)

Comptonite is a name proposed by Brewster in 1821 for the zeolite mineral of the thompsonite family, from Vesuvius, Italy (*Edinb. Phil. J.* 4:131(1821); (*Nat. Zeo.* 57(1985))).

CoSAPO-34 (CHA)
Union Carbide Corporation

Related Materials: chabazite

Structure

CHEMICAL COMPOSITION:
  $Co_{0.16}Al_{0.84}P_{0.86}Si_{0.16}O_4)*0.18i\text{-}prNH_2*0.20H_2O$
UNIT CELL CONSTANTS: (Å) $a = 13.797(7)$
                        $c = 14.849(7)$
SPACE GROUP: $R\bar{3}$
DENSITY (calc.) (g/cm^3):   1.69
PORE STRUCTURE: 8 member-rings

Atomic parameters and their estimated standard deviations in parentheses (*Zeo.* 11:192 (1991)):

atom	x	y	z
T1	0.2304(1)	0.2266(1)	0.1021(1)
T2	0.0046(1)	0.2288(1)	0.1092(1)
O1	0.0043(4)	0.2607(4)	0.0111(4)
O2	0.1156(4)	0.2378(4)	0.1343(4)
O3	0.1975(4)	0.0891(4)	0.1249(4)
O4	0.3212(4)	0.0124(4)	0.1689(4)
O5	0.000	0.000	0.270(2)
N	0.053	0.505	0.012
C1	0.171	0.571	0.041
C2	0.231	0.503	0.023
C3	0.232	0.684	−0.010

The structure of CoSAPO-34 is similar to that of chabazite, characterized by a three-dimensional arrangement of D6R units linked together by four-member rings producing the chabazite cage.

Cowlesite
(natural)

Cowlesite was first described in 1975 by Wise and Tschernich. This mineral occurs naturally in deposits at Goble, Oregon as well as Kuniga, Japan and Skye, Scotland (*Nat. Zeo.* 306(1985)).

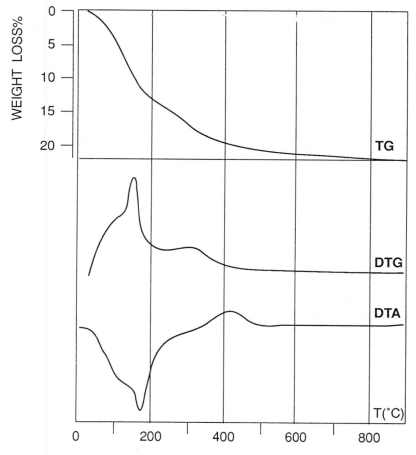

Cowlesite Fig. 1:   Thermal curves of cowlesite from Island Magee, County Antrim, Northern Ireland; in air, heating rate 20°C/min (*Nat. Zeo.* 311(1985)). (Reproduced with permission of Springer-Verlag Inc.)

Structure

(*Nat. Zeo.* 306(1985))

CHEMICAL COMPOSITION: $Ca_6(Al_{12}Si_{18}O_{60})*36H_2O$
SYMMETRY: orthorhombic
UNIT CELL CONSTANTS (Å): a = 11.27
                          b = 15.25
                          c = 12.61

X-Ray Powder Diffraction Data: (Nat. Zeo. 345(1988)) (d(Å) (I/I_0)) 15.2(100), 12.6(5), 11.3(5), 8.40(10), 7.62(15), 5.67(7), 5.52(5), 5.08(17), 4.70(7), 4.50(3), 4.16(5), 3.81(35), 3.75(15), 3.65(10), 3.25(5), 3.16(10), 3.052(20), 2.964(35), 2.934(25), 2.819(5).

Thermal Properties

Cowlesite has two main water losses, at 150 and 300°C, with a shoulder at 100°C.

CrGaAPO-17 (ERI)
(EP 158,976(1985))
Union Carbide Corporation

Related Materials: erionite

Structure

Chemical Composition: See FCAPO.

X-Ray Powder Diffraction Data: $(d(\text{Å})\ (I/I_0))$
11.5–11.4(vs), 6.61(s–vs), 5.72–5.70(s), 4.52–
4.51(w–s),   4.33–4.31(vs),   2.812–2.797
(w–s).

See AlPO$_4$-17 for a discussion of the framework topology of this chromium galloaluminophosphate.

Synthesis

ORGANIC ADDITIVE
quinuclidine

CrGaAPO-17 crystallizes from a reactive gel with a batch composition of: (1.0–2.0) R : (0.1–0.4) M$_2$O$_q$: (0.5–1.0) Al$_2$O$_3$: (0.5–1.0) P$_2$O$_5$: (40–100) H$_2$O (R refers to the organic additive, and q refers to the oxidation state of chromium and gallium).

Adsorption

Adsorption properties of CrGaAPO-17:

Adsorbate	Kinetic diameter, Å	Pressure, Torr	Temp., °C	Wt. % adsorbed*
O$_2$	3.46	100	–183	10
O$_2$		750	–183	12
n-Butane	4.3	100	24	4
H$_2$O	2.65	4.3	24	13
H$_2$O		20	24	14

*Typical amount adsorbed.

CSZ-1 (FAU/EMT)
(US4,309,313(1982))
W. R. Grace and Co.

Related Materials: faujasite/EMC-2

Structure

Chemical Composition: 0.78Na$_2$O : 0.22 Cs$_2$O :Al$_2$O$_3$: 4.9 SiO$_2$.

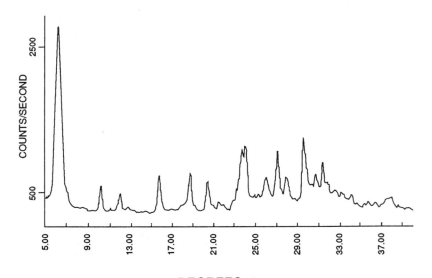

CSZ-1 Fig. 1:   X-ray powder diffraction pattern of CSZ-1 (*SSSC* 46:389(1989)).
(Reproduced with permission of Elsevier Publishing Company)

X-Ray Powder Diffraction Data: ($d(\text{Å})$ (I/I_0))
15.14(30), 14.24(90), 13.06(10), 8.66(15),
7.37(10), 5.61(8), 4.99(6), 4.71(21), 4.34(8),
4.14(15),   4.08(31),   3.682(65),   3.521(8),
3.397(100), 3.354(20), 3.331(8), 3.279(36),
3.166(5), 2.997(70), 2.966(15), 2.885(40),
2.780(25), 2.682(23), 2.614(7), 2.547(9),
2.503(10), 2.463(18), 2.369(12), 2.347(8),
2.302(5), 2.276(4), 2.168(10), 2.062(8),
2.026(15), 2.018(21).

The CSZ-1 structure consists of an intergrowth of zeolite Y and EMC-2.

Synthesis

ORGANIC ADDITIVES
none

The CSZ-1 structure is prepared from reactive gels containing cesium ions. Without the presence of cesium, the faujasite structure results. Gel compositions that successfully produce this material include: 10 SiO_2 : Al_2O_3 : 4 Na_2O : 0.5 CsCl : 1.4 NaCl : 150 H_2O. The range over which this material is claimed to form is Cs/(Cs + Na) of 0.02 to 0.15, Na_2O/SiO_2 of 0.2 to 0.4, SiO_2/Al_2O_3 of 7 to 16, and H_2O/SiO_2 of 15 to 30 (US4,309,313(1982)). If the cesium content is increased (Al_2O_3/Cs_2O = 0.4), pollucite is formed (US4,309,313(1982)). The morphology of CSZ-1 is similar to that of ZSM-3, crystallizing as hexagonal platelets. Crystal sizes range from 0.5 to 3 microns (*SSSC* 46:393(1989)). BET surface areas are reported between 750 and 760, compared to zeolite Y, which has a BET N_2 surface area of 830 to 860 m²/g.

CSZ-3 (FAU/EMT)
(US4,333,859(1982))
W. R. Grace and Co.

Related Materials: faujasite/EMC-2

Structure

Chemical Composition: (0.80–0.95) Na_2O : (0.02–0.20) Cs_2O : Al_2O_3 : (5.0–7.0) SiO_2 : (2–10) H_2O.

CSZ-3 is a cesium-containing zeolite with the intergrowth of the structures, faujasite and EMC-2. The SiO_2/Al_2O_3 ratio is higher than that of CSZ-1.

Synthesis

Synthesis of CSZ-3 utilizes the addition of seed slurries to encourage crystal formation. The seed slurry composition is: 13.3 Na_2O : Al_2O_3 : 12.5 SiO_2 : 267 H_2O. This composition was blended to form a clear solution, which gelled upon standing at room temperature overnight. The final batch composition used to crystallize CSZ-3 is: (1.8–7.0) Na_2O : (0.02–0.2) Cs_2O : Al_2O_3 : (6–20) SiO_2 : (90–400) H_2O. Crystallization occurs at temperatures around 100°C after 48 hours.

CuGS-2 (-STO)
(*SSSC* 49A:375(1989))

Structure

Chemical Composition: copper-germanium sulfide

X-Ray Powder Diffraction Data: ($d(\text{Å})$ (I/I_0))
8.43(100), 7.30(8), 6.53(3), 5.52(5), 5.16(11), 4.62(5), 4.40(27), 4.05(9), 3.65(8), 3.18(13), 3.04(8), 2.92(5), 2.86(5), 2.81(14), 2.76(4), 2.62(6), 2.23(5).

See CoMnGS-2 for a description of the GS-2 framework topology for this copper gallium sulfide.

Synthesis

ORGANIC ADDITIVE
tetramethylammonium$^+$

CuGS-2 crystallizes from a reactive gel with the composition: 0.8 $TMAHCO_3$: 0.27 TMACl : GeS_2 : 0.1 Cu acetate : 22 H_2O. Crystallization occurs after 18 hours at 150°C. The initial pH

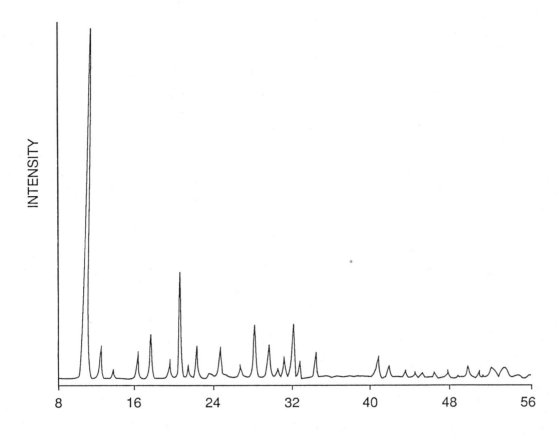

CuGS-2 Fig. 1:   X-ray powder diffraction pattern of CuGS-2 (*SSSC* 49A:375(1989)). (Reproduced with permission of Elsevier Publishing Company)

of the reactive slurry is 7 to 9, with a pH at 0.5 to 3 upon crystal formation.

CZH-5 (MTW)
(UK 2,079,735(1982))
Chevron Research Company

Related Materials: ZSM-12

Structure

Chemical Composition: (0.5–1.4) R_2O : 0.5 M_2O : Al_2O_3 : >5 SiO_2.

X-Ray Powder Diffraction Pattern: ($d(\text{Å})$ (I/I_0)) 11.85(50), 11.60(30), 9.97(25), 5.99(3), 5.81(4), 4.73(14), 4.68(5), 4.63(8), 4.43(5), 4.37(3), 4.25(100), 4.08(14), 4.06(7), 4.04(15), 3.96(6), 3.87(37), 3.83(28), 3.73(2), 3.64(3), 3.54(6), 3.46(16), 3.39(11), 3.33(13), 3.31(5), 3.19(7), 3.14(3), 3.05(5), 2.91(3), 2.89(6).

See ZSM-12 for a description of the framework topology.

Synthesis

ORGANIC ADDITIVE
choline

CHZ-5 crystallizes from a batch composition of: (12–200) SiO_2 : (1–17) M_2O : (5–25) R_2O : (50–150) MCl : (1500–15,000) H_2O. Crystallization occurs readily between 135 and 165°C, requiring more than 3 days of crystallization time.

D

D (CHA)

See Linde D.

D'achiardite

Incorrect spelling for dachiardite.

Dachiardite (DAC)
(natural)

Related Materials: svetlozarite

Dachiardite was first described by D'Achiardi in 1906 and named in honor of his father, who also was a distinguished mineralogist. Dachiardite occurs in hydrothermal deposits along with mordenite, epistilbite, stilbite, and heulandite. Sodium-rich dachiardite is found with mordenite, clinoptilolite, analcime, and heulandite (*Nat. Zeo.* 238(1985)).

Structure

(*Nat. Zeo.* 233(1985))

CHEMICAL COMPOSITION: $(Na,K,Ca_{0.5})_4$
  $(Al_4Si_{23}OO_{48})$* $18H_2O$
SYMMETRY: monoclinic
SPACE GROUP: C2/m
UNIT CELL CONSTANTS: (Å) $a = 18.69$
  $b = 7.50$
  $c = 10.26$
  $\beta = 107°54'$
PORE STRUCTURE: 3.7×6.7 Å (10 member-ring)
  3.6×4.8 Å (8 member-ring)

X-Ray Powder Diffraction Data: (*Nat. Zeo.* 337(1985)) (d(Å) (I/I_0)) 9.79(10), 8.90(50), 6.91(50), 6.00(35), 5.35(20), 4.97(50), 4.88(50), 4.61(10), 4.44(10), 4.23(10), 3.932(50), 3.848(10), 3.801(50), 3.773(20), 3.750(20), 3.834(20), 3.498(20), 3.452(100), 3.396(35), 3.375(20), 3.328(35), 3.204(100), 3.114(10),

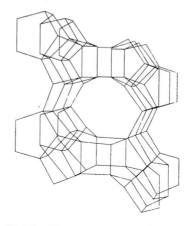

Dachiardite Fig. 1S: Framework topology.

3.077(10), 3.018(20), 2.964(50), 2.862(50), 2.712(50), 2.666(50), 2.607(10), 2.550(50), 2.517(20), 2.472(20), 2.449(20), 2.416(10), 2.387(20), 2.306(20), 2.273(10), 2.234(10), 2.216(10), 2.185(10), 2.170(10), 2.067(10), 2.040(20), 2.017(20), 1.992(35).

The framework structure of dachiardite can be thought of as cages of four pentagonal rings connected to build up a mordenite column. Five-member rings prevail in this framework. The main channel in this structure runs parallel to the c axis and is interconnected by channels parallel to the b axis, giving rise to a 10-member ring intersecting an eight-member-ring channel system. The pore openings are less than 4 Å. A cylindrical twinning is observed in crystals of this structure (*Z. Kristallogr.* 119:S53(1963)).

The SiO_2/Al_2O_3 ratios in this material are constant, with Si occupying ca. 83% of the tetrahedra. The content of exchangeable cations is subject to large variations, with the ratio of monovalent to divalent cations ranging between 1 and 114 and the ratio of sodium to potassium ranging from 0.43 to 2.31 (*Nat. Zeo.* 135(1985)).

Atomic coordinates and occupancy factors for dachiardite (*Z. Kristallogr.* 166:63(1984)):

	x	y	z	Occupancy
T(1)A	0.2905(4)	0.2084(11)	0.1496(7)	0.50
T(1)B	0.2846(4)	0.2053(10)	0.1660(7)	0.50
T(2)A	0.1914(3)	0.2901(8)	0.3371(5)	0.50
T(2)B	0.1929(3)	0.2978(7)	0.3714(5)	0.50
T(3)	0.0964(1)	0	0.7007(3)	1.00
T(4)	0.0816(1)	0	0.3793(3)	1.00
O(1)	0.3636(3)	0.3239(7)	0.2168(6)	1.00
O(2)	0.1162(3)	0.1770(7)	0.3265(5)	1.00
O(3)A	0.2188(6)	0.2642(16)	0.2070(12)	0.500
O(3)B	0.2382(7)	0.2370(19)	0.2652(14)	0.500
O(4)	0.1002(4)	0	0.5457(8)	1.00
O(5)	0.1688(5)	1/2	0.3487(9)	1.00
O(6)	0.3098(5)	0	0.1759(8)	1.00
O(7)	0.2335(6)	0.2452(19)	0.0131(13)	0.500
O(8)	0.2427(7)	0.2888(17)	0.5249(12)	0.500
O(9)	0.0103(3)	0	0.7080(7)	1.00
C(1)	−0.0091(4)	0.2598(11)	0.1297(9)	0.345(4)[a]
C(2)	0.0456(17)	1/2	0.5374(35)	0.156(3)[b]
W(1)	−0.0084(4)	1/2	0.2668(11)	1.00
W(2)	0.0884(15)	0	0.0330(30)	0.500
W(3)	0.0694(19)	0.1037(46)	0.0261(40)	0.250
W(4)	0.0724(18)	0.3931(42)	0.0261(40)	0.250
W(5)	0.0860(14)	1/2	0.0310(29)	0.500

[a]The occupancy refers to the scattering curve of Ca.
[b]The occupancy refers to the scattering curve of K.

Thermal Properties

A major water loss observed at 100°C continues at a slower rate to 200°C and up to 450°C. Lattice changes are observed after this temperature (*Sci. Rep. Niigata Univ. Ser. E. Geol. Miner.* 4:49(1977)). See Dachiardite Fig. 2 on the following page.

Infrared Spectrum

Mid-infrared Bands (cm⁻¹): 1210(sh), 1050(s), 775(w), 670(w), 558(w), 440(s) (*Zeo.* 4:369(1984)).

Danalite (SOD)
(*Am. Mineral.* 29:163(1944))

See sodalite.

Davyne (CAN)
(natural)

Related Material: cancrinite

Davyne or davynite was first reported in 1825 by Monticelli and Covelli. This name is used to describe the natural cancrinites that are potassium- and chlorine-rich. This mineral is found in deposits from the Monte Somma-Vesuvius area of Italy (*N. Jb. Miner. Mh.* 97(1990)).

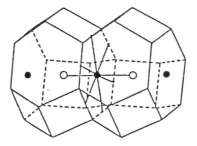

Davynite Fig. 1S: Framework topology.

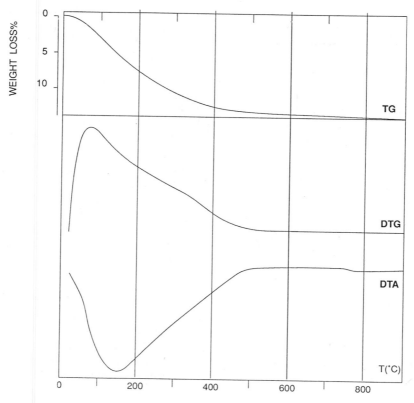

Dachiardite Fig. 2:   Thermal curves of dachiardite from Alpe di Siusi, Italy; in air, heating rate 20°C/min (*Nat. Zeo.* 238(1985)). (Reproduced with permission of Springer Verlag Inc.)

Structure

(N. Jb. Miner. Mh. 97(1990))

UNIT CELL CONTENTS: $Na_4K_2Ca_2(Si_6Al_6O_{24})$
$(SO_4)Cl_2$
SYMMETRY: hexagonal
SPACE GROUP: $P6_3/m$
UNIT CELL CONSTANTS: (Å) $a = 12.705(4)$
$c = 5.368(3)$

The aluminosilicate framework is related to that of cancrinite, with perfect ordering in the tetrahedral sites. The average bond distances are 1.728 Å and 1.604 Å for aluminum and silicon, respectively. The characteristic that distinguishes davyne from cancrinite is the position of the ions in the cancrinite cage. Ca–Cl–Ca chains occupy these channels, with the Ca^{+2} cations centered in the six-member ring with nearly equal bonds to the ring oxygens (Ca–O1 = 2.596 Å; Ca–O2 = 2.566 Å) and with coordination to two chloride ions (Ca–Cl = 2.704 Å). The chloride ions occupy sites in the center of the cancrinite cavity.

In the larger channels in the structure, the sites farthest from the framework are occupied mainly by potassium ions, with the sodium located closer to the framework wall. The sodium and potassium cations are linked to the framework at the O3 atom and to the sulfate through the two OB and OA atoms.

Site occupancies and fractional coordinates in davyne; M is the site multiplicity (*N. Jb. Miner. Mh.* 97(1990)):

Site	M	Occupancy	x	y	z
Si	1/2	1Si	0.3284(2)	0.4093(2)	3/4
Al	1/2	1Al	0.0691(2)	0.4086(2)	3/4
O1	1/2	1 O	0.2145(4)	0.4307(4)	3/4
O2	1/2	1 O	0.1002(4)	0.5567(4)	3/4
O3	1	1 O	−0.9987(3)	0.3230(3)	1.0084(8)
Ca	1/6	1 Ca	1/3	2/3	3/4
Cl	1/2	1/3 Cl	0.3150(14)	0.6372(6)	1/4
Na	1/2	1/2 Na	0.1510(9)	0.3097(13)	3/4
K	1/2	0.21Na + 0.29K	0.2217(9)	0.1136(7)	3/4
S	1/6	1/3 S	0	0	1/4
OA	1	1/18 O	0.0400	0.0200	0.0200
OB	1	1/6 O	0.0701	0.1132	0.3746
Cl′	1	1/18 O	0.0490	0.4000	−0.0750

Davynite (CAN)

See Davyne.

Deca-dodecasil 3R (DDR)
(clathrasil)

Related Materials: Sigma-1

Deca-dodecasil 3R is a member of the clathrasil family of silicates. Compositionally, it is a $120SiO_2*6M^{10}*9M^{12}*6M^{19}$, where M^{12} is N_2 and M^{19} is 1-aminoadamantane.

Structure

(*Z. Kristallogr.* 175:93(1986) (1987))

CHEMICAL COMPOSITION: SiO_2
SPACE GROUP: $R\bar{3}m$
UNIT CELL CONSTANTS (Å): $a = 13.860(3)$
$c = 40.891(8)$
PORE STRUCTURE: 4.4×3.6 Å, eight-member rings

Deca-dodecasil 3R is built by corner-sharing $[SiO_4]$ tetrahedra connected to pseudohexagonal layers of face-sharing pentagonal dodecahedra ($[5^{12}]$ cages). The stacking sequence for these layers is A̲B̲C̲ABC, and they are interconnected by additional $[SiO_4]$ tetrahedral-forming six-member rings between the layers. Such a stack-ing sequence produces two types of cages in addition to the pentagonal dodecahedron: a small decahedron, $[4^35^66^1]$, and a large 19-hedron, $[4^35^{12}6^18^3]$. A unique feature of this structure is the connection of the 19-hedron through eight-member-ring openings that are 4.5 Å in diameter. This opening is too small for the guest molecule to move.

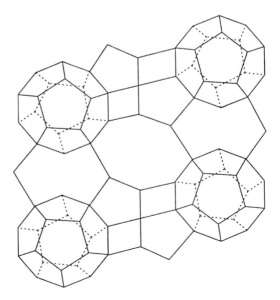

Deca-dodecasil 3R Fig. 1S: Framework topology

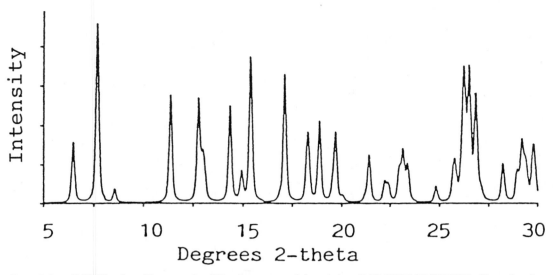

Deca-dodecasil 3R Fig. 2:   X-ray powder diffraction pattern of deca-dodecasil 3R (*SSSC* 37:57(1988)). (Reproduced with permission of Elsevier Publishing Company)

The polytypic series of deca-dodecasils (*Z. Kristallogr.* 175:93(1986)):

Polytype layer sequence	Lattice Parameters		Symmetry
	$a,b[Å]$	$c[Å]$	
deca-dodecasil AA	1H 13.96	13.63	hexagonal
deca-dodecasil ABAB	2H 13.86	27.26	hexagonal
deca-dodecasil ABCABC	3R 13.86	40.89	rhombohedral
deca-dodecasil AABAAB	3H 13.86	40.89	hexagonal
.			
.			
.			
deca-dodecasil	D 13.86	no periodicity	

Synthesis

GUEST MOLECULES
1-adamantane amine/ethylenediamine

Deca-dodecasil 3R crystallizes from aqueous silica-containing solutions prepared from the hydrolysis of tetramethoxysilane. Ethylenediamine is the base, with a molar ratio of en:SiO$_2$

Atomic parameters for Si and O in the silica framework with standard deviations
Z. Kristallogr. 175:93(1986)):

Atom	x	y	z
Si(1)	0.7267(2)	0.0511(2)	0.0700(0)
Si(2)	0.1264(1)	0.2527(3)	0.1095(1)
Si(3)	0.2020(2)	0.4039(3)	0.1711(1)
Si(4)	0.1227(2)	0.2454(3)	0.2325(1)
Si(5)	0.2256(2)	0	0
Si(6)	0	0	0.2039(2)
Si(7)	0	0	0.1280(2)
O(1)	0.3576(5)	0.3881(6)	0.6358(2)
O(2)	0.2792(6)	0.3749(7)	0.5768(2)
O(3)	0.4067(7)	0.2954(6)	0.5889(2)
O(4)	0.2277(4)	0.4554(8)	0.6589(3)
O(5)	0.1767(4)	0.3534(9)	0.7447(2)
O(6)	0.3233(10)	0.1616(5)	0.5391(2)
O(7)	0.5401(9)	0.2700(4)	0.5494(3)
O(8)	0.2713(4)	0.5426(8)	0.5529(3)
O(9)	0.1780(5)	0.3561(10)	0.5230(3)
O(10)	0.1870(8)	0	0.5
O(11)	0	0	0.1661(4)

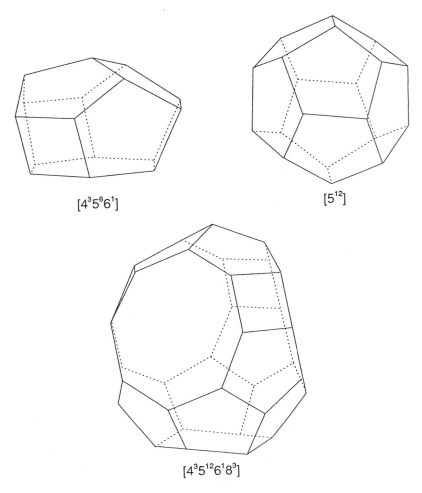

$[4^3 5^6 6^1]$ $[5^{12}]$

$[4^3 5^{12} 6^1 8^3]$

Deca-dodecasil 3R Fig. 3:   Three types of cages in deca-dodecasil 3R (*Z. Kristallogr.* 175:93(1986)). (Reproduced with permission of Wiesbaden Akademische Verlagsgesellschaft)

of 1:0.5. After hydrolysis, 1-adamantane amine is added, and the mixture crystallizes at 170°C after 6 days to several weeks.

Thermal Properties

Upon thermal treatment, the adamantane amine guest molecule decomposes, leaving the silicate framework. With a pore opening of 4.5 Å, the silicate begins to exhibit the molecular sieve properties of a small-pore material.

Desmine (STI)

This is an old name for stilbite.

Dodecasil 1H (DOH) (clathrasil)

Dodecasil 1H, $34SiO_2 * 3M^{12} * 2M^{12'} 1M^{20}$, is the simplest member of the polytypic series of dodecasils.

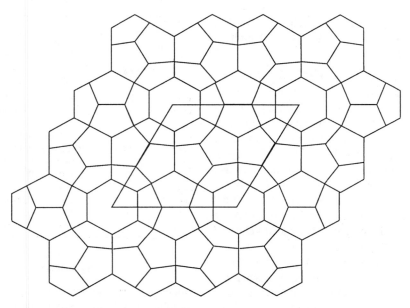

Dodecasil 1H Fig. 1S: Framework topology.

Structure

(*Z. Kristallogr.* 166:11(1984))

CHEMICAL COMPOSITION: SiO₂
SYMMETRY: hexagonal
SPACE GROUP: P6/mmm
UNIT CELL CONSTANTS (a): $a = 13.783(4)$
                          $c = 11.190(3)$
FRAMEWORK DENSITY: 18.5

X-Ray Diffraction Data: (*Z. Kristallogr.* 166:11(1984)) (d(Å) (I/I_0)) 12.011(2), 11.191(1), 8.194(2), 6290(8), 5.977(11), 5.892(25), 5.610(16), 5.283(100), 5.084(32), 4.529(14), 4.200(9), 4.014(2), 3.996(6), 3.762(27), 3.563(6), 3.524(48), 3.459(14), 3.322(28), 3.306(22), 3.293(20), 3.173(13), 2.995(2), 2.883(2), 2.860(3), 2.750(<1), 2.730(2), 2.641(1), 2.543(6), 2.468(7), 2.396(4), 2.341(2),

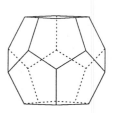

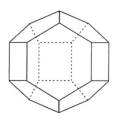

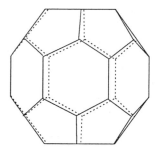

 $[5^{12}]$

$[4^3 5^6 6^3]$

$[5^{12} 6^8]$

Dodecasil 1H Fig. 2:   The three cage types present in dodecasil 1H (*Z. Kristallogr.* 166:11(1984)). (Reproduced with permission of Wiesbaden Akademische Verlagsgesellschaft)

2.307(2), 2.296(4), 2.215(3), 2.207(1), 2.180(1), 2.134(1), 1.965(2), 1.937(4), 1.882(2), 1.859(<1), 1.828(2).

Dodecasil 1H is a clathrate compound with an organic guest molecule, M, within three different types of cages. Corner-sharing [SiO$_4$] tetrahedra form a three-dimensional 4-connected net that is built up from hexagonal layers of face-sharing pentagondodecahedra ([5^{12}] cages), which form fundamental cages of the dodecasil series. Connecting these layers in an $\underline{AA}$ sequence are two new types of cages, the [4^{3}5^{6}6^3] cage and the [5^{12}6^8] cage. Guest molecules are trapped within these cages. The mean values of 170.6° for the Si–O–Si angles and 1.565 Å for the Si–O distances of the SiO$_2$ host framework of dodecasil 1H differ from other silica polymorphs but are comparable with the mean values found in the clathrasil melanophlogite (*Z. Kristallogr.* 166:11(1984)).

Atomic parameters with standard deviation for dodecasil 1H (*Z. Kristallogr.* 166:11(1984)):

Atom	No. of positions	x	y	z
Si(1)	12	0.4186(2)	0.2093(1)	0.2252(2)
Si(2)	12	0.3868(1)	0.0000(0)	0.3627(2)
Si(3)	6	0.2628(2)	0.1314(1)	0.0000(0)
Si(4)	4	0.3333(0)	0.6667(0)	0.1384(3)
O(1)	24	0.1052(3)	0.3933(4)	0.3031(4)
O(2)	12	0.5429(4)	0.2715(2)	0.8164(7)
O(3)	12	0.3405(5)	0.1702(2)	0.1139(5)
O(4)	6	0.3601(5)	0.0000(0)	0.5000(0)
O(5)	6	0.1865(6)	0.0000(0)	0.0000(0)
O(6)	6	0.5000(0)	0.0000(0)	0.3451(8)
O(7)	2	0.3333(0)	0.6667(0)	0.0000(0)

Synthesis

Dodecasil 1H can be prepared under hydrothermal conditions using hydrolyzed tetramethyloxysilane as the source of silica. Crys-

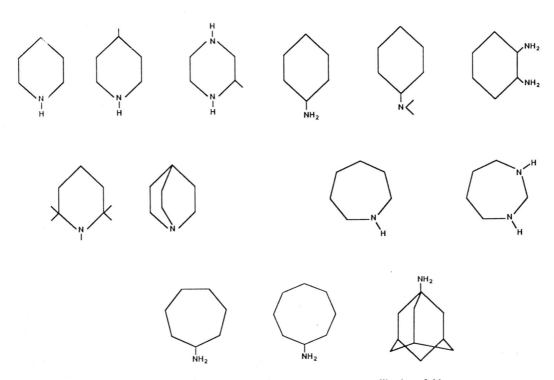

Dodecasil 1H Fig. 3:   Organic amines that have been shown to encourage crystallization of this structure.

Results of synthesis using characteristic guest species (*Ind. J. Chem.* 27:380(1988)):

Guest*	Molar ratio	Products† at temp (°C)			
		160	180	200	240
Pyrrolidine	—	D3C	D3C	D3C	D3C
1-Aminoadamantane	—	DD3R	D1H	D1H	D1H
Pyr, Adam	1:1	DD3R	DD3R	D1H	D1H
Pyr, Adam	1:10	DD3R	DD3R	D1H	D1H
Pyr, Adam	1:20	DD3R	DD3R	D1H	D1H
Pyr, Adam	10:1	D3C	D3C	D3C	D3C
Methylamine	—	Melano	Melano	Melano	D3C
MeNH₂, Adam	1:1	DD3R	D1H	D1H	D1H
MeNH₂, Pyr	1:1	D3C	D3C	D3C	D3C
Pyr, MeNH₂, Adam	1:1:1	D3C	D3C	D3C	D3C
Pyr, MeNH₂, Adam	10:1:1	D3C	D3C	D3C	D3C
Pyr, MeNH₂, Adam	10:1:10	DD3R	D1H	D1H	D1H
2-Aminopentane	—	Nonasil	Nonasil	—	—
Aminopen, Pyr	1:1	D3C	D3C	D3C	D3C

*Pyr = Pyrrolidine; Adam = 1-aminoadamantane; MeNH₂ = methylamine; and aminopen = 2-aminopentane.
†Melano = Melanophlogite; D3C = dodecasil 3C; D1H = dodecasil 1H; DD3R = deca-dodecasil 3R.

Effect of temperature on clathrasil formation (*Ind. J. Chem.* 27:380(1988)):

Guest	Product* at:			
	160°C	180°C	200°C	240°C
Methylamine	melano	melano	melano	D3C
Piperidine	D3C	D3C	D3C	D1H
1-Azabicyclo-octane	D3C	D3C	D1H	D1H
1-Aminoada-mantane	DD3R	D1H	D1H	D1H
Hexamethylene-imine	NS	NS	D1H	—
2-(Aminomethyl) tetrahydrofuran	NS	NS	D1H	—
1,2-Diamino-cyclohexane	NS	NS	D1H	—

*D3C = dodecasil 3C; melano = melanophlogite; D1H = dodecasil 1H; DD3R = deca-dodecasil 3R, and NS = nonasil.

tallization occurs with addition of the organic to this system and two weeks at elevated temperatures (*J. Incl. Phenom.* 4:85(1986)).

Dodecasil 3C (MTN) (clathrasil)

Related Materials: ZSM-39

This material is identified as $136SiO_2*16M^{12}*8M^{16}$

Structure

(*Nature* 294:340(1981))

CHEMICAL COMPOSITION: SiO_2
SYMMETRY: cubic
SPACE GROUP: Fd3
UNIT CELL CONSTANTS (Å): $a_0 = 19.402$

Dodecasil-3C is a clathrate compound with a silica framework. Guest molecules occupy two different types of cages. It is a cubic three-layer member of the polytypic series of dodecasils. [SiO$_4$] tetrahedra form a three-dimensional 4-connected net built up from pseudohexagonal layers of face-sharing pentagondodecahedra ([t^{12}] cages). This is the fundamental cage of the dodecasil series. The layers are connected in an ABCABC sequence, giving rise to two types of voids. In addition to the [5^{12}] cage there is the larger [5^{12}6^4] cage.

At room temperature, the dodecasil 3C crystals containing Kr, Xe, and N(CH$_3$)$_3$ are cubic, whereas those with tetrahydrofuran, tetrahydrothiophane, and piperidine are optically anisotropic.

Atomic parameters with standard deviations, (Z. *Kristallegr* 167:73(1984)):

Atom	x	y	z
Si(1)	0.0676(1)	0.0676(1)	0.3700(1)
Si(2)	0.2164(1)	0.2164(1)	0.2164(1)
Si(3)	0.125	0.125	0.125
O(1)	0.0932(2)	0.4065(2)	−0.0001(2)
O(2)	0.0433(2)	0.2995(3)	0.4505(3)
O(3)	0.125	0.3734(3)	0.125
O(4)	0.1704(2)	0.1704(2)	0.1704(2)
[5^{12}]	0	0	0
[5^{12}6^4]	0.625	0.625	0.625

Si–O distances in the range of 1.526–1.575 Å; mean value $d =$ 1.566 + Å.
Si–O–Si angles in the range of 169–180°; mean value is 174.5.

Synthesis

Dodecasil 3C was crystallized from silicate solutions prepared from the precipitation of tetraethylorthosilicate. The guest molecules, which were added prior to hydrothermal treatment, include CH$_4$, CO$_2$, N(CH$_3$)$_3$, N$_2$, and Ar. N$_2$, CO$_2$, and N(CH$_3$)$_3$ were detected as substantial constituents of the crystalline product.
ents of the crystalline product.

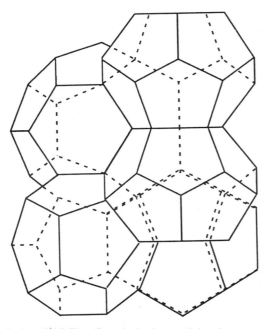

Dodecasil 3C Fig. 1S: Projection parallel to the pentagondodecahedra layer showing the [5^{12}] cage and the [5^{12}6^4] cage between the layers (*Z. Kristallogr.* 167:73(1984)). (Reproduced with permission of Wiesbaden Akademische Verlagsgesellschaft)

Thermal Properties

DSC and hot stage microprobe examination of dodecasil 3C crystals with tetrahydrofuran as guest molecules show a displacive phase transformation to the cubic high temperature phase taking place at temperatures varying from 100 to 120°C. The transformation energy is ca. 100 J/mol SiO$_2$. On cooling, the reverse transformation takes place between 95 and 105°C.

Doranite (ANA)
(natural)

Related Materials: analcime

Doranite was identified first by Greg and Lettsom in 1858 from deposits at the Knockagh escarpment near Carrickfergus, County Antrim, in Northern Ireland (*Zeo.* 7:286(1987)).

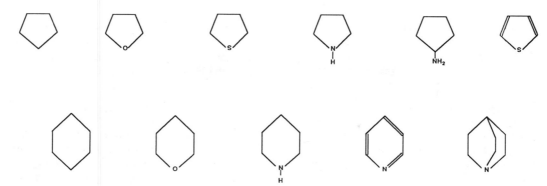

Dodecasil 3C Fig. 2:   Organics that have been used to encourage the crystallization of this structure.

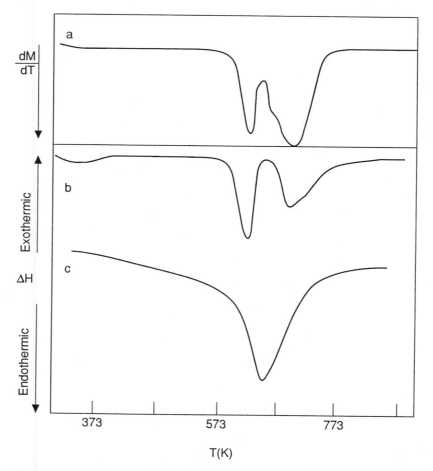

Doranite Fig.1:   Thermoanalytical curves for doranite and analcime (in static air). (a) DTG of doranite (20 K/min); (b) DSC of doranite (15 K/min); (c) DSC of analcime (15 K/min). (*Zeo.* 7:284 (1987)). (Reproduced with permission of Butterworth Publishing Company)

Structure

(*Zeo.* 7:286(1987))

CHEMICAL COMPOSITION: $Na_{10}Mg_3Al_{16}Si_{32}O_{96}$
  *$25H_2O$
PORE STRUCTURE: small-pore

X-Ray Powder Diffraction Data: (*Zeo.* 7:286(1987)) ($d(\text{Å})(I/I_0)$) 5.85(31), 3.52(100), 2.99(40), 2.92(10), 2.85(5), 2.72(17), 2.55(12), 2.47(6), 2.35(22), 2.25(5), 2.21(4), 2.03(34).

Doranite is structurally related to analcime, with about one-third of the cation positions filled by magnesium ions. See analcime for a description of the framework topology.

Thermal Properties

The thermal analysis confirms the magnesium ion sitting within the structural aluminosilicate framework of doranite and not the presence of a magnesium-rich second phase (*Zeo.* 7:286 (1987)).

E

E
(US2,962,355(1960))
Union Carbide Corporation

Structure

Chemical Composition: 0.4 Na_2O : 0.5 K_2O : 2.0 SiO_2 : Al_2O_3 : 3.3 H_2O.

X-Ray Powder Diffraction Data: (d(Å) (I/I_0)) 9.53(100), 7.13(16), 5.47(8), 4.23(18), 3.86(6), 3.54(4), 3.46(12), 3.41(4), 3.34(5), 3.14(14), 3.08(10), 3.00(18), 2.86(23), 2.63(7), 2.30(8), 2.23(9), 1.89(4), 1.67(5), 1.62(3), 1.48(3).

Zeolite E exhibits properties characteristic of a very small-pore molecular sieve.

Synthesis

ORGANIC ADDITIVE
none

Zeolite E crystallizes from a reactive sodium/potassium aluminosilicate gel, with equimolar ratios of potassium to sodium and with a SiO_2/Al_2O_3 of 2 and a water/Al_2O_3 ratio of 30. Crystallization occurs at 100°C after 64 hours, producing 2- to 4-micron-size cubic crystals.

Ion Exchange

The cations in zeolite E can be exchanged for calcium (89%), magnesium (52%), and lithium (47%).

Adsorption

Adsorbate	Temp., °C	Pressure, mm Hg	Wt% Adsorbed			
			K-Na-E	Ca-E	Mg-E	Li₂-E
H_2O	25	1	6.3	12.4	17.0	14.0
		4.5	15.9	15.5	20.4	18.1
		25	19.5	18.6	22.8	20.2

Adsorbate	Temp., °C	Pressure, mm Hg	Wt% Adsorbed			
			K-Na-E	Ca-E	Mg-E	Li₂-E
O_2	−196	100	0.4	1.8	0.6	0.4
CO_2	25	710	1.6			
C_2H_4	25	687	0.1			
SO_2	25	708	3.0			

ECR-1
(US4,657,748(1987))
Exxon Research and Engineering Company

Related Materials: mazzite/mordenite (proposed)

Structure

CHEMICAL COMPOSITION: (0.02–0.1)R_2O : (0.9–0.98)Na_2O : Al_2O_3 : (5–20) SiO_2 : x H_2O
SYMMETRY: primitive orthorhombic
SPACE GROUP: Pmmn
UNIT CELL CONSTANTS: (Å) a = 7.310(4)
(*Nature* 819(1987)) b = 18.144(6)
           c = 26.31(1)

X-Ray Powder Diffraction Data: (prepared from $(PrOH)_2Me_2N$)) d(Å) (I/I_0) 14.9(11), 10.6(18), 9.10(63), 7.87(36), 6.77(40), 6.56(9), 6.31(15), 5.91(22), 5.31(9), 4.99(14), 4.73(16), 4.44(17), 4.22(22), 4.11(3), 3.796(54), 3.724(11), 3.574(13), 3.500(79), 3.252(57), 3.172(100), 3.068(7), 3.008(27), 2.278(14), 2.603(7); (prepared from $(EtOH)_2Me_2N$) 14.8(20) 10.6(15), 9.10(44), 7.85(8), 6.76(45), 6.57(10), 6.29(19), 5.89(16), 5.32(9), 4.97(14), 4.73(6), 4.44(18), 4.23(35), 3.789(15), 3.730(17), 3.574(17), 3.497(40), 3.251(55), 3.170(100), 3.069(8), 3.006(27), 2.679(27), 2.588(15).

ECR-1 is proposed to be a 12-member-ring zeolite containing a regular intergrowth between sheets of mazzite and of mordenite. The repeat unit in this structure is 26.5 Å, representing the

157

two planes of alternating mazzite (15.5 Å) and mordenite (10.5 Å). The composition of this material resembles that of mazzite (Si/Al = 3.4) rather than that of mordenite (Si/Al > 5). The overall Si/Al ratio, determined by chemical analysis, is 3.35.

Synthesis

ORGANIC ADDITIVES
$(PrOH)_2Me_2N$ (DHPDM) (*ACS Sym. Ser.* 398:506(1989))
$(EtOH)_2Me_2N$ (DHEDM) (*ACS Sym. Ser.* 398:506(1989))

ECR-1 crystallizes in the presence of DHEDM and DHPDM. Without the addition of the or-

ganic, the faujasite (type Y) structure is formed. The gel composition used in crystallization of this material consists of: $0.6(DHEDM)_2O$: $1.35Na_2O : Al_2O_3 : 7.5 SiO_2 : 110 H_2O : 1.2NaCl$. Waterglass, N brand silicate of PQ Corporation, is used as the source of silica although colloidal silicas have been found equally effective. Hydrated alumina is used as the source of aluminum. Analcime and zeolite P (gismondine) generally are observed as the impurity phases present. Crystallization occurs between 120 and 160°C after 8 days (*ACS Sym. Ser.* 398:507(1989)).

Thermal Properties

See ECR-1 Fig. 2, p. 159.

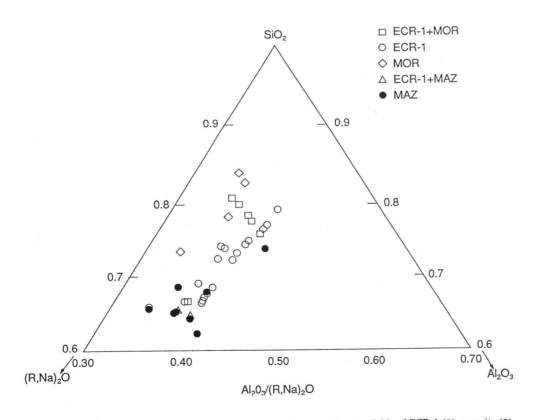

ECR-1 Fig. 1:   Crystallization composition diagram showing crystallization fields of ECR-1 (○), mazzite (●), mordenite (◇), ECR-1 + mazzite (△), and ECR-1 + mordenite (□) (*ACS Sym. Ser.* 398:506(1989)).

Thermal Properties

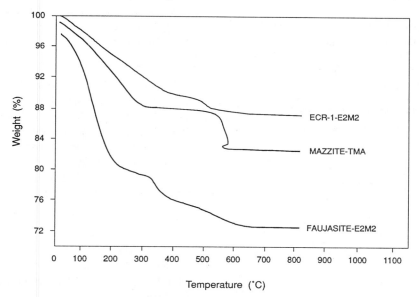

ECR-1 Fig. 2:   TGA of templated ECR-1, mazzite, and faujasite [E2M2 = (EtO4)$_2$ Me$_2$N] (*ACS Sym. Ser.* 398:506(1989)). (Reproduced with permission of the American Chemical Society)

Adsorption

Adsorption properties of ECR-1 (*ACS Sym. Ser.* 398:506(1989)):

Sample	Organic SiO$_2$/Al$_2$O$_3$	Unit Cell, Å			nC_6H_{12}[1],wt%	H$_2$O[2],wt%
		a	*b*	*c*		
A	DHEDM	6.70 18.15	26.31	7.31	1.6	10.8
B	DHEDM	6.88 18.13	26.18	7.31	2.3	11.5
C	DHEDM	6.54 18.15	26.09	7.31	2.0	9.4

1. 1 hr/45 Torr/20°C.
2. Wt. loss < 400°C.

ECR-2 (LTL)
(EP 142,350(1985))
Exxon Research and Engineering Company

Related Materials: Linde type L

Structure

CHEMICAL COMPOSITION: 0.9–1.1 M$_{2/n}$O : Al$_2$O$_3$: 2.5–5.1 SiO$_2$: x H$_2$O (M = K or mixture of K/Na)

SYMMETRY: hexagonal

UNIT CELL CONSTANTS: a = 18.49 Å
c = 7.52 Å

PORE STRUCTURE: 12-member ring, unidimensional large-pore

X-Ray Powder Diffraction Data: (d(Å) (I/I_0))
15.88(100), 7.49(20), 8.01(30), 5.81(17),
4.59(54), 4.41(15), 4.33(8.9), 3.92(72), 3.80(6),
3.78(2), 3.65(31), 3.47(60), 3.40(4), 3.28(34),
3.18(72), 3.06(47), 3.01(9), 2.91(54), 2.84(7),
2.79(6), 2.65(49), 2.62(20), 2.50(13), 2.47(11),
2.42(16), 2.37(5), 2.30(7).

See Linde type L for a description of the framework topology.

Synthesis

ORGANIC ADDITIVES
none

ECR-2 crystallizes from gel compositions in the range: (1–1.6) M_2O : Al_2O_3 : (2.5–5.1) SiO_2 : (80–140) H_2O, where M can be either K^+ or a mixture of K^+ and Na^+. Success of crystallization appears to be dependent on the presence of potassium. Crystallization occurs between 120 and 300°C within 2 to 8 days. Larger-size crystals are formed at the higher temperatures. Sources of silica and alumina include metakaolin and potassium silicate.

Thermal Properties

ECR-2 loses 10.2% of its weight between room temperature and 400°C.

ECR-4 (FAU/EMT)
(US4,714,601(1987))
Exxon Research and Engineering Company

Related Materials: faujasite/EMC-2

Structure

The difference between ECR-4 and type Y is based on the greater number of Si atoms with zero or one aluminum neighbor than with two or three neighbors in ECR-4, which is the reverse of what is observed in zeolite Y. EMC-2 is also present as an intergrowth.

Synthesis

ORGANIC ADDITIVES
($HOCH_2CH_2)_2(CH_3)_2N^+$
($CH_3CHOHCH_2)_2(CH_3)_2N^+$

ECR-4 is a faujasite-type zeolite prepared from a reactive mixture with a molar oxide ratio: (1.6–8) (Na^+,R) : (4–20) SiO_2 : (100–400) H_2O (R represents the organic additive). Nucleating seeds also were added to this system, amounting to 0.1 to 10 mol% relative to the Al_2O_3 content of the final material. Crystallization takes place between 90 and 120°C.

ECR-5 (CAN)
(EP 190,903(1986))
Exxon Research and Engineering Company

Related Materials: cancrinite

Structure

CHEMICAL COMPOSITION: (1.1–1.3) Na_2O : Al_2O_3 : (2.4–3.1) SiO_2
UNIT CELL CONSTANTS (A): a = 12.64 Å
c = 5.18 Å

X-Ray Powder Diffraction Data: d(Å) (I/I_0))
6.33(76), 5.47(14), 4.64(90), 4.13 (23),
3.65(97), 3.23(100), 3.04(6), 3.734(38),
3.595(37), 2.418(20), 2.392(9).

ECR-5 is structurally related to cancrinite. The main feature of the cancrinite structure is the single 12-ring channel parallel to the c axis. In the natural cancrinite, these 12-member rings are blocked, restricting adsorption. In ECR-5, however, the structure is open, with properties characteristic of the presence of large pores. The SiO_2/Al_2O_3 ratio is between 2.4 and 6, higher than that of the natural cancrinite.

Synthesis

ORGANIC ADDITIVES
none

ECR-5 is synthesized from reaction mixtures with composition ranges of:

Na_2O/Al_2O_3 : 2–17

SiO_2/Al_2O_3 : 2–25

H_2O/Al_2O_3 : 30–450

$(H_2O + NH_3)/Al_2O_3$: 50–600

The unique feature in this synthesis of the unblocked form of cancrinite is the utilization of ammonia in the synthesis gel. About 30% by weight ammonia is used. Crystallization occurs after five days at temperatures between 75 and 90°C. Phillipsite has been observed as an impurity phase in this system when stirring has been too vigorous. The sources of aluminum and silicon do not appear to strongly influence crystallization to the desired product.

Thermal Properties

Crystallinity is maintained when the as-synthesized material is heated to 500°C. TGA shows a weight loss of 13.5% between room temperature and 500°C.

Adsorption

Adsorption properties of ECR-5 (wt%):

n-Butane		n-Hexane	
0°C	−25°C	0°C	27°C
2.1%	4.2%	5.2%	2.2%

n-Butane at 100 Torr; n-hexane at 20 Torr.

ECR-10 (RHO)
(US4,960,578(1990))
Exxon Research and Engineering Company

Related Materials: rho

Structure

Chemical Composition: $(Na,Cs)_2O : (Al,Ga)_2O_3 : 2–4 SiO_2 : x H_2O$ (x = 0 to 6).

X-Ray Powder Diffraction Data: ($d(Å)$ (I/I_0))
10.55042(62), 6.0606(15), 5.2411(3), 4.6860(3), 4.2808(19), 3.9651(9), 3.4965(54), 3.3167(100), 3.1626(61), 3.0285(56), 2.9104(13), 2.7086(38), 2.5445(10), 2.4728(6), 2.4062(4), 2.3430(13), 2.2878(3), 2.2366(9), 2.1867(10), 2.1445(2),

2.0982(8), 2.0572(10), 2.0202(4), 2.0173(2), 1.9802(5), 1.9486(7), 1.8843(4), 1.8553(6), 1.8275(4), 1.7992(13), 1.7476(9), 1.7247(13), 1.6189(2), 1.5997(4).

ECR-10 is topologically related to zeolite rho.

Synthesis

ORGANIC ADDITIVES
none

ECR-10 crystallizes from a batch composition with a $Na_2O/(Ga,Al)_2O_3$ ratio between 1 and 3, $Cs_2O/(Ga,Al)_2O_3$ between 0.1 and 2, $Ga_2O_3/(Ga,Al)_2O_3$ between 0.7 and 1, $SiO_2/(Ga,Al)_2O_3$ of 2 to 4, and $H_2O/(Ga,Al)_2O_3$ between 30 and 250. Crystallization requires temperatures around 100°C after a few hours to several days.

ECR-17
(EP 259,526(1988))

See CSZ-1.

ECR-30 (EMT/FAU)
(EP 315,461(1981))
Exxon Research and Engineering Company

Related Materials: EMC-2/faujasite

Structure

Chemical Composition: 0.02 to 0.8 T_2O : 0.2 to 0.98 Na_2O : Al_2O_3 : 6-20 SiO_2 : x H_2O where T represents the organic amine cation, x represents 0 or an integer from 1 to 25.

X-Ray Powder Diffraction Data: ($d(Å)$ (I/I_0))
14.98(100), 14.39(43.20, 13.29(30.7), 8.65(14.6), 8.079(22), 7.491(4.2), 7.413(9.8), 7.29(1.3), 6.664(0.5), 6.485(2.1), 5.904(0.8), 5.622(8.0), 5.556(4.50, 5.373(2.7), 5.269(1.0), 5.189(8.8), 4.876(2.2), 4.718(1.5), 4.568(0.9), 4.564(0.9), 4.450(9.9), 4.430(0.2), 4.325(5.0), 4.040(2.2), 4.037(1.1), 3.965(2.3), 3.745(1.8), 3.714(1.0), 3.625(0.3), 3.604(0.6), 3.598(2.1), 3.498(0.8), 3.489(0.2), 3.459(0.2), 3.437(0.8), 3.369(1.4), 3.343(0.3), 3.322(1.9), 3.269(1.8), 3.236(0.9), 3.174(0.4), 3.141(0.2), 3.101(1.7),

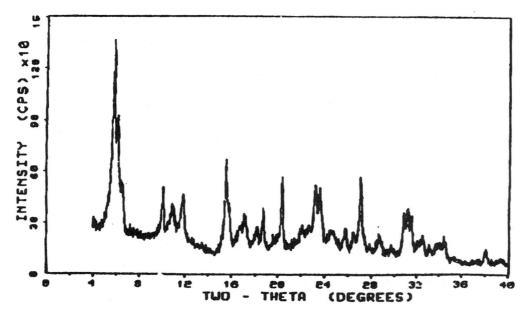

ECR-30 Fig. 1:   X-ray powder diffraction pattern.

3.094(0.2), 3.027(0.2), 2.996(0.2), 2.980(0.2), 2.941(0.3), 2.933(0.3), 2.919(1.4), 2.883(1.3), 2.878(1.9), 2.860(3.5), 2.287(0.4), 2.794(0.3), 2.789(0.5), 2.766(1.7)

ECR-30 represents a material which contains an intergrowth between EMC-2 (major phase) and faujasite (minor phase).

Synthesis

ORGANIC ADDITIVES
methyltriethylammonium +

The material crystallizes with the range of $Na+T/Na$ between 3 and 5; $(Na,T)_{20} : Al_2O_3$ between 1.6 and 10; SiO_2/Al_2O_3 between 14 and 50 and H_2O/Al_2O_3 between 150 to 600. Crystallization occurs in a temperature range of 70 to 160°C with crystallization times ranging up to 67 days.

ECR-32 (FAU/EMT)
Exxon Research and Engineering Company

ECR-32 represents a higher silica faujasite with EMC-2 as a minor component.

Edingtonite (EDI)
(natural)

Related Materials: K-F
                        Linde F

The natural barium aluminosilicate edingtonite was first described by Haidinger in 1825 in Scotland (*Nat. Zeo.* (1985)).

Structure

(*Zeo. Mol. Sieve.* 141(1974))

CHEMICAL COMPOSITION: $Ba_2(Al_4Si_6O_{20})*8H_2O$
SYMMETRY: tetragonal
SPACE GROUP: $P42_1m$
UNIT CELL CONSTANTS: $a = 9.58$ Å
                       $c = 6.52$ Å
VOID FRACTION (determined from water
  content): 0.36
FRAMEWORK DENSITY (g/cc): 1.68
PORE STRUCTURE: two-dimensional, 3.5 × 3.9 Å

X-Ray Powder Diffraction Data: (*Zeo. Mol. Sieve.* 215(1974)) (d(Å) (I/I_0)) 6.49(80), 5.37(80), 4.80(90), 4.64(90), 4.29(40), 3.86(20), 3.58(100), 3.39(60), 3.255(50), 3.078(60), 3.010(80), 2.934(70), 2.749(100), 2.655(40), 2.591(90), 2.461(50), 2.232(80), 2.260(80),

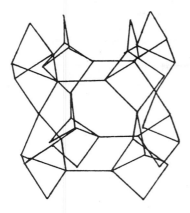

Edingtonite Fig. 1S: Framework topology.

2.201(60), 2.178(60), 2.143(20), 2.132(70),
2.062(70), 2.035(50), 1.9754(50), 1.933(60),
1.889(50), 1.834(70), 1.795(50), 1.766(50),
1.694(40), 1.678(20), 1.653(60), 1.644(60),
1.623(40), 1.606(40).

The framework structure of edingtonite consists of the simplest method of crosslinking the chains of 4–1 units of tetrahedra. Si/Al ordering very similar to that found in natrolite is observed in edingtonite, with each unit of five tetrahedra having an Si/Al ratio of 3:2 (*Acta Crystallogr.* B32:1623(1976)). Examples have been found

of disordered edingtonites (*N. Jb. Miner. Mh.* 373(1984)). The Ba^{+2} cations are located on the twofold rotation axis at the intersection of the channels and are surrounded by six framework oxygen atoms and four water molecules.

Atomic parameters for edingtonite (*N. Jb. Miner. Mh.* 373(1984)):

	occ.	x/a	y/b	z/c
Ba(1)	0.928	1/2	0	0.6393(1)
Ba(2)	0.048	1/2	0	0.5691(6)
T(1)	1	0	0	0
T(2)	1	−0.1734(1)	0.0934(1)	0.3812(1)
O(1)	1	0.1735(2)	0.3265	0.6232(4)
O(2–3)	1	−0.0447(2)	0.1962(2)	0.4651(2)
O(4–5)	1	−0.1372(2)	0.0382(2)	0.1443(3)
OW(1)	0.918	0.1753(3)	0.3247	0.1448(6)
OW(2)	0.933	0.3793(4)	0.1207	−0.211(8)
H(1)	0.918	0.2385	0.3693	0.0594
H(2)	0.933	0.4134	0.2020	0.0435

Thermal Properties

Four water losses are observed in this mineral between room temperature and 450°C. Structurally, there are only two independent water sites in this zeolite. Crystal destruction is observed at 400°C (*Nat. Zeol.* 65(1985)).

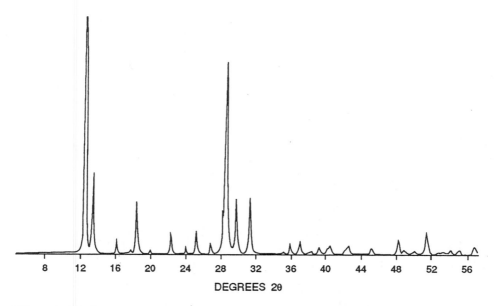

Edingtonite Fig. 2:   X-ray powder diffraction pattern for edingtonite (*ACS* 368:162(1988)).
(Reproduced with permission of the American Chemical Society)

Infrared Spectrum

Band frequencies (cm^{-1}) in the raman spectrum of edingtonite (in the region below 1200 cm^{-1}) (*Nat. Zeo.* 257(1988)):

A	B1	B2	B3
65	81		68
97			
104			
135			200
152			
		202	
235	342		292
278			
315			
359	357		

Band frequencies (cm^{-1}) in the raman spectrum of edingtonite (in the region below 1200 cm^{-1}) (*Nat. Zeo.* 257(1988)):

A	B1	B2	B3
407	380	350	338
	494		
432			
504			
530			
599			690
656		750	752
713			962
956			
993			
1056			1083
1103			
		1085	

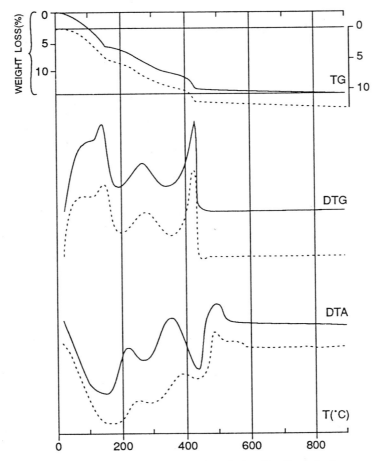

Edingtonite Fig. 3:   Thermal curves of the orthorhombic edingtonite from Bohlet Mine (solid line) and the tetragonal edingtonite from Ice River (broken line); in air, heating rate 20°C/min (*N. Jb. Miner. Mh.* 373(1984)).
(Reproduced with permission of E. Schweizerbart, Stuttgart)

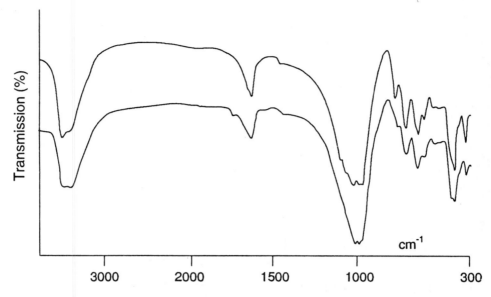

Edingtonite Fig. 4:   IR spectra of the orthorhombic edingtonite from Bohlet Mine (top) and the tetragonal edingtonite from Ice River (bottom) (*J. Jb. Miner. Mh.* 373(1984)). (Reproduced with permission of E. Schweizerbart, Stuttgart)

Ion Exchange

Fusion with sodium chlorate, potassium thiocyanate, silver, and thallium nitrate at 270°C results in ion exchange in this structure; however, exchange is not complete (*Mineral. Mag.* 23:483(1945)). Edingtonite is found to exhibit piezoelectric properties (*Rend. Soc. Miner. Ital.* 9:268(1953)).

Structure

(15th Congress of the IUCr, Bordeaux, July 1990, *pp* C-177-178)

CHEMICAL COMPOSITION: $(Na,H)_{16}Al_{16}Si_{60}O_{192}$
SPACE GROUP: $P6_3/mmc$
UNIT CELL CONSTANTS (Å) : $a = 17.384$
$c = 28.341$

EMC-1 (FAU)
(Fr 88 13,269(1988))
Ecole Nationale Superieure de Chimie de Mulhouse

Related Materials: faujasite

EMC-1 represents a high silica (SiO$_2$/Al$_2$O$_3$ = 3-10) faujasite prepared in the presence of crown-ethers.

EMC-2 (EMT)
(Fr 88 13,269(1988))
Ecole Nationale Superieure de Chimie de Mulhouse

Related Materials: Breck structure six hexagonal faujasite

Atomic parameters for EMC-2:

Atom	x	y	z
Si(1)	0.3729	0.0954	0.0167
Si(2)	0.4276	0.0340	0.1065
Si(3)	0.4881	0.1551	0.9276
Si(4)	0.4869	0.1525	0.1967
O(1)	0.4562	0.1282	0.9809
O(2)	0.2921	0.0000	0.0000
O(3)	0.3426	0.1713	0.0177
O(4)	0.4016	0.0932	0.0712
O(5)	0.4301	0.0714	0.1590
O(6)	0.3564	0.9262	0.1096
O(7)	0.5285	0.0571	0.0930
O(8)	0.4820	0.2410	0.9079
O(9)	0.5939	0.1878	0.9239
O(10)	0.4719	0.2359	0.1788
O(11)	0.5889	0.1779	0.1874
O(12)	0.4522	0.1196	0.2500

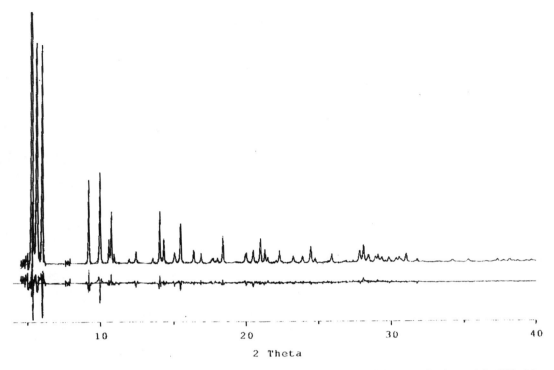

EMC-2 Fig. 1:   X-ray powder diffraction pattern for EMC-2 (*15th Congress of the IUCr,* Bordeaux, July 1990, *PP C-177-178*).

EMC-2 differs from the faujasite structure in the connection of the faujasite sheets.

Synthesis

ORGANIC ADDITIVES
1,4,7,10,13,16 hexaoxacyclooctadecane (18-crown-6)

EMC-2 crystallizes from a reactive gel with a batch composition of: 10 SiO_2 : Al_2O_3 : 2.4 Na_2O : 1 "crown ether" : 130 H_2O. Without the crown ether, crystallization of Y zeolite results (*PP 8th IZC* 127(1989)).

Encilite (MFI)
(EP 160,136(1985))

See ZSM-5.

Epidesmine (STI)

Obsolete synonym for stellerite.

Epistilbite (EPI)
(natural)

The name epistilbite first was proposed for a natural zeolite identified in 1826. The name is due to the zeolite's similarity to stilbite.

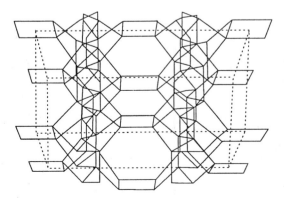

Epistilbite Fig. 1S: Framework topology.

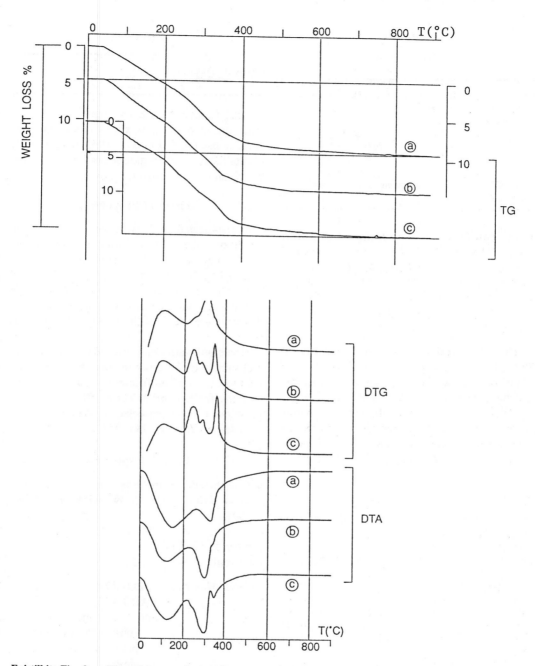

Epistilbite Fig. 2:   Thermal curves of epistilbite from (a) Elba, Italy, (b) Berufjord, Iceland, and (c) Yugawara, Japan; in air, heating rate 20°C/min for TG (and DTG) and 10°C/min for DTA (*Nat. Zeo.* 242(1985)). (Reproduced with permission of Springer Verlag Inc.)

Structure

(Zeo. Mol. Sieve. 216(1974))

CHEMICAL COMPOSITION: $Ca_{0.56} Na_{0.79} K_{0.04} Ba_{0.01}$
$(Al_{6.02}Si_{18}O_{48}) \cdot 15.46H_2O$
SPACE GROUP: C2
UNIT CELL CONSTANTS (Å): $a = 9.101(2)$
$b = 17.741(1)$
$c = 10.226(1)$
$\beta = 124°66'$
PORE STRUCTURE: intersecting 10- and 8-member
rings 3.4 × 5.6 Å and 5.2 × 3.7 Å

X-Ray Powder Diffraction Data: *(Zeo. Mol. Sieve.* 216(1974)) $(d(Å)$ $(I/I_0))$ 8.90(100), 6.90(33), 6.12(3), 4.92(55), 4.64(7), 4.49(9), 4.44(10), 4.33(20), 4.24(4), 4.01(8), 3.924(23), 3.870(73), 3.824(14), 3.802(16), 3.737(8), 3.446(92), 3.332(14), 3.270(21), 3.262(21), 3.208(83), 3.154(3), 3.114(2), 3.061(7).

The framework of epistilbite consists of four-, five-, and eight-member rings. Linked chains of five-member rings are present in this structure. The cations Ca^{+2} and Na^+ occupy sites near the periphery of the cavity. Coordination is to some of the framework oxygen atoms as well as water molecules *(Mineral. Mag.* 36:480(1967); *A. Kristallogr.* 173:257(1985)).

Atomic coordinates for epistilbite *(Mineral. Mag.* 36:480(1967)):

Atom	x	y	z
T(A)	0.001	0.088	0.161
T(B)	0.293	0.208	0.390
T(C)	0.707	0.197	0.097
Ca, Na	0.760	0	0.251
O(1)	0.021	0	0.215
O(2)	0	0.100	0
O(3)	0.812	0.117	0.132
O(4)	0.170	0.134	0.309
O(5)	1/2	0.180	0
O(6)	1/2	0.179	1/2
O(7)	3/4	1/4	0
O(8)	0.773	0.233	0.261
O(9)	1/4	1/4	1/2

Synthesis

ORGANIC ADDITIVE
none

This structure has been prepared from aluminosilicate reaction mixtures containing calcium ions *(JCS* London 1953:1879(1953)). Crystallization occurs at temperatures of around 250°C and at SiO_2/Al_2O_3 ratios of 7.

Thermal Properties

When subjected to dehydration under vacuum conditions, the epistilbite structure is thermally stable to at least 250°C. Readsorption of water is possible; however, no other gases or vapors will enter the pores. It is thought that, in its dehydrated state, the cations block the pores in this structure. See Epistilbite Fig. 2 on previous page.

The slight variations observed from one mineral sample to the other in epistilbite are attributed to dehydration differences due to possible stacking faults *(Am. Mineral.* 59:1055(1974)). Epistilbite exhibits piezoelectric properties *(Rend. Soc. Miner. Ital.* 9:268 (1953)).

Infrared Spectrum

Mid-infrared Vibrations (cm⁻¹): 1175sh, 1050sh, 795w, 690w, 563w, 455s *(Zeo.* 4:369 (1984)).

Erionite (ERI) (natural)

Related Materials: AlPO₄-17
SAPO-17
and related metal-substituted aluminophosphate molecular sieves
LZ-220

Erionite is a fibrous mineral first named in 1898; its name is derived from the Greek word for wool. It occurs as a common intergrowth with offretite *(Mineral. Mag.* 32:261(1959); *Nature* 214:1005(1967); *ACS* 101:230(1971); *Cryst. Res. Technol.* 15:869(1980)).

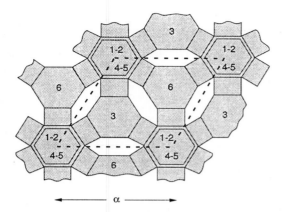

Erionite Fig. 1S: Framework topology.

Structure

(*Nat. Zeo.* 200(1985))

CHEMICAL COMPOSITION: $NaK_2MgCa_{1.5}$
($Al_8Si_{28}O_{72}$)*$28H_2O$
SYMMETRY: hexagonal
SPACE GROUP: $P6_3/mmc$
UNIT CELL CONSTANTS: (Å) $a = 13.15$
$c = 15.05$
PORE STRUCTURE: three-dimensional, eight-member
rings, 3.6 × 5.1 Å

X-Ray Powder Diffraction Data: (*Nat. Zeo.*
335(1985)) (d(Å) (I/I_0)) 11.56(76), 9.18(5),
7.55(9), 6.65(48), 6.31(6), 5.76(28), 5.39(20),
4.60(26), 4.58(30), 4.35(78), 4.18(28),
3.839(60), 3.771(94), 3.586(57), 3.423(5),
3.323(45), 3.291(17), 3.277(13), 3.193(13),
3.153(26), 3.122(11), 2.936(13), 2.880(76),
2,851(100), 2.826(65), 2.688(28), 2.513(33),
2.494(28), 2.215(19), 2.125(13), 2.091(7),
1.994(7).

The framework of erionite consists of columns formed by cancrinite cages joined by double six-ring units. Each cancrinite cage is connected to adjacent columns by single six-rings.
This gives rise to a 12-member-ring channel
system. However, the cancrinite cages in the
columns are alternately rotated by 60°, placing
a six-ring unit into the main channel every 15.2
Å.

Cations regularly occupy two types of sites
in the cavities of the framework and are, in part,

irregularly distributed in the large channels (*Bull.
Soc. Fr. Mineral. Cristallogr.* 92:250(1969)).

Atomic parameters for erionite (*Bull. Soc. Fr. Mineral.
Cristallogr.* 92:250(2969)):

Atom	x	y	z
Si(1)	0.120	0.120	0.120
Si(2)	0.120	0.120	0.606
Si(3)	0.000	0.239	0.106
Si(4)	0.000	0.239	0.606
Si(5)	0.210	0.120	0.250
Si(6)	0.166	0.255	0.750
Si(7)	0.046	0.378	0.750
O(1)	0.108	0.000	0.128
O(2)	0.126	0.000	−0.133
O(3)	0.054	0.162	0.628
O(4)	0.062	0.188	−0.633
O(5)	0.219	0.000	0.250
O(6)	0.109	0.330	0.750
O(7)	0.136	0.136	0.000
O(8)	0.000	0.273	0.000
O(9)	0.226	0.321	0.750
O(10)	0.048	0.500	0.750
O(11)	0.175	0.158	0.162
O(12)	0.167	0.185	0.662
O(13)	0.000	0.339	0.662

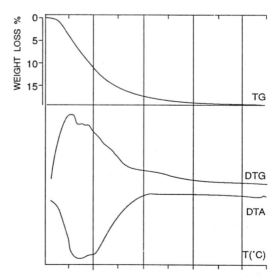

Erionite Fig. 2:   Thermal curves of erionite from
Agate Beach, Oregon; in air, heating rate 20°C/min (*Nat.
Zeo.* 206 (1985)). (Reproduced with permission of
Springer Verlag Inc.)

Synthesis

ORGANIC ADDITIVES
tetramethylammonium+ (TMA) (*JCS*
    (A)1970:1470(1970))
benzyltrimethylammonium+ (US3,699,139(1972))
benzyltriethylammonium+ (*SSSC* 28:429(1986))
1,4-diazobicyclo (2,2,2) octane monobasic and dibasic
    (*SSSC* 28:429(1986))

Erionite crystallizes from reactive aluminosilicate gels in the presence of TMA^+, Na^+,
and K^+ cations, with SiO_2/Al_2O_3 ratios around
15 (*JCS* (a)1970:1470(1970)). Common impurity phases reported include offretite and clinoptilolite. Similarities in the X-ray diffraction patterns of TMA-E (EAB), erionite, and offretite
made identification difficult in many early reports (*Nature* 213:1004(1967); *J. Solid State
Chem.* 37:204(1981)).

Thermal Properties

There are three peaks observed in the thermal
curves of erionite, at 100, 140, and 170°C, consistent with the presence of three structural types
of water molecules. See Erionite Fig. 2, p. 169.
Water loss is nearly complete at 300°C (*Nat.
Zeo.* 205(1985)). Dehydrated erionite has a very
stable framework structure, even upon exposure
to water vapor at 375°C.

Infrared Spectrum

Mid-infrared Vibrations (cm⁻¹):   1160bwsh,
1056bs, 782m, 725bw, 671vw, 634m, 580w,
533w, 470ms, 436m, 411w (*Zeo.* 9:104(1989)).

Ion Exchange

The main cation species in the natural mineral
are K^+ and Ca^{+2}. The K^+ cation shows considerable resistance to ion exchange, indicating
that it is locked within the structure. With an
increasing degree of ammonia exchange, initially the Na^+ and finally the K^+ cations are
replaced (*Zeo.* 9:224(1989)). Ammonia is desorbed from the ion-exchanged forms at inter-
vals from 267°C to 627°C. Dealumination occurs when the degree of ammonia exchange is
greater than 85%.

Cation exchange in erionite (*Zeo. Mol. Sieve.* 564(1974)):

Exchange	Conc (N)	T (°C)	x_{max}[a]
Na^+ to Cs^+	1.0	25	
K^+ to Na^+	1.0	25	[b]
Na^+ to 1/2 Ca^{+2}	1.0	25	1.0
Na^+ to 1/2 Sr^{+2}	1.0	25	1.0
Ca^{+2} to 1/2 Sr^{+2}	1.0	25	1.0

a. x_{max} = equiv. exchanged/g atoms of Al in the zeolite.
b. All of the K^+ is not exchangeable.

Adsorption

Adsorption properties of erionite (*Zeo. Mol. Sieve.* 622(1974)):

Adsorbate	T (°C)	P (Torr)	Wt% adsorbed
Argon	−196	0.1	14
		10	20.5
		100	24
Oxygen	−196	0.020	11
		3	18
		50	20.4
Xenon	−196	70	27
		100	32
		700	35
Nitrogen	−196	0.1	12
		10	14.5
		100	16
		700	19.4
	273	100	0.6
		700	1.2
		1400	5
H_2O	25	0.1	10
		1	13
		20	22
NH_3	25	0.4	5.8
		10	8.4
		100	11.4
		700	12.1
CO_2	25	20	9.5
		100	11.5
		300	13.5
		700	14.7
n-C_3H_8	25	10	5
		100	6.3
		700	7
n-C_4H_{10}	25	10	7.6
		100	9.0
		700	10.5
n-C_6H_{14}	98	25	1.8
		350	2.6

Adsorption properties of erionite (*Zeo. Mol. Sieve.* 622(1974)):

Adsorbate	T (°C)	P (Torr)	Wt% adsorbed
n-C_5H_{12}	25	10	9.3
		100	10.8
		400	14.5
Isobutane	25	none	
Isobutane		adsorbed	
1-Butene	25	10	8.3
		100	9.6
		700	12
Neopentane	25	None	
		adsorbed	
Benzene	25	None	
		adsorbed	

ETS-1
(US4,853,292(1989))
Engelhard Corporation

Structure

Chemical Composition: $1.0 \pm M_{2/n}O : TiO_2 : y\ SiO_2 : z\ H_2O$ (M cation with valence n; $y = 2.5$ to 25 and z from 0 to 100).

X-Ray Powder Diffraction Pattern: (d(Å) (I/I_0)) 7.80(100), 7.80(100) 7.16(5), 4.69(5), 4.47(5), 3.90(15), 3.39(5), 3.20(30), 3.03(5), 2.60(35), 2.45(15), 2.38(10), 1.97(10), 1.95(20), 1.90(15).

This material exhibits properties characteristic of a layered material.

Synthesis

ORGANIC ADDITIVE
none

This structure crystallizes from reaction mixtures with Si/Ti of 1, Na + K/Si of 3, OH/Si of 3, H_2O/SiO_2 of 60 and OH/H_2O of 10. The synthesis mixture employs sodium silicate, titanium sequioxide (Ti_2O_3), sodium and potassium hydroxide, KF*$2H_2O$. The pH initially was

around 12.2. Crystallization occurs after a week at 125°C.

ETS-2
(US4,853,292(1989))
Engelhard Corporation

Structure

Chemical Composition: titanium oxide phase with low levels of SiO_2.

X-Ray Powder Diffraction Pattern: (d(Å) (I/I_0)) 8.75(85), 3.70(40), 3.16(100).

This material exhibits properties characteristic of a low-thermal stability layered material.

Synthesis

ORGANIC ADDITIVE
none

This structure crystallizes from reaction mixtures similar to those of ETS-1 except that potassium is not used in the crystallization.

ETS-4
(US4,938,939(1990))
Engelhard Corporation

Related Material: Zorite

Structure

Chemical Composition: $1.0 \pm M_{2/n}O : TiO_2 : y\ SiO_2 : z\ H_2O$ (M cation with valence n; $y = 2.5$ to 25 and z from 0 to 100) (Si/Ti = 2.6).

X-Ray Powder Pattern: (d(Å) (I/I_0)) 11.6(45), 6.9(95), 5.27(35), 4.45(25), 3.61(25), 3.45(50), 3.38(35), 3.066(95), 2.979(100), 2.902(55), 2.763(20), 2.641(25), 2.587(60), 2.506(10), 2.426(20).

This material exhibits properties characteristic of a small-pore molecular sieve.

Synthesis

ORGANIC ADDITIVE
none

This structure crystallizes from reaction mixtures with a broad range of Si/Ti between 1 and 10, H_2O/SiO_2 between 2 and 100, and M/SiO_2 between 0.1 and 10. The most preferred range is between 2 and 3, 10 and 25, and 1 and 3. The synthesis mixture employs sodium silicate, titanium chloride (in HCl), KF*2H$_2$O, and sodium hydroxide. The pH initially is around 10.2. Crystallization occurs after a week at 150°C.

Adsorption

Adsorption properties of ETS-4:

Sorbate	As-synthesized, wt%	Sodium exchange, wt%	Calcium exchange, wt%
H_2O		13.1	16
n-Hexane		0.6	4.1
Benzene		0.5	0
SO_2		0.2	12

ETS-10
(US4,853,292(1989))
Engelhard Corporation

Structure

Chemical Composition: $1.0 \pm M_{2/n}O : TiO_2 : y\ SiO_2 : z\ H_2O$ (M cation with valence n; y = 2.5 to 25 and z from 0 to 100) (Si/Ti = 4.75).

X-Ray Powder Pattern: (d(Å) (I/I_o)) 14.7(15), 7.2(10), 4.93(5), 4.41(25), 3.74(5), 3.60(100), 3.45(25), 3.28(20), 2.544(10), 2.522(25), 2.469(10).

This material exhibits properties characteristic of a large-pore molecular sieve.

Synthesis

ORGANIC ADDITIVE
none

This structure crystallizes from reaction mixtures with a broad range of Si/Ti between 1 and 10, H_2O/SiO_2 between 2 and 100, and M/SiO_2 between 0.1 and 10. The most preferred range is between 2 and 3, 10 and 25, and 1 and 3. The synthesis mixture employs sodium silicate, titanium chloride (in HCl), KF*2H$_2$O, and sodium hydroxide. The pH initially is around 10.2. Crystallization occurs after a week at 150°C.

Thermal Properties

In the rare-earth-exchanged form this material exhibits crystal stability until 450°C, and the hydrogen form is stable to at least 500°C.

Adsorption

Adsorption properties of ETS-10:

Sorbate	As-synthesized, wt%	Rare earth exchange, wt%	Hydrogen form, wt%
H_2O	12.9	12.6	15
n-Hexane	8.2	7.2	7.8
1,3,5-Trimethyl-benzene	0.4–0.5	0.4–0.5	0.4–0.5
Triethylamine	8.4	4.7	11.1

EU-1 (EUO)
(EP 42,226(1981))
Imperial Chemical Industries

Related Material: ZSM-50
                    TPZ-3

Structure

(*Zeo.* 8:74(1988))

CHEMICAL COMPOSITION: (0.5–1.5) R_2O : Al_2O_3 : >10 SiO_2 : (0–100) H_2O
SYMMETRY: orthorhombic

SPACE GROUP: Cmma
UNIT CELL DIMENSIONS: (Å) $a = 13.74$
$b = 22.33$
$c = 20.18$
UNIT CELL VOLUME: (Å) ca. 6173.5
PORE STRUCTURE: (Å) medium-pore, 5.7 × 4.1°
(6.8 × 5.8 pocket, 8.1 Å deep)

X-Ray Powder Pattern: (as-synthesized) (EP
42,226(1981))   (d(Å)   (I/I_0))   11.03(s–vs),
10.10(m–s), 9.78(w), 6.84(w), 5.86(vw–w),
4.66(s–vs), 4.31(vs), 4.00(s–vs), 3.82(s–vs),
3.71(m–s), 3.44(m), 3.38(m), 3.26(s), 3.16(vw),
3.11(vw), 2.96(vw), 2.71(vw), 2.55(vw),
2.48(vw), 2.42(vw), 2.33(vw), 2.30(vw),
2.13(vw);   (calcined)   11.11(vs),   10.03(vs),
9.78(w–m), 7.62(w–m), 6.84(m), 6.21(vw–w),
5.73(w), 4.87(vw), 4.60(vs), 4.30(vs), 3.97(s–
vs), 3.77(s), 3.71(w–m), 3.63(vw–w), 3.42(m),
3.33(m), 3.27(s–vs), 3.23(m–s), 3.15(w–m),

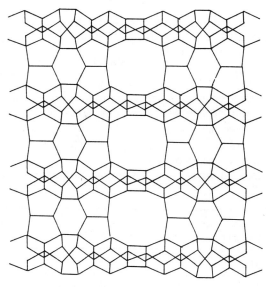

EU-1 Fig. 1S: Framework topology.

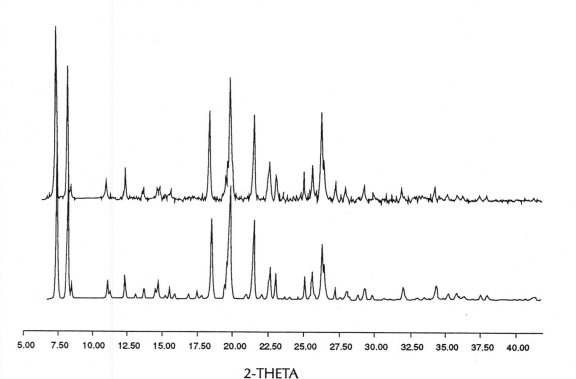

2-THETA

EU-1 Fig. 2:   X-ray powder diffraction pattern for EU-1 (*Zeo.* 8:74(1988)). (Reproduced with permission of Butterworth Publishing Company)

3.07(w–m), 2.93(w–m), 2.69(vw), 2.57(vw), 2.51(w), 2.45(vw), 2.41(vw), 2.32(vw), 2.29(vw), 211(vw).

Zeolite EU-1 is a high silica medium-pore zeolite containing a unidimensional channel system of 10-member rings in the [100] direction. Twelve-member rings off these channels define deep side pockets contained within the structure. The secondary building units defining this structure are the 5–1 and single four-ring units.

Atomic coordinates from Reitveld refinement of EU-1 in Cmma (*Zeo.* 8:74(1988)):

Atom	x	y	z	U	PP
Si(1)	0.115(1)	0.2500	0.123(2)	0.01500	1.0
Si(2)	0.187(1)	0.3703(9)	0.0630(9)	0.01500	1.0
Si(3)	0.289(1)	0.4315(7)	0.185(1)	0.01500	1.0
Si(4)	0.	0.448(1)	0.053(1)	0.01500	1.0
Si(5)	0	0.528(1)	0.181(2)	0.01500	1.0
Si(6)	0.117(1)	0.2500	0.277(2)	0.01500	1.0
Si(7)	0.191(1)	0.3739(8)	0.308(1)	0.01500	1.0
Si(8)	0.287(1)	0.4302(7)	0.431(1)	0.01500	1.0
Si(9)	0	0.450(1)	0.308(1)	0.01500	1.0
Si(10)	0	0.534(1)	0.432(1)	0.01500	1.0
O(1)	0	0.2500	0.113(4)	0.03000	1.0
O(2)	0.158(2)	0.3068(7)	0.089(2)	0.03000	1.0
O(3)	0.139(5)	0.2500	0.199(2)	0.03000	1.0
O(4)	0.2500	0.363(2)	0	0.03000	1.0
O(5)	0.245(3)	0.405(2)	0.118(2)	0.03000	1.0
O(6)	0.094(1)	0.407(1)	0.044(2)	0.03000	1.0
O(7)	0.2500	0.5000	0.196(3)	0.03000	1.0
O(8)	0.405(1)	0.431(1)	0.176(2)	0.03000	1.0
O(9)	0.260(3)	0.389(2)	0.246(2)	0.03000	1.0
O(10)	0	0.5000	0	0.03000	1.0
O(11)	0	0.477(2)	0.126(1)	0.03000	1.0
O(12)	0	0.497(2)	0.251(2)	0.03000	1.0
O(13)	0	0.2500	0.289(4)	0.03000	1.0
O(14)	0.164(2)	0.3065(7)	0.309(2)	0.03000	1.0
O(15)	0.094(1)	0.409(1)	0.303(2)	0.03000	1.0
O(16)	0.244(3)	0.391(2)	0.374(1)	0.03000	1.0
O(17)	0.2500	0.5000	0.424(4)	0.03000	1.0
O(18)	0.2500	0.403(3)	0.5000	0.03000	1.0
O(19)	0.404(1)	0.425(1)	0.427(2)	0.03000	1.0
O(20)	0	0.485(2)	0.376(1)	0.03000	1.0
O(21)	0	0.5000	0.5000	0.03000	1.0
X(51)	0.2500	0.2000	0.5000	0.05000	1.01(6)
X(52)	0.4000	0.2500	0.3820	0.05000	1.27(6)
X(53)	0.5000	0.1250	0.0670	0.05000	0.53(7)
X(54)	0.5000	0.1860	0.2980	0.05000	0.68(6)
X(55)	0.5000	0.2020	0.1760	0.05000	0.63(4)
X(56)	0	0.1830	0.4500	0.05000	0.40(5)

Synthesis

ORGANIC ADDITIVES
polymethylene diamines (EP 42,226(1981))
$(Me_3N(CH_2)_6Me_3)$ (EP 42,226(1981))

This structure crystallizes from a batch reaction composition of: 10 Na_2O : 10 $HexBr_2$: Al_2O_3 : 60 SiO_2 : 3000 H_2O at 200°C within six hours. Crystal formation is observed in the temperature range of 150 to 220°C. The pH range over which this structure forms is 11.6 (initial) to 11.8 (final) (*Zeo.* 3: 186(1983)). EU-1 also can form from an alkali-free system with a composition of: 5 $Hex.O.H_2O$: 2.5 $(NH_4)_2O$: Al_2O_3 : 60 SiO_2 : 3000 H_2O (*Zeo.* 5:153(1985)). Crystallization in this system occurs within 24 hours at 200°C. Co-crystallization of the EU-2 phase is observed when the SiO_2/Al_2O_3 ratio of the gel is above 120. Generally, EU-1 crystallizes at a SiO_2/Al_2O_3 between 30 and 120. "Soap bar"–shaped crystals of EU-1 form.

Adsorption

Adsorption properties of calcined EU-1 (*Proc. 6th IZC* 894:1983):

Sorbate	Kinetic diameter, nm	Sorbate pressure, p/mm Hg	Time, t/h	%Wt. increase
Water	0.265	4.7	2	6.9
			19	11.0
n-Hexane	0.430	45.8	2	9.5
p-Xylene	0.585	1.6	2	10.5
			18.5	10.8
Cyclohexane	0.600	27.0	2	1.1

EU-2
(BP 2,077,709(1981))
Imperial Chemical Industries

Related Material: ZSM-48

Structure

Chemical Composition: (0.5–1.5)R_2O : Al_2O_3 : >70 SiO_2 : (0–100) H_2O.

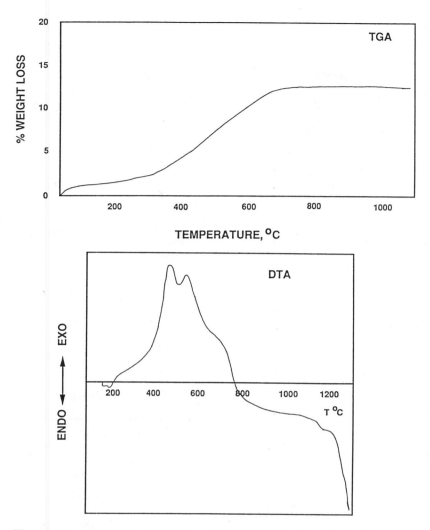

EU-1 Fig. 3:   TGA/DTA of EU-1 (*Proc. 6th IZC* 894:1983). (Reproduced with permission of Butterworth Publishing Company)

X-Ray Powder Data: $(d(Å)\ (I/I_0))$ 11.74(17), 10.13(14), 6.33(7), 5.85(7), 4.33(5), 4.18(86), 3.89(100), 3.69(7), 3.37(7), 3.08(5), 2.85(18), 2.09(5).

EU-2 exhibits properties related to the presence of medium pores in the structure. See ZSM-48 for a structural description.

Synthesis

ORGANIC ADDITIVES
polymethylene diamines (BP 2,077,709(1981))

EU-2 crystallizes between 180 and 200°C from reaction mixtures with a SiO_2/Al_2O_3 of 70 or

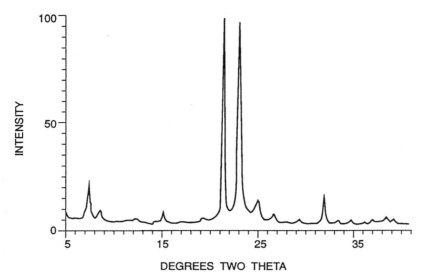

EU-2 Fig. 1:   X-ray powder diffraction pattern for EU-2 (*Zeo*. 5:153(1985)).
(Reproduced with permission of Butterworth Publishing Company)

higher. The OH/SiO_2 is in the range of 0.1 to 6.0, with water/SiO_2 between 1 and 100. The ratio of the organic amine cation ($(CH_3)_3N(CH_2)_nN(CH_3)_3^{+2}$, $n = 4$ and 9) to the total cation composition is between 0.1 and 1.0. A batch composition that results in highly crystalline EU-2 is: 60 SiO_2 : Al_2O_3 : 10 Na_2O : 10 RBr_2 : 3000 H_2O, where R is the diquaternary amine (*SSC* 28:215 (1986)). Crystallization is complete at 180°C after several days. Longer crystallization times in the diquaternary amine systems results in recrystallization to quartz or ferrierite. For EU-2 crystallization, at higher aluminum contents EU-1 appears as an impurity phase. Crystal agglomerates are observed for the hexamethonium and nonamethonium forms of EU-2, whereas tetramethonium cations produce 3- to 5-micron-size bars of this material. EU-2 can be crystallized from alkali-free systems with a batch composition of: 5 R_2O : 2.5 $(NH_4)_2O$: 60 SiO_2 : 3000 H_2O. Under these conditions and between 180 and 220°C, crystallization occurs in under 72 hours (*Zeo*. 5:153(1985)).

Adsorption

Adsorption properties of EU-2 (BP. 2,077,709(1981)):

Sorbate	Kinetic diameter, Å	Time, min	Wt % adsorbed, g/100 g	Voidage available, cc/100 g
Water	2.7	10	0.1	negl.
		120	0.2	
		1440	0.5	
n-Hexane		10	5.9	8.9
		60	6.4	9.7
		120	6.7	10.2
p-Xylene	5.85	10	5.7	7.6
		60	7.6	10.1
		120	7.6	10.1
m-Xylene	6.8	10	3.8	5.0
		60	5.7	7.6
		120	5.7	7.6
Cyclo-hexane	6.0	10	3.2	4
		60	3.5	4.4
		120	3.9	4.9
Symm. trimethyl-benzene	7.6	10	0.1	negl.
		1440	0.3	

EU-4
(EP 63,436(1982))
Imperial Chemical Industries

pyrrolidine (US4,528,171(1985))
hexamethonium^{+2} (HEX) (*Zeo.* 5:153(1985))

Structure

Chemical Composition: 0–1 M_2O : Al_2O_3 : at least 100 SiO_2 : 0–35 H_2O.

X-Ray Powder Data: $(d(\text{Å})$ $(I/I_0))$ 11.1(vs), 9.2(w–vs), 7.62(m), 6.87(m), 6.29(m), 5.98(w–m), 4.63(vs), 4.47(w–s), 4.29(vs), 3.98(vs), 3.80(w), 3.75(m), 3.68(s), 3.58(w–s), 3.42(m), 3.32(s), 3.28(s–vs), 3.23(m), 3.11(w).

Synthesis

ORGANIC ADDITIVES
propyltrimethylammonium$^+$ (EP 63,436(1982))

EU-4 crystallizes well at 180°C from a batch composition similar to that of EU-1: 60 SiO_2 : 10 Na_2O: 10 RBr_2 : 3000 H_2O, where the R is $(CH_3)_3N(CH2)_3N(CH_3)_3$ (*Proc. 7th IZC* 215 (1986)). The preferred range claimed for the crystallization of this material is a SiO_2/Al_2O_3 at least 40, OH/SiO_2 between 0.1 and 1.0, M + R/SiO_2 between 0.05 and 2.0, and R/M + R between 0.1 and 1, with H_2O/SiO_2 between 1 and 100.

Crystallization of this structure as a major phase occurs at SiO_2/Al_2O_3 ratios greater than 60. Under certain conditions, quartz appears as an impurity phase.

Adsorption

Adsorption properties at 25°C of EU-4 calcined at 450°C for 70 hours (EP63,436(1982)):

Adsorbate	Kinetic diameter Å	Pressure, mm Hg	Time, hours	Wt% ads., g/100 g	Voidage avail., cc/100 g
Water	2.7	6.0	17	0.78	0.78
n-Hexane	4.3	49	17	1.96	2.97
Cyclohexane	6.0	27.6	18.0	1.76	2.25

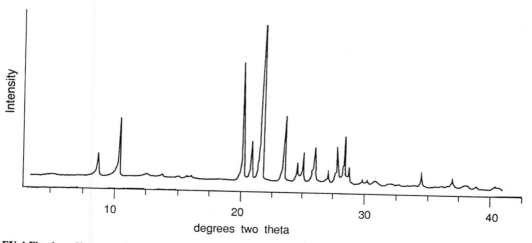

EU-4 Fig. 1:   X-ray powder diffraction pattern for EU-4 (*SSSC* 28:215 (1986)). (Reproduced with permission of Elsevier Publishing Company)

EU-7
(US4,581,212(1986))
Imperial Chemical Industries

Structure

Chemical Composition: 0.4–2.5 R_2O : Al_2O_3 : at least 20 SiO_2 : 0–600 H_2O.

X-Ray Powder Data: $(d(Å))$ (I/I_0) 7.96(12), 6.66(15), 4.23(100), 3.97(11), 3.65(9), 3.44(2), 3.33(98), 3.26(51), 3.12(7), 2.915(10), 2.868(1), 2.770(3), 2.691(1), 2.522(20), 2.502(13), 2.475(1), 2.370(10), 2.344(5).

Synthesis

ORGANIC ADDITIVES
butane 1,4-diamine (US4,581,212(1986))
hexamethylenediamine (US4,581,212(1986))
piperizine (US4,581,212(1986))

EU-7 is crystallized from a batch composition of: 60 SiO_2 : Al_2O_3 : 8–10 Cs_2O: 10–20 R : 3000 H_2O. Butane 1,4-diamine is the preferred organic under these conditions, producing the highest-crystallinity product. Crystallization occurs readily from cesium-containing solutions. Crystobalite, ferrierite, and ZSM-5 all have been observed as impurity phases under these conditions. Under conditions that produce ferrierite, the ferrierite phase is the first to appear, with EU-7 appearing as the ferrierite redissolves. Crystallization takes place around 150 to 180°C. When rubidium hydroxide is used in place of cesium, crystobalite is the major phase, with Nu-10 and EU-7 appearing as minor impurities. Rod-shaped crystals of EU-7 results from these syntheses, with dimensions of 0.2 micron by 1 to 4 microns in length.

Thermal Properties

EU-7 has been shown to be thermally stable to at least 550°C, with calcination to this temperature resulting in little change in the X-ray diffraction pattern.

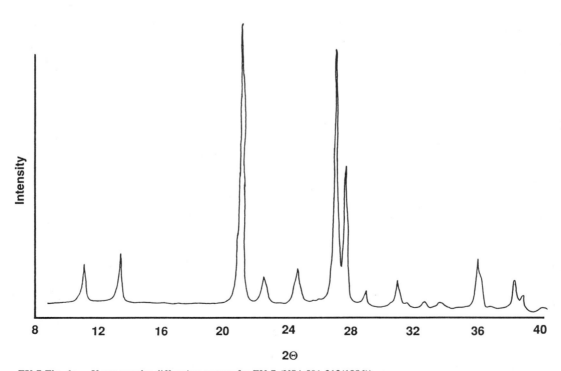

EU-7 Fig. 1:   X-ray powder diffraction pattern for EU-7 (US4,581,212(1986)).

EU-12
(US4,581,211(1986))
Imperial Chemical Industries

Structure

Chemical Composition: $0.5–2.0\ R_2O : Al_2O_3$: at least $5\ SiO_2 : 0–1000\ H_2O$.

X-Ray Powder Data: $(d(\text{Å})\ (I/I_0))$ 14.3(6), 11.2(40), 8.93(14), 7.58(19), 7.23(31), 6.70(30), 5.91(6), 5.62(21), 5.51(10), 4.93(19), 4.60(46), 4.48(34), 4.20(57), 3.81(52), 3.73(8), 3.62(49), 3.55(73), 3.44(9), 3.35(100), 3.21(37), 3.10(57), 2.980(10), 2.954(22), 2.914(17), 2.868(10), 2.816(7), 2.762(10), 2.646(8), 2.544(5),

Synthesis

ORGANIC ADDITIVES
tetramethylammonium$^+$ (TMA)

EU-12 is prepared from a typical batch composition of: $60\ SiO_2 : Al_2O_3 : 10\ Rb_2O : 5\ TMA_2O : 3013\ H_2O$. Crystallization occurs around 180°C after 100 hours in a stirred auto-

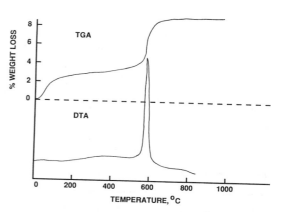

EU-12 Fig. 2:   TGA/DTA Traces of EU-12 (US4,581,211(1986)).

clave. A preferred range for crystallization is with the SiO_2/Al_2O_3 between 10 and 600, $RbOH/SiO_2$ between 0.15 and 0.65, H_2O/SiO_2 between 25 and 75, R/SiO_2 in the range of 0.05 to 0.25, and RbZ (Z = acid radical) between 0 and 0.25. Potassium ions have been used to crystallize this structure; however, in this system, ferrierite appears as an impurity phase. Other impurity phases observed in this system include zeolites P and HS and several unidentified minor products.

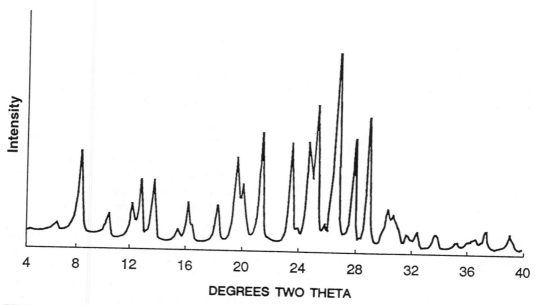

EU-12 Fig. 1:   X-ray powder diffraction pattern for EU-12 (US4,581,211 (1986)).

Thermal Properties

EU-12 is thermally stable to at least 600°C, with little change in the X-ray diffraction pattern.

EU-19
(JCS Dalton Trans. 2513(1988))

Structure

(Acta Crystallogr. B44;73(1988))

CHEMICAL COMPOSITION: piperazine*SiO$_2$
SYMMETRY: monoclinic
SPACE GROUP: C2/c
UNIT CELL PARAMETERS: (Å) a = 13.57
b = 4.90
c = 22.46
β = 91.67

D_x = 2.04 g/cm^3

X-Ray Powder Diffraction Data: (d(Å) (I/I_0))
11.43(100), 6.900(0.7), 5.977(3.2), 5.821(11.1), 5.708(1.4), 4.621(1.6), 4.499(14.3), 4.308(58.1), 3.978(64.3), 3.802(13.4), 3.650(4.1), 3.616(1.4), 3.435(13.4), 3.352(10.6), 3.282(30,4), 3.270(32.5), 3.118(0.9), 2.986(4.6), 2.912(0.7), 2.882(1.6), 2.856(3.0), 2.746(1.8), 2.690(1.4), 2.666(2.1), 2.592(1.7), 2.499(5.8), 2.453(5.3), 2.410(2.8), 2.378(0.7), 2.342(1.4), 2.302(0.7), 2.285(2.1).

EU-19 consists of double layers of Si$_6$O$_{13}$ units parallel to the [001] planes and held together by linkages through the strongly hydrogen-bonded piperazine cations. Each sheet is composed of eight Si atom rings. The structure

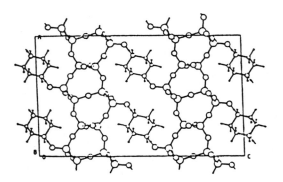

EU-19 Fig. 1S: Framework topology.

is more open than ZSM-39. Removal of the organic results in the generation of EU-20. The transformation takes place pseudomorphically, possibly involving crosslinking of the double sheets (Acta Crystallogr. B44:73(1988)).

Fractional coordinates for EU-19; estimated standard deviations in parentheses (Acta Crystallogr. B44:73(1988)):

	x	y	z
Si(1)	0.4671(4)	0.3329(10)	0.6835(2)
O(1)	0.3515(8)	0.2805(19)	0.6762(4)
O(2)	0.5247(8)	0.1419(20)	0.6407(4)
O(3)	0.4878(7)	0.6367(21)	0.6655(4)
O(4)	0.5000	0.2500	0.7500
Si(2)	0.5627(4)	−0.1677(10)	0.6335(2)
O(5)	0.5733(8)	−0.2217(24)	0.5653(4)
O(6)	0.1660(8)	0.3079(22)	0.6664(5)
Si(3)	0.2543(4)	0.2457(10)	0.0327(35)
O(7)	0.2535(8)	0.4464(22)	0.7668(4)
N(1)	0.6621(13)	0.3526(32)	0.5263(7)
C(1)	0.6885(17)	0.4093(46)	0.4651(9)
C(2)	0.7537(16)	0.3105(39)	0.5625(9)
H(11)	0.608(12)	0.498(35)	0.538(7)
H(12)	0.636(12)	0.149(33)	0.519(7)
H(21)	0.624(11)	0.444(33)	0.451(6)
H(22)	0.724(13)	0.568(32)	0.469(7)
H(31)	0.737(11)	0.265(30)	0.603(7)
H(32)	0.791(12)	0.521(33)	0.569(6)

Synthesis

ORGANIC ADDITIVES
piperazine (Zeo. 8:501(1988))
piperazine + trimethybenzylammonium$^+$ (Zeo. 8:501(1988))

The conditions used for synthesizing EU-19 are: 10 PIP : 20 SiO$_2$: 250 H$_2$O. Crystallization occurs readily at lower temperatures (120–150°C) and at pH around 11.8. At the higher temperature range, co-crystallization occurs with the clathrasil and ZSM-39. The addition of trimethylbenzylamine to this reaction mixture suppresses the formation of EU-19; however, it does not encourage formation of any other phase. Crystallization is slow in this system, requiring 70 days.

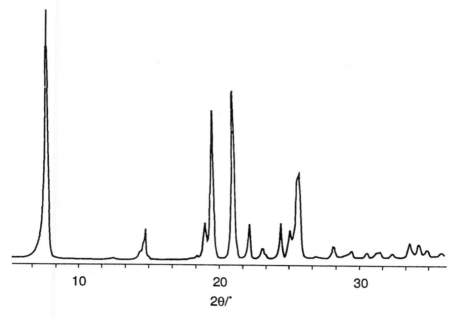

EU-19 Fig. 2:   X-ray powder diffraction pattern of EU-19 (*JCS Dalton Trans*. 2513 (1988)). (Reproduced with permission of the Chemical Society)

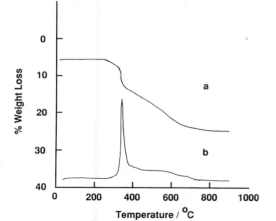

EU-19 Fig. 3:   TGA and DTA of EU-19 (*JCS Dalton Trans*. 2513 (1988)). (Reproduced with permission of the Chemical Society)

Thermal Properties

EU-19 converts to EU-20 upon thermal treatment to 400°C.

EU-20
(JCS Dalton Trans. 2513 (1988))

Structure

Chemical Composition: SiO_2.

X-Ray Powder Pattern Data: (d(Å) (I/I_0))
8.26(100) 6.88(18.5), 6.37(14.1), 5.32(15.90), 4.50(18.91),   4.33(7.10),   4.15(53.70), 4.08(34.40),   3.57(17.60),   3.48(15.90), 3.38(7.10), 3.31(16.70), 3.18(11.50), 3.12(6.2), 3.00(4.40), 2.89(0.90), 2.77(0.90), 2.70(2.60), 2.60(2.60), 2.47(2.60), 2.41(4.00), 2.39(4.80).

EU-20 is prepared through thermal treatment of EU-19 and is thought to be a crosslinked tectosilicate formed from thermal collapse of EU-19, consisting predominately of eight and

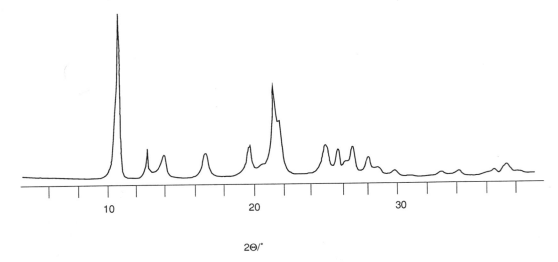

2Θ/°

EU-20 FIG. 1: X-ray powder diffraction pattern for EU-20 (*JCS Dalton Trans.* 2513 (1988)). (Reproduced with permission of the Chemical Society)

five silicon atom rings (*JCS Dalton Trans.* 2513(1988)).

Synthesis

See EU-19.

Thermal Properties

EU-20 appears to be thermally stable up to 900°C, with the X-ray diffraction pattern increasing with sharpness with the increase in thermal treatment of the material. Both the TGA and the DTA are featureless (*JCS Dalton Trans.* 2513(1988)).

F

F (EDI)

See Linde F.

Faroelite (THO)

Faroelite is an old and not widely used synonym for thompsonite.

Faujasite (FAU)*
(natural)

Related Materials: Linde type X
Linde type Y
SAPO-37
CSZ-3
LZ-210

The rare natural mineral faujasite first was described in 1842. Faujasite has the most open framework of all the natural zeolites, with a 51% void volume. It is one of the few magnesium-rich zeolites, along with offretite, mazzite, and ferrierite. Because of the industrial importance of the synthetic analogs types X and Y, the faujasite framework topology has been the focus of much study. Natural faujasite has been found in combination with phillipsite, gismondine, chabazite, gonnardite, natrolite, and analcime in deposits in Hawaii. In Germany, it has also been found in deposits along with chabazite and offretite (*Nat. Zeo.* 214 (1985)).

Structure

(*Zeo. Mol. Sieve.* 218 (1974))

CHEMICAL COMPOSITION: $Na_{20}Ca_{12}Mg_8$
$(Al_{60}Si_{132}O_{384})*235H_2O$
SPACE GROUP: Fd3m
UNIT CELL CONSTANT (A): $a = 24.60$

*The synthetic faujasites Linde types X and Y are discussed in this section when data on both compositional variants of the faujasite structure are handled together.

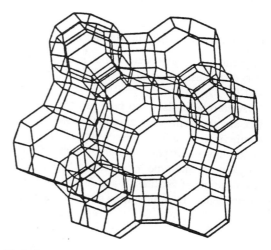

Faujasite Fig. 1S: Framework topology.

FRAMEWORK DENSITY (g/cc): 1.27
PORE STRUCTURE: large three-dimensional 12-member ring pores, 7.4 Å

X-Ray Powder Diffraction Data: ($d(Å)$ (I/I_0))
14.28(100), 8.74(19), 7.45(12), 7.14(5), 6.17(2), 5.67(78), 5.04(4), 4.75(44), 4.37(52), 4.18(3), 4.13(3), 3.903(18), 3.760(95), 3.562(8), 3.457(22), 3.301(67), 3.210(19), 3.083(3), 3.015(25), 2.910(36), 2.850(78), 2.760(28), 2.710(11), 2.632(32), 2.587(14), 2.518(4), 2.418(4), 2.374(23), 2.222(5), 2.182(13), 2.157(7), 2.095(8), 2.058(11), 2.002(2), 1.983(4).

The cubic unit cell of faujasite has a large cell dimension and contains 192 $(Si,Al)O_4$ tetrahedra. The framework of the faujasite structure can be described as a linkage of TO_4 tetrahedra in a truncated octahedron in a diamond-type structure. The truncated octahedron is referred to as the sodalite unit or sodalite cage. The octahedra are joined by hexagonal prisms forming the large voids in this structure. These large voids, or cages, are referred to as supercages and are connected to four other supercages through 12-member rings and to four sodalite units through six-member rings. Each supercage shares four-rings with six other sodalite units. Faujasite, Linde type X, and type Y topologically all have this framework structure.

Cation Sites in the Faujasite Structure

Six cation sites have been defined in the faujasite structure, designated as sites I, I', II, II', III, and IV.

A correlation is noted between the unit cell size and the framework aluminum content. Discontinuities have been observed near 77 and 62 aluminum atoms per unit cell, which differentiate type X and type Y zeolites. The chemical composition of zeolites X and Y can be traced to the method by which these structures are synthesized. In zeolites X and Y, the relation between the number of tetrahedral Al atoms, N_{Al}, and the Si/Al ratio is:

$$N_{Al} \frac{192}{(1 + R)}$$

where $R = N_{Si}/N_{Al}$. The aluminum ions (N_{Al}) in the unit cell of zeolite X vary from 96 to about 77. In zeolite Y, N_{Al} is between 76 and 48. The value of R then varies between 1 and 1.5 for zeolite X and between 1.5 and 3.0 for zeolite Y.

Discontinuities differentiating X from Y are also observed in the vapor phase adsorption of triethylamine, having a kinetic diameter of 7.8 Å. Because of the different location of Ca^{+2} in

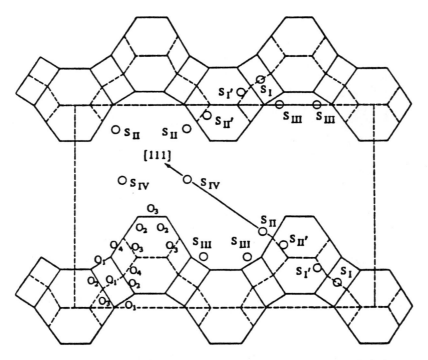

Faujasite Fig. 2:   The cation sites and their designation in zeolites X, Y, and faujasite. Starting at the center of symmetry and proceeding along the threefold axis toward the center of the unit cell, site I is the sixfold site located in the center of the double six-ring (hexagonal prism). Site I' is on the inside of the beta cage adjacent to the D6R. Site II' is on the inside of the sodalite unit adjacent to the single six-ring. Site II approaches the single six-ring outside the beta cage and lies within the large cavity opposite site II'. Site III refers to positions in the wall of the large cavity, on the fourfold axis in the large 12-ring aperture. The four different types of oxygens, O(1), O(2), O(3), O(4), also are indicated on their relative positions (*Mol. Sieves* 96 (1974)). (Reproduced with permission of John Wiley and Sons)

Cation distribution in natural faujasite (*Zeo. Mol. Sieve.* 98(1874):

Site		Na^{+a}	K^{+g}	Ca^{2+b}	Ba^{2+h}	Ni^{2+c}	Ce^{3+d}	La^{3+e}	Ca^{2+f},Na^+
16 I	De	10	8.6	14.2	7.3	10.6	3.4 Na$^+$	11.8	—
	Hy	—		—		—		—	
32 I'	De	9	12.9	2.6	5.0	3.2	11.5	2.6	—
	Hy	14		9.7		—	18 Na$^+$	3.3	17
32 II'	De	—		—		1.9	16,O$_x$	1.4,O$_x$	—
	Hy	15		11.5		—	32,O$_x$	28.3,O$_x$	32,O$_x$
32 II	De	32	31.7	11.4	11.3	6.4	10.7,Na$^+$	1.5	—
	Hy	11		23,4,O$_x$		—	26,O$_x$	14.2,O$_x$	11,O$_x$
48 III	De	—							
	Hy	—							
16 IV	De	—		—			—	—	
	Hy	—		2.2			6	10.3	
Reference		129	121	125	121	131	132	133	123
			130		130				

a. Data based on Na- and Ca-enriched faujasite from Kaiserstuhl. Dehydration in vacuo at 350°C. Si/Al = 2.27. Degree of exchange assumed to be > 80%.

b. Data on Ca^{+2} form from a completely exchanged crystal; composition: Ca$_{27}$[(AlO$_2$)$_{56}$(SiO$_2$)$_{133}$] • xH$_2$O. Dehydration in vacuo; temp. raised to 475°C over 4 days and held at 475°C for 12 hours before cooling to room temperature.

c. Ni^{+2} faujasite obtained by treatment with 1.0 M NiCl$_2$ at 90°C. Dehydrated in vacuo at 400°C for 7 hours, 1 × 10^{-6} Torr. Intensity data collected at room temperature. Composition: Ni$_{22}$Ca$_4$[(AlO$_2$)$_{58}$(SiO$_2$)$_{132}$]. Two site I' peaks assigned to Ni. Location of site II is farther from the six-ring than the usual site I.

d. Ce^{+3} faujasite uc contents given as: Ce$_{12}$Ca$_{7.6}$Na$_{7.8}$[(AlO$_2$)$_{59}$(SiO$_2$)$_{133}$] • 270H$_2$O. Data on hydrated crystal at room temperature. Dehydration in vacuo at 350°C. Six Ce^{+3} in hydrated crystal (site IV) located at random in supercages. Residual H$_2$O or OH shown as o$_x$.

e. La^{+3} faujasite completely exchanged except for possible hydrolysis. Residual Na, Mg, Ca <0.1%. Dehydration in vacuo at 475°C for 7 hours. Data collected at room temperature. 19 La atoms expected, but 15.9 found. The La in site I is displaced along the threefold axis, half-atoms 0.17 A from site I. Structure (420°C) showed same distribution of La^{+3}: 11.7 site I, 2.5 site I', 1.4 site II.

f. Structure determined on hydrated crystal of mineral; Na–Ca cations assumed to be dominant; 17 located. H$_2$O molecules shown by O$_x$.

g. 33 K$^+$ ions per unit call.

h. 23.1 Ba^{+2} ions per unit cell.

these two structures, (C$_2$H$_5$)N is not adsorbed in zeolite CaX with Si/Al ratio greater than 1.5, but it is readily adsorbed in CaY (*Soc. Chem. Ind.* London 47 (1968)).

Synthesis

See Linde type X and Linde type Y.

Thermal Properties

See Faujasite Fig. 4.

Infrared Spectrum

In the OH region of the infrared spectrum nearly all of the samples of the synthetic faujasites X and Y exhibit four OH bands, which occur at 3740, 3680 to 3705, 3640 (strong), and 3520 to 3610 (medium to weak) cm^{-1}. The band at 3740 cm^{-1} is attributed to amorphous silica in the samples. The bands around 3680 to 3705 cm^{-1} are assigned to undissociated adsorbed water or to the OH's of the metal (OH) complexes, which form from water dissociation in the zeolites. The bands at 3640 cm^{-1} and 3540 to 3610 cm^{-1} generally are attributed to acidic hydroxyls, with the more prominent former one being related to the oxygens 01 and vibration into the supercage. The latter bands are associated with the 03, 02, and 04 oxygens, where 04–H vibrates inside the large cavity, and 03–H and 02–H point into the less accessible beta cages.

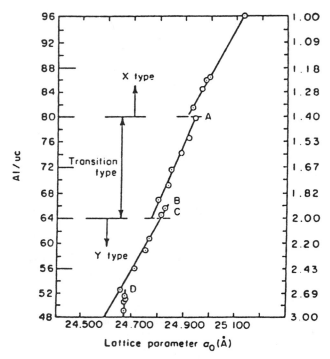

Faujasite Fig. 3:   Correlation between the lattice parameter (a_o) and the number of aluminum ions per unit cell for the synthetic faujasite types X and Y (*ACS Sym. Ser.* 101:171 (1971)). (Reproduced with permission of the American Chemical Society)

IR frequencies observed in cm^{-1} (*Zeo.* 3:13(1983))):

Zeolite	(SiO_2)	(MeOH?)	(01)	(02,03,04)	BPy	HPy,BPy,LPy	LPy
		OH stretching bands					
HNa-Y	3740w	—	3640s	3544s	1541s	1488s	1452w
HNa-Y*	3740w	—	—	—	—	1495m	1455s
BeNa-Y	3471 s	3705w	3638s	3540m	1540m	1488m	1450s
MgNa-Y	3741w	3685w	3641m	3580w	1548m	1490m	1448s
						1499m	
CaNa-Y	3739w		3640m	3590w	1549vw	1495m	1446s
SrNa-Y	3740w		3640vw	—	1547vw	1490-	1443s
BaNa-Y	3740w		—	—	1545vw	1490m	1442s
LaNa-Y	3740w	—	3640m	3520w	1547m	1490s	1447s
MgNa-X	3740w	3695m	3652s	3586w	1545m	1495m	1446s
						1488s	1440s
CaNa-X	3740w	3682w	3650vw	3610w	1545vw	1493w	1445s
SrNa-X	3740w	3682w	3652w	3610w	1545vw	1492m	1445s
BaNa-X	3740w	—	—	—	—	1490w	1442s
LaNa-X	3740w	3680w	3644s	3510m	1539s	1489s	1450w
			3600m				1444s

HNa-Y*: fully dehydroxylated at 923K under high vacuum; s: strong, m: medium, w: weak, vw: very weak. OH bands at 3740 cm^{-1} and around 3520–3580 cm^{-1} remained unchanged; bands around 3640 cm^{-1} disappeared upon pyridine adsorption; bands around 3700 cm^{-1} were changed in the case of La-X, eliminated with Be-Y, and weakened with Mg-Y.

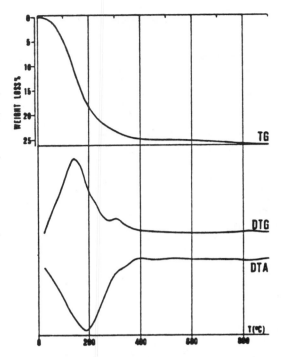

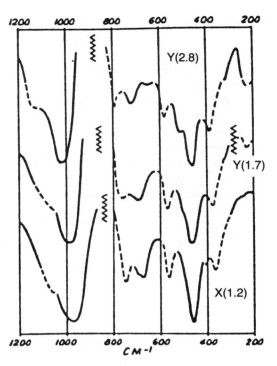

Faujasite Fig. 4:   Thermal curves of faujasite from Sasbach, Germany; in air, heating rate 20°C/min (*Nat. Zeo.* 219(1985)). (Reproduced with permission of Springer Verlag Inc.)

Faujasite Fig. 5:   Infrared spectra for zeolites X and Y with different Si/Al contents; numbers in parenthesis refer to Si/Al in the zeolite (*ACS Sym. Ser.* 101:201(1971)). (Reproduced with permission of the American Chemical Society)

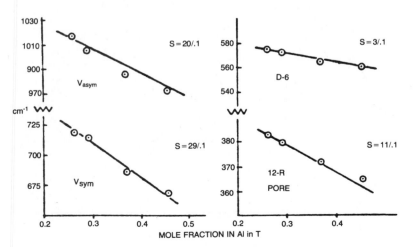

Faujasite Fig. 6:   Frequency vs. atom fraction of Al in the framework for zeolite X and Y for several infrared bands (*ACS Sym. Ser.* 101:201 (1971)). (Reproduced with permission of the American Chemical Society)

The effect of dehydration on the mid-range infrared spectrum is minor, with the frequency maximum usually increasing by 10 to 20 cm^{-1}. The effect of cation movement and framework distortion has been examined for Ca-exchanged Y zeolites with Si/Al of 2.5. Dehydration causes migration of the Ca^{+2} cation from inside the sodalite cage into a position near the center of the double six-ring (i.e., site I). A framework distortion and change in symmetry follow. The infrared spectrum also changes with this distortion. The band near 570 cm^{-1} shifts to 635 cm^{-1}, and the 390 cm^{-1} band shifts to 415 cm^{-1}. The character of the broad symmetric stretch at 710 to 750 cm^{-1} also changes.

Adsorption

Adsorption equilibrium data for natural faujasite (*Zeo. Mol. Sieve.* (1974)):

Adsorbate	T(K)	P(Torr)	g/g adsorbed
H_2O	298	22	0.33
O_2	90	700	0.29
N_2	77	700	0.24
Ar	90	730	0.28
SF_6	298	710	0.23
$(C_2F_5)_3N$	298	40	0.24
$(C_4F_9)_3N$	298	0.07	0.009

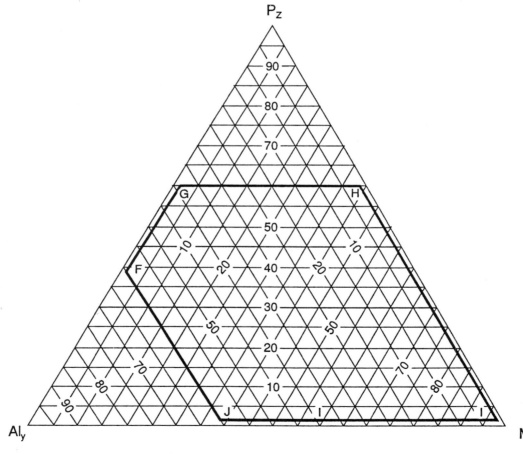

FCAPO Fig. 1: Composition range of reaction mixtures for the crystallization of FCAPO molecular sieves (EP 158,976(1986)).

FCAPO
(EP 158,976(1985))
Union Carbide Corporation

FCAPO is an acronym that denotes "framework constitutents" of elements in a framework of aluminum and phosphorus where all components are present as framework tetrahedral oxides. These components includes the elements arsenic, beryllium, boron, chromium, gallium, germanium, lithium, and vanadium. A number designation after the acronym identifies the aluminophosphate structural group. Compositional ranges for the reaction mixtures and products are shown in FCAPO Figs. 1 and 2.

FCAPO-5 (AFI)
(EP 158,976(1985))
Union Carbide Corporation

Related Materials: AlPO$_4$-5

X-Ray Powder Diffraction Data: $(d(\text{Å})\,(I/I_0))$
12.1–11.56(m–vs), 4.55–4.46(m–s), 4.25–4.17(m–vs), 4.00–3.93(w–vs), 3.47–3.40 (2–m).

FCAPO-5 materials exhibit properties characteristic of large-pore molecular sieves. See AlPO$_4$-5 for a description of the framework topology.

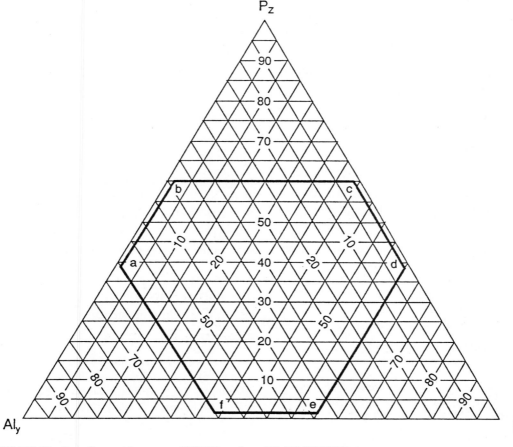

FCAPO Fig. 2:   Composition range of FCAPO products (EP 185,976(1985)).

FCAPO-11 (AEL)
(EP 158,976(1985))
Union Carbide Corporation

Related Materials: AlPO$_4$-11

X-Ray Powder Diffraction Data: (d(Å) (I/I_0))
9.51–9.17(m–s), 4.40–4.31(m–s), 4.25–4.17(s–vs), 4.04–3.95(m–s), 3.95–3.92(m–s), 3.87–3.80(m–vs).

FCAPO-11 materials exhibit properties characteristic of medium-pore molecular sieves. See AlPO$_4$-11 for a description of the framework topology.

FCAPO-14
(EP 158,976(1985))
Union Carbide Corporation

Related Materials: AlPO$_4$-14

X-Ray Powder Diffraction Data: (d(Å) (I/I_0))
10.3–9.93(vs), 6.81(w), 4.06–4.00(w), 3.51(w), 3.24(w), 3.01(w).

FCAPO-14 exhibits properties of small-pore molecular sieves.

FCAPO-16 (AST)
(EP 158,976(1985))
Union Carbide Corporation

Related Materials: AlPO$_4$-16

X-Ray Powder Diffraction Data: (d(Å) (I/I_0))
7.83–7.63(m–vs), 4.75–4.70(w–s), 4.06–3.99(m–vs), 3.363–3.302(2–m), 3.008–2.974(w–m).

FCAPO-16 materials exhibit properties of very small-pore molecular sieves.

FCAPO-17 (ERI)
(EP 158,976(1985))
Union Carbide Corporation

Related Materials: AlPO$_4$-17
                    erionite

X-Ray Powder Diffraction Data: (d(Å) (I/I_0))
11.5–11.4(vs), 6.61(s–vs), 5.72–5.70(s), 4.52–4.51(w–s), 4.33–4.31(vs), 2.812–2.797(w–s).

FCAPO-17 materials exhibit properties of small-pore erionite type molecular sieves.

FCAPO-18 (AEI)
(EP 158,976(1985))
Union Carbide Corporation

Related Materials: AlPO$_4$-18

X-Ray Powder Diffraction Data: (d(Å) (I/I_0))
9.21–9.16(vs), 5.72–5.70(m), 5.25–5.19(m), 4.41–4.39(m), 4.24–4.22(m), 2.814–2.755(m).

FCAPO-18 materials exhibit properties characteristic of small-pore molecular sieves.

FCAPO-20 (SOD)
(EP 158,976(1985))
Union Carbide Corporation

Related Materials: AlPO$_4$-20
                    sodalite

X-Ray Powder Diffraction Data: (d(Å) (I/I_0))
6.46–6.22(m–vs), 4.54–4.44(w–s), 3.70–3.63(m–vs), 2.614–2.564(vw–w), 2.127–2.103(vw–w).

FCAPO-20 materials exhibit properties characteristic of very small-pore sodalite sieves.

FCAPO-31 (ATO)
(EP 158,976(1985))
Union Carbide Corporation

Related Materials: AlPO$_4$-31

X-Ray Powder Diffraction Data: (d(Å) (I/I_0))
10.40–10.28(m–s), 4.40–4.37(m), 4.06–4.02 (w–m), 3.93–3.92(vs), 2.823–2.814(w–m).

FCAPO-31 materials exhibit properties characteristic of medium-pore molecular sieves.

FCAPO-33 (ATT)
(EP 158,976(1985))
Union Carbide Corporation

Related Materials: AlPO$_4$-33

X-Ray Powder Diffraction Data: (as-synthesized form) (d(Å) (I/I_0)) 9.56–9.26(w–m), 7.08–6.86(vs), 5.25–5.13(w–m), 4.34–4.25(w–m), 3.73–3.67(w–m), 3.42–3.38(w–m), 3.27–3.23(vs); (calcined form) (d(A) (I_0)) 6.73–6.61(vs), 4.91–4.83(m), 4.82–4.77(m), 3.36–3.34(m), 2.80–2.79(m).

FCAPO-33 materials exhibit properties characteristic of small-pore molecular sieves.

FCAPO-34 (CHA)
(EP 158,976(1985))
Union Carbide Corporation

Related Materials: SAPO-34
 chabazite

X-Ray Powder Diffraction Data: $(d(\text{Å})\,(I/I_0))$
9.41–9.17(s–vs), 5.57–5.47(vm–m), 4.97–4.82(w–s), 4.37–4.25(m–vs), 3.57–3.51(vw–s), 2.95–2.90(w–s).

FCAPO-34 materials exhibit properties characteristic of small-pore molecular sieves. See chabazite for a description of the framework topology.

FCAPO-35 (LEV)
(EP 158,976(1985))
Union Carbide Corporation

Related Materials: SAPO-35
 levyne

X-Ray Powder Diffraction Data: $(d(\text{Å})\,(I/I_0))$
8.19–7.97(m), 5.16–5.10(s–vs), 4.23–4.18(m–s), 4.08–4.04(vs), 2.814–2.788(m).

FCAPO-35 materials exhibit properties characteristic of small-pore molecular sieves. See levyne for a description of the framework topology.

FCAPO-36 (ATS)
(EP 158,976(1985))
Union Carbide Corporation

Related Materials: MAPO-36

X-Ray Powder Diffraction Data: $(d(\text{Å})\,(I/I_0))$
11.5–11.2(vs), 5.47–5.34(w–m), 4.70–4.60(m–s), 4.31–4.27(w–s), 4.08–4.04(m), 4.00–3.95(w–m).

FCAPO-36 materials exhibit properties characteristic of large-pore molecular sieves.

FCAPO-37 (FAU)
(EP 158,976(1985))
Union Carbide Corporation

Related Materials: SAPO-37
 faujasite

X-Ray Powder Diffraction Data: $(d(\text{Å})\,(I/I_0))$
14.49–14.03(vs), 5.72–5.64(w–m), 4.80–4.72 (w–m), 3.79–3.75(w–m), 3.31–3.29(w–m).

FCAPO-37 materials exhibit properties characteristic of large-pore molecular sieves. See faujasite for a description of the framework topology.

FCAPO-39 (ATN)
(EP 158,976(1985))
Union Carbide Corporation

Related Materials: AlPO$_4$-39

X-Ray Powder Diffraction Data: $(d(\text{Å})\,(I/I_0))$
9.41–9.21(w–m), 6.66–6.51(m–vs), 4.93–4.82(m), 4.19–4.12(m–s), 3.95–3.87(s–vs), 2.96–2.93(w–m).

FCAPO-39 molecular sieves exhibit small-pore adsorption properties.

FCAPO-40 (AFR)
(EP 158,976(1985))
Union Carbide Corporation

Related Materials: SAPO-40

X-Ray Powder Diffraction Data: $(d(\text{Å})\,(I/I_0))$
11.79–11.48(vw–m), 11.05–10.94(s–vs), 7.14–7.08(w–vs), 6.51–6.42(m–s), 6.33–6.28(w–m), 3.209–3.187(w–m).

FCAPO-40 materials exhibit properties characteristic of large-pore molecular sieves.

FCAPO-41 (AFO)
(EP 158,976(1985))
Union Carbide Corporation

Related Materials: SAPO-41

X-Ray Powder Diffraction Data: $(d(\text{Å})\,(I/I_0))$
6.51–6.42(w–m), 4.33–4.31(w–m), 4.21–4.17(vs), 4.02–3.99(m–s), 3.90–3.86(m), 3.82–3.80(w–m), 3.493–3.440(w–m).

FCAPO-41 materials exhibit properties characteristic of medium-pore molecular sieves.

FCAPO-42 (LTA)
(EP 158,976(1985))
Union Carbide Corporation

Related Materials: SAPO-42
                        Linde type A

X-Ray Powder Diffraction Data: $(d(\text{Å})\,(I/I_0))$
12.36–11.95(m–vs), 7.08–6.97(m–s), 4.09–4.06(m–s), 3.69–3.67(vs), 3.273–3.255(s), 2.974–2.955(m–s).

FCAPO-42 materials exhibit properties characteristic of small-pore molecular sieves.

FCAPO-44 (CHA)
(EP 158,976(1985))
Union Carbide Corporation

Related Materials: SAPO-44
                     chabazite

X-Ray Powder Diffraction Data: $(d(\text{Å})\,(I/I_0))$
9.41–9.26(vs), 6.81–6.76(w–m), 5.54–5.47(w–m), 4.31–4.26(s–vs), 3.66–3.65(w–vs), 2.912–2.889(w–s).

FCAPO-44 materials exhibit properties characteristic of small-pore chabazite molecular sieves.

FCAPO-46 (AFS)
(EP 158,976(1985))
Union Carbide Corporation

Related Materials: MgAPSO-46

X-Ray Powder Diffraction Data: $(d(\text{Å})\,(I/I_0))$
12.3–10.9(vs), 4.19–4.08(w–m), 3.95–3.87(vw–m), 3.351–3.278(vw–w), 3.132–3.079(vw–w).

FCAPO-46 materials exhibit properties characteristic of large-pore molecular sieves. See MgAPSO-46 for a description of the framework topology.

FCAPO-47
(EP 158,976(1985))
Union Carbide Corporation

Related Materials: MnAPO-47

X-Ray Powder Diffraction Data: $(d(\text{Å})\,(I/I_0))$
5.57–5.54(w–m), 4.33–4.31(s), 3.63–3.60(w), 3.45–3.44(w), 2.940–2.931(w).

FCAPO-47 materials exhibit properties characteristic of small-pore molecular sieves.

FeAPSO
(EP 161,491(1985))
Union Carbide Corporation

The FeAPSO molecular sieves are composed of iron, aluminum, phosphorus, and silicon oxides. The number associated with the FeAPSO name indicates the aluminophosphate framework structure type. See FeAPSO Figs. 1 and 2.

FeAPSO-5 (AFI)
(EP 161,491(1985))
Union Carbide Corporation

Related Materials: $AlPO_4$-5

Structure

Chemical Composition: see FeAPSO.

X-Ray Powder Diffraction Data: $(d(\text{Å})\,(I/I_0))$
(as-synthesized)   11.91(100),   11.05(4),* 7.03(13),* 6.81(7), 5.93(15),   5.54(<1),* 5.37(1),*   5.19(1),*   4.82(<1),*   4.48(33), 4.37(5),*   4.21(27),   4.04(sh),*   3.969(38), 3.934(sh),* 3.604(2), 3.548(1),* 3.440(15), 3.278(1),* 3.187(2),* 3.143(1),* 3.079(6), 2.979(19),   2.814(3),*   2.660(2),   2.600(9), 2.550(1),*   2.564(1),   2.380(4),   2.171(1), 2.137(2), 2.108(1), 2.076(1), 2.015(1), 1.985(1),* 1.907(3), 1.774(1), 1.653(1); (calcined 600°C) 11.95(100), 11.19(sh),* 10.46(35), 10.46(35),* 6.89(18), 5.99(8), 5.72(13),* 5.40(2),* 5.22(5),* 4.50(31),   4.40(14),* 4.21(33),   4.15(sh),* 4.04(sh),* 3.960(83), 3.739(1),* 3.59(2), 3.548(2),* 3.434(31), 3.302(2),* 3.198(3),* 3.074(14), 2.974(22), 2.840(29),* 2.827(5), 2.596(15), 2.564(3),* 2.488(1),* 2.430(4), 2.380(8),   2.356(2),* 2.298(2),* 2.151(2),* 2.137(2),   2.103(1),   2.067(2),   2.006(2), 1.949(2),* 1.907(4), 1.771(4), 1.653(2) (*peak contains, impurity).

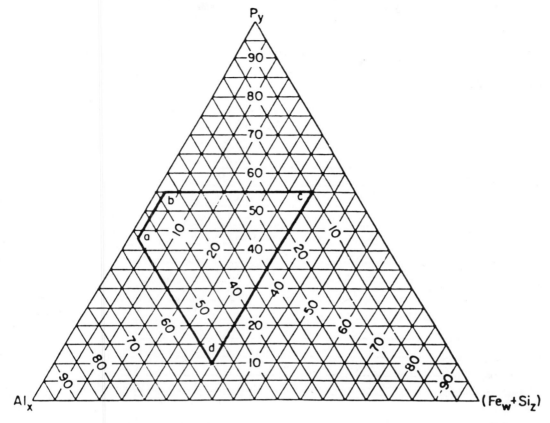

FeAPSO Fig. 1:   Ternary diagram indicating the parameters of the preferred compositions, given in mole fraction, for the iron silicoaluminophosphate molecular sieves (EP 161,491 (1985)).

FeAPSO-5 exhibits properties characteristic of a large-pore molecular sieve. See $AlPO_4$-5 for a description of the structure.

Synthesis

ORGANIC ADDITIVES
tetraethylammonium$^+$
tripropylamine
tetrabutylammonium$^+$
cyclohexylamine

FeAPSO-5 is prepared in pure form from reaction mixtures using tripropylamine as the organic additive. Aluminum isopropoxide is used as the source of aluminum, and phosphoric acid is the source of phosphorus. Iron acetate and colloidal silica (Ludox LS) provide the iron and silicon, respectively. The batch composition that successfully results in FeAPSO-5 includes: 0.9 Al_2O_3 : 0.9 P_2O_5 : 0.2 SiO_2 : 0.2 FeO : 1.0 tripropylamine : 50 H_2O. When TEAOH is used as the organic species, FeAPSO-34 is the major product, with FeAPSO-5 present as a minor phase. Crystallization occurs between 150 and 200°C within 2 to 7 days. TBA$^+$ can be used successfully to crystallize this structure when the batch composition is: 0.8 Al_2O_3 : 1.0 P_2O_5 : 0.5 SiO_2 : 0.4 FeO : 2.0 TBAOH : 83 H_2O.

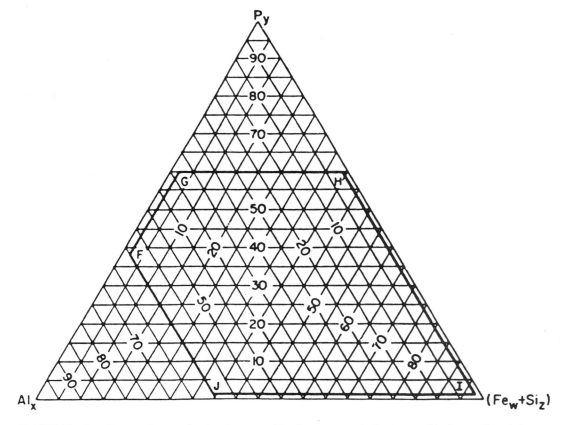

FeAPSO Fig. 2:   Ternary diagram showing the compositional ranges as mole fraction, used in the reaction mixtures producing the FeAPSO molecular sieves (EP 161,491 (1985)).

Adsorption

Adsorption properties of FeAPSO-5:

Adsorbate	Kinetic diameter Å	Pressure, Torr	Temperature, °C	Wt% adsorbed
O_2	3.46	100	−183	9.7
O_2		734	−183	11.6
Neopentane	6.2	100	24.5	3.8
O_2	6.0	59	23.7	5.7
H_2O	2.65	4.6	23.9	10.7
H_2O	2.65	20.0	23.6	19.2

FeAPSO-11 (AEL)
(EP 161,491(1985))
Union Carbide Corporation

Related Materials: $AlPO_4$-11

Structure

Chemical Composition: See FeAPSO.

X-Ray Powder Diffraction Data: ($d(\text{Å})$ (I/I_0)) (as-synthesized) 10.92(31), 9.36(47), 6.73(15), 5.64(34), 5.47(5), 4.67(6), 4.37(43), 4.23(100), 4.022(62), 3.952(sh),* 3.926(61), 3.850(86), 3.604(10), 3.376(25), 3.164(sh),** 3,121(17), 3.079(sh), 3.028(7), 2.840(9), 2.755(19), 2.667(2),** 2.629(9), 2.415(6), 2.386(14), 2.298(5), 2.108(5), 2.027(6), 1.804(5), 1.678(5), 1.656(3); (calcined 600°C) 10.98(60), 9.31(72), 6.86(sh),** 6.76(20), 6.46(3),** 6.03(3),** 5.57(55), 5.51(sh), 5.04(3),** 4.46(sh),** 4.37(28), 4.17(100), 4.06(sh),* 3.969(88), 3.867(sh),* 3.802(70), 3.708(3),** 3.648(5),** 3.562(4),* 3.453(7),* 3.363(20), 3.220(5),** 3.079(sh), 3.018(20), 2.940(7),* 2.814(10), 2.739(18), 2.629(5), 2.600(4),** 2.522(4),*

2.481(4),   2.368(10),   2.090(3),   2.023(5), 1.859(3),* 1.838(3),* 1.681(3), 1.650(3)* (*peak contains impurity; **impurity peak).

FeAPSO-11 exhibits properties characteristic of a medium-pore molecular sieve. See AlPO$_4$-11 for a description of the framework topology.

Synthesis

ORGANIC ADDITIVES
di-*n*-propylamine

FeAPSO-11 is prepared in pure form from reaction mixtures using di-*n*-propylamine as the organic additive. Aluminum isopropoxide is used as the source of aluminum, and phosphoric acid is the source of phosphorus. Iron acetate and colloidal silica (Ludox LS) provide the iron and silicon, respectively. The batch composition that successfully results in FeAPSO-11 includes: 0.9 Al$_2$O$_3$: 0.9 P$_2$O$_5$: 0.2 SiO$_2$: 0.2 FeO : 1.0 di-*n*-propylamine : 50 H$_2$O). FeAPSO-11 is the major product, with FeAPSO-31 appearing as a minor phase. Crystallization occurs between 150 and 200°C within 2 to 7 days.

Adsorption

Adsorption properties of FeAPSO-11:

Adsorbate	Kinetic diameter Å	Pressure, Torr	Temperature, °C	Wt% adsorbed
O$_2$	3.46	100	−183	7.6
O$_2$		734	−183	9.2
Neopentane	6.2	100	24.5	0.2
Cyclohexane	6.0	59	23.7	4.2
H$_2$O	2.65	4.6	23.9	10.8
H$_2$O	2.65	20.0	23.6	16.7

FeAPSO-16 (AST)
(EP 161,491(1985))
Union Carbide Corporation

Related Materials: AlPO$_4$-16

Structure

Chemical Composition: see FeAPSO.

X-Ray Powder Diffraction Data: (d(Å) (I/I_0))
(as-synthesized)   10.28(7),**   8.12(sh),** 7.83(58), 6.71(8),** 5.61(2),** 5.14(21),** 5.01(2),** 4.76(40), 4.37(sh),** 4.29(sh),** 4.21(sh),** 4.07(100), 3.883(10), 3.770(2), 3.562(1),** 3.453(1), 3.363(22), 3.290(sh),** 3.121(sh),** 3.089(9), 3.008(24), 2.797(10),** 2.747(4), 2.592(8),* 2.522(1),** 2.377(8), 2.270(3), 2.045(2), 1.879(7),* 1.845(2),** 1.778(1),** 1.746(1), 1.675(2)* (*peak may contain impurity; **impurity peak).

FeAPSO-16 exhibits properties characteristic of a very small-pore molecular. See AlPO$_4$-16 for a description of the framework topology.

Synthesis

ORGANIC ADDITIVES
quinuclidine
methylquinuclidine

FeAPSO-16 is prepared in pure form from reaction mixtures using quinuclidine or methylquinuclidine as the organic additive. Aluminum isopropoxide is used as the source of aluminum, and phosphoric acid is the source of phosphorus. Iron acetate and colloidal silica (ludox LS) provide the iron and silicon, respectively. The batch composition that successfully results in FeAPSO-16 includes: 0.9 Al$_2$O$_3$: 0.9 P$_2$O$_5$: 0.2–0.6 SiO$_2$: 0.4 FeO : 2.0 quin : 83 H$_2$O. Crystallization occurs at 150°C between 2 and 7 days.

FeAPSO-20 (SOD)
(EP 161,491(1985))
Union Carbide Corporation

Related Materials: sodalite
AlPO$_4$-20

Structure

Chemical Composition: See FeAPSO.

X-Ray Powder Diffraction Data: (d(Å) (I/I_0))
(as-synthesized) 6.32(59), 4.47(47), 3.998(4), 3.654(100), 3.164(16), 2.831(12), 2.584(16), 2.394(2), 2.240(4), 2.110(5), 1.909(4), 1.758(8); (calcined 600°C) 12.56(6), 11.82(6), 6.31(100),

4.43(28), 3.935(6), 3.733(5), 3.635(45), 3.143(11), 2.823(11), 2.578(9).

FeAPSO-20 exhibits properties characteristic of a very small-pore material. See $AlPO_4$-20 for a description of the framemwork topology.

Synthesis

ORGANIC ADDITIVES
tetramethylammonium$^+$

FeAPSO-20 is prepared in pure form from reaction mixtures using tetramethylammonium$^+$ as the organic additive. Aluminum isopropoxide is used as the source of aluminum, and phosphoric acid is the source of phosphorus. Iron acetate and colloidal silica (Ludox LS) provide the iron and silicon, respectively. The batch composition that successfully results in FeAPSO-20 includes: 0.9 Al_2O_3 : 0.9 P_2O_5 : 0.2 SiO_2 : 0.2 FeO : 1.0 TMAOH : 50 H_2O. Crystallization occurs at 150°C between 2 and 6 days.

Adsorption

Adsorption properties of FeAPSO-20:

Adsorbate	Kinetic diameter Å	Pressure, Torr	Temperature, °C	Wt% adsorbed
O_2	3.46	99	−183	1.5
O_2		749	−183	8.5
H_2O	2.65	4.6	23.2	22.7
H_2O	2.65	16.8	23.5	30.0

FeAPSO-31 (ATO)
(EP 161,491(1985))
Union Carbide Corporation

Related Materials: $AlPO_4$-31

Structure

Chemical Composition: See FeAPSO.

X-Ray Powder Diffraction Data: (d(Å) (I/I_0)) (as-synthesized) 10.41(64), 9.35(5),* 6.81(1), 6.07(1), 5.64(3), 5.20(6), 4.85(3), 4.39(49), 4.22(9),* 4.05(32), 3.936(100), 3.833(6), 3.546(4), 3.474(4), 3.372(2), 3.195(13), 3.110(1), 3.008(7), 2.821(20), 2.739(1), 2.555(9), 2.489(2), 2.418(2), 2.390(2), 2.358(3), 2.293(4), 2.275(3), 2.244(2), 2.006(2), 1.947(3), 1.871(2), 1.799(2), 1.770(4), 1.650(2); (calcined 600°C) 10.26(73), 9.04(3),* 6.83(1), 5.95(5), 5.46(4), 5.16(11), 4.80(4), 4.35(50), 4.016(44), 3.909(100), 3.795(3), 3.521(5), 3.449(9), 3.174(13), 2.990(12), 2.876(2),** 2.806(30), 2.739(2), 2.542(10), 2.475(5), 2.407(3), 2.378(2), 2.346(3), 2.282(4), 2.234(3), 2.052(2), 2.013(2),* 1.997(2), 1.940(5), 1.909(2), 1.863(3), 1.848(2), 1.793(2), 1.765(6), 1.653(3), 090(3), 2.023(5), 1.859(3),* 1.838(3),* 1.681(3), 1.650(3)* (*peak contains impurity; **impurity peak).

FeAPSO-31 exhibits properties characteristic of a large-pore molecular sieve. See $AlPO_4$-31 for a description of the framework topology.

Synthesis

ORGANIC ADDITIVES
di-*n*-propylamine

FeAPSO-31 is prepared in pure form from reaction mixtures using di-*n*-propylamine as the organic additive. Catapal (pseudoboehmite) is used as the source of aluminum and phosphoric acid as the source of phosphorus. Iron acetate and colloidal silica (ludox LS) provide the iron and silicon, respectively. The batch composition that successfully results in FeAPSO-31 includes: 0.9 Al_2O_3 : 0.9 P_2O_5 : 0.2 SiO_2 : 0.2 FeO : 1.0 di-*n*-propylamine : 50 H_2O. FeAPSO-31 is the major product, with FeAPSO-11 appearing as a minor phase. Crystallization occurs between 150 and 200°C within 2 to 7 days. At the lower temperatures, FeAPSO-46 is found to crystallize.

Adsorption

Adsorption properties of FeAPSO-31:

Adsorbate	Kinetic diameter Å	Pressure, Torr	Temperature, °C	Wt% adsorbed
O_2	3.46	99	−183	6.8
O_2		749	−183	11.6
Neopentane	6.2	100	23.4	3.6
Cyclohexane	6.0	57	23.4	6.9
H_2O	2.65	4.6	23.9	6.5
H_2O	2.65	16.8	23.5	21.3

FeAPSO-34 (CHA)
(EP 161,491(1985))
Union Carbide Corporation

Related Materials: chabazite
$AlPO_4$-34

Structure

Chemical Composition: See FeAPSO.

X-Ray Powder Diffraction Data: ($d(\text{Å})$ (I/I_0))
(as-synthesized) 12.1(5)8, 9.5(100), 7.0(10),
6.3(8), 5.99(2),* 5.57(32), 4.96(7), 4.53(3sh),*
4.35(50), 3.99(6), 3.88(2), 3.548(10), 3.466(11),
3.243(2), 3.164(2), 3.038(2sh), 2.940(19),
2.876(12), 2.607(4), 2.481(2), 2.281(2),
2.090(3), 1.914(2), 1.864(3), 1.791(2), 1.728(2),
1.684(1); (calcined 600°C) 11.79(7),* 9.21(100),
6.81(17), 5.47(9), 5.25(1), 4.93(5), 4.60(5),
4.46(2), 4.25(17), 3.943(7), 3.802(2), 3.678(2),
3.548(5), 3.401(7), 3.278(1), 3.164(2), 3.058(2),
2.885(16) (*impurity peak).

FeAPSO-34 exhibits properties characteristic
of a small-pore molecular sieve. See $AlPO_4$-34
for a description of the framework topology.

Synthesis

ORGANIC ADDITIVES
tetraethylammonium$^+$
di-*n*-propylamine

FeAPSO-34 is prepared in pure form from
reaction mixtures using di-*n*-propylamine as the
organic additive. Aluminum isopropoxide is used
as the source of aluminum, and phosphoric acid
is the source of phosphorus. Iron acetate and
colloidal silica (Ludox LS) provide the iron and
silicon, respectively. The batch composition that
successfully results in FeAPSO-34 includes: 0.9
Al_2O_3 : 0.9 P_2O_5 : 0.2–0.6 SiO_2 : 0.2 FeO : 1.0
TEAOH : 50 H_2O. FeAPSO-34 is the major
product at 150°C after 3 to 8 days of crystallization. FeAPSO-5 occurs as an impurity
phase over the temperature range of 150 to
200°C. Using di-*n*-propylamine as the organic,
FeAPSO-5 appears as the major phase.

FeAPSO-35 (LEV)
(EP 161,491(1985))

Related Materials: levyne
SAPO-35

Structure

Chemical Composition: See FeAPSO.

X-Ray Powder Diffraction Data: ($d(\text{Å})$ (I/I_0))
(as-synthesized) 10.19(5), 8.03(sh), 7.77(51),
6.57(9), 5.55(3), 5.09(28), 4.95(5), 4.76(50),**
4.22(15), 4.06(100), 3.885(9),** 3.793(8),
3.548(3), 3.365(25),** 3.285(7), 3.118(16),
3.091(sh),* 3.010(27),** 2.780(sh),
2.754(13),** 2.591(7), 2.381(10),* 2.053(3),**
1.886(8),* 1.774(1), 1.754(2),** 1.684(3)**
(*peak may contain impurity; **impurity peak).

FeAPSO-35 exhibits properties characteristic
of a small-pore molecula sieve. See levyne for
a description of the framework topology.

Synthesis

ORGANIC ADDITIVES
quinuclidine
methylquinuclidine

FeAPSO-35 is prepared in pure form from
reaction mixtures using quinuclidine and its
methyl derivative as the organic additives. Alu-

minum isopropoxide is used as the source of aluminum, and phosphoric acid is the source of phosphorus. Iron acetate and colloidal silica (Ludox LS) provide the iron and silicon, respectively. The batch composition that successfully results in FeAPSO-35 includes: 0.9 Al_2O_3 : 0.9 P_2O_5 : 0.2–0.6 SiO_2 : 0.2 FeO : 1.0 quin : 50 H_2O. FeAPSO-35 is the minor product at both 150°C and 200°C after 3 to 8 days of crystallization. FeAPSO-16 occurs as the major phase over this temperature range. Using the methyl derivative as the organic, FeAPSO-16 still appears as the major phase in this system.

FeAPSO-44 (CHA)
(EP 161,491(1985))
Union Carbide Corporation

Related Materials: SAPO-44
                          chabazite

Structure

Chemical Composition: See FeAPSO.

X-Ray Powder Diffraction Data: ($d(Å)$ (I/I_0)) (as-synthesized) 9.31(28), 6.83(21), 5.49(9), 5.10(2), 4.67(2), 4.23(100), 4.07(10), 3.850(2), 3.635(30), 3.427(58), 3.180(6), 2.966(11), 2.894(12), 2.714(2), 2.525(39), 2.137(7), 2.125(5), 2.072(5), 1.904(11).

FeAPSO-44 exhibits properties characteristic of a small-pore molecular sieve.

Synthesis

ORGANIC ADDITIVES
cyclohexylamine

FeAPSO-44 is prepared only as a minor phase with FeAPSO-5 in the crystallization mixture: 0.9 Al_2O_3 : 0.9 P_2O_5 : 0.2 SiO_2 : 1.0 cyclohexylamine : 50 H_2O. Aluminum isopropoxide is used as the source of aluminum, and phosphoric acid is the source of phosphorus. Iron acetate and colloidal silica (Ludox LS) provide the iron and silicon, respectively. FeAPSO-44 is the minor product at 220°C after 5 days of crystallization.

FeAPSO-46 (AFS)
(EP 161,491(1985))
Union Carbide Corporation

Related Materials: Co, Mg, Mn, Zn–
                          containing $AlPO_4$-based
                          molecular sieves

Structure

Chemical Composition: See FeAPSO.

X-Ray Powder Diffraction Data: ($d(Å)$ (I/I_0)) (as-synthesized) 13.42(3), 11.38(100), 7.11(2), 6.70(2), 6.41(1), 5.91(1), 5.77(1), 5.31(2), 5.13(<1), 4.47(1), 4.31(3), 4.11(7), 3.885(4), 3.660(3), 3.534(<1), 3.307(3), 3.206(2), 3.147(1), 3.093(3), 2.985(1), 2.959(<1), 2.889(<1), 2.885(2), 2.814(<1), 2.711(1), 2.606(1), 2.490(3), 2.448(<1), 2.259(<1), 2.188(<1), 2.049(1), 1.902(1), 1.811(<1), 1.768(<1), 1.743(<1); (calcined 500°C) 12.92(9), 11.04(100), 6.51(4), 5.76(3), 5.55(3), 5.17(3), 4.17(2), 4.006(2), 3.793(2), 3.575(2), 3.232(2), 2.797(2).

FeAPSO-46 exhibits properties characteristic of a large-pore molecular sieve. See MgAPSO-46 for a description of the framework topology.

Synthesis

ORGANIC ADDITIVES
di-n-propylamine.

FeAPSO-46 is prepared in pure form from reaction mixtures using di-n-propylamine as the organic additive. Catapal (pseudoboehmite) is used as the source of aluminum, and phosphoric acid is the source of phosphorus. Iron acetate and colloidal silica (Ludox LS) provide the iron and silicon, respectively. The batch composition that successfully results in FeAPSO-31 includes: 0.9 Al_2O_3 : 0.9 P_2O_5 : 0.2 SiO_2 : 0.2 FeO : 1.0 d-n-propylamine : 50 H_2O. FeAPSO-46 is the major product at 150°C after 7 to 8 days of crystallization.

Adsorption

Adsorption properties of FeAPSO-46:

Adsorbate	Kinetic diameter Å	Pressure, Torr	Temperature, °C	Wt% adsorbed
O_2	3.46	100	−183	2.6
O_2		749	−183	11.7
Neopentane	6.2	100	23.4	1.1
Cyclohexane	6.0	57	23.4	6.4
H_2O	2.65	4.6	23.2	7.2
H_2O	2.65	16.8	23.5	13.0

Adsorption

Adsorption properties of FeCoMgAPO-17:

Adsorbate	Kinetic diameter Å	Pressure, Torr	Temperature, °C	Wt% adsorbed*
O_2	3.46	100	−183	10
O_2		750	−183	12
n-Butane	4.3	100	24	4
H_2O	2.65	4.3	24	13
H_2O		20	24	14

*Typical amount adsorbed.

FeCoMgAPO-17 (ERI)
(EP 158,977(1985))
Union Carbide Corporation

Related Materials: AlPO$_4$-17
                     erionite

Structure

Chemical Composition: See XAPO.

X-Ray Powder Diffraction Data: $(d(Å) (I/I_0))$
11.5–11.4(vs), 6.61(s–vs), 5.72–5.70(s), 4.52–
4.51(w–s), 4.33–4.31(vs), 2.812–2.797(w–s).

FeCoMgAPO-17 exhibits properties characteristic of a small-pore molecular sieve. See AlPO$_4$-17 for a description of the topology.

Synthesis

ORGANIC ADDITIVES
quinuclidine

FeCoMgAPO-17 crystallizes from a reactive gel with a molar oxide composition of: (1.0–2.0) R : (0.05–0.2) M_2O_q : (0.5–1.0) Al_2O_3 : (0.5–1.0) P_2O_5 : (40–100) H_2O (R represents the organic additive, and q denotes the oxidation states of the metals (M)).

FeMgAPO-5 (AFI)
(EP 158,977(1985))
Union Carbide Corporation

Related Materials: AlPO$_4$-5

Structure

Chemical Composition: See XAPO.

X-Ray Powder Diffraction Data: $(d(Å) (I/I_0))$
12.1–11.56(m–vs),   4.55–4.46(m–s),   4.25–
4.17(m–vs),   4.00–3.93(w–vs),   3.47–3.40
(w–m).

FeMgAPO-5 is considered to be a large-pore molecular sieve, based on its physical properties. See AlPO$_4$-5 for the framework topology.

Synthesis

ORGANIC ADDITIVES
tetrapropylammonium$^+$

FeMgAPO-5 crystallizes from a reactive gel with a molar oxide composition: (1.0–2.0) R : (0.05–0.2) Fe_2O_q : (0.1–0.4) MgO : (0.5–1.0) Al_2O_3 : (0.5–1.0) P_2O_5 : (40–100) H_2O (R represents the organic additive in this system; q denotes the oxidation states of the iron).

Adsorption

Adsorption properties of FeMgAPO-5:

Adsorbate	Kinetic diameter Å	Pressure, Torr	Temper- ature, °C	Wt% adsorbed*
O_2	3.46	100	−183	7
O_2		750	−183	10
Neopentane	6.2	700	24	4
H_2O	2.65	4.3	24	12
H_2O		20	24	12

*Typical amount adsorbed.

Adsorption

Adsorption properties of FeMnAPO-11:

Adsorbate	Kinetic diameter Å	Pressure, Torr	Temper- ature, °C	Wt% adsorbed*
O_2	3.46	100	−183	5
O_2		750	−183	6
Cyclohexane	6.0	90	24	4
H_2O	2.65	4.3	24	6
H_2O		20	24	8

*Typical amount adsorbed.

FeMnAPO-11 (AEL)
(EP 158,977(1985))
Union Carbide Corporation

Related Materials: AlPO₄-11

Structure

Chemical Composition: See XAPO.

X-Ray Powder Diffraction Data: $(d(\text{Å})\ (I/I_0))$
9.51–9.17(m–s), 4.40–4.31(m–s), 4.25–4.17(s–vs), 4.04–3.95(m–s), 3.95–3.92(m–s), 3.87–3.80(m–vs).

FeMnAPO-11 exhibits properties characteristic of a medium-pore molecular sieve. See AlPO₄-11 for a detailed description of the topology.

Synthesis

ORGANIC ADDITIVES
di–n–propylamine

FeMnAPO-11 crystallizes from a reactive gel with a molar oxide composition of: (1.0–2.0) DPA : (0.05–0.2) Fe₂O₃ : (0.1–0.4) MnO : (0.5–1.0) Al₂O₃ : (0.5–1.0) P₂O₅ : (40–100) H₂O.

FeMnAPO-44 (CHA)
(EP 158,977(1985))
Union Carbide Corporation

Related Materials: SAPO-44
chabazite

Structure

Chemical Composition: See XAPO.

X-Ray Powder Diffraction Data: $(d(\text{Å})\ (I/I_0))$
9.41–9.26(vs), 6.81–6.76(w–m), 5.54–5.47(w–m), 4.31–4.26(s–vs), 3.66–3.65(w–vs), 2.912–2.889(w–s).

FeMnAPO-44 exhibits properties characteristic of a small-pore molecular sieve.

Synthesis

ORGANIC ADDITIVES
cyclohexylamine

FeMnAPO-44 crystallizes from a reactive gel composition with a molar oxide ratio of: (1.0–2.0) R : (0.05–0.2) M₂O_q : (0.5–1.0) Al₂O₃ : (0.5–1) P₂O₅ : (40–100) H₂O (R represents the organic additive, and q denote the oxidation state of the metal (M)).

Adsorption

Adsorption properties of FeMnAPO-44:

Adsorbate	Kinetic diameter Å	Pressure, Torr	Temperature, °C	Wt% adsorbed*
O_2	3.46	100	−183	13
O_2		750	−183	16
n-Hexane	4.3	100	24	2
H_2O	2.65	4.3	24	15
H_2O		20	24	17

*Typical amount adsorbed.

Ferrierite (FER)
(natural and synthetic)

Related Materials: FU-9
ISI-6
Nu-23
Sr-D
ZSM-35

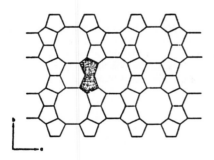

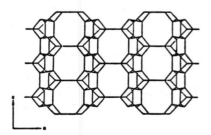

Ferrierite Fig. 1S: Framework topology.

Natural ferrierite first was identified in British Columbia and was named in 1918 in honor of W. F. Ferrier, the first person to discover this zeolite phase (*Nat. Zeo.* (1985)). The mineral ferrierite is rare and is one of the most siliceous naturally occurring zeolites.

Structure

(*Nat. Zeo.* (1985); *Z. Kristallogr.* 169:201(1984); *Zeo.* 7:442(1987); *Zeo.* 9:152(1989))

UNIT CELL CONTENTS: $(Na,K)Mg_2Ca_{0.5}$ $(Al_6Si_{30}O_{72})*20H_2O$
SYMMETRY: orthorhombic
SPACE GROUP: Immm
UNIT CELL CONSTANTS: (Å) (natural) $a = 19.18$
$b = 14.14$
$c = 7.50$
(TMA) $a = 18.88$
$b = 14.11$
$c = 7.45$
FRAMEWORK DENSITY (g/cc):   1.76
VOID FRACTION (determined from water content): 0.28
PORE STRUCTURE: 10-member ring intersecting 8-member ring 4.2 × 5.4 Å and 3.5 × 4.8 Å

X-Ray Powder Diffraction Data: (*Nat. Zoo.* 339 (1985)) $(d(Å)(I/I_o))$ 11.38(3), 9.60(100), 6.98(5), 6.63(3), 5.84(18), 4.97(2), 4.80(5), 4.58(2), 4.01(21), 3.974(38), 3.888(14), 3.797(20), 3.708(31), 3.562(14), 3.535(26), 3.493(22), 3.416(8), 3.318(7), 3.310(6), 3.199(9), 3.151(10), 3.130(4), 3.076(12), 2.977(13), 2.955(4), 2.901(4), 2.846(1), 2.726(5), 2.696(6), 2.643(3), 2.581(4), 2.572(3), 2.432(4), 2.372(9), 2.319(1), 2.255(2), 2.212(2), 2.190(2), 2.115(4), 2.109(3), 2.051(3), 2.021(3), 2.006(3), 1.969(4).

The aluminosilicate framework consists of chains of five-member rings that are parallel to the orthorhombic c-axis and are crosslinked by four-member rings. The main 10-member ring channels are along the [001] with the intersect-

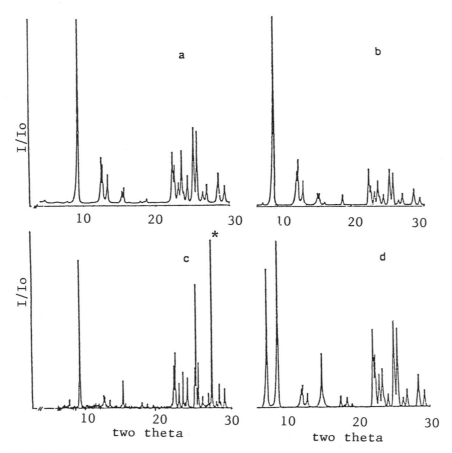

Ferrierite Fig. 2:   Observed and calculated X-ray diffraction profiles. (a) Observed pattern of as-synthesized ferrierite (*Zeo.* 9:152(1989)); (b) calculated pattern based on the ferrierite framework alone; (c) observed pattern of the natural ferrierite (line marked with asterisk due to NaCl); (d) calculated pattern for natural ferrierite based on the structure refinement by Vaughan (*Acta Crystallogr.* 21:983 (1966)). (Reproduced with permission of Butterworths Publishers)

ing eight-rings along the [010]. There are four unique T atom sites in this structure. The 5–1 secondary building unit is characteristic of the mordenite family of zeolites. Faults are observed on (010) and {110} (*Zeo.* 5:81(1985); *Am. Mineral.* 71:989(1986); *Z. Kristallogr.* 169:201(1984)). The (110) reflection at 11.38 is very weak in the natural and synthetic ferrierites, being significantly more intense for the calculated pattern for natural ferrierite. In the natural hydrated zeolite, the Mg^{+2} cations are completely coordinated to the adsorbed water, with a regular coordination sphere of 6 H_2O.

Synthesis

ORGANIC ADDITIVES *Zeo.*
TMA
pyridine
piperidine
C_3 diamines
C_4 diamines
ethylene diamine
choline
pyrrolidine
2,4-pentanedione
N-methylpyridinium[+]
alkyl piperidine

Unit cell dimensions and silica contents of different orthorhombic ferrierites (*Zeo.* 7:442(1987); *Zeo.* 9:152(1989)):

Ferrierite phase	SiO_2, wt%	a, Å	b, Å	c, Å
Silver Mt., CA	65	19.224(3)	14.142	7.510(2)
Francois Lake, BC	66.1	19.218(5)	14.153(5)	7.509(3)
Pinaus Lake, BC	67.8	19.205(4)	14.151(5)	7.496(3)
Rodope Mt., Bulgaria	70.0	19.161(4)	14.156(5)	7.500(3)
Kamloops Lake, BC	71.4	19.156(5)	14.127(3)	7.489(3)
Santa Monica Mts., CA	73.1	18.973(7)	14.140(6)	7.248(9)
TMA	—	19.173(7)	14.129(2)	7.485(1)
Calcined	—	18.863(4)	14.102(1)	7.449(1)
H form	—	18.869(2)	14.135(2)	7.459(1)
H (after 1215K)	—	18.765(4)	14.089(3)	7.430(1)
Al-free ferrierite	86.2	18.557(6)	13.889(3)	7.249(9)

*This is the theoretical value calculated from the unit cell contents. The experimental value is 85.6%.

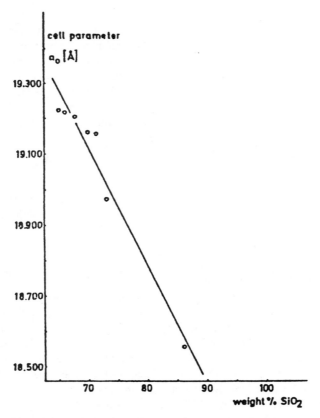

Ferrierite Fig. 3:   Variation of unit cell axis *a* of ferrierites with their silica contents (*Zeo.* 7:442(1987)). (Reproduced with permission of Butterworths Publishers)

Atomic coordinates for natural ferrierite (Z. *Kristallogr.* 169:201(1984)):

Atom	PP	x	y	z
T(1)	*	0.1552(2)	0	0
T(2)		0.0843(2)	0.2931(3)	0
T(3)		0.2718(2)	0	0.2920(6)
T(4)		0.3229(1)	0.2024(2)	0.2072(3)
O(1)		0	0.216(1)	0
O(2)		0.249(1)	0	0.500
O(3)		0.1020(6)	0.0901	0
O(4)		0.2021(7)	0	0.179(2)
O(5)		0.2500	0.2500	0.2500
O(6)		0.3428(8)	0.220(1)	0
O(7)		0.1157(4)	0.2501(8)	0.182(1)
O(8)		0.3211(5)	0.0914(6)	0.248(1)
Mg		0	0	0.500
I	1.00	0	0	0.233(2)
II	0.40	0.403(1)	0.429(3)	0
III	0.67	0.450(3)	0.366(3)	0
IV	1.06	0.4286(6)	0	0
V	0.32	0.440(3)	0	0.23(1)
VI	0.76	0.500	0.1216(7)	0.20(1)
VII	0.5	0.38(1)	0.5000	0

*Population parameters are given for the nonframework atoms I–VII only and are referred to neutral oxygen. Estimated standard deviations are 0.05.

Early crystallization studies produced ferrierite from a reaction mixture containing strontium cations (*Zeo.* 1:130(1981)). Aluminum-free ferrierite has been prepared (*Zeo.* 7:442(1987)) by using H_2BO_3 and ethylene diamine after 8 weeks of crystallization at 180°C. Both the boric acid and the amine are found in the structure after crystallization. This crystallization gives a platy morphology. The ferrierite structure has been claimed to be prepared in the presence of fluoride ion (EP 269,503(1988)). The major difference between ferrierite and the other synthetic materials such as ZSM-35, ZSM-38, and FU-9 lies in the intensity of the (110) reflection.

Ferrierite can be synthesized in the presence of TMA from the gel: 0.62 (TMA) OH : 1.5 Na_2O : 1.0 Al_2O_3 : 15.2 SiO_2 : 300 H_2O. Crystallization occurs at 250°C for 8 hours (*Zeo.* 9:152(1989)). Crystallization under less alkali conditions occurs from gels containing (0.06–0.14) (alkali)$_2$O : SiO_2 : (0.028–0.06) Al_2O_3 : (0.05–1) organic : (5–500) H_2O, where the or-

ganic is either pyridine or piperidine (JP 87,153,115(1987)).

Natural ferrierite has a tendency to grow in platelet-shaped crystallites as well as needles. The synthetic materials produced in the presence of TMA form ellipsoidal aggregates (*Zeo.* 9:152 (1989)). Structural decomposition is observed for the hydrogen exchanged form at temperatures above 1300K, characteristic of high silica zeolites.

Thermal Properties

For the TMA-prepared materials, the loss in weight initially reflects the loss of water from the pores. Loss of 5.3 H_2O per formula unit was observed between 300 and 700K. The TMA decomposes at around 770K (*Zeo.* 9:152(1989)). For the aluminum-free ferrierite, there are three distinct regions of weight loss. About 2% initial loss of the physically adsorbed water is followed by a 9% further decrease between 350 and 800°C, attributable to the breakdown of the organic and occluded boric acid. Between 800 and 900°C, a further loss is observed due to further decomposition of the boric acid. Decomposition of the structure occurs at 1000°C with conversion to cristobalite (*Zeo.* 7:442(1987)). See Ferrierite Figs. 4 and 5.

Infrared Spectrum

Mid-Infrared Vibrations (cm^{-1}): (*Zeo.* 4:369 (1984)) 1218(sh), 1060(s), 780(w), 695(w), 563(w), 445(s).

Ion Exchange

Ammonium and sodium ion-exchanged natural ferrierite has been examined for efficacy in exchanging lead ions from solution. The exchange is not complete. Starting with the sodium form, the exchange is not reversible; but it is reversible starting from the NH_4^+ form. This behavior is attributed to the presence of impurity ions in this natural materials (*Zeo.* 7:153(1987)). Ferrierite is highly selective for cesium and strontium, and this selectivity has made it desirable for applications in radioactive waste treatment (*Zeo.* 4:215(1984)).

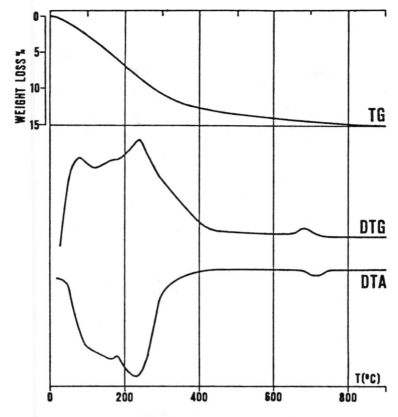

Ferrierite Fig. 4: Thermal curves of ferrierite from Albero Bassi, Italy; in air, heating rate 20°C/min (*Nat. Zeo*. 244 (1985)). (Reproduced with permission of Springer Verlag Publishers)

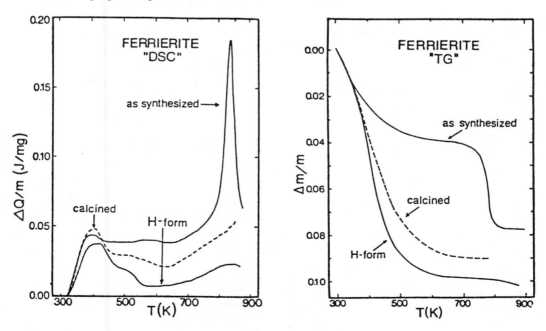

Ferrierite Fig. 5: Observed d.s.c.-signal (left) and weight loss (right) upon heating for as-synthesized, calcined, and H-form ferrierite (*Zeo*. 9:152(1989)). (Reproduced with permission of Butterworths Publishers)

Adsorption

Adsorption equilibrium data for natural ferrierite from Kamloops, BC (*Zeo. Mol. Sieve.* 625(1974)):

Adsorbate	T, K	P, Torr	g/g adsorbed
O_2	90	700	0.105
C_2H_4	298	750	0.041
n-C_4H_{10}	298	NA	
O_2	90	700	0.064*
C_2H_4	298	700	0.025*

*Sedimentary ferrierite from Nevada.

Adsorption

Adsorption properties of FeTiCoAPO-34:

Adsorbate	Kinetic diameter Å	Pressure, Torr	Temperature, °C	Wt% adsorbed
O_2	3.46	100	−183	13
O_2		750	−183	18
n-Hexane	4.3	100	24	6
H_2O	2.65	4.3	24	15
H_2O		20	24	21

FeTiCoAPO-34 (CHA)
(EP 158,977(1985))
Union Carbide Corporation

Related Materials: chabazite
                    $AlPO_4$-34

Structure

Chemical Composition See XAPO

X-Ray Powder Diffraction Data: $(d(\text{Å})\ (I/I_0))$
9.41–9.17(s–vs),   5.57–5.47(vw–w),   4.97–4.82(w–s),   4.37–4.25(m–vs),   3.57–3.51(vw–s),   2.95–2.90(w–s).

FeTiCoAPO-34 exhibits properties characteristic of a small-pore molecular sieve. For a detailed description of the framework topology, see $AlPO_4$-34.

Synthesis

ORGANIC ADDITIVES
tetraethylammonium$^+$

FeTiCoAPO-34 crystallizes from a reactive gel composition with a molar oxide ratio of: 1.0 $R : 0.4\ M_2O_q : 0.8\ Al_2O_3 : P_2O_5 : 0.8\ CH_3COOH : 50\ H_2O$ (R represents the organic additive, and q denotes the oxidation state of the metal (M)).

Flokite (MOR)

Obsolete name for mordenite.

Foresite (STI)

Obsolete name for stilbite.

Franzinite (CAN)

Franzinite is a cancrinite-related salt-bearing tectosilicate (*Hydrothermal Chem. of Zeolites,* 307(1982))

FU-1
(US4,300,913(1981))
Imperial Chemical Industries

Structure

Chemical Composition: 0.6–$1.4\ R_2O : Al_2O_3 :$ greater than $5\ SiO_2 : 0$–$40\ H_2O$.

X-Ray Powder Diffraction Data: $(d(\text{Å})\ (I/I_0))$
9.51(31), 8.35(8), 6.92(28), 6.61(9), 6.26(9), 5.25(16), 4.61(63), 4.48(6), 4.35(13), 4.07(19), 4.00(9.4),   3.89(13),   3.73(23),   2.68(3), 3.44(100).

Synthesis

ORGANIC ADDITIVE
TMA$^+$

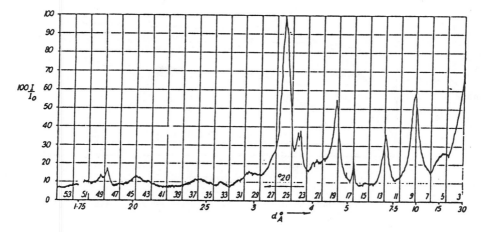

FU-1 Fig. 1:   X-ray powder diffraction pattern for FU-1 (US4,300,013(1981)).

FU-1 crystallizes from a reactive aluminosilicate gel in the presence of TMA cations. Crystallization occurs over a range of SiO_2/Al_2O_3 between 10 and 200 and an OH/SiO_2 between 0.4 and 0.8. Crystallization occurs readily when the organic amine has a concentration between 0.2 and 0.9 (organic/organic + inorganic cation). Sodium is the preferred inorganic cation. Temperatures that induce crystallization of this structure are 175 and 200°C after 24 hours. Common impurities include TMA-sodalite, ZSM-4, TMA analcime, and Nu-1. Nu-1 can be generated in greater proportion in this system if it is seeded with Nu-1 crystals. Cristobalite is also a noteworthy impurity. Stirring of the crystallization mixture has been employed.

Infrared Spectrum

Mid-infrared Vibrations for FU-1 (cm^{-1}):
(1230–1240(m), 1055–1065(s), 950(w), 780–790(w), 590(w), 430–440(m–s).

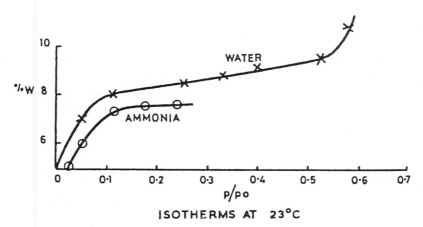

FU-1 Fig. 2:   Full isotherms for ammonia and water at 23°C for HFU-1 (US4,300,013(1981)).

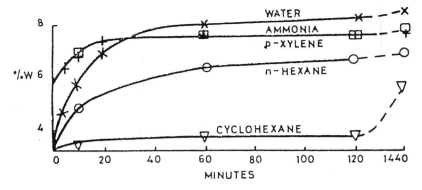

FU-1 Fig. 3:   Adsorption rates of HFU-1 for cyclohexane, *n*-hexane, *p*-xylene, ammonia, and water (US4,300,013(1981)).

Adsorption

Adsorption properties of FU-1 (Ger. 2,748,276(1978)):

Adsorbate	Diameter, Å	Wt% adsorbed	Apparent void filled (cc/100 g)
Triethylbenzene	8	0.2	0.28
Cyclohexane	6	3.7 (10 min)	4.7
		5.7 (17 hr)	7.2
p-Xylene	5.85	7.7	8.9
n-Hexane	4.3	6.8	10.4
Methanol	3.8	6.0	10.0
Water	2.65	8.5	8.5
Ammonia	2.6	7.7	10.0

FU-9 (FER)
(EP 58,529(1982))
Imperial Chemical Industries

Related Materials: ferrierite

Structure

CHEMICAL COMPOSITION: $(0.5–1.5) R_2O : Al_2O_3 :$ $(15–30) SiO_3 : (0–500) H_2O$
SYMMETRY: orthorhombic
UNIT CELL CONSTANTS (Å): $a = 19.0$
$b = 14.1$
$c = 7.5$

X-Ray Powder Diffraction Data: $(d(Å)) (I/I_0))$
11.3(7), 9.5(100), 7.05(21), 6.99(22), 6.61(19), 5.77(13), 5.67(3), 4.97(8), 4.84(1), 4.75(2), 4.57(2), 3.99(52), 3.94(37), 3.85(18), 3.78(32), 3.66(14), 3.56(40), 3.53(55), 3.49(52), 3.38(10), 3.31(15), 3.14(21), 3.05(8), 2.950(9), 2.898(5), 2.713(3), 2.643(4), 2.617(1), 2.575(3), 2.545(1), 2.477(3), 2.414(2), 2.347(3), 2.308(1), 2.260(1), 2.150(2), 2.109(2), 2.207(2), 1.998(6).

The structure of zeolite FU-9 is proposed to be that of an unfaulted ferrierite with properties similar to those of Sr-D.

Synthesis

ORGANIC ADDITIVES
TMA + trialkylamine

FU-9 crystallizes in the presence of TMA and trialkylamine from reactive mixtures with a molar composition of: $4.5 Na_2O : 3 QCl : Al_2O_3 :$ $25 SiO_2 : 900 H_2O$, where Q is 90% TMACl and 10% trimethylamine*HCl. Crystallization occurs after 24 hours at 180°C. At shorter crystallization times, FU-1 is observed, whereas using 100% trimethylamine*HCl causes zeolite ZSM-3 to form. Small-pore ferrierite is obtained when TMA only is used in the crystallization.

Thermal Properties

FU-9 can be heated to 550°C in the presence of wet air and still maintain x-ray crystallinity.

Adsorption

Adsorption capacity of FU-9 with SiO_2/Al_2O_3 of 20 (EP 55,529(1982)):

n-Hexane	Cyclohexane	p-Xylene	m-Xylene
6.2	<0.2	3.3	2.5

Voidage, in cc/100 g, for FU-9 at 25°C:

Water	Methanol	n-Hexane	Cyclo-hexane	p-Xylene	m-Xylene
9.1	13.8	9.4	<0.3	3.8	2.9

FZ-1 (MFI)
(EP B31,255(1981))

See ZSM-5.

G

G

See K-G.

GaAPO-5 (AFI)
(EP 158,976(1985))
Union Carbide Corporation

Related Materials: AlPO$_4$-5

Structure

Chemical Composition: See FCAPO.

X-Ray Powder Diffraction Data: (d(Å) (I/I_0))
12.1–11.56(m–vs), 4.55–4.46(m–s), 4.25–4.17 (m–vs), 4.00–3.93(w–vs), 3.47–3.40(w–m).

GaAPO-5 exhibits properties characteristic of a large-pore molecular sieve. See AlPO$_4$-5 for a description of the framework topology.

Synthesis

ORGANIC ADDITIVE
tripropylamine

GaAPO-5 crystallizes from a reactive gel with a batch composition of: (1.0–2.0) R : (0.05–0.2) Ga$_2$O$_3$: (0.5–1.0) Al$_2$O$_3$: P$_2$O$_5$: (40–100) H$_2$O.

Adsorption

Adsorption properties of GaAPO-44:

Adsorbate	Kinetic diameter Å	Pressure, Torr	Temperature, °C	Wt% adsorbed*
O$_2$	3.46	100	−183	13
O$_2$		750	−183	16
n-Hexane	4.3	100	24	2
H$_2$O	2.65	4.3	24	15
H$_2$O		20	24	17

*Typical amount adsorbed.

GaAPO-44 (CHA)
(EP 158,976(1985))
Union Carbide Corporation

Related Materials: SAPO-44
chabazite

Structure

Chemical Composition: See FCAPO.

X-Ray Powder Diffraction Data: (d(Å) (I/I_0))
9.41–9.26(vs), 6.81–6.76(w–m), 5.54–5.47(w–m), 4.31–4.26(s–vs), 3.66–3.65(w–vs), 2.912–2.889(w–s).

GaAPO-44 exhibits properties characteristic of a small-pore molecular sieve.

Synthesis

ORGANIC ADDITIVE
cyclohexylamine

GaAPO-44 crystallizes from a reactive gel with a batch composition of: (1.0–2.0) R : (0.05–0.2) Ga$_2$O$_3$: (0.5–1.0) Al$_2$O$_3$: (0.5–1.0) P$_2$O$_5$: (40–100) H$_2$O (R represents the organic additive).

Adsorption

Adsorption properties of GaAPO-44:

Adsorbate	Kinetic diameter Å	Pressure, Torr	Temperature, °C	Wt% adsorbed*
O$_2$	3.46	100	−183	13
O$_2$		750	−183	16
n-Hexane	4.3	100	24	2
H$_2$O	2.65	4.3	24	15
H$_2$O		20	24	17

*Typical amount adsorbed.

GaAsO$_4$-2
(RR 8th IZC 47(1989))

Structure

CHEMICAL COMPOSITION: Ga$_{24}$Al$_{24}$O$_{104}$F$_8$*16DMA
SYMMETRY: monoclinic
SPACE GROUP: P2$_1$/n
UNIT CELL CONSTANTS (Å): a = 18.0114
                             b = 10.4664
                             c = 19.0354
                             β = 113.976°
PORE STRUCTURE: 10-member rings with zigzag
  8-member-ring channels

X-Ray Powder Diffraction Data: (d(Å) (I/I_0))
8.75(100), 8.18(42), 6.80(9), 6.28(2), 5.49(2),
5.24(2), 5.03(9), 4.86(3), 4.73(4), 4.48(3),
4.10(8), 3.75(12), 3.59(4), 3.31(10), 3.23(15),
3.14(7), 3.01(8), 2.89(9), 2.78(9), 2.74(4),
2.61(9), 2.44(5), 2.34(4), 2.19(4), 2.02(3),
1.66(2), 1.63(4).

In the GaAsO$_4$-2 structure, all of the As atoms are tetrahedrally coordinated. Of the 24 Ga
atoms, one-third are 6-coordinated and two-thirds

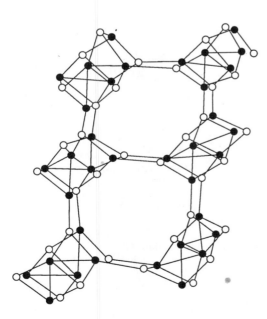

GaAsO$_4$-2 Fig. 1S: Framework topology.

are 5-coordinated. Nonframework oxygen atoms (from water) and F ions participate in the
coordination to the gallium atoms. The dimethylamine is located in the 10-member-ring
channels.

Synthesis

ORGANIC ADDITIVES
dimethylamine (DMA)

GaAsO$_4$-2 is prepared from a mixture of gallium hydroxide and pyroarsenic acid, dimethylamine, and water. The molar ratio of the reactive gel is: 3.6 DMA : 1.0 Ga$_2$O$_3$: 1.2
As$_2$O$_5$: 40 H$_2$O. This gel was aged at 180°C
for 5 days, after which hydrofluoric acid was
added (HF/Ga$_2$O$_3$ = 1.2). Crystallization then
occurred after 10 days at 200°C.

GaGS-1
SSSC 49A:375(1989))

Structure

Chemical Composition: gallium germanium
sulfide

X-Ray Powder Diffraction Data: (d(Å) (I/I_0))
9.63(100), 6.06(33), 4.82(13), 3.91(25), 3.30(6),
3.02(23), 2.39(7), 2.35(5), 2.21(3), 1.91(6),
1.80(6).

See GaGS-1 Fig. 1 on p. 212.

Synthesis

ORGANIC ADDITIVE
tetramethylammonium$^+$ (TMA)

Galaktite (NAT)

Galaktite is an obsolete synonym for natrolite,
found frequently in early papers.

GaPO$_4$-14
(JCS Chem. Commun. 606(1985))

Related Materials: AlPO$_4$-14

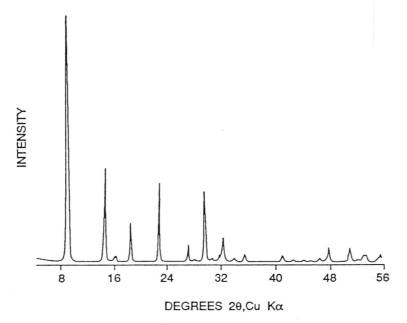

GaGS-1 Fig. 1: X-ray powder diffraction pattern for GaGS-1 (*SSSC* 49A:375 (1989)). (Reproduced with permission of Elsevier Publishing Company)

Structure

CHEMICAL COMPOSITION: (*i*-prNH₃)
($Ga_4(PO_4)_4OH$)*H_2O
SYMMETRY: triclinic
SPACE GROUP: P1
UNIT CELL CONSTANTS (Å): $a = 9.601$
$b = 9.757$
$c = 10.701$

The structure consists of crosslinked GaO_4 and PO_4 tetrahedra that produce voids in which the isopropylamine moiety is located. The organic amine is charged, resulting in a $C_3H_{10}N^+$ unit within the structure. The amine is hydrogen-bonded to the framework via N–H(1) . . . O(10) with a H(1) . . . O(10) distance of 2.32 Å. N–H(2) . . . O(6) has a distance of 1.95 A, and N–H(3) . . . O(17) has a distance of 2.09 Å. Phosphorus in the framework is 4-coordinate; however, the aluminum is 4-, 5-, and 6-coordinate. The building unit of GaPO₄-14 is edge-shared octahedra connected via corners with two trigonal bipyramids of $GaO_4(OH)$.

Selected bond lengths (Å) are: P–O, 1.53; Ga(1)–O, 1.92; Ga(2)–O, 1.82; Ga(3)–O, 1.81; Ga(4)–O, 1.97; Ga(1)–O(9), 2.38(3); Ga(4)–O(9), 2.053(3); Ga(4)–O(9), 2.084(3).

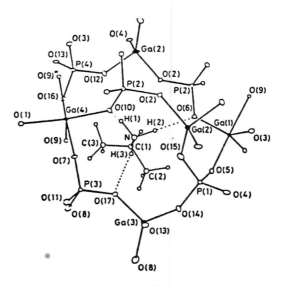

GaPO₄-14 Fig. 1: Atom position.

Synthesis

ORGANIC ADDITIVE
isopropylamine

GaPO$_4$-14 crystallizes from a reactive gel with a batch composition of: 1.0 R : Ga$_2$O$_3$: P$_2$O$_5$: 40 H$_2$O. Crystallization occurs after 142 hours at 200°C.

GaPO$_4$-21
(Acta Crystallogr. C42:144(1986))

Related Materials: AlPO$_4$-21

Structure

CHEMICAL COMPOSITION: Ga$_3$(PO$_4$)$_3$*C$_3$H$_9$N*H$_2$O
SYMMETRY: monoclinic
SPACE GROUP: P2$_1$/n
UNIT CELL CONSTANTS (Å): a = 8.700
b = 18.146(2)
c = 9.087
β = 107.28°
DENSITY: D$_x$ = 2.77 g/cm^3
PORE STRUCTURE: intersecting eight-member rings with Ga–OH–Ga bridges

The structure of GaPO$_4$-21 is essentially the same as that of AlPO$_4$-21. The framework consists of crosslinked corrugated sheets containing edge-shared three- and five-member rings with a 5-coordinated gallium polyhedron and tetrahedral phosphorus. Four-member rings are arranged in a crankshaft manner, consisting of alternating GaO$_4$ and PO$_4$ tetrahedra. The organic amine trapped within the structure is disordered over two sites. Thermal treatment to remove the organic results in a transformation to GaPO$_4$-25.

Unit Cell Parameters for GaPO$_4$-25 (Acta Crystallogr. C42:144(1986)):

SYMMETRY: orthorhombic
UNIT CELL CONSTANTS (Å): a = 8.40
b = 18.5
c = 15.0

Atomic positions ($\times$ 10^4) for GaPO$_4$-21 (*Acta Crystallogr.* C42:144(1986)):

Atom	x	y	z
Ga(1)	818(1)	2952(1)	2911(1)
Ga(2)	1364(1)	1079(1)	1610(1)
Ga(3)	5291(1)	3209(1)	1966(1)
P(1)	3502(2)	4268(1)	3596(2)
P(2)	3148(2)	2070(1)	−180(2)
P(3)	2670(2)	1694(1)	4965(2)
O(1)	395(6)	2050(2)	1800(6)
O(2)	1983(5)	3812(2)	3019(6)
O(3)	2606(5)	2500(2)	4425(5)
O(4)	−356(6)	3060(3)	4323(6)
O(5)	−778(6)	3439(2)	1242(6)
O(6)	4712(6)	4045(2)	2732(6)
O(7)	3473(6)	2680(2)	1048(6)
O(8)	6197(6)	3483(2)	500(6)
O(9)	6721(6)	2665(3)	3460(6)
O(11)	2627(6)	1151(2)	3664(5)
O(12)	−714(6)	856(2)	307(6)
O(13)	3015(5)	5064(2)	3235(6)
O(14)	2716(6)	1352(2)	467(6)
H(1)	−364(98)	2131(55)	963(97)
C(1)	1381(54)	4557(32)	8175(51)
C(2)	993(55)	4948(19)	6546(27)
C(3)	3204(28)	4587(18)	9035(31)
N	291(21)	4943(11)	9046(21)
C(1*)	1459(37)	4431(15)	7714(34)
C(2*)	1740(44)	5160(13)	7069(38)
C(3*)	2026(71)	4287(22)	9420(32)
N*	1872(23)	3753(9)	6852(22)

Second molecule of isopropylamine designated C(1), C(2*), etc.; each molecule has occupancy 0.5 fixed during refinement.

Synthesis

ORGANIC ADDITIVE
isopropylamine (ip$_4$NH$_2$)

GaPO$_4$-21 crystallizes from reactive gels with a composition of: 1.0 ip$_4$NH$_2$: Ga$_2$O$_3$: P$_2$O$_5$: 40 H$_2$O. Crystallization occurs at 200°C after 86 hours. Upon thermal treatment to 600°C, after 48 hours GaPO$_4$-21 converts to GaPO$_4$-25, which is isostructural with AlPO$_4$-25.

GaPO$_4$-25 (ATV)
(Acta Crystallogr. C42:144(1986))

See GaPO$_4$-21 and AlPO$_4$-25.

Garronite (GIS)
(natural)

Related Materials: gismondine

Garronite first was described by Walker in 1962. Initially it was thought to be similar to zeolite P. Further study confirmed its similarity to gismondine (*Nat. Zeo.* 122 (1985)). Garronite has been found along with deposits of phillipsite, chabazite, thomsonite, analcime, levyne, and natrolite and sometimes with gmelinite and mesolite.

Structure

(*Nat. Zeo.* 122 (1985))

CHEMICAL COMPOSITION: $NaCa_{2.5}$ $(Al_6Si_{10}O_{32})$
  *$13H_2O$
SPACE GROUP: $I4_1/amd(?)$
UNIT CELL CONSTANTS (Å): $a = 9.85$
                              $c = 10.32$
PORE STRUCTURE: intersecting eight-member rings
  3.1 × 4.5 Å and 2.8 × 4.8 Å

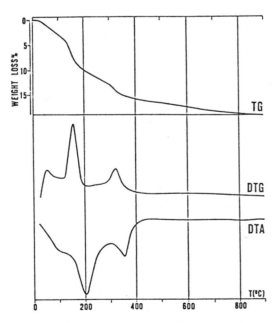

Garronite Fig. 1:   Thermal curves for garronite from Goble, Oregon; in air, heating rate 20°C/min (*Nat. Zeo.* 130(1985)). (Reproduced with permission of Springer Verlag Publishers)

X-Ray Powder Diffraction Data: (d(Å) (I/I_0))
7.15(75), 4.95(59), 4.15(78), 4.07(27), 3.576(2), 3.499(4), 3.244(50), 3.144(100), 2.992(1), 2.895(4), 2.744(3), 2.710(10), 2.674(72), 2.573(11), 2.415(2), 2.337(7), 2.212(4), 2.124(4), 2.072(7), 2.031(3), 2.015(3), 1.987(7)

Garronite is more silica-rich than gismondine, with a disordered (Si,Al) distribution. Garronite is also high in sodium. The members of the gismondine group exhibit variable channel dimensions due to considerable flexibility of the framework.

Thermal Properties

Garronite shows three water loss peaks, occurring at 60, 170, and 320°C. See Garronite Fig. 1.

GeAPO
(US 4,888,167(1989))

GeAPO-5 (AFI)
(EP 158,976(1985))
Union Carbide Corporation

Related Materials: $AlPO_4$-5

Structure

Chemical Composition: See FCAPO.

X-Ray Powder Diffraction Data: (d(Å) (I/I_0))
12.1–11.56 (m–vs), 4.55–4.46(m–s), 4.25–4.17(m–vs), 4.00–3.93(w–vs), 3.47–3.40(w–m).

GeAPO-5 exhibits properties characteristic of a large-pore molecular sieve. See $AlPO_4$-5 for a description of the framework topology.

Synthesis

ORGANIC ADDITIVE
tripropylamine

GeAPO-5 crystallizes from a reactive gel with a batch composition of: (1.0–2.0) R : (0.05–0.2) GeO_2 : (0.5–1.0) Al_2O_3 : (0.5–1.0)

Adsorption

Adsorption properties of GeAPO-5:

Adsorbate	Kinetic diameter Å	Pressure, Torr	Temperature, °C	Wt% adsorbed*
O_2	3.46	100	−183	7
O_2		750	−183	10
Neopentane	6.3	700	24	4
H_2O	2.65	4.3	24	4
H_2O		20	24	12

*Typical amount adsorbed.

Adsorption

Adsorption properties of GeAPO-34:

Adsorbate	Kinetic diameter Å	Pressure, Torr	Temperature, °C	Wt% adsorbed*
O_2	3.46	100	−183	13
O_2		750	−183	18
n-Hexane	4.2	100	24	6
H_2O	2.65	4.3	24	15
H_2O		20	24	21

*Typical amount adsorbed.

P_2O_5 : (40–100) H_2O (R represents the organic additive.

GeAPO-34 (CHA)
(EP 158,976(1985))
Union Carbide Corporation

Related Materials: $AlPO_4$-34
chabazite

Structure

Chemical Composition: See FCAPO.

X-Ray Powder Diffraction Data: (d(Å) (I/I_0))
9.41–9.17(s–vs), 5.57–5.47(vw–m), 4.97–4.82(w–s), 4.37–4.25(m–vs), 3.57–3.51(vw–s), 2.95–2.90(w–s).

GeAPO-34 exhibits properties characteristic of a small-pore molecular sieve similar to chabazite.

Synthesis

ORGANIC ADDITIVE
tetraethylammonium$^+$

GeAPO-34 crystallizes from a reactive gel with a batch composition of: (1.0–2.0) R : (0.05–0.2) GeO$_2$: (0.5–1.0) Al$_2$O$_3$: (0.5–1.0) P$_2$O$_5$: (40–100) H$_2$O.

GeBAPO-44 (CHA)
(EP 158,976(1985))
Union Carbide Corporation

Related Materials: chabazite

Structure

Chemical Composition: See FCAPO.

X-Ray Powder Diffraction Data: (d(Å) I/I_0))
9.41–9.26(vs), 6.81–6.76(w–m), 5.54–5.47(w–m), 4.31–4.26(s–vs), 3.66–3.65(w–vs), 2.912–2.889(w–s).

The framework topology of GeBAPO-44 is similar to that of SAPO-44 and is considered to be chabazite-like.

Synthesis

ORGANIC ADDITIVES
cyclohexylamine

GeBAPO-44 crystallizes from a reactive gel containing the molar oxide ratio: 1.0–2.0 R : 0.05–0.2 M$_2$O$_q$: 0.5–1.0 Al$_2$O$_3$: 0.5–1.0 P$_2$O$_5$: 40–100 H$_2$O (R represents the organic additive, and q the oxidation state of the gemanium and boron).

Adsorption

Adsorption properties of GeBAPO-44:

Adsorbate	Kinetic diameter Å	Pressure, Torr	Temper- ature, °C	Wt% adsorbed*
O_2	3.46	100	−183	13
O_2		750	−183	16
n-Hexane	4.3	100	24	2
H_2O	2.65	4.3	24	15
H_2O		20	24	17

*Typical amount adsorbed.

Genthelvite (SOD)
(Feldspars and Feldspathoids (Brown, W. L., ed., Reidel, Dordrecht, Netherlands) 480(1984))

See sodalite.

Gismondine (GIS)
(natural)

Related Materials: amicite
garronite
gobbinsite
MAPSO-43
Na-P1
Na-P2
P_c
P_t

Gismondine first was ascribed to a mineral sample found near Rome, Italy in 1817 and was named after its discoverer, Gismondi, in that same year (*Nat. Zeo.* 122(1985)).

Structure

(*Zeo. Mol. Sieve.*149 (1974))

CHEMICAL COMPOSITION: $Ca_4(Al_8Si_8O_{32})*16H_2O$
SYMMETRY: monoclinic
SPACE GROUP: $P2_1/c$
UNIT CELL CONSTANTS (Å): $a = 10.02$
$b = 10.62$
$c = 9.84$
$\beta = 92°25'$

VOID FRACTION (determined from water content): 0.46
FRAMEWORK DENSITY: 1.52 g/cc
PORE STRUCTURE: intersecting eight-member rings 4.5 × 3.1 and 2.8 × 4.8 A

X-Ray Powder Diffraction Data: (*Zeo. Mol. Sieve.*[221] 1974) (d(Å) (I/I_0)) 10.00(3), 7.30(63), 5.94(7), 5.77(15), 5.32(4), 5.00(17), 4.91(52), 4.68(17), 4.46(10), 4.33(6), 4.27(100), 4.21(51), 4.18(34), 4.05(30), 4.02(6), 3.642(8), 3.606(8), 3.587(5), 3.431(16), 3.383(8), 3.338(47), 3.186(90), 3.132(71), 3.065(4), 3.022(5), 2.993(16), 2.955(1), 2.873(4), 2.825(5), 2.782(11), 2.744(76), 2.714(59), 2.693(78), 2.662(69), 2.658(72), 2.624(17), 2.607(11), 2.567(7), 2.521(15), 2.492(7), 2.475(8), 2.467(9), 2.458(8), 2.407(13), 2.389(7),

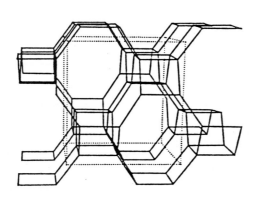

Gismondine Fig. 1S: Framework topology.

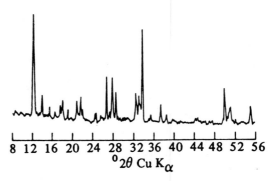

Gismondine Fig. 2:   X-ray powder diffraction pattern for gismondine from Capo Di Bove, Italy (*Zeo. Mol. Sieves* 206(1974)). (Reproduced with permission of John Wiley and Sons, Inc.)

2.340(16),   2.293(4),   2.265(6),   2.242(5),   2.193(15),   2.135(8),   2.102(6),   2.080(6),   2.046(10), 2.033(10), 2.011(12), 1.967(8).

The structure of gismondine was determined for a hydrated mineral sample (*Naturwissenschaften* 45:4888(1958); *Am. Mineral.* 48:664(1962)). It later was shown to be structurally related to the synthetic zeolite P. Single crystal neutron diffraction studies at 15K show that the mean tetrahedral T–O distances are consistent with an essentially ordered distribution of silicon and aluminum with minor substitution of excess silicon in the Al(1) sites (*Zeo.* 6:361(1986)). Full calcium occupancy was observed, and coordination around the calcium involves two framework oxygens and 4.3 water oxygens.

Synthesis

ORGANIC ADDITIVES
tetramethylammonium$^+$ (TMA)

This structure has been crystallized in the presence of Na, (Na,TMA), (Na,Li), (Na,K), (Na,Ba), and (Li,Cs,TMA). The preferred cation appears to be sodium (*Zeo.* 1:130(1981)). For further discussion of the synthetic analogs of gismondine, see gismondine(TMA) and Na-P1.

Atomic coordinates ($\times$ 10^5) and occupancy factors (*Zeo.* 6:361(1986)):

Atom	x/A	y/B	z/C	G
Si(1)	41,701(27)	11,342(27)	18,365(29)	
Si(2)	90,696(27)	87,029(26)	16,244(27)	
Al(1)	9,783(33)	11,445(31)	16,605(35)	
Al(2)	58,866(33)	86,492(33)	14,799(34)	
O(1)	8,594(23)	16,268(21)	99,653(23)	
O(2)	26,410(21)	8,016(21)	21,490(23)	
O(3)	43,831(21)	14,684(22)	2,648(22)	
O(4)	24,580(21)	40,345(21)	30,058(22)	
O(5)	99,985(24)	98,818(21)	20,919(25)	
O(6)	4,313(20)	24,651(21)	25,317(21)	
O(7)	47,039(21)	22,764(20)	28,039(22)	
O(8)	51,143(22)	99,265(20)	22,922(20)	
Ca	71,243(22)	7,174(33)	35,977(30)	
OW1	25,843(26)	10,659(23)	49,653(24)	
H11	20,848(61)	31,767(54)	1,194(60)	
H21	26,257(52)	9,735(54)	39,871(49)	
OW2	59,508(23)	13,096(22)	55,060(24)	
H12	53,278(52)	29,830(49)	3,790(55)	
H22	56,213(49)	41,867(47)	12,553(53)	
OW3	90,921(26)	11,782(25)	49,859(28)	
H13	97,987(321)	17,930(323)	47,756(339)	0.384(0.197)
H23	94,105(60)	42,852(63)	7,771(75)	
H33	96,046(167)	30,710(188)	27(148)	0.616(0.197)
OW4	77,553(35)	23,209(38)	23,491(48)	0.697(0.024)
H14	72,229(219)	28,728(243)	17,445(204)	0.613(0.189)
H24	86,970(117)	25,243(152)	22,989(145)	0.697(0.024)
H34	65,855(1025)	27,923(663)	19,458(756)	0.084(0.189)
OW5	73,447(85)	32,137(81)	41,318(107)	0.303(0.024)
H15	76,015(281)	15,794(206)	480(256)	0.303(0.024)
H25	74,585(206)	39,881(170)	35,918(262)	0.303(0.024)
OW6	77,947(92)	19,590(112)	16,474(123)	0.303(0.024)
H16	73,394(594)	26,797(578)	14,126(617)	0.303(0.024)
H26	87,648(477)	22,894(340)	19,538(424)	0.303(0.024)

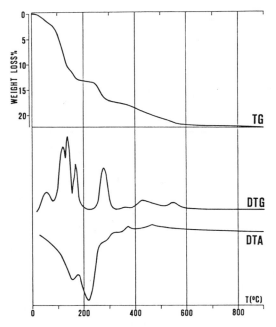

Gismondine Fig. 3:   Thermal curves of gismondine from Montalto di Castro, Italy; in air, heating rate 10°C/min for TG (and DTG), 20°C/min for DTA (*Nat. Zeo.* 130(1985)). (Reproduced with permission of Springer Verlag Publishers)

Thermal Properties

The thermal behavior of gismondine is not straightforward. There are five main dehydra-tion steps between room temperature and 300°C. In addition, there is some minor water loss between 300 and 600°C. This behavior is consistent with the six different water sites in this structure (*Nat. Zeo.* 129(1985)).

Gismondine (TMA) (GIS)
(*Helv. Chim. Acta* 53:1285(1970))

Related Materials: gismondine

Structure

CHEMICAL COMPOSITION: $(CH_3)_4NAlSi_3O_8*H_2O$
SYMMETRY: tetragonal
SPACE GROUP: $I4_1/amd$
UNIT CELL CONSTANTS (Å): $a = 9.84$
                                                      $c = 10.02$
PORE STRUCTURE: intersecting eight-member rings

X-Ray Powder Diffraction Data: $(d(Å)\ (I/I_0))$
7.13(80),   5.216(5),   4.216(35),   4.065(91),
3.695(6),   3.281(100),   3.098(11),   2.775(1),
2.735(24),   2.666(3),   2.614(1),   2.434(1),
2.302(4), 2.198(4), 2.107(3), 2.045(1), 2.031(1),
1.998(12),   1.960(2),   1.904(1),   1.889(1),
1.780(6), 1.758(7), 1.743(6).

The structural framework is that of gismondine. The ratio of Si/Al in this material is higher than that found in the natural mineral.

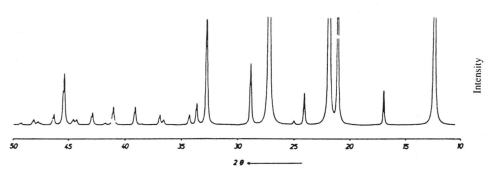

Gismondine (TMA) Fig. 1:   X-ray powder diffraction pattern for TMA-gismondine (*Helv. Chim. Acta* 53:1285(1970)). (Reproduced with permission of Springer Verlag Publishers)

Atomic coordinates for TMA-gismondine (*Helv. Chim. Acta* 53:1285 (1970)):

Atom	Position	x	y	z
T	16g	0.1485(5)	0.6015(5)	1/8
O(1)	16h	0	0.5820(10)	0.1795(10)
O(2)	16f	0.1785(10)	1/2	0
N	4b	0	1/4	3/8
1/4C(1)	32i	0.1485(45)	0.1915(20)	0.3510(40)
14C(2)	32i	0.0780(20)	0.2175(30)	0.2615(20)

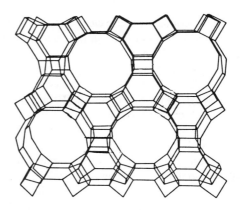

Gmelinite Fig. 1S: Framework topology.

Synthesis

ORGANIC ADDITIVE
tetramethylammonium$^+$ (TMA)

The gismondine structure is prepared in the presence of TMA cations after 8 to 10 days at 130°C. Thermal treatment to remove the organic from the pores of the structure results in the generation of "cubic" Na-P1.

Gismondite (GIS)

Little-used synonym for gismondine.

Giuseppettite (CAN)
(natural)

Related Materials: cancrinite

Giuseppettite is a cancrinite-type phase with unit cell parameters of $a = 12.85$ and $c = 42.22$ Å with an ABABABACBABABABC stacking sequence (*Proc. NATO Study Institute of Feldspars and Feldspathoids*, Rennes, France (1983)).

Gmelinite (GME)
(natural/synthetic)

The mineral gmelinite was first named by Brewster in 1825 to honor the chemist G. C. Gmelin.

Structure

(*Zeo. Mol. Sieve.* 150 (1974))

CHEMICAL COMPOSITION: $Na_8[(AlO_2)_8 (SiO_2)_{16}]$ *$24H_2O$
SYMMETRY: hexagonal
SPACE GROUP: $P6_3/nnc$
UNIT CELL CONSTANTS (Å): a = 13.75; c = 10.05
VOID FRACTION (determined from water content): 0.44
FRAMEWORK DENSITY: 1.46 g/cc
PORE STRUCTURE: 12-ring intersecting 8-ring, 3.6 × 3.9 Å and 7.0

X-Ray Powder Diffraction Pattern: (*Zeo. Mol. Sieve.* 222 (1974)) (d(Å) (I/I_0)) 11.91(100), 7.68(5), 5.12(18), 5.02(53), 4.50(9), 4.11(42), 3.44(7), 3.22(25), 2.98(27), 2.92(8.4), 2.85(16), 2.69(21), 2.60(2.3), 2.56(2.0), 2.40(2.7), 2.29(3.5), 2.08(11), 2.05(3.2), 1.83(3.1), 1.81(3.3), 1.79(3.1), 1.73(2.0), 1.72(5.3), 1.67(2.5), 1.58(2.1).

The gmelinite structure is similar to that of chabazite. In fact, chabazite–gmelinite intergrowths are commonly found in many samples. The gmelinite framework can be produced by joining D6R units through tilted four-member rings. There are two cation sites within the D6R unit. Six monovalent cations can be located near the eight-member rings.

Synthesis

ORGANIC ADDITIVES
tetramethylammonium$^+$ (TMA)
polymeric quaternary amines

The gmelinite structure can be crystallized in the presence of sodium, strontium, and mixtures of calcium and TMA, as well as mixtures of sodium and TMA, with the preferred cation being sodium (*Zeo.* 1:130(1981); *JCS* London 1961: 983(1961)). With SiO_2/Al_2O_3 of 9 and Na_2O/Al_2O_3 of 4, crystalline mixtures of chabazite and gmelinite are produced (*JCS* London 1959: 195(1959)).

A fault-free gmelinite can be prepared by utilizing polymeric quaternary amines in the synthesis. Batch compositions with SiO_2/Al_2O_3 between 10 and 60, Na_2O/R_2O between 0.5 to 0.95, and OH/SiO_2 between 0.3 to 1.3, with H_2O/OH of 15 to 40, will result after 3 to 15 days at 90°C in the crystallization of fault-free gmelinite (US4,061,717(1977)).

Thermal Properties

Water is lost in three steps from the gmelinite structure, an initial shoulder at 100°C and two

peaks at 175 and 300°C, and can be related to the presence of three water sites in the structure.

Apparent activation energies of the dehydration of gmelinite was estimated from the TGA for the temperature range between 17 and 692°C.

Activation energies of the dehydration of gmelinite *Z. Phys. Chem.* 48:257(1966)):

Peak temperature, °C	No. H_2O lost	E_A (kcal/mol)
17	8	3.8
197	8	9.2
314	3	10.7
419	2	34.2
588	1.5	49.6
692	CO_2 (from calcite)	56.3

The potassium exchanged form of gmelinite exhibits distinctly different water desorption/adsorption properties from the parent mineral. Essentially all of the water is given off below 300°C in overlapping steps, with maxima at 80 and 150°C. The material adsorbs very slowly at ambient temperature, and the loosely bound water is released at a relatively low temperature. The adsorption of water causes the gmelinite lattice to expand in the C direction.

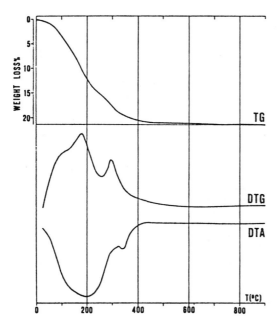

Gmelinite Fig. 2:   Thermal curves of gmelinite from Flinders, Australia; in air, heating rate 20°C/min (*Nat. Zeo.* 173 (1985)). (Reproduced with permission of Springer Verlag Publishers)

Adsorption

Adsorption equilibrium data for gmelinite (Nova Scotia) (*Zeo. Mol. Sieve.* 625:(1974)):

Adsorbate	T, °C	P, Torr	g/g adsorbed
H_2O	25	4	0.23
NH_3	25	700	0.23
CO_2	25	700	0.14
SO_2	25	700	0.39
O_2	−298	85	0.087
N_2	−298	710	0.062
Ar	−298	93	0.085
C_2H_4	25	760	0.037
C_2H_6	25	700	0.030

(a)

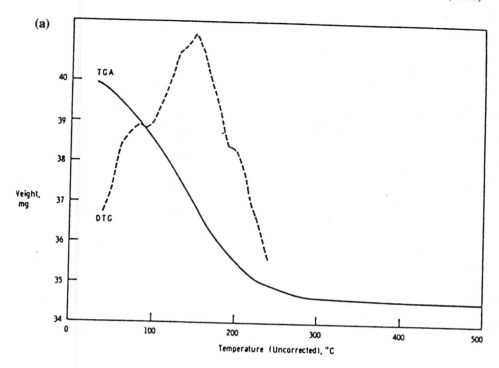

(b)

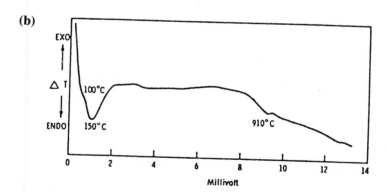

Gmelinite Fig. 3:   (a) Differential thermal analysis of K-gmelinite (13.84% H_2O); heating rate 30°C/min; He atmosphere (b) Thermogravimetric analysis of K-gmelinite (13.84% H_2O); heating rate: 20°C/min; flowing He (*Nat. Zeo.* 421(1976)). (Reproduced with permission of Pergamon Press)

Gobbinsite (GIS)
(natural)

Related Materials: gismondine

In 1982, Nawaz and Malone found a deposit of gobbinsite in the Gobbins, Island Magee, Northern Ireland (*Mineral. Mag.* 46:365(1982)). Based on the X-ray powder diffraction pattern, it was considered to be the sodium equivalent of gismondine. Natural gobbinsite is generally found as aggregates of fibrous crystals sometimes associated with gmelinite.

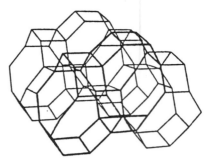

Gobbinsite Fig. 1S: Framework topology.

Structure

(*Nat. Zeo.* 122 (1985))

CHEMICAL COMPOSITION: $Ca_{0.6}Na_{2.6}K_{2.2}$
 $(Al_6Si_{10}O_{32})*12H_2O$
SPACE GROUP: $Pmn2_1$

UNIT CELL CONSTANTS (Å): $a = 10.108(1)$
                            $b = 9.766(1)$
                            $c = 10.171(1)$
PORE STRUCTURE: intersecting eight-member rings

X-Ray Powder Diffraction Data: (*Nat. Zeo.* 326 (1985)) (d(Å) (I/I_0)) 7.11(100b), 5.78(20),

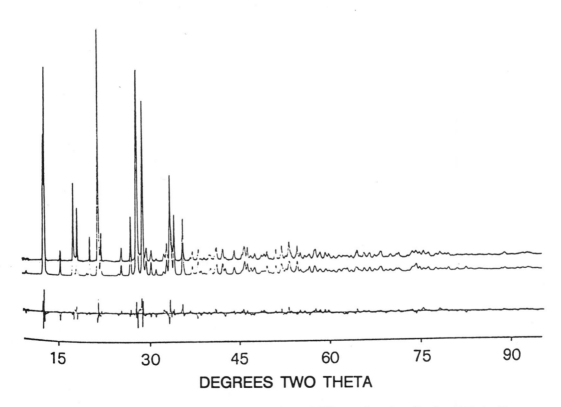

Gobbinsite Fig. 2:   Observed (upper), calculated (middle), and difference (lower) profiles for gobbinsite (*Z. Kristallogr.* 171:281(1985)). (Reproduced with permission of Wiesbaden Akademische Verlagsgesellschaft Publishers)

5.06(50),   4.89(30),   4.41(25),   4.12(100),
4.03(20), 3.515(10), 3.326(30), 3.201(100),
3.106(80), 3.040(10), 2.968(10b), 2.887(5),
2.757(5), 2.699(80b), 2.651(40), 2.539(25),
2.435(20), 2.379(20), 2.317(10), 2.256(15),
2.206(25), 2.153(25), 2.057(20), 1.986(25).

The gobbinsite structure is related to the gis-mondine-type framework, containing two-dimensional intersecting eight-member-ring channels. It also can be considered as a stacking of two-dimensional arrays of double crankshaft chains parallel to [100] and [010], inducing a flexible framework. The sodium ions are located in the nine-rings perpendicular to the a-axis and coordinated to three framework oxygens and two nonframework (water) oxygens. The symmetry is a distorted trigonal bipyramid around the sodium ion. Water also acts as a bridge between the sodiums. The calcium and potassium ions were not located. Because there is no crystallographically required order in these cations, the

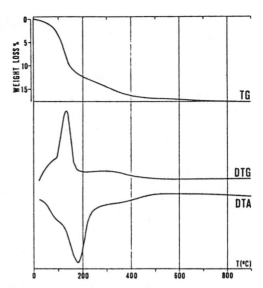

Gobbinsite Fig. 3:   Thermal curves of gobbinsite from Gobbins, Northern Ireland; in air, heating rate 20°C/min (*Nat. Zeo.* 122 (1985)). (Reproduced with permission of Springer Verlag Publishers)

Positional and population parameters for gobbinsite (*Z. Kristallogr.* 171:281(1985)):

Atom	Point symmetry	x	y	z	Population	Multiplicity
T(1)	1	0.156(1)	0.432(2)	−0.191(2)	4	4
T(2)	1	0.154(2)	0.110(2)	−0.242(2)	4	4
T(3)	1	0.345(2)	0.073(2)	0	4	4
T(4)	1	0.348(2)	0.389(2)	0.048(3)	4	4
O(1)	1	0.186(3)	0.275(2)	−0.241(3)	4	4
O(2)	m	0	0.441(5)	−0.162(5)	2	2
O(3)	1	0.200(4)	0.538(3)	−0.309(3)	4	4
O(4)	1	0.254(3)	0.459(3)	−0.061(3)	4	4
O(5)	m	0	0.074(3)	−0.197(5)	2	2
O(6)	1	0.198(4)	0.048(3)	−0.394(3)	4	4
O(7)	1	0.254(3)	0.046(3)	−0.138(2)	4	4
O(8)	1	0.312(3)	0.226(2)	0.059(3)	4	4
O(9)	m	0.500	0.073(5)	−0.026(4)	2	2
O(10)	m	0.500	0.414(5)	0.009(5)	2	2
Na(1)	1	0.247(5)	0.228(6)	0.284(8)	2.6(2)	4
K(1)	m	0.500	−0.071(4)	−0.615(4)	1.7(1)	2
H_2O(1)	m	0.500	0.243(8)	0.321(8)	2.2(1)	2
H_2O(2)	m	0	0.395(5)	0.146(5)	2.9(2)*	2
H_2O(3)	m	0	0.314(!)	0.306(1)	1.8(2)	2
H_2O(4)	1	0.341(6)	0.271(5)	0.559(6)	5.1(1)*	4
H_2O(5)	m	0.500	0.338(6)	0.600(7)	2.9(2)*	2

Numbers in parentheses are the esd's in the units of the least significant digit given. Those parameters without esd's were held fixed in least-squares refinement.

Population parameters are given in the number or atoms or ions per unit call.

*High values are probably due to the presence of K^+ or Ca^{+2} ions in addition to H_2O at these positions.

Na^+ and Ca^{+2} are thought to occupy the eight-rings randomly.

Thermal Properties

Gobbinsite exhibits three water losses and one shoulder in the DTG, which appears at 80°C, a peak at 150°C, and a continuous loss between 200 and 400°C (*Nat. Zeo.* 122 (1985)).

Gonnardite (NAT)
(natural)

Related Materials: natrolite

Gonnardite first was identified by Lacroix in 1896. Significant confusion over the identification of the structure of this mineral has led to some wrongly deduced identifications. It first was thought to be a disordered thomsonite, as well as being related to tetranatrolite.

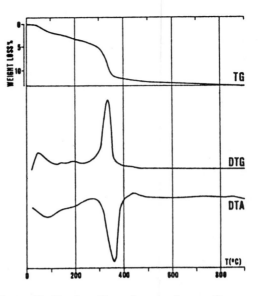

Gonnardite Fig. 1:   Thermal curves of gonnardite from Kloch, Styria, Austria; in air, heating rate 20°C/min (*Nat. Zeo.* 71 (1984)). (Reproduced with permission of Springer Verlag Publishers)

Structure

(*Nat. Zeo.* 71 (1985))

CHEMICAL COMPOSITION: $Na_5Ca_2(Al_9Si_{11}O_{40})$
  $*12H_2O$
SYMMETRY: tetragonal or orthorhombic
UNIT CELL CONSTANTS: (Å) $a = 13.35$
                              $b = 13.35$
                              $c = 6.65$

X-Ray Powder Diffraction Data: (*Nat. Zeo.* 321 (1985)) ($d(Å)$ (I/I_0)) 6.64(76), 5.90(39), 4.70(61), 4.41(37), 4.20(24), 3.675(3), 3.580(4), 3.213(31), 3.109(12), 2.954(27), 2.898(100), 2.608(11), 2.468(17), 2.351(7), 2.306(5), 2.268(5), 2.211(17), 2.178(3), 2.102(3), 2.067(4), 1.981(4).

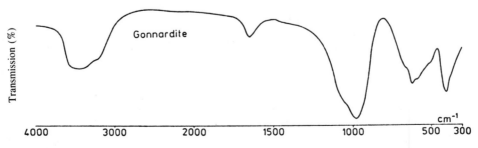

Gonnardite Fig. 2:   Infrared spectrum of gonnardite (*Nat. Zeo.* 349 (1984)). (Reproduced with permission of Springer Verlag Publishers)

Little is known about the exact structure of gonnardite. Based on the X-ray powder diffraction pattern and the infrared spectrum, it is more similar to tetranatrolite than to any other fibrous zeolite.

Thermal Properties

There is one major water loss peak in gonnardite, which occurs at 330°C, with a few minor losses occurring at 50, 130, 200, and 420°C. A lattice contraction is observed for *a* and *b* between the temperatures of 20 and 50°C. Crystallinity is destroyed at 300°C in this mineral.

Infrared Spectrum

See Gonnardite Figure 2.

Goosecreekite (GOO) (natural)

Goosecreekite first was described in 1980 by Dunn, Peacor, Newberry, and Ramik, and was named after the locality in Loudon County, Virginia in which it was found (*Canad. Mineral.* 18:323(1980)).

Structure

(*Am. Mineral. 71*, 1494(1986))

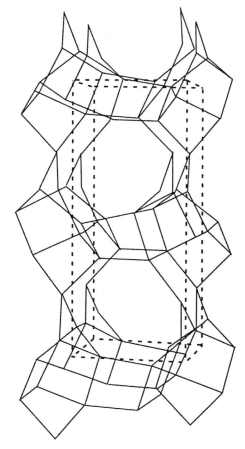

Goosecreekite Fig. 1S: Framework topology.

CHEMICAL COMPOSITION: $CaAl_2Si_6O_{16} \cdot 5H_2O$
SYMMETRY: monoclinic
SPACE GROUP: $P2_1$
UNIT CELL CONSTANTS: (Å) $a = 7.401(3)$
$b = 17.439(6)$
$c = 7.293(3)$
$\beta = 105.44(4)°$
PORE STRUCTURE: intersecting eight-member rings
4.0 × 2.8 × 3.5 and 4.1 × 2.7 × 3.3 Å

This structure consists of three distinct sets of intersecting channels parallel to the *a*, *b*, and *c** axes. The TO_4 tetrahedral form four-, six-, and eight-member rings forming distorted layers parallel to (010). The (010) layers resemble those in brewsterite [$(Sr,Ba)Al_2Si_6O_{16} \cdot 5H_2O$]. These structures differ in the crosslinking to form the three-dimensional framework. T_9O_{18} subunits, which can describe this structure, form vertex-sharing chains in the *b* direction linked in the *c* direction by four-member rings. Some Si–Al ordering is observed (*Am. Mineral.* 71: 1494(1986)).

The channel system and the exchangeable cations in goosecreekite, when compared with brewsterite, differ in that both minerals have channels bounded by eight-rings oriented parallel to the *a* and *c** axes, but in brewsterite there is no channel in the *b* direction. In brewsterite, the larger Sr^{+2} and Ba^{+2} cations have a higher coordination number and bond to five water molecules and four framework oxygens. In goosecreekite, the Ca^{+2} is coordinated to only two framework oxygens.

Atomic coordinates ($\times 10^4$) for goosecreekite (*Am. Mineral*. 71:1494(1986)):

	x	y	z
Ca	6557(3)	2721	1989(3)
Si(1)	3374(3)	4732(2)	6282(3)
Si(2)	3168(3)	1294(2)	5997(3)
Si(3)	0219(3)	2599(2)	5249(3)
Si(4)	1044(3)	0578(2)	8612(3)
Si(5)	0892(3)	3877(2)	8325(3)
Si(6)	2394(3)	0045(2)	2768(3)
Al(1)	7482(3)	1285(2)	5887(4)
Al(2)	0682(4)	3982(2)	2632(4)
O(1)	7228(9)	0612(4)	4105(9)
O(2)	5336(9)	1478(4)	6346(9)
O(3)	8372(8)	2097(4)	5064(9)
O(4)	9065(9)	1001(4)	7936(9)
O(5)	2607(9)	4392(4)	8023(9)
O(6)	2666(9)	4226(4)	4408(9)
O(7)	5594(9)	4662(4)	7062(9)
O(8)	2017(9)	2065(4)	5155(9)
O(9)	2657(9)	1069(4)	7949(9)
O(10)	2515(9)	0596(4)	4581(9)
O(11)	0811(9)	3051(4)	7261(9)
O(12)	9698(9)	3179(4)	3431(9)
O(13)	9052(10)	4718(4)	2174(9)
O(14)	1728(9)	0585(4)	0904(9)

	x	y	z
O(15)	1202(9)	3741(4)	0505(9)
O(16)	9022(9)	4346(4)	7315(9)
H20(1)	3512(12)	2234(6)	0964(14)
H20(2)	6565(15)	2989(6)	8717(13)
H20(3)	5102(11)	3063(5)	4504(10)
H20(4)	7574(13)	1530(5)	1048(11)
H20(5)	5650(10)	4074(5)	1595(13)

Note: Esd's are given in parentheses. The y coordinate of Ca was held constant to fix the origin.

GS-7
(*SSSC* 49A:375(1989))

Structure

Chemical Composition: germanium sulfide

X-Ray Powder Diffraction Data: (d(Å) (I/I_0))
12.37(100), 9.02(8), 8.48(9), 8.06(20), 6.88(7), 5.33(24), 3.86(8), 3.84(7), 3.78(4), 3.25(10), 3.18(7), 2.67(7).

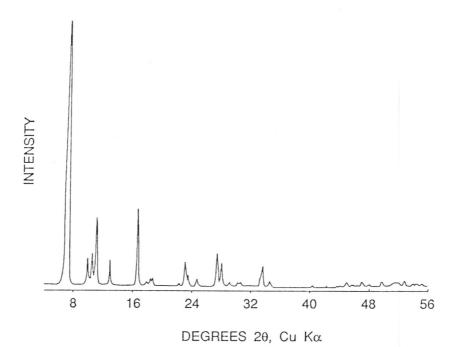

GS-7 Fig. 1: X-ray powder diffraction pattern for GS-7 (*SSSC* 49A: 375(1989)). (Reproduced with permission of Elsevier Publishers)

Synthesis

ORGANIC ADDITIVE
dipropylamine

GS-7 crystallizes from a reactive gel with a composition of: 0.8 DPA : GeS_2 : 15 H_2O. Crystallization occurs after 66 hours at 100°C. The initial pH of the reactive slurry is 7 to 9, with a pH of 0.5 to 3 upon crystal formation.

GS-9
(*SSSC* 49A:375(1989))

Structure

Chemical Composition: germanium sulfide

X-Ray Powder Diffraction Data: (d(Å) (I/I_0))
12.25(61),   11.08(7),   9.04(100),   8.73(18), 5.59(20), 5.50(6), 4.86(19), 4.79(7), 4.37(5), 4.25(5), 3.22(25), 2.93(6).

Synthesis

ORGANIC ADDITIVE
cyclohexylamine

GS-9 crystallizes from a reactive gel with a composition of: 0.8 $RHCO_3$: GeS_2 : 15 H_2O. Crystallization occurs after 162 hours at 100°C. The initial pH of the reactive slurry is 7 to 9, with a pH of 0.5 to 3 upon crystal formation.

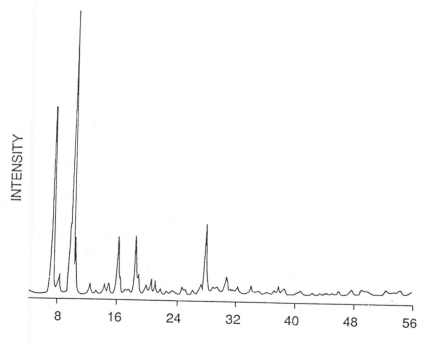

GS-9 Fig 1:   X-ray powder diffraction pattern for GS-9 (*SSSC* 49A: 375(1989)). (Reproduced with permission of Elsevier Publishing Company)

H

H

See Linde H.

H1

See AlPO4-H1.

H2

See AlPO4-H2.

H3

See AlPO4-H3.

H4

See AlPO4-H4.

H5

See variscite and AlPO4-H5.

H6

See AlPO4-H6.

Harmotome (PHI) (natural)

Related Materials: phillipsite

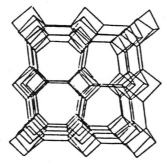

Harmotome Fig. 1S: Framework topology.

Harmotome was one of the early zeolites identified. It was not named until 1801.

Structure

(Nat. Zeo. 134(1985))

CHEMICAL COMPOSITION: $Ba_2(Ca_{0.5},Na)$
$(Al_5Si_{11}O_{32})*12H_2O$
SPACE GROUP: $P2_1/m$
UNIT CELL CONSTANTS (Å): $a = 9.88$
$b = 14.14$
$c = 8.69$
$\beta = 124°49'$
PORE STRUCTURE: intersecting eight-member rings
3.6 Å, 3.0 × 4.3 Å, and 3.2 × 3.3 Å
VOID FRACTION   (determined from water content):
0.31
FRAMEWORK DENSITY: 1.59 g/cc

X-Ray Powder Diffraction Data: (*Nat. Zeo.* 329(1985)) (*d*(Å) (*I/I₀*)) 8.12(60), 7.16(65), 7.05(19), 6.39(95), 5.03(32), 4.31(33), 4.30(35), 4.11(56), 4.07(67), 4.05(59), 3.895(32), 3.667(10), 3.573(5), 3.528(10), 3.466(18), 3.410(5), 3.241(62), 3.195(28), 3.169(70), 3.126(100), 3.075(31), 2.918(29), 2.897(13), 2.847(10), 2.747(27), 2.730(54), 2.697(61), 2.678(54), 2.671(64), 2.628(15), 2.561(9), 2.5229(33), 2.515(20), 2.470(11), 2.464(11), 2.369(18), 2.343(15), 2.320(23), 2.299(17), 2.261(10), 2.241(15), 2.151(18), 2.147(19), 2.127(7), 2.066(17), 2.059(18), 2.053(16), 2.025(10), 2.005(8), 1.964(8).

The distinguishing features between phillipsite and harmotome are the overall symmetry and the chemical composition. Harmotome is consistent in composition with 11 Si, 2 Ba, and (1/2Ca + Na) near 1 in the unit cell. The Ba^{+2} cations coordinate with six of the framework oxygens and four H_2O molecules, and are located at the channel intersections. Eight of the 12 H_2O molecules in the unit cell are coordinated with the Ba^{+2} cations (*Dokl. Akad. Nauk SSR* 188:670(1969)).

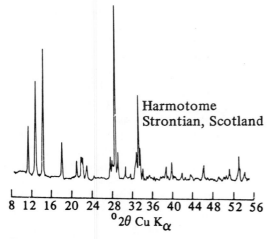

Harmotome Fig. 2:   X-ray powder diffraction pattern for natural harmotome from Strontian, Scotland (*Zeo. Mol Sieve.* 206 (1974)). (Reproduced with permission of John Wiley and Sons, Inc.)

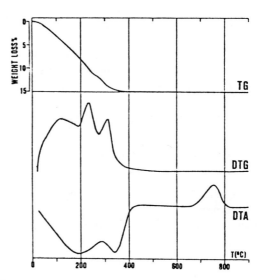

Harmotome Fig. 3:   Thermal curves of harmotome from Andreasberg, Germany; in air, heating rate 20°C/min (*Nat. Zeo.* 145(1985)). (Reproduced with permission of Springer Verlag Publishers)

Positional and population parameters for harmotome (*Acta Crystallogr.* B30:2426(1974)):

Atom	Population	x/a	y/b	z/c
T(1)	4	0.7367(1)	0.0248(1)	0.2840(1)
T(2)	4	0.4214(1)	0.1410(1)	0.0136(1)
T(3)	4	0.0577(1)	0.0075(1)	0.2898(1)
T(4)	4	0.1216(1)	0.1390(1)	0.0375(1)
O(1)	4	0.1042(4)	0.0896(2)	0.1958(4)
O(2)	4	0.6470(3)	0.5726(2)	0.1679(4)
O(3)	4	0.6163(3)	0.1186(2)	0.1792(4)
O(4)	4	0.0050(3)	0.9083(2)	0.1711(4)
O(5)	4	0.9057(3)	0.0515(2)	0.2955(3)
O(6)	4	0.3137(3)	0.3709(2)	0.1017(4)
O(7)	4	0.7808(3)	0.4856(2)	0.4976(4)
O(8)	2	0.5885(5)	3/4	0.0573(5)
O(9)	2	0.0661(5)	1/4	0.0256(6)
Ba	2	0.86290(5)	1/4	0.19441(5)
Ca	0.60(3)	0.5869(20)	0.6286(13)	0.4799(25)
W(1)	2	0.8004(8)	3/4	0.4889(7)
W(2)	2	0.1148(8)	3/4	0.4593(7)
W(3)	4	0.3027(5)	0.8628(3)	0.1324(5)
W(4)	2	0.4611(17)	1/4	0.5134(13)
W(5)	2	1/2	1/2	1/2

Thermal Properties

There are five different water sites contained within the harmotome structure, but water loss is observed in four steps, 50 (shoulder), 120,

230, and 320°C. Complete dehydration is observed at 400°C.

Adsorption

Dehydrated harmotome adsorbs only small polar molecules such as ammonia.

Hauyn (SOD)
(*N. Jb. Min. Abh.* **109:201(1968)**)

See sodalite. Hauyn represents a salt-bearing tectosilicate with the sodalite structure containing occluded salt.

Haydenite (CHA)

Obsolete synonym for chabazite.

Helvin (SOD)
(*Am. Mineral.* **20:163(1944)**)

See sodalite. Helvin represents a salt-bearing tectosilicate of the sodalite family.

Herschelite (CHA)

Herschelite is a sodium-rich chabazite exhibiting a special pseudohexagonal habit (*Am. Mineral.* 74:1337(1989)).

Heulandite (HEU)
(natural)

Related Materials: clinoptilolite
LZ-219

The name heulandite first was proposed in 1822 for a monoclinic mineral species that was considered to be the same as stilbite.

Structure

(*Nat. Zeo.* 256(1985))

CHEMICAL COMPOSITION: $(Ca,Sr,K_2)_{1.2}$ $(Al_{2.4}Si_{6.6}O_{18})$
SYMMETRY: monoclinic
SPACE GROUP: Cm or C2
UNIT CELL CONSTANTS (Å): $a = 17.73$
$b = 17.82$
$c = 7.43$
beta $= 116°$
VOID FRACTION   (determined from water content): 0.39
PORE STRUCTURE: intersecting 8- and 10-member rings 2.6 × 4.7 Å, 3.0 × 7.6 Å, 3.3 × 4.6 Å

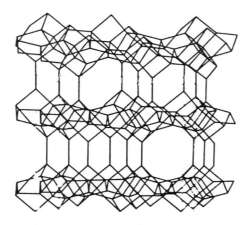

Heulandite Fig. 1S: Framework topology.

X-Ray Powder Diffraction Data: (*Cryst. Res. Technol.* 24:10(1989)) (*d*(Å)) (natural) 9.0216, 7.9677, 6.8072, 6.3684, 5.2907, 4.6447, 4.5503, 4.5046, 4.2667, 4.0568, 4.0026, 3.7370, 3.4916, 3.3934, 3.1853, 3.1092, 3.0317, 2.9968, 2.8790, 2.7745, 2.7172, 2.6281, 2.4937, 2.3520, 2.1650, 2.1261, 2.1025, 2.0660, 2.0286, 1.8983, 1.8233; (dehydrated at 250°C) 8.8416, 8.7113, 8.1508, 7.8272, 7.6915, 7.3114, 7.1930, 6.8597, 6.6295, 6.1911, 6.1062, 6.0235, 5.7699, 5.7144, 5.5715, 5.5198, 5.2595, 4.9124, 4.7786, 4.5852, 4.5046, 4.4709, 4.3222, 4.2687, 4.1697, 3.6912, 3.3934, 3.2653, 3.0798, 2.9968, 2.9677, 2.9081, 2.8433, 2.7172, 2.6281, 2.5841, 2.4870; (dehydrated at 350°C) 9.0677, 8.0766, 5.3064, 4.6447, 4.5503, 4.4597, 4.2891, 4.0844, 4.0476, 3.4916, 3.3934, 3.1853, 3.0317, 3.0067, 2.8835, 2.7745, 2.5626, 2.3520, 2.1025, 2.0286, 1.8983, 1.7866; (dehydrated at 400°C) 8.5847, 6.6543, 6.1146, 5.7699, 5.5717, 5.2595, 5.0088, 4.9124, 4.6811, 4.5872, 4.5046, 4.2087, 4.1218, 3.2653, 3.0798, 2.9773; (dehydrated at 450°C) 8.1508, 4.4158, 4.0844, 3.8652, 2.9203, 2.9018, 1.9844.

There are two main channels parallel to the *c* direction, one formed by a 10-member ring and the other by an eight-member ring. Connecting the layers containing these channels are six-, five-, and four-member rings. Parallel to the *a* direction are channels formed by eight-member rings. Some Si,Al ordering is found in the heulandite framework. There are two cation sites in this structure; each is in one of the two main channels which are parallel to *c*. The cations located at these sites coordinate both water molecules and framework oxygens on one side only.

Synthesis

ORGANIC ADDITIVE
none

Heulandite and clinoptilolite encompass a solid solution series. Heulandite has been synthesized from gel compositions of $Li_2O : Al_2O_3 : 8 SiO_2 : 8.5 H_2O$ within 2 to 3 days at 250 to 300°C

Atomic coordinates for heulandite (*SSSC* 28:449(1986)):

Atomic site	x	y	z
T(1)	0.17914(8)	0.17033(6)	0.0961(2)
T(2)	0.21308(7)	0.40985(6)	0.5014(2)
T(3)	0.20729(7)	0.19126(6)	0.7149(2)
T(4)	0.06465(7)	0.29962(6)	0.4112(2)
T(5)	0.0	0.21483(9)	0.0
O(1)	0.1976(4)	0.5	0.4522(8)
O(2)	0.2299(2)	0.1211(2)	0.6098(5)
O(3)	0.1827(2)	0.1552(2)	0.8827(5)
O(4)	0.2371(2)	0.1067(2)	0.2532(5)
O(5)	0.0	0.3278(3)	0.5
O(6)	0.0817(2)	0.1605(2)	0.0634(5)
O(7)	0.1261(3)	0.2348(2)	0.5484(6)
O(8)	0.0089(2)	0.2678(2)	0.1860(5)
O(9)	0.2116(2)	0.2539(2)	0.1794(5)
O(10)	0.1171(2)	0.3732(2)	0.4001(5)

Occupancies of Al atoms in the tetrahedral sites; the Al contents, per 18 oxygen atoms in the structure, are listed together (*SSSC* 28:449(1986)):

Tetrahedral site	RB(ex)-HEU	Faroer HEU*
Al contents	2.11	2.34
T(1)	24%	22%
T(2)	35	42
T(3)	17	27
T(4)	18	12
T(5)	22	28

*Tschermaks Min. Petr. Mitt. 18:129(1972).

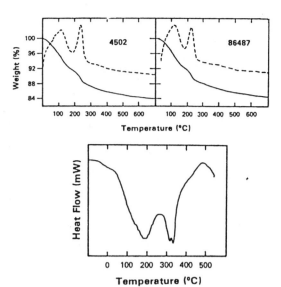

Heulandite Fig. 2:   (top) TGA curves for heulandite; solid lines are weight-loss curves, and dashed lines are derivative curves (%/min). (bottom) DSC curve for heulandite; weight loss ranges between 15 and 16% (*Nat. Zeo.* 565(1988)).

Thermal Properties

The low bond density between the layers in one direction results in structural changes in heulandite during dehydration. At temperatures below 130°C, heulandite will adsorb H_2O and NH_3 (*Zeo. Molec. Sieve.* (1974)). After dehydration at higher temperatures, no adsorption occurs. Argon diffusion rates are minimally affected upon dehydration (*Surf. Sci.* 14:77(1969)).

Upon heating to 200°C, 12 water molecules per unit cell are lost with a concomitant contraction of the lattice. Rehydration results in the restoration of the lattice dimensions. Between 230 and 260°C there is a sharp contraction of the unit cell, which amounts to 15% of the unit cell volume. This phase, which is referred to as heulandite B or phase B, does not rehydrate immediately. With time, however, the water content and cell dimensions will return to approximately those of the original parent material. After prolonged heating at 450°C the lattice is destroyed (*Cryst. Res. Technol.* 24:1027 (1989)). This behavior is not observed in clinoptilolite.

(*Am. Mineral.* 48:1372(1963)). This material also was successfully prepared without the presence of seeds from a batch composition of CaO : Al_2O_3 : 7 SiO_2 : 5 H_2O under high water vapor pressure (15,000 psi) and temperatures ranging from 250 to 360°C (*J. Geol.* 68:41(1960)). Strontium and barium cations also have been used in the synthesis of this structure. In all cases the SiO_2/Al_2O_3 ratios range between 5 and 8 (*Mat. Res. Bull.* 2:951(1967)).

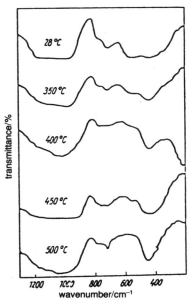

Heulandite Fig. 3:   Infrared spectra of heulandite samples fully hydrated and dehydrated at different temperatures in the range 1200 to 400 cm⁻¹ (*Cryst. Res. Technol.* 24:1027 (1989)). (Reproduced with permission of Akademie-Verlag Berlin)

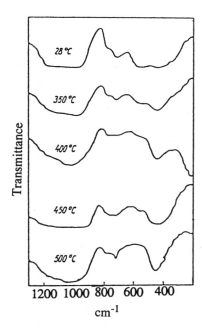

Heulandite Fig. 4:   Water bands of heulandite samples fully hydrated and dehydrated at different temperatures (*Cryst. Res. Technol.* 24:1027 (1989)). (Reproduced with permission of Akademie-Verlag Berlin)

Infrared Spectrum

Mid-infrared Vibrations (cm⁻¹):   1060, 760sh, 710w, 550, 425. With thermal treatment the bands near 760 and 550 cm⁻¹ decrease with increasing temperature, disappearing after dehydration at 500°C.

Hexagonal Faujasite (EMT) (*Zeo.* 11:98(1991))

The Proper Nomenclature is EMC-2.

Structure

Chemical Composition: $SiO_2/Al_2O_3 = 6.8$.

X-Ray Powder Diffraction Data: (d(Å) (I/I_0)) 15.02(100), 14.23(57), 13.29(26), 8.68(38), 8.02(43), 7.51(12), 7.41(27), 5.68(31), 5.57(17), 5.27(11), 4.87(7), 4.72(7), 4.43(6), 4.34(28), 3.76(18), 3.60(10), 3.44(9), 3.36(8), 3.28(28), 3.10(12), 2.89(16), 2.78(6).

Synthesis

See EMC-2.

Thermal Properties

See Hexagonal fau Figure 2.

Adsorption

Adsorption data for FAU and hexagonal faujasite (*Zeo.* 11:98(1991)):

Molecule	Kinetic diameter, Å	Pressure, Torr	T, K	Wt% adsorbed FAU	EMT
H_2O	2.65	5	300	0.26	0.23
H_2O		15	300	0.33	0.32
O_2	3.46	100	77	0.37	0.32
N_2	3.64	100	77	0.27	0.25
n-Hexane	4.30	40	300	0.20	0.18
Cyclohexane	6.00	50	300	0.23	0.22
Neopentane	6.20	700	300	0.17	0.17

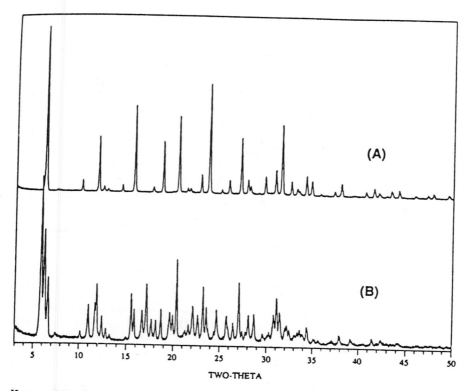

Hexagonal fau Fig. 1:   X-ray powder diffraction pattern of (A) FAU(c) and (B) hexagonal faujasite (*Zeo*. 11:98 (1991)). (Reproduced with permission of Butterworths Publishers)

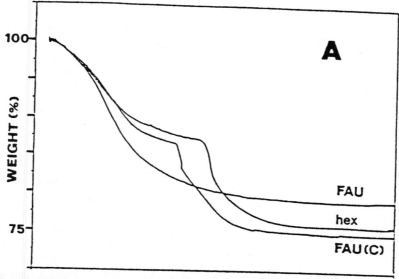

Hexagonal fau Fig. 2:   Thermal decomposition patterns for FAU, FAU(C), and hexagonal faujasite: (A) thermogravimetric analyses and (B) differential thermal analyses (*Zeo*. 11:98 (1991)). (Reproduced with permission of Butterworths Publishers)

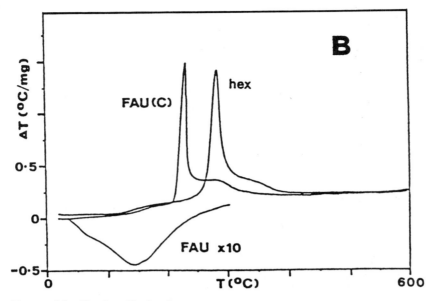

Hexagonal fau Fig. 2:   Continued.

Holdstite (MTN)
(*Nature* 303:223(1983))

Natural clathrasil with the ZSM-39 topology.

HP (FAU)
(US4,407,782(1983))
Texaco

Related Materials: faujasite

Structure

CHEMICAL COMPOSITION: 21.7 Na$_2$O : 37.8 Al$_2$O$_3$:
   40.1 SiO$_2$
SYMMETRY: cubic
UNIT CELL CONSTANT (Å): $a = 25.07$

For an aluminum-rich faujasite structure with a Si/Al ratio of 1, the unit cell constant is predicted to be 25.13 Å. The normally prepared type X material has a SiO$_2$/Al$_2$O$_3$ of 2.45 and a lattice constant of 24.996 Å. Though structurally related to faujasite, zeolite HP represents a higher aluminum-containing material.

Synthesis

ORGANIC ADDITIVE
none

Crystallization of this molecular sieve occurs from a batch composition with a Si/Al between 0.25 and 1.00, which under normal crystallization conditions (i.e., atmospheric pressure and 20 hours of aging at room temperature) results in zeolite type A. However, when aging occurs at 93°C at 50,000 psig with a subsequent 6 hours of crystallization time, zeolite type HP is produced. Impurities in this system include type A.

Adsorption data for high pressure zeolite, HP (US4,407,782(1983)):

Compound	Temp(°C)	Zeolite HP		Zeolite X	
In flow of helium		Behavior	Time	Behavior	Time
Benzene	350	Desorbed	10 min	Desorbed	25 min
Decahydronaphthalene	350	Desorbed	24 min	Desorbed	50 min
Diphenylmethane	350	Not desorbed	24 min	Not desorbed	26 min
Triisopropylbenzene	300	Not desorbed	24 min	Not desorbed	24 min
	350	Not desorbed	16 min	Not desorbed	20 min
$(C_4F_9)_3N$	350	Desorbed with air		Desorbed with air	
	350	Desorbed with air		Desorbed with air	

HS (SOD)

An abbreviation commonly used to designate hydroxysodalite.

Hsianghualite (ANA)
(natural)

Hsianghualite is a beryllosilicate with the analcime structure. Compositionally, it contains $Li_{16}Ca_{24}(Be_{24}Si_{24}O_{96})*F_{16}$, with a space group $I2_13(a = 12.8\ 8A)$ *Am. Mineral.* 44:1327(1958)).

Hydroxysodalite (SOD)

Also HS; see sodalite.

I

ISI-1 (TON)
(EP 87,017(1983))
Research Association for Petroleum Alternatives Development

Related Materials: theta-1

Structure

CHEMICAL COMPOSITION: $p(M_{2/n}O)(Al_2O_3)q(SiO_2)$ ($0.3 \leq p \leq 3.0$; $q \geq 10$).

X-Ray Powder Diffraction Data: (d(Å) (I/I_0))
10.89(s), 8.74(m), 6.94(m), 5.43(w), 4.58(w), 4.37(vs), 4.11(w), 3.68(vs), 3.62(s), 3.46(s), 3.34(w), 3.30(w), 3.22(w), 2.98(w), 2.94(w), 2.90(w), 2.78(w), 2.74(w), 2.71(w), 2.52(m), 2.43(w), 2.37(w).

ISI-1 exhibits properties characteristic of a medium-pore molecular sieve. See ZSM-22 for a discussion of the framework topology.

Synthesis

ORGANIC ADDITIVE
methanol

Batch compositions that result in the crystallization of ISI-1 include silica/alumina ratios greater than 10, methanol/water ratios between 0.1 and 10, methanol/silica between 5 and 100, and hydroxyl group/silica between 0.01 and 0.5. Alkali or alkaline earth metal/silica is between 0.1 and 3. Crystallization occurs at an initial pH of around 8.5 after 20 hours at 170°C. The crystals resulting are thermally stable to calcination at 550°C.

ISI-4 (MTT)
(EP 102,497(1984))
Research Association for Petroleum Alternatives Development

Related Materials: ZSM-23

Structure

X-Ray Powder Diffraction Data: (d(Å) (I/I_0))
11.31(s), 10.92(vs), 10.03(w), 7.83(m), 6.08(w), 5.24(w), 4.90(w), 4.51(vs), 4.42(w), 4.24(s), 4.14(w), 3.89(s), 3.73(m), 3.69(vs), 3.61(vs), 3.53(m), 3.43(s), 3.32(w), 3.16(w), 2.79(w), 2.83(w), 2.52(m).

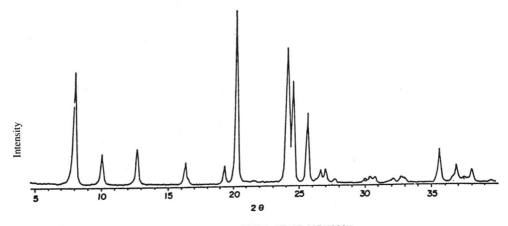

ISI-1 Fig. 1: X-ray powder diffraction pattern of ISI-1 (EP 87,017(1983)).

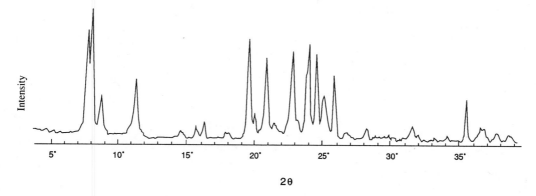

ISI-4 Fig. 1:   X-ray powder diffraction pattern of ISI-4 (EP 102,497(1984)).

ISI-4 exhibits properties characteristic of a medium-pore molecular sieve. See ZSM-23 for a description of the framework topology.

Synthesis

ORGANIC ADDITIVES
ethylene glycol
monoethanolamine

ISI-4 crystallizes from a reactive mixture with a molar oxide ratio of SiO_2/Al_2O_3 greater than 10, ethylene glycol/H_2O of 0.5 to 10, and ethylene glycol/SiO_2 of 0.01 to 0.5. Crystallization occurs between 120 and 200°C after 20 hours. The pH of the initial mixture is generally between 7.5 and 12.0.

Thermal Properties

ISI-4 does not lose crystallinity upon heating to 500°C.

ISI-6 (FER)
(US4,578,259(1986))
Research Association for Petroleum Alternatives Development

Related Materials: ferrierite

Structure

CHEMICAL COMPOSITION: $pM_{2/n}O : Al_2O_3 : qSiO_2$
(p = 0.05–3.0, q = 5–500).

X-Ray Powder Diffraction Data: (d(Å) (I/I_0))
9.44(100), 7.07(4.40), 6.92(4.30), 6.59(4.30), 5.74(4.30), 3.97(20–100), 3.92(20–70), 3.83(5–50), 3.77(5–50), 3.64(20–100), 3.53(20–100), 3.46(20–80), 3.36(4–20), 3.30(4–20), 3.12(4–30), 3.04(4–20).

ISI-6 exhibits properties characteristic of a medium-pore molecular sieve. See ferrierite for a description of the framework topology.

Synthesis

ORGANIC ADDITIVE
pyridine or pyridine derivatives with:
   ethylene glycol
   morpholine
   monoethanolamine
   ethylene diamine

ISI-6 crystallizes from a batch composition with these molar oxide ratios: SiO_2/Al_2O_3 greater than 5, pyridine/SiO_2 of 0.01 to 100, second organic/SiO_2 of 0.01 to 100, hydroxyl/SiO_2 of 0.001 to 0.5, and H_2O/SiO_2 of 5 to 1000. Crystallization occurs around 170°C after 20 hours.

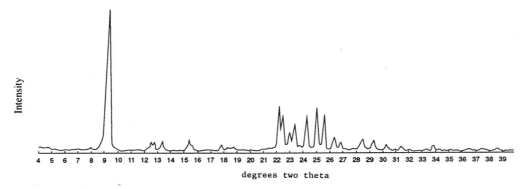

ISI-6 Fig. 1:   X-ray powder diffraction pattern of ISI-6 (US4,578,259(1986)).

J-K

J

See Linde J.

K-A
(*Rend. Accad. Sci. Fis. Mat. Naples* 35:165(1968))

K-A is thought to represent a mixture of K-E and synthetic kalsilite prepared from a potassium aluminosilicate gel (*ACS* Symposium series 40:30(1977)).

K-E (ANA)
(*JCS* 2882(1956))

This aluminosilicate was prepared from a potassium gel system and has properties related to those of analcime.

Kehoeite (ANA)
(natural)

Kehoeite is a phosphate mineral with the analcime framework.

K-F (EDI)
(*JCS* 2882(1956); *JCS* (A) 2475(1968))

Related Materials: edingtonite
Linde F
Z

K-F is an earlier designation of (zeolite) Z reported in 1956, as well as of Linde F reported in 1968. The zeolite Z and Linde F have diffraction patterns related to that of edingtonite. The designation K-F has been used to identify both materials.

The structure was refined in 1974 (*Z. Kristallogr.* 140:10(1974)).

K-G (CHA)
(JCS, 2882(1956))

Related Materials: chabazite

Structure

Chemical Composition: $K_2O : Al_2O_3$: ca. $3.9SiO_2$

X-Ray Powder Diffraction Data: $(d(\text{Å}) (I/I_0))$
[SiO_2/Al_2O_3 = 2.3] 9.47(ms), 6.90(m), 5.22(m), 4.32(s), 3.97(ms), 3.70(w), 3.46(w), 3.11(mw), 2.93(vvs), 2.80(w), 2.59(s), 2.29(s), 2.19(ms), 2.09(m), 1.90(w), 1.84(m), 1.75(vw), 1.71(s), 1.63(m), 1.57(ms), 1.48(w), 1.45(w)

[SiO_2/Al_2O_3 = 2.72] 9.45(ms), 6.94(m), 5.55(vvw), 5.24(m), 4.78(vvw), 4.32(s), 3.95(ms), 3.68(mw), 3.45(mw), 3.18(mw), 2.93(vvs), 2.79(mw), 2.59(ms), 2.29(ms), 2.17(m), 2.10(mw), 1.89(vw), 1.84(mw), 1.71(ms), 1.64(m), 1.57(mw), 1.52(mw), 1.48(w)

[SiO_2/Al_2O_3 = 4.15] 9.38(ms), 8.02(m), 7.00(mw), 5.60(mw), 5.11(m), 4.66(w), 4.34(s), 3.92(ms), 3.61(m), 3.47(mw), 3.21(mw), 2.93(vvs), 2.61(m), 2.31(mw), 2.10(mw), 1.86(w), 1.82(m), 1.73(m), 1.65(mw), 1.56(mw), 1.52(mw), 1.49(w)

The synthetic zeolite K-G is related to the natural zeolite mineral chabazite.

Synthesis

ORGANIC ADDITIVES
none

Zeolite K-G crystallizes from a reactive potassium containing gel with a typical batch composition of 2.5 K_2O : 5 SiO_2 : Al_2O_3 and x H_2O. Crystallization occurs at temperatures of 150°C. Silica gel and colloidal silica as well as potas-

potassium silicate are used as the sources of silica. Hydrous aluminum hydroxide is the source of aluminum.

K-H
(Zeolites, Their Synthesis, Properties and Applications, 2:10(1964)(Russian))

See Linde J.

K-I
(JCS (A) 2475(1968))

This name was proposed in 1968 for a material that is structurally similar to Linde Q.

K-L (LTL)
(JCS Dalton 12:1254(1972))

This aluminosilicate was prepared in a potassium-containing gel system and is related to Linde type L.

K-M (MER)
(JCS 2882(1956))

Related Material: merlinoite

K-M was designated for an apparently synthetic phillipsite. This material was later found structurally to be merlinoite.

K-W (MER)
(US3,012,853 (1961))
Union Carbide Corporation

See Linde W.

Krymskite (PHI)
(natural)

Krymskite is a name for a variety of phillipsite-wellsite type zeolites showing minor morphological and chemical deviations (*N. Jb. Miner. Abh.* 128:312(1977)).

Kurcite (PHI)
(natural)

Kurcite is a name for a variety of phillipsite-wellsite type zeolites showing minor morphological and chemical deviations (*N. Jb. Miner. Abh.* 128:312(1977)).

KZ-1 (MTT)
(Zeo. 3:8(1983))

Related Materials: ZSM-23

Structure

X-Ray Powder Diffraction Data: (d(Å) (I/I_0))
10.92(100), 9.94(25), 7.83(48), 6.07(6), 5.57(7), 5.41(16), 4.90(7), 4.91(90), 4.24(76), 4.13(19), 3.88(94), 3.71(74), 3.60(86), 3.52(36), 3.43(48), 3.29(6), 3.15(10), 3.03(5), 2.82(9), 2.63(3), 2.52(27), 2.45(13), 2.38(5), 2.33(8).

KZ-1 exhibits properties consistent with the presence of medium pores in the structure. See ZSM-23 for a description of the topology.

Synthesis

ORGANIC ADDITIVES
pyrrolidine
2-aminopropane
dimethylamine

KZ-1 crystallizes from reactive organic-containing (alumino)silicate gels. Crystallization occurs between 120 and 160°C after 40 hours. ZSM-39 and cristobalite both have been observed as impurity phases in this system. The morphology varies with the amine. Plates are formed in the presence of pyrrolidine approximately 0.1 micron thick. Needle-shaped crystals less that 0.1 micron in diameter to 10 microns long form in the dimethylamine system, whereas 2-aminopropane produces larger 0.5 × 20-micron needles.

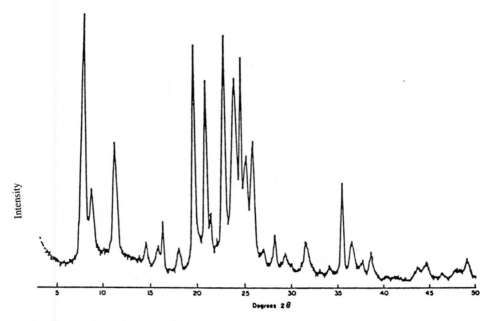

KZ-1 Fig. 1:   X-ray powder diffraction pattern for KZ-1. The sample was calcined at 600°C in air for 1 hour, then allowed to equilibrate to room temperature in air (ca. 70% RH) (*Zeo.* 3:8(1983)). (Reproduced with permission of Butterworths Publishers)

Range of reactant compositions that produced KZ-1 (given in mole ratios relative to H_2O) (*Zeo.* 3:8(1983)):

Reactant	Low	Normally used	High
H_2O		1	
SiO_2		0.0022	
NaOH	0.004	0.009	0.018
Amine	0.007	0.01	0.02
Al_2O_3	0	0.0002	0.0004
H_2SO_4	0	0.0075	

4.55(14), 4.35(100), 4.11(1), 3.66(85), 3.60(67), 3.45(46), 3.34(7), 3.29(7), 3.21(3), 2.97(3), 2.93(5), 2.90(5), 2.78(5), 2.73(13), 2.71(4), 2.52(22), 2.43(14), 2.40(5), 2.36(12), 2.28(2), 2.23(3).

KZ-2 exhibits properties consistent with the presence of medium pores in the structure. See theta-1 for a description of the topology.

KZ-2 (TON)
(*Zeo.* 3:8(1983))

Related Materials: ZSM-22

Structure

X-Ray Powder DIffraction Data: ($d(\text{Å})$ (I/I_0))
10.78(80),   8.67(19),   6.92(29),   5.41(15),

Synthesis

ORGANIC ADDITIVES
diethylamine
1-aminobutane
1,4-diaminohexane
2,2'-diaminodiethylamine

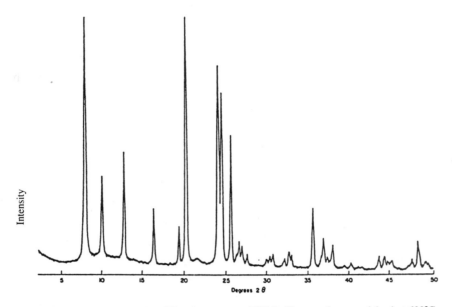

KZ-2 Fig. 1:   X-ray powder diffraction pattern of KZ-2. The sample was calcined at 600°C in air for 1 hour, then allowed to equilibrate to room temperature in air (ca. 70% RH) (*Zeo.* 3:8(1983)). (Reproduced with permission of Butterworths Publishers)

Range of reactant compositions that produced KZ-2 (given in mole ratios relative to H_2O) (*Zeo.* 3:8(1983)):

Reactant	Low	Normally used	High
H_2O		1	
SiO_2		0.0022	
NaOH		0.009	
Amine		0.01	
Al_2O_3	0	0.0002	0.0004
H_2SO_4	0	0.0075	

KZ-2 crystallizes from reactive organic-containing (alumino)silicate gels. Crystallization occurs under reaction conditions similar to those of KZ-1. The morphology does not vary with the amine. Rectangular rod-shaped crystals, 0.5–0.5 × 2–3 microns, are formed.

L

L

See Linde type L.

Large Port (Pore) Mordenite (MOR)

Large port mordenite is a term used to identify mordenite material that does not contain pore-restricting blockages. See mordenite.

Laumonite (LAU)

An earlier spelling of the mineral laumontite.

Laumontite (LAU) (natural)

Related Materials: leonhardite

Laumontite first was identified by Hauy in 1801, and was named in honor of Gillet de Laumont, a coworker of Hauy. Several changes in the spelling occurred from the original lomonite. Laumontite reversibly transforms to leonhardite upon partial dehydration. In nature, calcite and natrolite are found associated with this mineral. Other coexisting phases include quartz, anal-cime, and montmorillonite. The fully hydrated laumonite is not normally found. A laboratory synthesis of laumonite has not been reported.

Structure

(*ACS* 101:259(1971))

CHEMICAL COMPOSITION: $Ca_4(Al_8Si_{16}O_4)*16\ H_2O$
SYMMETRY: monoclinic
SPACE GROUP: Am (or A2)
UNIT CELL CONSTANTS (A): $a = 7.6$
$\qquad\qquad\qquad\qquad\qquad b = 14.8$
$\qquad\qquad\qquad\qquad\qquad c = 13.1$
$\qquad\qquad\qquad\qquad\qquad \gamma = 112$
PORE STRUCTURE: single 10-member rings 4.0 × 5.3 Å

X-Ray Powder Diffraction Data: (*Nat. Zeo.* 100(1985)) (d(Å) (I/I_0)) 9.43(78), 6.83(56), 6.18(9), 5.04(18), 4.72(16), 4.49(32), 4.15(100), 3.762(8), 3.657(42), 3.506(94), 3.404(26), 3.358(34), 3.265(63), 3.196(45), 3.091(4), 3.031(45), 2.947(8), 2.876(38), 2.876(38), 2.644(2), 2.628(3), 2.571(34), 2.537(3), 5.517(5), 5.515(9), 2.452(7), 2.437(43), 2.388(2), 2.358(23), 2.267(9), 2.215(8), 2.178(8), 2.150(28), 2.087(4), 1.990(5).

The framework of laumontite consists of 4-, 6-, and 10-member rings. The channels form parallel to [100]. A twisted six-ring is considered to be the secondary building unit of this structure. The Si/Al distribution is perfectly ordered in the framework of this structure. The calcium ions and water molecules are located in the 10-ring channels (*N. Jb. Miner. Monat.* 33(1967); *ACS Sym. Ser.* 101:259(1971)). There is one fourfold Ca site on a mirror with coordination approximating a trigonal prism with two vertices occupied by water and four coordinated to the framework. This cation is not expected to change its position upon change in hydration because the framework oxygens are on both sides of the cation.

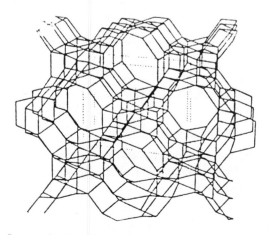

Laumontite Fig. 1S: Framework topology.

243

Atomic coordinates for laumontite (*N. Jb. Miner. Monat.* 33(1967)):

Atom	x/a_o	y/b_o	z/c_o
Ca(1)	0.729(5)	0.224(2)	0.500
Ca(2)	0.242(5)	0.768(2)	0.500
Si(1)	0.332(5)	0.260(2)	0.384(1)
Al(II)	0.751(5)	0.365(2)	0.312(1)
Si(III)	0.162(5)	0.419(2)	0.385(1)
Si(IV)	0.813(5)	0.582(2)	0.382(1)
Al(V)	0.220(5)	0.623(2)	0.307(1)
Si(VI)	0.646(5)	0.738(2)	0.379(1)
O(1)	0.276(6)	0.233(3)	0.500
O(2)	0.554(6)	0.281(3)	0.379(2)
O(3)	0.924(6)	0.347(3)	0.381(2)
O(4)	0.285(6)	0.354(3)	0.346(1)
O(5)	0.745(6)	0.333(3)	0.183(1)
O(6)	0.225(6)	0.441(2)	0.500
O(7)	0.760(6)	0.479(3)	0.317(1)
O(8)	0.206(6)	0.503(3)	0.304(1)
O(9)	0.750	0.543	0.500
O(10)	0.032(6)	0.638(3)	0.381(2)
O(11)	0.707(6)	0.649(3)	0.337(2)
O(12)	0.210(6)	0.666(3)	0.185(2)
O(13)	0.416(6)	0.701(3)	0.373(2)
O(14)	0.738(6)	0.753(3)	0.500
O(15)	0.222(7)	0.380(4)	0.122(3)
O(16)	0.835(7)	0.620(3)	0.105(2)

Thermal Properties

Water is lost from samples of laumontite in three steps. The first step has a peak at 100°C, which corresponds to the loss of three water molecules; the second water loss is observed at 240°C and related to five further waters lost; the third loss of water occurs at 400°C, and it too is a loss of five water molecules.

Infrared Spectrum

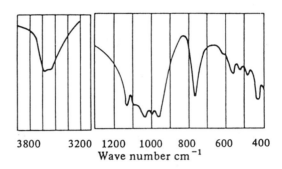

3800   3200 1200   1000   800   600   400
Wave number cm^{-1}

Laumontite Fig. 3:   Infrared spectrum of laumontite (*Zeo. Mol. Sieve* (1974)). (Reproduced with permission of John Wiley and Sons, Inc.)

Adsorption

Equilibrium adsorption capacity data for laumonite (*Zeo. Mol. Sieve.* 626(1974)):

Adsorbate	T, °C	P, Torr	g/g adsorbed
CO_2	25	700	<0.01
NH_3	25	700	15.2

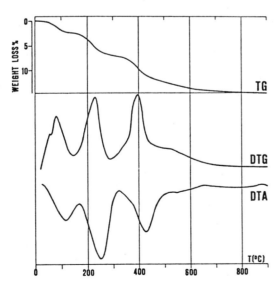

Laumontite Fig. 2:   Thermal curves of laumontite from Nasik, India; in air, heating rate 20°C/min (*Nat. Zeo.* 100 (1985)). (Reproduced with permission of Springer Verlag Publishers)

Leonhardite (LAU) (natural)

Related Materials: laumontite

The name leonhardite was coined by Blum and Delffs in 1843 for the partially dehydrated form of laumontite.

Structure

(*Nat. Zeo.* 100(1985))

CHEMICAL COMPOSITION: $Ca_4(Al_8Si_{16}O_{48})*14H_2O$
SPACE GROUP: $C2/m$
UNIT CELL CONSTANTS (A): $a = 14.75$
$b = 13.07$
$c = 7.60$
$\beta = 111°54'$

X-Ray Powder Diffraction Data: (*Am. Mineral.* 48:683(1963)) ($d(Å)$ (I/I_0)) 9.49(100), 6.86(36),   6.54(2),   6.19(2P),   5.19(1.7), 5.052(6P),   4.731(19.6P),   4.500(8P), 4.314(2.6P), 4.156(61), 3.768(2.4), 3.667(13), 3.510(31.7P),   3.411(7.7P),   3.367(3.7P), 3.272(21), 3.205(7.8), 3.152(16P), 3.033(27), 2.950(3.4), 2.881(13.7), 2.798(2.7), 2.629(3), 2.575(14),   2.521(4),   2.463(3),   2.439(14P), 2.361(12.3), 2.278(2.7), 2.268(6.5), 2.217(4.6), 2.180(5.4), 2.153(18.6), 2.082(1.5), 2.060(1.5), 2.042(1.5), 1.991(4.8), 1.955(11.6) (P = preferred orientation).

The principal difference between leonhardite and laumontite is the degree of hydration, with leonhardite being the less hydrous phase. These differences are characterized by a decrease in

Atomic coordinates ($\times 10^5$) for leonhardite (*Zeo.* 9:377(1989)):

Atom	x/A	y/B	z/C
Si(1)	23810(13)	38177(15)	15515(24)
Si(2)	8281(13)	38285(14)	32651(14)
Al*	12982(16)	30997(17)	73198(20)
O(1)	25981(15)	50000	22658(28)
O(2)	21108(10)	37669(11)	92717(19)
O(3)	14973(10)	38130(11)	55330(18)
O(4)	14659(10)	33896(10)	20664(20)
O(5)	33598(10)	31686(11)	26834(19)
O(6)	5002(24)	50000	26109(28)
O(7)	1051(10)	31005(11)	71791(20)
Ca	27540(19)	50000	75776(36)
OW1	41286(547)	50000	3570(674)
OW2	41489(91)	45514(101)	5147(177)
OW3	50000	43199(39)	50000
OW4	38445(52)	42560(63)	62407(113)
OW5	37628(111)	38162(112)	65890(248)
OW6	39527(106)	37909(120)	77087(212)
OW7	40290(126)	38065(147)	84035(249)
OW8	41917(100)	37515(120)	98301(220)

*Al site assumed to be $Al_{0.95}Si_{0.05}$.

the reaction indices and by small lattice variations. Leonhardite is the more common form. The dehydration–hydration cycle is reversible, however, and this reversibility is sensitive to the presence of Na or K substituting for the calcium ions in the structure (*Am. Mineral.* 37:812(1952)). The minimum amount of water that provides stability in the calcium form is around 11 molecules per unit cell. Increasing the water produces distortion of the calcium coordination, which increases the occurrence of 7-coordination around the calcium. The ease of dehydration of laumontite to leonhardite is related to the presence of physically adsorbed water, which is not coordinated to the cations, and to the ability of the calcium atom to vary its coordination toward water (*Zeo.* 9:377(1989)).

Leucite (ANA)
(natural)

Chemical Composition: $K(AlSi_2O_6)$. See analcime.

Leucite represents the potassium form of analcime. It exhibits no adsorption properties due to the size of the K^+ cations.

Leucophosphite
(*Am. Mineral.* 57:397(1972))

See $AlPO_4$-15.

Levyne (LEV)
(natural)

Related Materials: SAPO-35
ZK-20
and related metal-
substituted
aluminophosphates

The natural calcium aluminosilicate levyne was named in 1825. Intergrowths of the natural mineral are common with erionite and/or offretite.

Structure

(*Nat. Zeo.* 192(1985))

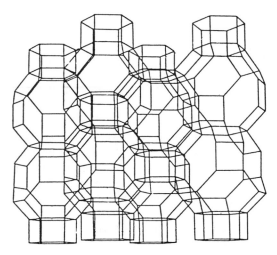

Levyne Fig. 1S: Framework topology.

CHEMICAL COMPOSITION: $NaCa_{2.5}(Al_6Si_{12}O_{36})$
*$18H_2O$
SPACE GROUP: $R\bar{3}m$
UNIT CELL CONSTANTS (Å): $a_{rh} = 10.84$
$\alpha_{rh} = 75°57'$
or:
$a_{hex} = 13.35$
$c_{hex} = 22.90$

PORE STRUCTURES: eight-member rings, 4.8 × 3.6 Å

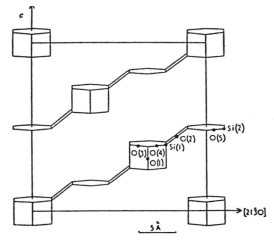

Levyne Fig. 2: The sequence of layers in levyne, showing alteration between hexagonal prisms and hexagonal rings (*Trans. Farad. Soc.* 55:1915(1959)). (Reproduced with permission of the Chemical Society)

X-Ray Powder Diffraction Data: (*Nat. Zeo.* (334(1985))) (d(Å) (I/I_0)) 10.32(28), 8.15(78), 7.66(17), 6.67(20), 5.16(42), 5.03(3), 4.27(46), 4.08(100), 3.850(26), 3.591(7), 3.475(23), 3.438(9), 3.332(19), 3.156(52), 3.084(25), 2.865(16), 2.855(15), 2.800(85), 2.714(8), 2.623(40), 2.581(6), 2.521(14), 2.445(3), 2.395(14), 2.293(8), 2.250(3), 2.223(11), 2.173(5), 2.144(7), 2.133(12), 2.129(14), 2.103(5), 2.064(10), 2.039(4), 1.972(3).

Approximate atom positions for levyne (*Trans. Faraday Soc.* 55:1915(1959)):

Atom	x	y	z	No. positions
Si,Al(1)	0.31(5)	0.82(5)	0.07	12
Si,Al(2)	0.25	−0.25	0.50	6
O(1)	0.28	−0.28	0	6
O(2)	0.45(5)	0.76	0.11	12
O(3)	0.03	0.03	0.68	6
O(4)	0.20	0.20	0.85	6
O(5)	0.37	0.37	0.72	6

The framework of the levyne structure can be described as alternate layers of hexagonal prisms and single six-member rings, alternating such that every seventh layer is superimposable upon the first.

Thermal Properties

Two prominent water losses are observed for the natural mineral levyne, occurring at 70 and 300°C.

Adsorption

Adsorption equilibrium data for levyne (*Zeo. Mol. Sieve.* 626(1974)):

Adsorbate	Temperature, °C	Pressure, Torr	Wt% adsorbed
CO_2	25	700	12
O_2, N_2	−186	700	Not adsorbed

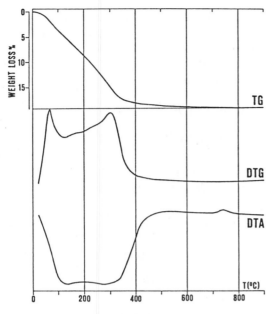

Levynite (LEV)

Common name for levyne.

Li-A (ABW)

Li-A was the first name reported by Barrer and White for a new crystalline aluminosilicate. It was later changed to LiA(BW) so that it would not be confused with Linde type A. See A(BW).

LiAPO-5 (AFI)
(EP 158,976(1985))
Union Carbide Corporation

Related Materials: AlPO₄-5

Structure

Chemical Composition: See FCAPO.

X-Ray Powder Diffraction Data: (d(Å) (I/I_0)) 12.1–11.56(m–vs), 4.55–4.46(m–s), 4.25–4.17(m–vs), 4.00–3.93(w–vs), 3.47–3.40(w–

m). See AlPO₄-5 for a description of the framework topology.

Synthesis

ORGANIC ADDITIVE
tripropylamine

LiAPO-5 crystallizes from a reactive gel with a batch composition of: 1.0–2.0 R : 0.05–0.2 As₂O₃ : 0.5–1.0 Al₂O₃ : 0.5–1.0 P₂O₅ : 40–100 H₂O.

Adsorption

Adsorption properties of LiAPO-5:

Adsorbate	Kinetic diameter, Å	Pressure, torr	Temperature, °C	Wt% adsorbed*
O₂	3.46	100	−183	7
O₂		750	−183	10
Neopentane	6.2	700	24	4
H₂O	2.65	4.3	24	4
H₂O		20	24	12

*Typical amount adsorbed.

LiAPO-11 (AEL)
(EP 158,976(1985))
Union Carbide Corporation

Related Materials: AlPO₄-11

Structure

Chemical Composition: See FCAPO.

X-Ray Powder Diffraction Data: (d(Å) (I/I_0)) 9.51–9.17(m–s), 4.40–4.31(m–s), 4.25–4.17(s–vs), 4.04–3.95(m–s), 3.95–3.92(m–s), 3.87–3.80(m–vs).

See AlPO₄-11 for a description of the framework topology.

Synthesis

ORGANIC ADDITIVE
di-*n*-propylamine

LiAPO-11 crystallizes from a reactive gel with a batch composition of: 1.0–2.0 R : 0.05–0.2 Li_2O : 0.5–1.0 Al_2O_3 : 0.5–1.0 P_2O_5 : 40–100 H_2O (R refers to the organic additive).

Adsorption

Adsorption properties of LiAPO-11:

Adsorbate	Kinetic diameter, Å	Pressure, torr	Temperature, °C	Wt% adsorbed*
O_2	3.46	100	−183	5
O_2		750	−183	6
Cyclo-hexane	6.0	90	24	4
H_2O	2.65	4.3	24	6
H_2O		20	24	8

*Typical amount adsorbed.

Li-H
(*JCS* 1167(1951))

Structure

Chemical Composition: Si/Al = 4.

X-Ray Powder Diffraction Data: (d(Å) (I/I_0))
9.84(m), 8.39(mw), 8.10(w), 6.68(ms), 5.83(w), 4.88(ms), 4.55(vw), 4.27(s), 3.97(vw), 3.78(w), 3.58(mw), 3.49(w), 3.40(mw), 3.32(mw), 3.08(vw), 2.97(w), 2.80(w), 2.66(w), 2.52(m), 2.38(vvw), 2.02(vw), 1.87(w).

Synthesis

ORGANIC ADDITIVE
none

Li-H was synthesized from a lithium aluminosilicate gel with a batch composition of: > 1.0 Li_2O : 8.0 SiO_2 : Al_2O_3 : excess H_2O. Crys-

tals formed after 36 to 60 hours at 220°C and pH of 11. Thirty-micron rod-shaped crystals were produced.

Linde A (LTA)

See Linde type A.

Linde B (GIS)

A disused term for synthetic gismondine. See Na-P1 for a description of synthetic gismondine.

Linde D (CHA)
(BP 868,846(1961))
Union Carbide Corporation

Related Materials: chabazite

Structure

Chemical Composition: 0.9 (x Na_2O : (1 − x) K_2O) : Al_2O_3 : 4.5–4.9 SiO_2 : ca. 7 H_2O.

X-Ray Powder Diffraction Data: (d(Å) (I/I_0))
9.42(66), 6.89(67), 5.54(15), 5.03(62), 4.33(62), 3.98(27), 3.89(23), 3.60(12), 3.45(39), 3.19(15), 2.94(100), 2.69(9), 2.61(38), 2.30(16), 2.09(22), 1.81(29), 1.73(23); (Ca^{+2}-exchanged zeolite D) 9.21(78), 6.81(83), 4.93(100), 4.27(89), 3.41(56), 2.92(95), 2.81(44), 2.58(39), 2.50(28), 2.06(28), 1.79(22), 1.72(22).

Linde D exhibits properties indicating a small-pore chabazite molecular sieve.

Synthesis

ORGANIC ADDITIVES
none

Linde D is prepared from reactive gel mixtures with molar oxide ratios in the following ranges: Na_2O + K_2O/SiO_2 of 0.45 to 0.65; Na_2O/Na_2O + K_2O of 0.74 to 0.92; SiO_2/Al_2O_3 of 28; H_2O/Na_2O + K_2O of 18 to 45. Crystallization occurs well at a temperature of around 100 to 120°C. At 100°C, crystallization results after 61 hours. Crystals are approximately 4 microns in diameter, with a density of 2.035 g/cm³ when fully hydrated.

Adsorption

Adsorptive properties of Linde D (BP. 868,846(1961)):

Adsorbate	Pressure, mm Hg	Temperature, °C	Grams adsorbed per 100 g of adsorbent
Propane	1	25	3.7
	100	25	7.2
	700	25	8.8
	1	25	4.0*
	100	25	7.1*
	700	25	8.1*
Nitrogen	0.1	−196	9.4
	10	−196	11.6
	100	−196	12.8
	700	−196	15.1
	0.1	−196	8.6+
	10	−196	11.6+
	100	−196	12.8+
	700	−196	15.4+
n-Pentane	1	25	3.5
	50	25	7.9
	400	25	10.2
	1	25	6.0*
	50	25	9.5*
	400	25	12.2*
Butene-1	10	25	9.5
	100	25	10.6
	400	25	——
	700	25	11.1
Cyclopropane	700	25	2.4
Water	0.1	25	10.3
	1	25	16.2
	4.5	25	20.1
	24	25	25.8
Argon	0.1	−196	12.3
	10	−196	17.2
	150	−196	22.4
Isobutane	700	25	1.5
	700	25	0.9

* 0.92 Ca-exchanged zeolite D.
+ 0.71 Zn-exchanged zeolite D.

Ion Exchange

Slight variations in the X-ray pattern between the various cation-exchanged forms are observed. The patterns show substantially all of

Cation exchange in Linde D (BP 868,846(1961)):

Exchanging salt	Moles salt per mole zeolite D*	Composition of exchanged form
Sodium chloride	100	0.95 Na₂O : 0.05 K₂O
Magnesium chloride hexahydrate	50	0.72 MgO : 0.24 K₂O : 0.04 Na₂O
Zinc nitrate hexahydrate	50	0.96 ZnO : 0.29 K₂O (tr. Na₂O)
Strontium chloride	50	0.96 SrO : 0.04 K₂O (tr. Na₂O)
Lithium sulfate	50	0.66 Li₂O : 0.34 K₂O

*One mole is equivalent to 544 grams based on the formula: 0.47 Na_2O : 0.53 K_2O : 1.00 Al_2O_3 : 4.65 SiO_2 : 4.7 H_2O.

the same lines, and all meet the requirements of a unit cell of approximately the same size.

Linde F (EDI)
(US2,996,358(1961))
Union Carbide Corporation

Related Materials: zeolite K-F
edingtonite

Structure

CHEMICAL COMPOSITION: $K_{20}Al_{20}Si_{20}(O_{80}*25H_2O)$
SYMMETRY: orthorhombic primitive (K^+ form)
tetragonal primitive ($Na(ex)K^+$ form)
UNIT CELL CONSTANTS (A)   for orthorhombic:
$a = 13.921$
$b = 14.011$
$c = 13.136$
for tetragonal:
$a = 10.056$
$c = 6.680$
PORE STRUCTURE: [110] 2.8 × 3.8 Å eight-member rings; [001] variable eight-rings (Z. *Kristallogr.* 140:S10(1974))

X-Ray Powder Diffraction Data: (US 2,996,358(1961)) (d(Å) (I/I_0)) 6.95(100), 6.51(11), 3.48(21), 3.09(56), 2.96(72), 2.81(39), 2.25(8), 1.74(6), 1.69(6), 1.64(5).

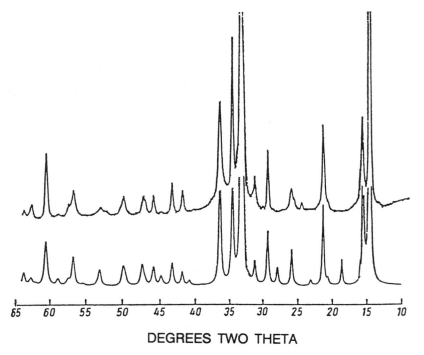

DEGREES TWO THETA

Linde F Fig. 1:   X-ray powder diffraction pattern (upper) and calculated pattern (lower) for Na(ex) K-F (*Z. Kristallogr.* 140:S10(1974)). (Reprinted with permission of Wiesbaden Akademische Verlagsgesellschaft Publishers)

The structure of synthetic edingtonite (F) is made up of four-member-ring chains similar to those found in natrolite and thomsonite. The chains are joined to form boat-shaped eight-member rings. There are two different (Si,Al) positions and three different oxygen atom sites in this structure (*Z. Kristallogr.* 140:S10(1974)). An approximate effective pore diameter determined for this zeolite, based on adsorption data, is 3.6 Å.

Parameters of hydrated Na(ex) K–F (*Z. Kristallogr.* 140:S10(1974)):

Atom	x	y	z
Si,Al(1)	0	0	0
Si,Al(2)	0.127(2)	0.134(2)	0.134(2)
O1	0.099(5)	0.093(5)	0.133(5)
O2	−0.018(6)	0.184(4)	0.466(9)
O3	0.240(4)	0.2600	0.389(9)

Synthesis

ORGANIC ADDITIVES
none

Zeolite F crystallizes in the K_2O : Al_2O_3 : 2 SiO_2 : H_2O system at temperatures between 50 and 120°C. K_2O/SiO_2 ratios range between 1.4 and 4.0, and the H_2O/K_2O ratios range from about 10 to 20. Linde F crystallizes with SiO_2/Al_2O_3 ratios in the gel between 1 and 3. The best crystallization of this material occurs at 100°C (US2,996,358(1961)). Impurities include synthetic kaliophilite. The mixture of F and kaliophilite is covered in the patent literature as Linde M (US2,995,423(1961)).

Thermal Properties

Linde F is thermally stable to at least 350°C.

Adsorption

Adsorption properties of Linde F (US2,996,358(1961)):

Adsorbate	Temperature, °C	Pressure, mm Hg	Wt% adsorbed
H_2O	25	0.010	2.0
		1.0	3.6
		4.5	12.0
		25	15.5
CO_2	25	0.05	2.3
		7	4.7
		35	6.2
		133	6.9
		312	7.5
		682	7.9
NH_3	25	0.06	1.8
		12	6.3
		89	7.2
		312	7.6
		707	7.6
SO_2	25	0.15	7.0
		14	14.9
		53	15.6
		152	16.4
		703	17.0
C_2H_4	25	560	1.9
CH_3OH	25	0.035	0.9
		0.55	7.1
		13	9.4
		30	9.8
		120	10.9
O_2	-196	130	1.4

Linde H
(US3,020,789(1961))
Union Carbide Corporation

Related Materials: Q + K-G (*ACS*
102:30(1974))
K-G + K-I (JCS A
2475(1968))

Structure

(*Zeo. Mol. Sieve.* 152 1974)

UNIT CELL CONTENTS: $K_{14}[(AlO_2)_{14}(SiO_2)_{14}*28H_2O$
SYMMETRY: hexagonal
UNIT CELL CONSTANTS (Å): a = 13.4
c = 13.2
FRAMEWORK DENSITY (g/cc):   2.18
VOID VOLUME (g/cc): 0.22

X-Ray Powder Diffraction Data: (d(Å) (I/I_0))
13.4(64), 11.6(100), 10.6(9), 9.50(24), 6.86(62),
6.02(11), 5.27(38), 4.74(8), 4.46(9), 4.41(8),
4.31(16), 4.19(23), 3.95(48), 3.72(17), 3.36(16),
3.26(18), 3.16(25), 3.14(22), 3.00(31), 2.92(94),
2.68(17), 2.66(25), 2.63(8), 2.59(41), 2.55(15),
2.37(7), 2.34(7), 2.28(60), 2.19(28), 2.10(21),
1.90(8), 1.85(29), 1.76(7), 1.73(11), 1.71(33).

Synthesis

ORGANIC ADDITIVE
none

Zeolite H crystallizes from reactive aluminosilicate mixtures with the following molar ratios of the gel: K_2O/SiO_2 between 1 and 4,
SiO_2/Al_2O_3 between 1.5 and 2.4, and H_2O/K_2O
between 23 and 34. Crystallization occurs at
100°C. Crystals grow as hexagonal plates (Linde
Q) and round spherulites (zeolite K-G).

Adsorption

Adsorption properties of zeolite H (US3,020,789(1961)):

Adsorbate	Temperature, °C	Pressure, mm Hg	Wt% adsorbed
H_2O	25	0.001	2.0
		0.02	5.2
		1.0	7.4
		4.5	17.4
		25	21.5
CO_2	25	10	0.8
		36	1.4
		160	2.8
		346	3.8
		710	4.3
SO_2	25	1	3.1
		10	4.6
		44	6.1
		129	7.6
		305	8.4
		708	9.3
O_2	-196	101	1.2
		687	0.1

Linde J
(US3,011,869(1961))
Union Carbide Corporation

Related Materials: K-H

Structure

(Zeo. Mol. Sieve. 360 1974)

CHEMICAL COMPOSITION: 0.9 K_2O : Al_2O_3 : 2.1
  SiO_2 : x H_2O
SYMMETRY: tetragonal
UNIT CELL CONSTANTS (Å): $a = 9.45$
                                          $c = 9.92$
DENSITY: 2.22 g/cc
VOID VOLUME: 0.08 cc/g
PORE STRUCTURE (approx. based on adsorption): 2.6

X-Ray Powder Diffraction Data: $(d(\text{Å})\ (I/I_0))$
6.86(54), 5.57(11), 4.77(32), 4.72(16), 4.27(15),
4.00(51), 3.23(46), 3.18(4), 3.13(93), 3.04(40),
3.00(41), 2.97(36), 2.89(100), 2.87(61),
2.68(14), 2.66(20), 2.64(25), 2,.61(11),
2.58(23), 2.33(14), 2.30(8), 2.19(14), 2.15(6),
2.09(4), 2.00(6), 1.95(11), 1.91(11), 1.86(7),
1.83(7), 1.79(12), 1.72(8), 1.72(6), 1.70(4),
1.65(11), 1.63(6).

Linde J exhibits behavior characteristic of a
very small-pore molecular sieve.

Synthesis

ORGANIC ADDITIVES
none

Linde J crystallizes from a reactive potas-
sium-ion-containing aluminosilicate gel with
K_2O/SiO_2 around 4, SiO_2/Al_2O_3 of 4, and
H_2O/K_2O around 10. Crystallization requires 89
hours at 100°C.

Adsorption

Adsorption properties of Linde J (US3,011,869(1961)):

Adsorbate	Temperature, °C	Pressure, mm Hg	Wt% adsorbed
H_2O	25	25	8.4
NH_3	25	675	3.9
SO_2	25	711	2.2
CO_2	25	691	1.0
C_2H_4	25	681	0.7

Linde L (LTL)

See Linde type L.

Linde M
(US2,995,423(1961))
Union Carbide Corporation

Related Materials: Linde F + synthetic
                                    kaliophilite

Structure

Chemical   Composition: K_2O : 2.1
SiO_2 : Al_2O_3 : 1.7 H_2O.

X-Ray Powder Diffraction Data: $(d(\text{Å})\ (I/I_0))$
7.02(18), 6.55(6), 4.50(8), 4.25(21), 3.98(3),
3.50(6), 3.10(100), 2.98(16), 2.82(15), 2.60(27),
2.26(5), 2.23(5), 2.12(14).

Synthesis

ORGANIC ADDITIVES
none

Linde M crystallizes from a reactive potassium aluminosilicate gel with a batch composition: 49 K_2O : 10 SiO_2 : Al_2O_3 : 420 H_2O.
Crystallization occurs at 100°C after 66 hours.
Sheaflike crystals 1 to 1.5 microns in size with
a density of 2.34 are produced.

Adsorption

Adsorption properties of Linde M (US2,995,423(1961)):

Adsorbate	Temperature, °C	Pressure, mm Hg	Wt% adsorbed
H_2O	25	24	10.3
Methanol	25	100	2.5

Linde N (LTN)

See Linde type N.

Linde Q
(US2,991,151(1961))
Union Carbide Corporation

Related Materials: K-I

Structure

Chemical Composition: Si/Al 1 to 1.2

X-Ray Powder Diffraction Data: ($d(Å)$ (I/I_0))
(K-Q) 13.6(47), 11.8(100), 6.96(3), 6.75(20),
6.02(11), 5.86(6), 5.30(4), 4.77(4), 4.44(8),
4.21(17), 3.72(17), 3.69(5), 3.37(16), 3.28(21),
3.24(8), 3.14(26), 3.01(39), 2.92(27), 2.67(31),
2.63(5), 2.55(17), 2.49(4), 2.34(10), 2.30(4),
2.22(7), 2.14(7), 2.10(7); (Na-Q) 13.4(52),
11.6(100), 6.96(4), 6.70(20), 5.98(17), 5.79(4),
4.74(7), 4.39(11), 4.17(20), 3.70(15), 3.67(13),
3.35(15), 3.25(22), 3.22(15), 3.13(22), 3.00(50),
2.91(24), 2.66(24), 2.62(7), 2.53(9), 2.48(7),
2.32(9), 2.28(2), 2.20(7), 2.14(7), 2.09(9); (Li-
Q) 13.4(46), 11.6(100), 6.91(5), 6.65(12),
6.02(22), 5.82(5), 5.37(2), 4.74(12), 4.44(7),
4.17(24), 3.72(24), 3.69(22), 3.36(17), 3.26(24),
3.23(12), 3.14(27), 3.00(46), 2.91(24), 2.67(29),
2.63(5), 2.54(15), 2.49(5), 2.33(7), 2.29(5),
2.21(5), 2.14(7), 2.09(10); (Ca-Q) 13.4(42),
11.6(100), 6.86(9), 6.65(36), 5.94(15), 5.75(12),
5.27(6), 4.72(18), 4.44(18), 4.13(36), 3.70(18),
3.67(12),   3.32(13),   3.225(10),   3.23(30),
3.11(15), 2.99(60), 2.89(18), 2.65(24), 2.60(12),
2.52(12), 2.48(6), 2.35(6), 2.31(9), 2.18(6),
2.14(6), 2.11(12).

Synthesis

ORGANIC ADDITIVES
none

Linde Q crystallizes from reactive aqueous
potassium aluminosilicate mixtures with the following mole ratios of oxides K_2O/SiO_2 of about
4, SiO_2/Al_2O_3 of about 4, and H_2O/K_2O of about
12. Crystallization occurs in the temperature range
of 25 to 50°C after 162 hours.

Infrared Spectrum

Assignment of mid-infrared framework vibrations (*Zeo.*
11:116(1991)):

Band frequency, cm^{-1}	Assignment
1150–900	Asymetric stretch (ext. + int.)
745.5	Symmetric stretch (ext.)
673 sh	Symmetric stretch (int.)
666.5	
627.5 sh	
598.2	"Double ring" (ext.)
499.2	O–T–O bend (int.)
437	
414.5	
349.7	S8R "pore opening vibrations"
313.2	S12R (ext.)

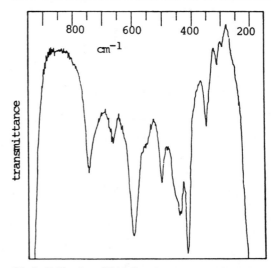

Linde Q Fig. 1:   Mid-infrared spectrum (framework
vibrations, 900–300 cm^{-1}, of as-synthesized Linde Q
(*Zeo.* 11:116 (1991)). (Reproduced with permission of
Butterworths Publishers)

Adsorption

Adsorption properties of Linde Q (K^+ form)
(US2,991,151(1961)):

Adsorbate	Temperature, °C	Pressure, mm Hg	Wt% adsorbed
H_2O	25	0.011	3.9
		0.2	6.3
		4.5	11.5
		25	18.3
CO_2	25	0.013	0.9
		3	4.6
		12	5.8
		97	7.6
		385	8.5
		720	9.7
NH_3	25	0.15	1.3
		12	4.1
		57	4.7
		142	5.3
		338	6.2
		694	6.9
SO_2	25	0.09	2.2
		4	2.9
		12	3.9
		23	4.4
		96	4.8
		314	5.5
		721	6.3
N_2	−196	24	0.6
		400	1.4
		710	2.5
O_2	−196	85	1.6
		120	1.4
Ar	−196	93	1.6
C_2H_6	25	700	0
C_2H_4	25	700	0
C_3H_8	25	700	0

Ion Exchange

Linde Q undergoes ion exchange readily. Starting from the potassium form, 65% sodium exchanged, 58% lithium, 81% calcium, 45% magnesium, 64% strontium, and 66% barium could be prepared.

Linde R (CHA) (US3,030,181(1962)) Union Carbide Corporation

Related Materials: chabazite

Structure

Chemical Composition: $Na_2O : Al_2O_3 : 3.45–3.65 SiO_2 : 0–7 H_2O$.

X-Ray Powder Diffraction Data: ($d(Å)$ (I/I_0)) 9.51(88), 6.97(35), 5.75(16), 5.61(26), 5.10(45), 4.75(12), 4.37(78), 4.13(12), 4.02(14), 3.92(35), 3.80(16), 3.63(41), 3.48(25), 3.34(12), 3.21(18), 3.13(12), 2.95(100), 2.89(16), 2.80(14), 2.71(14), 2.66(10), 2.62(25), 2.53(22), 2.39(10), 2.14(6), 2.10(14), 1.93(10), 1.89(10), 1.82(18), 1.76(6), 1.73(16), 1.69(4).

Linde R exhibits properties characteristic of a small-pore chabazite molecular sieve. See chabazite for a description of the framework topology.

Adsorption properties of ion-exchanged Linde Q (US2,991,151(1961)):

Adsorbate	Temp. °C	Pressure, mm Hg	Wt% adsorbed						
			NaQ	LiQ	CaQ	MgQ	ZnQ	SrQ	BaQ
H_2O	25	25	22.3	27.0	20.7	21.9	25.4	19.8	17.9
CO_2	25	700	4.7	12.1	11.8	9.8	12.5	11.3	7.9
NH_3	25	700	5.8	9.1	12.2	11.3	15.8	—	—
C_2H_4	25	700	0	0	5.1	2.3	5.8	4.9	0.4
O_2	−196	120	1.1	3.0	6.9	2.8	6.3	4.1	1.4
C_2H_6	25	700	0	0	7.8	1.1	7.4	—	—
SO_2	25	700	0	4.5	26.5	20.8	22.3	—	—
N_2	−196	700	0	2.3	3.2	2.6	4.8	—	—

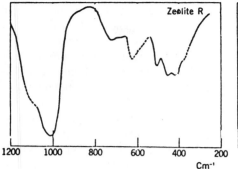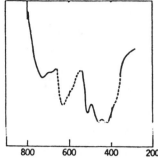

Linde R Fig. 1:   Infrared spectra of zeolite R (*Advances in Chemistry Series* 101, *Molecular Sieve Zeolites*, Flanigan, E. M, Sand, L. B., eds., 201(1971). (Reproduced with permission of the American Chemical Society)

Synthesis

ORGANIC ADDITIVE
none

Linde R crystallizes from a reactive sodium aluminosilicate gel with a batch composition typically in the range of 3.2 Na_2O : 4 SiO_2 : Al_2O_3 : 260 H_2O. Crystallization occurs at 100°C after 16 hours. The crystals resulting from this synthesis are irregularly shaped between 0.6 and 7 microns in size with a density of 1.98.

Infrared Spectrum

Mid-infrared Vibrations (cm⁻¹): (SiO_2/Al_2O_3 = 3.25) 11.36mwsh, 1007s, 738w, 678w, 625m, 508m, 426m, 370wsh (*ACS* 101:201 (1971).

Ion Exchange

Linde R can be ion-exchanged with soluble salts of lithium, magnesium, zinc, potassium, strontium, and barium. In exchange with calcium chloride, 52% of the original sodium ions in the structure are replaced.

Adsorption

Adsorptive properties of dehydrated synthetic zeolite R (US3,030,181(1961)):

Adsorbate	Pressure, mm Hg	Temper-ature, °C	Wt% adsorbed
H_2O	0.1	25	17.2
	5	25	22.8
	24	25	25.9
CO;	0.3	25	2.1
	50	25	13.6
	700	25	17.6
	700	25	13.9*
Argon	0.1	−196	0
	5	−196	0
	150	−196	2.8
	150	−196	2.3*
Propane	0.5	25	0
	50	25	0
	700	25	1.2
O_2	700	−183	16.0
	700	−183	4.3*
N_2	700	−196	5.5
	700	−196	3.6*
Cyclopropane	700	25	2.3

*0.52 Ca-exchanged synthetic zeolite R.

Linde T (ERI/OFF)
(US2,950,952(1960))
Union Carbide Corporation

Related Materials: erionite/offretite
intergrowth

See also T.

Structure

(*Zeo. Mol. Sieve.* 173(1974))

CHEMICAL COMPOSITION: $(Na_{1.2}K_{2.8})[(AlO_2)_4$
$(SiO_2)_{44}]*14H_2O$
FRAMEWORK DENSITY: 1.50 g/cc
VOID FRACTION (determined from water
content): 0.40
PORE STRUCTURE: three-dimensional, 3.6 × 4.8 Å

X-Ray Powder Diffraction Pattern: (*Zeo. Mol. Sieve.* 367 (1974)) (d(Å) (I/I_0)) 11.45(100), 9.18(4), 7.54(13), 6.63(54), 6.01(2), 5.74(6), 4.99(2), 4.57(8), 4.34(45), 4.16(3), 4.18(2), 3.82(16), 3.76(56), 3.67(1), 3.59(30), 3.42(2), 3.31(16), 3.18(12), 3.15(18), 2.93(11), 2.87(38), 2.85(45), 2.68(11), 2.61(2), 2.51(8), 2.49(13), 2.30(1), 2.21(6), 2.12(5), 2.09(3), 1.99(2), 1.96(2), 1.89(8), 1.87(2), 1.84(4), 1.78(8), 1.77(5), 1.75(2), 1.71(3), 1.66(9), 1.59(5), 1.52(1), 1.51(2), 1.47(3), 1.41(1), 1.39(3).

The X-ray powder diffraction patterns of zeolite T indicate an intergrowth of erionite and offretite. The X-ray reflections with (201) and (211) odd l lines, indicative of erionite, are very weak in zeolite T, suggesting that the material consists mainly of the offretite structure.

The cation positions are within the double six-ring units; the gmelenite cages and the six-rings are within the main cavities (*Izv. Akad. Nauk SSSR Ser. Khim* 6:116(1965)). In both offretite and erionite there are two different T atom sites, in the hexagonal prism (T1 site) and the six-member ring (T2 site). Their location relative to the large cage differs. The cation density is larger near the T2 sites. To maintain short-range electrical neutrality, the aluminum density is larger on the T2 sites. The relative displacement of aluminum from T1 to T2 sites is related to the fraction R_1/R_2, where R_1 and R_2 are the Si/Al ratios of the two sites. The magnitude of this ratio is dependent on the cation filling, which is larger for erionite than for offretite. The cations in the hexagonal prisms and those in the cancrinite cages compensate for aluminum on T1 sites, whereas cations in the gmelinite cages compensate for aluminum on T2 sites, and the cations in the offretite channels contribute equally to both T1 and T2. The cations that are located in the erionite supercages are assumed to compensate for aluminum on T2 sites only. this nonrandom Si/Al distribution has been investigated by NMR techniques.

Synthesis

ORGANIC ADDITIVES
none

Zeolite T crystallizes in the range of molar oxide gel compositions containing both sodium and potassium with a ratio of $Na_2O/(Na_2O + K_2O)$ between 0.7 and 0.8. The ratio of SiO_2/Al_2O_3 is from 20 to 28, and the M_2O/SiO_2 (where M = Na + K) ranges between 0.4 and 0.5. The water content is between 40 and 42 for $H_2O/(Na_2O + K_2O)$. Other phases that crystallize, depending on the amounts of sodium and hydroxide, are Linde type L and chabazite. In general, crystallization is sensitive to changes in the hydroxide content of the reactive gel. Crystallization occurs after 166 hours at 100°C.

2θ (Cu–rad)

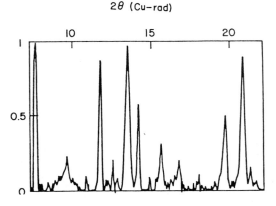

Linde T Fig. 1: High-angle part of the X-ray powder diffraction pattern of zeolite T containing 30% erionite (*Zeo.* 4:474(1986)). (Reproduced with permission of Butterworths Publishers)

Synthesis of T-type zeolites (*Zeo.* 6:474(1986)):

Sample	Si/Al (gel)	Si/Al (crystal)	Na/K (gel)	Na/K (crystal)	Crystallized offretite(%)	Erionite intergrowth(%)
T/1	7.5	3.0	3.0	0.4	30	30
T/2	7.5	3.0	3.0	0.4	75	30
T-exl	13.3	3.2	5.0	0.4	45	30
Jenkins*	9.0	2.3	3.0	0.4	6	30

*US3,578,398(1971).

Infrared Spectrum

Mid-infrared Vibrations (cm^{-1}): [SiO_2/Al_2O_3 = 7.0] 1156wsh, 1059s, 1010s, 771w, 718w, 623mw, 575w, 467ms, 433ms, 410wsh, 366wsh (*ACS* 101:201(1971))

Adsorption

Equilibrium adsorption capacities for zeolite T (*Zeo. Mol. Sieve.* 621(1974)):

Adsorbate	T, K	P, Torr	Wt %	
Argon	77	0.1	8	
		10	13.5	
Krypton	90	1	17	
		10	23	
		18	24	
Oxygen	77	0.02	4	
		3	12.5	
		50	15	
		100	16	
Xeron	195	3.8	20	(slow rate)

Adsorbate	T, K	P, Torr	Wt %
Nitrogen	77	0.1	7
		10	9.8
		100	10.7
		700	12
H_2O	298	0.1	7.5
		4	16
		20	18.6
	373	16	11
		22	12
NH_3	298	0.4	1.5
		10	4.7
		100	6.6
		700	9.8
CO_2	298	20	4.8
		100	12
		300	14
		700	15
H_2O	298	1	4.9
		50	8.5
		100	4.3
		150	10
CCl_2F_2	298	NA	

(continued)

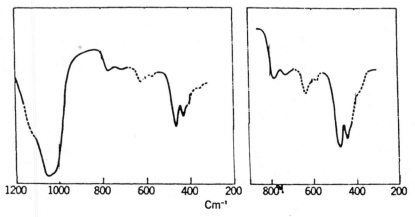

Linde T Fig. 2:   Mid-infrared spectrum of zeolite T (*Advances in Chemistry Series 101, Molecular Sieve Zeolites,* Flanigan, E. M, Sand, L. B., eds., 201(1971). (Reproduced with permission of the American Chemical Society)

Adsorbate	T, K	P, Torr	Wt %
n-C_3H_8	298	10	1.8
		100	2.5
		700	2.7
n-C_3H_6	298	10	4.0
		100	4.7
		700	5.5
n-C_5H_{12}	298	10	6.6
		100	8.1
		700	11.2
Isobutane	298	NA	
Cyclohexane	298	NA	
Benzene	298	NA	

Zeolite T does not adsorb molecules with a kinetic diameter greater than 4.3 at 400K.

Linde Type A
US2,882,243(1959))
Union Carbide Corporation

Related Materials: alpha
LZ-215
N-A
SAPO-42
ZK-4
ZK-21
ZK-22

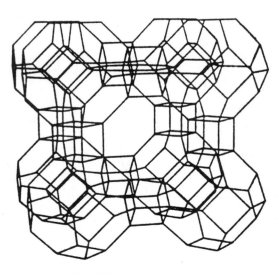

Linde type A Fig. 1S: Framework topology.

Structure

Na-A (*Zeo.* 7:148(1987))

CHEMICAL COMPOSITION: $Na_{12}[(AlO_2)_{12} (SiO_2)_{12}]^*$
27H_2O (pseudo cell) pseudo cell × 8 for true cell
SYMMETRY: cubic
SPACE GROUP: Pm3m (pseudo cell)
Fm3c (true cell)
UNIT CELL CONSTANTS: a = 24.64 Å (12.32 Å for pseudo cell)
DENSITY: 1.99 cc/g
UNIT CELL VOLUME: 1870 Å³ (pseudo cell)
FRAMEWORK DENSITY: 1.27 g/cc
PORE STRUCTURE: three-dimensional eight-ring parallel to ⟨100⟩; 4.1 Å

X-Ray Powder Diffraction Data: $(d(\text{Å}))$ (I/I_0)
17.7(40), 14.5(3), 8.67(18), 7.25(16), 6.22(21), 5.80(8), 5.47(3), 5.07(13), 4.33(40), 1.17(5), 3.66(5), 3.384(100), 2.903(2), 2.522(11), 2.181(3), 1.843(20), 1.631(2).

Positional parameters for type A framework (*Zeo.* 7:148 (1987)):

	x	y	z
T	0	0.18616(8)	0.37708(7)
O1	0	0.2447(3)	0.5
O2	0	0.2844(2)	0.2844(2)
O3	0.1122(2)	0.1122(2)	0.3583(2)

LTA Framework Structure

Two types of polyhedra make up the aluminosilicate framework of zeolite A. One is a simple cubic arrangement of eight tetrahedra (double four-rings), and the other is an octahedron of 24 tetrahedra, which is referred to as the beta cage and is similar to the sodalite structure. The free diameter in zeolite A is 6.6 Å for the beta cage and 11.4 Å for the large cavity or the alpha cage, with a pore window opening of 4.1 Å. Synthesized in the absence of organic additives, this zeolite contains equal amounts of silicon and aluminum in the tetrahedral oxide framework. The ordering of the Si and Al in the framework has been confirmed by numerous methods, including structural refinement (Z. *Kristallogr.* 133:134(1971)). The true unit cell contains 192

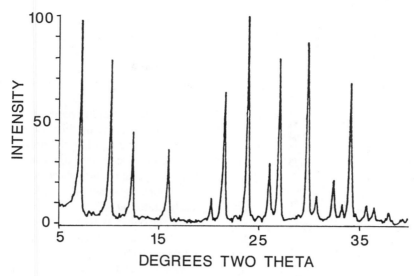

Linde type A Fig. 2:   X-ray powder diffraction pattern of Na-A. (Reproduced with permission of author)

tetrahedra and has a lattice constant of 24.6 Å, whereas the pseudo cell has 24 tetrahedra and a constant of 12.3 Å. A number of notations have been used to describe these sites, and the ones used in this work include: site I, the six-ring sites; site II, the eight-ring sites; and site III, the four-ring sites. Generally the exchangeable divalent cations occupy sites I and II. Site II has a weak affinity for all of the cations (*J. Phys. Chem.* 82:1655(1978)). Bond angles in type A are: Si–O(I)–Al of 145.5°, Si–O(II)–Al of 159.5°, and Si–O(III)–Al of 144.1°, with representative O–T–O angles of 108.0 to 110.9° for SiO_4 and 107.0 to 112.0° for AlO_4, giving a mean value of 109.47° (*Z. Kristallogr.* 133:134 (1971)). In the dehydrated state Na-A has bond angles of 145.1°, 165.6°, and 145.5°, respectively (*J. Phys. Chem.* 77:805(1973)).

The six-ring site is the most important site for the location of cations. The deviation of the cations from the center of the ring, which is defined as the intersection of the threefold axis with the plane determined by the 3-coordinated O(3) oxygens, changes with the type of cation in the structure.

In general, the water content increases with decreasing ionic radius of the cations. For example, Tl^+ ($r = 1.49$ Å) contains 22 waters, and Na^+ ($r = 0.98$ Å) contains 27. The calcium-exchanged A has the highest with 30, as there are half the number of cations present. Discussion of hydrated and dehydrated zeolite A will be dealt with by cation type.

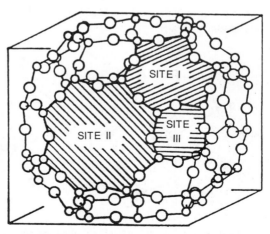

Linde type A Fig. 3:   Unit cell structure of zeolite A (*5th IZC*, 291 (1980)).

Cation-exchange forms of zeolite A (*JACS* 78:5963(1956)): *JACS* 78:5972(1956)):

Cation content/ unit cell	Hydrated density, g/cc	a, Å	No. H₂O's
Na₁₂	1.99	12.32	27
K₁₂	2.08	12.31	24
Na₄.₂Li₇.₈	1.91	12.04	24
Na₈.₂Cs₃.₈	2.26	12.30	—
Na₂.₄Tl₉.₆	3.36	12.33	20
Na₄Mg₄	2.04	12.29	—
Ca₆	2.05	12.26	30
NaSr₅.₅	—	12.32	—
Ag₁₂	2.76	12.38	24
Mn₃.₄Na₅.₂*	—	12.19	22.2
CO₃.₉₅Na₄.₁*	—	12.24	24.4
Ni₁.₉Na₈.₂*	—	12.25	22.8
Cu₁.₄Na₉.₂*	—	12.17	21.5
Cu₃.₁Na₅.₈*	—	12.24	23.0
Cu₄Na₄*	—	12.17	—
Zn₄.₇₅Na₂.₅*	—	12.17	20.5

*8th IZC 1053(1989).

Zeolite Na A

There are 12 sodium ions in the hydrated zeolite Na-A with eight located near the center of the six-rings on the threefold axis inside of the alpha cage (*JACS* 82:1041 (1960)). The remaining four ions are located in the eight-rings. The water molecules in this hydrated material form a pentagonal dodecahedron in the alpha cage (*Z. Kristallogr.* 133:134 (1971)). Such an arrangement is not unusual and is observed in clathrate compounds as well.

In dehydrated zeolite A, eight sodium ions are displaced 0.4 Å into the alpha cages from the center of the six-rings (*JACS* 78:5972 (1956); *Z. Kristallogr.* 126:135 (1968)). Three of the sodium ions are located in the eight-rings about 1.2 Å from the center (*J. Phys. Chem.* 77:805 (1973)). These cations partially block the aperture, thus influencing the adsorption of gases and vapors, and regulate the pore size. The remaining ions are located opposite the four-ring. When dehydration is performed at 350°C, the lattice remains apparently cubic. If the zeolite is dehydrated at 400°C, then the symmetry is no longer cubic at room temperature, and orthorhombic distortion is suggested. With hydration at room temperature, complete recovery to the original structure is observed (*Zeo.* 3:99 (1983)).

When solvated with methanol (Na₁₂Si₁₂ Al₁₂O₄₈*9CH₃OH, Pm3m; a = 12.298(2)Å), the 12 Na⁺ cations occupy three sites: six are on the threefold axes near the center of the six-rings, three are associated with the eight-ring oxygens, and three are in the large cavity opposite the four-rings. The nine methanol oxygens are all in the large cavity, associated with the 12 Na⁺ ions in two ways: six oxygens bridge between six-ring and four-ring Na ions, and three coordinate to eight-ring Na ions. As compared with the hydrated or the dehydrated zeolite, two Na ions have moved from the six-ring to four-ring sites. When acetonitrile is absorbed on zeolite Na-A, occupancies of the cations are the same; however, the disorder in the acetonitrile

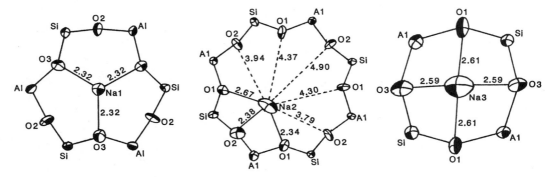

Linde type A Fig. 4:   Plot of Na atoms and adjacent rings in dehydrated zeolite Na-A. Displacement ellipsoids at 50% probability level (*J. Am. Chem. Soc.* 102:4708 (1980)). (Reproduced with permission of the American Chemical Society)

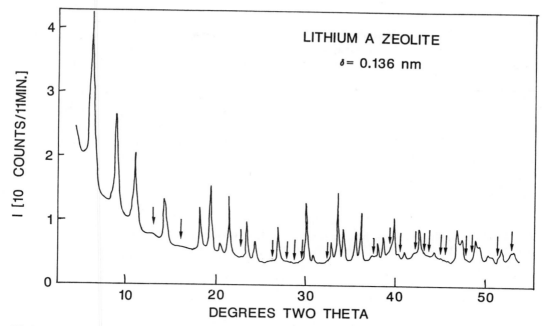

Linde type A Fig. 5:   Powder neutron diffraction pattern of Li-A zeolite. The arrows mark the positions of weak lines (*Zeo.* 3:255 (1983)). (Reproduced with permission of Butterworths Publishers)

precludes any definitive assignment of positions (*Zeo.* 3:348 (1983)).

The structure of the ammonia complex of Na-A has been determined. Twelve NH$_3$ molecules are found in two sites in the beta cage, and eight are coordinated to the threefold axis. The remaining four are hydrogen-bonded to those eight and to the framework oxygens. Twenty ammonia molecules occupy the larger cavity (alpha cage), and many are hydrogen-bonded to other NH$_3$ molecules as well as to the framework oxygen. The sodium cations in this material are near the center of the eight-ring windows (*J. Phys. Chem.* 76:2597(1972)).

Zeolite Li A

The crystal structure determined for dehydrated Li-exchanged A (Li$_{9.7}$Na$_{2.3}$Al$_{12}$Si$_{12}$O$_{48}$, based on neutron diffraction; Li$_9$Na$_3$Al$_{12}$Si$_{12}$O$_{48}$, based on X-ray diffraction) indicates that the majority of the Li cations are located at the six-ring centers (*JCS Faraday Trans.* 1,75:898(1979); *Zeo.* 3:255(1983)). Framework distortion is observed

Change in the unit cell of zeolite A with exchange of Na for Li (*Proc. 6th IZC* 684 (1984)):

Zeolite mole fraction			Solution mole fraction		
Li	Na	Li + Na/Al	Li	Na	Unit cell, Å
0	1.0	1.0			12.305
0.05	0.95	0.96	0.38	0.62	12.298
0.08	0.92	0.98	0.45	0.55	12.285
0.10	0.90	0.96	0.51	0.49	12.253
0.12	0.88	0.94	0.54	0.46	12.238
0.16	0.84	0.96	0.59	0.41	12.216
0.19	0.81	0.93	0.62	0.38	12.209
0.26	0.74	0.93	0.67	0.33	12.155
0.28	0.72	0.92	0.69	0.31	12.177
0.29	0.71	1.01	0.70	0.30	12.188
0.36	0.64	0.90	0.80	0.20	12.159
0.40	0.60	0.97	0.85	0.15	12.121
0.45	0.55	1.0	0.88	0.12	12.177
0.60	0.40	0.98	0.92	0.08	12.082
0.76	0.24	0.96	0.95	0.05	12.063
0.87	0.13	0.95	0.97	0.03	12.054
0.94	0.06	0.95	mxch*	mxch*	12.020

*mxch = multiple exchange

from K-A to Na-A to Li-A, following the charge-to-radius ratio of the exchangeable cations (*Zeo.* 3:255 (1983)). The size of the eight-ring, based on the smaller cross-distances, increases in the order:

$$d\text{-KA} < d\text{LiA} < d\text{TlA} < d\text{NaA}$$

with the shape of the eight-rind depending on the cation location in the framework six-ring.

Zeolite K A

The crystal structure of hydrated and dehydrated potassium-exchanged A (KA) has been examined (*Zeo.* 7: 423 (1987); *J. Phys. Chem.* 79:2157 (1975)). The symmetry of this zeolite depends

upon the exchangeable cations. Thermal treatment of the sample also is important. A representation of the dependence on the phase transitions of KA with temperature and pressure is shown in Linde type A Fig. 7.

The cations in this zeolite can be located in the eight-ring or the six-ring sites: in the sodalite cage by the six-ring, in the 26 hedron by the six-ring, against the four-rings in the 26 hedron, or in the center of the sodalite cage (*Zeo.* 4: 365 (1984)).

Zeolite NH₄A

In dehydrated NH_4A, the cations occupy all of the four- and eight-ring sites (*JACS* 103:3441 (1981)).

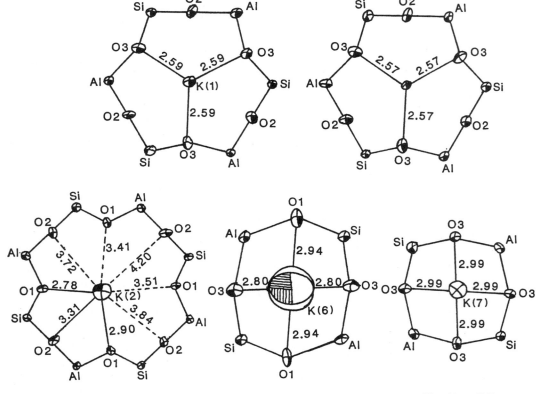

Linde type A Fig. 6:   Plot of nearest neighbors to K atoms in dK-A zeolite. Displacement ellipsoids at 50% probability level (*J. Am. Chem. Soc.* 102:4708(1980)). (Reproduced with permission of the American Chemical Society)

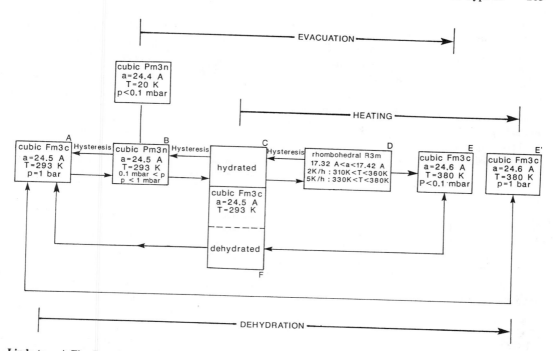

Linde type A Fig. 7:   Schematic representation of the pressure and temperature–dependent phase transitions of KA (*Zeo.* 7:423 (1987)). (Reproduced with permission of Butterworths Publishers)

Zeolite CaA

Calcium-exchanged zeolites are especially useful industrially for drying and purifying natural gas, carbon dioxide removal, separation of nitrogen and oxygen from air, separation of refinery gases, production of pure hydrogen from waste gases, and separation of *n*- and isoparaffins. These zeolites have calcium contents of 60 to 80%. The adsorptive effect of the calcium depends critically on the number and the location of these cations. The water desorption and the sensitivity of the structure to thermal treatments will depend on the coordination and the distribution of the calcium ions. For the nearly fully exchanged dehydrated Ca_6A zeolite, none of the Ca^{2+} ions are located in the eight-ring sites (*JACS* 105:1191 (1983); Proc. 5th IZC 214(1980)). In the hydrated material, the calcium positions are along the threefold axis on either side of the six-ring. Eighty percent of the water in the structure is localized (*Z. Kristallogr.* 192:142 (1975)). With partial dehydration, fourfold coordination in the large cavity is preferred, with a small amount of Ca^{+2} detected in the center of the sodalite cages. For the partially exchanged dehydrated material, Ca_4Na_4-A, both of the cations occupy six-ring positions, with sodium inside the beta cage and calcium on the opposite side (*Acta Crystallogr.* 22:162 (1967)). This arrangement results in apertures that are completely open and capable of admitting molecules with diameters of around 4.3 Å. In the hydrated state the water molecules are arranged in a distorted dodecahedron formed by 12 localized water molecules. Fifteen delo-

Deviations of atoms (Å) from the (111) plane at 0(3)[a] (*JACS* 100:3091(1978)):

$Ca_6 - A$		$Sr_6 - A$	
Ca(1)	0.51	Sr(1)	0.28
Ca(2)	0.06	Sr(2)	−0.78
Ca(3)	−0.63	Sr(3)	−0.04
O(2)	0.11	0(2)	−0.04

a. A negative deviation indicates that the atom lies on the same side of the plane as the origin.

calized waters connect the clusters through the eight-ring channels. After rehydration of Ca_4Na_4-A, neither the cations nor the water molecules are observed to return to their original location (*Zeo.* 7:148 (1987)). With calcium exchange levels at Ca_5Na_2-A, the sodium is thought to be in the large cavity at the six-ring and the calcium slightly displaced at the six-ring into the beta cage (*J. Solid State Chem.* 51:83 (1984)).

Zeolite Sr A

Dehydrated Sr_6 A is very similar to d-Ca_6-A with three Sr ions distributed over two nonequivalent, threefold axis positions. As with Ca-A, the distribution of the five cations minimizes intercationic repulsions. The sixth ion is associated with the eight-ring. The Sr cation at the eight-ring does not lie on the plane of the eight-ring, but lies 0.9 Å from it (*JACS* 100:3091 (1978)).

Zeolite Tl A

In the dehydrated Tl^+-exchanged A, the larger Tl^+ ions are located off the plane of the six-rings and extend from the plane of the ring into the alpha cage (*JACS* 78:5972(1956)). Some also are positioned inside the beta cages. The remaining four Tl^+ ions are near the center of the eight-rings (*J. Phys. Chem.* 76:2593(1972)). Only small changes are observed in cation positions in the zeolite framework upon dehydration.

Deviations of atoms from 111 planes at O(3) (in Å)[a]
J. Phys. Chem. 76:2593(1972)):

	Dehydrated	Hydrated
T1(2)	−1.82	−1.71
O(3)	0	0
O(2)	0.37	0.14
T1(1)	1.54	1.53

a. A negative deviation indicates that the atom lies on the same side of the plane as the origin.

Zeolite Ba A

Fully dehydrated Ba_6A is not crystalline; however, partially ion-exchanged forms have been

studied (*ACS* 135:137(1980)). Such instability arises from the inability of Ba^{+2} to adjust to lower coordination numbers.

		# Ba ions near		
		6-ring		
Composition		Large cavity	Sodalite cavity	8-ring
Ba_6-A	12.288	$2\frac{1}{2}$	$1\frac{1}{2}$	2
d-Ba_6-A*	12.189	2	1	3
d-$Ba_{3.5}Na_5$-A	12.267	$2\frac{1}{2}$	0	1
d-Ba_1Na_{10}-A	12.262	$\frac{1}{2}$	0	$\frac{1}{2}$

*Partially dehydrated.

Deviation of atoms (Å) from the (111) plane at O(3) (*ACS* 135:137 (1980)):

	h-Ba_6-A	part d-Ba_6-A	d-$Ba_{3.5}Na_5$-A	d-Ba_1Na_{10}-A
O(2)	0.07	0.14	0.09	0.08
Ba(1)	1.23	1.45	0.95	1.38
Ba(2)	−1.22	−1.19	—	—
Ba(3)	3.01	2.77	—	—
Na or Na(1)	—	—	−0.12	0.24

Other Cation-Exchange Forms of Zeolite A

The crystal structures of dehydrated $Mg_{1.2}Ag_2Na_{6.6}$-A, $Ca_{1.2}Ag_3Na_{6.6}$-A, $Ba_{1.2}Ag_3Na_{6.6}$-A, $Zn_{1.2}Ag_3N1_{6.6}$-A, and Ag_3Na_9-A have been examined. At 380°C the eight-ring windows are not fully occupied by the cations. For the six-ring sites as well as the eight-ring sites, a mixed occupancy with Ag, Na, and divalent cations is probable. Charged clusters are observed for dehydrated AgNa-A (*Zeo.* 3:149 (1983)). Cobalt ions, in dehydrated partially exchanged zeolite A, are located in sites with perfect trigonal coordination with the oxygens of the six-ring. There are four cobalt ions and four sodium ions per unit cell. When hydrated, the cobalt ions occupy two positions: one at the center of the beta cage and three off the six-rings in the large alpha cage on the threefold axis (*JCS Chem. Com-*

mun. 1287 (1972)). In general, incomplete exchange of NaA with divalent cations shows all divalent ions to occupy six-ring sites upon dehydration (*JACS* 3091(1978)).

A theoretical investigation of site selectivities for transition metal cations such as Mn^{+2}, Co^{+2}, Fe^{+2}, Ni^{+2}, and Zn^{+2} ions, based on the cation–lattice interaction energy, reveals that these cations preferentially occupy the six-ring oxygen sites, but that the last incoming ions prefer the eight-member oxygen ring sites ((*Proc. 5th IZC*, 291(1980)).

Zeolite N-A
(See also ZK-4 and N-A)
(US3,306,922(1967))

CHEMICAL COMPOSITION: $(Na_4,TMA_3)[(AlO_2)_7$ $(SiO_2)_{17}]$ $21H_2O$ (pseudo cell) variations Na/Al up to 0.9; Si/Al between 1.25 and 3.75 pseudo cell × 8 for true cell
SYMMETRY: cubic
SPACE GROUP: Pm3m (pseudo cell)
                       Fm3c (true cell)
UNIT CELL CONSTANTS: 12.12 Å for pseudo cell
UNIT CELL VOLUME: 1780 Å^3 (pseudo cell)
FRAMEWORK DENSITY: 1.27 g/cc

This zeolite is a silica-rich zeolite containing TMA trapped within the pores as a countercation. The Si/Al ratio is claimed to vary from 1.25 to 3.75 (US3,306,922(1967)). The lattice constant is 12.12 A, with the unit cell typically containing five fewer AlO_4 tetrahedra and five fewer cations then Na-A. Thermal decomposition is the only way to remove the TMA cation, as this ion cannot be replaced by sodium or calcium.

Synthesis

ORGANIC ADDITIVES
generally none
TMA (US3,306,922(1967))

Zeolite A generally crystallizes in the Na^+ form from a composition of 2 Na_2O : Al_2O_3: 2 SiO_2 : 35 H_2O, using sodium aluminate and sodium silicate or colloidal silica and sodium

hydroxide at 25 to 175°C. This structure has been prepared in the presence of Na, (Na,TMA), (Na,K), (Na,Ba), (Na,Ba,TMA), and (Li,Cs, TMA), but the preferred cation is sodium (*Zeo.* 1:130(1981)). Silica-rich forms of this zeolite have been prepared by using gels of higher silica and NaOH content and using TMA in place of some of the alkali metal. The silica-rich gels that crystallize type A appear to crystallize well within an hour, after which hydroxysodalite is observed to form rapidly. The presence of chloride or sulfate inhibits the formation of zeolite A. When nitrate salts are used, cancrinite is crystallized instead. Triethanol amine addition retards the rate of crystallization; the amine is thought to act as a suspending agent. Changes in the morphology, from cubic crystals around 2 to 3 microns to crystals with beveled edges around 5 to 6 microns in size, have been observed (*Zeo.* 7:387(1987)). However, for applications such as detergent use, synthesis of small crystals of this zeolite is desirable (US4,222,995(1980); JP 83,213,627(1983)). With direct hydrothermal synthesis of calcined kaolin, extrudates can be formed producing pure zeolite pellets of type A (*Zeo. Mol. Sieve.* 315(1974)). Clear homogeneous solutions can produce Na-A, however, with time, and Na-X and HS will form with extended crystallization times (*7th IZC* 177(1986)).

In the crystallization of type A zeolite in the Na^+ system, very high concentrations of hydroxide result in the formation of zeolite P or hydroxysodalite (HS) (*Zeo. Mol. Sieve.* 278 (1974)). It has been observed that zeolite X can form an overgrowth onto zeolite A (*PP 8th IZC* 3(1989)). With the addition of $NaAlO_2$ and sufficient NaOH, both mordenite and clinoptilolite will transform into type A (JP 83,213,626 (1983)). In the commercial process excess carbonate and chloride contaminants can be removed by allowing some of the type A to recrystallize to the sodalite phase, which traps the contaminants so that the mother liquor can be recycled (US4,330,518 (1982)).

Thermal Properties

Type Na-A shows a continuous loss of water with increasing temperature, with a total weight

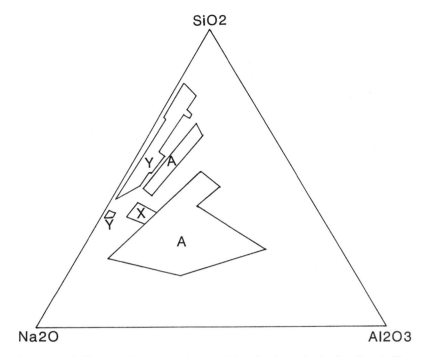

Linde type A Fig. 8:   Patented batch compositions for the synthesis of zeolites A, X, and Y. Coordinates expressed as mole percent (*ACS Symposium Series* 218:3(1983), *Intrazeolite Chemistry*, Stuckey, G. D. Dwyer F. G., eds.). (Reproduced with permission of the American Chemical Society)

loss of 22.5%. The DTA is endothermic between 25 and 300°C, become exothermic at 860 and 910°C. Thermal stability is limited to 700°C, with the formation of cristobalite occurring around 800°C. The DTA curves for zeolite Na-A, Ca-A, Ag-A and NH_4-A, are shown in the accompanying figures. Increasing the NH_4^+ content decreases the thermal stability in this zeolite. Highly ammonium-ion-exchanged NH_4-A decomposes to a sillimanite phase, whereas mixed forms result in recrystallization to a carnegite phase.

Infrared Spectrum

The Raman frequency around 500 cm^{-1} is sensitive to the framework T–O–T angle for several high-aluminum-containing zeolites, including zeolite A. Raman frequency is observed at 490 cm^{-1} for Na-A, with an average T–O–T angle of 148.3°, and this value shifts to 497 cm^{-1} in the Li-exchanged A with a decrease in T–O–T angle to 141.2° (*Zeo*. 8:306(1988)).

Infrared spectral data for A type zeolites (*ACS* 101:201(1971)):

Zeolite	SiO$_2$/Al$_2$O$_3$	Infrared bands, cm^{-1}
Na-A	1.88	1090vwsh, 995s, 660vw, 550ma, 464m, 378ms, 260vwb?
Ca-A	1.9	1130vwsh, 1050vwsh, 742wsh, 705vwsh, 665vw, 542ms, 460m, 376m
N-A	3.58	1131vwsh, 1030vwsh, 1030s, 750vwsh, 675vw, 572ms, 474m, 385m
N-A	6.01	1151vwsh, 1044s, 750vwsh, 698vw, 581ms, 475m, 393m

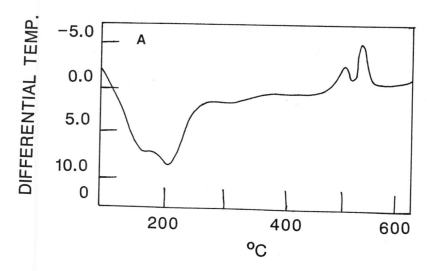

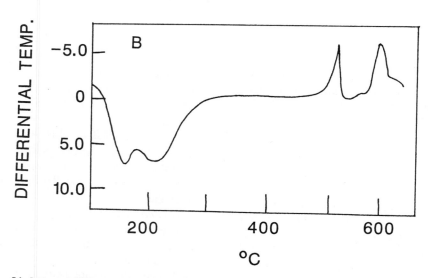

Linde type A Figs. 9a and 9b: DTA curves for (9a) zeolite Na-A and (9b) calcium-exchanged A (*JCS* 3811(1958)). (Reproduced with permission of the Chemical Society)

Adsorption

The adsorption properties of this zeolite have been examined in detail because of its prime utility as a desiccant and an adsorbent. The stan-dard method of adsorption analysis is the ad-sorption capacity measured gravimetrically (or volumetrically). Other methods, such as NMR and IR, also have been used. In the following

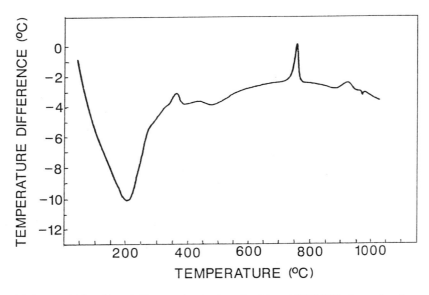

Linde type A Fig. 10:   DTA curve for Ag-A zeolite (*Zeo.* 3:271(1983)). (Reproduced with permission of Butterworths Publishers)

Near-infrared combination and overtone bands of water adsorbed on Na-A (*Zeo.* 4:244(1984)):

cm^{-1}	Assignment
7190	v_o-2OH residual OH
6930	v_o-2OH NaII(8 rings) . . . H_2O
6892	v_o-2OH NaI(6 rings) . . . H_2O
6670	
5192	$v_;zo$-1OH + δ O-1HOH H_2O . . . NaII
5160	v_o-1OH + δ O-1HOH H_2O . . . NaI
4780	δ O-3HOH
4455	v_o-1OH + δ O-1 OH residual OH groups

tables, the pressure P is expressed in millimeters of mercury or Torr. P/P_0 is the pressure P of a vapor below its critical temperature and P_0 is the equilibrium vapor pressure. Capacities are given in grams per gram of dehydrated zeolite unless otherwise noted. Prior to adsorption the zeolite is outgassed at 350 to 400°C in vacuum.

Neutron diffraction studies of methane and acetylene adsorbed in zeolite 4A show that these molecules sit in front of the NaII and NaIII cation sites. At temperatures of 150K for methane and 300K for acetylene the molecules be-

come highly mobile (*Zeo.* 2:260 (1982)). For the potassium form, ethane is adsorbed to 6.9 wt% at 4.14×10^8 Pa at 350°C (*Zeo.* 4:22(1984)). At high temperatures (300–500°C) and high pressure (51–101 MPa), Kr is adsorbed into various ion-exchanged type A zeolites. Under these conditions, the Kr is encapsulated into the alpha cages of the Na, K, and Rb forms of A. The larger the cation is, the slower the desorption of Kr (*Zeo.* 4:291(1984)).

In xenon adsorbed in zeolite CaA, there is a nearly direct correlation between the chemical shift and the density of the adsorbed xenon with no change in the shape of the NMR peak (*5th IZC* 510(1980)). Methane is slightly preferred over CD_4 in Ca-A with the heats of adsorption differing by about 100 cal/mol (ΔH_0) kcal/mol(CH_4 = 5100/CD_4 = 5000) (*Zeo.* 9:159(1989)). The infrared spectrum of methane adsorbed in Na-A changes with temperature. The growth of the v_1 forbidden band and the degeneracy splitting of v_3 are related to the electric field in the zeolite cavity; $v_1 = 2,882$ cm^{-1}, $v_3 = 3002$ cm^{-1}, $v_3^2 = 2982$ cm^{-1} (*J. Physique Lett.* 45(1984)255)). NH_3 and CO_2 both preferentially adsorb at the cation sited at the eight-

Adsorption capacities for zeolite Na A (*Zeo. Mol. Sieve.* 607(1974)):

Adsorbate	T, K	P, Torr	g/g	Comments
Argon	77	100	<0.01	b
	195	100	5	a
		300	15	a
		700	30	a
	273	100	1	a
		300	1.5	a
		700	3.7	a
Oxygen	90	0.2	0.11	b
	195	40	0.003	b
		150	0.01	b
		700	0.044	b
		100	6	a
		300	18	a
		700	34	a
Nitrogen	77	700	<0.01	
	195	100	0.065	b
		300	0.085	b
		700	0.115	b
		100	30	a
		300	42	a
		700	49	a
H_2O	298	0.025	0.16	b
		0.20	4	b
		0.25	20	b
	373	1	0.06	b
		4	0.13	b
		12	0.17	b
		20	0.19	b
NH_3	298	3	0.090	b
		10	0.11	b
		100	0.15	b
		700	0.175	b
CO_2	198	10	0.25	b
		700	0.30	b
	298	2	0.07	b
		10	0.12	b
		100	0.165	b
	423	100	0.034	b
		700	0.105	b
CO	198	15	0.070	b
		100	0.091	b
		700	0.11	b
	273	150	0.024	b
		700	0.055	b
SO_2	298	0.1	0.16	b
		10	0.28	b
		100	0.30	b
		700	0.35	b
CH_4	198	150	0.023	b
		700	0.058	b
	273	150	0.007	b
		700	0.022	b

Adsorbate	T, K	P, Torr	g/g	Comments
C_2H_2	298	10	0.072	b
		50	0.094	b
		100	0.10	b
		700	0.11	b
	423	10	0.015	b
		50	0.033	b
		100	0.042	b
		700	0.060	b
C_2H_4	298	15	0.073	b
		100	0.095	b
		700	0.11	b
C_2H_6	298	10	0.010	b
		100	0.048	b
		700	0.080	b
n-C_4H_{10}	298	N.A.		
n-C_6H_{14}	298	N.A.		
Isobutane	298	N.A.		
Neopentane	298	N.A.		
C_3H_6	298	10	0.11	b
		100	0.135	b
		700	0.14	b
Benzene	298	N.A.		
CH_3OH	298	0.01	0.066	b
		0.1	0.13	b
		100	0.19	b
H_2S	298	10	0.16	b
		100	0.22	b
		400	0.24	b
SF_6	298	N.A.		
CCl_2F_2	298	N.A.		

a. cc STP/g.
b. g/g
N.A.: Zeolite A does not adsorb molecules with a kinetic diameter greater than 3.6 Å at 77K and 4.0 Å at 300K.

Adsorption capacities for zeolite Ca A (*Zeo. Mol. Sieve.* 607(1974)):

Adsorbate	T, K	P, Torr	g/g	Comments
Argon	77	0.1	0.28	b
		1	0.33	b
		100	0.35	b
	195	100	12	a
		300	30	a
		700	55	a
	273	100	0.5	a
		300	15	a
		700	3.3	a

(continued)

Adsorbate	T, K	P, Torr	g/g	Comments		Adsorbate	T, K	P, Torr	g/g	Comments
Oxygen	77	0.1	0.16	b		n-C$_6$H$_{14}$	298	0.2	0.13	b
		0.2	0.26	b				10	0.145	b
		100	0.31	b				100	0.145	b
	195	100	0.035	b		n-Neptane	278	0.003	0.085	b
		300	0.065	b				0.01	0.11	b
		700	0.095	b				0.1	0.14	b
Nitrogen	77	3	0.21	b				10	0.175	b
		10	0.22	b			423	0.01	0.024	b
		700	0.24	b				0.1	0.055	b
	195	100	0.07	b				10	0.12	b
		300	0.085	b				10	0.12	b
		700	0.11	b		Isobutane	298	N.A.		
Krypton	273	11000	90	a,c		n-Decane	457	100	0.14	b
		45000	120	a,c		n-Dodecane	488	300	0.15	b
H$_2$O	298	0.025	0.19	b		Neopentane	298	N.A.		
		0.1	0.22	b		C$_2$H$_2$	298	5	0.055	b
		4	0.27	b				50	0.080	b
		20	0.30	b				100	0.088	b
	373	1	0.11	b				700	0.11	b
		4	0.17	b		C$_2$H$_4$	195	0.01	0.020	b
		12	0.19	b				0.1	0.085	b
		20	0.20	b				10	0.12	b
NH$_3$	298	3	0.10	b		Benzene	298	<0.006		b
		10	0.135	b		CH$_3$OH	298	0.01	0.085	b
		100	0.175	b				0.1	0.16	b
		700	0.19	b				1	0.20	b
CO$_2$	298	1	0.066	b				10	0.22	b
		10	0.13	b						
		100	0.20	b						
		700	0.24	b						
	373	15	0.60	b						
		100	0.12	b						
		700	0.20	b						
CO	198	15	0.077	b						
		100	0.11	b						
		700	0.15	b						
	273	100	0.038	b						
		700	0.070	b						
SO$_2$	298	0.1	0.17	b						
		10	0.30	b						
		100	0.34	b						
		700	0.36	b						
H$_2$S	298	10	0.21	b						
		100	0.28	b						
SF$_6$	298	N.A.								
I$_2$	393	90	0.86	b						
S	324	50	0.34	b						
CH$_4$	195	19	0.013	b						
C$_2$H$_6$	298	25	0.025	b						
		100	0.053	b						
		700	0.085	b						
n-C$_4$H$_{10}$	298	2	0.07	b						
		10	0.10	b						
		700	0.13	b						

a. cc STP/g.
b. g/g
N.A.: Zeolite A does not adsorb molecules with a kinetic diameter greater than 3.6 Å at 77K and 4.0 Å at 300K.

member ring in Li-A and Ca-A. CO$_2$ interacts with Na-A at the cations in the six-rings (*5th IZC* 476(1980)).

Ion Exchange

Ion exchange in this zeolite can be successfully accomplished in aqueous solution. Type A has an exchange capacity of 7.0 milliequiv/g anhydrous and 5.5 milliequiv/g hydrated. Under acid conditions, however, destruction of the framework is observed. Use of ammonium-exchanged forms with subsequent thermal treatment to decompose the ammonium cation also results in structural collapse. Exchange of lithium and magnesium ions occurs with consid-

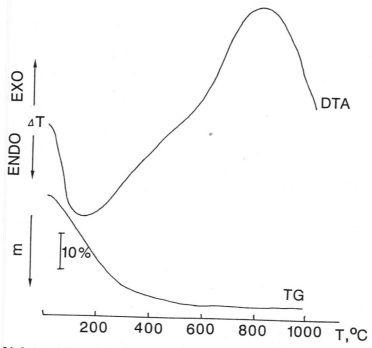

Linde type A Fig. 11:   Thermal analysis of $Na_{4.4}(NH_4)_{7.6}$-A (*Thermochim. Acta* 93:753(1985)). (Reproduced with permission of Elsevier Publishers).

erable difficulty. Exchange of sodium ions with barium has been achieved, but crystal decomposition occurs upon dehydration. Cesium exchange is not complete (*Trans. Faraday Soc.* 54:1074 (1958)). Fully Cs^+-exchanged zeolite A can be prepared, not by traditional aqueous methods but by treating dehydrated type A directly with cesium vapor. Zeolite A in the sodium, potassium, or calcium form reacts with gaseous Cs at 350°C such that all of the sodium, potassium, or calcium ions are reduced, and the metal atoms leave the zeolite. Treatment of the material by this method results in excess cesium in the zeolite, as adsorbed Cs_4^{+3} clusters and a composition of $Cs_{12.5}$-A (*J. Phys. Chem.* 91:15 (1987); *JCS Chem. Commun.* 1225(1987); *JACS* 109:26,7986(1987); *ACS Symp. Sev.* 368:177 (1988); *J. Phys. Chem.* 89:4420(1985)).

Complete ion exchange through solution ex-

Summary of reactions, products, and crystallographic data for metal vapor exchange in zeolite A (*PP 7th IZC* 4A-7(1986)):

Zeolite	Metal	Product	Color	$E°$aq, V	a, Å
Ca_6A	Cs	Cs_6Ca_3-A	black	0.05	12.240(4)
Ca_6-A	Na	Ca_6-A	colorless	−0.16	12.270(2)
Zn_6-A $\cdot 3H_2O$	Na	Na_9-A	black	1.95	12.286(7)
Ag_{12}-A $\cdot 3H_2O$	Na	Na_4Ag_4-A	black	3.51	12.267(5)

change of a hydrated NH_4^+-A has been found to contain an additional molecule of CsOH in the sodalite cavity. Avoidance of close Cs^+– Cs^+ contact through the six-rings in the structure results in a siting of four Cs^+ ions per large cavity at nearly opposite four-rings rather than

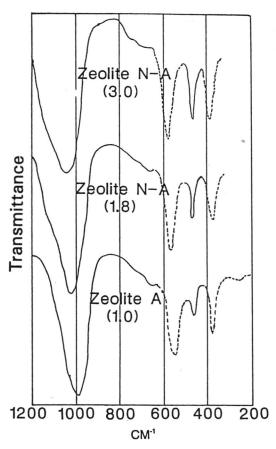

Transmittance

Zeolite N–A
(3.0)

Zeolite N–A
(1.8)

Zeolite A
(1.0)

1200 1000 800 600 400 200
CM⁻¹

Linde type A Fig. 12:   Infrared spectra of zeolites A
and N-A; numbers of parenthesis are Si/Al values
(*Advances in Chemistry Series* 101, *Molecular Sieve
Zeolites,* Flanigan, E. M, Sand, L. B., eds., 201(1971).
(Reproduced with permission of the American Chemical
Society)

Calcium ion exchange for sodium exhibits a
nonuniform dependence on temperature. Per-
colation theory appears to be obeyed in this system, where it can be considered as a cubic array
of cages interconnected by open or closed windows. True equilibrium is predicted to be at-
tained after several months at 85°C (*Proc. 7th
IZC* 601(1986)).

Solvents other than water also have been examined, and the efficiency of exchange has been
determined. These solvents include methanol,
ethanol, ethylene glycol, dioxane, and dimeth-
ylsulfoxide (*J. Inorg. Nucl. Chem.* 34:1069
(1972); ibid. 32:2389(1070); ibid. 33:1927(1971);
J. Phys. Chem. 75:85(1971)). Exchange using
alcohol as the solvent does proceed, but it is a
slower process. At higher alcohol concentrations in mixed alcohol/water systems, the selectivity increases (*J. Phys. Chem.* 75:85(1971)).
With lithium the equilibrium constant decreases
with increasing alcohol concentration (*J. Inorg.
Nucl. Chem.* 33:1927(1971)). When DMSO is
present in high concentration, the ion exchange
capacity is found to be reduced by 85%. In
general the rate of exchange decreases with decreasing dielectric constant (*Zeo. Mol. Sieve.*
392(1974)).

For hydrated monovalent cations, a linear
relationship is found between the infrared frequency of selected framework vibrations and the

opposite five-rings. The hydroxide is expected
to be between the two Cs^+ ions in the sodalite
cage (*Zeo.* 9:146(1989)).

The order of decreasing selectivity for univalent ions in zeolite A is (*JACS* 78:5963(1956)):
Ag > Tl > Na > K > NH4 > Rb > Li > Cs;
and for the divalent ions the order of decreasing
selectivity is: Zn > Sr > Ba > Ca > Co >
Ni > Cd > Hg > Mg. For the organic amine
exchanged forms it is: Na > methylamine >
ethylamine > propylamine > butylamine (*Zeo.*
8:423(1988)).

Raman frequencies of hydrated zeolite A(cm⁻¹) (*J. Phys.
Chem.* 89:1861(1985)):

Li⁺	Na⁺	K⁺	Tl⁺	NH₄⁺	Assignment
356,381	355	331	333	324	
440	405	405	400	406	
497	490	487	482	490	Si–O
					deformation
				660	
730	699	694	665	697	Al–0
					stretch
	738	735	738	732	
845	850	840	840	850	
937	971	960	952	975	
996	1000	998			Si–O
	(vw)	(vw)			stretch
1059	1039	1030	1030	1043	
1084	1099	1095	1084	1103	

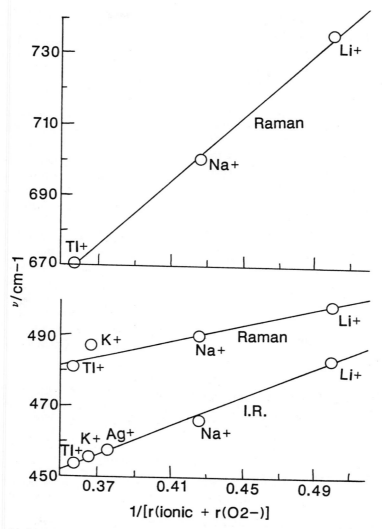

Linde type A Fig. 13: Frequency shifts of lattice vibrations for monovalent exchanged Linde A as observed by Raman and infrared spectroscopy. (The Raman spectra of K^+ and Ag^+ forms were so weak that only the absorption at 487 cm^{-1} for the K^+ form was observable.) (*JCS Chem. Commun.* 733 (1972)). (Reproduced with permission of the Chemical Society)

reciprocal of the sum of the cation and framework oxygen ionic radii (*JSC Chem. Commun.* 733(1972)).

Salts have been known to occlude within the pore structure of this zeolite. Such occlusion considerably changes the properties of the zeolite, including conductivity and thermal prop-

erties. Zeolite A occludes 9 $AgNO_3$ molecules per unit cell from silver nitrate melts (*JCS* 299(1958)) and 10 $NaNO_3$ from molten $NaNO_3$ (*J. Phys. Chem.* 72:2885(1968)). Because of the large size of the sulfate anion, $NaSO_4$ does not occlude in type A. $CaNO_3^+$ ion pairs exchange for sodium, but there is no evidence of exchange by $SrNO_3^+$ ion pairs (*J. Phys. Chem.* 72:4704(1968)). Movement of the lithium ion in lithium salt inclusion complexes was examined using Li^+ NMR (*Proc. 5th IZC,* 327(1980)). Three cation sitings exist, differing in both their mobility and local structure. $Li_3NO_3^{+2}$ is planar and lies on the 111 axis in the alpha cage.

Ag$^+$-exchanged type A has been examined for its electrical conductivity. The conductivity changes linearly over a wide temperature range from room temperature up to 650°C, where it becomes constant (*Zeo.* 3:271(1983)).

Linde Type L (LTL)
(US3,216,789)
Union Carbide Corporation

Related Materials: Perlialite
LZ-212
Ba-G

Structure

(*Zeo. Mol. Sieve.* (1974))

CHEMICAL COMPOSITION: $K_9[(AlO_2)_9(Si_2)_{27}]*22H_2O$
SYMMETRY: hexagonal
SPACE GROUP: P6/mmm
UNIT CELL CONSTANTS (Å): a = 18.4 a
c = 7.5 a
FRAMEWORK DENSITY: 1.61 g/cc
VOID FRACTION (determined from water content): 0.32
PORE STRUCTURE: unidimensional 12-member rings 7.1 Å

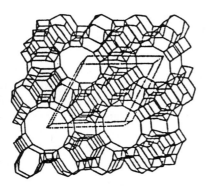

Linde type L Fig. 1S: Framework topology.

X-Ray Powder Diffraction Data: (d(Å) (I/I_0))
15.8(100),   7.89(14),   7.49(15),   5.98(25), 5.75(11), 4.57(32), 4.39(13), 4.33(13), 3.91(30), 3.78(13), 3.66(19), 3.48(23), 3.26(14), 3.17(34), 3.07(22), 3.02(15), 2.91(23), 2.65(19), 2.62(8), 2.53(8), 2.45(9), 2.42(11), 2.19(11).

The LTL structure is based on polyhedral cages formed by five six-member and six four-member rings. Such an arrangement is similar to that found in erionite, cancrinite, and offretite. This arrangement results in the formation of a unidimensional channel system. In the hydrated form of zeolite L, the structure has four cation positions. Only the cations sitting within the 12-member-ring channel system will readily exchange. The other three cations are located outside the main channels and occupy sites in close proximity to the framework oxygen atoms. The preferred Si/Al ratio is 3.0, with ordering of Si and Al. Upon dehydration, the cations sited at the wall of the 12-member ring channel migrate into the six-member rings.

Parameters of hydrated zeolite L at room temperature (*Z. Kristallogr.* 128,S:352(1969)):

Atom	Fractional occupancy*	x	y	z
Si,Al(1)		0.0946(30)	0.3595(33)	1/2
Si,Al(2)		0.1662(22)	0.4989(26)	0.2137(43)
O(1)		0	0.2674(65)	1/2
O(2)		0.1646(28)	0.3292	1/2
O(3)		0.2620(19)	0.5240	0.260(10)
O(4)		0.1004(30)	0.4078(27)	0.330(6)
O(5)		0.4261(28)	0.8522	0.275(10)
O(6)		0.1360(43)	0.4691(43)	0
Na,K	0.7	1/3	2/3	0
K(1)	1.0	1/3	2/3	1/2
K(2)	0.9	0	1/2	1/2
Na(2)	0.6	0	0.303(7)	0
H₂O(1)	0.7	0.0700(42)	0.1400	1/2
H₂O(2)	0.25	0	0.09(2)	1/2
H₂O(3)	0.5	0	0	1/2
H₂O(4)	0.25	0	0.135(21)	0.243(61)
H₂O(5)	0.25	0.092(8)	0.184	0.203(34)
H₂O(6)	0.7	0	0	0.189(46)
H₂O(7)	0.7	0.116	0.232	0
H₂O(8)	0.5	0	0.195(13)	0

*Estimated standard deviation on the order of 0.05.

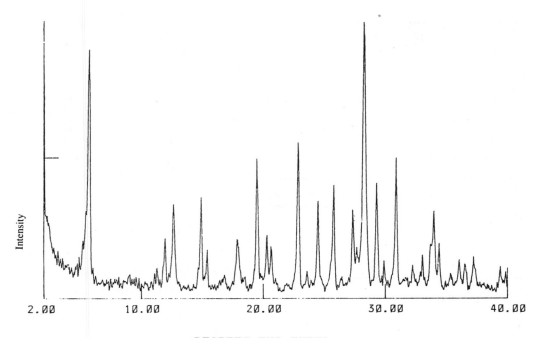

DEGREES TWO THETA

Linde type L Fig. 2:   X-ray powder diffraction pattern, (Reproduced with permission of E. Seugling, George Tech)

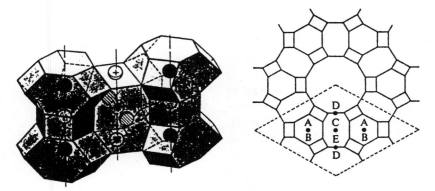

Linde type L Fig. 3:   Cation positions in zeolite L. (left) Projection of framework of zeolite L parallel to c showing main c-channels 7.1 to 7.8 A in diameter (*Zeo. Mol. Sieve*. 115 (1974)). (Reproduced with permission of John Wiley and Sons, Inc.)

Cation positions in zeolite L (Z. *Kristallogr.* 128:352(1969)):

Site	Location	Number of sites/uc	Occupancy observed
A	center of D6R	2	1.4
B	center of e-cage	2	2.0
C	between adjacent e-cages	3	2.7
D	wall of main channel	6	3.6
	total	13	9.7*

*The calculated value of 9.7 is based on partial occupancy of site A by Na^+. If K^+ is used, a value of 9.1 is obtained, in good agreement with 9 as found by analysis.

Synthesis

ORGANIC ADDITIVES
tetraalkylammonium$^+$
TMA$^+$

Zeolite L first was prepared from a batch composition containing a mixture of potassium and sodium oxides. The ratio of $K_2O/(K_2O + Na_2O)$ ranges from 0.33 to 1.0, and the M_2O/SiO_2 is between 0.35 and 0.5. The SiO_2/Al_2O_3 ratio falls between 10 and 28, and water to alklai oxide ranges from 15 to 41. Crystallization takes place at 100°C after 64 hours (US 3,216,789(1965)). A cesium-containing zeolite L can be prepared from a synthesis mixture with a composition of: K_2O/Cs_2O between 3 and 10, H_2O/K_2O from 40 to 100, and SiO_2/Al_2O_3 ranging between 7 and 13 (EP 280,513(1988)). Alkaline earth metals in conjunction with K^+ also have been claimed to form this structure (EP 142,355(1985); EP 142,353(1985)) at temperatures of at least 150°C. The preferred cations appear to be potassium and barium (*Zeo.* 1:130(1981)). Seeding has been shown to aid in the crystallization (EP 142,347(1985)). Rhyolitic pumice can be used as a source of starting materials, requiring 6 days of crystallization time at 120 to 140°C in the presence of high amounts of K^+. Phillipsite is a common impurity phase at the lower temperatures, whereas the higher temperatures, at lower K^+ content, will produce mordenite, chabazite, and erionite (*Zeo.* 2:290(1982)).

Infrared Spectrum

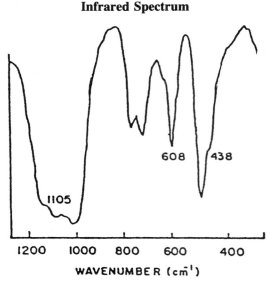

Linde type L Fig. 4:   Mid-infrared spectrum of zeolite L (*Zeo.* 10:598 (1990)). (Reproduced with permission of Elsevier Publishers)

Adsorption

Adsorption equilibrium values for adsorbates in zeolite (Na,K)L (*Zeo. Mol. Sieve.* 616(1974)):

Adsorbate	T, K	P, Torr	Wt% adsorbed
Argon	88	0.1	3.4
		1	8
		10	12.5
		150	19.5
Krypton	90	0.5	17.5
		2	26.5
		10	30
Oxygen	90	10	12.4
		100	14.9
		500	18.3
		750	21.4
Nitrogen	88	0.1	5.4
		1	10
		100	13.5
		700	18
H_2O	298	4	13.2
		10	14.7
		22	19.7
NH_3	298	10	4.0
		100	7.0
		500	8.0
		700	8.0

Adsorbate	T, K	P, Torr	Wt% adsorbed
CO₂	298	10	2.4
		100	7.1
		500	10.5
		750	11.1
SO₂	298	10	15
		100	18.0
		500	22
		700	23
n-C₄H₁₀	298	10	6.5
		100	7.5
		300	8.5
		700	9.0
Isobutane	298	10	5.3
		100	6.7
		300	7.4
		750	8.1
Neopentane	298	10	5.2
		100	6.6
		500	7.9
		750	8.6
Benzene	298	0.2	3.0
		1	6.0
		10	10.0
		80	17
(C₃H₇)N	323	2	11.9
(C4H9)3N	323	0.065	0.05
(C₄H₉)₃N	323	0.5	0.05

Linde Type N (LTN)
(US3,414,602(1968))
Union Carbide Corporation

Related Materials: Z-21

Structure

(*Zeo. Mol. Sieve*. 163(1974))

CHEMICAL COMPOSITION: NaAlSiO₄*1.35H₂O
SYMMETRY: cubic
SPACE GROUP: Fd3
UNIT CELL CONSTANT (Å): $a = 36.93$
PORE STRUCTURE: very small six-member-ring
   openings, 2.6 Å
VOID VOLUME: 0.16 cc/g

X-Ray Powder Diffraction Data: (d(Å) (I/I_0))
21.655(41), 13.173(68), 11.191(73), 9.299(50),
7.138(14), 6.569(100), 6.268(36), 5.864(23),
5.655(16), 4.826(27), 4.522(14), 4.358(20),

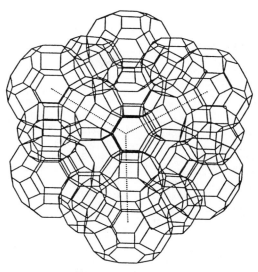

Linde type N Fig. 1S: Framework topology.

4.056(89), 3.776(57), 3.718(80), 3.574(23),
3.482(20), 3.377(20), 3.347(41), 3.139(41),
3.051(20), 2.991(61), 2.827(25), 2.763(59),
2.668(30), 2.452(32), 2.178(25), 1.81(23).

Linde type N consists of TO₄ units arranged
as single four- or single six-member rings. The
structure can be thought of as an interpenetration
of two simpler structures, those of sodalite and
ZK-5 (*Z. Kristallogr*. 1´0:313(1982)). Complete ordering of Si and Al is observed. The
bond distances of Al–O and Si–O vary between
1.70(2) and 1.75(2) and 1.58(2) and 1.63(2),
respectively.

Synthesis

ORGANIC ADDITIVE
TMA

Linde type N crystallizes from an organic and
sodium-containing reactive gel with Na₂O +
TMA₂O/SiO₂ between 0.9 and 1.1, Na₂O/SiO₂
between 0.4 and 0.55, and SiO₂/Al₂O₃ around
2. The water/alkali and organic oxide is from
30 to 50. Crystallization occurs after aging with
agitation for 24 hours, followed by heating with
stirring to 100°C for 2 hours. Without agitation,
Linde type A crystallizes.

Atomic coordinates for Linde type N (Z. *Kristallogr.* 160:313 (1982)):

Atom	X	y	z
Al1	0.3733(2)	0.2497(2)	0.3111(2)
Al2	0.3737(2)	0.2470(2)	0.4318(2)
Al3	0.4546(2)	0.3287(2)	0.3912(2)
Al4	0.3974(2)	0.3396(2)	0.5398(2)
Si1	0.3120(2)	0.2496(2)	0.3728(2)
Si2	0.4356(2)	0.2466(2)	0.3708(2)
Si3	0.3911(2)	0.3291(2)	0.4536(2)
Si4	0.5399(2)	0.3393(2)	0.4014(2)
O1	0.3388(4)	0.2610(4)	0.3403(4)
O2	0.3834(4)	0.2877(4)	0.2848(4)
O3	0.3607(4)	0.2132(4)	0.2843(4)
O4	0.4132(4)	0.2388(4)	0.3340(4)
O5	0.3333(4)	0.2374(4)	0.4086(4)
O6	0.4117(4)	0.2388(4)	0.4055(4)
O7	0.4695(4)	0.2198(4)	0.3724(4)
O8	0.4514(4)	0.2876(4)	0.3702(4)
O9	0.3713(4)	0.2901(4)	0.4490(4)
O10	0.4246(4)	0.3346(4)	0.4265(4)
O11	0.3613(4)	0.3593(4)	0.4436(4)
O12	0.4065(4)	0.3316(4)	0.4944(4)
O13	0.4970(4)	0.3342(4)	0.4088(4)
O14	0.5601(4)	0.3406(4)	0.4402(4)
O15	0.5486(4)	0.3760(4)	0.3807(4)
O16	0.5536(5)	0.3042(4)	0.3781(4)

Thermal Properties

Linde type N converts to beta-cristobalite at temperatures around 900°C.

Adsorption

The adsorption properties of this aluminosilicate are limited, as only 4.3 wt% water is adsorbed after 6 hours with a total water adsorption of 16 wt% after 4 days at 18 Torr. Neither oxygen nor nitrogen is adsorbed in this structure.

Linde Type X (FAU)
(US2,882,244(1959))
Union Carbide Corporation

Related Materials: faujasite

See Linde type Y and faujasite for further discussion of this material.

Structure

CHEMICAL COMPOSITION: $0.9 \pm 0.2 \, M_{2/n}O$: $Al_2O_3 : 2.5 \pm 0.5 \, SiO_2$: up to 8 H_2O
FRAMEWORK DENSITY (g/cc): 1.31
VOID FRACTION (determined from water content): 0.5
PORE STRUCTURE: three-dimensional, 7.4 Å

X-Ray Powder Diffraction Data: $(d(Å) \, (I/I_0))$

14.465(100), 8.845(18), 7.538(12), 5.731(18), 4.811(5), 4.419(9), 4.226(1), 3.946(4), 3.808(21), 3.765(3), 3.609(1), 3.500(1), 3.338(8), 3.253(1), 3.051(4), 2.944(9), 2.885(19), 2.794(8), 2.743(2); (Ca-X) $(d(Å) \, (I/I_0))$ 14.371(100), 8.792(9), 7.506(4), 5.709(16), 4.793(5), 4.405(11), 3.936(2), 3.800(20), 3.754(2), 3.593(2), 3.486(2), 3.328(12), 3.241(3), 3.041(4), 2.934(8), 2.875(6), 2.783(7), 2.732(4).

See faujasite for a description of the framework topology.

Li X

Site populations of extra framework sites found in the hydrated and dehydrated LiNaX zeolite (*Zeo.* 10:61 (1990)):

Site	h-LiNax	d-LiNaX
I	5 H_2O	9 H_2O or (5 H_2O + 3 Na)
I'	17 Li, 19 H_2O or (7 Na + 9 H_2O)	25 Li, 5 Na
II'	7 H_2O	—
II	30.7 or (28 Na + 5 H_2O)	16 Li, 11 Na
III	23 H_2O	9 Na

Na X

There are 85 sodium ions in the unit cell of Na-X: 16 in either the sodalite cages or the hexagonal prisms (site I' and site I, respectively), 32 in the large cavities in the plane of the six tetrahedra connecting the supercages and the sodalite cages (site II), and 37 in either crystal-

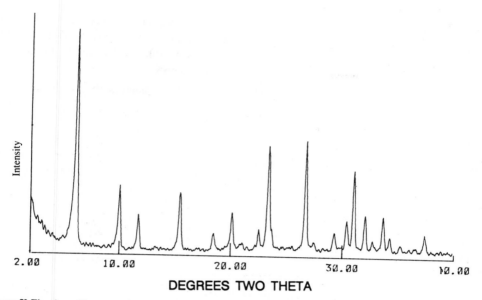

Linde type X Fig. 1:   X-ray powder diffraction pattern of Linde type X. (Reproduced with permission of author)

lographically equivalent sites or constant motion within the supercages. The distribution of ions over the different zeolite sites is influenced by the affinity of the sites for the ions as well as by the hydration properties of the ions (*Proc. 6th IZC* 49(1984)). Higher H_2O loadings lead to a higher occupancy of H_2O sites generally and a lower Na occupancy of site II than in slightly dehydrated samples. There are 20 mobile $Na(H_2O)_6^+$ ions per unit cell in the fully hydrated samples but only 14 in a less hydrated sample (*Zeo.* 2:167(1982)).

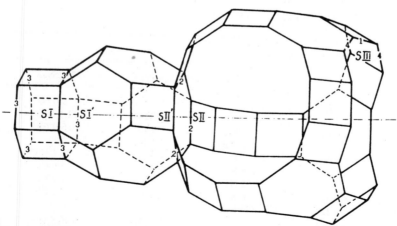

Linde type X Fig. 2:   Location of the cation sites I, I′, II′, II, and III (*Zeo.* 11:287 (1991)). (Reproduced with permission of Elsevier Publishers)

K, Rb, and Cs X

Effect of salt treatment on the unit cell size:

Crystal	Treatment	$a(\text{Å})$
h-KNaX	0.2 KCl, 26 hr. 20°C	25.07(1)
d-KNaX	measured at 400°C	24.86(1)
h-RbNaX	0.2 RbCl, 25 hr. 20°C	25.12(1)
d-RbNaX	dehydrated at 400°C	24.83(1)
h-CsNaX	0.1 CsCl, 46 hr. 20°C	25.11(1)
d-CsNaX	measured at 400°C	24.98(1)

Synthesis

ORGANIC ADDITIVE
none

Zeolite X with a SiO_2/Al_2O_3 molar ratio of 2 to 3 was prepared from reaction mixtures having SiO_2/Al_2O_3 of between 3.0 and 5.0 and NaOH as the mineralizing agent. The lowest attainable SiO_2/Al_2O_3 ratio crystals are for a 2.3 ratio (US2,882,244(1959)). SiO_2/Al_2O_3 ratios between 1.95 and 2.25 are crystallized when a mixture of NaOH and KOH is used (*Zeo.* 7:451(1987))

Preformed monolithic shapes of self-bonded type X have been prepared, co-crystallized with type A.

Hydroxysodalite is an impurity phase produced in the crystallization of type X. Other impurity phases include S and P (*Zeo.* 2:271(1982); *Zeo.* 7:451(1987)). Aging of the reaction mixture prior to crystallization from room temperature to 80°C before crystallization at 100°C can improve the purity of the type X phase.

Thermal Properties

See Linde type X Figures 3 and 4.

Infrared Spectrum

Vibrational frequencies for sodium X zeolite (*Zeo.* 5:188 (1985)):

(Si/Al = 1.12)

Raman (cm^{-1}) Temperature (K) 293	IR (cm^{-1}) 83	IR (cm^{-1}) 283	Assignments
External modes:			
14			Translational unit cell
16			modes (acoustic
20			modes)
29			Probably two phonon
32			transitions
41,46,52.5			
57.2		76	Translational Na/Na
79			lattice modes—Site
			III sites
90		95	Rotational SiO$_4$ and
100		104	AlO$_4$ lattice modes
107.5		125	
175		160	Translational Na/Na
195		185	lattice modes—SI',
218		206	SII', and SII sites
244			Translational Si/Si
250			and Al/Al lattice
268			modes
274			*(continued)*

Distribution of nonframework atoms over sites (*Zeo.* 11:287 (1991)):

	h-KNaX	d-KNaX	h-RbNaX	h-CsNaX	d-CsNaX
Site I	5.5 K	10.8 Na	3.5 Na	3.0 Na	1.0 Cs
	9.0 H$_2$O	—	10.0 H$_2$O	12.5 H$_2$O	12.5 Na
Site I'	13.6 K	6.0 K	2.7 Rb	16.9 Na	5.6 Cs
	5.6 Na	10.2 Na	22.4 Na	—	5.0 Na
Site II'	25.8 H$_2$O	9.2 K	24.6 H$_2$O	26.1 H$_2$O	1.6 Cs
Site II	24.7 K	18.4 K	22.7 Rb	12.6 Cs	—
	—	11.3 Na	8.8 Na	11.7 Na	32.0 Na
Site III +	6.2 K	23.6 K	9.2 Rb	27.2 CFs	37.9 Cs
supercage	7.6 Na	10.5 Na	—	—	—
	83.4 H$_2$O	—	95.4 H$_2$O	138.7 H$_2$O	46.1 Cs
Summarized	49.9 K	57.2 K	34.6 Rb	39.8 Cs	49.5 Na
content	13.2 Na	42.8 Na	34.7 Na	31.6 Na	—
	118.2 H$_2$O	—	130.0 H$_2$O	177.3 H$_2$O	—

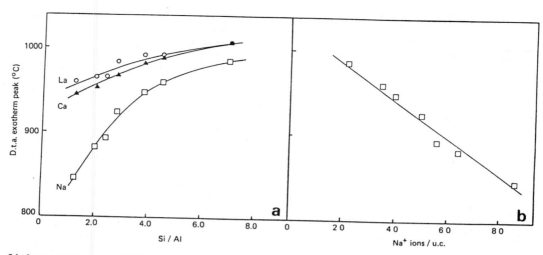

Linde type X Fig. 3:   (a) Thermal stability of faujasites in the sodium, calcium, and lanthanum forms as a function of Si/Al ratio. (b) Thermal stability of faujasites in the sodium form as a function of the number of Na^+ ions per unit cell (*Zeo.* 6:60(1986)). (Reproduced with permission of Butterworths Publishers)

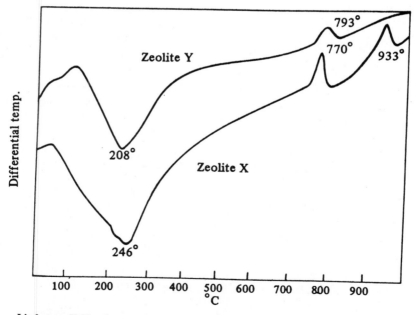

Linde type X Fig. 4:   DTA curves for a typical zeolite X and zeolite Y. The dehydration endotherm peak for zeolite X is about 40° higher than in zeolite Y (*Zeo. Mol. Sieve.* 445 (1974)). (Reproduced with permission of John Wiley and Sons, Inc.)

Temperature (K) 293	IR (cm^{-1}) 83	283	Assignments

(Si/Al = 1.12)
Raman (cm^{-1})

Internal modes:

293	83	283	Assignments
290		315vw	Bending modes
350			
368		364m	
380		395w	
463	460,445	461s	
503	500	500shvw	
528			
560	563	562m	Symmetric stretch of Si–O–Al chain
645	615	612vw	
730	692	700m	
740	760	752m	
	865vw		Probably combination transition observed at 83K
	975sh		Asymmetric stretch of zeolite Si–O–Al chain
932	990	975vs	
	1070		
	1095		
1075	1145	1078w	
	1180		

vs = very strong; s = strong; m = medium; w = weak; vw = very weak; sh = shoulder.

See faujasite for infrared spectrum of Linde type X.

Near-infrared combination and overtone bands of water adsorbed on Na X (*Zeo.* 4:244 (1984)):

Wavenumber, cm^{-1}	Assignment
7320	v O-2OH	terminal OH groups
7250	v O-2OH	NaIII . . . HOH . . .
7160	v O-2OH	O1H
7020	v O-2OH	bulk water
6950	v-O-2OH	H_2O, special adsorbed
	overlapping	
6800	v O-2OH	NaIII . . . H_2O
5325	v O-1OH + δ O-1H_2O	NaIII . . . H_2O
5220	v O-1OH + δ O-1H_2O	bulk water
5210	v O-1OH + δ O-1H_2O	special adsorbed
5060	v O-1OH + δ O-1H_2O	NaIII . . . H_2O . . .
4850	δ O-3 H_2O	H_2O
4650	v O-1OH + δ O-10H	terminal OH groups

Ion Exchange

Organic ion exchange follows the order: NH_4 > methylamine > sodium > ethylamine > propylamine > butylamine, dimethylamine > diethylamine.

The exchange capacity for Pb(II) has been examined, and it has been found that the higher the silica/aluminia ratio is, the lower the selectivity toward Pb(II) ions (*Zeo.* 5:158(1985)).

Maximum exchange levels for zeolite X, Y, and dealuminated Y expressed in fractions of the sodium cation exchange capacity as determined by radio tracer methods (*Zeo.* 6:51 (1986)):

Al/UC	Temp. °C	Cs$^+$	NH$_4^+$	Ca^{+2}	La^{+3}	K$^+$
86	25	0.65(2)	0.83(2)	1.00(1)	0.86(2)	0.99(1)
64	25	0.70(1)	0.88(2)	0.88(1)	0.76(1)	0.99(2)
	65	0.73(2)	0.92(1)	0.91(1)	—	—
50	25	0.71(1)	0.74(2)	0.74(2)	0.71(1)	1.00(1)
	65	0.73(1)	0.86(2)	0.84(2)	0.72(1)	—
40	25	0.74(1)	0.74(1)	0.74(1)	0.72(2)	1.00(2)
	65	0.76(1)	—	—	0.74(1)	—
25	25	0.78(1)	0.76(2)	0.76(2)	0.74(1)	1.00(2)
	65	0.78(2)	—	0.86(1)	0.74(2)	—
22	25	0.88(1)	0.89(1)	0.81(1)	0.82(1)	0.95(1)
	65	0.90(1)	0.92(2)	0.85(2)	0.88(2)	0.95(2)
	80	0.91(1)	0.94(2)	0.91(1)	0.91(1)	—

Adsorption

Adsorption of Tertiary Amines by X and Z Zeolite (*Zeo. Mol. Sieves* 606(1974)):

	$(C_2F_5)_2$ NC_3F_7	$(C_3H_7)_3$ N	$(C_4H_9)_3$ N	$(C_4F_9)_3$ N^a
	x/m = g/g Activated Zeolite			
Critical dimension (A)	7.7	8.1	8.1	10.2
Pressure (torr)	43	3	1	0.55
Temp (°C)	25	25	25	25
Zeolite X (SI/Al = 1.25)				
LiX (59)[b]	0.542	0.236	0.191	—
Zeolite X	0.521	0.229	0.227	0.031
CsX (51)	0.296	0.147	0.153	—
CaX (84)	0.487	0.018	0.012	0.016
BaX (93)	0.296	0.035	0.074	—
Zeolite Y (Si/Al = 2.4)				
Zeolite Y	—	—	0.228	0.018
CaY (85)	0.529	0.216	0.210	0.008
"Columbia" activated carbon	—	—	—	0.64

[a] Adsorption measured after four hours
[b] Numbers in parentheses indicate mole % exchange of the Na form.

Adsorption properties for NaX (Si/Al ca. 1.25) (*Zeo. Mol. Sieve.* (1974)):

Adsorbate	T, K	P, Torr	x/m
Argon	78	1	0.33
		10	0.40
		100	0.42
	90	1	0.30
		10	0.34
		100	0.40
		700	0.41
	195	50	0.0043
		100	0.0083
		700	0.047
Oxygen	90	20	0.30
		400	0.31
	195	60	0.005
		100	0.083
		700	0.054
Nitrogen	77	0.001	0.125
		0.1	0.16
		10	0.25
		700	0.26

Adsorbate	T, K	P, Torr	x/m
	195	40	0.039
		100	0.060
		700	0.12
Krypton	195	10	0.01
		50	0.05
		100	0.10
		600	0.41
H_2O	298	0.4	0.24
		1	0.26
		20	0.34
	373	0.5	0.095
		1	0.11
		10	0.18
NH_3	298	2	0.10
		10	0.14
		100	0.175
		700	0.190
	373	2	0.059
		10	0.175
		100	0.115
		700	0.155
CO_2	195	0.1	0.06
		1	0.20
		6	0.33
		100	0.39
		700	0.26
CO	198	15	0.062
		100	0.11
		700	0.165
SO_2	298	0.05	0.14
		1	0.31
		10	0.39
		700	0.44
H_2S	298	10	0.2224
		100	0.294
		400	0.34
SF_6	298	10	0 04
		100	0.23
		300	0.33
CCl_2F_2	298	700	0.36
Hg	596	50	0.002
		200	0.010
I_2	393	90	1.11
P	597	60	0.44
S	593	60	0.41
CH_4	195	20	0.010
		100	0.041
		700	0.080
	273	700	0.020
C_2H_6	298	50	0.025
		200	0.07
		700	0.09
C_3H_8	298	15	0.090
		100	0.13
		700	0.14

(continued)

Adsorbate	T, K	P, Torr	x/m
n-C_4H_{10}	298	2	0.12
		10	0.15
		700	0.18
n-C_6H_{14}	298	0.2	0.17
		10	0.19
		100	0.20
n-Heptane	298	0.01	0.175
		1	0.21
		40	0.21
	423	0.01	0.015
		0.1	0.076
		1	0.11
		30	0.15
Isobutane	298	10	0.081
		100	0.16
		700	0.18
n-Pentane	298	1	0.16
		100	0.19
Neopentane	298	0.1	0.082
		1	0.11
		50	0.14
		700	0.156
Isooctane	298	2	0.18
		40	0.21
C_2H_2	298	2	0.062
		10	0.093
		100	0.128
		700	0.147
C_2H_4	298	10	0.050
		100	0.085
		700	0.105
Benzene	298	0.04	0.16
		0.1	0.19
		1	0.24
		10	0.25
m-Xylene	298	0.01	0.035
		0.1	0.115
		1	0.23
		10	0.24
CH_3OH	298	0.01	0.074
		0.1	0.15
		1	0.20
		10	0.23

x/m is in g/g.

Adsorption properties for CaX (Si/Al ca. 1.25, Ca/Al$_2$ ca. 0.85) (*Zeo. Mol. Sieve.* (1974)):

Adsorbate	T, K	P, Torr	x/m
Krypton	273	11,000	90
		45,000	120
Oxygen	90	20	0.32
		700	0.38
Nitrogen	77	30	0.26
		700	0.27
	238	20	0.016
		100	0.031
		700	0.056
H_2O	300	3	0.31
		20	0.36
NH_3	298	2	0.10
		10	0.14
		100	0.185
		700	0.220
	373	2	0.065
		10	0.080
		100	0.12
		700	0.180
CO_2	298	20	0.125
		100	0.20
		700	0.29
	373	100	0.072
		700	0.16
CO	273	10	0.026
		100	0.044
		700	0.070
SF_6	298	10	0.090
		100	0.23
		300	0.30
CH_4	273	20	0.005
		100	0.016
		700	0.029
Neopentane	209	1.5	0.07
		50	0.138
		700	0.153

Linde Type Y (FAU) (US,130,007(1964)) Union Carbide Corporation

Related Materials: faujasite

See Linde type X and faujasite for further discussion of the structure and properties of this material.

Structure

CHEMICAL COMPOSITION: 0.9 ± 0.2 M$_{2/n}$O :
  Al$_2$O$_3$: 3–6 SiO$_2$: up to 9 H$_2$O
FRAMEWORK DENSITY (g/cc):   1.25–1.29
VOID FRACTION (determined from water
  content):   0.48
PORE STRUCTURE: three-dimensional, 7.4 Å

X-Ray Powder Diffraction Data: (Na-Y) (d(Å))
(I/I_0)) 14.29(100) 8.75(9), 7.46(29), 5.68(44),

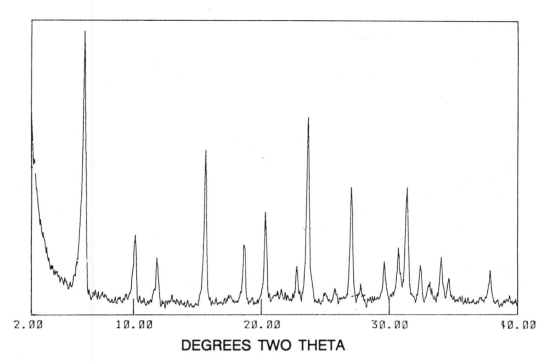

Linde type Y Fig. 1:   X-ray powder diffraction pattern of zeolite type Y. (Reproduced with permission of author)

4.76(22),   4.38(35),   3.91(12),   3.775(47), 3.573(4),   3.466(9),   3.308(37),   3.222(8), 3.024(16),   2.917(21),   2.858(48),   2.767(20), 2.717(7),   2.638(19),   2.595(11),   2.382(13), 2.188(8), 2.162(4), 2.100(8), 2.063(7), 1.933(4), 1.910(5), 1.772(5), 1.750(9), 1.704(12).

See faujasite for a description of the framework topology.

In dehydrated Li Y, the lithium cations are shown to locate in the center of the framework six-rings at the sites I' and II (*Zeo.* 3:255(1983)). The framework aluminum content can be determined by using the unit cell parameters (see faujasite for description). In the hydrated state, both water and framework oxygen complete the ligands for the cations. The water controls the population of the ion at various sites in the structure (*Zeo.* 7:528(1987)). In the Na form, only water induces serious migration of the ions, whereas the other extreme is observed for CaY,

with the best-coordinated sites increasing in occupancy with rising temperature (*J. Phys. Chem.* 88:1916(1984); *Zeo.* 4:41(1984); *Zeo.* 5:257 (1985)).

Synthesis

ORGANIC ADDITIVE
none

Zeolite Y crystallizes from a preferred range of Na_2O/SiO_2 between 0.4 and 0.6, with SiO_2/Al_2O_3 between 15 and 25 and Na_2O/H_2O between 20 and 50. Crystallization occurs readily between 80 and 125°C. The temperature affects the size of the crystals, with the larger crystals forming at lower temperatures (US 3,130,007(1964)). Other phases identified to crystallize when the Na_2O/SiO_2 and H_2O/Na_2O

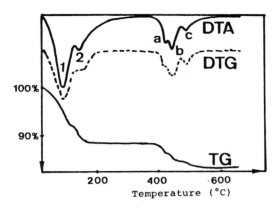

Linde type Y Fig. 2: Thermal decomposition of NH₄Y zeolite under vacuum. Heating rate 0.2°C/min. (*PP Proc. 7th IZC* 2B-8(1986)). (Reprinted with permission of Butterworths Publishers)

were changed include herschelite, noselite, and zeolites P and A (*Zeo.* 4:453(1988)). Rapid addition of sodium aluminate to a sodium silicate solution that has been modified by addition of a small quantity of aluminum has been claimed to efficiently produce the type Y structure commercially (US4,264,562(1981)).

Thermal Properties

The decomposition range for NH₄-Y is between 473 and 863K (*Zeo.* 3:249(1983)).

Infrared Spectrum

See table on p. 287.

Ion Exchange

Organic ion exchange follows the order: NH₄ > methylamine > ethylamine > propylamine > butylamine > Na > dimethylamine > diethylamine.

See Linde type X for further description of ion-exchange properties of Linde type Y.

Adsorption

Adsorption properties for NaY (Si/Al ca. 2.4–2.5) (*Zeo. Mol. Sieve.* 606 (1974)):

Adsorbate	T, K	P, Torr	x/m
Krypton	273	11,000	105
		45,000	150
Oxygen	90	20	0.29
		100	0.32
		700	0.34
	195	100	0.005
		700	0.033
Nitrogen	77	10	0.25
		100	0.27
		700	0.28
	195	50	0.015
		100	0.024
		700	0.072
H_2O	298	0.1	0.080
		1	0.25
		20	0.35
NH_3	298	2	0.080
		10	0.11
		100	0.15
		700	0.18
	373	2	0.040
		10	0.060
		100	0.095
		700	0.135
CO_2	298	10	0.032
		30	0.062
		100	0.14
		700	0.23
n-Pentane	298	1	0.145
		100	0.180
		400	0.180
	373	1	0.035
		10	0.095
		100	0.140
n-C_6H_{14}	298	0.3	0.175
		10	0.190
		100	0.190
n-C_7H_{16}	298	2	0.16
		10	0.20
		40	0.21
Isooctane	298	2	0.17
		40	0.20
Neopentane	298	10	0.116
		100	0.126
		700	0.14
Benzene	298	0.1	0.17
		1	0.23
		10	0.25
		70	0.26
$(C_4F_9)_3N$	323	0.5	0.028

Vibrational frequencies for sodium Y zeolite (*Zeo.* 5:188 (1985)):

(Si/Al = 2.43) Raman cm^{-1}	IR cm^{-1}		Assignments
Temperature (K) 293	83	293	
External modes:			
33		49	Probably two phonon transitions
41, 50			
60			Translation Na/Na lattice modes—SiIII sites
71		71	
76			
		92	Rotational SiO$_4$ and AlO$_4$ lattice modes
		100	
99.5		108	
104		115	
		169	Translational Na/Na lattice modes—SI', SII', and SII sites
175		188	
195		201	
220		224	
257			
268			Translational Si/Si and Al/Al lattice modes
276			
Internal modes:			
305		318vw	Pending modes
350			
367			
380		380m	
		409w	
460		460s	
505		510shw	
526			
566		570m	
660		650w	
720	720	720m	Symmetric stretch of zeolite Si–O–Al chain
740	795	720m	
	900w		Probably combination transition observed at 83K
	1000sh	995sh	Asymmetric stretch of zeolite Si–O–Al chain
975	1030	1012vs	
	1070	1045sh	
1055	1145	1080m	
1125	1180	1130m	

vs = very strong; s = strong; m = medium; w = weak; vw = very weak; sh = shoulder.

Adsorption properties for CaY (Si/Al ca. 2.4–2.5) (*Zeo. Mol. Sieve*. 606 (1974)):

Adsorbate	T, K	P, Torr	x/m
Oxygen	90	1	0.085
		10	0.25
		700	0.37
	195	60	0.010
		100	0.016
		700	0.040
Nitrogen	77	3	0.22
		10	0.25
		700	0.30
	195	70	0.015
		100	0.023
		700	0.060
H_2O	298	0.2	0.12
		1	0.24
		20	0.34
	373	0.2	0.07
		1	0.09
		10	0.16
		20	0.20
NH_3	298	2	0.090
		10	0.13
		100	0.18
		700	0.21
CO_2	298	10	0.026
		30	0.042
		100	0.070
		700	0.15
	373	30	0.026
		100	0.044
		700	0.095
n-Pentane	298	1	0.135
		10	0.160
		100	0.175
	373	3	0.055
		10	0.082
		100	0.130
		400	0.15
n-C_7H_{16}	298	2	0.16
		10	0.20
		40	0.20
Isooctane	298	2	0.16
		40	0.19
Benzene	298	0.1	0.16
		1	0.22
		10	0.23
		70	0.25

Linde W (MER)
(US3,012,853 (1961))
Union Carbide Corporation

Related Material: merlinoite
                       K-M

Structure

(*Zeo. Mol. Sieve*. 175(1974))

CHEMICAL COMPOSITION: $K_{42}[(AlO_2)_{42}) (SiO_2)_{76}]*$
   $107H_2O$
SYMMETRY: cubic
UNIT CELL CONSTANTS (Å) $a = 20.1$
DENSITY: 2.18 g/cc
VOID VOLUME: 0.22 cc/cc
PORE STRUCTURE: small-pore eight-member rings

X-Ray Powder Diffraction Data: (US 3,012,853(1961)) (d(Å) (I/I_0)) 9.99(20), 8.17(49), 7.09(54), 5.34(28), 5.01(56), 4.45(21), 4.28(35), 3.64(20), 3.25(100), 3.17(75), 2.95(71), 2.72(53), 2.67(12), 2.60(8), 2.54(26), 2.43(9), 2.40(8), 2.18(10), 1.91(5), 1.86(7).

Zeolite W exhibits properties characteristic of a small-pore molecular sieve.

Synthesis

ORGANIC ADDITIVE
none

Linde W crystallizes from a reactive aluminosilicate gel prepared from metakaolin and silica gel with a SiO_2/Al_2O_3 ratio between 4 and 10.

Adsorption

Adsorption properties of Linde W (*Zeo. Mol Sieve*. 627 (1974)):

Adsorbate	Activation temp., °C	T, K	P, Torr	Wt% adsorbed
Ar	400	673	544atm	22
NH_3	250	298	700	0.11
CO_2	250	298	700	0.057
N_2	250	77	700	NA

Linde X (FAU)

See Linde type X.

Linde Y (FAU)

See Linde type Y.

Liottite (LIO)
(natural)

Liottite first was identified in 1977, and was named in honor of Liotti, who first located the material.

Structure

(*Am. Mineral.* 321(1977))

CHEMICAL COMPOSITION: $(Ca,N1_2,K_2)$
  $9[Al_{18}Si_{18}O_{72}] *(SO_4,CO_3)_6CL_3(OH)_3*2H_2O$
SYMMETRY: hexagonal
SPACE GROUP: P6m2
UNIT CELL CONSTANTS (Å) $a = 12.8$
                               $b = 16.1$
PORE STRUCTURE: six-member-ring openings only

X-Ray Powder Diffraction Data: (*Am. Mineral.* 321(1977) (d(Å) (I/I_0)) 11.12(2), 9.17(1), 6.52(1), 5.37(1), 4.84(36), 4.210(2), 4.114(2), 4.077(2), 3.790(8), 3.715(100), 3.375(2), 3.217(2), 3.095(1), 3.058(1), 3.033(3), 2.885(4), 2.787(9), 2.760(2), 2.745(5), 2.686(10), 2.678(7), 2.630(3), 2.610(<1), 2.456(2),

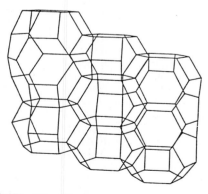

Liottite Fig. 1S: Framework topology.

2.427(2), 2.290(2), 2.225(2), 2.174(1), 2.155(2), 2.141(24), 2.102(4), 2.060(5), 1.999(1), 1.957(2), 1.932(6), 1.872(3), 1.816(<1), 1.801(10), 1.782(1), 1.768(1), 1.753(1), 1.732(2), 1.646(1), 1.618(1), 1.606(6), 1.581(<1), 1.559(1), 1.525(2), 1.505(1), 1.478(<1), 1.474(1), 1.391(1), 1.378(4), 1.346(4), 1.314(1).

Liottite exhibits a close relationship to the cancrinite–vishneivite–davyne series of minerals. Liottite contains calcium as the dominant cation, with SO_4^{-2} the dominant anion outside the framework. The framework is made up of a six-layer sequence of six-member rings with a stacking sequence of ABABAC (*Am. Mineral.* 321(1977)).

Lomonite (LAU)

Early name for laumontite.

Lovdarite (LOV)
(natural)

Lovdarite was first found by Men'schikov in the Lovozero pluton on Mount Karnasurt in Russia (*Dokl. Akad. Nauk SSSR* 221:154(1974)).

Structure

(*Eur. J. Mineral.* 2:809(1990))

CHEMICAL COMPOSITION: $Na_{12}K_4Be_8Si_{28}O_{72}*18H_2O$
SYMMETRY: orthorhombic
SPACE GROUP: Pma2
UNIT CELL CONSTANTS (Å): $a = 38.58$
                                     $b = 6.93$
                                     $c = 7.15$
PORE STRUCTURE: nine-member ring 3.7 × 3.2 × 4.4 Å
                                 eight-member ring 3.7 × 3.6 Å

X-Ray Powder Diffraction Data: (d(Å) (I/I_0)) 9.895(19.5), 7.150(17.9), 6.826(0.9), 6.725(100), 6.597(2), 6.541(30), 6.137(6), 5.795(84), 5.676(2.3), 5.214(0.8), 4.976(54.7), 4.947(6.7), 4.937(4.0), 4.848(2.2), 4.826(1.7),

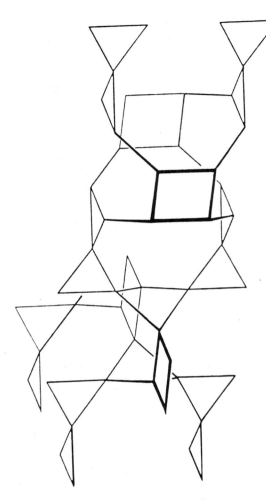

Lovdarite Fig. 1S: Framework topology.

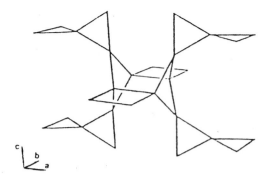

Lovdarite Fig. 2:   Basic configuration of lovdarite framework (*PP 7th IZC* 23 (1986)). (Reprinted with permission of Butterworths Publishers)

2.469(1.7), 2.466(5.1), 2.458(4.8), 2.445(1.7), 2.424(2.1), 2.413(0.6), 2.389(1.6), 2.383(0.6), 2.382(3.6), 2.347(3.1), 2.366(1.1), 2.356(1.3), 2.338(1.3), 2.331(0.8), 2.328(0.8), 2.310(0.7), 2.294(3.1), 2.288(17.3), 2.287(5), 2.275(5.7), 2.266(0.8) (from calculated pattern in *Zeo.* 10:426S(1990)).

The lovdarite structure is composed of a three-dimensional structure containing silicon and beryllium tetrahedra as a part of the structure. The alkali cations, K and Na, occupy sites within the cavities.

4.778(1.8), 4.656(1.3), 4.446(7.4), 4.381(1.7), 4.068(0.9), 3.973(0.8), 3.958(0.6), 3.736(1.1), 3.713(3.1), 3.518(14.6), 3.509(6.9), 3.465(3.8), 3.463(7.6), 3.452(2.1), 3.437(11.5), 3.413(9.7), 3.362(10.9), 3.351(2.5), 3.298(55.6), 3.295(7.3), 3.270(3.3), 3.193(30.9), 3.177(19.7), 3.174(2.1), 3.143(35.1), 3.137(55.3), 3.118(27.4), 3.109(11.2), 28.82(5.1) 3.080(62), 3.068(24.8), 3.035(13.7), 3.025(7.9), 2.995(1.8), 2.974(6.1), 2.954(10.9), 2.949(2.8), 2.916(8), 2.901(16.8), 2.862(7.6), 2.827(1.7), 2.819(2.9), 2.770(0.6), 2.749(22.9), 2.730(2.5), 2.673(24), 2.653(1.9), 2.638(5.9), 2.629(2.5), 2.607(5.5), 2.597(2.1), 2.575(0.6), 2.544(6.9), 2.488(9), 2.478(12.2), 2.474(5.8),

Atomic positional parameters for lovdarite (*Eur. J. Mineral.* 2:809 (1990)):

	x	y	z
Si1	0.2113(1)	0.0504(7)	0.2770
Si2	0.2113(1)	0.0497(8)	0.7179(9)
Si3	0.1952(1)	0.3671(7)	−0.0008(9)
Si4	0.1248(1)	0.5026(7)	−0.0269(9)
Si5	0.0582(1)	0.5267(7)	0.1399(9)
Si6	0.0333(1)	0.2593(7)	0.4469(9)
Si7	0.0458(1)	0.9202(7)	0.4588(9)
Be1	0.1805(4)	0.7820(30)	0.0050(33)
Be2	0.0670(5)	0.5415(36)	0.7334(34)
K	0.3860(1)	0.0417(7)	0.1828(9)
Na1	1/4	0.6098(19)	0.2437(18)
Na2	1/4	0.6217(17)	0.7683(17)
Na3	0.1283(2)	0.8542(13)	0.6990(13)
Na4	0.0069(4)	0.2840(24)	0.8654(26)
O1	1/4	−0.0319(24)	0.2723(22)
O2	0.2103(3)	0.2603(20)	0.1857(19)
O3	0.1858(3)	−0.0960(18)	0.1901(17)

	x	y	z
O4	0.2023(2)	0.0905(17)	0.5007(19)
O5	1/4	−0.0324(26)	0.7287(24)
O6	0.2106(3)	0.2624(19)	0.8147(18)
O7	0.1858(3)	−0.1014(20)	0.8030(18)
O8	0.2065(2)	0.5870(16)	0.0011(18)
O9	0.1540(2)	0.3356(16)	0.0027(18)
O10	0.1409(3)	0.7122(17)	−0.0041(19)
O11	0.0979(26)	0.4563(18)	0.1463(18)
O12	0.1057(29)	0.4735(19)	0.7794(19)
O13	0.0355(3)	0.3608(19)	0.2493(19)
O14	0.0563(3)	0.7205(22)	0.2606(22)
O15	0.0448(3)	0.5539(20)	−0.0671(20)
O16	0.0514(3)	0.0501(19)	0.4226(17)
O17	0.0493(3)	0.3782(19)	0.6145(20)
O18	−0.0060(3)	0.2198(19)	0.5039(20)
O19	0.0678(3)	−0.2501(17)	0.6341(18)
W1	0.2115(3)	0.5643(19)	0.5123(22)
W2	0.1438(3)	0.6958(20)	0.4247(19)
W3	0.1290(3)	0.1820(21)	0.5459(20)
W4	0.0744(4)	0.0616(25)	0.8790(25)
W5	0	0	0.0134(60)

LZ-105 (MFI)
(US4,257,885(1981))
Union Carbide Corporation

LZ-105 identifies a molecular sieve material with the ZSM-5 topology, prepared in the absence of organic amine additives.

LZ-132 (LEV)
(EP 91,048(1983))
Union Carbide Corporation

Related Materials: levyne

Structure

CHEMICAL COMPOSITION: x $M_{2/n}O$: Al_2O_3 : y SiO_2 : z H_2O (x = 0–3.5; y = 10–80; z = 0–15)
PORE STRUCTURE: eight-member rings 3.7 × 4.8 Å

X-Ray Powder Diffraction Data: ($d(Å)$ (I/I_0))
(as-synthesized) 10.2(12), 7.97(41), 7.56(2), 6.51(21), 5.47(11), 5.07(84), 4.93(15), 4.46(2), 4.19(58), 4.00(100), 3.75(32), 3.52(7), 3.40(4), 3.26(18), 3.11(50), 3.02(9), 2.805(6), 2.757(39), 2.578(9), 2.536(3), 2.475(3), 2.356(2), 2.254(4),

2.217(1), 2.181(2), 2.122(3), 2.090(7), 2.023(2), 1.998(4), 1.910(1), 1.884(4), 1.852(5), 1.762(9), 1.719(1), 1.647(3); (calcined) 10.05(26), 7.97(100), 7.56(7), 6.51(71), 5.47(3), 5.04(23), 4.40(sh), 4.17(27), 3.97(40), 3.75(14), 3.51(4), 3.38(3), 3.23(12), 3.09(4), 2.788(5), 2.730(20), 2.564(5), 2.515(3), 2.455(2), 2.332(1), 2.254(1), 2.171(1), 2.118(2), 2.080(2), 2.014(1), 1.989(1), 1.888(1), 1.877(2), 1.838(2), 1.801(3), 1.755(3), 1.707(1).

LZ-132 exhibits properties characteristic of a small-pore molecular sieve.

Synthesis

ORGANIC ADDITIVE
methylquinuclidine

LZ-132 crystallizes from a batch composition of: Al_2O_3 : 20–50 SiO_2 : 4.5–4.0 Na_2O : 420 H_2O : 5.6 R (where R is methylquinuclidine iodide). Crystallization occurs at 150°C after 167 hours. The replacement of Cs_2O for Na_2O also produces this structure, but the crystallization time is lengthened. Increasing the silica and organic concentrations also results in longer crystallization times.

Adsorption

Equilibrium adsorption properties of LZ-132 (EP 91,048 (1983)):

	Kinetic diameter, Å	Pressure, Torr	Temp., °C	Wt. % adsorbed
O_2	3.46	102	−183	26.0
O_2		755	−183	30.1
n-Butane	4.3	101	24	1.4
n-Butane		301	24	1.9
n-Butane		303	50	2.7
n-Hexane	4.3	11.5	24	13.1
n-Hexane		83	24	15.0
Isobutane	5.0	762	24	0.42
H_2O	2.65	4.6	24	14.1
H_2O		20.0	24	26.1

LZ-133 (LEV)
(EP 91,049(1983))
Union Carbide Corporation

Related Materials: levyne

Structure

CHEMICAL COMPOSITION: $x M_{2/n}O : Al_2O_3 : y$ $SiO_2 : z H_2O$ ($x = 0-3.5; y = 10-80; z = 0-15$)
PORE STRUCTURE: eight-member rings, levyne-type structure

X-Ray Powder Diffraction Data: ($d(Å)$ (I/I_0))
(as-synthesized) 11.5(6), 9.03(9), 7.31(9), 6.56(31), 6.42(26), 6.24(24), 6.07(10), 5.47(11), 5.16(32), 4.98(6), 4.90(8), 4.80(18), 4.60(100), 4.48(34), 4.25(9sh), 4.31(9), 4.15(20), 3.90(24), 3.80(13), 3.66(62), 3.58(11), 3.45(28), 3.28(23), 3.21(29), 3.14(26), 3.12(27), 3.02(27), 2.905(6), 2.885(2), 2.780(4), 2.722(3), 2.652(8), 2.571(2), 2.321(11); (calcined 600°C) 11.5(20), 10.9(18), 9.03(33), 7.31(28), 6.46(100), 6.37(sh), 6.24(67), 6.03(10), 5.47(7), 5.13(14), 4.96(sh), 4.90(9), 4.77(12), 4.58(91), 4.46(37), 4.29(8), 4.13(16), 3.87(20), 3.75(11), 3.63(65), 3.58(sh), 3.44(30), 3.26(27), 3.18(33), 3.13(29), 3.11(30), 3.01(36), 2.940(7), 2.885(5), 2.772(6), 2.730(6), 2.667(sh), 2.637(11), 2.564(6), 2.495(32).

LZ-133 exhibits properties characteristic of a small-pore molecular sieve.

Synthesis

ORGANIC ADDITIVE
methylquinuclidine

LZ-133 crystallizes from a reactive gel with a molar oxide composition of: Al_2O_3 : 27–40 SiO_2 : 4.5 Na_2O : 420 H_2O : 5.6 R (where R refers to methylquinuclidine iodide). Crystallization is readily achieved at 200°C after 24 hours. Increasing the concentration of SiO_2 and the organic increases the time of crystallization. Using the mixed alkali system Na/Cs also lengthens the time required before crystallization is achieved.

Adsorption

Equilibrium adsorption data for LZ-133 (EP 91,049 (1983)):

	Kinetic diameter, Å	Pressure, Torr	Temp., °C	Wt. % adsorbed
O_2	3.46	97.0	−183	12.0
O_2		502	−183	14.4
n-Hexane	4.3	99.0	23	4.8
n-Butane	4.3	760	23	1.42*
Isobutane	5.0	504	23	0.67
H_2O	2.65	4.6	23	10.8
H_2O		19.0	23	19.4

LZ-135
(US4,857,288(1989))
UOP

Structure

Chemical Composition: $x M_{2/n}O : Al_2O_3 : y$ SiO_2 (M = cation of valence n; $x = 0-3.5$; $y = 3.5-6.5$).

X-Ray Powder Diffraction Data: ($d(Å)$ (I/I_0))
15.78(20), 13.60(7), 10.28(27), 9.12(10), 7.90(64), 7.38(17), 6.81(17), 6.71(16), 6.24(30), 5.95(13), 5.83(9), 5.47(4), 4.87(15), 4.72(13), 4.40(14), 4.29(10), 4.15(9), 3.93(34), 3.82(100), 3.69(48), 3.60(37), 3.453(31), 3.351(11), 3.153(31), 3.132(31), 3.038(61), 2.912(29), 2.867(25), 2.823(43), 2.739(16), 2.691(8), 2.571(9), 2.536(10), 2.442(9), 2.398(6), 2.254(11), 2.141(10), 2.118(6), 2.067(10), 1.957(7), 1.918(17), 1.855(9), 1.811(6), 1.774(10), 1.759(7), 1.704(7), 1.664(7).

This material is a small-pore molecular sieve with a pore diameter of 4.3 Å.

Synthesis

ORGANIC ADDITIVES
tetraethylammonium/tetramethylammonium cation

LZ-135 crystallizes from a reactive gel with a molar composition of: 0.1–6 R_2O : 0.1–6 Q_2O : 1–10 $M_{2/n}O$: Al_2O_3 : 4–6 SiO_2 : 50–250 H_2O. R = tetraethylammonium cation; Q is tetramethylammonium cation. Sodalite and P_c occur as common impurities.

Thermal Properties

LZ-135 is stable to at least 500°C, the temperature required to remove the trapped organic from the pores.

Adsorption

Adsorption properties of LZ-135:

Adsorbate	Kinetic diameter, Å	Pressure, Torr	Temp., °C	Wt. % adsorbed
O_2	3.46	108	−183	13.7
O_2		730	−183	14.7
n-Hexane	4.3	50	23.9	4.3
n-Hexane		102	24.0	5.0
Isobutane	5.0	104	23.7	0.8
H_2O	2.65	4.6	23.6	17.1
H_2O		19.5	23.2	19.6

LZ-200
(US4,348,369(1982))
Union Carbide Corporation

Structure

X-Ray Powder Diffraction Data: $(d(Å)\,(I/I_0))$
12.1(vs), 6.99(m), 4.75(m), 4.57(m), 4.42(m), 3.49(m), 3.03(s), 2.89(m), 2.81(m), 2.64(m).

Analysis of the adsorption properties of this material indicate the presence of small pores slightly smaller than those of potassium zeolite A.

Synthesis

ORGANIC ADDITIVE
none

LZ-200 is prepared from an aluminosilicate gel with a composition of: 1.76 SiO_2 : Al_2O_3 : 2.06 Na_2O : 56 H_2O. Crystallization occurs after 10 hours at 100°C. Other phases observed as impurities or transient species in this synthesis include hydroxysodalite, cancrinite, and zeolite A.

Adsorption

Summary of adsorption properties of LZ-200 (US 4,348,369 (1982)):

Adsorbate	Temp., °C	Pressure, Torr	Gas adsorbed cc/g LZ-200
O_2	−183	10–750	not adsorbed
O_2	0	300–750	not adsorbed
N_2	−196	100–750	not adsorbed
CH_3OH	23	80	0.015–0.045
CO_2	23	100	0.089–0.124
		750	0.155–0.162
H_2O	23	4.6	0.174–0.184
		19.5	0.203–0.206

LZ-202 (MAZ)
(US4,503,023(1985))
Union Carbide Corporation

LZ-202 represents high-silica mazzite, a material that has dealuminated using $(NH_4)_2SiF_6$.

LZ-Designation of Zeolites: Secondary Synthesis (US 4,503,023 (1985)):

Designation	Zeolite Type
LZ-210	Si substituted Y zeolite
LZ-211	Si substituted mordenite
LZ-212	Si substituted L zeolite
LZ-213	Si substituted omega zeolite
LZ-214	Si substituted Rho
LZ-215	Si substituted N-A zeolite
LZ-216	Si substituted W zeolite
LZ-217	Si substituted offretite
LZ-218	Si substituted chabazite
LZ-219	Si substituted clinoptilolite
LZ-220	Si substituted erionite

(continued)

Designation	Zeolite Type	Designation	Zeolite Type
LZ-221	Si substituted ferrierite	LZ-248	Zn substituted Y zeolite
LZ-222	Si substituted LZ-105	LZ-249	Cr substituted mordenite
LZ-223	Si substituted ZSM-5	LZ-250	Cr substituted LZ-202 (omega type)
LZ-224	Fe substituted Y zeolite	LZ-251	Cr substituted L zeolite
LZ-225	Ti substituted Y zeolite	LZ-252	Sn substituted mordenite
LZ-226	Fe substituted mordenite	LZ-253	Sn substituted LZ-202 (omega type)
LZ-227	Ti substituted mordenite	LZ-254	Sn substituted L zeolite
LZ-228	Fe substituted L zeolite	LZ-255	Si-Ti substituted LZ-202 (omega type)
LZ-229	Ti substituted L zeolite	LZ-256	Si-Ti-Fe substituted LZ-202 (omega type)
LZ-230	Si substituted W zeolite	LZ-257	Ti-Fe substituted LZ-202 (omega type)
LZ-231	Ti substituted clinoptilolite	LZ-258	Si-Fe substituted LZ-202 (omega type)
LZ-232	Ti substituted erionite	LZ-259	Si-Ti-Fe substituted Y zeolite
LZ-233	Ti substituted offretite	LZ-260	Si-Fe substituted Y zeolite
LZ-234	Fe substituted erionite	LZ-261	Ti-Fe substituted Y zeolite
LZ-235	Fe substituted chabazite	LZ-262	Si-Fe substituted mordenite
LZ-236	Ti substituted chabazite	LZ-263	Ti-Fe substituted mordenite
LZ-237	Fe substituted clinoptilolite	LZ-264	Si-Ti-Fe substituted mordenite
LZ-238	Sn substituted Y zeolite	LZ-265	Ti-Fe substituted L zeolite
LZ-239	Cr substituted Y zeolite	LZ-266	Si-substituted NH4-beta
LZ-240	2r substituted Y zeolite	LZ-267	Si-substituted ZSM-20
LZ-241	Ti substituted ZSM-5	LZ-268	Fe-substituted NH4-beta
LZ-242	Ti substituted LZ-105	LZ-269	Fe-substituted LZ-105
LZ-243	Si-Ti substituted Y zeolite	LZ-270	Ti-substituted NH4-beta
LZ-244	Si-Ti substituted mordenite	LZ-271	Fe-substituted LZ-202 (omega type)
LZ-245	Co substituted Y zeolite	LZ-273	Sn4 + − substituted zeolite L
LZ-246	Si substituted LZ-202 (omega type)	LZ-274	Sn4 + − substituted zeolite mordenite
LZ-247	Ti substituted LZ-202 (omega type)	LZ-275	Sn4 + − substituted zeolite Y

M

M

See Linde M.

MAPO-5 (AFI)
(US4,567,029(1986))
Union Carbide Corporation

Related Materials: $AlPO_4$-5

Structure

Chemical Composition: Magnesium alumi-nophosphate. See MeAPO.

X-Ray Powder Diffraction Data: $(d(\text{Å})(I/I_0))$ 12.1(83), 6.92(10), 5.99(21), 4.53(56), 4.23(60), 3.99(100), 3.60(6), 3.47(34), 3.09(19), 3.00(20), 2.683(5), 2.614(16), 2.442(3), 2.386(12), 2.146(3), 1.918(5), 1.661(3).

For a discussion of the framework topology of MAPO-5 see $AlPO_4$-5.

Synthesis

ORGANIC ADDITIVES
N,N-diethanolamine
tripropylamine
tetrapropylammonium$^+$
2-methylpyridine
tetraethylammonium$^+$
diisopropylamine
cyclohexylamine
triethylamine
N-methylbutylamine
N-ethylbutylamine
dibutylamine

MAPO-5 crystallizes from reactive gels with varying amounts of magnesium (from magnesium acetate) and alumina. A general batch composition is: 1.0 R : a MgO : b Al$_2$O$_3$: P$_2$O$_5$: ca. 40 H$_2$O : $2a$ CH$_3$COOH, with Mg and Al varied so that Mg + Al = 1. Crystallization takes place between 150 and 200°C,

depending on the organic chosen, with crystallization times between 24 (using diisopropylamine, cyclohexylamine, N-methylbutylamine, N-ethylbutylamine, and dibutylamine) and 168 hours (using triethylamine).

Adsorption

Adsorption properties of MAPO-5:

Adsorbate	Kinetic diameter, Å	Pressure, Torr	Temp., °C	Wt. % adsorbed
O_2	3.46	102	−183	15.4
O_2		706	−183	18.0
Cyclohexane	6.0	74	25	10.5
n-Butane	4.3	705	24	7.4
Neopentane	6.2	701	24	7.2
H_2O	2.65	4.6	24	17.3
H_2O		22	24	26.3

MAPO-11 (AEL)
(US4,567,029(1986))
Union Carbide Corporation

Related Materials: $AlPO_4$-11

Structure

Chemical Composition: Magnesium alumi-nophosphate. See MeAPO.

X-Ray Powder Diffraction Data: $(d(\text{Å})(I/I_0))$ 10.78(44), 9.21(65), 6.71(23), 5.61(44), 4.65(7), 4.35(71), 4.19(69), 4.00(71), 3.95(57), 3.90(67), 3.82(100), 3.63(8), 3.59(13), 3.36(41), 3.16(12), 3.11(25), 3.03(11), 2.83(14), 2.73(29), 2.62(11), 2.47(8), 2.39(17), 2.03(8), 1.80(6).

For a description of the framework topology of MAPO-11 see $AlPO_4$-11.

Synthesis

ORGANIC ADDITIVES
diisopropylamine
tripropylamine
N-methylbutylamine
dibutylamine
N-ethylbutylamine
di-n-propylamine

MAPO-11 crystallizes from reactive aluminophosphate gels containing magnesium acetate replacing some of the aluminum in the system. Crystallization occurs readily at temperatures around 150 to 200°C after 24 hours or longer, depending on the organic employed in the crystallization. An impurity phase was reported for some systems. A typical composition includes: 1.0 $C_6H_{15}N$: 2x MgO : (1 − x) Al_2O_3 : 1.0–1.25 P_2O_4 : ca. 40 H_2O.

Adsorption

Adsorption properties of MAPO-11:

Adsorbate	Kinetic diameter, Å	Pressure, Torr	Temp., °C	Wt. % adsorbed
O_2	3.46	102	−183	6.0
O_2		706	−183	7.2
Xenon	4.0	751	24	8.7
n-Butane	4.3	705	24	2.9
Cyclohexane	6.0	74	25	1.5
Neopentane	6.2	701	24	0.3
H_2O	2.65	4.6	24	12.6
H_2O		22	24	14.3

MAPO-12
(US4,567,029(1986))
Union Carbide Corporation

Related Materials: $AlPO_4$-12

Structure

Chemical Composition: magnesium aluminum phosphate. See MeAPO.

X-Ray Powder Diffraction Data: $(d(\text{Å})(I/I_0))$
14.5(37), 7.97(16), 7.25(25), 6.77(68), 6.33(6), 5.19(25), 4.82(27), 4.27(100), 4.13(17), 4.00(44), 3.73(49), 3.56(33), 3.38(37), 3.10(16), 2.998(25), 2.912(44), 2.714(14), 2.644(21), 2.585(19).

See $AlPO_4$-12 for a description of the framework topology for MAPO-12

Synthesis

ORGANIC ADDITIVE
ethylenediamine

MAPO-12 crystallizes from a reaction composition of: 1.0 $C_2H_8N_2$: 0.167 MgO : 0.917 Al_2O_3 : 1.0 P_2O_5 : 39.8 H_2O : 0.33 CH_3COOH. Crystallization occurs after 24 hours at 200°C.

MAPO-14
(US4,567,029(1986))
Union Carbide Corporation

Related Materials: $AlPO_4$-14

Structure

Chemical Composition: magnesium aluminophosphate. See MeAPO.

X-Ray Powder Diffraction Data: $(d(\text{Å})(I/I_0))$
10.1(100), 8.19(11), 6.81(21), 5.99(5), 5.61(15), 5.01(15), 4.82(2), 4.27(7), 4.06(22), 3.97(10), 3.51(19), 3.36(4), 3.24(18), 3.09(13), 3.05(15), 3.01(23), 2.90(12), 2.75(8), 2.698(1), 2.529(4), 2.361(1), 2.265(4).

For a detailed description of the topology of MAPO-14 see $AlPO_4$-14.

Synthesis

ORGANIC ADDITIVES
ethylenediamine
isopropylamine

MAPO-14 is crystallized from a batch composition with a molar oxide ratio of: 1.0 $C_2H_8N_2$: 0.4 MgO : 0.8 Al_2O_3 : 1.0 P_2O_5 : 39.6 H_2O : 0.8 CH_3COOH. Magnesium acetate hydrate is used as the source of magnesium. Crystallization occurs after 24 hours at temperatures between 150 and 200°C. Using isopropylamine in the reactive gel results in a mixture of phases.

MAPO-16 (AST)
(US4,567,029(1986))
Union Carbide Corporation

Related Materials: $AlPO_4$-16

Structure

Chemical Composition: magnesium aluminophosphate. See MeAPO.

X-Ray Powder Diffraction Data: $(d(\text{Å})(I/I_0))$ 7.69(55), 4.72(49), 4.02(100), 3.87(13), 3.35(19), 3.07(13), 2.99(23), 2.37(6), 1.87(6).

For a description of the framework topology for MAPO-16 see $AlPO_4$-16.

Synthesis

ORGANIC ADDITIVE
quinuclidine

MAPO-16 is prepared from a reactive gel with a molar oxide ratio of: 1.0 quin : 0.4 MgO : 0.8 Al_2O_3 : 1.0 P_2O_5 : 40 H_2O. Crystallization occurs after 24 hours at 150°C. Both magnesium oxide and magnesium acetate have been used successfully in this synthesis.

Thermal Properties

MAPO-16 maintains its crystalline framework structure upon heating to 600°C.

Adsorption

Adsorption properties of MAPO-16:

Adsorption	Kinetic diameter, Å	Pressure, Torr	Temp., °C	Wt. % adsorbed
O_2	3.46	701	−183	4.6
O_2		101	−183	2.8
n-Butane	4.3	701	24	0.6
H_2O	2.65	4.6	24	19.9
H_2O		20	24	27.5

The adsorption properties were measured on the calcined form of this material.

MAPO-17 (ERI)
(US4,567,029(1986))
Union Carbide Corporation

Related Materials: $AlPO_4$-17
erionite

Structure

Chemical Composition: magnesium aluminophosphate. See MeAPO.

X-Ray Powder Diffraction Data: $(d(\text{Å})(I/I_0))$ 11.4(100), 9.05(50), 6.60(51), 6.21(33), 6.05(14), 5.72(70), 5.34(36), 4.93(39), 4.53(88), 4.33(96), 4.15(83), 3.81(55), 3.73(48), 3.52(87), 3.25(23), 3.10(41), 2.92(19), 2.861(10), 2.812(28), 2.677(37), 2.604(14), 2.575(25).

For a description of the framework topology see $AlPO_4$-17.

Synthesis

ORGANIC ADDITIVES
quinuclidine
cyclohexylamine

MAPO-17 crystallizes from a reactive magnesium-containing aluminophosphate gel with a molar oxide ratio of: 1.0 R : 0.167 MgO : 0.917 Al_2O_3 : 1 P_2O_5 : 39.8 H_2O : 0.33 CH_3COOH.

Crystallization occurs readily at 200°C after 168 hours. MAPO-17 obtained from this system has a minor impurity phase.

MAPO-20 (SOD)
(US4,567,029(1986))
Union Carbide Corporation

Related Materials: AlPO$_4$-20
                    sodalite

Structure

Chemical Composition: magnesium aluminophosphate. See MeAPO.

X-Ray Powder Diffraction Data: $(d(Å)(I/I_0))$
6.37(47), 4.53(42), 4.04(8), 3.69(100), 3.21(15), 2.86(12), 2.61(15), 2.42(1), 2.26(4), 2.13(4), 1.93(4), 1.78(8).

For a description of the framework topology of MAPO-20 see sodalite.

Synthesis

ORGANIC ADDITIVES
tetramethylammonium$^+$
tetraethylammonium$^+$
quinuclidine

MAPO-20 crystallizes from a reactive magnesium aluminophosphate mixture containing: 1 R : 0.33 MgO : 0.83 Al$_2$O$_3$: P$_2$O$_5$: 40 H$_2$O. Crystallization occurs readily at 200°C after 24 hours, using tetramethylammonium cation as the organic additive. Other organic amines also result in the crystallization of this framework structure, but not as a major component. Both magnesium oxide and magnesium acetate have been used as a source of magnesium ion in this system.

Adsorption

Adsorption properties of MAPO-20:

Adsorbate	Kinetic diameter, Å	Pressure, Torr	Temp., °C	Wt. % adsorbed
O$_2$	3.46	101	− 183	0.5
O$_2$		701	− 183	2.1
n-Butane	4.3	703	24	0.4
H$_2$O	2.65	4.6	24	23.8
H$_2$O		22	24	31.5

MAPO-34 (CHA)
(US4,567,029(1986))
Union Carbide Corporation

Related Materials: AlPO$_4$-34
                    chabazite

Structure

Chemical Composition: magnesium aluminophosphate. See MeAPO.

X-Ray Powder Diffraction Data: $(d(Å)(I/I_0))$
9.31(100), 6.92(14), 6.28(17), 5.54(39), 4.90(23), 4.31(91), 4.00(5), 3.85(6), 3.52(25), 3.45(17), 3.23(4), 3.14(4), 3.03(5), 2.93(31), 2.86(23), 2.61(7), 2.47(3), 2.28(3), 2.09(4), 1.92(3), 1.87(5), 1.79(4).

MAPO-34 exhibits properties characteristic of a small-pore material. See chabazite for description of the framework topology of the aluminophosphate-based series.

Synthesis

ORGANIC ADDITIVES
tetraethylammonium$^+$
isopropylamine

MAPO-34 crystallizes from a reactive gel with a molar oxide ratio of : 1 TEA OH : 0.4 MgO : 0.8 Al$_2$O$_3$: P$_2$O$_5$: 40 H$_2$O. Crystallization occurs readily after 24 hours at temperatures ranging between 150 and 200°C. Magnesium oxide and magnesium acetate hydrate both have been utilized for this crystallization.

Thermal Properties

MAPO-34 is thermally stable with little change in the X-ray powder diffraction pattern when heated to 600°C.

Adsorption

Adsorption properties of MAPO-34:

Adsorbate	Kinetic diameter, Å	Pressure, Torr	Temp., °C	Wt. % adsorbed
O_2	3.46	117	−183	25.4
O_2		708	−183	30.9
n-Butane	4.3	712	23	13.1
Isobutane	5.0	717	24	0.8
H_2O	2.65	4.6	24	29.0
H_2O		16	23	33.2

MAPO-35 (LEV)
(US4,567,029(1986))
Union Carbide Corporation

Related Materials: levyne
SAPO-35

Structure

Chemical Composition: magnesium aluminophosphate. See MeAPO.

X-Ray Powder Diffraction Data: $(d(\text{Å})(I/I_0))$ 10.2(21), 7.98(41), 6.61(28), 5.54(9), 5.10(100), 4.96(17), 4.53(11), 4.19(62), 4.04(100), 3.90(9), 3.82(21), 3.75(9), 3.55(4), 3.44(26), 3.12(34), 3.11(sh), 3.07(sh), 2.78(47), 2.59(14), 2.54(4), 2.14(4), 2.10(4), 1.88(6), 1.84(4).

MAPO-35 exhibits properties of a small-power molecular sieve structure.

Synthesis

ORGANIC ADDITIVE
quinuclidine

MAPO-35 crystallizes from a batch composition with a molar oxide ratio of: 1 quin : 0.4 MgO : 0.8 Al_2O_3 : P_2O_5 : 40 H_2O. Crystallization occurs after 146 hours at 100°C. Magnesium oxide is used as the source of magnesium.

MAPO-36 (ATS)
(US4,567,029(1986))
Union Carbide Corporation

Related Materials: SAPO-36

Structure

Chemical Composition: magnesium aluminophosphate. See MeAPO.

X-Ray Powder Diffraction Data: $(d(\text{Å})(I/I_0))$ 11.2(100), 10.8(sh), 6.56(5), 5.61(12), 5.40(18), 4.67(55), 4.27(39), 4.15(sh), 4.08(sh), 4.04(33), 3.97(32), 3.90(12), 3.74(8), 3.28(13), 3.21(sh), 3.16(9), 3.10(9), 2.97(7), 2.814(8), 2.714(3), 2.592(11).

MAPO-36 exhibits properties characteristic of a large-pore material.

Synthesis

ORGANIC ADDITIVES
tripropylamine
TPAOH

MAPO-36 crystallizes from a reactive gel with a molecular oxide ratio of: 1 R : 0.167–0.40 MgO : 0.8–0.917 Al_2O_3 : P_2O_5 : 40 H_2O. Crystallization occurs after 24 hours at 150°C to 200°C. Both magnesium oxide and magnesium acetate hydrate have been used successfully as sources of magnesium in this system. An impurity phase was noted in the system.

Thermal Properties

The X-ray powder diffraction pattern does not significantly change upon thermal treatment to 600°C.

Adsorption

Adsorption properties of MAPO-36:

Adsorbate	Kinetic diameter, Å	Pressure, Torr	Temp., °C	Wt% adsorbed
O_2	3.46	101	−183	18.9
O_2		703	−183	23.5
n-Butane	4.3	712	23	7.3
n-Hexane	4.3	105	23	10.5
Cyclohexane	6.0	73	23	9.1
Neopentane	6.2	745	23	6.4
H_2O	2.65	4.6	24	22.2
H_2O		22	24	31.1
CO_2	3.3	297	−82	22.0

MAPO-39 (ATN)
(US4,567,029(1986))
Union Carbide Corporation

Related Materials: other metal-substituted molecular sieves
$AIPO_4$-39

Structure

CHEMICAL COMPOSITION: magnesium aluminophosphate. See MeAPO.
UNIT CELL CONSTANTS (Å): $a = 13.09$
                          $c = 5.18$
PORE STRUCTURE: unidimensional eight-member rings

X-Ray Powder Diffraction Data: $(d(Å)(I/I_O))$
9.41(28), 6.66(49), 4.93(44), 4.19(82), 3.95(100), 3.31(4), 3.12(10), 3.02(17), 2.96(39), 2.75(14), 2.65(10), 2.46(3), 2.368(10), 2.206(3), 1.973(3), 1.873(4), 1.852(4), 1.781(4); (after calcination to 500°C) $(d(Å)(I/I_O))$ 9.12(12), 6.46(100), 4.77(50), 4.11(60), 3.85(87), 3.25(10), 3.07(14), 2.96(22), 2.91(33), 2.698(16), 2.585(9), 2.404(3), 2.332(8), 2.176(2), 1.957(1), 1.848(3), 1.759(4).

The MAPO-39 framework consists of cages similar to those found in cancrinite, differing

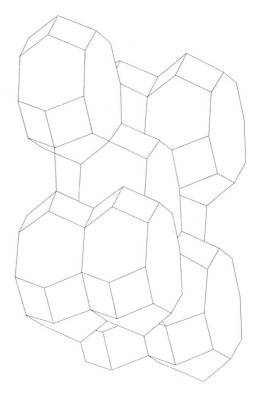

MAPO-39 Fig. 1S: Framework topology.

only in ring size, with MAPO-39 having eight-member rings and cancrinite six-member rings. This gives a channel system that is unidimensional with eight-member-ring openings. (IUC Conference, Bordeaux (July 1990)).

Synthesis

ORGANIC ADDITIVES
dipropylamine
isopropylamine

MAPO-39 crystallizes from a reactive gel with a molar composition of : R : 0.167–0.4 MgO : 0.8–0.917 Al_2O_3 : P_2O_5 : 19.8 H_2O : 0.33 CH_3COOH. Crystallization occurs at 150 to 200°C after 24 hours. The presence of a minor impurity phase was reported in this system.

Thermal Properties

See MAPO-39 Figures 3 and 4.

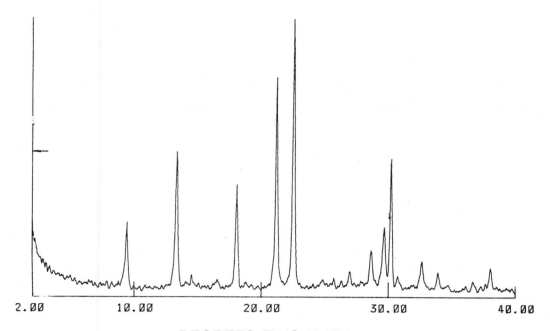

DEGREES TWO THETA

MAPO-39 Fig. 2: X-ray powder diffraction pattern of MAPO-39 prepared in the presence of di-*n*-propylamine. (Reproduced with permission of F. Testa, U. della Calabria)

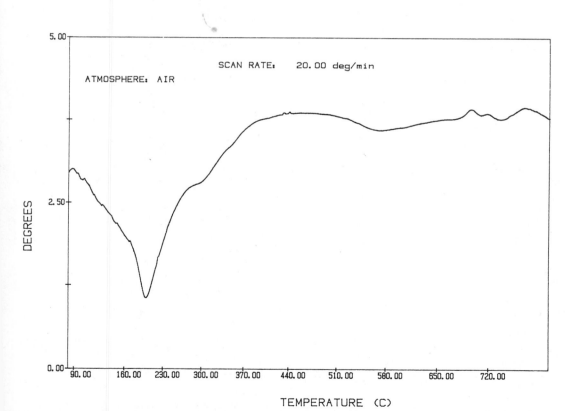

MAPO-39 Fig. 3: DTA of MAPO-39 from di-*n*-propylamine-containing system. (Reproduced with permission of B. Duncan, Georgia Tech)

301

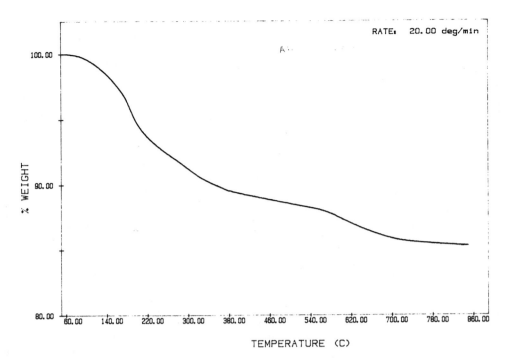

MAPO-39 Fig. 4: TGA of MAPO-39 from di-*n*-propylamine-containing system. (Reproduced with permission of B. Duncan, Georgia Tech)

Infrared Spectrum

See MAPO-39 Figure 5.

Adsorption

Adsorption properties of MAPO-39:

Adsorbate	Kinetic diameter, Å	Pressure, Torr	Temp., °C	Wt. % adsorbed
O$_2$	3.46	102	− 183	8.2
O$_2$		706	− 183	10.0
n-Butane	4.3	705	24	0.43
Cyclohexane	6.0	74	25	0.44
Xenon	4.0	750	24	9.3
Neopentane	6.2	701	24	0
H$_2$O	2.65	4.6	24	17.6
H$_2$O		22	24	22.8

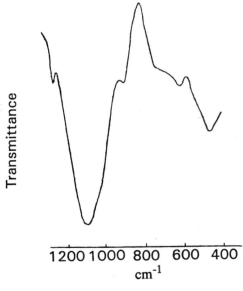

MAPO-39 Fig. 5: Mid-infrared spectrum of MAPO-39, as synthesized. (Reproduced with permission of B. Duncan, Georgia Tech)

MAPO-44 (CHA)
(US4,567,029(1986))
Union Carbide Corporation

Related Materials: SAPO-44
                    chabazite

Structure

Chemical Composition: magnesium alumino-phosphate. See MeAPO.

X-Ray Powder Diffraction Data: $(d(\text{Å})(I/I_0))$
94.1(86), 6.81(17), 6.42(3), 5.50(32), 5.10(3), 4.67(9), 4.29(100), 4.10(35), 3.93(8), 3.87(14), 3.66(56), 3.41(17), 3.24(10), 3.02(sh), 2.98(17), 2.90(54), 2.76(4), 2.73(4), 2.51(9), 2.344(1), 2.304(1), 2.265(1), 2.249(4), 2.233(3), 2.076(3), 1.929(1), 1.895(6), 1.873(5), 1.817(8), 1.707(5).

MAPO-44 exhibits properties characteristic of a small-pore molecular sieve chabazite.

Synthesis

ORGANIC ADDITIVE
cyclohexylamine

MAPO-44 crystallizes from a reactive gel composition: R : 0.6 MgO : 0.7 Al_2O_3 : P_2O_5 : 1.2 CH_3COOH: 39 H_2O. Crystallization occurs after 24 hours at temperatures between 150 and 200°C. Rhombohedral crystals are obtained.

Adsorption

Adsorption properties of MAPO-44:

Adsorbate	Kinetic diameter, Å	Pressure, Torr	Temp., °C	Wt. % adsorbed
O_2	3.46	12	−183	14.0
O_2		704	−183	16.2
Xenone	4.0	751	23	12.4
n-Butane	4.3	705	23	1.4
H_2O	2.65	20	23	19.2

MAPO-47 (CHA)
(US4,567,029(1986))
Union Carbide Corporation

Related Materials: chabazite

Structure

Chemical Composition: magnesium alumino-phosphate. See MeAPO.

X-Ray Powder Diffraction Data: $(d(\text{Å})(I/I_0))$
9.41(100), 6.86(8), 6.37(3), 5.54(18), 5.04(4), 4.67(4), 4.31(55), 4.08(5), 3.87(7), 3.60(13), 3.44(11), 3.23(4), 3.04(4), 2.93(24), 2.90(sh), 2.61(3), 2.515(2), 1.873(5), 1.811(3), 1.725(3); (500°C calcined materials) $(d(\text{Å})(I/I_0))$ 9.21(100), 6.81(21), 6.28(6), 5.47(21), 4.90(14), 4.62(3), 4.29(54), 3.99(3), 3.83(4), 3.53(12), 3.43(12), 3.21(3), 3.01(3), 2.91(22), 2.87(13), 2.592(4), 1.855(3).

MAPO-47 exhibits properties characteristic of a small-pore chabazite-type molecular sieve.

Synthesis

ORGANIC ADDITIVE
N,N-diethylethanolamine

MAPO-47 crystallizes from reactive gels with a molar oxide ratio of: 2 R : 0.167 MgO : 0.917 Al_2O_3 : P_2O_5 : 0.33 CH_3COOH : 5.5 C_3H_7OH : 40 H_2O. Crystallization occurs at 150° C after 144 hours. Magnesium acetate is used as the source of magnesium and aluminum isopropoxide as the source of aluminum.

Adsorption

Adsorption properties of MAPO-47:

Adsorbate	Kinetic diameter, Å	Pressure, Torr	Temp., °C	Wt. % adsorbed
O_2	3.46	12	−183	17.6
O_2		706	−183	21.5
Xenon	4.0	751	24	17.0
n-Butane	4.3	705	24	4.0
H_2O	2.65	20	23	27.4

MAPO-50 (AFY)
(US4,853,197(1989))
UOP

Structure

Chemical Composition: magnesium aluminophosphate. See MeAPO.

X-Ray Powder Diffraction Data: $(d(\text{Å})(I/I_0))$
11.40(21), 11.00(100), 9.01(16), 6.35(4), 5.20(2), 4.14(4), 3.91(2), 3.78(22), 3.68(14), 3.40(6), 3.40(6), 3.06(3), 3.00(7), 2.89(5), 2.40(2).

See CoAPO-50 for a description of the framework topology.

Synthesis

ORGANIC ADDITIVE
di-*n*-propylamine

MAPO-50 crystallizes from a reactive mixture with a molar oxide ratio of: 2.5 R : 0.3 MgO : 0.85 Al_2O_3 : 1.0 P_2O_5 : 0.7 CH_3COOH : 50 H_2O. Crystallization occurs at 150°C after 144 hours.

Adsorption

Adsorption properties of MAPO-50:

Adsorbate	Kinetic diameter, Å	Pressure, Torr	Temp., °C	Wt. % adsorbed
O_2	3.46	700	−183	7.2
n-Hexane	4.30	88	24	3.4

MAPSO

See MgAPSO.

Mazzite (MAZ)
(natural)

Related Materials: LZ-202
omega
ZSM-4

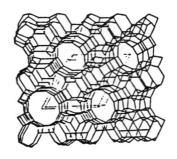

Mazzite Fig. 1S: Framework topology.

Mazzite has the distinction of having been the first natural mineral identified after its synthetic analogs were known. The mineral was named after Professor F. Mazzi, a mineralogist at Pavia University in Italy (*Nat. Zeo.* 160(1985)).

Structure

(*AZST* (1987))

CHEMICAL COMPOSITION: $K_3Ca_{1.5}Mg_2$
$(Al_{10}Si_{26}O_{72})* 28H_2O$
SYMMETRY: hexagonal
SPACE GROUP: P6₃/mmc
UNIT CELL CONSTANTS: (Å) $a = 18.39$
$c = 7.65$

DENSITY (calc): 2.108
PORE STRUCTURE: unidimensional 12 member rings

X-Ray Powder Diffraction Data: (*Nat. Zeo.* (1985)) $(d(\text{Å})(I/I_0))$ 15.93(35), 9.20(60), 7.96(35), 6.89(25), 6.02(53), 5.53(12), 5.31(17), 4.729(50), 4.423(12), 3.986(20), 3.824(95), 3.717(25), 3.655(47), 3.531(90), 3.474(12), 3.452(10), 3.185(100), 3.102(30), 3.065(38), 3.010(40), 2.941(100), 2.865(10), 2.681(12), 2.643(16), 2.552(20), 2.511(5), 2.446(1), 2.422(9), 2.393(1), 2.302(22), 2.298(22), 2.210(9), 2.147(10), 2.123(17), 2.037(10), 2.006(7), 1.9910(10), 1.9708(3), 1.9100(18), 1.8860(1), 1.8386(17), 1.7768(6), 1.6967(12), 1.6703(20), 1.6181(18), 1.5820(7), 1.5160(1), 1.5036(1), 1.4438(10).

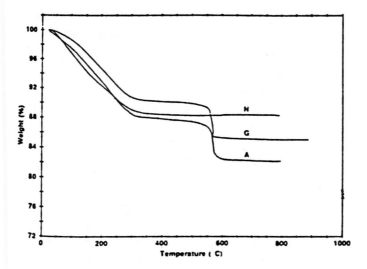

Mazzite Fig. 2:   Thermogravimetric analysis (TGA) scans of synthetic aluminum mazzite (A), gallium mazzite (G), and gallium mazzite after prior calcination at 600°C followed by NaCl/NaOH exchange (N). The heating rates were 10°C/min. The organic additive used in the synthesis of these materials was TMA$^+$ (*Mat. Res. Bull.* 20:125(1985)). (Reproduced with permission of Pergamon Press)

Atomic parameters with their standard deviations for the natural mineral mazzite (*Cryst. Struct. Commun.* 3:339 (1974)):

Atom	x/a	y/b	z/c	Occupancy (%)
Si1	0.4902(2)	0.1583(2)	1/4	100
Si2	0.0933(1)	0.3536(1)	0.0444(3)	100
O1	0.5178(6)	0.2589(6)	1/4	100
O2	0.5752(6)	0.1504(6)	1/4	100
O3	0.1004(4)	0.3821(5)	1/4	100
O4	0.1116(3)	0.4353(3)	−0.0723(7)	100
O5	0.1612(4)	0.3224(4)	0.0005(9)	100
O6	0	0.2740(4)	0	100
K, Na, Ca	1/2	0	0	46(2)
Mg	1/3	2/3	1/4	100
Ca	0	0	0.068(12)	21(2)

In the mazzite framework the characteristic feature is the presence of gmelinite cages. Each unit cell contains two cages shifted by $c/2$. The main channels in this zeolite are large 12-member rings with an average free aperture of ~7.5 Å. Two types of smaller channels are present: the first consists of stacked gmelinite cages surrounded by six-member rings; the second is between two crosslinked rows of cages and is sur-rounded by eight-member rings. The cations K, Na, and Ca are located in the smaller channels between the rows of gmelinite cages. The Mg in this mineral is at the center of the gmelinite cage and coordinates only to water molecules. Ca cations, to a lesser extent, also are located in these large cages. No Si/Al ordering is observed (*Cryst. Struct. Commun.* 3:339(1974)).

The gallosilicate analog also has been prepared and the structure studied. No Si/Ga ordering within the structure has been observed (*Mat. Res. Bull.* (20:125(1985)).

Thermal Properties

See Mazzite Figure 2.

MCH (CHA)
(UK 2,061,900(1981))
Imperial Chemical Industries

Related Materials: herschelite
                          chabazite

Structure

Chemical Composition: $1.0 \pm R_2O : Al_2O_3$ $: 4–7 \ SiO_2 : 0–8 \ H_2O$.

X-Ray Powder Diffraction Data: $(d(\text{Å})(I/I_0))$ 9.5(33), 5.4(12), 4.22(66), 3.2(23*), 2.9(100*), 2.6(13*), (* indicates incompletely resolved peaks).

Synthesis

ORGANIC ADDITIVE
none

MCH crystallizes from a reactive gel with molar oxide ratios of: SiO_2/Al_2O_3 of 5 to 20, OH^-/SiO_2 of 0.4 to 1.6, H_2O/OH^- between 10 and 30, and X^-/Al_2O_3 of 0 to 10. X is the anion of the strong acid. Crystallization occurs between 2 and 72 hours at temperatures ranging between 80 and 130°C. Gmelinite is a common impurity phase in this system.

Adsorption

Comparison of adsorption properties of MCH and natural herschelite:

	Adsorption wt% at 25°C and $P/P_o = 0.5$	
Sorbate	MCH	Natural herschelite
H_2O	23.5	24
n-Hexane	3.6	under 0.1
p-Xylene	2.0	under 0.1

Chemical Composition: $M_{x/m}^{m+} : (AlO_2)_{1-x}^- : (PO_2)_{1-x}^+ : (SiO_2)_{x+y}: N_{y/n}^{n-}$. M is a cation of valence *m;* N is an anion of valence *n; x* and *y* are numbers between -1 and 1.

MCM-1
(EP 146,384(1984))
Mobil Oil Corporation

Related Materials: $AlPO_4$-H3

Structure

Chemical Composition: $M_{x/m}^{m+} : (AlO_2)_{1-x}^- : (PO_2)_{1-x}^+ : (SiO_2)_{x+y}: N_{y/n}^{n-}$. M is a cation of valence *m;* N is an anion of valence *n; x* and *y* are numbers between -1 and 1.

Diffraction Lines: $(d(\text{Å})(I/I_0))$ 9.6677(29.97), 6.8569(100), 6.4870(77), 5.600(5), 4.8729(51), 4.8414(50), 4.2482(85), 3.9700(5), 3.6099(9), 3.4401(7), 3.3900(42), 3.0597(77), 2.9312(15), 2.8927(17), 2.7777(11), 2.6781(41); (after thermal treatment to 450°C) 6.8339(95), 4.8477(11), 4.7900(10), 4.5300(8), 4.2731(100), 4.1441(43), 3.6246(10), 3.5438(2), 3.4200(10), 3.2900(5), 3.0823(12), 3.0427(14), 2.6961(3).

Synthesis

ORGANIC ADDITIVE
tetrapropylammonium$^+$

MCM-1 is prepared via a two-phase synthesis using an organic phase that consists of hexanol and $Si(OC_2H_4)$ and a water phase consisting of the aluminum source, phosphoric acid, and the organic additive. The reaction mixture contains 9.3 Si/38.8 P and 51.9 Al. Crystallization occurs at 150°C after 65 hours of crystallization with a starting pH between 5 and 7 (EP 146,384(1984)). The resulting crystals are dried at 80°C.

Thermal Properties

Upon thermal treatment of this material, a change in the X-ray powder diffraction pattern is observed.

MCM-2 (CHA)
(EP 146,384(1984))
Mobil Oil Corporation

Related Materials: chabazite
SAPO-34

Structure

Chemical Composition: $M_{x/m}^{m+} : (AlO_2)_{1-x}^- . (PO_2)_{1-x}^+ : (SiO_2)_{x+y}: N_{y/n}^{n-}$. M is a cation of valence *m;* N is an anion of valence *n; x* and *y* are numbers between -1 and 1.

X-Ray Powder Diffraction Data: $(d(\text{Å})(I/I_0))$ 9.2412(99), 6.86(25), 6.4868(7), 6.2515/915),

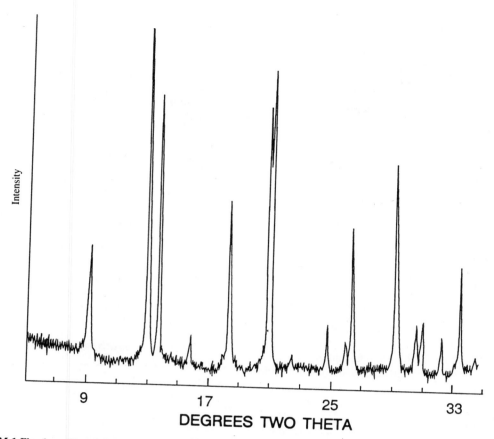

MCM-1 Fig. 1: X-ray powder diffraction pattern of MCM-1. (Reproduced with permission of B. Duncan, Georgia Tech)

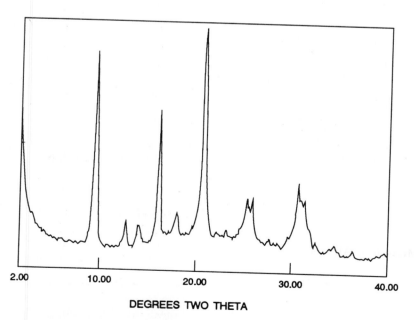

MCM-2 Fig. 1: X-ray powder diffraction pattern of MCM-2. (Reproduced with permission of B. Duncan, Georgia Tech)

5.524(73), 4.8868(21), 4.8257(12), 4.3030(100), 4.2584(24), 4.000(5), 3.84(5), 3.5075(21), 3.4376(23), 3.3947(7), 3.1239(2), 3.0495(14), 3.0160(6), 2.9190(34), 2.8492(25); (after thermal treatment to 450°C) 9.2476(100), 6.8414(23), 6.28(2), 5.505(17), 4.9465(5), 4.6200(2), 4.2923(36), 3.8415(5), 3.54(7), 3.4266(8), 3.2100(2), 3.16(2), 2.9086(15), 2.8621(9).

Synthesis

ORGANIC ADDITIVES
tetrapropylammonium$^+$
tetraethylammonium$^+$

MCM-2 is synthesized from a two-phase synthesis mixture containing an organic layer (hexanol and $Si(OC_2H_5)_4$) and an aqueous layer consisting of phosphoric acid, a source of alumina, and the organic additive. The composition of the reaction mixture that crystallizes this structure is: 10.8 Si : 45 P : 44.2 Al. Crystallization occurs after 168 hours at 150°C at a pH between 5 and 7.

Thermal Properties

MCM-2 appears to be thermally stable, with the X-ray diffraction pattern showing little decrease in overall peak intensity between room temperature and 450°C.

Adsorption

Adsorption properties of MCM-2:

	Temperature, °C	Wt% adsorbed
Water	60	10.3
n-Hexane	90	9.3
p-Xylene	90	2.7
2-Methylpentane	90	1.1
Cyclohexane	90	0.8
o-Xylene	120	0.9

MCM-3
(EP 146,384(1985))
Mobil Oil Corporation

Related Materials: variscite

Structure

Chemical Composition: $M_{x/m}^{m+}$: $(AlO_2)_{1-x}^-$: $(PO_2)_{1-x}^+$: $(SiO_2)_{x+y}$: $N_{y/n}^{n-}$. M is a cation of valence m; N is an anion of valence n; x and y are numbers between -1 and 1.

X-Ray Powder Diffraction Data: $(d(\text{Å})(I/I_0))$
5.1933(44), 4.8933(16), 4.5500(1), 4.2976(100), 3.9617(23), 3.9145(63), 3.8453(14), 3.6884(29), 3.5868(49), 3.3852(5), 3.0406(20), 2.9094(13), 2.8588(14), 2.7565(12).

Synthesis

ORGANIC ADDITIVE
tetrapropylammonium$^+$

A two-phase synthesis is used in preparing the SAPO analog to the aluminophosphate variscite. The organic phase contains hexanol and $Si(OC_2H_5)_4$, and the aqueous phase comprises the aluminum source, phosphoric acid, and the organic quaternary amine TPAOH. The inorganic composition of the mixture is: 12.4 Si : 52 P : 35.6 Al. Crystallization occurs after 65 hours at 150°C at a pH range between 5 and 7. Crystals are dried at 80°C.

MCM-4
(EP 146,384(1985))
Mobil Oil Corporation

Structure

Chemical Composition: $M_{x/m}^{m+}$: $(AlO_2)_{1-x}^-$: $(PO_2)_{1-x}^+$: $(SiO_2)_{x+y}$: $N_{y/n}^{n-}$. M is a cation of valence m; N is an anion of valence n; x and y are numbers between -1 and 1.

X-Ray Powder Diffraction Data: $(d(\text{Å})(I/I_0))$
4.4626(14), 4.3539(36), 4.2694(100), 4.0690(16), 3.9834(3), 3.74(5), 3.6516(31), 3.3698(62), 3.0467(8), 2.9447(7).

Synthesis

ORGANIC ADDITIVE
tetrapropylammonium$^+$

MCM-4 is prepared from a two-phase synthesis mixture with the organic phase being composed of hexanol and $Si(OC_2H_5)_4$ and the aqueous phase consisting of the aluminum source, phosphoric acid, and the quaternary amine TPAOH. Crystallization occurs at 150°C after 168 hours with the pH initially between 5 and 7.

The samples were dried at 80°C before the X-ray diffraction pattern was run.

MCM-5
(EP 146,384(1985))
Mobil Oil Corporation

Structure

Chemical Composition: $M_{x/m}^{m+}$: $(AlO_2)_{1-x}^-$: $(PO_2)_{1-x}^+$: $(SiO_2)_{x+y}$: $N_{y/n}^{n-}$. M is a cation of valence m; N is an anion of valence n; x and y are numbers between -1 and 1.

X-Ray Powder Diffraction Data: $(d(\text{Å})(I/I_0))$
8.5984(100), 6.7810(3), 4.7545(5), 4.6389(6), 4.5429(2), 4.4200(2), 4.3500(3), 4.2206(3), 4.1134/92), 3.8541(2), 3.7092(7), 3.6724(4), 3.4981(1), 3.3886(4), 3.3331(0.5), 3.2150(5), 3.1616(3), 3.0206(2), 2.9090(1), 2.8887(2), 2.7450(1), 2.7005(2), 2.6774(1), 2.6472(1), 2.5890(1), 2.5760(1).

Synthesis

ORGANIC ADDITIVE
tetrapropylammonium$^+$

MCM-5 is prepared from a two-phase synthesis mixture with the organic phase being composed of hexanol and $Si(OC_2H_5)_4$ and the aqueous phase consisting of the aluminum source, phosphoric acid, and the quaternary amine TPAOH. The inorganic composition consists of 9.3 Si : 38.6 P : 52.1 Al. Crystallization occurs after 168 hours at 150°C with an initial pH in the range of 5 to 7. After isolation of the crystalline product, MCM-5 is dried at 80°C.

MCM-6 (AFI)
(EP 146,384(1985))
Mobil Oil Corporation

Related Materials: SAPO-5

Structure

Chemical Composition: $M_{x/m}^{m+}$: $(AlO_2)_{1-x}^-$: $(PO_2)_{1-x}^+$: $(SiO_2)_{x+y}$: $N_{y/n}^{n-}$. M is a cation of valence m; N is an anion of valence n; x and y are numbers between -1 and 1.

X-Ray Powder Diffraction Data: $(d(\text{Å})(I/I_0))$
11.81(93), 6.84(12), 5.92(32), 4.48(81), 4.20(79), 3.96(100), 3.67(5), 3.42(42), 3.07(19), 2.96(22), 2.66(7), 2.59(22).

MCM-6 exhibits properties characteristic of a large-pore $AlPO_4$-5 molecular sieve. See $AlPO_4$-5 for a discussion of the framework topology.

Synthesis

ORGANIC ADDITIVE
tetrapropylammonium$^+$

MCM-6 is prepared from a two-phase synthesis mixture with the organic phase being composed of hexanol and $Si(OC_2H_5)_4$ and the aqueous phase consisting of the aluminum source, phosphoric acid, and the quaternary amine TEAOH. Small amounts of the inorganic base CsOH are used. The reaction mixture is cured for 24 hours at 130°C prior to crystallization at 180°C. Crystal formation is observed after 144 hours. The pH does not change after crystallization is complete, remaining at 6.5.

MCM-7 (AEL)
(EP 146,384(1985))
Mobil Oil Corporation

Related Materials: SAPO-11

Structure

Chemical Composition: $M_{x/m}^{m+}$: $(AlO_2)_{1-x}^-$: $(PO_2)_{1-x}^+$: $(SiO_2)_{x+y}$: $N_{y/n}^{n-}$. M is a cation of valence m; N is an anion of valence n; x and y are numbers between -1 and 1.

X-Ray Powder Diffraction Data: $(d(\text{Å})(I/I_0))$
10.85(18), 9.32(30), 6.70(12), 5.64(26),

4.37(86), 4.21(100), 4.00(60), 3.93(65), 3.90(71), 3.83(77), 3.11(9), 3.00(10), 2.72(12).

See AlPO$_4$-11 for a discussion of the framework topology.

Synthesis

ORGANIC ADDITIVE
di-n-propylamine

MCM-7 is prepared from a two-phase synthesis mixture with the organic phase composed of hexanol and Si(OC$_2$H$_5$)$_4$ and the aqueous phase consisting of the aluminum source, phosphoric acid, and the neutral amine di-n-propylamine. The reaction mixture is cured for 24 hours at 130°C prior to crystallization at 200°C. Crystal formation is observed after 24 hours. The pH of the reacting gel initially is 5.5, increasing to 9 when crystallization is complete.

MCM-8
(EP 146,384(1985))
Mobil Oil Corporation

Structure

Chemical Composition: $M_{x/m}^{m+}$: (AlO$_2$)$_{1-x}^-$: (PO$_2$)$_{1-x}^+$: (SiO$_2$)$_{x+y}$: $N_{y/n}^{n-}$. M is a cation of valence m; N is an anion of valence n; x and y are numbers between -1 and 1.

X-Ray Powder Diffraction Data: (d(Å)(I/I_0))
6.31(68), 4.47(48), 3.65(100), 3.16(15), 2.83(12), 2.59(13), 3.99(5).

Synthesis

ORGANIC ADDITIVE
tetrapropylammonium$^+$

MCM-8 is prepared from a two-phase synthesis mixture with the organic phase composed of hexanol and Si(OC$_2$H$_5$)$_4$ and the aqueous phase consisting of the aluminum source, phosphoric acid, and the quaternary amine TMAOH. Small amounts of the inorganic base KOH are used. The reaction mixture is cured for 24 hours at

130°C prior to crystallization at 180°C. Crystal formation is observed after 144 hours. The pH changes from an initial value of 5.5 to a final 7 after 144 hours.

MCM-9 (VFI)
(EP 146,384(1985))
Mobil Oil Corporation

Related Materials: VPI-5

Structure

X-Ray Powder Diffraction Data: (d(Å)(I/I_0))
16.41(62), 10.85(18), 9.66(32), 9.32(27), 8.20(21), 6.86(76), 6.71(12), 6.50(54), 6.16(11), 5.64(27), 5.46(4), 4.85(60), 4.74(15), 4.67(6), 4.36(49), 4.23(93), 4.14(37),   4.08(100), 4.01(28), 3.93(47), 3.83(41), 3.76(14), 3.61(14), 3.39(31), 3.34(9), 3.27(25), 3.16(14), 3.06(65), 2.94(18), 2.89(21), 2.84(7), 2.78(10), 2.73(16), 2.68(33), 2.62(8); (after thermal treatment to 450°C).   16.36(47),   14.05(21),   10.92(6), 8.94(41), 8.19(10), 6.90(13), 5.5(30), 4.47(23), 4.36(7), 4.08(100), 3.94(60), 3.78(34), 3.54(18), 3.45(10), 3.33(8), 3.28(13), 3.22(7), 3.01(33), 2.95(11), 2.82(8), 2.74(18).

See VPI-5 for a discussion of the framework topology.

Synthesis

ORGANIC ADDITIVE
d-n-propylamine

MCM-9 is prepared from a two-phase synthesis mixture with the organic phase composed of hexanol and Si(OC$_2$H$_5$)$_4$ and the aqueous phase consisting of the aluminum source, phosphoric acid, and the natural diamine d-n-propylamine. The reaction mixture is cured for 24 hours at 130°C prior to crystallization at 200°C between 24 and 48 hours. The initial pH for the system is between 5 and 7. The crystals are filtered, washed, and dried at 80°C. Impurity phases of SAPO-11 and MCM-1, silicon-containing analogs of AlPO$_4$-H6, have been observed in this system (*Appl. Catal.* 51:L13 (1989)).

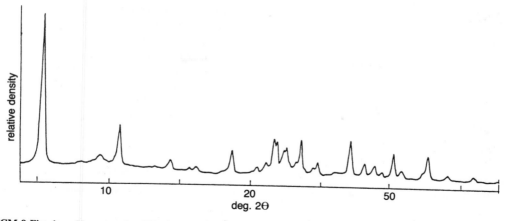

relative density

deg. 2Θ

MCM-9 Fig. 1:   X-ray powder diffraction pattern of MCM-9 (*Applied Catalysis* 51:213 (1989)). (Reproduced with permission of Elsevier Publishers)

Adsorption

Adsorption properties of MCM-9 (*Appl. Catal.* 51:L13 (1989)):

	Wt%
n-hexane	15.35
Cyclohexane	14.41
Triisopropylbenzene	14.29

MCM-10
(EP 146,384(1985))
Mobil Oil Corporation
Structure

Chemical Composition: $M_{x/m}^{m+}$: $(AlO_2)_{1-x}^{-}$: $(PO_2)_{1-x}^{+}$: $(SiO_2)_{x+y}$: $N_{y/n}^{n-}$. M is a cation of valence *m;* N is an anion of valence *n;* x and y are numbers between -1 and 1.

X-Ray Powder Diffraction Data: $(d(Å)(I/I_0))$
11.84(17),   10.20(78),   7.65(85),   6.86(24), 5.93(13), 5.69(36), 5.11(57), 5.01(64), 4.49(11), 4.37(86),   4.09(100),   3.95(12),   3.80(43), 3.77(14), 3.42(27), 3.22(81), 3.18(9), 2.99(21), 2.96(21), 2.93(32), 2.85(37), 2.78(7), 2.68(42), 2.59(19); (after thermal treatment to 450°C) 11.75(21),   10.10(85),   7.56(100),   6.81(72), 5.65(25), 5.08(29), 4.94(78), 4.46(17), 4.35(99), 4.28(26), 4.08(83), 3.85(16), 3.75(21), 3.41(42), 3.18(68), 2.96(30), 2.92(51), 2.88(11), 2.83(37), 2.66(43), 2.58(20).

Synthesis

ORGANIC ADDITIVE
Diquat-7(OH)$_2$

MCM-10 is prepared from a two-phase synthesis mixture with the organic phase consisting of hexanol and Si(OC$_2$H$_5$)$_4$ and the aqueous phase containing the source of aluminum, phosphoric acid, and the organic quaternary amine Diquat-7(OH)$_2$. This reaction mixture has a pH of 6 and is digested at 130°C for 24 hours before crystallization at 180°C. Crystals are observed after 144 hours.

Thermal Properties

Little change is observed in the X-ray powder diffraction pattern for MCM-10 after thermal treatment to 450°C.

MCM-20
(US4,968,652(1990))
Mobil Oil Corporation

Kenyaite-type layered material

MCM-25
(US4,968,652(1990))
Mobil Oil Corporation

Pillared material

MCM-21
(US, 696,807(1987))
Mobil Oil Corporation

Structure

Chemical Composition: iron silicoal-uminophosphate.

X-Ray Powder Diffraction Data: $(d(Å)(I/I_0))$
9.83(vs), 6.09(w), 5.41(m), 4.91(w), 4.74(w), 4.60(w), 4.17(w), 4.00(w), 3.55(w), 3.35(w), 3.28(w), 3.22(w), 3.07(w), 3.05(w), 2.95(w), 2.82(s), 2.74(vs), 2.65(m), 2.57(w), 2.48(w), 2.45(w), 2.36(w), 2.34(w), 2.29(w), 2.12(w), 2.09(w), 2.03(w), 1.96(w), 1.94(w), 1.92(w), 1.91(w), 1.87(w), 1.82(w), 1.75(w), 1.72(w), 1.71(w), 1.70(w), 1.64(m), 1.61(w), 1.58(w), 1.55(w).

Synthesis

ORGANIC ADDITIVES
tetrapropylammonium$^+$
tetramethylammonium$^+$
ethylenediamine
quinuclidine

MCM-21 is prepared from solutions containing phosphoric acid and iron nitrate. Tetrapropylammonium hydroxide or quinuclidine is used as the organic amine additive, and sodium hydroxide is the inorganic base. Crystallization occurs after 6 days at 150°C. If other organic additives such as TMAOH or ethylenediamine are used, crystallization occurs after 5 days at 200°C.

MCM-22
(US4,954,325(1990))
Mobil Oil Corporation

Related Materials: SSZ-25
PSH-3

Structure

Chemical Composition: Si/Al at least 10

X-ray Powder Diffraction Data: $(d(Å)(I/I_0))$:
30.0 (w-m), 22.1 (w), 12.36 (m-vs), 11.03 (m-s), 8.86 (w-m), 6.18 (m-vs), 6.00 (w-m), 5.54 (w-m), 4.92 (w), 4.64 (w), 4.41 (w-m), 4.25 (w), 4.10 (w-s), 4.06 (w-s), 3.91 (m-vs), 3.75 (w-m), 3.56 (w-m), 3.42 (vs), 3.30 (w-m), 3.20 (w-m), 3.14 (w-m), 3.07 (w), 2.99 (w), 2.82 (w), 2.78 (w), 2.68 (w), 2.59 (w).

Synthesis

ORGANIC ADDITIVES
hexamethyleneimine

Crystallization occurs at 150°C after 5 to 7 days.

MeAPO
(US4,567,029(1986))
Union Carbide Corporation

"MeAPO" represents the short-hand notation used to describe the metal aluminosilicates patented by Union Carbide in 1986. The "Me" elements include cobalt (Co), manganese (Mn), magnesium (M), and zinc (Z). A number placed after

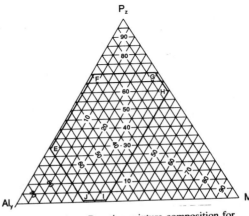

MeAPO Fig. 1:   Reaction mixture composition for MeAPO's (US4,567,019(1986)).

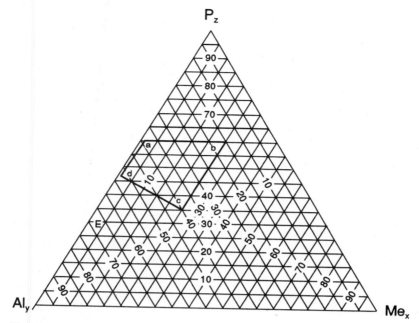

MeAPO Fig. 2:   Preferred product compositions for MeAPO's (US4,567,029 (1986)).

this notation, as for example, in ZAPO-5, indicates the aluminophosphate framework topology. The preferred ranges of compositions (reaction mixture and product) are shown in the ternary diagrams (MeAPO Figs. 1 and 2).

Melanophlogite is a rare silica mineral containing up to 8% occluded organics. This mineral first was described in 1876 (*N. Jb. Miner. Mh.* 119(1982)). The natural mineral is tetragonal at room temperature, whereas the synthetic form is cubic.

MeAPSO-43 (GIS)
SSSC28:103(1986)
Union Carbide Corporation

A metal containing aluminophosphate with the Gismondine topology. (Me = Fe,Mg,Mn,Co or Zn)

Melanophlogite (MEP)
(natural)

Related Materials: cubic gas hydrates of
type I

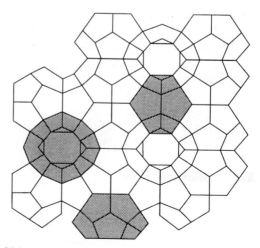

Melanophlogite Fig. 1S: Framework topology.

Structure

(*Z. Kristallogr* 164:247(1983))

CHEMICAL COMPOSITION: $46SiO_2*6M^{14}*2M^{12}$
(M^{12} = Ch_4,N_2; M^{14} = CO_2,N_2).
SYMMETRY: cubic
SPACE GROUP: Pm3n
UNIT CELL CONSTANTS (Å): a = 13.78
D_f = 19.0
CAGE TYPE: $2[5^{12}]$;$6[5^{12}6^2]$

X-Ray Powder Diffraction Data: (*Z. Kristallogr.* 164:247(1983)) $(d(Å)(I/I_0))$ (natural)
9.47(3), 6.70(3), 6.00(8), 5.47(6), 4.238(1), 3.870(7), 3.717(5), 3.579(10), 3.350(1), 3.251(5), 3.159(4); (synthetic) 6.71(3), 6.01(80), 5.47(6), 3.87(7), 3.72(5), 3.59(10), 3.26(5), 3.18(4).

Melanophlogite is a clathrate compound that consists of a three-dimensional framework of $[SiO_2]$ tetrahedra containing trapped small guest molecules. At room temperature the natural clathrasil is tetragonal and microtwinned. A displacive phase transformation occurs at 65°C. This transformation is dependent on the locality from which the samples were obtained.

The host framework consists of tetrahedra that are corner-linked to give a three-dimensional 4-connected net that is isotypic with the $[OH_4]$ framework of the cubic gas hydrates. There are two types of cages per unit cell: two pentagondodecahedra, $[5^{12}]$ cages, and six tetrakaidecahedra, $[5^{12}6^2]$ (*Z. Kristallogr.* 164:247

Atomic parameters for melanophlogite (*Z. Kristallogr.* 164:247 (1983)):

Atom	x	y	z
Si(1)	0	0.3098(1)	0.1142(1)
Si(2)	0.1826(1)	0.1826(1)	0.1826(1)
Si(3)	0.25	0	0.5
O(1)	0.0963(2)	0.2465(2)	0.1360(2)
O(2)	0	0.4056(2)	0.1813(2)
O(3)	0.3423(3)	0	0
O(4)	0.25	0.25	0.25
$[5^{12}6^2]$	0.25	0.5	0
$[5^{12}]$	0	0	0

Guest molecule distribution and population densities for melanophlogite; results obtained by combination of mass spectroscopic analysis and refinement of site occupation factors and temperature factors of guest molecules (*Z. Kristallogr.* 164:247 (1983)):

Cage type	Guest molecule	Guest molecule distribution [%]	e^-/cage, calc.	e^-/cage, exp.	Population density [%]
$[5^{12}]$	CH_4	100	10	9	90
$[5^{12}6^2]$	N_2	78	14	12	59
	CO_2	22	22		17

(1983)). The CH_4 and N_2 are thought to be located in the $[5^{12}]$ cages and Co_2 and N_2 in the $[5^{12}6^2]$ cages.

Synthesis

ORGANIC GUESTS
N_2, Kr, Xe, H_2O, CO_2, CH_4, CH_3NH_2

This structure has been prepared in the presence of methylamine at temperatures between 160 and 200°C (*Ind. J. Chem.* 27a:380(1988)). This material has been reported to transform to quartz with grinding (*N. Jb. Miner. Mh.* 119(1982)). Though the guest molecules are removed before 500°C, the framework is thermally stable at temperatures up to 1000°C.

Infrared Spectrum

Mid-infrared Vibrations cm^{-1}: 1118(s), 795(w), 456(s) (*Zeo.* 4:369(1984))

Merlinoite (MER) (natural)

Related Materials: K-M
                     Linde W

A recently discovered mineral (1977), merlinoite was found in Italy and named in honor of Professor Merlino of the University of Pisa.

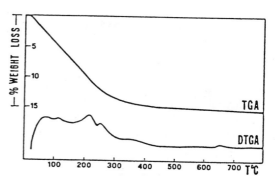

Merlinoite Fig. 2: TG and DTGA curves of merlinoite. Heating rate 10°C/min (*N. Jb. Miner. Mh.* 1977:355(1977)). (Reproduced with permission of E. Schweizbart'sche Verlagsbuchhandlung)

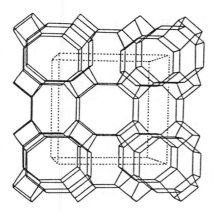

Merlinoite Fig. 1S: Framework topology.

Structure

(AZST (1977))

CHEMICAL COMPOSITION: $(K,Na)_5(Ba,Ca)_2$
$(Al_5Si_{23}O_{64})*24H_2O$
SYMMETRY: orthorhombic
SPACE GROUP: Immm
UNIT CELL CONSTANTS: (Å) $a = 14.12$
$b = 14.23$
$c = 9.95$
$D_{measured}$: 2.14(4)g/cm³
PORE STRUCTURE: eight-member rings 5.1 × 3.4 and 3.1 × 3.5 Å

X-Ray Powder Diffraction Data: (*Nat. Zeo.* 330 (1985)) (d(Å)(I/I_0)) 10.02(12), 8.15(12), 7.12(90), 7.08(88), 5.36(40), 5.03(35), 4.98(20), 4.48(37), 4.29(28), 4.07(8), 3.655(19), 3.556(5), 3.526(4), 3.336(2), 3.258(44), 3.241(41), 3.226(38), 3.176(100), 2.935(34), 2.770(16), 2.730(27), 2.720(30), 2.670(11), 2.552(16), 2.540(12), 2.535(12), 2.507(11), 2.435(8), 2.428(8), 2.391(4), 2.386(3), 2.354(2), 2.181(6), 2.065(4), 2.005(4), 1.977(4).

The structure of merlinoite contains double eight-member rings joined by four-member rings in an s-shape configuration along the c direction, giving rise to pore openings of 3.1 × 3.5/2.7 × 3.6 and 3.4 × 5.1/3.3 × 3.3 Å. There are eight framework oxygens around each (K,Ba)

site, and the other three cations (Ca,K,Na) are in the center of the largest cage, located at the niche in the "double crankshaft." The structure is very similar to that of phillipsite, and the similar X-ray powder diffraction patterns led to confusion in assignment of structures. The synthetic phases K-M and Linde W first were both identified as phillipsite-type structures, and later were shown to be more closely associated with the merlinoite structure (*JCS* 1956:2882(1956); *ACS Sym. Ser.* 40:30(1977)).

Thermal Properties

The DTA curve shows a broad peak between 20 and 180°C, with peaks also occurring at 215 and 250°C. The total loss in weight for merlionite is 15.57%. (*N. Jb. Miner. Mh.* 1977:355(1977)).

Mesole (THO)

Obsolete name for thomsonite.

Mesoline (LEV)

Mesoline is an obsolete synonym for levyne.

Mesolite (NAT)
(natural)

Related Materials: natrolite

Mesolite represents the intermediate member of the natrolite family of structures, which was named from the Greek meaning "middle stone" in 1816 (*Nat. Zeo.* 36(1985)).

Structure

(*Nat. Zeo.* 35(1985))

CHEMICAL COMPOSITION: $Na_{16}Ca_{16}(Al_{48}Si_{72}O_{240})^*$
    64H_2O
SPACE GROUP: Fdd2
UNIT CELL CONSTANTS: (Å) $a = 18.41$
                               $b = 56.65$
                               $c = 6.55$
FRAMEWORK DENSITY: 1.75 g/cc
PORE STRUCTURE: eight-member rings, 2.6 × 3.9 Å

X-Ray Powder Diffraction Data: (*Nat. Zeo.* 318(1985))   (d(Å) (I/I_0))   6.59(100), 6.12(4), 5.86(38), 5.41(4), 4.72(49), 4.60(29), 4.46(22),

4.40(3), 4.35(24), 4.20(32), 4.14(16), 3.917(2), 3.864(3), 3.373(2), 3.294(4), 3.218(14), 3.163(19), 3.091(7), 3.084(7), 2.975(10), 2.931(16), 2.885(55), 1.857(56), 2.598(2), 2.582(4), 2.574(5), 2.568(5), 2.472(8), 2.443(1), 2.421(9), 2.361(2), 2.312(4), 2.270(3), 2.237(4), 2.198(17), 2.178(4), 2.130(2), 2.081(1), 2.069(1), 2.044(2), 2.022(1), 2.000(1), 1.986(1).

Atomic coordinates with estimated standard deviations for mesolite (*Acta Crystallogr.* C42:937 (1986)):

Atom	x	y	z
Si(1)	0.75000	0.75000	0.00000
Si(2)	0.15499(2)	0.76532(1)	0.87210(9)
Si(3)	0.90188(2)	0.81979(1)	0.65040(9)
Si(4)	0.08794(2)	0.84688(1)	0.65213(9)
Si(5)	0.00435(2)	0.83180(1)	0.27740(9)
Al(1)	0.78689(2)	0.78025(1)	0.61310(9)
Al(2)	0.03630(2)	0.80211(1)	0.89287(9)
Al(3)	0.95155(2)	0.86233(1)	0.89294(9)
O(1)	0.76818(5)	0.77289(2)	0.86482(17)
O(2)	0.02298(5)	0.80800(2)	0.15228(17)
O(3)	0.98498(6)	0.85319(2)	0.12809(18)
O(4)	0.82365(4)	0.80837(1)	0.59682(16)
O(5)	0.07262(4)	0.77385(1)	0.86525(17)
O(6)	0.16644(5)	0.85935(2)	0.67096(18)
O(7)	0.84335(5)	0.75866(2)	0.51529(18)
O(8)	0.09631(5)	0.92339(2)	0.79447(17)
O(9)	0.88933(5)	0.84183(2)	0.80185(17)
O(10)	0.95568(5)	0.80062(2)	0.75319(18)
O(11)	0.20947(5)	0.78462(2)	0.95950(17)
O(12)	0.02193(5)	0.86358(2)	0.71805(18)
O(13)	0.93735(5)	0.82799(2)	0.43383(17)
O(14)	0.18020(5)	0.75695(2)	0.64166(17)
O(15)	0.07638(5)	0.83892(2)	0.41273(17)
Na	0.96736(3)	0.76089(1)	0.63172(14)
Ca	0.22807(1)	0.92737(1)	0.88068(9)
OW(1)	0.05347(7)	0.76839(2)	0.37684(22)
OW(2)	0.78167(10)	0.81598(3)	0.07685(25)
OW(3)	0.20069(8)	0.84820(3)	0.18338(23)
OW(4)	0.81793(6)	0.85687(2)	0.36556(28)
H(11)	0.0450(13)	0.7813(4)	0.322(4)
H(12)	0.0881(14)	0.7679(4)	0.420(5)
H(21)	0.7773(13)	0.8037(5)	0.018(5)
H(22)	0.8085(17)	0.8232(5)	0.004(6)
H(31)	0.2274(16)	0.8560(5)	0.247(6)
H(32)	0.1685(14)	0.8438(4)	0.260(5)
H(41)	0.8072(13)	0.8702(4)	0.387(5)
H(42)	0.8615(16)	0.8525(5)	0.359(6)

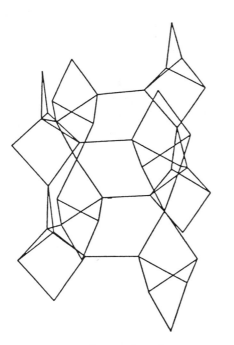

Mesolite Fig. 1S: Framework topology.

The mesolite structure is completely ordered with respect to Si/Al, having a similar framework topology to natrolite and scolecite with an

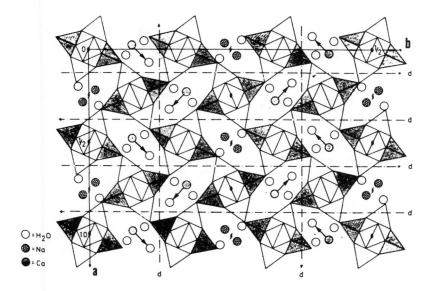

Mesolite Fig. 2:   The structure of mesolite projected along c, symmetry Fdd2 with tripled b; 2 and 2_1 only are shown; $H_2O(3)$ corresponds to the circle that appears as the farthest from Ca. The arrows show the Ca–H_2O vector (*Nat. Zeo.* 39 (1985)). (Reproduced with permission of Springer Verlag Publishers)

intermediate composition. In mesolite there are two types of chains, one surrounded by two Na and two Ca channels and the second surrounded by one Na and three Ca channels. This structure separates mesolite from natrolite or scolecite, as there is a large difference in the T–O–T angles between these three structure. The distortion is most evident in the angles associated with O atoms bridging two adjacent chains. The Si(3)–O(4)–Al(1), Si(2)–O(5)–Al(2), and Si(4)–O(6)–Al(3) angles are similar to those of scolecite.

Each Na channel has twice the number of cations as a Ca-containing channel in this structure. The sodium cations coordinate to four framework oxygen atoms on one side of the tetrahedral spiral forming the channel and to two water molecules. Each water is shared by two adjacent Na atoms along the channel, causing the Na coordination polyderon to become a distorted trigonal prism. The Ca cations also are coordinated to four framework O atoms of the silicate, but are slightly more centered in the channel than the Na cations. This arrangement produces four similar Ca–O distances instead of the two long, two short distances around the Na cation.

Thermal Properties

The thermal analysis of mesolite has been found to be sample-dependent. The major dehydration steps, however, occur at 247, 327, and 397°C. The peak around 327°C is similar to that of natrolite and may be related to the loss of water in the Na-containing channels. The peak at 397°C

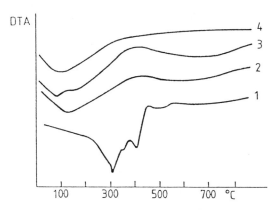

Mesolite Fig. 3:   DTA curves for mesolite: (1) original hydrated samples; (2) rehydrated after evacuation at 300°C; (3) at 400°C; (4) at 600°C (*Nat. Zeo.* 265 (1988)). (Reproduced with permission of Springer Verlag Publishers)

may correspond to the loss of the two remaining water molecules in the Ca-containing channels (*Acta Crystallogr*. C42:937(1986)).

Infrared Spectrum

Ion Exchange

No exchange is observed with Na- or Ca-containing solutions, but high exchange has been observed with fused salts (180–270°C) for Na, K, Li, NH_4, Ag, and Tl). Upon exchange, complete removal of Na occurs, but only partial

Vibrational spectra of mesolite (*Cryst. Res. Technol*. 18:1045 (1983)):

Type of vibration	IR cm^{-1}	FIR cm^{-1}	REF cm^{-1}	R cm^{-1}
Antisymmetric + symmetric	3620 VS	—	—	—
stret. O–H	3520 VS			
	3420 VS			
Combination band H_2O	3250 S	—	—	—
	2520 VW			
Bending H_2O	1640 M	—	—	1620 W
external vibrations				
between TO$_4$	1410 M	—	1105 M	1100 VW
antisymmetric stret. T–O	1020 S	—	1070 W	1080 VW
	960 VS		1020 M	1010 W
			980 W	990 M
			940 W	960 W
			900 W	910 W
				860 W
Symmetric stret. T–O	750 VW	—	805 VW	805 W
	710 VW		715 W	770 W
			690 W	720 VW
				700 VW
				680 W
Bending O–T–O	420 W	—	420 M	480 VW
			405 M	460 M
				405 S
"Pore opening" oxygen	520 W	—	—	640 W
translation mode of lattice	380 VW	370 VW	360 M	360 VW
	320 VW	330 VW	340 W	340 VW
		310 VW	315 VW	330 W
Ca–O	—	280 VW	—	290 VW
		205 M		260 W
				240 W
Na–O	—	180 VW	—	180 VW
		160 W		
Complex Ca–H_2O	—	105 M		105 VW
				90 W
Complex Na–H_2O		130 VW		148 W
optical mode of lattice		110 M		70 W
		52 VW		50 W

IR—infrared absorption spectra; REF—infrared reflectance spectra; R—Raman scattering spectra; VS—very strong; W—weak; VW—very weak; S—strong; M—medium; W—weak; FIR—far infrared absorption spectra.

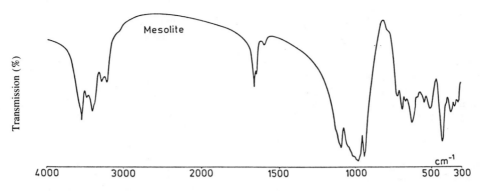

Mesolite Fig. 4:   Infrared spectrum of mesolite (*Nat. Zeo*. 347 (1985)). (Reproduced with permission of Springer Verlag Publishers)

exchange of the calcium ions is noted (*Miner. Mag.* 23:421(1933)).

Mesotype (NAT)

Obsolete synonym used in early papers to describe natrolite.

Metaheulandite II

Metaheulandite II was coined for the first transition of heulandite, which is observed as the heulandite structure contracts by 8%. This also was called the heulandite B phase. This contraction occurs at 230°C (*Nat. Zeo.* 267(1985)).

Metanatrolite

Metanatrolite was the name given in 1890 to describe dehydrated natrolite (*Nat. Zeo.* 36 (1985)).

Metavariscite
(natural)

Related Materials: lucinite

Metavariscite, which is dimorphous with variscite, first was reported in 1925. A natural

aluminophosphate mineral, metavariscite occurs as fine microcrystals in cavities of variscite and has been found in deposits in Utah, Nevada, and Pacific Ocean islands (*Phosphate Minerals* 74(1984)).

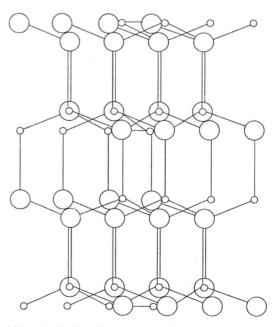

Metavariscite Fig. 1S: Framework topology.

Structure

(Acta Crystallogr. B29:2292(1973))

CHEMICAL COMPOSITION: $AlPO_4*2H_2O$
SYMMETRY: monoclinic
SPACE GROUP: $P2_1/n$
UNIT CELL CONSTANTS (Å): $a = 5.178(2)$
$b = 9.514(2)$
$c = 8.454(2)$
$\beta = 90.35°$

DENSITY: (measured) 2.54

X-Ray Powder Diffraction Data: (from author) (synthetic) (d(Å) (I/I_0)) 6.30(m), 4.78(s), 4.55(s), 4.43(w), 4.23(s), 4.01(w), 3.50(s), 3.24(w), 3.10(w), 2.70(vs), 2.57(w), 2.50(m), 2.42(w), 2.39(w), 2.28(mw), 2.20(m), 2.16(w), 2.094(w), 2.060(w), 2.002(w), 1.950(s), 1.925(w), 1.853(w), 1.810(w), 1.780(w), 1.758(w), 1.733(w), 1.710(m), 1.662(w), 1.637(w), 1.620(w), 1.582(w), 1.530(w), 1.510(w), 1.484(m).

The metavariscite structure is composed of PO_4 tetrahedra that share vertices with four $AlO_4(OH_2)_2$ octahedra and form six-member T-atom rings. The sequence of linkages is an up-down-down-up-down-down (or boat) conformation. The two waters coordinate the aluminum in *cis*-positions and donate two short and two longer hydrogen bonds to the phosphate oxygens. None of these hydrogen bonds is along the octahedron edge. Both waters are true H_2O species and are not coordinated to the structure as OH^- or H_3O^+ species (*Acta Crystallogr.* B29:2292(1973)).

Atomic Parameters for metavariscite (*Acta Crystallogr.* B29:2292 (1973)):

	x	y	z
Al	0.40309(9)	0.32545(5)	0.30626(5)
P	−0.09105(8)	0.14688(4)	0.18371(4)
O1	0.16505(22)	0.17902(11)	0.27036(13)
O2	−0.09291(23)	0.21677(12)	0.02094(13)
O3	−0.31481(22)	0.20439(12)	0.28127(13)
O4	−0.11458(21)	−0.01392(11)	0.17227(13)
OW1	0.40410(30)	0.44767(15)	0.32202(18)
OW2	0.40410(31)	0.36239(14)	0.07903(15)
H11	0.139(5)	0.530(3)	0.288(4)
H12	−0.037(7)	0.432(3)	0.331(5)
H21	0.274(7)	0.334(5)	0.023(4)
H22	0.539(8)	0.369(5)	0.023(4)

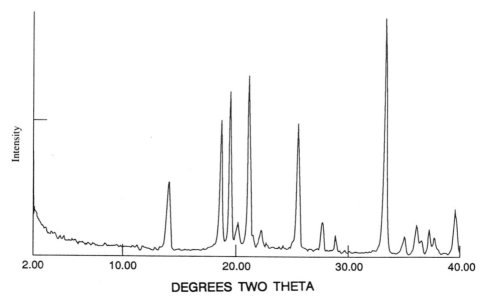

Metavariscite Fig. 2: X-ray powder diffraction pattern of metavariscite. (Reproduced with permission of B. Duncan, Georgia Tech)

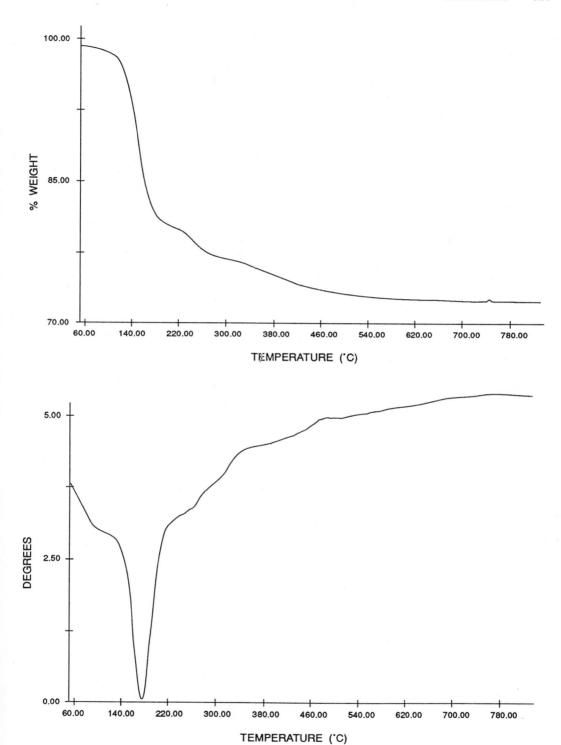

Metavariscite Fig. 3:   TGA(top)/DTA(bottom) of metavariscite run in air. (Reproduced with permission of B. Duncan, Georgia Tech)

Synthesis

ORGANIC ADDITIVES
none

Synthetic metavariscite is prepared from reactive mixtures containing a P_2O_5/Al_2O_3 ratio of 2.73. It crystallizes with or without the presence of potassium in the crystallization mixture. Metavariscite is prepared in pure form with relative ease after 1.5 hours at 100°C. It also occurs with other $AlPO_4$ hydrate phases including $AlPO_4$-H1, -H2, -H3, and -H4, as well as variscite. In the presence of organic additives, the temperature generally will dictate which structure is formed, with metavariscite being observed at temperatures below 125° (*Bull. Chem. Soc. Fr.* 1762(1961); from author).

Thermal Properties

Metavariscite reversibly transforms at room temperature upon dehydration over the desiccant P_2O_5 to $AlPO_4$-A, or upon thermal treatment to between 70 and 120°C. With further heating, $AlPO_4$-A transforms to tridymite (*Bull. Chem. Soc. Fr.* 1762(1961)).

Infrared Spectrum

MgAPSO
(EP 158,348(1985))
Union Carbide Corporation

MgAPSO molecular sieves represent the magnesium silicoaluminophosphate molecular sieves.

MgAPSO-5 (AFI)
(EP 158,348(1985))
Union Carbide Corporation

Related Materials: $AlPO_4$-5

Structure

Chemical Composition: magnesium silicoaluminophosphate. See MgAPSO.

X-Ray Powder Diffraction Pattern: (d(Å) (I/I_0)) 12.28–11.95 (69–100), 7.00–6.86 (8–12), 6.07–5.93 (15–35), 4.58–4.48 (38–73), 4.26–4.21 (sh–100), 4.013–3.969 (48–100), 3.648–3.583 (0–14), 3.480–3.434 (23–44), 3.110–3.069 (12–20), 3.013–2.964 (15–21), 2.683–2.656 (2–11), 2.622–2.589 (11–19), 2.455–2.430 (0–4), 2.405–2.380 (5–11), 2.222–2.217(0–1), 2.196–2.171 (0–3), 2.159–2.132 (3–4), 2.122–2.101 (0–3), 2.094–2.080 (0–2), 2.032–2.015(0–2), 2.002–1.989 (0–2), 1.969–1.959 (0–1), 1.926–1.905 (4–6), 1.811(0–1), 1.794–1.787(0–3), 1.771–1.762 (0–4), 1.752–1.746 (0–1), 1.664–1.648 (0–4).

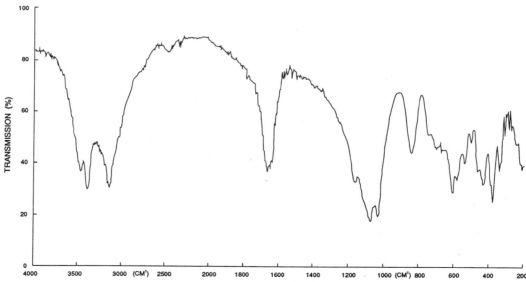

Metavariscite Fig. 4:   Infrared spectrum of metavariscite. (Reproduced with permission of B. Duncan, Georgia Tech)

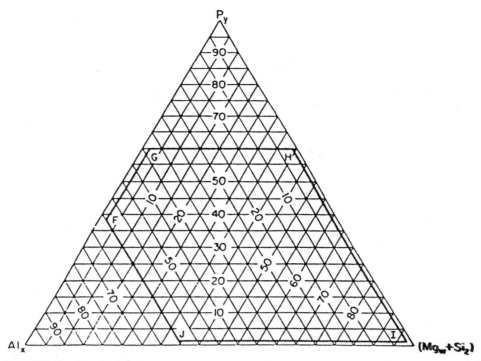

MgAPSO Fig. 1:   Preferred composition field for the MgAPSO molecular sieves (EP 158,348 (1985)).

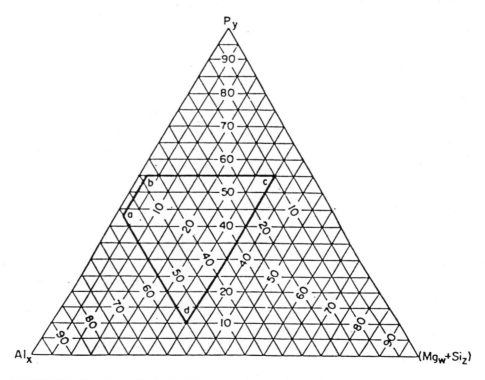

MgAPSO Fig. 2:   Crystallization field for the reactive gels that produce the MgAPSO molecular sieves (EP 158,348 (1985)).

The MgAPSO-5 structure exhibits properties characteristic of a large-pore molecular sieve. See AlPO$_4$-5 for a description of the framework topology.

Synthesis

ORGANIC ADDITIVES
tripropylamine
tetrapropylammonium$^+$
di-*n*-propylamine
diethylethanolamine
isopropylamine
tetraethylammonium$^+$
tetrabutylammonium$^+$
cyclohexylamine

MgAPSO-5 crystallizes from a reactive gel with a molar oxide ratio range as shown for MgAPSO. Crystallization occurs over a range of times, but generally begins after 40 hours. This structure occurs in the temperature range of 150 to 200°C, depending on the organic additive utilized. Co-crystallizing phases with MgAPSO-5 include MgAPSO-26, -47, -11, -39, -34, and -44.

Adsorption

Adsorption properties of MgAPSO-5:

Adsorbate	Kinetic diameter, Å	Pressure, Torr	Temp., °C	Wt. % adsorbed*
O$_2$	3.46	99	−183	13.2
O$_2$		749	−183	15.5
Cyclohexane	6.00	57	23.4	7.9
Neopentane	6.2	100	23.4	5.0
H$_2$O	2.65	4.6	23.2	16.0
H$_2$O		16.8	23.5	21.3

---

*Calcined in air to 600°C for 2.25 hr.

MgAPSO-11 (AEL)
(EP 158,348(1985))
Union Carbide Corporation

Related Materials: AlPO$_4$-11

Structure

Chemical Composition: magnesium silicoaluminophosphate. See MgAPSO.

X-Ray Powder Diffraction Data: (d(Å) (I/I_0)) 11.33–10.85(sh–35), 9.83–9.21(6–60), 6.86–6.71(sh–22), 5.75–5.57(sh–30), 5.56–5.42(sh–3), 4.75–4.65(0–4), 4.44–4.33(sh–38), 4.27–4.19(100), 4.040–3.969(sh–72), 3.969–3.900(sh–90), 3.900–3.850(21–48), 3.810(0–4), 3.648–3.576(0–9), 3.401–3.339(0–21), 3.143–3.100(sh–17), 3.048–3.028(0–6), 3.018–2.979(0–17), 2.867–2.823(0–15), 2.763–2.730(0–18), 2.652–2.600(9–13), 2.515(0–3), 2.488–2.442(0–11), 2.398–2.374(0–17), 2.301–2.276(0–3), 2.241–2.214(0–1), 2.191–2.181(0–1), 2.161–2.146(0–4), 2.113–2.094(0–5), 2.036–2.019(0–4), 1.814–1.801(0–3), 1.687–1.681(0–3).

MgAPSO-11 exhibits properties of a medium-pore molecular sieve. See AlPO$_4$-11 for a description of the topology.

Synthesis

ORGANIC ADDITIVES
di-*n*-propylamine
diisopropylamine

MgAPSO-11 crystallizes from a reactive gel with a composition in the range shown for MgAPSO. Crystallization generally occurs between 150 and 200°C. Crystals begin to form after 40 hours or longer. Common co-crystallizing phases include MgAPSO-5, -33, -34, -39, and -46.

MgAPSO-16 (AST)
(EP 158,348(1985))
Union Carbide Corporation

Related Materials: AlPO$_4$-16

Structure

Chemical Composition: magnesium silicoaluminophosphate. See MgAPSO.

X-Ray Powder Diffraction Data: $(d(\text{Å})\ (I/I_0))$
7.76–7.69(sh–64), 4.75–4.72(10–45), 4.07–
4.00(100), 3.900–3.818(15–26), 3.332–3.267
(16–40), 3.084–3.079(sh–17), 3.008–2.988(9–
45), 2.730–2.968(sh–3), 2.592–2.573(4–10),
2.380–2.368(4–7), 2.287–2.256(2–5), 2.044–
2.036(2–10),   1.876–1.873(7–10),   1.746–
1.743(1–2).

MgAPSO-16 exhibits properties characteristic of a very small-pore material.

Synthesis

ORGANIC ADDITIVES
quinuclidine

MgAPSO-16 crystallizes from a reactive gel
within the molar composition range shown for
MgAPSO. Crystallization generally begins after
48 hours at temperatures ranging between 150
and 200°C. MgAPSO-35 is a common co-crystallizing phase in this system.

MgAPSO-20 (SOD)
(EP 158,348(1985))
Union Carbide Corporation

Related Materials: AlPO$_4$-20
                       sodalite

Structure

Chemical Composition: magnesium silicoaluminophosphate. See MgAPSO.

X-Ray Powder Diffraction Data: $(d(\text{Å})$
$(I/I_0))$ 6.42–6.23(42–100), 4.55–4.41(22–43),
4.050–3.964(3–7),   3.695–3.603(56–100),
3.198–3.121(11–15),   2.861–2.791(10–12),
2.610–2.601(10–16), 2.417–2.408(1–2), 2.260–
2.253(3–4), 2.130–2.124(5), 1.927–1.921(4–
5), 1.772–1.768(8).

MgAPSO-20 exhibits properties characteristic of a very small-pore material.

Synthesis

ORGANIC ADDITIVES
tetramethylammonium$^+$

MgAPSO-20 crystallizes from a reactive gel
within a molar oxide range as shown for
MgAPSO. Crystallization begins after 18 hours
at temperatures between 100 and 200°C.

Adsorption

Adsorption properties of MgAPSO-20:

Adsorbate	Kinetic diameter, Å	Pressure, Torr	Temp., °C	Wt. % adsorbed*
O$_2$	3.46	99	−183	0.8
O$_2$		750	−183	2.7
H$_2$O	2.65	4.6	23.2	16.5
H$_2$O		16.8	23.5	19.9

*Calcined in air to 600°C for 1.5 hr.

MgAPSO-34 (CHA)
(EP 158,348(1985))
Union Carbide Corporation

Related Materials: chabazite
                       AlPO$_4$-34

Structure

Chemical Composition: magnesium silicoaluminophosphate. See MgAPSO.

X-Ray Powder Diffraction Data: $(d(\text{Å})\ (I/I_0))$
(as synthesized) 9.32(100), 6.91(15), 6.30(15),
5.55(52), 4.94(21), 4.32(92), 4.002(4), 3.864(5),
3.540(23),   3.455(18),   3.243(3),   3.151(4),
3.029(4), 2.932(33), 2.866(22), 2.833(5*),
2.775(2), 2.611(7), 2.332(2), 2.480(8), 2.277(4),
2.100(3), 1.915(4), 1.862(6), 1.795(4), 1.727(4),
1.684(2), 1.649(4); (after 500°C calcination)
9.12(100), 6.76(22), 6.24(1), 5.44(15), 4.90(10),
4.60(3), 4.24(31), 4.11(sh), 3.969(3), 3.809(3),
3.513(11),   3.389(10),   3.132(4),   2.979(sh),
2.885(23),   2.652(2),   2.564(3),   2.455(1),
2.071(1), 1.845(2), 1.781(2), 1.752(1), 1.725(1),
1.698(2) (*indicates impurity peak).

MgAPSO-34 exhibits properties characteristic of a small-pore molecular sieve.

Synthesis

ORGANIC ADDITIVES
diisopropylamine
tetraethylammonium$^+$

MgAPSO-34 crystallizes from the range of gel compositions outlined for MgAPSO. Crystallization occurs after 40 hours at temperatures ranging between 100 and 200°C. Other molecular sieve phases observed in this system include MgAPSO-5 and -11.

Adsorption

Adsorption properties of MgAPSO-34:

Adsorbate	Kinetic diameter, Å	Pressure, Torr	Temp., °C	Wt. % adsorbed*
O_2	3.46	100	−183	21.7
O_2		734	−183	33.6
Isobutane	5.00	300	23	1.3
n-Hexane	4.3	51	24	10.4
H_2O	2.65	4.6	23	27.1
H_2O		18.5	24	32.9

*Calcined in air to 600°C for 1.5 hr.

MgAPSO-35 (LEV)
(EP 158,348(1985))
Union Carbide Corporation

Related Materials: levyne

Structure

Chemical Composition: magnesium silicoaluminophosphate. See MgAPSO.

X-Ray Powder Diffraction Pattern: (d(Å) (I/I_0)) 10.65–10.05(10–21), 8.35–7.97(36–100), 6.76–7.46(17–100), 5.64–5.54(0–9), 5.22–5.04(25–80), 5.01–4.98(0–sh), 4.31–4.19(sh–54), 4.11–4.00(40–100), 3.867–3.754(sh–27), 3.619–3.534(5–8), 3.453–3.376(0–8), 3.351–3.267(10–16), 3.290(sh–10), 3.175–3.100 (24–44), 3.089–3.008(5–23), 2.844–2.840 (sh–7), 2.805–2.763(19–37), 2.763–2.739(sh), 2.629–2.585(5–9), 2.522–2.488(0–4), 2.404–2.368(0–6), 2.287–2.259(0–4), 2.212–2.206(0–1), 2.166–2.141(0–5), 2.132–2.118(0–6), 2.036–2.023(0–7), 2.014–2.010(0–1), 1.914–1.907(0–2), 1.888–1.873(0–4), 1.870–1.859(0–4), 1.848–1.834(0–5), 1.797–1.774(0–7), 1.664–1.653 (0–4).

MgAPSO-35 exhibits properties characteristic of a small-pore levyne molecular sieve.

Synthesis

ORGANIC ADDITIVES
quinuclidine
methylquinuclidine

MgAPSO-35 crystallizes from reactive gels in the range of compositions shown for MgAPSO. Crystallization occurs after 40 hours at temperatures ranging between 150 and 200°C. MgAPSO-16 is a common co-crystallizing phase when quinuclidine is used as the organic additive.

Adsorption

Adsorption properties of MgAPSO-35:

Adsorbate	Kinetic diameter, Å	Pressure, Torr	Temp., °C	Wt. % adsorbed*
O_2	3.46	100	−183	6.7
O_2		734	−183	9.2
Isobutane	5.0	100	24	0.3
n-Hexane	4.3	51	24	1.1
H_2O	2.65	4.6	23	11.5
H_2O		19.5	23	17.7

*Calcined in nitrogen at 500°C for 2 hr.

Adsorption properties of MgAPSO-35:

Adsorbate	Kinetic diameter, Å	Pressure, Torr	Temp., °C	Wt. % adsorbed*
O_2	3.46	100	−183	11.2
O_2		744	−183	14.0
Isobutane	5.0	100	22.8	0.2
n-Hexane	4.3	49	22.3	5.7
H_2O	2.65	4.6	23.1	16.1
H_2O		17.8	22.9	20.5

*Calcined in air at 500°C for 6.7 hr.

The adsorption data indicated incomplete removal of the organic with nitrogen calcination.

MgAPSO-36 (ATS)
(EP 158,348(1985))
Union Carbide Corporation

Related Materials: other metal-substituted aluminophosphate molecular sieves

Structure

Chemical Composition: magnesium silicoaluminophosphate. See MgAPSO.

X-Ray Powder Diffraction Pattern: (d(Å) (I/I_0)) 11.3–11.05(100), 10.92–10.78(0–sh), 6.58(6–8), 5.64–5.57(sh), 5.44–5.37(24–31), 4.70–4.60(38–41), 4.29–4.27(25–49), 4.23 (0–sh), 4.12–4.08(sh), 4.077–4.058(sh), 3.978(25–42), 3.900–3.867(sh), 3.739–3.723(5–9), 3.290–3.278(14–16), 3.178–3.148(7–10), 3.100–3.069(10–12), 2.988–2.969(3–7), 2.814–2.797(8–11), 2.714–2.698(1–3), 2.592–2.564(7–16), 2.515–2.495(3–4), 2.392–2.386(2–3), 2.293–2.281(1), 2.249–2.238(2–3), 2.186(4), 2.103–21.080(1–2), 2.060–2.045(1–2), 2.006–1.998(1), 1.949(0–1), 1.922–1.910(3), 1.867(0–1), 1.787–1.784(1–2), 1.707(0–2), 1.661–1.656(1–3).

MgAPSO-36 exhibits properties characteristic of a large-pore molecular sieve.

Synthesis

ORGANIC ADDITIVES
tripropylamine
tetrapropylammonium$^+$

MgAPSO-36 crystallizes from reactive gels within the composition range given for MgAPSO. Crystallization occurs between 150 and 200°C after a minimum of 48 hours at these temperatures. MgAPSO-5 is a common co-crystallizing phase.

Adsorption

Adsorption properties of MgAPSO-36:

Adsorbate	Kinetic diameter, Å	Pressure, Torr	Temp., °C	Wt. % adsorbed*
O$_2$	3.46	100	−183	12.9
O$_2$		734	−183	15.4
Isobutane	5.0	100	24	5.2
Cyclohexane	6.0	59	23.7	9.0
Neopentane	6.2	100	24.5	5.5
H$_2$O	2.65	4.6	23	16.8
H$_2$O		20	23.6	23.5

*Calcined in nitrogen at 500°C for 2 hr and in air at 600°C for an additional 2 hr.

MgAPSO-39 (ATN)
(EP 158,348(1985))
Union Carbide Corporation

Related Materials: AlPO$_4$-39

Structure

Chemical Composition: magnesium silicoaluminophosphate. See MgAPSO.

X-Ray Powder Diffraction Pattern: (d(Å) (I/I_0)) 9.61–9.21(20–53), 6.76–6.56(25–53), 4.98–4.85(23–34), 4.27–4.17(70–100), 4.004–3.900(97–100), 3.326–3.296(3–4), 3.191–3.175(0–4), 3.121–3.100(sh–17), 3.038–2.998 (13–20), 2.979–2.950(17–29), 2.763–2.740(10–16), 2.644–2.622(sh–11), 2.448–2.439(0–2), 2.380–2.362(5–9), 2.217–2.201(0–5), 2.103–2.085(0–2), 2.014(0–1), 1.966–1.961(0–2), 1.926–1.922(0–1), 1.877–1.864(4–5), 1.859–1.841(0–3), 1.791–1.778(3–5), 1.755–1.746(0–4), 1.692–1.681(0–2).

MgAPSO-39 exhibits properties consistant with those of a small-pore molecular sieve. See MAPO-39 for a description of the framework topology.

Synthesis

ORGANIC ADDITIVES
di-*n*-propylamine
diisopropylamine

MgAPSO-39 crystallizes from a reactive gel in a composition range shown for MgAPSO. Crystallization begins after 48 hours in a range of temperatures between 150 and 200°C. Co-crystallizing phases include MgAPSO-5, -11, -31, and -46.

MgAPSO-43 (GIS)
(EP 158,348(1985))
Union Carbide Corporation

Related Materials: gismondine

Structure

Chemical Composition: magnesium silicoaluminophosphate. See MgAPSO.

X-Ray Powder Diffraction Pattern: (d(Å) (I/I_0)) 7.20–6.83(35–100), 5.85–5.37(2–4), 5.13–5.09(12), 4.51–4.47(3–5), 4.15–4.12(49–100), 3.653–3.635(2), 3.336–3.319(7–9), 3.232–3.215(39–50), 3.182–3.165(18–25), 3.126–3.107(5–6), 2.710–2.699(8–12), 2.510–2.502(3–4), 1.997–1.991(3), 1.788–1.785(4), 1.759–1.750(1–2), 1.728–1.725(3–4), 1.707–1.700(2).

MgAPSO-43 exhibits properties characteristic of a small-pore molecular sieve.

Synthesis

ORGANIC ADDITIVES
di-*n*-propylamine

MgAPSO-43 crystallizes from a reactive gel within the composition range shown for Mg-APSO. Crystallization occurs when catapal is used as the source of aluminum at temperatures of 150°C after 122 hours. MgAPSO-46 is a common co-crystallizing phase in this system.

MgAPSO-44 (CHA)
(EP 158,348(1985))
Union Carbide Corporation

Related Materials: chabazite
                        SAPO-44

Structure

Chemical Composition: magnesium silicoaluminophosphate. See MgAPSO.

X-Ray Powder Diffraction Pattern: (d(Å) (I/I_0)) 9.61–9.37(100), 6.92–6.81(11–35*), 6.51–6.33(2–4), 6.11–6.07(0–5), 5.57–5.50(20–36), 5.16–4.93(7), 4.72–4.67(7–53), 4.33–4.28(58–100), 4.10–4.08(0–18), 3.986–3.934(sh), 3.867–3.817(8), 3.663–3.562(17–58*), 3.453–3.406 (10–18*), 3.243–3.209(1–12), 3.175(0–3), 3.018–2.998(0–sh), 3.008–2.931(4–15*), 2.931–2.885(16–48), 2.849–2.831(sh–1), 2.780–2.755(1–5), 2.739–2.698(sh–3), 3.578(0–1), 2.543–2.495(3–6), 2.442(0–1), 2.344–2.338(0–1), 2.309–2.304(0–1), 2.276–2.254(0–1), 2.151–2.141(0–6*), 2.090–2.076(0–2), 1.953(0–1), 1.929–1.914(0–5), 1.895–1.888(0–8*), 1.877–1.866(0–5), 1.824–1.797(4–10), 1.771–1.765(0–1), 1.728–1.704(4–10), 1.689–1.681(0–2) (*peak may contain impurity).

MgAPSO-44 exhibits properties characteristic of a small-pore chabazite-type molecular sieve.

Synthesis

ORGANIC ADDITIVE
cyclohexylamine

MgAPSO-44 crystallizes from a reactive gel within the composition range shown for Mg-APSO. Crystallization occurs after 40 hours at 150 to 200°C. MgAPSO-5 is a common phase, co-crystallizing with MgAPSO-44.

MgAPSO-46 (AFS)
(EP 158,348(1985))
Union Carbide Corporation

Related Materials: other metal-substituted
                        aluminophosphate
                        molecular sieves

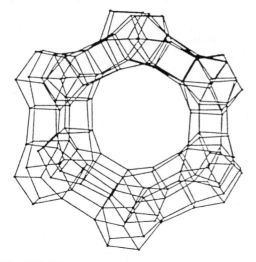

MgAPSO-46 Fig. 1S: Framework topology.

Structure

CHEMICAL COMPOSITION: See MgAPSO.
SPACE GROUP: P3c1
UNIT CELL CONSTANTS: (Å) $a = 13.2251$
                             $b = 13.2251$
                             $c = 26.8922$
                             $\gamma = 120$
UNIT CELL VOLUME: 4080 Å^3

X-Ray Powder Diffraction Pattern: (d(Å) (I/I_0)) 13.44(3), 11.48(100), 8.76(<1), 7.15(2), 6.71(2), 6.44(3), 5.95(1), 5.79(2), 5.33(3), 5.10(<1), 4.48(1), 4.34(4), 4.29(sh), 4.13(12), 3.906(6), 3.682(3), 3.534(<1), 3.320(4), 3.219(3), 3.163(2), 3.109(4), 3.000(1), 2.873(2), 2.823(<1), 2.722(<1), 2.622(1), 2.505(2), 2.462(<1), 2.417(<1), 2.344(<1), 2.276(<1), 2.201(<1), 2.141(<1), 2.062(1), 1.977(<1), 1.914(<1), 1.845(<1), 1.821(<1), 1.778(<1), 1.752(<1).

MgAPSO-46 contains a 12-member ring pore system with interconnecting eight-member rings (*SSSC* 37:269(1988)).

Refined parameters for the framework atoms of MgAPSO-46 (*SSSC* 37:269 1988)):

Atom[1]	x	y	z	B (Å^2)
P1	0.66667	0.33333	0.06039	1.22
P2	0.19862	0.18082	0.13165	1.62
P3	0.46965	0.10767	0.19050	1.02
P4	0.20093	0.96592	0.30722	1.73
P5	0.46960	0.31952	0.36555	1.16
P6	0.00000	0.00000	0.43771	1.60
Al1	0.00000	0.00000	0.06457	1.55
Al2	0.46967	0.31584	0.13168	2.55

(continued)

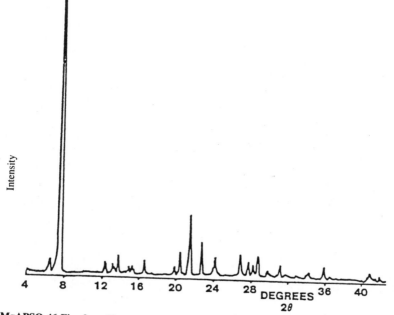

MgAPSO-46 Fig. 2:   X-ray powder diffraction pattern for MgAPSO-46 (*Abstracts, BZA Meeting,* 1989).

Atom[1]	x	y	z	B (Å^2)
Al3	0.19875	0.96524	0.18925	1.04
Al4	0.46726	0.11576	0.30775	1.42
Al5	0.19352	0.17290	0.36246	1.46
Al6	0.66667	0.33333	0.43950	1.39
O12	0.9846	0.1174	0.5852	1.75
O16	0.0000	0.0000	0.9957	2.49
O21	0.6875	0.2336	0.0819	1.36
O22	0.7350	0.0649	0.1160	1.02
O23'	0.9192	0.4727	0.1661	2.29
O23	0.7826	0.2713	0.1759	1.91
O32	0.1515	0.2587	0.1616	2.65
O32'	0.1910	0.0927	0.1748	1.96
O33	0.3389	0.9936	0.1796	2.58
O34	0.1638	0.2199	0.7555	3.21
O43	0.8976	0.5052	0.7477	1.74
O44	0.3218	0.0036	0.3158	1.95
O45	0.5189	0.4365	0.3483	1.15
O45'	0.4884	0.2442	0.3316	2.85
O54	0.1405	0.2616	0.3398	1.76
O54'	0.1867	0.0777	0.3187	2.29
O55	0.7279	0.0750	0.3741	2.09
O56	0.9997	0.8889	0.4222	2.66
O61	0.6666	0.3333	0.0081	2.85
O65	0.7979	0.4676	0.4151	2.54

Synthesis

ORGANIC ADDITIVE
di-*n*-propylamine

MgAPSO-46 crystallizes from a reactive gel within the composition range shown for Mg-APSO. Crystallization occurs after 122 hours at 150°C, and MgAPSO-43 appear as a co-crystallized phase in this system.

Adsorption

Adsorption properties of MgAPSO-46:

Adsorbate	Kinetic diameter, Å	Pressure, Torr	Temp., °C	Wt. % adsorbed*
O$_2$	3.46	100	−183	20.7
O$_2$		734	−183	24.7
Neopentane	6.2	100	24.5	8.4
Isobutane	5.0	100	24	7.8
Cyclohexane	6.0	59	23.7	11.9
H$_2$O	2.65	4.6	23	22.0
H$_2$O		20.0	23	27.4

*Calcined in nitrogen at 500°C for 1.75 hr.

MgAPSO-47 (CHA)
(EP 158,348(1985))
Union Carbide Corporation

Related Materials: chabazite

Structure

Chemical Composition: magnesium silicoaluminophosphate. See MgAPSO.

X-Ray Powder Diffraction Pattern: (d(Å) (I/I$_0$)) 9.29(100), 6.84(9), 6.36(5), 5.52(22), 5.03(9), 4.66(2), 4.30(53), 4.06(7), 3.961(2), 3.859(7), 3.598(21), 3.432(12), 3.222(5), 3.190(3), 3.126(1*), 3.022(3), 2.919(21), 2.893(sh), 2.837(2), 2.763(1), 2.695(2), 2.597(4), 2.567(1), 2.510(3), 2.338(2), 2.305(1), 2.270(2), 2.126(2), 1.085(1), 1.907(2), 1.870(4), 1.810(3), 1.768(1), 1.745(1), 1.719(2), 1.695(1), 1.681(1), 1.645(2) (*impurity peak).

MgAPSO-47 exhibits properties characteristic of a small-pore chabazite-type molecular sieve.

Synthesis

ORGANIC ADDITIVE
diethylethanolamine

MgAPSO-47 crystallizes from a reactive gel within the composition range shown for Mg-APSO. Crystallization occurs at 100°C after 22 hours, and MgAPSO-5 is a common impurity in this system. Ludox is used as the source of silica and catapal as the source of aluminum.

Adsorption

Adsorption properties of MgAPSO-47:

Adsorbate	Kinetic diameter, Å	Pressure, Torr	Temp., °C	Wt% adsorbed*
O$_2$	3.46	99	−183	14.1
O$_2$		725	−183	29.2
Isobutane	5.0	100	22.8	0.2
n-Hexane	4.3	49	23.3	4.2
H$_2$O	2.65	4.6	23.1	18.5
H$_2$O		17.8	22.9	28.7

*Calcined in air at 500° for 1.75 hr.

MnAPO-5 (AFI)
(US4,567,029(1986))
Union Carbide Corporation

Related Materials: AlPO$_4$-5

Structure

Chemical Composition: manganese aluminophosphate. See XAPO.

X-Ray Powder Diffraction Data: (d(Å) (I/I_0))
12.11(100), 6.86(110), 5.99(33), 4.53(47),
4.25(32), 3.99(57), 3.62(5), 3.45(27), 3.09(11),
2.99(19), 2.681(4), 2.614(15), 2.442(3),
2.398(7), 1.918(4).

MnAPO-5 exhibits properties characteristic of a large-pore molecular sieve. For a detailed description of the framework topology of MnAPO-5 see AlPO$_4$-5.

Synthesis

ORGANIC ADDITIVES
tetraethylammonium$^+$
diethylethanolamine
tripropylamine
diisopropylamine

MnAPO-5 crystallizes from a reactive gel composition with a molar oxide ratio of: 1.0 R : 0.167 MnO : 0.917 Al$_2$O$_3$: P$_2$O$_5$: 0.33 CH$_3$COOH : 5.5 C$_3$H$_7$OH : 45 H$_2$O. Manganese acetate or manganese sulfate is an effective source of manganese, and hydrated alumina or aluminum isopropoxide is the source of aluminum. Crystallization occurs readily after 24 hours at temperatures between 150 and 200°C.

Adsorption

Adsorption properties of MnAPSO-5:

Adsorbate	Kinetic diameter, Å	Pressure, Torr	Temp., °C	Wt% adsorbed*
O$_2$	3.46	100	-183	14.1
O$_2$		750	-183	19.6
Neopentane	6.2	12	24	6.5

Adsorbate	Kinetic diameter, Å	Pressure, Torr	Temp., °C	Wt% adsorbed*
Neopentane		710	24	7.9
H$_2$O	2.65	4.3	23	13.3
H$_2$O		20.6	23	26.5

*Typical amount adsorbed.

MnAPO-11 (AEL)
(US4,567,029(1986))
Union Carbide Corporation

Related Materials: AlPO$_4$-11

Structure

Chemical Composition: manganese aluminophosphate. See XAPO.

X-Ray Powder Diffraction Data: (d(Å) (I/I_0))
11.05(41), 9.51(56), 6.81(27), 5.68(43), 5.47(6),
4.70(9), 4.40(57), 4.21(63), 4.04(49), 3.97(46),
3.93(50), 3.87(100), 3.66(9), 3.62(14), 3.39(35),
3.18(10), 3.12(25), 3.04(9), 2.849(9), 2.747(25),
2.629(11), 2.481(8), 2.392(11), 2.298(5),
2.113(5), 2.032(8), 1.804(6), 1.687(5).

MnAPO-11 exhibits properties characteristic of a medium-pore molecular sieve. For a detailed description of the framework topology of MnAPO-11 see AlPO$_4$-11.

Synthesis

ORGANIC ADDITIVES
diisopropylamine
di-*n*-propylamine

MnAPO-11 crystallizes from a reactive gel composition with a molar oxide ratio of: 1.0 R : 0.4 MnO : 0.8 Al$_2$O$_3$: P$_2$O$_5$: 0.8 CH$_3$COOH : 50 H$_2$O. Manganese acetate or manganese sulfate is an effective source of manganese, and hydrated alumina or aluminum isopropoxide is the source of aluminum. Crystallization occurs readily after 24 hours at temperatures between 150 and 200°C.

MnAPO-16 (AST)
(US4,567,029(1986))
Union Carbide Corporation

Related Materials: AlPO$_4$-16

Structure

Chemical Composition: manganese aluminophosphate. See XAPO.

X-Ray Powder Diffraction Data: (d(Å) (I/I_0))
7.83(100), 4.75(28), 4.06(88), 3.87(23), 3.35(26), 3.08(7), 2.998(15), 2.747(3), 2.585(5), 2.374(7), 2.270(2), 2.045(2), 1.877(6), 1.746(2), 1.678(2).

MnAPO-16 exhibits properties characteristic of a very small-pore molecular sieve.

Synthesis

ORGANIC ADDITIVES
quinuclidine

MnAPO-16 crystallizes from a reactive gel composition with a molar oxide ratio of: 1.0 R : 0.167 MnO : 0.917 Al$_2$O$_3$: P$_2$O$_5$: 0.33 CH$_3$COOH : 40 H$_2$O. Manganese acetate or manganese sulfate is an effective source of manganese, and hydrated alumina or aluminum isopropoxide is the source of aluminum. Crystallization occurs readily after 24 hours at a temperature of 150°C.

MnAPO-34 (CHA)
(US4,567,029(1986))
Union Carbide Corporation

Related Materials: chabazite
AlPO$_4$-34

Structure

Chemical Composition: magnesium silicoaluminophosphate. See MgAPSO.

X-Ray Powder Diffraction Data: (d(Å) (I/I_0))
9.31(100), 6.86(18), 6.28(18), 5.51(49), 4.90(26), 4.31(89), 3.99(5), 3.85(5), 3.52(26), 3.45(22), 3.23(5), 3.14(5), 3.028(8), 2.934(32),

2.858(27), 2.607(8), 2.468(5), 1.918(5), 1.910(3), 1.855(6).

MnAPO-34 exhibits properties characteristic of a small-pore molecular sieve.

Synthesis

ORGANIC ADDITIVES
tetraethylammonium$^+$

MnAPO-34 crystallizes from a reactive gel composition with a molar oxide ratio of: 1.0 R : 0.16–0.4 MnO : 0.8–0.917 Al$_2$O$_3$: P$_2$O$_5$: 0.33 CH$_3$COOH : 40 H$_2$O. Manganese acetate or manganese sulfate is an effective source of manganese, and hydrated alumina or aluminum isopropoxide is the source of aluminum. Crystallization occurs readily after 24 hours at a temperature of 150°C.

MnAPO-35 (LEV)
(US4,567,029(1986))
Union Carbide Corporation

Related Materials: levyne
SAPO-35

Structure

Chemical Composition: manganese aluminophosphate. See XAPO.

X-Ray Powder Diffraction Data: (d(Å) (I/I_0))
10.28(15), 8.04(47), 6.66(29), 5.61(7), 5.13(73), 4.98(13), 4.23(53), 4.08(100), 3.85(20), 3.77(6), 3.56(6), 3.34(23), 3.13(26), 3.11(24), 3.08(sh), 2.797(46), 2.592(10), 2.515(6), 1.881(6), 1.848(6), 1.781(6), 1.664(6).

MnAPO-35 exhibits properties characteristic of a small-pore levyne-type molecular sieve.

Synthesis

ORGANIC ADDITIVES
quinuclidine

MnAPO-35 crystallizes from a reactive gel composition with a molar oxide ratio of: 1.0

R : 0.5 MnO : 0.75 Al_2O_3 : P_2O_5 : CH_3COOH : 50 H_2O (R represents the organic additive in this system). Manganese acetate or manganese sulfate is an effective source of manganese, and hydrated alumina or aluminum isopropoxide is the source of aluminum. Crystallization occurs readily after 24 hours at a temperature of 150°C.

MnAPO-36 (ATS)
(US4,567,029(1986))
Union Carbide Corporation

Structure

Chemical Composition: manganese aluminophosphate. See XAPO.

X-Ray Powder Diffraction Data: (d(Å) (I/I_0)) 11.19(100), 10.78(30), 6.56(6), 5.57(14), 5.37(33), 4.67(45), 4.27(*), 4.10(24), 4.04(37), 3.88(10), 3.74(11), 3.29(21), 3.16(12), 2.959(10sh), 2.805(13), 2.585(*), 2.501(4) (*contribution to peak by shared impurity peak).

MnAPO-36 exhibits properties characteristic of a large-pore molecular sieve.

Synthesis

ORGANIC ADDITIVES
tripropylamine

MnAPO-36 crystallizes from a reactive gel composition with a molar oxide ratio of: 1.5 R : 0.167 MnO : 0.917 Al_2O_3 : P_2O_5 : 0.33 CH_3COOH : 40 H_2O (R represents the organic additive in this system). Manganese acetate or manganese sulfate is an effective source of manganese, and hydrated alumina is the source of aluminum. Crystallization occurs readily after 72 hours at a temperature of 150°C.

MnAPO-44 (CHA)
(US4,567,029(1986))
Union Carbide Corporation

Related Materials: chabazite
SAPO-44

Structure

Chemical Composition: manganese aluminophosphate. See XAPO.

X-Ray Powder Diffraction Data: (d(Å) (I/I_0)) 9.41(100), 6.81(7), 6.42(2), 5.54(15), 5.13(2), 4.70(5), 4.31(41), 4.11(16), 3.97(3), 3.87(3), 3.65(22), 3.41(7), 3.21(5), 3.018(3), 2.969(8), 2.912(29), 2.763(1), 2.714(3), 2.536(3), 1.899(3), 1.892(2), 1.877(4), 1.821(4), 1.707(3).

MnAPO-44 exhibits properties characteristic of a small-pore molecular sieve.

Synthesis

ORGANIC ADDITIVES
cyclohexylamine

MnAPO-44 crystallizes from a reactive gel composition with a molar oxide ratio of: 1.0 R : 0.4 MnO : 0.8 Al_2O_3 : P_2O_5 : 0.8 CH_3COOH : 50 H_2O (R represents the organic additive in this system). Manganese acetate is an effective source of manganese, and hydrated alumina is the source of aluminum. Crystallization occurs readily after 168 hours at a temperature of 150°C.

MnAPO-47
(US4,567,029(1986))
Union Carbide Corporation

Related Materials: related metal-substituted materials.

Structure

Chemical Composition: manganese aluminophosphate. See XAPO.

X-Ray Powder Diffraction Data: (d(Å) (I/I_0)) 9.41(100), 6.92(*), 5.57(14), 5.07(4), 4.33(49), 3.62(*), 3.45(*), 2.94(19), 2.903(13) (*represents a peak shared with an impurity).

MnAPO-47 exhibits properties characteristic of a small-pore molecular sieve.

Synthesis

ORGANIC ADDITIVES
diethylethanolamine

MnAPO-47 crystallizes from a reactive gel composition with a molar oxide ratio of: 1.0 R : 0.167 MnO : 0.917 Al_2O_3 : P_2O_5 : 0.33 CH_3COOH : 45 H_2O (R represents the organic additive in this system). Manganese acetate is an effective source of manganese, and aluminum isopropoxide is the source of aluminum. Crystallization occurs readily after 96 hours at a temperature of 150°C.

MnAPSO
(EP 161,490(1985))
Union Carbide Corporation

MnAPSO compositionally represents a family of manganese silicoaluminophosphates with a preferred range of compositions for Al/P/(Mn + Si). See MnAPSO Figs. 1 and 2.

MnAPSO-5 (AFI)
(EP 161,490(1985))
Union Carbide Corporation

Related Materials: AlPO4-5

Structure

Chemical Composition: See MnAPSO.

X-Ray Powder Diffraction Data: (as-synthesized) $d(\text{Å})$ (I/I_0)) 12.81(13)*, 11.79(100), 11.05(5)*, 9.72(4)*, 9.51(4)*, 6.81(14), 6.46(3)*, 5.91(27), 5.37(3)*, 4.80(7)*, 4.48(43), 4.23(58), 3.99(75), 3.60(6), 3.440(42), 3.079(18), 2.979(34), 2.667(8), 2.600(21), 2.436(4), 2.386(10), 2.176(5), 2.146(5), 2.141(5), 2.122(5), 2.080(3), 2.019(3), 1.914(7), 1.778(5), 1.762(3), 1.656(5); (calcined 500°C)

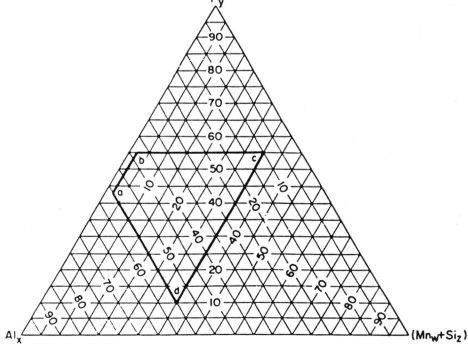

MnAPSO Fig. 1:   Ternary diagram of the preferred composition, in mole fractions, for MnAPSO molecular sieves (EP 161,590(1985)).

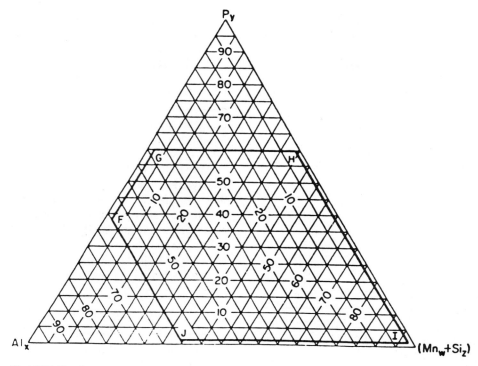

MnAPSO Fig. 2:   Ternary diagram of the reaction mixture, in mole fractions, for MnAPSO molecular sieves (EP 161,590(1985)).

11.95(100), 11.33(4)*, 6.86(25), 5.91(21), 5.37(3)*, 5.31(3)*, 5.07(5)*, 4.48(40), 4.19(40), 3.95(43), 3.427(30), 3.069(11), 2.969(35), 2.660(5), 2.592(19), 2.423(4), 2.374(6), 2.127(4), 2.099(3), 1.973(3), 1,899(5), 1.647(4) (*peaks may contain impurities).

MnAPSO-5 exhibits properties of a large-pore molecular sieve. (See AlPO$_4$-5 for a description of the framework structure.)

Synthesis

ORGANIC ADDITIVES
tetraethylammonium$^+$
tripropylamine
tetrabutylammonium$^+$
cyclohexylamine
tetrapropylammonium$^+$
dipropylamine

MnAPSO-5 crystallizes from the reaction mixture: 1.0 R : 0.2 MnO : 0.9 Al$_2$O$_3$: 0.9 P$_2$O$_5$: 0.2–0.6 SiO$_2$: 50 H$_2$O. It appears as a stable phase at 150 to 200°C when R = TEAOH is used in the synthesis mixture at 150°C, with MnAPSO-34 crystallizing first. Pure phase crystallization was not identified in this organic-containing system. When tripropylamine is used, MnAPSO-5 crystallizes between 4 and 11 days, at temperatures ranging between 150 and 200°C. In TBAOH, at 200°C, MnAPSO-5 crystals are formed; however, at lower temperatures no crystalline phases were identified. This structure will crystallize in the presence of cyclohexylamine at 200°C, but MnAPSO-44 appears as a minor product. TPAOH in the crystallizing mixture provides very pure MnAPSO-5 crystals. Small amounts of MnAPSO-5 are observed when dipropylamine is used. Generally, crystallization times required to produce this structure occur between 2 days (in the presence of TPAOH) and 25 days (in the presence of dipropylamine). Temperatures are generally between 150 and 200°C.

Catapal or aluminum isopropoxide has been used as a source of aluminum, phosphoric acid as a source of phosphorus, Ludox LS as a source of silica, and manganese acetate as a source of Mn cations.

Adsorption

Adsorption properties of MnAPSO-5:

Adsorbate	Kinetic diameter, Å	Pressure, torr	Temp., °C	Wt% adsorbed*
O_2	3.46	102	−183	8.9
O_2		750	−183	10.8
n-Butane	4.3	504	23.0	4.4
Cyclohexane	6.0	65	23.4	5.4
H_2O	2.65	4.6	23.0	8.1
H_2O	2.65	19.5	23.0	17.1

MnAPSO-11 (AEL)
(EP 161,490(1985))
Union Carbide Corporation

Related Materials: AlPO$_4$-11

Structure

Chemical Composition: See MnAPSO.

X-Ray Powder Diffraction Data: (d(Å) (I/I_0))
(as-synthesized) 10.92(36), 9.31(61), 6.76(19), 4.65(36), 5.47(10), 4.65(13), 4.33(45), 4.21(100), 4.00(55), 3.95(52), 3.92(61), 3.83(71), 3.63(13), 3.59(16), 3.5629(13), 3.38(26), 3.153(13), 3.121(23), 3.028(13), 2.84(16), 2.730(23), 2.622(16), 2.54(10), 2.508(10), 2.475(10), 2.398(13), 2.370(16), 2.287(10), 2.108(10), 2.023(10), 1.866(3), 1.804(10), 1.681(10); calcined 600°C) 10.92(33), 9.03(60), 7.50(13), 6.92(27), 6.56(13), 5.99(sh), 5.51(67), 4.55(27), 4.46(40), 4.35(33), 4.13(73), 4.08(100), 4.00(73), 3.97(80), 3.79(73), 3.66(27), 3.453(33), 3.339(27), 3.267(33), 3.209(33), 3.132(27), 3.028(33), 2.998(40), 2.940(27), 2.814(20), 2.747(33), 2.637(20), 2.529(27), 2.423(20), 2.404(20), 2.356(20), 2.332(27), 2.201(20).

MnAPSO-11 exhibits properties of a medium-pore molecular sieve. (See AlPO$_4$-11 for a description of the framework topology.)

Synthesis

ORGANIC ADDITIVES
dipropylamine

MnAPSO-11 has been crystallized from reactive mixtures with the composition: 1.0 dPA : 0.2 MnO : 0.9 Al$_2$O$_3$: 0.9 P$_2$O$_5$: 0.2–0.6 Si$_2$: 50 H$_2$O. Crystallization generally occurs between 150 and 200°C within 2 to 25 days. Other phases observed to co-crystallize with MnAPSO-11 include MnAPSO-39 (minor), MnAPSO-31 (major), and MnAPSO-5 (minor).

Adsorption

Adsorption properties of MnAPSO-11:

Adsorbate	Kinetic diameter, Å	Pressure, Torr	Temp., °C	Wt% adsorbed
O_2	3.46	106	−183	7.0
O_2		744	−183	11.1
Neopentane	6.2	741	25.3	2.5
Isobutane	5.0	740	24.2	3.6
Cyclohexane	6.0	82	23.9	10.7
H_2O	2.65	4.6	24.9	5.1
H_2O	2.65	19	24.8	14.9

MnAPSO-16 (AST)
(EP 161,490(1985))
Union Carbide Corporation

Related Materials: AlPO$_4$-16

Structure

Chemical Composition: See MnAPSO.

X-Ray Powder Diffraction Data: (d(Å) (I/I_0))
(as-synthesized) 10.28(8)*, 8.04(23)*, 7.76(48), 6.66(11)*, 5.57(5)*, 5.13(24)*, 5.01(8)*, 4.75(40), 4.21(19)*, 4.06(100)**, 3.87(13), 3.83(10)*, 3.75(5)*, 3.548(5), 3.351(26)**,

3.339(sh)*, 3.209(5), 3.100(15)*, 3.079(15), 2.998(24), 2.797(16)*, 2.747(7)*, 2.585(10)**, 2.515(5)*, 2.380(11), 2.270(5), 2.151(5)*, 2.049(5), 1.877(10)**, 1.845(5)*, 1.746(5), 1.678(5); (calcined 600°C) 7.69(100), 6.66(9)*, 4.77(25), 4.37(44)*, 4.33(41)*, 4.13(66)*, 4.06(72)**, 3.88(31), 3.79(13)*, 3.363(31)**, 3.198(13), 3.079(19), 3.008(14), 2.747(13), 2.585(13)**, 2.522(16)*, 2.380(13), 1.888(9)** (*impurity peak; **peak may contain impurity).

MnAPSO-16 exhibits properties of a very small-pore material. (See $AlPO_4$-16 for a description of the framework topology.)

Synthesis

ORGANIC ADDITIVES
quinuclidine
methylquinuclidine

MnAPSO-16 has been crystallized from reactive mixtures with the composition: 1.0 quin : 0.2 MnO : 0.9 Al_2O_3 : 0.9 P_2O_5 : 0.2–0.6 SiO_2 : 50 H_2O. Crystallization generally occurs between 150 and 200°C within 2 to 11 days. Other phases observed to co-crystallize with MnAPSO-16 include MnAPSO-35 (minor) when quinuclidine is used as the organic additive. When methylquinuclidine is used, MnAPSO-16 is the minor product and MnAPSO-35 the major product isolated.

MnAPSO-20 (SOD)
(EP 161,490(1985))
Union Carbide Corporation

Related Materials: $AlPO_4$-20
                    sodalite

Structure

Chemical Composition: See MnAPSO.

X-Ray Powder Diffraction Data: ($d(Å)$ (I/I_0)) (as-synthesized) 6.35(49), 4.49(43), 4.02(3), 3.75(1)*, 3.67(100), 3.117(13), 2.842(11), 2.595(16), 2.400(2), 2.247(4), 2.118(4),

1.917(4), 1.764(7); (calcined 500°C) 12.51(2), 6.33(100), 4.48(40), 4.00(4), 3.66(99), 3.168(17), 2.835(15), 2.589(17), 2.243(3), 2.116(4), 1.913(4) (*impurity peak).

MnAPSO-20 exhibits properties of a very small-pore material. (See $AlPO_4$-20 for a description of the framework topology.)

Synthesis

ORGANIC ADDITIVES
tetramethylammonium$^+$

MnAPSO-20 has been crystallized from reactive mixtures with the composition: 1.0 TMA : 0.2 MnO : 0.9 Al_2O_3 : 0.9 P_2O_5 : 0.2–0.6 SiO_2 : 50 H_2O. Crystallization generally occurs between 150 and 200°C within 4 days. No other phases were observed.

Adsorption

Adsorption properties of MnAPSO-20:

Adsorbate	Kinetic diameter, Å	Pressure, Torr	Temp., °C	Wt% adsorbed
O_2	3.46	102	−183	0.7
O_2		744	−183	1.2
H_2O	2.65	4.6	23.3	9.0
H_2O	2.65	20	23.2	13.7

MnAPSO-31 (ATO)
(EP 161,490(1985))
Union Carbide Corporation

Related Materials: $AlPO_4$-31

Structure

Chemical Composition: See MnAPSO.

X-Ray Powder Diffraction Data: ($d(Å)$ (I/I_0)) (as-synthesized) 11.22(4), 10.27(61), 5.17(5), 4.81(4), 4.36(49), 4.19(4), 4.04(30), 4.02(32),

3.92(100), 3.526(5), 3.450(3), 3.181(12),
2.995(6), 2.812(22), 2.548(9), 2.482(3),
2.411(3), 2.382(3), 2.353(3), 1.2346(3),
2.285(3), 2.266(3), 2.241(3), 1.942(3), 1.866(2),
1.766(5), 1.654(2); (calcined 500°C) 10.31(58),
5.98(4), 5.18(9), 4.81(4), 4.36(52), 4.03(44),
3.92(100), 3.526(7), 3.460(8), 3.181(15),
2.998(11), 2.879(3), 2.811(33), 2.546(11),
2.477(6), 2.409(3), 2.383(3), 2.348(3), 2.289(4),
2.236(3), 2.000(3), 1.942(5), 1.909(4), 1.864(3),
1.767(6).

MnAPSO-31 exhibits properties of a medium-pore molecular sieve. (See $AlPO_4$-31 for a description of the structure.)

Synthesis

ORGANIC ADDITIVES
dipropylamine

MnAPSO-31 has been crystallized from reactive mixtures with the composition: 1.0 DPA: 0.2 MnO : 0.9 Al_2O_3 : 0.9 P_2O_5 : 0.2–0.6 SiO_2 : 50 H_2O. Crystallization generally occurs between 150 and 200°C within 2 to 10 days. Other phases observed to co-crystallize with MnAPSO-31 include MnAPSO-46 (minor) and MnAPSO-11 (minor).

Adsorption

Adsorption properties of MnAPSO-31 (EP 161,490 (1985)):

Adsorbate	Kinetic diameter, Å	Pressure, Torr	Temp., °C	Wt% adsorbed
O_2	3.46	105	−183	5.6
O_2		741	−183	9.7
Neopentane	6.2	739	23.5	4.6
H_2O	2.65	4.6	23.8	5.8
H_2O	2.65	18.5	24.0	15.5

MnAPSO-34 (CHA)
(EP 161.490(1985))
Union Carbide Corporation

Related Materials: chabazite
                        $AlPO_4$-34

Structure

Chemical Composition: See MnAPSO.

X-Ray Powder Diffraction Data: ($d(Å)$ (I/I_0))
(as-synthesized) 9.21(100), 6.86(17), 6.24(15),
5.51(33), 4.90(23), 4.31(69), 3.99(10), 3.85(8),
3.543(25), 3.453(19), 3.243(10), 3.143(10),
3.028(10), 2.931(27), 2.867(23), 2.652(8),
2.614(12), 2.475(8), 2.103(6), 2.080(6),
1.914(6), 1.863(8), 1.794(6), 1.728(6), 1.650(6);
(calcined 425°C) 9.21(100), 6.86(25), 6.28(5),
5.47(15), 4.96(15), 4.65(5), 4.27(37), 4.00(5),
3.97(5), 3.83(7), 3.534(15), 3.427(12), 3.220(4),
3.153(5), 3.008(4), 2.912(17), 2.849(11),
2.763(3), 2.592(5), 2.481(4), 2.321(3), 2.265(3),
2.099(3), 2.076(3), 1.903(1), 1.859(3), 1.791(3),
1.719(4), 1.681(3).

MnAPSO-34 exhibits properties of a small-pore molecular sieve.

Synthesis

ORGANIC ADDITIVES
tetraethylammonium $^+$

MnAPSO-34 has been crystallized from reactive mixtures with the composition: 1.0 TEAOH : 0.2 MnO : 0.9 Al_2O_3 : 0.9 P_2O_5 : 0.2–0.6 SiO_2 : 50 H_2O. Crystallization generally occurs between 100 and 200°C within 2 to 14 days. MnAPSO-34 appears to be a low-temperature product, as it can be prepared in pure form after 7 days at 100°C. Other phases observed to co-crystallize with MnAPSO-34 include MnAPSO-5 (major). The purity of the product, either MnAPSO-5 or MnAPSO-34, is dependent on the temperature and the time of crystallization.

Adsorption

Adsorption properties of MnAPSO-34:

Adsorbate	Kinetic diameter, Å	Pressure, Torr	Temp., °C	Wt% adsorbed
O_2	3.46	103	−183	11.4
O_2		731	−183	15.6
Isobutane	5.0	741	24.5	0.8
n-Hexane	4.3	103	24.4	4.6
H_2O	2.65	4.6	24.4	15.2
H_2O	2.65	18.5	23.9	24.4

MnAPSO-35 (LEV)
(EP 161,490(1985))
Union Carbide Corporation

Related Materials: SAPO₄-35
levyne

Structure

Chemical Composition: See MnAPSO.

X-Ray Powder Diffraction Data: $(d(Å) (I/I_0))$
(as-synthesized) 10.28(14), 8.12(45), 6.61(23),
5.57(11), 5.10(80), 4.98(16), 4.25(57),
4.06(100), 3.83(34), 3.59(9), 3.466(7),
3.314(21), 3.153(50), 3.069(11), 2.849(9),
2.788(41), 2.614(14), 2.571(7), 2.543(5),
2.508(7), 2.386(5), 2.281(5), 2.156(7), 2.118(7),
2.032(5), 1.910(7), 1.884(7), 1.841(7), 1.791(9),
1.670(5), 1.658(7); (calcined 500°C) 10.28(27),
8.12(96), 7.76(14), 6.61(41), 5.61(14), 5.13(68),
5.01(sh), 4.27(64), 406(100), 3.82(32), 3.59(23),
3.466(18), 3.314(27), 3.153(59), 3.069(23),
2.849(18), 2.780(46), 2.622(18), 2.578(14),
2.508(9), 2.156(9), 2.127(9), 2.032(9), 1.918(9),
1.888(9), 1.845(9), 1.791(14), 1.664(9),
1.650(9).

MnAPSO-35 exhibits properties of a small-pore levyne molecular sieve.

Synthesis

ORGANIC ADDITIVES
quinuclidine
methylquinuclidine

MnAPSO-35 has been crystallized from reactive mixtures with the composition: 1.0 quin : 0.2 MnO : 0.9 Al_2O_3 : 0.9 P_2O_5 : 0.2–0.6 SiO_2 : 50 H_2O. Crystallization generally occurs between 100 and 200°C within 2 to 11 days. Other phases observed to co-crystallize with MnAPSO-35 include MnAPSO-16, a major component at the higher temperatures in the presence of quinuclidine and a minor component when methylquinuclidine is used as the organic additive.

Adsorption

Adsorption properties of MnAPSO-35:

Adsorbate	Kinetic diameter, Å	Pressure, Torr	Temp., °C	Wt% adsorbed
O_2	3.46	103	−183	1.8
O_2		731	−183	2.6
n-Hexane	6.2	103	24.4	0.8
H_2O	2.65	4.6	24.4	9.9
H_2O	2.65	18.5	23.9	15.9

MnAPSO-44 (CHA)
(EP 161,490(1985))
Union Carbide Corporation

Related Materials: SAPO-44
chabazite

Structure

Chemical Composition: See MnAPSO.

X-Ray Powder Diffraction Data: $(d(Å) (I/I_0))$
(as-synthesized) 9.39(100), 6.83(20), 6.45(4),
5.52(43), 5.12(5), 4.68(7), 4.29(84), 4.09(21),
3.94(8), 3.86(9), 3.65(58), 3.409(22), 3.205(10),
3.012(5), 2.969(16), 2.900(50), 2.753(4),
2.721(6), 2.577(3), 2.528(9), 2.336(2), 2.299(2),
2.255(2), 2.143(3), 2.125(3), 2.076(2), 1.922(2),

1.890(7), 1.870(4), 1.814(7), 1.701(6); (calcined 500°C) 9.21(100), 6.79(26), 6.26(3), 5.46(12), 4.93(18), 4.60(3), 4.25(28), 3.99(3), 3.80(3), 3.526(13), 3.387(9), 3.137(3), 3.123(4), 2.990(2), 2.976(2), 2.921(3), 2.875(7), 2.811(2), 2.791(2), 2.560(3).

MnAPSO-44 exhibits properties characteristic of a small-pore molecular sieve. (See CoAPSO-44 for a description of the structure.)

Synthesis

ORGANIC ADDITIVES
cyclohexylamine

MnAPSO-44 has been crystallized from reactive mixtures with the composition: 1.0 cycloC$_6$NH$_2$: 0.2 MnO : 0.9 Al$_2$O$_3$: 0.9 P$_2$O$_5$: 0.2–0.6 SiO$_2$: 50 H$_2$O. Crystallization generally occurs between 150 and 200°C within 3 to 9 days. Other phases observed to co-crystallize with MnAPSO-44 include an unidentified phase, MnAPSO-31 (minor), and MnAPSO-5 (major). Crystallization of pure MnAPSO-44 occurs at 200°C after 4 days of crystallization.

Adsorption

Adsorption properties of MnAPSO-44:

Adsorbate	Kinetic diameter, Å	Pressure, Torr	Temp., °C	Wt% adsorbed
O$_2$	3.46	102	−183	5.6
O$_2$		744	−183	9.7
n-Hexane	4.3	95	23.6	1.3
Isobutane	5.0	746	24.1	22.7
H$_2$O	2.65	4.6	24.8	22.7
H$_2$O	2.65	19	29.8	27.7

MnAPSO-47 (CHA)
(EP 161,490(1985))
Union Carbide Corporation

Related Materials: chabazite

Structure

Chemical Composition: See MnAPSO.

X-Ray Powder Diffraction Pattern: (d(Å) (I/I_0)) (as-synthesized) 18.44(1), 9.38(100), 6.89(5), 6.40(3), 5.56(9), 5.06(4), 4.69(3), 4.32(30), 4.08(4), 3.98(1), 3.88(3), 3.61(11), 3.445(7), 3.234(2), 3.199(1), 3.033(2), 2.930(10), 2.901(7), 2.845(1), 2.700(1), 2.604(2), 2.576(1), 2.516(20), 2.343(1), 2.297(1), 2.277(1), 2.132(1), 2.091(1), 1.911(1), 1.874(5), 1.813(2), 1.722(1), 1.698(1); (calcined 500°C) 17.80(1), 9.12(100), 8.85(1), 6.75(5), 6.23(1), 5.45(2), 4.92(2), 4.58(3), 4.24(7), 3.98(1), 3.80(1), 3.521(2), 3.385(2), 3.176(10), 3.125(1), 2.977(1), 2.876(3), 2.837(2), 2.645(1), 2.562(1), 1.838(1).

MnAPSO-47 exhibits properties of a small-pore molecular sieve.

Synthesis

ORGANIC ADDITIVES
diethylaminoethanol

MnAPSO-47 has been crystallized from reactive mixtures with the composition: 2.0 DEEA : 0.2 MnO : 0.9 Al$_2$O$_3$: 0.9 P$_2$O$_5$: 0.2–0.6 SiO$_2$: 50 H$_2$O. Crystallization generally occurs at 150°C within 9 to 18 days. No other phases are observed to co-crystallize with MnAPSO-47 in this system.

MnGS-3
(SSSC 49A:375(1989))

Structure

Chemical Composition: manganese germanium sulfide.

X-Ray Powder Diffraction Data: (d(Å) (I/I_0)) 7.93(100), 7.13(3), 6.75(7), 4.90(39), 4.08(17), 3.96(24), 3.17(45), 3.04(9), 2.85(4), 2.77(11), 2.30(10), 1.86(9).

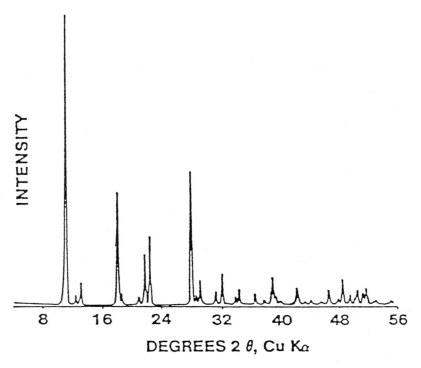

MnGS-3 Fig. 1:   X-ray powder diffraction pattern for
MnGS-3 (*SSSC* 49A:375(1989)). (Reproduced with permission of Elsevier Publishers)

Synthesis

ORGANIC ADDITIVE
tetramethyammonium$^+$

MnGS-3 crystallizes from a reactive gel with
the composition: 0.8 TMAHCO$_3$: 0.27 TMACl
: GeS$_2$: 0.1 Mn acetate : 22 H$_2$O. In this preparation, crystalline GeS$_2$ is used. Crystallization
occurs after 90 hours at 150°C. The initial pH
of the reactive slurry is 7 to 9, with a pH of 0.5
to 3 upon crystal formation.

"Modified" Dachiardite (DAC) (natural)

Related Materials: dachiardite

Minato first identified a dachiardite mineral in
the Onoyama gold mine in Japan. Characteristic
of these samples are diffuse and streaked diffraction maxima, initially interpreted as being
due to structural disorder. Mineral samples from
Hokiya-dake, Nagano, Japan exhibited sharp
diffraction spots ("Annual Meeting of the Mineralogical Society of Japan" (1964); *Contrib.
Mineral. Petrol.*, 49:63 (1975); *Natural Zeolites*, 105 (1978); *Eur. J. Mineral.* 2:187 (1990)).

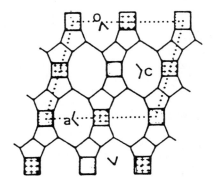

"Mod" Dachiardite Fig. 1S: Framework topology.

Structure

(*Eur. J. Mineral.* 2:187(1990))

CHEMICAL COMPOSITION: $Mg_{0.01}Ca_{1.76}Na_{0.39}$ $K_{0.51}Si_{19}Al_{4.55}Fe_{0.01}O_{48}*13.13H_2O$
SPACE GROUP: C2/m
UNIT CELL CONSTANTS (Å): $a = 18.625$
$b = 7.508$
$c = 10.247$
$\beta = 108.056°$

VOLUME: $1362.34 \ A^3$

The refinement of the structure of Hokiya-dake dachiardite shows the presence, in 16% abundance, of domains referred to as "modified" dachiardite in association with domains of "normal" dachiardite. There is significant partial ordering of Si and Al. The T3 and T4 tetrahedra of the four-member rings are the richest in aluminum. There are eight extra-framework sites; seven are very close to those found in dachiardite in Elba, Italy, and one is not found in the Italian dachiardite (*Eur. J. Mineral.* 2:187 (1990)).

Atomic coordinates and occupancy factors (%) for dachiardite from Hokiya-dake, Japan (*Eur. J. Mineral.* 2:187(1990)):

Atom	x/a	y/b	z/c	Occupancy
T1	0.2872(2)	0.2063(4)	0.1578(4)	92(4)
T1M	0.288(3)	0.266(6)	0.168(5)	8(4)
T2	0.1917(2)	0.2959(4)	0.3554(4)	92(4)
T2M	0.188(3)	0.230(7)	0.356(5)	8(4)
T3	0.0964(2)	0	0.6995(3)	92(2)
T3M	0.089(3)	1/2	0.691(5)	8(2)
T4	0.0809(2)	0	0.3792(4)	92(2)
T4M	0.078(3)	1/2	0.372(5)	8(2)
O1	0.3636(4)	0.3204(10)	0.2181(7)	92(4)
O1M	0.376(5)	0.152(13)	0.226(9)	8(4)
O2	0.1152(4)	0.1719(10)	0.3284(7)	92(4)
O2M	0.113(6)	0.314(14)	0.316(10)	8(4)
O3	0.2264(4)	0.2563(12)	0.2338(8)	100
O4	0.0991(5)	0	0.5440(10)	92(2)
O4M	0.083(9)	1/2	0.528(15)	8(2)
O5	0.1669(6)	1/2	0.3461(12)	92(2)
O5M	0.171(7)	0	0.356(14)	8(2)
O6	0.3107(6)	0	0.1769(11)	92(2)
O6M	0.326(7)	1/2	0.199(14)	8(2)
O7	1/4	1/4	0	100
O8	1/4	1/4	1/2	100
O9	0.0102(4)	0	0.7079(9)	92(2)

Atom	x/a	y/b	z/c	Occupancy
O9M	0.004(9)	1/2	0.611(17)	8(2)
C1*	0.4902(4)	0.2391(14)	0.1335(9)	32(1)
C2*	0.5422(16)	0	0.5361(30)	24(1)
C3*	1/2	0.252(11)	0	14(1)
W1	0.4908(11)	0	0.2565(18)	92(1)
W2	0.0835(16)	0	0.0336(29)	54(3)
W3	0.0869(27)	0.1334(83)	0.0217(51)	23(3)
W4	0.0743(22)	0.6055(94)	0.0262(45)	26(5)
W5	0.0919(26)	1/2	0.0324(44)	48(5)

*Cation sites in the structure.

Percent of aluminum in the tetrahedral sites calculated according to the methods of Alberti and Gotardi (*Z. Kristallogr.* 175:249(1988)) and Jones (*Acta Crystallogr.* B24:355(1968)) for Hokiya-dake and the Elba dachiardites (*Eur. J. Mineral.* 2:187(1990)):

	Hokiya-dake		Elba	
	A.G.	Jones	A.G.	Jones
T1	5	0	12	0
T2	2	0	6	0
T3	28	18	24	15
T4	31	22	40	31
Average	12.3	7.2	16.5	7.6
From chemical analysis	18.9		20.4	

Monoclinic Ferrierite (FER) (*Am. Mineral.* 70:619(1985))

See ferrierite.

Monophane (EPI)

Obsolete synonym for epistilbite.

Montesommaite (MON) (natural)

Montesommaite is a rare mineral that occurs with dolomite, calcite, chabazite, and natrolite around Pollena, Mt. Somma-Vesuvius, Italy. It was named after Monte Somma, the high ridge of the volcanic cone that preceded Vesuvius. It occurs as transparent, colorless, euhedral crystals up to 0.1 mm in length.

Structure

(Am. Mineral. 75:1415(1990))

CHEMICAL COMPOSITION: $(K, Na)_9Al_9Si_{23}O_{64}*$ $10H_2O$
SYMMETRY: orthorhombic
SPACE GROUP: Fdd2
UNIT CELL CONSTANTS(Å): $a = b = 10.099$ (1)
$c = 17.307(3)$
DENSITY: 2.34 g/cm^3
PORE STRUCTURE: 8-member rings

X-Ray Powder Diffraction Data: $(d(Å)$ $(I/I_0))$
6.589(75), 4.488(0.7), 4.334(43), 3.299(100), 3.130(100), 2.797(30), 2.513(18), 2.347(22), 2.256(11), 2.178(16), 2.005(2), 1.959(5), 1.865(1), 1.784(22), 1.720(16), 1.655(9), 1.595(0.7), 1.571(11), 1.558(18), 1.546(13), 1.497(0.7), 1.442(2), 1.420(5), 1.397(3), 1.381(3), 1.320(11), 1.291(6), 1.262(3), 1.253(8), 1.237(4), 1.227(0.7), 1.212(3), 1.174(5).

Several levels of pseudosymmetry are found in montesommaite, and the powder data have been indexable on the pseudotetragonal cell. DLS modeling of the framework structure suggests that montesommatie can be related to a 4.8^2 net as proposed by Smith *(Am. Mineral. 63:960 (1978))*. Such a net also occurs in merlinoite and gismondine. The (100) projection is a $(5.^28)_2(5.8^2)_1$ net with five-member-ring sheets separating the eight-member rings.

Refined atomic coordinates for the idealized montesommaite substructure *(Am. Mineral. 75:1415(1990))*:

Atom	x	y	z	U(iso) (Å^2)
K	0	0	0/2	0.025
T	0	0.463(4)	0.090(2)	0.005
O(1)	0	1/4	0.116(8)	0.015
O(2)	0.186(7)	0.436	7/8	0.015
O(3)	0	0	0	0.015
H$_2$O	0	1/4	0.342(15)	0.015

Note: Calculations are based upon a cell having $A = 7.141$ Å, $C = 17.307$ Å, and symmetry Ī4$_1$/amd. Atom T represents (4.7 Al + 11.3 Si). The K and H$_2$O sites contain 4.3 K and 4 O, respectively. Esd's are in parentheses.

Mordenite (MOR) (natural)

Related Materials: Na-D
Ca-Q
Zeolon
LZ-211

The silica-rich mineral zeolite mordenite first was named by How in 1864 after the location Morden, King's Co., Nova Scotia, where samples were located *(Nat. Zeo. 223(1985))*. Commercial synthetic mordenite prepared by the Norton Company is referred to as Zeolon.

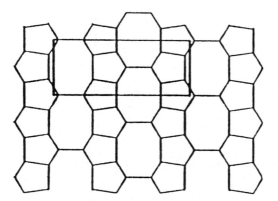

Montesommaite Fig. 1S: Framework topology.

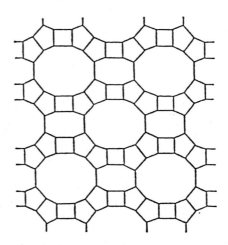

Mordenite Fig. 1S: Framework topology.

Structure

(*Nat. Zeo.* 223(1985))

CHEMICAL COMPOSITION: Na$_3$KCa$_2$
  (Al$_8$Si$_{40}$O$_{96}$)*28H$_2$O
SYMMETRY: orthorhomic
SPACE GROUP: Cmcm
UNIT CELL CONSTANTS: (Å) $a = 18.11$
                          $b = 20.46$
                          $c = 7.52$
PORE STRUCTURE: intersecting 8- and 12-member
  rings 6.5 × 7.0 Å and 2.6 × 5.7 Å

X-Ray Powder Diffraction Data: (*Zeo. Mol. Sieve.* 231(1974) (*d*(Å) (*I/I$_0$*)) 13.58(18), 10.26(5), 9.06(100), 6.59(14), 6.40(17), 6.07(4), 5.80(18), 4.88(3), 4.60(2), 4.53(31), 4.46(2), 4.15(8), 4.00(70), 3.842(7), 3.765(4), 3.629(3), 3.568(4), 3.532(2), 3.476(43), 3.4217(11), 3.394(33), 3.291(3), 3221(40), 3.201(34), 3.155(2), 3.101(4br), 3.028(1), 3.017(2), 2.942(5), 2.895(13), 2.741(2), 2.715(2), 2.701(5), 2.633(3), 2.588(1), 2.565(10), 2.521(7), 2.459(4), 22.436(2br), 2.294(1), 2.279(1), 2.263(1), 2.232(2), 2.166 (2br), 2.117(1), 2.052(7), 2.035(2), 2.019(2), 1.997(2).

Mordenite exists with a nearly constant SiO$_2$/Al$_2$O$_3$ ratio of 10, and is considered to be the most siliceous of the zeolite minerals. In the distribution of Si and Al in the tetrahedral framework sites, there is a small Al-enrichment in the four-member rings that connect hexagonal sheets (*Z. Kristallogr.* 115:439(1961); *Nat. Zeo.* 53(1978); *Zeo.* 9:170(1989)).

The structure consists of chains that are crosslinked by the sharing of neighboring oxygens. The building units of this structure consist of four- and five-member rings. Each of the framework tetrahedran belongs to one or more five-member rings in the framework. The zeolite contains a two-dimensional channel system, but for practical applications it is considered to be a unidimensional, large-pore molecular sieve with side pockets because of the constraint on the eight-member-ring openings for diffusion. The main cage is defined by two four-rings and two hexagonal sheets connected by them. This case has two exits, through two eight-member rings.

Natural mordenite contains crystal stacking faults in the *c* direction or amorphous material or cations in the channels that severely restrict diffusion into the large channel system. Many

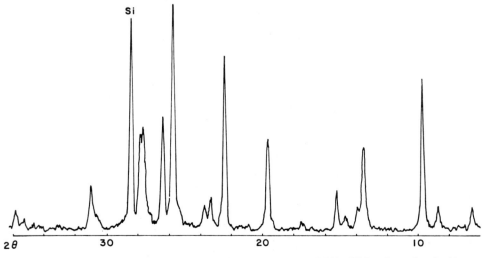

Mordenite Fig. 2:   X-ray powder diffraction pattern for mordenite (*Zeo.* 5:251 (1985)). (Reproduced with permission of Butterworths Publishers)

of the synthetic mordenites do not exhibit such pore restriction.

The cations are located mainly in three sites. Site A is near the center of the eight-member ring that connects the main cage, described above, with the smaller channel; site D is near the center of the eight-member ring that connects the cage with the larger channel; and site E is in the larger channel far from the four-member rings. These three sites are preferred by most cations except Rb and Ba, which are found in sites normally

Change in the unit cell parameters (A) and asymmetric infrared stretch as a function of aluminum content (*Zeo.* 7:427 (1987)):

SiO_2/A_2O_3	a	b	c	v_{as} T–O (cm^{-1})
12.6	18.25	20.51	7.50	1074
17.6	18.18	20.41	7.48	1080
19.5	18.14	20.31	7.47	1084
35	18.23	20.43	7.47	1085
66	18.20	20.41	7.46	1088
165	18.22	20.44	7.47	1090

Atomic coordinates and occupany factors (%) for natural mordenite from Elba, Italy (*Z. Kristallogr.* 175:249(1986)):

Atom	x	y	z	Occupancy
T1	0.3007(1)	0.0724(1)	0.0412(3)	100*
T1P	0.6993(1)	0.9276(1)	0.9588(3)	100*
T2	0.3037(1)	0.3088(1)	0.0461(3)	100*
T2P	0.6963(1)	0.6912(1)	0.9539(3)	100*
T3	0.0870(2)	0.3941(1)	1/4	100*
T4	0.0867(2)	0.2278(1)	1/4	100*
O1	0.1243(3)	0.4170(3)	0.4287(7)	100
O1P	0.8757(3)	0.5830(3)	0.5743(7)	100
O2	0.1226(3)	0.1954(3)	0.4254(7)	100
O2P	0.8774(3)	0.8046(3)	0.5746(7)	100
O3	0.2362(3)	0.1224(2)	0.9865(8)	100
O3P	0.7638(3)	0.8776(2)	0.0135(8)	100
O4	0.0958(4)	0.3064(4)	1/4	100
O5	0.1696(4)	0.1960(5)	3/4	100
O6	0.1778(5)	0.4208(4)	3/4	100
O7	0.2321(4)	1/2	1/2	100
O8	1/4	1/4	1/2	100
O9	0	0.4067(6)	1/4	100
O10	0	0.2050(6)	1/4	100

*The occupancy refers to a scattering curve of Si 80% and Al 20%.

Comparison of site occupancies for mordenite (*Nat. Zeo.* 53(1976)):

Type		hNa*	hCa	hK*	dCa	dK	dH
I	4b	4Na	1.70Ca	1.92K	1.66Ca		0.65Na
II	4c	4W	3.95W				
	8f**			4.5W			
	8d**					3.34K	
III	8g**	4W		4.7W	0.62Ca		
	4c		4.7W				
IV	8f*			3.02K	0.45Ca		
	4c	4?	3.4W				
	8d**					3.04K	
V	16h		10.3W				
	8f	8W		8.4W			
VI	8g	8W	8.7W	5.8W,K	0.63Ca		
	8d					0.91K	
VII	4c	1Na	1.68Ca				
VIII	8c			6.5W,K			

*Cation and water assignments are uncertain, and mixed occupancy may be important

**These sites are effectively fourfold because of close approach if fully occupied.

occupied by water. All Ca ions are localized in an 8-coordinated site; K ions alternate with water molecules in a 6-coordination site (*Z. Kristallogr.* 175:249(1986)). A comparison of hydrated sodium exchanged with dehydrated sodium and hydrated calcium and dehydrated H$^+$ mordenites shows only small changes in the dimensions of the channels.

Sites I and II are close, as are II and IV. The formal charge distribution for dCa and dK is similar when these pairs are considered.

Synthesis

ORGANIC ADDITIVE
benzyltrimethylammonium$^+$
alkyl phenol/alkyl sulfonates
N-ethylpyridinium$^+$
trioctylamine

Early investigations into the synthesis of this known mineral utilized high temperature and high pressures. Mordenite synthesis first was claimed in 1927, and it was prepared from feldspars and alkali carbonates at 200°C over a period of 7 days (*Econ. Geol.* 22:843(1927)). Later a more reproducible method was reported requiring 300°C (*JCS* 2158(1948)). At 100 to 175°C using diatomite and sodium aluminate this zeolite was formed after 7 to 16 days (*Mol. Sieves* 71(1968)). Crystallization generally occurs in the range of 8 to 15 days. A wide variety of silica and alumina sources have been used to produce this structure, generally in the presence of the sodium of the cation (*Zeo.* 6:2(1986)). A higher-silica mordenite can be prepared without the additions of organics. A SiO_2/Al_2O_3 of 19.1 was prepared from a crystallization mixture having a mole ratio of 22.9 (*Zeo.* 6:30(1986)). Generally, mordenite will begin to appear at a SiO_2/Al_2O_3 of 8, with impurities of phillipsite being observed. At the higher silica contents quartz is a general impurity. Crystallization occurs in a pH range of 8 to 10, but analcime will begin to appear at higher ph (12.8 (*Mol. Sieves.* 78(1968)). Calcium-containing mordenite was synthesized at very high pressures and temperatures (1000–5000 atm and 330–500°C) (*Am. Mineral.* 43:476(1958); *Geochim. Cosmochim. Acta* 17:53(1959)). Strontium-containing mordenite was prepared in a SiO_2/Al_2O_3 of 9 at temperatures between 110 and 450°C (*JCS* London 485(1964)). Lithium mordenite was prepared at temperatures between 150 and 200°C (*ACS* 101:127 (1971)). The preferred cations include sodium and the alkaline earth ions (*Zeo.* 1:130(1981)).

Much of this synthesis produced a small-pore mordenite similar to that of the natural material, based on adsorption studies. X-ray diffraction cannot distinguish between large-port and small-port materials. Generally, small-port materials are prepared at the higher temperatures, whereas low temperatures (100–260°C) will result in the large-port material (*Mol. Sieves* 71(1968)). The small-port materials exhibit adsorption properties of a material with only a 4 A opening, whereas the large-pore mordenite exhibits characteristic adsorption properties related to the structurally determined presence of the open 12-member-ring channel system. Erionite has been transformed to mordenite under hydrothermal conditions at 300°C and 30 MPa (*Zeo.* 6:95(1986)). Large port mordenite also has been

prepared using 1 wt% H-mordenite as seed crystals in the presence of potassium. In the mixed cation system, the mordenite was found to preferentially incorporate potassium into the structure over sodium. The K/Na ratios obtained during synthesis and during ion exchange were found to be the same (*Zeo.* 5:309(1985)).

Solid solutions of Be,Al-mordenites also have been prepared (*Clay Science* 7:49(1987)). These materials crystallize in the presence of benzyltrimethylammonium base. The batch composition of the gels for these solid solutions are in the range: $(4.87 + 0.33x) Na_2O : (1.65 - 0.33x) K_2O : (0.44 - 0.08x) BTMA_2O : 0.06x Al_2O_3 : (0.30 - 0.60x)BeO : 20SiO_2 : 600H_2O$ $(0 < x < 5)$. Crystallization occurs after 3 to 5 days at 150°C.

Mordenite has been crystallized from clear aqueous solutions at 100°C at 1 atm pressure.

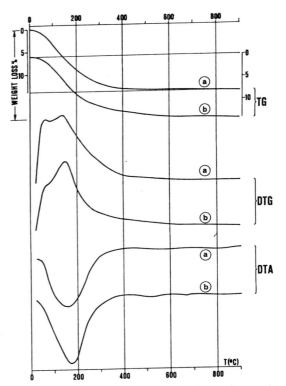

Mordenite Fig. 3:   Thermal curves of mordenite from (a) Morden, Canada and (b) Poona, India, in air, heating rate 20°C/min (*Natl. Zeo.* 223(1985)) (Reproduced with permission of Springer Verlag Publishers)

The batch composition is in the range: 10 Na$_2$O : 0.075–0.4Al$_2$O$_3$: 28–40 SiO$_2$: 360 H$_2$O. Crystals obtained had a SiO$_2$/Al$_2$O$_3$ ratio of 15 (*Am. Mineral.* 65:1012(1980)).

Thermal Properties

Temperatures (K) of the DTA and DTG maxima for mordenite and dealuminated mordenites (*J. Phys. Chem.* 88:1130(1984)):

Sample	n-Butylamine DTA	n-Butylamine DTG	Ammonia DTA	Ammonia DTG
HM	393(20.01)	373	393(11.32)	373
	573(shoulder)		853(shallow)	
	623(shoulder)			
	678(6.53)	668		
	843(shallow)	683		
Deal. HM(18)	358(6.56)	423	363(13.28)	353
	478(shoulder)			
	678(9.64)	663	>773(shallow)	
	718(0.77)	693		
	843(shallow)			
Deal. HM(26)	348(5.65)	333	353(20.88)	348
	478(3.77)	473	>773(shallow)	
	678(8.28)	563		
	713(0.75)	693		
	843(shallow)			

Numbers in parentheses indicate peak heights (10$_2$ mm/g).

Infrared Spectrum

Evaluation of the studied vibrational spectra of mordenite (*Zeo.* 3:329 (1983)):

Type of vibration	i.r. (cm⁻¹)	FIR (cm⁻¹)	REF (cm⁻¹)	R (cm⁻¹)
1	3620s	—	—	3590w
	3480vs	—	—	3530w
				3520w
2	2950w	—	—	—
3	1640m	—	—	—
4	910vs	—	1090s	1080w
			1050m	1010w
			1010m	980w
5	780vs	—	740w	815w
	705w			760w
	720w			700vw
6	1210w	—	1220w	860w
			1180w	
7	440vs	405	480m	460vw
			440m	450w
			430m	440w
				410s
				405s
8	660vs	—	690w	680w
9	620w	—	—	595w
10	—	280w	—	268w
		270w		248w
		260w		220w
		240vw		
11	—	225w	—	210w
		205m		200w
		190w		195w
				190vw
12	—	95w	—	105vw
				98vw
13	—	170w	—	170m
		160w		150m
14	—	370m	—	340w
		355w		320w
		290w		295w
		150w		68w
		115w		52vw
		70w		
		50vw		

i.r. = infrared absorpiton spectra, FIR = Far-infrared absorption spectra, REF = reflectance infrared spectra, R = Raman spectra, vs = very strong, s = strong, m = medium, w = weak, vw = very weak.
1 = antisymmetric and symmetric stretching O–H
2 = combination band H$_2$O
3 = bending H–O–H
4 = antisymmetric stretching Si(Al)–O
5 = symmetric stretching Si(Al)–O
6 = external Si(Al)–O
7 = bending 0–Si (A1)–O
8 = "pore opening"
9 = vibration H$_2$O
10 = Ca–O
11 = Na–O
12 = Ca–H$_2$O
13 = Na–H$_2$O
14 = optical translational mode of lattice
Group vibrational representation: G$_{vib}$ = 69A$_g$(R) + 69B$_{1g}$ + 69B$_{2g}$(R) + 69B$_{3g}$(R) + 69A$_u$ + 69BN$_{1u}$(IR) + 69B$_{1u}$(IR) + 69B$_u$(IR).

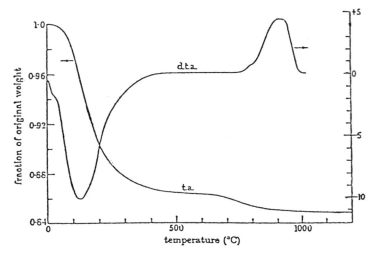

Mordenite Fig. 4:   TGA/DTA of the hydrogen form of mordenite (*JCS* 466(1964)). (Reproduced with permission of the Chemical Society)

Ion Exchange

Rates of ion exhange in mordenite (60°C) (*Zeo. Mol. Sieve.* 574 (1974)):

System	D (cm^2/s)
$Na^+ \rightarrow K^+$	1.3×10^{-12}
$K^+ \rightarrow Na^+$	1.8×10^{-13}
$Na^+ \rightarrow Rb^+$	7.5×10^{-12}
$Rb^+ \rightarrow Na^+$	8.1×10^{-13}
$Na^+ \rightarrow 1/2Ca^{+2}$	2.3×10^{-12}
$Cs^+ \rightarrow Na^+$	1.0×10^{-13}

Mordenite is highly selective for cesium and strontium. These properties make this material desirable for application in radioactive waste treatment (*Zeo.* 4:215(1984)). Cooper ions can be exchanged in this zeolite, using solid CuO, CuF_2, or $Cu_3(PO_4)_2$ in a physical mixture with mordenite at temperatures between 550 and 800°C (*Zeo.* 6:175(1986)). Tin can ion-exchange into mordenites with up to 3.6 molecules per unit cell upon heating of H-mordenite mixed with tin(II) chloride dihydrate in an oxygen atmosphere (*Solid State Ionics* 35:51(1989)).

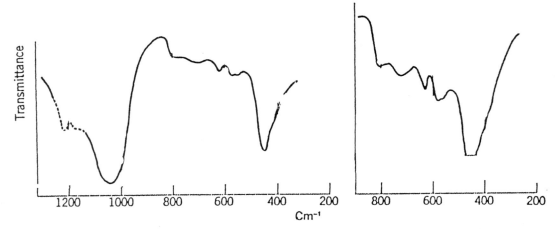

Mordenite Fig. 5:   Infrared spectrum in the mid-IR regions for synthetic mordenite (Zeolon) (*Advances in Chemistry Series* 101, *Molecular Sieve Zeolites*, Flanigan, E. M, Sand, L. B., eds., 201(1971). (Reproduced with permission of the American Chemical Society)

Adsorption

Adsorption properties of natural mordenite* from Nova Scotia (*Zeo, Mol. Sieve.* 618 (1974)):

Adsorbate	T, K	P, Torr	g/g
Argon	90		NA
Oxygen	90	10	0.11
		100	0.123
		700	0.13
	195	700	0.063
Nitrogen	77	100	NA
	195	700	0.061
H_2O	298	1	0.095
		5	0.136
		20	0.15
CO_2	298	10	0.075
		100	0.107
		600	0.126
CO	298	25	0.085
		100	0.107
		600	0.125
CH_4			NA
C_2H_4	298	700	0.034 (slow)
C_2H_6	298		NA
C_3H_8	298		NA

*The typical mordenite mineral does not adsorb molecules with a critical kinetic diameter greater than 4.8 A at 300°K.

Adsorption properties of synthetic mordenite (H-Zeolon) (*Zeo. Mol. Sieve.* 619 (1974)):

Adsorbate	T, K	P, Torr	x/m
Argon			NA
Oxygen	90	10	0.182 (g/g)
		100	0.198
		700	0.212
Nitrogen	213	100	0.012
		700	0.029
CO_2	298	100	0.035
		700	0.070

Adsorbate	T, K	P, Torr	g/g
SO_2	273	50	0.083
		200	0.18
		700	0.24
HCl	293	40	0.0835
		120	0.104
		200	0.117
n-C_4H_{10}	298	10	0.045
		100	0.053
		500	0.058
		700	0.060
Isobutane	298	50	0.040
		100	0.043
		500	0.048
		700	0.049
Neopentane	298	50	0.044
		100	0.046
		500	0.052
		700	0.054

Adsorption properties for the sodium form of Zeolon (*Zeo. Mol. Sieve.* 619 (1974)):

Adsorbate	T, K	P, Torr	x/m
Argon			NA
Krypton	273	11000	60
			(cc(STP)/g
		45000	70
Oxygen	90	10	0.154 (g/g)
		100	0.165
		700	0.171
Nitrogen	77	10	0.154
		100	0.165
		700	0.138
	213	100	0.039
		700	0.055
CO_2	298	100	0.10
		700	0.12
SO_2	ca. 300	10.7	0.17
		34	0.175
		83	0.187

N

N (LTN)

See Linde type N.

N-A (LTA)
US3,306,922(1967)
Union Carbide Corporation

Related Structures: Linde type A

Structure

For a description of the structure see Linde type A.

Chemical Composition: 1.0 ± 0.1 $[(CH_3)_4 N]_2O + x$ $Na_2O : Al_2O_3 : 4.25 \pm 1.75$ $SiO_2 : y$ H_2O

X-Ray Powder Diffraction Data: $(d(\text{Å})$ $(I/I_0))$
12.08(100), 8.55(60), 6.99(45), 6.05(5), 5.41(8), 4.94(6), 4.28(11), 4.03(59), 3.83(7), 3.65(68), 3.36(24), 2.93(36), 2.85(11), 2.710(5), 2.644(5), 2.583(15), 2.474(5), 2.423(3), 2.375(0.6), 2.333(2), 2.250(0.5), 2.211(4), 2.141(6), 2.111(1.3), 2.078(2), 2.049(1), 2.019(8), 1.992(0.2), 1.966(1), 1.892(5).

Synthesis

ORGANIC ADDITIVES
TMA

Zeolite N-A is prepared from a gel composition of: 1.5 [TMA$_2$O] : Al$_2$O$_3$: 6 SiO$_2$: 200 H$_2$O after 8 days at 100°C. The SiO$_2$/Al$_2$O$_3$ ratio of the product can vary between 2 and 7 (US3,306,922(1967)). As crystallization is done in a glass vessel, significant amounts of sodium dissolved from the glass also are present and incorporated into this structure. Without the addition of sodium, no type A structure is formed. The rate at which N-A crystallizes is dependent on the amount of sodium present.

Infrared Spectrum

Infrared spectrum of N-A (*ACS* 101:201 (1971)):

SiO$_2$/Al$_2$O$_3$ = 3.58	SiO$_2$/Al$_2$O$_3$ = 6.01
1131vwsh	1151vwsh
1030s	1044s
750vwsh	750vwsh
675vw	698vw
572ms	581ms
474m	475m
385m	393m

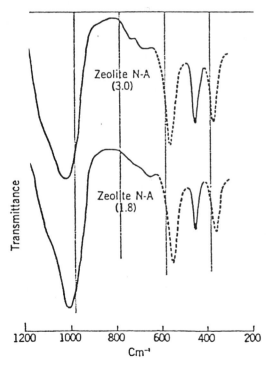

N-A Fig. 1: Infrared spectra in the mid-infrared region of N-A with two different SiO$_2$/Al$_2$O$_3$ ratios (*Advances in Chemistry Series* 101, *Molecular Sieve Zeolites,* Flanigan, E. M, Sand, L. B., eds., 201(1971). (Reproduced with permission of the American Chemical Society)

Na-A

See Linde type A.

Na-B (ANA)
(JCS 1561(1952))

Related Material: analcime

Structure

Chemical Composition: Si/Al = 2.

Synthesis

ORGANIC ADDITIVES
none

Na-B crystallizes from a batch composition of: >1.0 Na_2O : 4.0 SiO_2 : Al_2O_3 : xs water. Crystallization occurs at 180°C.

Na-D (MOR)
(JCS 1561(1952))

Related Material: mordenite

Structure

Chemical Composition: Na_2O : 10 SiO_2 : Al_2O_3 : 6.7 H_2O.

X-Ray Powder Diffraction Data: (*Zeo. Mol. Sieve.* 356(1974)) (d(Å) (I/I_0)) 13.53(40), 10.24( ), 9.06(50), 6.57(55), 6.39(—), 6.08(—), 5.80(15), 5.05(—), 4.83(—), 4.52(25), 4.15(—), 4.00(60), 3.84(—), 3.76(—), 3.53(—), 3.47(100), 3.39(—), 3.29(—), 3.22(—), 3.16(—).

Synthesis

ORGANIC ADDITIVES
none

Na-D crystallizes from a batch composition of: Na_2O : 8.2–12.3 SiO_2 : xs H_2O. Crystallization occurs after 2 to 3 days at temperatures between 265 and 295°C. Silicic acid and sodium aluminate are used as the sources of silica and alumina in this synthesis.

Na-J (JBW)
(JCS 1561(1952))

See nepheline hydrate I. This material was prepared in the sodium cation system over a temperature range of 270 to 450°C.

Na-K
(JCS 1561(1952))

This material is called nepheline hydrate II.

Na-M (MOR)
(Geokhimija 9:820(1963))

Aluminosilicate mordente prepared by Senderov.

Na-P1 (GIS)
(JCS 1959:195(1959))

Related Materials: gismondine

Na-P1 is a commonly crystallizes phase from sodium aluminosilicate gels. This zeolite was thought to be related to phillipsite, but was later

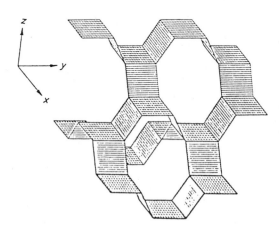

Na-P1 (gismondine-type) Fig. 1S: Framework topology.

shown to contain a gismondine framework structure. A multitude of names have added to the confusion. These include P_c or cubic P and P_t or tetragonal P, as well as P(B).

Structure

(Zeo. Mol. Sieve. 168(1974))

CHEMICAL COMPOSITION: $Na_6Al_6Si_{10}O_{32}*12H_2O$
SPACE GROUP: pseudosymmetry I4₁/a
UNIT CELL CONSTANTS (Å): $a = c = 10.04$
PORE STRUCTURE: intersecting eight-member rings
  4.5 × 3.1 Å and 2.8 × 4.8 Å

X-Ray Powder Diffraction Data: (d(Å) (I_{rel}))
([Na-P1]$_{calc}$) 7.101(515), 5.021(314), 4.491(4), 4.100(559), 3.551(8), 3.348(13), 3.176(1000), 3.028(15), 2.899(65), 2.785(8), 2.684(534), 2.511(51), 2.436(8), 2.367(88), 2.304(1), 2.246(21), 2.192(12), 2.141(12), 2.050(46), 2.009(1), 1.967(148), 1.933(1), 1.865(7), 1.834(26), 1.775(69), 1.748(6), 1.722(108), 1.698(4), 1.674(101), 1.651(3), 1.629(70), 1.588(38), 1.568(8), 1.550(42), 1.532(2), 1.514(13), 1.497(5), 1.481(75), 1.450(5),

1.435(7), 1.420(50), 1.406(3), 1.393(40), 1.380(4), 1.367(92); ([P_c]) 7.10(55), 5.01(35), 4.10(55), 3.16(100), 2.67(55), 2.52(5), 2.36(7), 2.054(5), 1.965(10), 1.771(7), 1.719(7), 1.667(7); ([P_t]) 7.132(85), 7.047(83), 5.776(5), 5.048(51), 4.914(26), 4.420(8), 4.108(94), 4.049(22), 3.527(4), 3.328(18), 3.194(100), 3.117(64), 3.036(10), 2.979(5), 2.776(3), 2.750(5), 2.694(46), 2.679(28), 2.653(21), 2.531(6), 2.435(5), 2.387(4), 2.257(2), 2.206(2), 2.159(2), 2.055(1), 1.982(2), 1.966(2), 1.929(1).

The structure of Na-P1 is based on a gismondine aluminosilicate framework. This structure contains feldspar-like chains that run parallel to the x and y axes. Based on the average T–O distance, the distribution of aluminum is uniform throughout the unit cell.

The structural coordinates were obtained with a pseudosymmetry of I4¹/a. The true symmetry is expected to be lower than I4̄.

The variations observed in the different cation-exchanged forms of P due to hydration and so forth are attributed to the flexibility of the framework structure. This flexibility is proposed to be the reason for the considerable twinning observed in crystals of this material.

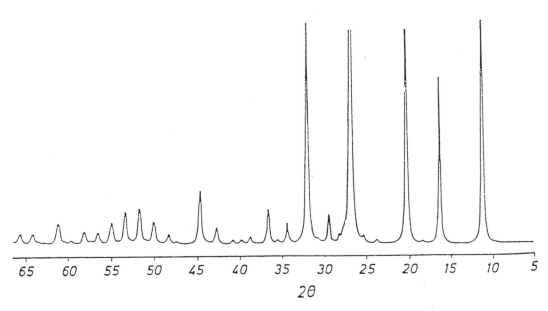

Na-P1 Fig. 2:   X-ray powder diffraction pattern for zeolite Na-P1 (*Z. Kristallogr.* 135:339 (1972)). (Reproduced with permission of Oldenbourg Verlag Publishers)

Data on various cation-exchanged forms of zeolite P(B) (*Zeo. Mol. Sieve.* (1974)):

Cation form	Unit cell dimensions, Å	H_2O, wt%	Density, g/cc	Refractive index
Li^+	$a = 9.99, c = 9.91$	18.7	2.02	1.489
Na^+, cubic[a]	$a = 10.05$	19.3	2.01–2.08	1.40
Na^+, tetr.[b]	$a = 10.11, c = 9.83$	16.9	2.15	1.482
K^{+b}	$a = 9.93, c = 9.67$	10.8	2.28	1.504
Ca^{+2b}	$a = 9.89, c = 10.30$	18	2.16	1.505
Ba^{+2b}	$a = 9.92, c = 10.36$	14.6	2.57	1.531
Mg^{+2}, cubic[a]	$a = 10.03$	14		
Zn^{+2}, cubic[a]	$a = 10.04$	14		

a. Ratio of si/Al of original B = ca. 1.5.
b. Unit cell contents of original P = $Na_{5.7}(AlO_2)_{5.7}(SiO_2)_{10.3} \cdot 12H_2O$.

Final parameters based on space group $I\bar{4}$ for Na-P1 (*Z. Kristoallogr.* 135:339 (1972)):

Atom	Position	Occupancy factor	x	y	z
T(1)	8g	1	0.1438(9)	0.1692(9)	−0.0181(8)
T(2)	8g	1	0.1683(8)	0.3579(9)	0.2329(7)
O(1)	8g	1	0.1890(25)	0.0215(15)	0.0410(20)
O(2)	8g	1	0.1885(20)	0.2895(20)	0.0880(20)
O(3)	8g	1	0.0090(25)	0.3460(30)	0.2985(20)
O(4)	8g	1	0.2830(20)	0.2845(25)	0.3300(25)
I	4e	0.42(4)	0	0	0.3250(45)
II	8g	0.70(5)	0.2195(25)	0.0240(35)	0.3515(30)
III	8g	0.50(5)	0.2930(45)	0.0020(60)	0.3520(45)
IV	8g	0.32(2)	0.0140(55)	0.1895(50)	0.4940(50)
V	8g	0.42(2)	0.3590(40)	0.0825(35)	0.2210(35)

Synthesis

ORGANIC ADDITIVES
none

Zeolite P1 is synthesized from a sodium aluminosilicate system. Batch compositions include: $4Na_2O : Al_2O_3 : 10 SiO_2 : 160 H_2O : 6$ NaCl at 120°C (*5th Int. Conf. on Zeolites,* Napoli, 58(1980)). Clear solutions with batch compositions of $15 Na_2O : Al_2O_3 : 32 SiO_2 : 1100 H_2O$ will transform into cubic P after several days at 135°C (*JCS* A:2909(1971); *Zeo.* 8:166(1988)). In gel compositions that result in the formation of the faujasite structure, with time it will convert to zeolite P1. Zeolite A also will transform into zeolite P in highly caustic media (*Zeo.* 2:135(1982)).

Ion Exchange

Cation exchange in zeolite P1 (25°c, 0.1 N solution) (*JCS* A:2909 (1971)):

Exchange	x_{max}
Na^+ to Li^+	0.5
Na^+ to K^+	1.0
Na^+ to Rb^+	0.9
Na^+ to Cs^+	1.0
Na^+ to $\frac{1}{2} Sr^{+2}$	1.0
Na^+ to $\frac{1}{2} Ba^{+2}$	1.0

Na-P2 (GIS)
(*JCS* 195(1959))

A variant of the synthetic analog of gismondine. See Na-P1 for a description of the framework topology.

Na-P3 (GIS)
(*JCS* 195(1959))

A variant of the synthetic analog of gismondine.

Na-F (FAU)
(*Am. Mineral.* 49:656(1964))

Aluminosilicate faujasite prepared by Taylor & Roy.

Na-K

See nepheline hydrate II.

Na-Q
(*JCS* 195(1959))

Aluminosilicate identified by Barrer et al.

X-Ray Powder Diffraction Data: (d(Å) (I/I_0))
12.36(vs), 8.77(s), 7.15(s), 6.35(w), 5.52(s), 5.03(w), 4.36(m), 4.102(s), 3.899(w), 3.712(s), 3.413(m), 3.287(ms), 3.072(vw), 2.982(vs), 2.900(m), 2.751(m), 2.681(mw), 2.623(s), 2.512(mw), 2.410(w), 2.369(mw), 2.247(w), 2.173(m), 2.143(mw), 2.109(mw), 2.078(mw), 2.050(m).

Na-R (FAU)
(*JCS* 195(1959))

Aluminosilicate faujasite prepared by Barrer et al. Also called species R.

X-Ray Powder Diffraction Data: (d(Å) (I/I_0))
14.52(vs), 8.84(ms), 7.56(ms), 5.74(s), 4.81(m), 4.42(ms), 4.20(vvw), 3.954(w), 3.813(ms), 3.769(mw), 3.621(vvw), 3.515(vvw), 3.048(m), 2.941(m), 2.881(s), 2.790(m), 2.745(w),
2.725(vw), 2.662(m), 2.617(mw), 2.545(mw), 2.403(m), 2.206(m), 2.181(m).

Na-S (GME)
(*JCS* 195(1959))

Aluminosilicate with the gmelinite topology reported by Barrer et al. Also called species S.

X-Ray Powder Diffraction Data: (d(Å) (I/I_0))
11.9(ms), ~9.5(vw), 6.85(s), 5.01(s), 4.46(mw), 3.962(mw), 3.431(s), 3.280(vvw), 3.167(vw), 2.970(m), 2.900(mw), 2.829(mw), 2.595(s), 2.284(w),   2.113(w),   2.083(s),   1.901(m), 1.800(s),   1.716(s),   1.687(m),   1.667(m), 1.646(w), 1.636(w).

Na-T (JBW)
(*JCS* 195(1959))

Aluminosilicate related to nepheline hydrate I. Also called species T.

X-Ray Powder Diffraction Data: (d(Å) (I/I_0))
8.2(ms), 7.6(mw), 5.6(m), 4.85(w), 4.66(mw), 4.46(ms), 3.25(m), 3.17(m), 2.93(s), 2.84(ms), 2.72(w), 2.58(w), 2.39(w), 2.34(w), 2.27(w), 2.23(w), 2.02(w), 1.89(w).

Na-V (THO)
(*JCS* 195(1959))

Aluminosilicate with the thomsenite topology synthesized by Barrer et al.

Natrolite (NAT)
(natural)

Related Materials: gonnardite
                     mesolite
                     scolecite

Naturalite originally was known mainly as "fibrous zeolite." The name natrolite first was proposed in 1803 (*Nat. Zeo.* 35(1985)). Natrolite occurs in nature in association with analcime but has not been observed with albite.

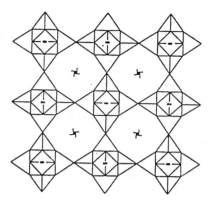

Natrolite Fig. 1S: Framework topology.

Structure

(*Zeo. Mol. Sieve.* 164(1974))

CHEMICAL COMPOSITION: $Na_{16}(Al_{16}Si_{24}O_{80})*16H_2O$
SYMMETRY: orthorhombic
SPACE GROUP: Fdd2
UNIT CELL CONSTANTS (Å): $a = 18.285$
                           $b = 18.630$
                           $c = 6.585$
FRAMEWORK DENSITY: 1.76 g/cc
PORE STRUCTURE: two-dimensional eight-member
rings 2.6 × 3.9 Å

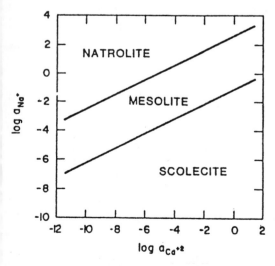

Natrolite Fig. 2: Stability fields of natrolite, mesolite, scolecite at 25°C, 1 atm, $a_{H_2O} = 1$ (*Am. Mineral.* 68:1134 (1983)). (Reproduced with permission of *American Mineralogist*)

X-Ray Powder Diffraction Data: (d(Å) (I/I_0))
6.53(74), 5.88(36), 4.66(35), 4.59(30), 4.39(58), 4.35(70), 4.15(42), 4.11(37), 3.622(2), 3.261(12), 3.192(42), 3.151(52), 3.098(29), 2.939(36), 2.897(9), 2.863(80), 2.844(74), 2.582(43), 2.570(71), 2.552(16), 2.448(88), 2.410(86), 2.331(15), 2.318(37), 2.288(18), 2.260(36), 2.239(7), 2.222(7), 2.194(58), 2.177(100), 2.076(11), 2.053(27), 1.933(23), 1.883(21).

The unit cell composition and symmetry of natrolite, scolecite, and mesolite differ, but all three zeolites possess the same type of framework structure. The natrolite framework is composed of linked chains of tetrahedra. The total number of cations and water molecules in the unit cell is 32, which does not vary among the three structures. The channel dimensions are small, and the space is occupied by the cations. Each sodium ion has for its nearest neighbors four framework oxygens and two water molecules. Ordering of silica and alumina in the framework of this structure is well known. Structural changes are associated with dehydration and ion exchange; therefore, this material is not considered as a molecular sieve adsorbent.

The lattice constants change linearly with the $Na_2O/(Na_2O + K_2O + CaO + MgO)$ ratio. The transformation from natrolite to mesolite occurs via a rotation of the secondary building units two-dimensionally along the a–b plane.

Fractional atomic coordinates with esd values in parentheses; Si(1) in special position (0,0,0) (*Zeo.*, 4: 140 (1984)):

	x	y	z
Na	2207(1)	306(1)	6177(3)
Al	374(1)	936(1)	6152(2)
Si(2)	1533(1)	2113(1)	6232(2)
O(1)	227(1)	687(1)	8662(4)
O(2)	702(1)	1817(1)	6097(4)
O(3)	984(1)	350(1)	5000(4)
O(4)	2063(1)	1528(1)	7259(4)
O(5)	1805(1)	2273(1)	3900(4)
O(w)	563(1)	1897(1)	1112(5)
H(1)	573(27)	1577(25)	531(77)
H(2)	842(26)	1854(28)	1801(79)

The *a* and *b* axis elongates with increasing Ca content. The conversion between mesolite and scolecite is a three-dimensional rotation of the SBU with an increase in the beta angle and elongation of the *c* axis (*PP 7th IZC* 1B-7(1986)).

Synthesis

ORGANIC ADDITIVE
none

Synthetically, natrolite has been observed to form from sodium aluminosilicate gels. Crystallization occurred between 100 and 200°C, and the natrolite species was readily observed when the gels were seeded. Analcime was a persistent impurity (*ACS* 101:149(1971)).

Thermal Properties

Dehydration causes a substantial change in the local charge balance as the cations move into different sites for coordination with the framework oxygens. The result is a change in the lattice dimensions. The dehydrated form of naturalite shows no tendency to readsorb molecules other than water and ammonia.

Infrared Spectrum

The state of the water in natrolite is characterized by absorption bands at 1640, 3230, 3350, and 3550 cm^{-1}, with a combined absorption band observed around 2150 cm^{-1}. All the bands disappear after vacuum treatment between 130 and 180°C. This type of dehydration is reversible.

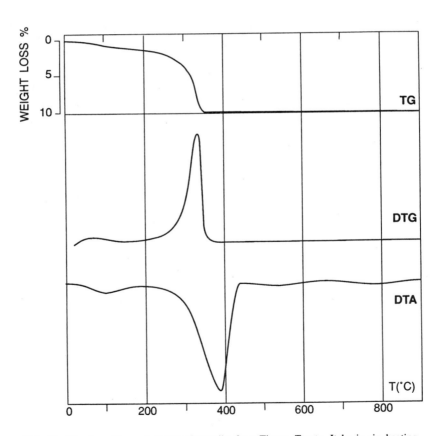

Natrolite Fig. 3:   Thermal curves of natrolite from Tierno, Trento, Italy; in air, heating rate of 20°C/min (*Nat. Zeo.* 49(1985)). (Reproduced with permission of Springer Verlag Publishers)

Lattice parameters of natrolite; dependence on heating temperature and time (*Coll. Czech. Chem. Commun.* 52:1211 (1987)):

Temp., °C	Time, hr	a, nm	b, nm	c, nm	b–a, nm
310	0.5	1.816(4)	1.868(6)	0.658(3)	0.052
310	32	1.836(3)	1.864(3)	0.658(2)	0.028
310	4905	1.823(2)	1.859(3)	0.657(1)	0.036
350	0.5	1.823(5)	1.873(5)	0.655(3)	0.050
350	32	1.831(7)	1.868(7)	0.665(6)	0.037
350	4102	1.832(6)	1.878(9)	0.659(3)	0.046
400	0.5	1.826(4)	0.864(6)	0.659(1)	0.038
400	32	1.803(10)	1.858(10)	0.638(10)	0.055
400	4097	1.816(10)	1.862(10)	0.669(7)	0.040
500	0.5	1.833(3)	1.872(6)	0.661(1)	0.039
500	32	1.832(3)	1.859(5)	0.659(1)	0.027
500	4096	1.804(7)	1.883(10)	0.663(3)	0.079

Vacuum treatment at 600°C results in a shift toward lower frequency by 10 cm^{-1} with an increasing halfwidth. The readsorbed water can be removed at a relatively low temperature (100°C). The combination of an increase of the amount of readsorbed water found by thermal analysis and its removal at relatively low temperatures suggests the formation of secondary porosity with dehydration. This secondary porosity is enhanced by anisotropic compression

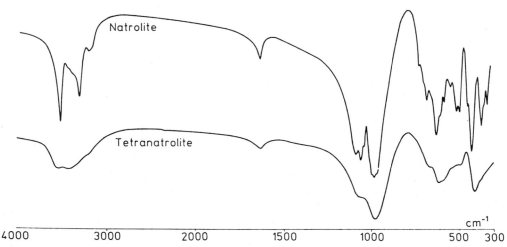

Natrolite Fig. 4:   Infrared spectrum of natrolite (*Nat. Zeo.* 347 (1985)). (Reproduced with permission of Springer Verlag Publishers)

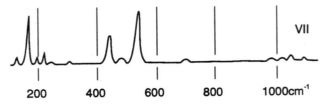

Natrolite Fig. 5:   Raman spectra of natrolite (*Acta Physica Hung.* 61:27(1987)). (Reproduced with permission of the Publishing House of the Hungarian Academy of Sciences)

and subsequent expansion of the natrolite crystal during dehydration and subsequent heating. These two opposite changes lead to the formation of microfaults along the *C* axis (*Nat. Zeo.* 265 (1988)).

The changes in the infrared absorption spectra in vibration bands, fine structure, and splitting are caused by variations of the local force fields with changes in symmetry and by other chemical and physical factors.

Evaluation of infrared spectra of natrolites (*Acta Montana* 70:75 (1985)):

Natro-lite	Stretching vibration O–H	Bending vibration H_2O	Stretching vibration T–O	Bending vibration O–T–O
1	3580vs 33220m	1620m	1180–1050vs fine structure	720m
2	3570vs 3310m	1625m	1180–1050vs fine structure	730–700m fine structure
3	3570vs	1630m	1190–1020vs fine structure	780w 720
4	3580vs	1620m	1040vs	720–700w fine structure
5	3590w 3560w 3390m 3310w	1625m	1150–1040vs wide band	720–710m doublets
6	3580s 3500w 3400m 3310w	1630m	1180–1140vs fine structure	790–690m fine structure
7	3580s 3450w 3360m	1620m	1140–1040vs fine structure	790–690m fine structure
8	3600vs 3590vs 3400w	1620m	1130–1040vs fine structure	790–700m fine structure
9	3620–3570vs 3380w 3605–3580vs doublets	1630m	1160–1100vs doublets	730m
10	3460 inflection 3370m	1620m	1180–1040vs fine structure	760–690m fine structure

Band frequencies (in cm^{-1}) in the Raman spectra of natrolite (in the region below 1200 cm^{-1}) (*Acta Montana* 70:75 (1985)):

A_1	A_2	B_1	B_2
	99		93
75	129	115	106
105	141	121	123
141	170	182	135
155			144
160	184		162
196	211	216	183
206			222
242	336	273	277
272		289	297
307			326
361			
415	413	331	358
434		358	
445	495		415
520		514	495
534	588	576	
598	633	619	604
703	708	967	722
727			964
965			990
	996		
993			1009
1024			
1047	1075		
1071			
1083	1092		
			1063
			1088

Na-X (FAU)

See Linde type X.

NaZ-21 (LTN)

See Z-21.

N-B (GIS)
(US3,306,922(1967))
Union Carbide Corporation

Related Materials: gismondine

Structure

Chemical Composition: 1.0 ± 0.2 [$(CH_3)_4$ N]$_2$O + x Na$_2$O : Al$_2$O$_3$: 3.5 ± 1.5 SiO$_2$: y H$_2$O

X-Ray Powder Diffraction Data: (d(Å) (I/I_0))
7.09(vs), 5.00(ms), 4.00(vs), 3.51(w), 3.34(vw), 3.17(vs), 2.677(s), 2.511(w), 2.238(vw), 2.136(w), 2.017(vw), 1.962(m), 1.829(vw), 1.772(mw), 1.717(m), 1.669(mw), 1.624(w), 1.479(mw), 1.390(w), 1.363(mw), 1.313(vw), 1.274(mw).

N-B represents the gismondine, or type B, structure prepared in the presence of organic amine cations.

Synthesis

ORGANIC ADDITIVE
tetramethylammonium$^+$

Zeolite N-B crystallizes from a tetramethylammonium ion–containing aluminosilicate gel with a batch composition of: 1.5TMA$_2$O : Al$_2$O$_3$: 3 SiO$_2$: 200 H$_2$O. Crystallization occurs at temperatures of 200°C after 6 days.

Nepheline Hydrate I (JBW)
(JCS 1561(1952))

Nepheline hydrate first was identified by Barrer and White in 1952 when they synthesized these aluminosilicates from a sodium oxide system (*JCS* 1561(1952)). This structure constitutes a link between the anhydrous tektosilicates and the zeolites.

Structure

(*Zeo.* 2:162(1982))

CHEMICAL COMPOSITION: Na$_3$AL$_3$Si$_3$O$_{12}$*2H$_2$O
SYMMETRY: orthorhombic
SPACE GROUP: Pna2$_1$
UNIT CELL CONSTANTS: (Å) $a = 16.426$
$b = 15.014$
$c = 5.2235$
DENSITY (calc): 2.38 g/cm^3
PORE STRUCTURE: eight-ring apertures

X-Ray Powder Diffraction Data: (*JCS* 1561(1952)) (d(Å) (I/I_0)) 8.19(60), 7.49(50), 5.53(20), 4.397(100), 3.432(15), 3.408(100),

2.960(35), 2.850(10), 2.607(10), 2.445(10), 2.420(5), 2.303(5).

Nepheline hydrate I consists of a tetrahedral aluminosilicate framework with alternating aluminum and silicon ions. There is a set of parallel two-repeat chains with single and double chains alternating. The largest channels are bound by eight-member rings. Two of the sodium ions are coordinated by seven framework oxygen atoms, whereas the coordination of the third sodium ion is with three framework oxygens and three water oxygens.

Atomic populations, positions, and displacements (*Zeo.* 2:162 (1982)):

Atom	x	y	z
Na(1)	0.0929(2)	0.2500(4)	0.0774(1)
Na(2)	0.4156(2)	0.2518(4)	0.273(1)
Na(3)	0.2086(4)	0.5044(6)	0.135(2)
Al(1)	0.2022(1)	0.2509(3)	0.25
Al(2)	0.4511(3)	0.3923(2)	0.734(1)
Al(3)	0.4511(2)	0.1068(2)	0.738(1)
Si(1)	0.3075(1)	0.2482(2)	0.7470(7)
Si(2)	0.0574(2)	0.1090(2)	0.232(1)
Si(3)	0.0559(2)	0.3890(2)	0.236(1)
O(1)	0.2252(3)	0.2495(6)	0.583(2)
O(2)	0.2878(3)	0.2498(6)	0.055(1)
O(3)	0.1411(7)	0.1606(6)	0.173(3)
O(4)	0.1436(6)	0.3447(5)	0.180(2)
O(5)	0.3618(6)	0.3339(6)	0.674(3)
O(6)	0.3592(6)	0.1606(6)	0.674(3)
O(7)	0.0228(6)	0.1392(6)	0.509(2)
O(8)	0.0230(5)	0.3563(5)	0.519(2)
O(9)	0.4886(5)	0.3583(5)	0.031(2)
O(10)	0.4932(6)	0.1447(6)	0.028(2)
O(11)	0.0736(4)	0.0048(7)	0.221(2)
O(12)	0.0689(3)	0.4954(7)	0.225(2)
W(1)	0.337(1)	0.492(2)	0.226(4)
W(2)	0.237(1)	0.497(1)	0.686(6)
Na(31)	0.216(2)	0.005(3)	0.12(1)
W(11)	0.168(4)	0.503(7)	0.76(2)

Nepheline Hydrate II
(*JCS* 1561(1952))

Identified as Na-K in the $Na_2O-Al_2O_3-SiO_2-H_2O$ system. Crystallization occurs over a temperature range between 270 and 450°C.

Nonasil (NON)
(clathrasil)

Related Materials: ZSM-51

Nonasil-$[4^1 5^8]$ is a member of the clathrasil family of materials with an $88SiO_2 * 8M^8 8M^9 * 4M^{20}$ composition. This material has no structural analog in the clathrate hydrates.

Structure

(*J. Incl. Phenom.* 4:339(1986))

CHEMICAL COMPOSITION: SiO_2
SPACE GROUP: Fmmm
UNIT CELL CONSTANTS (Å): $a = 22.232(6)$
$b = 15.058(4)$
$c = 13.627(4)$

X-Ray Powder Diffraction Data: (*J. Incl. Phenom.* 4:339(1986)) (d(Å) (I/I_0)) 11.1(23), 9.2(35), 7.56(3), 6.83(3), 6.25(3), 5.98(5), 5.82(4), 4.61(66), 4.47(28), 4.30(100) 4.26(32), 3.97(50), 3.752(14), 3.704(28), 3.567(27), 3.406(5), 3.329(24), 3.256(31), 3.231(16), 3.107(4), 3.068(3), 3.028(3), 2.989(7), 2.834(4), 2.778(2), 2.732(5), 2.684(10), 2.635(2), 2.565(3), 2.526(4), 2.509(15).

Nonasil-$[4^1 5^8]$ contains cagelike voids occupied by guest species. The three-dimensional 4-connected silica host framework has three types of cagelike voids, the 8-hedra, $[5^4 6^4]$, the 9-hedra, $[4^1 5^8]$ (fundamental cage), and the 20-hedra, $[5^8 6^{12}]$. It is in the last of these cages that the structure-controlling guest molecules reside.

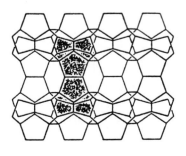

Nonasil Fig. 1S: Framework topology.

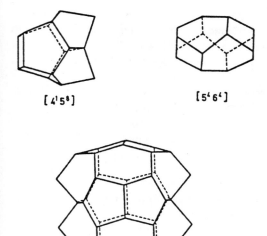

[4¹5⁸]   [5⁴6⁴]

[5⁸6¹²]

Atom	x	y	z
Si(3)	0.2950(7)	0.3270(13)	0
Si(4)	0.5	0.1051(9)	0.6156(12)
Si(5)	0.3173(7)	0.5	0.2831(13)
O(1)	0.3356(13)	0.3282(19)	0.9056(20)
O(2)	0.4427(8)	0.1462(15)	0.3374(16)
O(3)	0.3520(15)	0.4050(50)	0.2525(39)
O(4)	0.2702(28)	0.4329(39)	0
O(5)	0.3197(17)	0.5	0.5937(29)
O(6)	0.3667(37)	0.25	0.25
O(7)	0	0	0.1392(38)
O(8)	0.5	0.1155(28)	0.5
O(9)	0.25	0	0.25
O(10)	0.25	0.25	0
C(1)	0.5	0.4489	0.5700
C(2)	0.4659	0.3547	0.5
C(3)	0	0	0.5

Nonasil Fig. 2:   The three types of cages in nonasils-[4¹-5⁸] (*J. Incl. Phenom.* 4:339 (1986)). (Reproduced with permission of Reidel Publishers)

Comparison of lattice parameters in nonasils-[4¹5⁸] with different guest molecules and degassed nonasil; standard deviations are less than 0.1 Å (*J. Incl. Phenom.* 4:339 (1986)):

Fractional coordinates for the nonasil-[4¹5⁸] 88SiO₂* 4M₂ O*, M₂O = CH₃CH(NH₂)C₃H₇, refined in space group Fmmm; the guest species is represented by the positions C(1) to C(3) (*J. Incl. Phenom.* 4:339 (1986)):

Atom	x	y	z
Si(1)	0.3745(4)	0.1635(7)	0.3103(8)
Si(2)	0.2798(10)	0.5	0.5

Lattice parameters	Guest Molecules			
	1,2-diamino-cyclohexane	2-Amino-pentane	2-Amino-butane	No guest
a_o (Å)	22.21	22.23	22.17	22.17
b_o (Å)	15.04	15.06	14.99	15.01
c_o (Å)	13.64	13.63	13.63	13.63

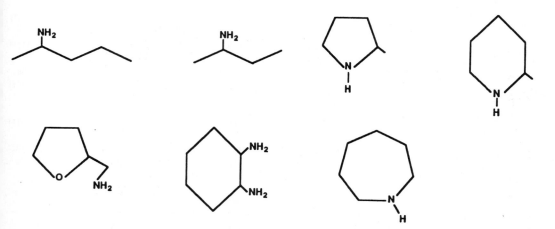

Nonasil Fig. 2:   Organics used in the preparation of nonasil.

Synthesis

ORGANIC ADDITIVES: See Nonasil Fig. 2.

Nonasil crystallizes from aqueous silicate solutions using different organic additives that become trapped within structural cages. Crystallization generally occurs at temperatures between 150 and 200°C, depending on the organic used in the synthesis. This clathrasil requires several weeks of crystallization time. When hexamethyleneimine is used, the best yield of product nonasil is obtained (*J. Incl. Phenom.* 4:339 (1986)).

Thermal Properties

Three distinct exotherms are observed in the DTA of nonasil: a very weak one in the range of 250 to 350°C, a stronger one at 530 to 610°C, and one in the range between 520 and 750°C.

Weight loss in the TGA begins slowly at around 150°C and becomes faster in the temperature ranges 520 to 610°C and 800 to 1000°C. A total of 5.75% weight loss is observed up to 700°C. A total weight loss of 7.2% occurs between room temperature and 1000°C. This indicates that the organic guest (hexamethyleneimine) occupies only the larger cage, and that the smaller cages are essentially empty (*J. Incl. Phenom.* 4:339(1986)).

Nosean (SOD)
(*Tscherm. Min. Petr. Mitt.* 10:225(1965))

See sodalite.

N-R (FAU)
(*JCS* 983(1961))

Aluminosilicate faujasite prepared in the presence of amine cations by Barrer et al.

Clathrasils synthesized with different guest substances at various temperatures; when two phases were obtained under the same conditions, the main product is given on the upper line (*J. Incl. Phenom.* 4:339 (1986)):

Guest	Reaction Temperature (°C)					
	150	160	170	180	190	200
2-Methyl pyrrolidine	Ns	Ns	Ns	Ns	Ns	Ns
Hexamethyleneimine	—	—	Ns	Ns	D1H	D1H
				Ns	Ns	
2-(Aminomethyl)-tetrahydrofuran	Ns	Ns	Ns	Ns	—	D1H
2-Methyl- piperidine	Ns	Ns	Ns	Ns	Ns	Ns
					D1H	D1H
1,2-Diamino-cyclohexane	Ns	Ns	Ns	Ns	Ns	D1H
			D1H	D1H	D1H	Ns
2-Methyl- piperazine	Ns	D1H	D1H	D1H	D1H	D1H
	D1H					
2-Aminobutane	Ns	Ns	Ns	Ns	D3C	D3C
		D3C	D3C	D3C	Ns	Ns
1-Aminobutane	—	—	—	Ns	—	—
				D3C[a]		
2-Aminopentane	Ns	Ns	Ns	Ns	Ns	Ns

Ns = nonasils; D1H = dodecasils 1H; D3C = dodecasils 3C; — = no experimental data.
[a]150 barr Ar pressure.

Nu-1
(US4,060,590(1977))
Imperial Chemical Industries

Structure

Chemical Composition: 0.9–1.3 R_2O : Al_2O_3 : 20–150 SiO_2 : 0–40 H_2O.

X-Ray Powder Diffraction Data: (d(Å) (I/I_0)) 8.87(18), 8.23(69), 6.53(43), 6.19(75), 4.43(52), 4.30(51), 4.08(37), 4.03(100), 3.965(73), 3.845(74), 3.81(22), 3.687(16), 3.508(29), 3.256(27), 2.858(15).

Zeolite Nu-1 exhibits properties characteristic of a small-pore and of a medium-pore molecular sieve.

Synthesis

ORGANIC ADDITIVE
TMA

Nu-1 crystallizes in the presence of TMA and sodium cations with a ratio of 0.05 to 1.0 (R_2O/ Na_2O + R_2O). It crystallizes over a range of SiO_2/Al_2O_3 between 20 and 200 with Na_2O/SiO_2 in the range of 0.05 to 0.25. Crystallization occurs at temperatures between 150 and 200°C from a static autoclave after 1.5 to 3.0 days. The reproducibility of this synthesis appears low.

Adsorption

Adsorption properties of NU-1 (*Catal. Highly Siliceous Zeolites* 343 (1986)):

	Water	*n*-Hexane	Isobutane	*p*-Xylene	Pyridine
Wt%	7.8	4.0	3.0	3.8	4.0

Nu-2 (BEA*)
(EP 55,046(1981))
Imperial Chemical Industries

Related Materials: beta

Structure

Chemical Composition: 0.5–1.8 R_2O : Al_2O_3 : at least 10 SiO_2 : 0–100 H_2O.

X-Ray Powder Diffraction Data: (d(Å) (I/I_0)) 11.33(23), 9.04(3), 7.56(4), 6.61(3), 6.03(3), 5.37(5), 4.51(2), 4.14(23), 3.96(100), 3.51(12), 3.46(3), 3.38(2), 3.31(21), 3.10(7), 3.02(21), 2.93(8), 2.91(5), 2.68(6), 2.59(3).

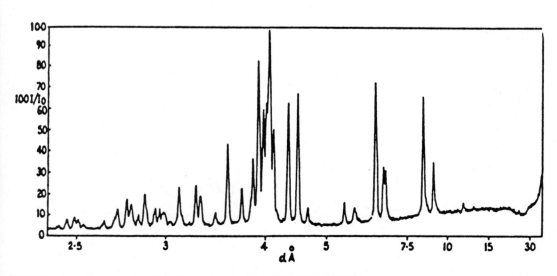

NU-1 Fig. 1:   X-ray powder diffraction pattern for Nu-1 (US4,060,590(1977)).

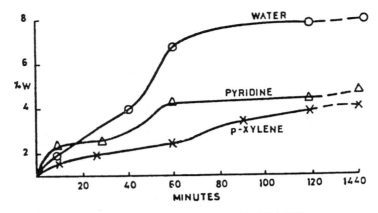

NU-1 Fig. 2:   Adsorption rates on HNu-1 (US4,060,590(1977)).

Synthesis

ORGANIC ADDITIVE
tetraethylammonium$^+$

Nu-2 crystallizes from a reactive aluminosilicate gel with a molar oxide ratio between 10 and 3000 and an OH/SiO_2 ratio between 0.1 and 2.0. The ratio of the organic amine cation to SiO_2 is low, generally falling between 0.02 and 0.4. Crystallization occurs readily at 150°C, with product being formed after 3 days. At lower temperatures (ca. 95°C) nearly one month is required before crystalline product is observed.

Zeolites ZSM-12 and ZSM-8 both have been identified as impurity phases in the synthesis of NU-2, crystallizing preferentially at low TEAOH/SiO_2 ratios. The organic amine in this synthesis undergoes significant Hofmann degradation during the crystallization, which consumes hydroxide and releases ethylene and triethylamine.

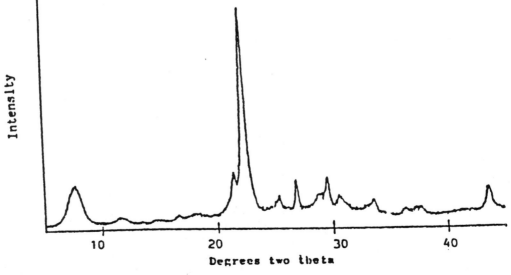

Nu-2 Fig. 1:   X-ray powder diffraction pattern for Nu-2 (*SSSC* 49A:151(1989)). (Reproduced with permission of Elsevier Publishers)

This is directly related to the temperature and the base level used in the synthesis (*SSSC* 49A:151(1989)).

Adsorption

Adsorption properties of zeolite Na,H-Nu-2:

Adsorbate	Kinetic* diameter A	Time, min.	Wt% adsorbed	Voidage available, cc/100 g
Water	2.7	10	7.6	7.6
		60	12.8	12.8
		1440	12.8	12.8
n-Hexane	4.3	10	11.8	17.9
		60	12.1	18.3
		1440	12.2	18.5
p-Xylene	5.85	10	12.7	14.6
		60	16.3	18.7
		120	16.3	18.7
		1440	16.2	18.6
m-Xylene	5.95	1440	3.5	4.0
Cyclohexane	6.0	10	0.2	0.26
		60	0.5	0.6
		120	0.5	0.6
		1440	1.0	1.3

*Lennard Jones kinetic diameter.

Adsorption properties of zeolite H-Nu-2:

Adsorbate	Kinetic* diameter, Å	Time, min	Wt% adsorbed	Voidage available, cc/100 g
Water	2.7	1440	13.6	13.6
p-Xylene	5.85	1440	17.0	19.5
m-Xylene	5.95	1440	15.9	18.3
Cyclohexane	6.0	1440	14.6	19.0
*Sym*trimethyl benzene	7.8	1440	0.8	0.9

*Lennard Jones kinetic diameter.

Nu-3 (LEV)
(EP 40,016(1981))
Imperial Chemical Industries

Related Materials: levyne

Structure

Chemical Composition: $0.5-1.5 \ R_2O : Al_2O_3$: at least $5 \ SiO_2 : 0-400 \ H_2O$.

X-Ray Powder Diffraction Data: ($d(\text{Å})$ (I/I_0))
10.11(8), 8.01(33), 7.56(1), 6.56(19), 5.50(10), 5.07(79), 4.94(14), 4.69(6), 4.62(2), 4.39(3.5), 4.21(56), 4.01(100), 3.78(35), 3.54(6), 3.42(3), 3.27(18), 3.18(2), 3.12(48), 3.03(9), 2.81(5), 2.75(38).

Lattice parameters for organic-containing Nu-3 (*BZA Meeting Abstracts* 1293 (1990)):

SPACE GROUP: R$\bar{3}$m

ORGANIC	a	c
1-Aminoadamantane	13.231 Å	22.290 Å
N-Methyl-quinuclidinium iodide	13.062 Å	22.601 Å
Calcined and H$^+$-exchanged	13.121 Å	22.558 Å

The levyne structure contains cages accessible via eight-member-ring openings. The organics are found to be ordered with well-defined orientations in this structure. The iodine, however, in the N-methyl-quinuclidinium iodide–prepared sample was not incorporated into the structure. Residual fragments of the organic are thought to be present in the calcined sample.

Synthesis

ORGANIC ADDITIVE
N-methyl-quinuclidine
substituted adamantanes

Zeolite Nu-3 is prepared from the reactive aluminosilicate gel with SiO_2/Al_2O_3 ranging between 10 and 300, inorganic/organic ratio between 0 and 2.0, H_2O/organic between 8 and 30, and OH/SiO_2 between 0.01 and 2.0. Sodium cations generally are used as the inorganic cations in this synthesis. Crystallization occurs readily after 3 days at 180°C (EP 40,016(1981)). This zeolite also has been observed as the more stable phase in the transformation of cubic P in a gel-free system (*Zeo.* 9:3(1989)).

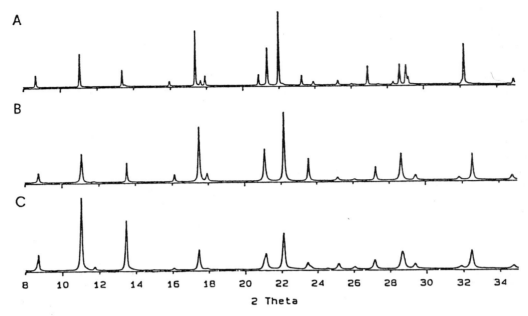

A

B

C

2 Theta

Nu-3 Fig. 1:   X-ray powder diffraction patterns of NU-3: (A) synthesized in the presence of 1-aminoadamantane, (B) synthesized in the presence of N-methyl-quinuclidinium iodide and (C) calcined and H$^+$-exchanged (*BZA Meeting Abstract* 129 (1990))

Adsorption

Adsorption properties of Nu-3 at 25°C and $P/P_o = 0.5$ (*EP* 40,016 (1981)):

Adsorbate	Kinetic* diameter, Å	Time, min	Wt% adsorbed	Voidage available, cc/100 g
Water	2.7	10	6.5	6.5
		60	9.3	9.3
		120	9.9	9.9
n-Hexane	4.3	10	1.5	2.3
		60	4.9	7.4
		120	5.9	8.9
		1440	6.0	9.9
p-Xylene	5.8	10	0.1	0.15
		1440	0.2	0.3

Structure

Chemical Composition: 0.1–7.5 M$_2$O : 0.2–2.5 Al$_2$O$_3$: 100 SiO$_2$: 0–50 H$_2$O

X-Ray Powder Diffraction Data: (d(Å) (I/I_0))
11.3(16), 11.1(20), 10.08(15), 9.90(8), 9.77(6), 9.05(1), 7.50(2), 7.09(1), 6.78(2), 6.44(5), 6.07(4), 6.05(5), 5.97(1), 5.75(8), 5.65(6), 5.63(2), 5.41(2), 5.19(2), 5.07(2), 5.01(4), 4.915(1), 4.629(9), 4.000(1), 4.495(2), 4.475(2), 4.386(13), 4.291(10), 4.124(13), 4.104(4), 4.039(6), 3.950(1), 3.880(100), 3.850(69), 3.743(51), 3.730(50), 3.678(27), 3.649(22), 3.629(5), 3.466(12), 3.364(6), 3.332(9), 3.329(4), 3.267(4), 3.260(4).

Nu-4 exhibits properties characteristic of a medium-pore ZSM-5 type molecular sieve.

Nu-4 (MFI)
(EP 65,401(1982))
Imperial Chemical Industries

Related Materials: ZSM-5

Synthesis

ORGANIC ADDITIVES
N,N-diethylethylene diamine
tetraethylene pentamine

Nu-4 crystallizes from a reactive mixture with a batch composition of: 18 tetraethylenepentamine : 11.1 Na$_2$O: 96.3 SiO$_2$: 3600 H$_2$O : 50 NaCl (*Zeo* 4:275(1984)). The temperature required for crystallization of this material is 160°C, with 1 day of crystallization time.

Thermal Properties

Nu-4 is stable to temperatures required to remove the organic from the pores of the material.

Adsorption

Adsorption properties of HNu-4 at 25°C:

HNu-4 monoclinic sample #	% sorption w/w			
	n-Hexane $P/P_o = 0.33$	*p*-Xylene $P/P_o = 0.5$	*m*-Xylene $P/P_o = 0.5$	Cyclohexane $P/P_o = 0.4$
A	10.5	8.5	1.9	—
B	11	9.7	4.2	—
C	11.2	9.8	6.8	0.5
D	12.0	10.3	10.0	2.5

Nu-5 (MFI)
(EP 54,386(1982))
Imperial Chemical Industries

Related Materials: ZSM-5

Structure

Chemical Composition: 0.5–1.5 R$_2$O : Al$_2$O$_3$: at least 10 SiO$_2$: 0–2000 H$_2$O

X-Ray Powder Diffraction Data: (d(Å) (I/I_0)) 11.1(70), 10.02(41), 9.96(37), 9.74(18), 9.00(3), 8.04(1), 7.44(6), 7.08(3), 6.71(7), 6.36(14), 5.99(15), 5.70(12), 5.59(13), 5.13(4), 5.03(6), 4.984(8), 4.623(7), 4.371(15), 4.266(15), 4.095(14), 4.014(11), 3.859(100), 3.821(70), 3.749(39), 3.725(54), 3.643(31), 3.598(4), 3.484(7), 3.358(10), 3.315(12), 3.054(12), 2.994(13), 2.979(13), 2.015(8), 1.996(8).

Nu-5 exhibits properties characteristic of a medium-pore ZSM-5 type molecular sieve.

Synthesis

ORGANIC ADDITIVE
pentaerythritol

Crystallization of Nu-5 occurs over a SiO$_2$/Al$_2$O$_3$ ratio between 50 and 200 with an alkali metal hydroxide/SiO$_2$ ratio between 0.1 and 0.25. The ratio of pentaerythritol to Al$_2$O$_3$ ranges between 1 and 50. Crystallization occurs after 48 hours at 180°C.

Adsorption

Adsorption properties of HNu-5:

	n-Hexane	*p*-Xylene	*m*-Xylene	Cyclohexane
Wt%	9.0	8.0	0.1	0.1

Nu-6
(EP 54,346(1982))
Imperial Chemical Industries

Structure

Chemical Composition: 0.5–1.5 R$_2$O : Al$_2$O$_3$: at least 10 SiO$_2$: 0–2000 H$_2$O.

X-Ray Powder Diffraction Data: (Nu-6(1)) (d(Å) (I/I_0)) 13.4(89), 11.3(6), 6.89(3), 5.46(13), 4.52(17), 4.48(15), 4.29(84), 4.23(19), 3.998(100), 3.683(34), 3.478(40), 3.382(91), 3.335(61), 3.107(13), 3.109(11), 2.986(3), 2.964(3), 2.484(17); (Nu-6(2)) (d(Å) (I/I_0)) 8.41(45), 6.67(42), 6.09(15), 4.61(27.5), 4.33(100), 4.19(sh), 4.10(sh), 3.94(2), 3.76(11), 3.65(15), 3.44(27), 3.33(76), 3.17(15), 3.05(9).

Zeolite Nu-6(2) exhibits properties of a medium-pore molecular sieve.

Synthesis

ORGANIC ADDITIVE
4,4'-bipyridyl/ethanol

Zeolite Nu-6(1) crystallizes from a reactive aluminosilicate gel with a SiO$_2$/Al$_2$O$_3$ ratio between 10 and 5000 and an inorganic (generally sodium) hydroxide/SiO_2 between 0.01 and 0.3. The ratio of the alcohol to Al_2O_3 is between 0 and 1000, and the ratio of the organic amine to Al_2O_3 is between 1 and 500. Crystallization occurs readily after 3 days at 140°C. Thermal treatment at 450°C of Nu-6(1) for 48 hours produces the X-ray powder diffraction pattern for Nu-6(2).

Adsorption Properties

Adsorption properties of Nu-6(2) at 25°C and P/P_o of 0.5 after 2 hours:

	n-Hexane	p-Xylene	m-Xylene	Cyclo-hexane	Water
Wt%	8.0	7.3	5.9	4.8	3.9

Nu-10 (TON)
(EP 77,624(1983))
Imperial Chemical Industries

Related Materials: theta-1

Structure

CHEMICAL COMPOSITION: 0.5–1.5 R_2O : Al_2O_3 : at least 20 SiO_2 : 0–4000 H_2O
SYMMETRY: orthorhombic
SPACE GROUP: Cmcm
UNIT CELL CONSTANTS: (Å) a = 13.853
b = 17.434
c = 5.04

X-Ray Powder Diffraction Data: (d(Å) (I/I_0))
10.64(36),   8.58(17),   6.91(17),   5.37(12),
4.55(17),   4.35(98),   3.64(100),   3.60(74),
3.45(40), 3.32(10), 2.51(24), 2.43(10), 2.35(7),
1.87(7).

Nu-10 exhibits properties characteristic of a medium-pore molecular sieve.

Synthesis

ORGANIC ADDITIVES
N,N'-diethylethylene diamine
N,N-diethylethylene diamine
diethylamine
triethyltetramine
1,3-propane diamine
1,2-propane diamine

Zeolite Nu-10 crystallizes from the reaction mixture: x DEA : x NAOH: 20 SiO_2 : w H_2O (w = 250, 1000; x = 0, 1, 3, 5; y = 20, 50, 100). The formation of this zeolite favors low DEA and Na^+ concentrations and temperatures between 150 and 180°C. Crystallization occurs within 5 to 10 days at pH ranges between 12.4 and 13.4. Morphologically the resulting Nu-10 crystals are lath- and fused lath-shaped balls, as well as platelets. The DEA appears to remain in the structure with 1.3 DEA per 24 SiO_2, but lesser amounts were observed when the crystallization mixture was stirred (*Zeo.* 8:508(1988)). When 1,2-propane diamine is used as the organic additive, a mixture of ferrierite and Nu-10 is observed to form. From the composition 18 R : x Na_2O : 96.3 SiO_2 : 3600 H_2O : 50 NaCl (where R = tetraethylenepentamine), Nu-10 crystallizes when x is between 0.48 and 7.22 and temperatures are between 150 and 180°C. These crystals are rod-shaped. When x = 11.1, the zeolite Nu-4 crystallizes.

This zeolite also can be prepared from a batch composition of: 5 M_2O : x Al_2O_3 : 60 SiO_2 : 3000 H_2O : 20 R (where R is $NH_2(CH_2)_nNH_2$, and M is an alkali metal cation). The pH ranges from 12.3 to 12.6 with highly crystalline materials prepared in both the potassium and the rubidium systems, with n falling between 6 and 10 for the carbon number of the diamine. ZSM-5 mixed with cristobalite forms when the inorganic cation is lithium. High SiO_2/Al_2O_3 ratios (greater than 170) also produce good-quality crystalline Nu-10. Pure silica systems result in the formation of ZSM-48 (*Zeo.* 4:280(1984)). The addition of excess NaCl also will change the crystallization products. Using tetraethylenepentamine (TEPA), ZSM-5 is formed, along with Nu-10 and cristobalite (*SSSC* 28:255(1986)).

Thermal Properties

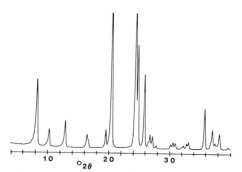

NU-10 Fig. 1:   X-ray powder diffraction pattern for DEA-silica-Nu-10 (*Zeo*. 8:508(1988)). (Reproduced with permission of Butterworths Publishers)

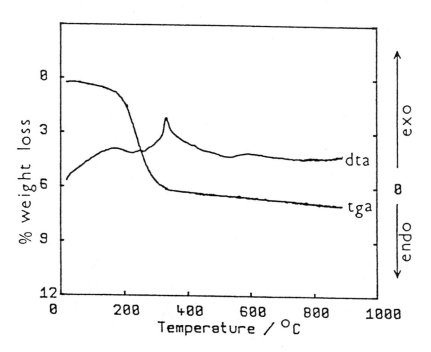

NU-10 Fig. 2:   Thermal analysis of DEA-silica-Nu-10 (*Zeo*. 8:508(1988)). (Reproduced with permission of Butterworths Publishers)

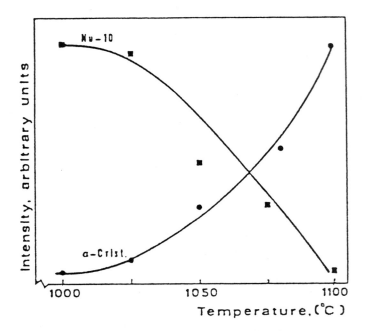

NU-10. Fig. 3:   Thermal breakdown of zeolite Nu-10 and
recrystallization to alpha-cristobalite as a function of treatment temperature
(*SSSC* 28:255(1986)). (Reproduced with permission of Elsevier Publishers)

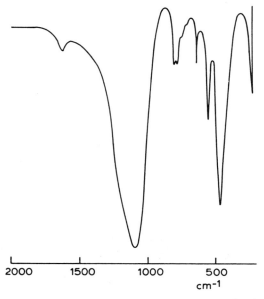

NU-10 Fig. 4:   KBr infrared spectrum of Nu-10 (*Zeo*. 8:60(1988)). (Reproduced with permission of Butterworths
Publishers)

Infrared Spectrum

The infrared spectrum is similar to that of other zeolites having the same main adsorption regions. The stretch appears near 1100 cm^{-1} (1209, 1117, 1081 cm^{-1}), and the deformation is situated near 500 cm^{-1} (637, 547, 463 cm^{-1}). See NU-10 Fig. 4.

Adsorption

Difficulties with the adsorption properties of this material were noted when the larger templates were used. Total organic removal was found to be difficult. The adsorption properties reflect the presence of residual carbon (*Zeo.* 4:275(1984)).

Adsorption properties of HNu-10 (*Zeo.* 4:275 (1984)):

Time, min	Wt% adsorbed at 25°C			Adsorbed at 0°C	
	m-Xylene	*p*-Xylene	pyridine	*n*-hexane	water
HNu-10 containing 0.9 wt% carbon:					
10	0.5	1.7	2.4	1.8	3.8
60	0.6	1.9	2.6	2.2	4.0
120	0.7	2.0	2.7	2.4	4.0
300	0.8	6.4	6.9	4.8	4.3
Hnu-10 containing 0.2 wt% carbon:					
10	0.5	5.6	6.9	3.5	2.0
60	0.7	6.0	7.0	6.1	2.5
120	0.8	6.6	7.6	6.4	2.4
300	0.8	6.5	7.7	6.5	4.2

Adsorbate (T, °K): N$_2$ (77) C$_2$H$_6$ (176) *n*-C$_4$H$_{10}$ (273) *i*-C$_4$H$_{10}$ (273):

Sample	Pore Volume, V_m (cm^3 (STP) g^{-1})			
Nu-10*	2.1	—	—	—
NaNu-10	70	36	27	17
HNu-10	72	36	28	20
Knu-10	67	34	24	22
RbNu-10	62	40	29	—

*As-synthesized.

Summary of the gravimetric adsorption data recorded at 296K (*Zeo.* 7:28 (1987)):

Sorbate	Kinetic diameter, Å	Sorbate Uptake (t = 90 min), del. V (mul/g)					
		Nu-10A		Nu-10B		Nu-10C	
		Na	H	K	H	Rb	H
Water	2.66	22.9	24.4	21.9	22.7	19.2	20.2
Methanol	3.35	59.7	56.6	55.8	51.0	49.0	53.1
n-Hexane	4.30	81.0	79.5	84.0	83.4	79.0	76.7
Benzene	5.85	37.7	34.9	41.9	39.2	37.1	39.7
p-Xylene	5.85	50.2	48.6	47.9	47.9	43.3	48.8
m-Xylene	6.80*	22.0	15.9	6.8	6.2	7.4	6.8
o-Xylene	6.80	9.2	5.8	4.3	3.2	3.5	4.9
2,3-DMB	6.10	20.8	17.4	8.4	15.0	5.8	16.7

*Taken to be the same as *o*-oxylene.

Nu-13 (MTW)
(EP 59,059(1982))
Imperial Chemical Industries

Related Materials: ZSM-12

Structure

Chemical Composition: 0–4 M$_2$O : 0.1–2.5 Al$_2$O$_3$: 100 SiO$_2$: 0–35 H$_2$O.

X-Ray Powder Diffraction Data: (d(Å) (I/I_0)) 11.8(19), 10.05(9), 4.79(16), 4.26(100), 4.08(46), 3.834(23), 3.648(3), 3.541(10), 3.395(11), 3.320(12), 3.198(6), 3.143(6), 3.043(5), 2.894(8), 2.515(12), 2.495(13).

See ZSM-12 for a description of the framework topology.

Synthesis

ORGANIC ADDITIVE
piperazine

Nu-13 crystallizes from a reactive gel with a batch composition of: 2.32 Na$_2$O : 92.7 piperazine : Al$_2$O$_3$: 96.3 SiO$_2$: 3371 H$_2$O : 54.8 NaCl. Crystallization occurs at 177°C after 3 days in a stirred autoclave. Minor impurities in some batches include quartz, ferrierite, and cristobalite (EP 59,059(1982)).

Thermal Properties

Nu-13 is thermally stable to the removal of the organic from the pores of the material at 450°C after 3 hours.

Nu-15 (TON)
(Collect. Czech. Chem. Commun. 55: 1750(1990))

Related Materials: Nu-10

Nu-15 is part of series of materials that include the hydrothermal decomposition products of glasses with variable Al_2O_3, SiO_2, V_2O_5 content. This zeolite first was prepared by Dubanska for the Institute of Geology and Geotechnique, Czechoslovak Academy of Sciences in 1990 (*Collect. Czech. Chem. Commun.* 55:1750 (1990)).

Structure

Chemical Composition: $2 Na_2O : 2 Al_2O_3 : 6 SiO_2 : V_2O_5$.

X-Ray Powder Diffraction Data: $(d(\overset{\circ}{A})$ $(I/I_0))$ 10.40(60), 8.27(40), 6.42(40), 5.54(25), 4.41(100), 3.69(65), 3.60(70), 3.41(90), 3.26(90), 2.76(45), 2.54(5), 2.44(5), 2.36(5), 1.89(10).

See theta-1 for a description of the framework topology.

Synthesis

ORGANIC ADDITIVE
none

Nu-15 is prepared by decomposing glass containing 15–17% Na_2O, 15–17% Al_2O_3, 25–40% SiO_2, and 30–45% V_2O_5 for 4 to 8 days at 120 to 160°C in strongly basic solutions of NaOH (1–4 mol/l).

Nu-23 (FER)
(EP 103,981(1984))
Imperial Chemical Industries

Related Materials: ferrierite

Structure

Chemical Composition: 0.5–$1.5 M_{2/n}O : Al_2O_3$: at least $5 SiO_2$.

X-Ray Powder Diffraction Data: $(d(\overset{\circ}{A})$ $(I/I_0))$ 11.3(5), 9.4(53), 7.00(21), 6.89(13), 6.58(12), 5.70(14), 4.94(6), 4.72(3), 4.25(1), 3.966(66), 3.931(39), 3.822(35), 3.771(49), 3.628(24), 3.535(100), 3.467(75), 3.355(21), 3.317(28), 3.246(4),   3.127(27),   3.038(18),   2.939(8), 2.887(8), 2.818(4), 2.702(5), 2.645(8), 2.575(5).

Analysis of the adsorption properties indicates the presence of a medium-pore system in Nu-23. See ferrierite for a description of the framework topology.

Synthesis

ORGANIC ADDITIVE
cyclohexylamine
2-aminocyclohexanol
methylaminocyclohexane

Nu-23 crystallizes from a reactive gel system having a molar batch composition of: $2.74 Na_2O$: 22.1 organic : Al_2O_3 : $728 SiO_2$: $659 H_2O$: 15.6 NaCl. Crystallization occurs readily after 4 days at 150°C, using cyclohexylamine as the organic additive. When 2-aminocyclohexanol was used, crystallization appeared more facile, requiring 48 hours for completion (EP 103,981(1984)).

Thermal Properties

Nu-23 is stable to temperatures of 450°C after the 3 hours required to remove the organic amine from the pores of the structure.

Adsorption

Adsorption properties for Nu-23 (EP 103,981 (1984)):

Adsorbate	Kinetic diameter, Å	P/P_o	Time, min	Wt adsorbed g/100 g	Voidage available, cc/100 g
Water	2.7	0.3	10	10.4	10.4
			60	11.8	11.8
			120	12.1	12.1
n-Hexane	4.3	0.33	10	5.8	6.7
			60	6.5	7.5
			120	6.7	10.1
p-Xylene	5.9	0.5	10	1.9	2.2
			60	2.8	3.2
			120	3.2	3.7
m-Xylene	6.2	0.5	10	1.9	2.2
			60	2.5	2.9
			120	3.2	3.7
Cyclohexane	6.3	0.5	10	0.8	1.0
			60	2.1	2.7
			120	2.6	3.3

Nu-32
(UK 2,158,056(1985))
Imperial Chemical Industries

Structure

Chemical Composition: 1.0 ± 0.3 $M_{2/n}O$: Al_2O_3 : 15–70 SiO_2 : 5–25 H_2O.

X-Ray Powder Diffraction Data: ($d(Å)$ (I/I_0))
6.99(6), 6.56(6), 4.96(6), 4.75(5), 4.64(22), 4.50(6), 4.214(6), 4.072(27), 3.919(7), 3.822(21), 3.504(10), 3.423(100), 3.354(8), 3.176(33), 3.070(15), 2.905(4), 2.778(33), 2.708(17), 2.645(19), 2.590(2), 2.325(13), 2.206(2), 2.021(4), 1.890(2), 1.862(21).

Analysis of the adsorption properties of Nu-32 indicates that this material is comprised only of small pores.

Synthesis

ORGANIC ADDITIVE
4,4'-bipyridyl

Nu-32 crystallizes from a reactive aluminosilicate gel system with the following molar batch composition: 7.95 Na_2O : 4.6 (4,4'-bipyridyl) : Al_2O_3 : 30 SiO_2 : 834 H_2O. Crystallization occurs in a stirred autoclave at temperatures of 180°C after 3 days (UK 2,158,056(1985)).

Thermal Properties

Calcination to 550°C for 17 hours to remove the organic additive does not substantially change the framework of this material.

Adsorption

Adsorption properties of Nu-32:

Sorbate	Time, min	% w/w sorbed
Water	10	6.4
	60	7.6
	120	7.8
Methanol	10	0.2
	60	0.4
	120	0.8
n-Hexane	10	0.2
	60	0.3
	120	0.4
p-Xylene	120	0.2

Nu-87
EP 377,291(1989))
Imperial Chemical Industries

Structure

Chemical Composition: 100 SiO_2 : > 10 Al_2O_3 : ≥ 20 $R_{2/n}O$ (R = cations of valency n).

X-Ray Powder Diffraction Data: ($d(Å)$ (I/I_0))
12.52(w), 11.06(s), 10.50(m), 8.31(w), 6.81(w), 4.62(m–s), 4.39sh(m–s), 4.31(vs), 4.16(m),

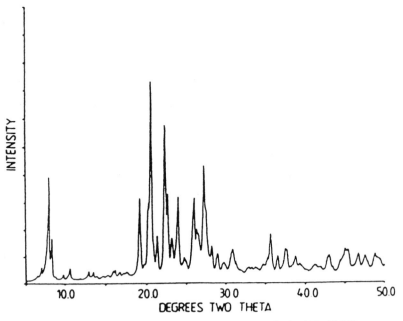

NU-87 Fig. 1:   X-ray powder diffraction pattern for as-synthesized Nu-87 (EP 377,291(1989)).

3.98(s–vs), 3.92(s), 3.83(w–m), 3.70(m–s), 3.61(w), 3.41(m–s), 3.37sh(m), 3.26(s–vs), 3.15(w), 3.08(w), 2.89(w–m), 2.52(w–m).

Nu-87 differs from EU-1 in that EU-1 does not contain an X-ray line at about 12.5 Å; EU-1 contains an X-ray line at about 10.1 Å that is absent from the pattern of Nu-87.

Synthesis

ORGANIC ADDITIVES
$R_1R_2R_3N(CH_2)_mNR_1R_2R_3$

Crystallization of Nu-87 occurs within a SiO_2/Al_2O_3 range of preferably 20 to 200, with ROH/SiO_2 between 0.1 and 0.5, H_2O/SiO_2 between 25 and 75, and Q/SiO_2 between 0.05 and 0.5. R represents the alkali or ammonium cations, and Q represents a nitrogen-containing organic cation. Salts can be added to accelerate crystallization. Nu-87 crystallizes in the presence of $[(CH_3)_3N(CH_2)_{10}N(CH_3)_3]^{+2}$. Analcime has been identified as an impurity phase in this system.

Adsorption

Adsorption data for H-Nu-87:

Sorbate	Adsorption temperature, °C	Relative pressure	Uptake, wt%	Apparent voidage filled, cm³/g*
Water	25.4	0.07	5.4	0.054
		0.28	8.3	0.083
		0.46	10.0	0.100
Methanol	25.2	0.10	10.3	0.130
		0.29	11.6	0.147
		0.50	12.3	0.156
n-Hexane	26.7	0.13	11.0	0.167
		0.31	11.5	0.175
		0.52	12.0	0.182
Toluene	26.7	0.11	12.3	0.142
		0.32	13.2	0.152
		0.48	13.6	0.157
Cyclohexane	26.7	0.12	11.6	0.149
		0.37	12.2	0.157
		0.49	12.5	0.160
Neopentane	0.0	0.11	3.29	0.05
		0.32	5.55	0.09
		0.54	8.54	0.14

*The apparent voidage filled was calculated by assuming the liquids maintain their normal densities at the adsorption temperature.

N-X (FAU)
(US3,306,922(1967))
Union Carbide Corporation

Related Materials: faujasite
Linde type X

Structure

Chemical Composition: 1.0 ± 0.1 [(CH$_3$)$_4$ N]$_2$O + x Na$_2$O : Al$_2$O$_3$: 2.5 ± 0.55 SiO$_2$: y H$_2$O.

X-Ray Powder Diffraction Data: (d(Å) (I/I_0))
14.47–14.37(vs), 8.85–8.80(m), 7.54–7.50(m), 5.73–5.71(s), 4.81–4.79(m), 4.41–4.46(m), 4.22–4.29(w), 3.95–3.93(w), 3.80–3.79(s), 3.68–3.66(m), 3.34–3.33(s), 3.05–3.01(m), 2.94–2.93(m), 2.89–2.87(s), 2.79–2.78(m), 2.74–2.73(w), 2.66–2.65(m), 2.55–2.54(vw), 2.50–2.49(vw), 2.45–2.44(vw), 2.40–2.39(m), 2.32–2.29(vw), 2.21–2.20(w), 2.19–218(vw), 2.12–21(w), 2.08–2.07(vw), 2.06–2.05(vw), 1.93–1.92(vw), 1.92–1.91(vw), 1.87–1.88(vw), 1.83–1.82(vw), 1.80–1.81(vw), 1.79–1.78(vw), 1.77–1.76(w), 1.72–1.71(w).

Zeolite N-X exhibits properties characteristic of a large-pore faujasite molecular sieve.

Synthesis

ORGANIC ADDITIVE
TMA$^+$

Zeolite N-X crystallizes from a tetramethylammonium ion–containing aluminosilicate gel with R$_2$O/Al$_2$O$_3$ between 1.5 and 4.2 and SiO$_2$/Al$_2$O$_3$ ratio between 2 and 3. Crystallization occurs readily after 8 days at 100°C.

N-Y (FAU)
(US3,306,922(1967))
Union Carbide Corporation

Related Materials: faujasite
Linde type Y

Structure

Chemical Composition: 1.0 ± 0.1 [(CH$_3$)$_4$ N]$_2$O + x Na$_2$O : Al$_2$O$_3$: (>3–6) SiO$_2$: y H$_2$O.

X-Ray Powder Diffraction Data: (d(Å) (I/I_0))
14.28(100), 8.77(18), 7.48(17), 5.69(11), 4.78(10), 4.39(7), 4.20(4), 4.135(1), 3.924(7), 3.786(20), 3.661(9), 3.474(1), 3.316(17), 3.032(6), 2.923(8), 2.866(17), 2.775(6), 2.721(2), 2.646(7), 2.533(2), 2.478(1), 2.430(1), 2.388(4), 2.301(1), 2.238(1), 2.193(2), 2.167(2), 2.101(1), 2.064(1), 2.023(1), 1.915(2), 1.897(1), 1.853(1), 1.841(1), 1.791(1), 1.777(1), 1.754(2), 1.707(2).

Zeolite N-Y exhibits properties characteristic of a large-pore faujasite molecular sieve.

Synthesis

ORGANIC ADDITIVE
TMA$^+$

Zeolite N-Y crystallizes from a tetramethylammonium ion–containing aluminosilicate gel with R$_2$O/Al$_2$O$_3$ between 1.5 and 2.5 and SiO$_2$/Al$_2$O$_3$ ratio between > 3 and > 4. Crystallization occurs after 13 days at 100°C.

Adsorption

Adsorption properties of zeolite N-Y:

Adsorbate	Temp, °C	P, Torr	Wt% adsorbed
H$_2$O	25	20	18.7
N$_2$	−196	25	13.0
		100	14.2
		500	15.3
		700	15.6

O

See Offretite. Has also be referred to as TMA-O.

Octadecasil (AST)
(Eur. J. Solid State Inorg. Chem. 28:345(1991))

Silica version of AlPO$_4$-16.

OE (OFF/ERI)
(EP 106,643(1983))
Toyo Soda

Related Materials: offretite/erionite

Structure

Chemical Composition: x M$_2$O * Al$_2$O$_3$ * 5–10 SiO$_2$ * y H$_2$O (M = Na, K; x = 0.8–2; y = 0–10).

X-Ray Powder Diffraction Data: (d(Å) (I/I_0)) 11.5(s), 9.2(w), 7.59(ms), 6.65(s), 6.32(w), 5.75(w), 5.40(w), 4.57(ms), 4.35(s), 4.17(w), 3.83(ms), 3.77(100), 3.59(s), 3.41(w), 3.31(ms), 3.23(w), 3.17(ms), 3.15(ms), 2.928(w), 2.864(s), 2.847(s), 2.680(ms), 2.586(ms), 2.354(w).

Synthesis

ORGANIC ADDITIVE
none

Zeolite OE crystallizes from reactive mixtures in the range: SiO$_2$/Al$_2$O$_3$ = 6–40; OH/SiO$_2$ = 0.3–1; K/K + Na = 0.1–0.9; H$_2$O/SiO$_2$ = 10–70. Crystallization occurs around 150°C after 40 hours with a stirred autoclave. Increasing the size of the autoclave and using a static autoclave suppress the crystallization of this structure. Small quantities of erionite are observed as intergrowths in the crystal. At higher crystallization temperatures, zeolite P appears simultaneously.

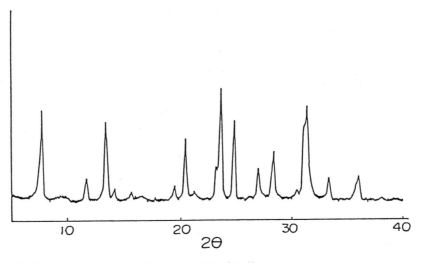

OE Fig. 1: X-ray powder diffraction pattern of zeolite OE (EP 106,643(1983)).

Adsorption

The adsorption capacity for cyclohexane is 1.9 wt% (25°C, 48 mm Hg).

Offretite (OFF)
(natural)

Related Materials: TMA-O
LZ-217

Structure

(*Nat. Zeo.* 209(1985))

CHEMICAL COMPOSITION: $KCaMg(A_5Si_{13}O_{36})$* $15H_2O$
SYMMETRY: hexagonal
SPACE GROUP: P6̄m2
UNIT CELL CONSTANTS (Å): $a = 13.29$
$c = 7.58$
PORE STRUCTURE: large 12-member ring channels 6.7 Å, intersecting 8-member rings 3.6 × 4.9 Å

X-Ray Powder Diffraction Data: (*Nat. Zeo.*
335 (1985)) ($d(Å)$ (I/I_0)) 11.61 (70), 7.60(13), 6.67(17), 6.36(7), 5.77(4), 4.59(30), 4.36(78), 3.842(50), 3.784(100), 3.607(52), 3.431(8), 3.329(19), 3.195(10), 3.172(17), 3.046(2), 2.046(17), 2.881(72), 2.862(94), 2.694(18), 2.644(3), 2.516(32), 2.499(16), 2.307(3), 2.218(22), 2.204(6), 2.129(7), 2.114(7), 2.090(8), 1.996(8).

The offretite structure is related to erionite and often is found intergrown with that zeolite.

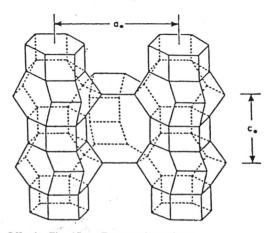

Offretite Fig. 1S: Framework topology.

The c spacing of offretite is half that of erionite, and all X-ray reflections with l odd for erionite are absent in the X-ray patterns of offretite (*Mineral. Mag.* 29:773(1951)). The framework structure consists of the AABAAB sequence of six-member rings. It is this sequencing that differentiates offretite from erionite. The framework consists of columns formed by cancrinite cages joined together by double six-ring units, and leads to the generation of 12-ring channels in the c direction. Some of the potassium ions are located in the gmelinite cages, with the remaining ions occupying sites in the large channels (*ACS* 101:230(1971); *Acta Crystallogr.* B28:825(1972)).

Fractional atomic coordinates for offretite (*Acta Crystallogr.* B28:825(1972)):

Atom	x/a	y/b	z/c	Occupancy
K	0	0	1/2	1
Ca(1)	2/3	1/3	0.377(5)	0.39(3)
Ca(2)	0	0	0.13(1)	0.07(3)
Mg	1/3	2/3	0	0.82(6)
Si(1)	0.0027(5)	0.2342(4)	0.2085(7)	1
Si(2)	0.0930(6)	0.4251(5)	1/2	1
O(1)	0.029(1)	0.351(1)	0.329(2)	1
O(2)	0.101(2)	0.202(2)	0.257(4)	1
O(3)	0.255(2)	0.127(3)	0.293(4)	1
O(4)	0.012(2)	0.267(2)	0	1
O(5)	0.230(3)	0.460(3)	1/2	1
O(6)	0.075(2)	0.537(2)	1/2	1
H₂O(7)	1/3	2/3	0.261(5)	0.90(8)
H₂O(8)	0.243(6)	0.486(6)	0	0.34(6)
H₂O(9)	0.16(1)	0.52(1)	0	0.14(5)
H₂O(10)	0.485(6)	0.242(8)	1/2	0.58(8)
H₂O(11)	0.562(7)	0.438(7)	0.172(9)	0.47(5)
H₂O(12)	0.53(3)	0.35(3)	0	0.17(4)
H₂O(13)	2/3	1/3	0.24(2)	0.30(8)

Synthesis

ORGANIC ADDITIVES
TMA⁺
monoethanolamine
diaminoethane
BTMA⁺
choline chloride
BTEA⁺
DABCO

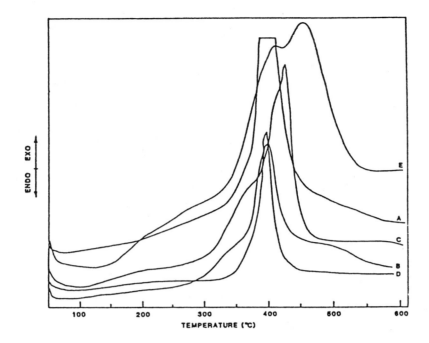

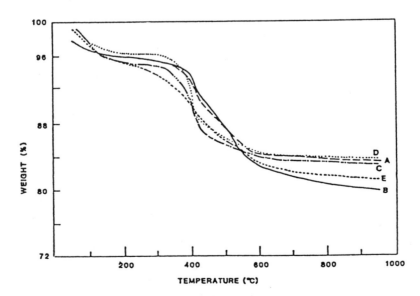

Offretite Fig. 2:   (top) DTA profiles for: (A) BTEA; (B) BTMA; (C) DABCO(I); (D) DABCO(II); (E) CC-offretite. Heating rate: 10°C/min; oxygen flow rate: 55 cm³/min. (bottom) Thermograms for: (A) BTEA; (B) BTMA; (C) DABCO(I); (D) DABCO(II); (E) CC-offretite. Heating rate: 10°C/min; oxygen flow rate: 180 cm³/min (*Zeo.* 7:265(1987)). (Reproduced with permission of Butterworths Publishers).

The synthesis of the pure offretite structure is difficult because of the recurring presence of the erionite as an intergrowth within the crystal. Offretite synthesized in the presence of TMA (*Zeo.* 9:104(1989); JP 83,1355,123(1983)) cation has a high degree of offretite character; however, based on analysis of the mid-infrared vibrations, it still contains a small amount of erionite impurity. The use of choline chloride (CC) and benzyl trimethylammonium chloride (BTMA) also produces offretite phases with small erionite impurities within the structure. Excel-

lent results in obtaining higher-quality offretite have been found with the use of benzyl-N,N,N-triethylammonium chloride (BTEA) as the organic amine additive to this system (*Zeo.* 9:104(1989)). Hydrogel compositions that have successfully resulted in the crystallization of offretite materials involve: Al_2O_3 : 12 SiO_2 : 2.2 Na_2O : 0.64 K_2O : 2.5 (organic amine) : 200 H_2O, with temperatures ranging between 150 and 200°C for 7 days. This structure also has been prepared in the presence of lithium and cesium (*JCS Dalton Trans.* 1020(1977)) or ru-

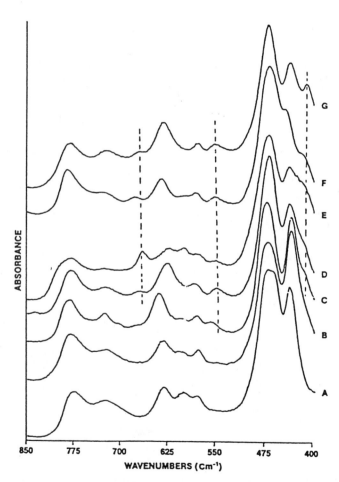

Offretite Fig. 3: Mid-infrared spectrum of several zeolites of the offretite-erionite family, crystallized from gels containing: (A) TMA, (B) CC, (C) BTMA, (D) DABCO(I), and (E) DABCO(II) ions as templating agents. Spectra for reference materials such as ZSM-34 and natural erionite are shown in (F) and (G), respectively (*Zeo.* 9:104(1989)). (Reproduced with permission of Butterworths Publishers)

bidium (*ACS* 218:21(1983)). In the lithium and cesium system, ZK-5 appears as a major impurity phase. The use of K/TMA in preparing offretite results in a very narrow distribution of particle sizes, around 1 micron, whereas the K/Na/TMA-prepared offretites will range in size from 1 micron to 6 microns (*Zeo.* 7:63(1987)).

Thermal Properties

The loss of TMA cations in offretite occurs at 300°C. Decomposition of choline occurs at 500°C (*Zeo.* 7:255(1987)). The decomposition range for NH_4 offretite is between 370 and 780°C (*Zeo.* 3:249(1983)).

Infrared Spectrum

Mid-infrared Spectrum (cm^{-1}): For TMA-offretite bands are reported at 438(ms), 471(ms), 581(w), 602(w), 633(mw), 725(bw), 777(m), 953(wsh), 1063(s), 1150(bwsh) (*Zeo.* 9:104 (1989)). The shoulder near 412 cm^{-1} and the weak band near 560 cm^{-1} suggest the presence of erionite. For CC-offretite bands are observed at 436(ms), 472(ms), 579(mw), 610(bw), 633(mw), 725(bw), 783(mw), 954(vwsh), 1064(s), 1159(bwsh). For BTMA-offretite bands are observed at 437(ms), 469(ms), 580(vw), 610(vw), 647(w), 745(bvw), 794(mw), 1070(s), 1187(bwsh). For BTEA-offretite bands are reported at 421(wsh), 443(ms), 471(ms), 594(bw), 647(bw), 746(bvw), 798(mw), 955(vwsh), 1077(s), 1187(bwsh). For DABCO-offretite bands are reported at 415(wsh), 438(ms), 470(ms), 555(w), 582(w), 635(mw), 677(vw), 728(bw), 785(m), 1057(s), 1152(bwsh) (*Zeo.* 9:104(1989)).

Ion Exchange

Protons can be exchanged into offretite by using an acid ion exchange resin (*Zeo.* 3:72(1983); *PP 8th IZC* 37(1989)).

Adsorption Properties

Adsorption properties for synthetic offretite (TMA, Na, K) (*Zeo. Mol. Sieve.* 623(1974)):

Adsorbate	T, K	P, Torr	Wt% adsorbed
Oxygen	90	10	18
		100	21
		700	23
n-Butane	298	10	10
		100	11.5
		700	12.2
Isobutane	298	10	5.8
		100	6.7
		700	8.0
Neopentane	298	10	5.5
		100	5.8
		700	6.9

Omega (MAZ)
(US4,241,036(1980); BP 1,178,186(1970))
Union Carbide Corporation

Related Materials: mazzite

Structure

(*Zeo. Mol. Sieve.* 167(1974))

CHEMICAL COMPOSITION: $Na_{6.8}TMA_{1.6}$
    $[(AlO_2)_8(SiO_2)_{28}]21H_2O$
SYMMETRY: hexagonal
SPACE GROUP: P6$_3$/mmcs
UNIT CELL CONSTANTS (Å) $a = 18.4$
                                        $c = 7.6$
DENSITY: 1.65 g/cc
VOID FRACTION (determined from water content): 0.38
PORE STRUCTURE: unidimensional 12-member rings 7.5 Å

X-Ray Powder Diffraction Data: (*Zeo. Mol. Sieve.* 364(1974)) (d(Å) (I/I_0)) 15.95(20), 9.09(86), 7.87(21), 6.86(27), 5.94(32), 5.47(6), 5.25–5.19(8), 4.695(32), 3.909(11), 3.794(58), 3.708(30), 3.620(25), 3.516(53), 3.456(20), 3.13(38), 3.074–3.02(21), 2.911(36), 2.640(6), 2.488(6), 2.342(17), 2.272(6), 2.139(5), 2.031(17), 1.978(5), 1.909(10), 1.748(6).

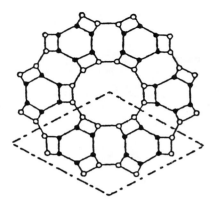

Omega Fig. 1S: Framework topology.

See mazzite for a description of the framework topology.

Synthesis

ORGANIC ADDITIVES
TMA$^+$
choline-chloride
pyrrolidine
DABCO

Omega will crystalize from reaction mixtures containing TMA at 100°C after faujasite first appears. The faujasite can be eliminated as an impurity if crystallization is at 135°C. Sodalite, analcime, mordenite, and cristabolite also can be formed as impurity phases in this system (*Zeo.* 7:203(1987)). Crystallization at the lower temperatures occurs after 6 days. The batch composition that produces omega is: 4–7 Na$_2$O

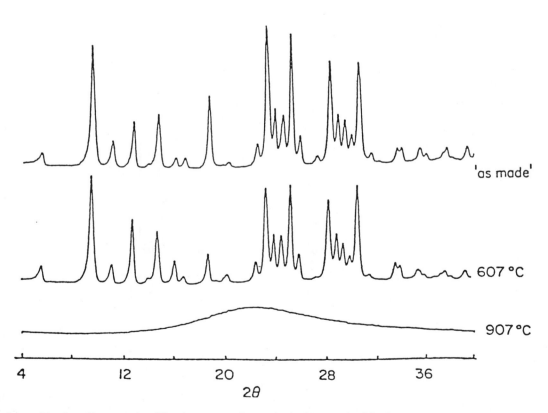

Omega Fig. 2:   X-ray powder diffraction pattern of as-synthesized (top) and calcined omega (*Zeo.* 4:263(1984)). (Reproduced with permission of Butterworths Publishers)

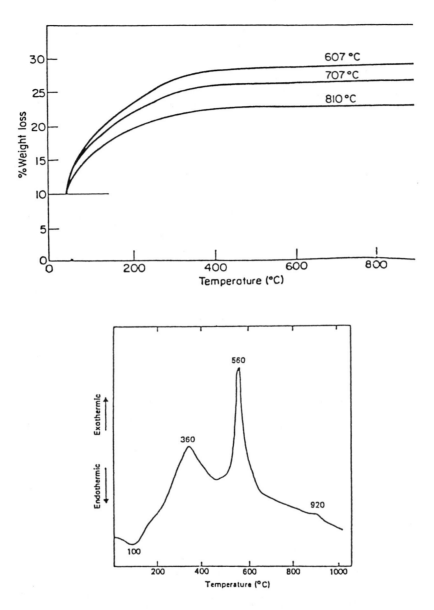

Omega Fig. 3:   DTA profile in air of zeolite omega prepared in the presence of TMA (*Zeo*. 7:203(1987)). (Reproduced with permission of Butterworths Publishers)

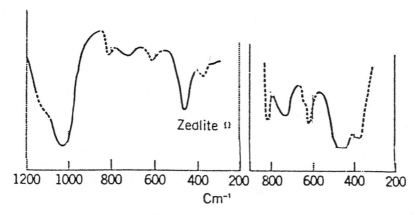

Zeolite Ω

1200 1000 800 600 400 200 800 600 400 200
Cm⁻¹

Omega Fig. 4: Mid-infrared vibrations of zeolite omega (*Advances in Chemistry Series* 101, *Molecular Sieve Zeolites*, Flanigan, E. M, Sand, L. B., eds., 201(1971). (Reproduced with permission of the American Chemical Society)

: Al_2O_3 : 15 SiO_2 : 300 H_2O : 2–4 TMABr (*Zeo.* 4:263(1984)). The zeolite S, or gmelinite phase, initially forms in this system. The pH range for crystallization is between 11.8 and 13. This zeolite generally crystallizes in a spherulite morphology and appears as 3.5- to 7.5-micron agglomerates. Fibrous crystals also have been prepared and are found to be more thermally stable than spherulitic ones.

Thermal Properties

Three exotherms are observed upon decomposition of the organic amine contained within the structure of omega. They occur at 360 and 560°C, with a small exotherm at 920°C resulting from decomposition of the zeolite structure (*Zeo.* 7:203(1987)). The temperature range over which NH_4-omega decomposes to form NH_3 and H-omega is between 310 and 780°C (*Zeo.* 3:249(1983)).

Infrared Spectrum

Mid-infrared Vibrations (cm⁻¹): (SiO_2/Al_2O_3 = 7.7) 1130(wsh), 1024(s), 805(mw), 722(mw), 610(mw), 451(ms), 372(m) (*ACS* 101:201 (1971)).

Adsorption

Adsorption properties of zeolite omega (US4,241,035(1980)):

Adsorbate	Cation form	Act. temp., °C	Temp., °C	Pressure, Torr	Wt% adsorbed
O_2	Na	400	−183	100	13.6
N_2	Na	350	−196	760	10.7
N_2	NH_4	250	−196	760	13.2
H_2O	Na	200	25	18	21.3
CO_2	Na	200	78	760	16.9
i-C_4H_{10}	Na	200	25	760	4.4
	Ca	400	25	760	4.8
	K	400	25	760	5.9
	NH_4	400	25	760	5.0
Neopentane	Na	400	25	700	7.5
$C_8F_{16}O$	Na	350	29	30	23.0
$(C_4F_9)_3N$	Na	350	50	0.7	20.3*
	Na	400	50	0.7	16.9*
	Ca	400	50	0.7	15.0*
	K	400	50	0.7	17.4*
	NH_4	250	50	0.7	15.6*
	Na	350	28	0.07	18.7*
	Na	400	28	0.07	13.0*

*After 5 hours.

Orizite (EPI)

Obsolete synonym for epistilbite.

P

General term for synthetic zeolites with the gismondine structure. P_c, P_t and P_o represent the cubic, tetragonal and orthorhombic phases of gismondine. See Na-Pl for a description of the framework topology.

Another material referred to as P also is known as species P. This material is related to the ZK-5 framework structure (see species P).

P-A (LTA)
(ACS 101:76(1971))

Related Materials: Linde type A

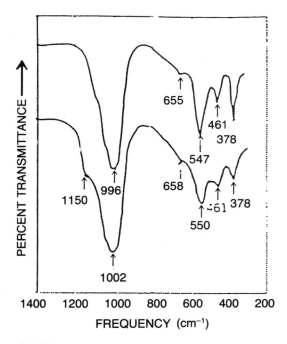

P-A Fig. 1: Infrared spectra of P-A (upper curve) and type A (lower curve (*Advances in Chemistry Series* 101, *Molecular Sieve Zeolites,* Flanigan, E. M, Sand, L. B., eds., 76(1971). (Reproduced with permission of the American Chemical Society)

Structure

Chemical Composition: P-A represents the phosphorus-containing analog of zeolite type A.

SYMMETRY: cubic
UNIT CELL CONSTANT (Å) $a = 12.24$ (pseudo cell)
TETRAHEDRA/U.C.: 24
D (measured): 2.11 g/cm^3

X-Ray Powder Diffraction Data: $(d(\text{Å})\,(I/I_0))$
12.2(100), 8.6(89), 7.07(57), 5.48(35), 4.99(5), 4.33(16), 4.08(57), 3.87(8), 3.69(81), 3.54(5), 3.40(32), 3.28(68), 2.96(92), 2.89(22), 2.74(16), 2.68(11), 2.62(49), 2.50(11), 2.45(8), 2.36(5), 2.24(3), 2.17(14), 2.14(8), 2.10(5), 2.07(5), 2.05(14), 2.02(2), 1.92(12), 1.89(7), 1.85(5), 1.83(5), 1.73(18).

Synthesis

The silicoaluminophosphate P-A crystallizes from a molar reactant composition of: ≥ 1.8 Na$_2$O : Al$_2$O$_3$: 1.6 SiO$_2$: 1.1 P$_2$O$_5$: ≥ 110 H$_2$O. Crystallization occurs at 125°C after 45 hours.

Infrared Spectrum

See P-A Figure 1.

Ion Exchange

P-A with an initial cation wt% of 15.1 exchanges 88% with K$^+$ and 93% with Ca^{+2}.

Adsorption

Adsorption properties of ion-exchanged forms of P-A:

Exchange form	Adsorbate	P, Torr	T, °C	Wt% adsorbed
Na	H$_2$O	20	25	26.7
K				20.7
Ca				27.8

Exchange form	Adsorbate	P, Torr	T, °C	Wt% adsorbed
Na	CO_2	750	25	—
K				13.2
Ca				21.4
Na	O_2	750	− 196	23.8
K				3.0
Ca				26.7
Na	N_2	750	− 196	—
K				—
Ca				22.6
Na	$n\text{-}C_4H_{10}$	750	25	—
K				—
Ca				11.5

P(B) (GIS)

See Na-P1

Pahasapaite (RHO)
(natural)

Related Materials: rho

Pahasapaite is a crystalline beryllophosphate first identified in 1987 by Rouse, Peacor, Dunn, Campbell, Roberts, Wicks, and Newbury (*N. Jb. Miner. Mh.* 433(1987)). This mineral was named after the Lakota Sioux word for Black Hills. Pahasapaite was discovered at the Tip Top Mine in Custer, South Dakota, associated with tiptopite, tinsleyite, fransoletite, and ehrleite.

Structure

(*N. Jb. Miner. Mh.* 433(1987))

CHEMICAL COMPOSITION: $(Ca_{5.5}Li_{3.6}K_{1.2}Na_{0.2\text{--}13.5})$, (-- represents absences) $Li_8Be_{24}P_{24}O_{96} * 38H_2O$
SYMMETRY: cubic
SPACE GROUP: I23
UNIT CELL CONSTANTS (Å): 13.781(4)
DENSITY: 2.28(4) g/cm³
PORE STRUCTURE: eight-member rings

X-Ray Powder Diffraction Data: (d(Å) (I/I_0))
9.60(100), 5.61(30), 4.86(30), 4.35(40), 3.97(20), 3.684(90), 3.439(20), 3.248(90), 3.081(30), 2.935(90), 2.809(20), 2.702(60), 2.516(30), 2.361(10), 2.297(20), 2.237(40), 2.180(2), 2.124(10), 2.075(10), 2.033(1), 1.991(5), 1.950(10), 1.878(1), 1.843(5), 1.810(2), 1.750(5), 1.723(5), 1.698(10), 1.672(5), 1.604(10), 1.564(5), 1.524(1), 1.488(2), 1.455(5), 1.423(10), 1.409(10), 1.395(15), 1.316(1), 1.292(2), 1.270(10), 1.233(5), 1.192(5), 1.166(5), 1.151(2), 1.142(2), 1.098(5).

The pahasapaite framework contains a three-dimensional network of large cavities connected to one another through narrow apertures. The cavities contain all of the water plus 10.5 (Ca, Li, K, Na) ions. The eight Li atoms lie on a separate eightfold axis and are not considered to be readily exchangable.

Paranatrolite (NAT)
(natural)

Related Materials: natrolite

Paranatrolite is the name given to the natrolite mineral from Mt. St. Hilaire, Quebec (*Nat. Zeo.* 36(1985)). This material has a higher water content than the other natrolite minerals have.

Structure

(*Nat. Zeo.* 35(1985))

CHEMICAL COMPOSITION: $Na_{16}(Al_{16}Si_{24}O_{80})* 24H_2O$
SYMMETRY: pseudo-orthorhombic
UNIT CELL CONSTANTS (Å): $a = 19.07$
$b = 19.13$
$c = 6.58$
PORE STRUCTURE: eight-member rings
2.6 × 3.9 Å

Parastilbite (EPI)

Obsolete synonym for epistilbite.

Partheite (-PAR)
(natural)

Partheite is a proposed name for a mineral found in the Taurus Mountains of Turkey (*Bull. Suisse Min. Petrogr.* 59:5(1979)).

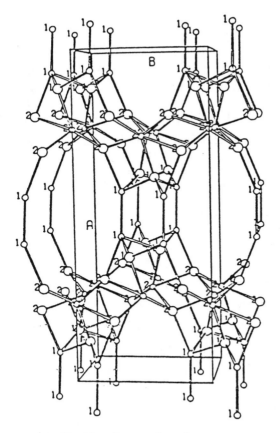

Partheite Fig. 1S: Framework topology.

3.532(15), 3.400(15), 3.190(40), 3.046(30),
3.000(10), 2.950(20), 2.900(30), 2.790(10),
2.752(15), 2.710(10), 2.682(5), 2.600(5),
2.531(10), 2.454(10), 2.427(5), 2.332(10),
2.265(5), 2.231(20), 2.200(5), 2.161(10),
2.1334(5), 2.094(5), 2.051(10), 2.010(5),
1.974(15).

Partheite consists of a three-dimensional aluminosilicate framework with hydroxyl groups associated with the framework at every second AlO_4 tetrahedron. Water and the calcium ions occupy sites in the large 10-member-ring zigzag channels. The framework contains eight-member rings and two four- and six-member rings as well.

Atom positions for partheite (Z. *Kristallogr.* 169:165(1984)):

Atom	x	y	z
Si(1)	0.06729(9)	0.1832(3)	0.2896(2)
Si(2)	0.23983(9)	0.0077(3)	0.4621(2)
Al(1)	0.1161(1)	0.0844(3)	0.6006(3)
Al(2)	0.1999(1)	0.3162(3)	0.2858(2)
Ca	0.35586(7)	0.1991(2)	0.0444(2)
O(1)	0.0695(2)	0.0181(7)	0.2162(6)
O(2)	0.0725(2)	0.1719(7)	0.4626(6)
O(3)	0.1222(2)	0.2883(7)	0.2295(6)
O(4)	0.1722(2)	0.0363(6)	0.0250(6)
O(5)	0.2081(2)	0.4669(6)	0.4096(6)
O(6)	0.2345(3)	0.1550(6)	0.3605(6)
O(7)	0.2340(2)	0.3599(6)	0.1221(6)
O(8)	0	0.2632(9)	1/4
OH	0.3523(3)	0.2673(7)	0.2918(6)
$H_2O(1)$	0.0712(3)	0.5050(8)	0.0159(9)
$H_2O(2)$	0.4541(3)	0.3070(8)	0.0800(7)

Structure

(*Nat. Zeo.* 315(1985))

CHEMICAL COMPOSITION: $Ca_8[Al_{16}Si_{16}O_{60}(OH)_8]*16H_2O$
SYMMETRY: monoclinic
SPACE GROUP: C2/c
UNIT CELL CONSTANTS: (Å) a = 21.59
b = 8.78
c = 9.31
β = 91°28′
DENSITY: 2.39 g/cm³
FRAMEWORK DENSITY: 18.2 T atoms/1000 Å³
PORE STRUCTURE: 10-member rings 3.5 × 6.9 Å

X-Ray Powder Diffraction Data: (*Nat. Zeo.* 346(1985)) (d(Å) (I/I_0)) 10.79(100), 8.12(80), 6.10(70), 5.39(5), 4.65(5), 4.38(10), 4.31(5), 4.05(20), 3.870(10), 3.740(50), 3.600(40),

Thermal Properties

Three endothermic peaks are observed in the DTA curves, at 255, 430, and 590°C (*Schweiz Miner. Petr. Mitt.* 49:5(1979)). No major structural change was observed after heating to 150°C for 64 hours. Structural transformation to anorthite was observed when partheite was heated at 400°C for 40 hours (Z. *Kristallogr.* 19:165 (1984)).

Paulingite (PAU)
(natural)

Paulingite is a natural zeolite mineral first identified in 1960 and named in honor of Nobel laureate Linus C. Pauling (*Am. Mineral.* 45:79(1960)).

Structure

(*Zeo. Mol. Sieve.* 169(1974))

CHEMICAL COMPOSITION: $(K_2,Na_2,Ca,Ba)_{76}$ $[(AlO_2)_{152}(SiO_2)_{520}]$ * 700 H_2O
SYMMETRY: cubic
SPACE GROUP: Im3m
UNIT CELL CONSTANT (Å): $a = 35.10$
FRAMEWORK DENSITY: 1.54 g/cc
VOID FRACTION (determined by water content): 0.49
PORE STRUCTURE: intersecting eight-member rings, 3.8 A × 3.8 Å

X-Ray Powder Diffraction Data: (*Nat. Zeo.* 331(1985)) (*d*(Å)) (*I/I₀*)) 12.42(5), 10.14(4), 9.39(10), 8.79(6), 8.29(60), 7.85(6), 7.49(5), 7.16(12), 6.89(67), 6.21(35), 5.85(21), 5.70(32), 5.42(17), 5.30(5), 5.18(6), 4.96(33), 4.78(76), 4.69(23), 4.39(28), 4.32(8), 4.26(21), 4.08(25), 3.974(4), 3.877(32), 3.700(11), 3.620(2), 3.583(60), 3.444(7), 3.377(12), 3.349(69), 3.261(89), 3.231(8), 3.204(6), 3.179(18), 3.129(58), 3.081(100), 2.991(39), 2.969(31), 2.944(7), 2.907(6), 2.869(11), 2.848(11), 2.828(7), 2.794(13), 2.726(39), 2.631(17), 2.618(45), 2.574(11), 2.522(12), 2.485(7), 2.449(10), 2.306(5), 2.286(7), 2.259(7), 2.178(4), 2.137(5), 2.106(6), 2.063(8), 2.049(12), 2.034(5), 2.008(10), 1.954(4).

The structure is composed of an extremely complex framework with eight different structural units: two double eight-member rings, three types of cages, two horseshoe-like configurations, and one toroidal channel. The A cage is a truncated cubo-octahedron; this cage is connected with a set of six double eight-rings. The B cage has two planar and four nonplanar eight-ring openings and 32 corners. One of the planar openings is connected with one double eight-ring, and the other with the other double eight-ring. These double eight-rings are alike but differ crystallographically. This part of the structure contains all crystallographically different Si,Al sites. The final (or C) cage has two planar six-ring openings and six nonplanar eight-ring opening. This cage plus two horseshoe-like configurations is not a center of symmetry. The whole framework topology can be described by these aggregates (*Sci.* 154:1004(1966)). The channels in paulingite are similar to those in Linde type A and ZK-5.

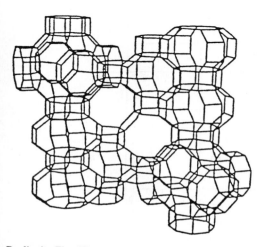

Paulingite Fig. 1S: Framework topology.

Atomic positions for paulingite (*Sci.* 154:1004(1966)):

Atom	No. per cube	x	y	z
(Si,Al)1	48	0.3137	1/4	1/2 − x
(Si,Al)2	48	0.4021	1/4	1/2 − x
(Si,Al)3	96	0.3132	0.2498	0.0979
(Si,Al)4	96	0.4558	0.1072	0.0443
(Si,Al)5	96	0.4019	0.1782	0.0448
(Si,Al)6	96	0.3126	0.1785	0.0446
(Si,Al)7	96	0.2592	0.1073	0.0445
(Si,Al)8	96	0.1708	0.1076	0.0441
O1	48	0.1635	0.0933	0
O2	48	0.2679	0.0968	0
O3	48	0.3041	0.1886	0
O4	48	0.4092	0.1900	0
O5	48	0.4484	0.0952	0
O6	48	0.4489	0.3794	0
O7	48	0.0713	x	0.1610

(continued)

Atom	No. per cube	x	y	z
O8	48	0.4308	x	0.2322
O9	48	0.1437	x	0.0545
O10	48	0.2860	x	0.1965
O11	48	0.2868	x	0.0890
O12	48	0.4299	x	0.0548
O13	96	0.2152	0.1211	0.0496
O14	96	0.2870	0.1414	0.0582
O15	96	0.3573	0.1672	0.0525
O16	96	0.4278	0.1417	0.0573
O17	96	0.3002	0.2163	0.0693
O18	96	0.4148	0.2142	0.0710
O19	96	0.3576	0.2623	0.0909
O20	96	0.3080	0.2351	0.1419
M1	16	0.1788	x	x
M2	24	0.2543	x	0
M3	48	0.3975	x	0.1445
(H_2O)1	16	0.140	x	x

Atom	No. per cube	x	y	z
(H_2O)2	16	0.217	x	x
(H_2O)3	48	0.209	x	0.053
(H_2O)4	48	0.348	x	0.200
(H_2O)5	48	0.350	x	0.082
(H_2O)6	48	0.220	x	0.140
(H_2O)7	48	0.139	x	0.217
(H_2O)8	48	0.334	0.275	0
(H_2O)9	48	0.422	0.285	0
(H_2O)10	24	0.292	1/2	0
(H_2O)11	24	0.367	x	0
(H_2O)12	12	0.178	0	0
(H_2O)13	12	0.270	0	0
(H_2O)14	12	0.458	0	0

Thermal Properties

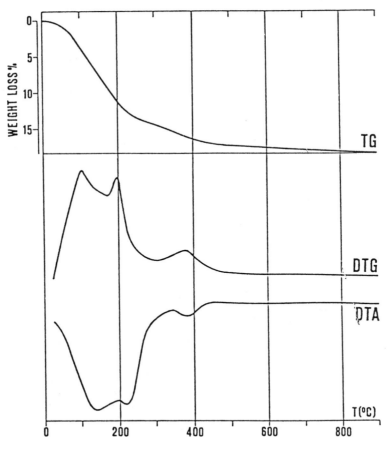

Paulingite Fig. 2:   Thermal curves of paulingite from Ritter, Oregon; in air, heating rate 20°C/min (*Nat. Zeo.* 167(1985)). (Reproduced with permission of Springer Verlag Publishers)

Adsorption

During dehydration, the structure appears to undergo irreversible distortion.

P-B(P) (GIS)
(*ACS* 101:76(1971))

Related Materials: gismondine

Structure

Chemical Composition: P-B(P) represents the phosphorus-containing analog of zeolite type B (B or P structure type).

SYMMETRY: tetragonal
UNIT CELL CONSTANT (Å): $a = 10.1$
                                           $c = 9.8$
TETRAHEDRA/U.C.: 16

X-Ray Powder Diffraction Data: (d(Å) (I/I_0))
7.08(58), 5.01(58), 4.96(45), 4.44(7), 4.10(57), 4.04(17), 3.53(5), 3.34(6), 3.20(100), 3.13(57), 2.71(37), 2.68(38), 2.66(22), 2.53(8), 2.20(5), 1.98(6), 1.68(7).

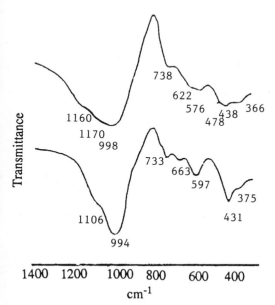

P-B Fig. 1:   Infrared spectra of P-B (upper curve) and type B (lower curve) (*Advances in Chemistry Series* 101, *Molecular Sieve Zeolites*, Flanigan, E. M, Sand, L. B., eds., 76(1971). (Reproduced with permission of the American Chemical Society)

Synthesis

The silicoaluminophosphate P-B crystallizes from a molar reactant composition of: $\geqslant 0.4$ Na_2O : Al_2O_3 : 0.6 SiO_2 : 0.7 P_2O_5: $\geqslant 110$ H_2O. Crystallization occurs at 200°C after 70 hours.

Infrared Spectrum

See P-B Figure 1.

Ion Exchange

P-B with an initial cation wt% of 12.1 exchanges 100% with K^+.

Adsorption

Adsorption properties of ion-exchanged forms of P-B:

Exchange form	Adsorbate	P, Torr	T, °C	Wt% adsorbed
Na	H_2O	20	25	—
K				4.1
Na	CO_2	750	25	—
K				2.4
Na	O_2	750	−196	—
K				—

P-C (ANA)
(*ACS* 101:76(1971))

Related Materials: analcime
                              type C

Structure

Chemical Composition: P-C represents the phosphorus-containing analog of zeolite type C or analcime.

SYMMETRY: cubic
UNIT CELL CONSTANT (Å): $a = 13.73$
TETRAHEDRA/U.C.:   48
D (measured):   2.30 g/cm³

X-Ray Powder Diffraction Data: (d(Å) (I/I_0))
5.64(80), 4.87(16), 3.68(6), 3.44(100), 3.27(3),

2.93(45), 2.81(6), 2.70(14), 2.51(13), 2.37(8), 1.91(8), 1.87(5), 1.75(12), 1.72(3), 1.70(3).

Synthesis

The silicoaluminophosphate P-C crystallizes from a molar reactant composition of: $\geqslant$0.5 Na$_2$O : Al$_2$O$_3$: 0.6 SiO$_2$: 0.5 P$_2$O$_5$: $\geqslant$55 H$_2$O. Crystallization occurs at 210°C after 160 hours.

Infrared Spectrum

See P-C Figure 1.

Adsorption

Adsorption properties of ion-exchanged forms of P-C:

Exchange form	Adsorbate	P, Torr	T, °C	Wt% adsorbed
Na	H$_2$O	20	25	4.4

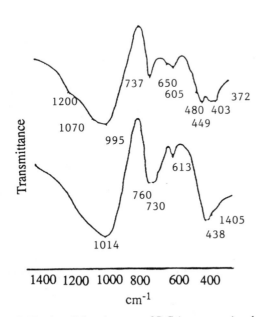

P-C Fig. 1: Infrared spectra of P-C (upper curve) and analcime (lower curve) (*Advances in Chemistry Series* 101, *Molecular Sieve Zeolites*, Flanigan, E. M, Sand, L. B., eds., 76(1971). (Reproduced with permission of the American Chemical Society)

P$_c$ (GIS)

A name used to describe a dimensionally cubic synthetic zeolite with the gismondine structure. See Na-Pl for a description of the framework topology. Unit cell constant: $a = b = c = 10.04$ Å (from J. Higgins).

P-(Cl)

See species P.

Perlialite
(*Proc. 6th IZC* 570(1983))

(See Linde type L.) Perlialite is a hydrated silicate proposed to be similar to the synthetic zeolite Linde L.

P-G (CHA)
(*ACS* 101:76(1971))

Related Materials: chabazite
                      type G

Structure

Chemical Composition: P-G represents the phosphorus-containing analog of synthetic chabazite.

SYMMETRY: rhombohedral
UNIT CELL CONSTANTS (Å) $a = 9.44$
                         $\alpha = 94°28'$
TETRAHEDRA/U.C.: 12

X-Ray Powder Diffraction Data: (d(Å) (I/I_0))
9.46(100), 6.97(21), 5.61(17), 5.10(21), 4.72(10), 4.53(5), 4.36(74), 4.15(7), 4.02(9), 3.90(43), 3.62(28), 3.48(16), 3.25(12), 3.14(10), 2.95(95), 2.92(53), 2.71(9), 2.64(19), 2.54(14), 2.33(9), 2.11(7), 1.89(8), 1.82(14), 1.73(12).

Synthesis

The silicoaluminophosphate P-G crystallizes from a molar reactant composition of: $\geqslant$0.5 K$_2$O : Al$_2$O$_3$: 1.0 SiO$_2$: 0.5 P$_2$O$_5$: $\geqslant$110 H$_2$O. Crystallization occurs at 150°C after 116 hours.

Infrared Spectrum

See P-G Figure 1.

Ion Exchange

P-G with an initial cation wt% of 13.9 exchanges 83% with Na^+ and 99% with Ca^{+2}.

Adsorption

Adsorption properties of ion-exchanged forms of P-G:

Exchange form	Adsorbate	P, Torr	T, °C	Wt% adsorbed
Na	H_2O	20	25	26.1
K				23.3
Ca				26.2
Na	CO_2	750	25	9.6
K				11.0
Ca				—
NH_4				—
Na	O_2	750	−196	9.8
K				11.2
Ca				28.4

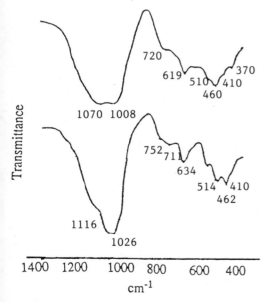

P-G Fig. 1: Infrared spectra of P-G (upper curve) and chabazite type G (lower curve) (*Advances in Chemistry Series* 101, *Molecular Sieve Zeolites,* Flanigan, E. M, Sand, L. B., eds., 76(1971). (Reproduced with permission of the American Chemical Society)

Exchange form	Adsorbate	P, Torr	T, °C	Wt% adsorbed
NH_4				4.0
Na	N_2	750	−196	8.8
K				9.1
Ca				24.6
Na	$n\text{-}C_4H_{10}$	750	25	1.8
K				1.6
Ca				3.5

Phacolite (CHA)

A variety of chabazite, phacolite forms triclinic crysals with a high degree of twinning (*PP* 4B-4, *7th IZC* (1986))

Phakolite (CHA)

See phacolite.

Phase A
(*Soviet Phys. Crystallogr.* 16:1035(1972))

Structure

CHEMICAL COMPOSITION: $Ba_2[X]BaCl_2[(Si,Al)_8O_{16}$
SPACE GROUP: I4/mmm
UNIT CELL CONSTANTS (Å): $a = 14.194$
$c = 9.234$
PORE STRUCTURE: eight member rings
X is the tetrahedral group [(Si,Al)(O,OH)$_4$ or (OH,Cl)].

Atomic coordinates for Phase A (*Sov. Phys. Crystallogr.* 16:1035(1972)):

Atom	x	y	z
Si,Al	0.2386(7)	0.3917(7)	0.157(1)
O(1)	0.193(2)	0.390(2)	0
O(2)	0.350	0.350	0.156(3)
O(3)	0.158	0.342	1/4
O(4)	0.257(3)	1/2	0.213(3)
Ba(1)	0.1816(4)	1/2	1/2
Ba(2)	1/2	1/2	0.288(1)
Cl	0.375	0.375	1/2
X*	0	1/2	1/4

*O^{-2} was used for f_x.

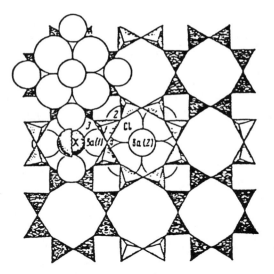

Phase A Fig. 1S: Framework topology.

PHI
(US4,124,686(1978))
Union Carbide Corporation

Structure

Chemical Composition: 0.97–1.1 $M_{2/n}O$: Al_2O_3 : 4–7 SiO_2 : 0–6 H_2O.

X-Ray Powder Diffraction Pattern: $(d(Å)(I/I_0))$
11.63(m), 9.51(s), 7.69(vw), 6.97(s), 5.61(s), 5.04(s), 4.31(s), 3.97(vw), 3.43(vs), 2.92(vs), 2.69(vw), 2.61(w), 2.51(vw), 2.09(w), 1.90(w), 1.81(w), 1.74(vw), 1.72(w).

Synthesis

ORGANIC ADDITIVE
TMA$^+$

Zeolite phi crystallizes from a reactive aluminosilicate gel system with a SiO_2/Al_2O_3 ratio between 9 and 15. The ratio of base (Na_2O + TMA_2O)/SiO_2 is between 0.5 and 0.6 and the ratio of Na_2O/Na_2O + TMA_2O between 0.8 and 0.95. Crystallization occurs after 68 hours at 100°C.

Thermal Properties

See Phi Figure 2.

Infrared Spectrum

See Phi Figure 3.

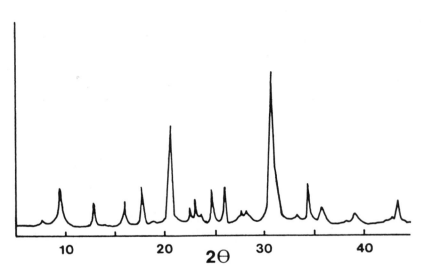

Phi Fig. 1:   X-ray powder diffraction pattern of zeolite phi (*Zeo.* 11:349(1991)). (Reproduced with permission of Butterworths Publishers)

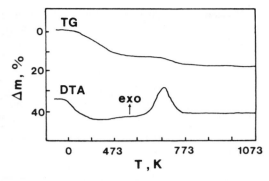

Phi Fig. 2: TG/DTA pattern in air of zeolite phi (Si/Al = 2.5) (*Zeo.* 11:349(1991)). (Reproduced with permission of Butterworths Publishers)

Adsorption

Adsorption properties of zeolite phi (US4,124,686(1978)):

	Pressure, Torr	Temperature, °C	Wt% adsorbed
O_2	100	−183	12.1
O_2	750	−183	18.4
n-Butane	750	25	8.1
Isobutane	750	25	2.7
Neopentane	750	25	3.5
$(C_4F_9)_3N$	0.5	50	8.0

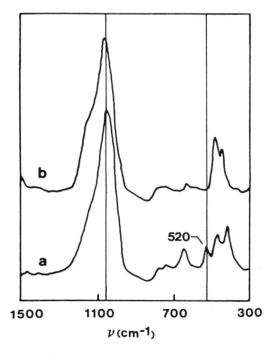

Phi Fig. 3: Infrared spectra of zeolite phi (Si/Al = 2.5) (a) and offretite (b) (*Zeo.* 11:349(1991)). (Reproduced with permission of Butterworths Publishers)

Phillipsite (**PHI**)
(natural)

Related Materials: harmotome
ZK-19
wellsite

The name phillipsite first was introduced in 1825 by Levy for samples of the harmotome mineral from Sicily, Italy (*Nat. Zeo.* 134(1985)).

Structure

(*Nat. Zeo.* 134(1985))

CHEMICAL COMPOSITION: $Ba_2(Ca_{0.5},Na)$ $(Al_5Si_{11}O_{32})*12H_2O$
SYMMETRY: monoclinic
SPACE GROUP: $P2_1/m$
UNIT CELL CONSTANTS (Å): $a = 9.88$
$b = 14.30$
$c = 8.67$
$\beta = 124°49'$
PORE STRUCTURE: eight-member rings 3.6, 3.0 × 4.3, 3.2 × 3.3 A

X-Ray Powder Diffraction Data: (*Nat. Zeo.* 326(1985)) ($d(Å)$ (I/I_0)) 7.11(100), 5.78(20), 5.06(50), 4.89(3), 4.41(25), 4.12(100), 4.03(20), 3.515(10), 3.326(30), 3.201(100), 3.106(80), 3.040(10), 2.968(10br), 2.887(5), 2.757(5), 2.699(80br), 2.651(40), 2.539(25), 2.435(20), 2.379(20), 2.317(10), 2.256(15), 2.206(25), 2.153(25), 2.057(20), 1.986(25).

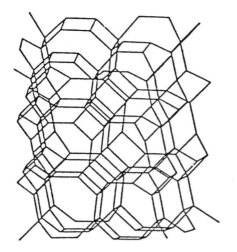

Phillipsite Fig. 1S: Framework topology.

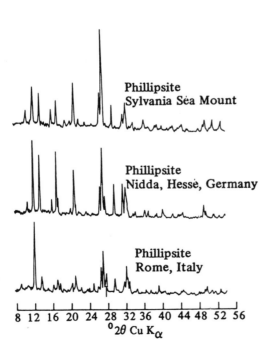

Phillipsite
Sylvania Séa Mount

Phillipsite
Nidda, Hessè, Germany

Phillipsite
Rome, Italy

$8\ 12\ 16\ 20\ 24\ 28\ 32\ 36\ 40\ 44\ 48\ 52\ 56$
$^{\circ}2\theta$ Cu K$_{\alpha}$

Phillipsite Fig. 2:   X-ray powder diffraction pattern of natural phillipsite from three locations (*Zeo. Mol. Sieve.* 206(1974)). (Reproduced with permission of John Wiley and Sons, Inc.)

Phillipsite contains a two-dimensional channel system like that of harmotome, differing only in the chemical composition and symmetry (see harmotome for a description of the topology). The channels intersect in a plane perpendicular to the c axis. It is thought that the channels are subject to blockages, which, in the dehydrated mineral, are thought to be the cations. No Si,Al ordering is observed. There are two cation sites outside the framework that are fully occupied by Ba or K, resulting in a lowering of the symmetry of the structure.

Stacking faults, if extensively present, would transform a phillipsite crystal into one with domains of phillipsite and merlinoite.

Phillipsite can be complexely twinned to give a diffraction pattern displaying tetragonal symmetry with the same unit cell dimensions as those of the untwinned crystal.

Atomic coordinates for phillipsite (*Acta Crystallogr.* 15:655(1962)):

Atom	x	y	z
Si(1)	0.153	0.018	0.892
Si(2)	0.833	0.017	0.891
Si(3)	0.335	0.861	0.238
Si(4)	0.650	0.860	0.237
O(1)	0.815	0.083	0.156
O(2)	0.773	0.104	0.840
O(3)	0.235	0.102	0.849
O(4)	0.178	0.075	0.174
O(5)	0.992	0.048	0.885
O(6)	0.509	0.874	0.205
O(7)	0.203	0	0
O(8)	0.796	0	0
O(9)	0.303	0.250	0.740
O(10)	0.707	0.250	0.750
Na	0.010	0.250	0.865

Synthesis

ORGANIC ADDITIVES
NH_4^+
$NMeH_3^+$
$NMe_2H_2^+$
NMe_3H^+
TMA^+

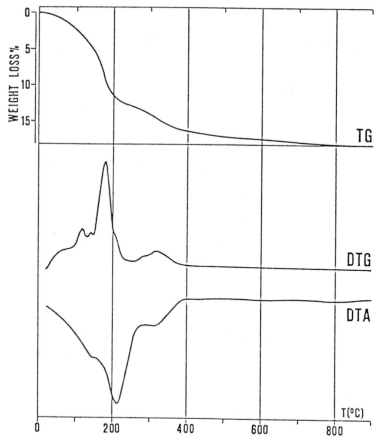

Phillipsite Fig. 3:   Thermal curves of phillipsite from Casal Brunori, Rome, Italy: in air, heating rate 20°C/min (*Nat. Zeo.* 146(1985)). (Reproduced with permission of Springer Verlag Publishers)

The phillipsite structure appears as a regular impurity phase, and in many crystallization systems many zeolites will recrystallize to form phillipsite. It has been prepared in the K, Ba, Ca, NH$_4$, (Na,K), (Na,TMA), (Ca,TMA) systems.

Thermal Properties

When dehydrated, phillipsite decomposes at temperatures above 250°C. See Phillipsite Figure 3.

Ion Exchange

See Phillipsite Figure 4.

P-L (LTL)
(*ACS* **101:76(1971)**)

Related Materials: Linde type L

Structure

Chemical Composition: P-L represents the phosphorus-containing analog of zeolite type L.

SYMMETRY: hexagonal
UNIT CELL CONSTANT (Å): $a = 18.75$; $c = 15.03$
TETRAHEDRA/U.C.: 72
D (measured): 2.21 g/cm^3

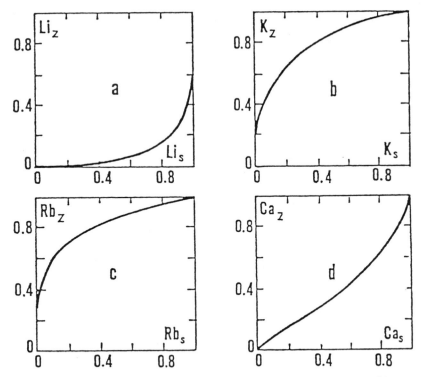

Phillipsite Fig. 4:   Exchange isotherms for Pine Valley, Nevada phillipsite at 25°C, total normality 0.1 N (*JCS* 2904(1971)). (Reproduced with permission of the Chemical Society)

X-Ray Powder Diffraction Data: $(d(\text{Å})\ (I/I_0))$
16.0(100), 8.0(4), 7.55(8), 6.09(18), 5.86(4), 4.65(29), 4.47(9), 4.37(6), 3.96(28), 3.68(9), 3.51(19), 3.32(15), 3.22(31), 3.09(24), 3.05(5), 2.94(28), 2.88(4), 2.82(4), 2.69(24), 2.64(8), 2.53(9), 2.50(4), 2.46(4), 2.44(5), 2.32(3), 2.30(3), 2.22(9), 2.06(3), 1.96(2), 1.88(6).

Synthesis

The silicoaluminophosphate P-L crystallizes from a molar reactant composition of: $\geq 1.0\ K_2O$: Al_2O_3 : 1.5 SiO_2 : 1.0 P_2O_5 : $\geq 110\ H_2O$. Crystallization occurs at 175°C after 166 hours.

Infrared Spectrum

See P-L Figure 1.

Ion Exchange

P-L with an initial cation wt% of 17.8 exchanges 40% with Na^+, 90% with NH_4, 42% with Ca^{+2}, and 5 wt% with La.

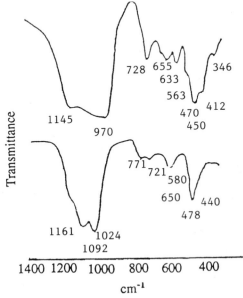

P-L Fig. 1:   Infrared spectra of P-L (upper curve) and Linde type L (lower curve) (*Advances in Chemistry Series* 101, *Molecular Sieve Zeolites*, Flanigan, E. M, Sand, L. B., eds., 76(1971). (Reproduced with permission of the American Chemical Society)

Adsorption

Adsorption properties of ion-exchanged forms of P-L:

Exchange form	Adsorbate	P, Torr	T, °C	Wt% adsorbed
Na	H_2O	20	25	15.9
K				14.4
Ca				16.4
NH_4				14.8
La				15.6
Na	CO_2	750	25	—
K				—
Ca				—
Na	O_2	750	−196	5.3
K				8.9
Ca				14.0
NH_4				12.1
La				11.6
Na	N_2	750	−196	—
K				—
Ca				—
Na	n-C_4H_{10}	750	25	—
K				—
Ca				—
Na	iso-C_4H_{10}	750	25	2.2
K				3.3
Ca				3.1
NH_4				1.7
La				1.7
Na	Neopentane	750	25	1.4
K				2.2
Ca				1.1
NH_4				0.6
Na	$(C_3H_7)_3N$	2	50	1.4
K				2.2
Ca				0.8
NH_4				0.6
La				1.2

P_o

Orthorhombic form of gismondine.

Pollucite (ANA)
(natural)

Pollucite, $CsAl_2Si_4O_{12}$, appears to be the only naturally occurring cesium-containing feldspathoid. It is structurally related to analcime.

P-R (CHA)
(ACS 101:76(1971))

Related Materials: chabazite
zeolite R

Structure

Chemical Composition: P-R represents the phosphorus-containing analog of chabazite.

SYMMETRY: rhombohedral
UNIT CELL CONSTANTS (Å): $a = 9.44$
$\alpha = 94°28'$

TETRAHEDRA/U.C.:   12

X-Ray Powder Diffraction Data: $(d(\text{Å})\ (I/I_0))$
9.46(100), 6.94(29), 5.59(10), 5.09(32), 4.72(9), 4.35(76), 4.11(5), 4.00(8), 3.90(24), 3.63(32), 3.47(16), 3.25(10), 2.94(82), 2.71(5), 2.62(11), 2.53(8), 2.32(5), 2.10(8), 1.89(8), 1.82(9), 1.73(6).

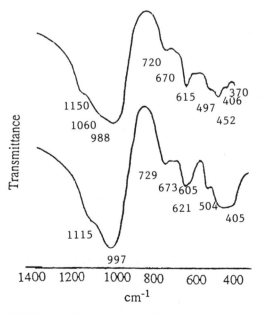

P-R Fig. 1:   Infrared spectra of P-R (upper curve) and chabazite (type R) lower curve (*Advances in Chemistry Series* 101, *Molecular Sieve Zeolites*, Flanigan, E. M, Sand, L. B., eds., 76(1971). (Reproduced with permission of the American Chemical Society)

Synthesis

The silicoaluminophosphate P-R crystallizes from a molar reactant composition of: $\geqslant 1.2$ Na_2O : Al_2O_3 : 1.8 SiO_2 : 0.9 P_2O_5 : $\geqslant 110$ H_2O. Crystallization occurs at 125°C after 94 hours.

Infrared Spectrum

See P-R Figure 1.

Ion Exchange

P-R with an initial cation wt% of 13.2 exchanges 91% with K^+, 87% with NH_4^+, and 96% with Ca^{+2}.

Adsorption

Adsorption properties of ion-exchanged forms of P-R:

Exchange form	Adsorbate	P, Torr	T, °C	Wt% adsorbed
Na	H_2O	20	25	19.0
K				18.0
Ca				11.1
NH_4				0.8
Na	CO_2	750	25	11.6
K				11.2
Ca				9.0
NH_4				—
Na	O_2	750	−196	6.6
K				0
Ca				9.6
NH_4				0.2
Na	N_2	750	−196	—
K				—
Ca				—
Na	n-C_4H_{10}	750	25	6.6
K				1.6
Ca				—
NH_4				0.2

Pseudonatrolite (MOR)

Obsolete synonym for mordenite.

PSH-3
(EP 64,205(1982))
Bayer AG

Related Materials: MCM-22
  SSZ-25

Structure

Chemical Composition: $M_{2/n}O$: Al_2O_3 : 20–150 SiO_2.

X-Ray Powder Diffraction Data: $(d(\text{Å})\,(I/I_0))$
12.63 (vs), 10.92(m), 8.84(vs), 6.86(w), 6.15(s), 5.50(w), 4.91(w), 4.60(w), 4.39(w), 4.09(w), 3.91(m), 3.75(w), 3.56(w), 3.41(vs), 3.30(w), 3.19(w),   3.11(w),   2.836(w),   2.694(w), 2.592(w),   2.392(w),   2.206(w),   2.122(w), 2.036(w), 1.973(w), 1.873(w), 1.855(w).

Synthesis

ORGANIC ADDITIVE
$(CH_2)_6NH$

PSH-3 crystallizes to from a gel with batch compositions in the range: SiO_2/Al_2O_3 of 10 to 200, OH/SiO_2 of 0.05 to 10, R/Na_2O of 0.5 to 5.0. R is hexamethylenimine. Crystallization occurs at temperatures between 80 and 180°C after 12 to 144 hours.

Adsorption

Adsorption in g/100 g for PSH-3:

H_2O	7.7
n-Hexane	6.9
Cyclohexane	6.1

P_t (GIS)

The tetragonal form of synthetic gismondine.

Ptilolite (MOR)

Member of the mordenite group of minerals found in Jefferson Co., Colorado, USA. This name is not used today (*Mineral. Mag.* 31:887(1958)).

P-W
(ACS 101:76(1971))

Related Materials: type W

Structure

Chemical Composition: P-W represents the phosphorus-containing analog of zeolite W.

SYMMETRY: tetragonal
UNIT CELL CONSTANTS (Å): $a = 20.17$; $c = 10.03$
TETRAHEDRA/U.C.: 64

X-Ray Powder Diffraction Data: $(dl(Å)\ (I/I_0))$
10.2(18), 8.3(42), 7.2(61), 5.40(17), 5.07(29), 4.51(27), 4.31(23), 4.11(19), 3.68(18), 3.23(71), 3.19(100), 2.96(45), 2.80(16), 2.75(39), 2.69(18), 2.57(23), 2.46(10), 2.20(8), 2.08(4), 1.79(5), 1.78(6), 1.73(8).

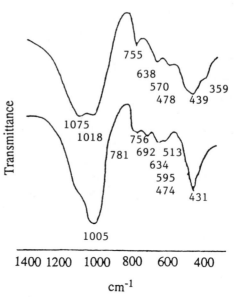

P-W Fig. 1:   Infrared spectra of P-W (upper curve) and type W (lower curve) (*Advances in Chemistry Series* 101, *Molecular Sieve Zeolites*, Flanigan, E. M, Sand, L. B., eds., 76(1971). (Reproduced with permission of the American Chemical Society)

Synthesis

The silicoaluminophosphate P-W crystallizes from a molar reactant composition of: $\geqslant 0.5\ K_2O$: Al_2O_3 : $1.6\ SiO_2$: $0.5\ P_2O_5$: $\geqslant 110\ H_2O$. Crystallization occurs at 150°C after 68 hours.

Infrared Spectrum

See P-W Figure 1.

Adsorption

Adsorption properties of ion-exchanged forms of P-W:

Exchange form	Adsorbate	P, Torr	T, °C	Wt% adsorbed
Na	H₂O	20	25	23.1
K				19.1
Ca				10.6

PZ-1
(EP 91,537(1983))
Degussa

Related materials: magadiite

Structure

Chemical Composition: high-silica aluminosilicate.

X-Ray Powder Diffraction Data: (PZ-1) $(d(Å)$ $(I/I_0))$ 15.49(100), 9.97(7), 9.18(1), 7.31(2), 5.15(17), 4.84(13), 4.55(7), 4.23(5), 3.68(6), 3.56(14), 3.44(59), 3.30(30), 3.15(28), 2.15(18), 1.88(3), 1.83(6); (PZ-2) $(d(Å)$ $(I/I_0))$ 13.48(100), 7.34(24), 3..55(55), 3.40(85), 1.85(20).

PZ-1 represents a hydrous layer silicate structure related to the known mineral magadiite.

Synthesis

ORGANIC ADDITIVE
$R_3P-(CH_2)_n-PR_3X_2$

PZ-1 crystallizes from an aluminosilicate gel with a SiO_2/Al_2O_3 ratio between 20 and 250 and an R/SiO_2 ratio between 0.01 and 0.2. The organic used in this synthesis is $Bu_3P-(CH_2)_3-PBu_3Br_2$. The preferred inorganic cation is sodium, where the Na_2O/SiO_2 ratio is between 0.01 and 0.5. Crystallization occurs after 14 hours at 175°C. After treatment with either HCl or NH_4Cl followed by thermal treatment, PZ-2 is obtained.

Adsorption

Adsorption properties reported for PZ-1 and PZ-2 (EP 91,537(1983)):

	n-Hexane	Benzene	Water
PZ-1:			
wt%	1.0	3.0	7.0
PZ-2:			
wt%	6.0	4.0	5.0

PZ-2

See PZ-1.

Q–R

Q

This term designates two different species; it can refer to Linde Q or to species Q.

Q-[Br]

See species Q.

R (FAU)
(JCS London 1959:195((1959))

Related Materials: faujasite

Zeolite R, identified in 1959, was identified as the synthetic faujasite related to Linde X.

See also (Type) R (CHA)

Outdated designation for a synthetic chabazite as coined by Union Carbide in the early 1970s.

RbAlSiO₄ (ABW)

See A (BW).

Rb-D
(JCS, Dalton Trans. 2534(1972))

Related Materials: K-F
Z

Synthesis

Rb-D crystallizes from a reactive rubidium-containing aluminosilicate gel at a SiO_2/A_2O_3 of 2, using metakaolin, silica, and RbOH as the starting materials and temperatures of 80°C.

Rb-M (MER)
(JCS Dalton Trans. 2534(1972))

Related Materials: Linde W
merlinoite

Synthesis

Rb-M crystallizes from a rubidium-containing aluminosilicate gel with a SiO_2/Al_2O_3 ratio between 4 and 6, using metakaolin, silica, and RbOH. Crystallization occurs after 7 days at 80° C.

Reissite (EPI)

Obsolete synonym for epistilbite.

Rho (RHO)
(US3,720,750(1973))
Exxon

Related Materials: pahasapaite
LZ-214
beryllophosphate-R

Structure

(JACS 112:4821(1990))

UNIT CELL COMPOSITION:
$(Na,Cs)_{12}[Al_{12}Si_{36}O_{96}]*44H_2O$
SYMMETRY: cubic
SPACE GROUP: Im3m
UNIT CELL CONSTANTS (Å): $a = 15.0$
(dehydrated NH_4^+ form) $a = 14.821$
(dehydrated H^{+4} form) $a = 14.982$
(Li^+ form) $a = 14.3390(5)$
(Li,Cs^+ form) $a = 14.4925(5)$

(Ag^+ form) $a = 14.2251(6)$
(CaH^+ form, 400°C) $a = 13.970(5)$
(Sr^{++} form, 250°C) $a = 14.045(1)$
(Ba^{++} form, 200°C) $a = 14.184(2)$
(Cd^{++} form, 350°C) $a = 14.488(3)$
(Na,Cs^+ form, 25°C) $a = 15.031(1)$
PORE STRUCTURE: eight-member rings, 3.6 Å diameter

X-Ray Powder Diffraction Data: $(d(Å) (I/I_0))$
10.33(100), 7.25(3), 5.95(30), 5.14(1), 4.61(2), 4.20(13), 3.90(16), 3.64(1), 3.44(51), 3.26(52), 3.11(24), 2.979(32), 2.862(8), 2.667(22), 2.578(1), 2.501(4), 2.433(4), 2.368(2), 2.309(7),

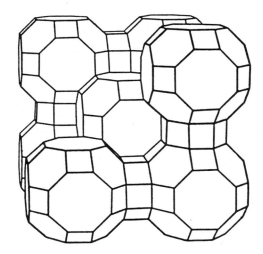

Rho Fig. 1S: Framework topology.

2.254(3), 2.201(9), 2.154(2), 2.067(6), 2.025(3), 1.987(2), 1.951(4), 1.918(2), 1.855(3), 1.824(3), 1.798(4), 1.771(5), 1.746(0.5), 1.722(4), 1.698(6), 1.675(0.5), 1.853(2), 1.633(0.5), 1.613(2), 1.594(1), 1.575(2), 1.556(1), 1.540(1), 1.506(2), 1.490(3), 1.475(4).

The rho structure consists of a body-centered cubic arrangement of alpha cages joined together through double eight-member rings. The three-dimensional topology contains intersecting eight-member-ring channels. Distortion of the alpha cages is observed in the NH_4^+ and Na^+/Cs^+ forms of this structure. The eight-member rings in the H^+ form are completely open, whereas the NH_4^+ cations are all located at the center of the eight-member ring (*Zeo.* 4:51(1984)). The Ca,D-rho shows the largest reported deviation from Im3m symmetry for this structure ($a = 13.9645(7)$A). The calcium atoms are located in the center of the double eight-ring, distorting the framework to generate a tetrahedral environment (*JACS* 112:4821(1990)). In the Li^+-exchanged form, the lithium cations occupy sites in the six-member ring, leaving the eight-ring pores open (*JACS* 112:4821(1990)).

Synthesis

ORGANIC ADDITIVES
none

Summary of rho structural data (*JACS* 112:4821(1990)):

Form	Temp., K	a	<T–O>	<T–O–T>
CaD	298	13.9645	1.651	131.5
Sr	473	14.045		
$CaND_4$(II)	298	14.110	1.642	135.6
Ba	473	14.184		
Ag	298	14.2251	1.649	133.4
Li	298	14.3390	1.66	
$LiND_4$	298	14.372		
Rb	298	14.375	1.648	136.5
$CaND_4$(I)	298	14.410	1.642	135.6
$(ND_4)_6$	11	14.4245	1.640	137.5
$(ND_4)_6$	11	14.425	1.637	137.9
Tl	298	14.461	1.644	138.9
Na	298	14.4848	1.655	136.3
Cd	623	14.488		
LiCs	298	14.4925	1.645	137.1
$(ND_4)_6$	298	14.5264	1.631	140.7
D	11	14.601	1.635	140.0
K	298	14.613	1.648	136.5
D	13	14.620		
CsD	298	14.6536	1.640	139.8
$NaCs(H_2O)$	298	14.6566	1.636	140.8
CsD	298	14.6652	1.638	140.8
NaCs	298	14.678	1.683	132.8
dD	294	14.694	1.632	143.0
CsD	493	14.7014	1.642	139.8
D	295	14.7237	1.630	143.5
$K(H_2O)$	298	14.7412		
D	423	14.7580	1.632	143.8
$(NH_4)_{12}$	373	14.821	1.632	143.8
D	13	14.850	1.610	145.8
D	573	14.8680	1.645	141.0
D	298	14.8803	1.632	144.5
CD_3OD	11	14.969		
H	773	14.982	1.610	149.2
Ca		14.995		
H	293	15.00	1.610	145.7
D_2O	423	15.027	1.655	142.5
$NaCs(H_2O)$	298	15.031	1.660	139.6
D	298	15.0387	1.624	146.8
D	623	15.0620	1.625	147.5
D	298	15.0686	1.627	146.8
D	298	15.0696	1.626	147.1
D	623	15.0799	1.627	147.0
D	298	15.0976	1.629	146.8

Zeolite rho crystallizes from reactive sodium/cesium aluminosilicate gels at 100° after 6 days. Prior to the hydrothermal synthesis, the gel is aged at room temperature for 6 days.

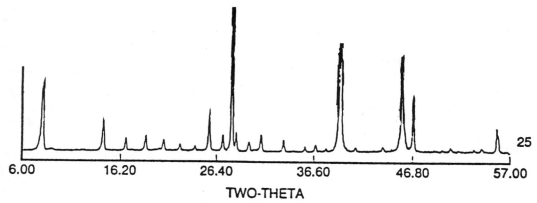

Rho Fig. 2:   X-ray powder diffraction pattern for zeolite Ca,ND₄-rho (*JACS* 112:4821(1990)). (Reproduced with permission of the American Chemical Society)

Positional parameters for NH_4^+ and H^+ forms of zeolite rho (*Zeo.* 4:51(1984)):

	Point symmetry	x	y	z	Population	Multiplicity
(a) NH₄-rho ($I\bar{4}3m$)						
T	1	0.2648(2)	0.1135(2)	0.4103(2)	48	48
O(1)	1	0.2109(3)	0.0200(4)	0.3868(3)	48	48
O(2)	m	0.1936(4)	0.1936(4)	0.3932(7)	24	24
O(3)	m	0.3623(3)	0.1191(5)	0.3623(3)	24	24
N	mm	0	0	0.3842(8)	12	12
X	$\underline{4}2m$	0	0	1/2	2.4(6)	6
(b) H-rho (Im3m)						
T	2	1/4	0.1030(2)	0.3970(2)	48	48
O(1)	m	0.2191(3)	0	0.3831(3)	48	48
O(2)	m	0.1668(2)	0.1668(2)	0.3783(3)	48	48

Numbers in parentheses are the esd values in the units of the least significant digit given. Population parameters are given as the number of atoms or ions per cell.

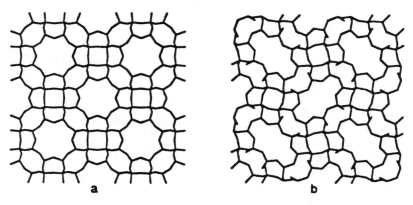

Rho Fig. 3:   Framework structural changes from (a) dehydrated H-rho to (b) Ca,D-rho 112:4821(1990)). (Reproduced with permission of the American Chemical Society)

Thermal Properties

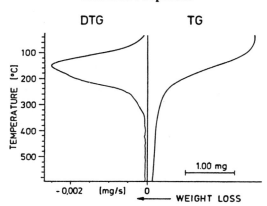

Rho Fig. 4:   Thermogravimetric (TG) and differential thermogravimetric (DTG) curves for Na,Cs-rho (*Z. Kristallogr.* 187:253(1989)). (Reproduced with permission of Oldenbourg Verlag Publishers)

Rhodesite
(natural)

Related Materials: delhagelite

Rhodesite is a natural aluminosilicate of the delhagelite group of minerals. This structure was successfully synthesized by Babaev, Ganbarov, and Ragimov of the Academy of Sciences of the Azerbaijan SSR.

Structure

(*RR 8th IZC* 46(1989))

SYMMETRY: orthorhombic (synthetic)
SPACE GROUP: Pman
UNIT CELL CONSTANTS (Å): $a = 23.42$
                          $b = 6.54$
                          $c = 7.05$

Rhodesite is structurally related to the dehayelite group of minerals, which have eight-member-ring cavities approximately 4 A in diameter. The large cations that fill the channels for this mineral are potassium. These cations also are found in delhayelite, hydrodelhayelite, and mauntainite. Ba^{+2} are found in macdonaldite. (*Z. Kristallogr.* 149:155(1979)

Synthesis

ORGANIC ADDITIVES
none

This mineral is found in deposits in South Africa as well as California. Rhodesite can be synthesized from a batch composition of: x KOH : 7.64 Na_2O : Al_2O_3 : 32.26 SiO_2 : 5.36 CaO : 2.34 MgO, over a temperature range of 160 to 200°C for 200 hours. The crystals appear as aggregates of radiant white fibers (*RR 8th IZC* 46(1989)).

Thermal Properties

Thermally, dehydration occurs in three steps, at 20 to 150°C, 240 to 305°C, and 605 to 820°C, with decomposition of the structure occurring between 605 and 820°C.

Adsorption

Synthetic rhodesite adsorbs 3.75 mmol/g of CO_2 (*RR 8th IZC* 46(1989)).

Roggianite (ROG)
(natural)

Roggianite is a natural hydrated calcium aluminosilicate found in deposits in Italy. Thomsonite is generally observed to be associated with this material (*Proc. 5th IZC* 205(1980)).

Structure

(*Proc. 5th IZC* 205(1980))

CHEMICAL COMPOSITION: $Ca_{16}[Al_{16}Si_{32}O_{88}(OH)_{16}]$ $(OH)_{16}$*26 H_2O
SYMMETRY: tetragonal
SPACE GROUP: I4/mcm
UNIT CELL CONSTANTS (Å): $a = 18.3$
                          $c = 9.2$
PORE STRUCTURE: 12-member ring with diameter of 4.2 Å

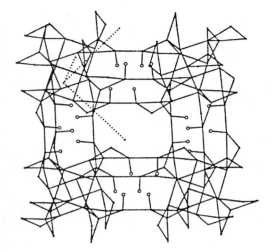

Roggianite Fig. 1S: Framework topology.

X-ray Powder Diffraction Data: ($d(\text{Å})$ (I/I_0))
13.040(100), 9.230(49), 6.525(5), 6.150(42), 5.834(20), 4.589(2), 4.472(2), 4.339(10), 4.111(9), 3.752.(7), 3.605(41), 3.411(68), 3.246(12), 3.198(34), 3.154(43), 2.905(14), 2.869(24), 2.833(23), 2.740(3), 2.652(17), 2.623(6), 2.5985(10), 2.5481(21), 2.4549(10), 2.3530(6), 2.2962(13), 2.2790(8), 2.2600(5), 2.2281(9), 2.1660(4), 2.1348(6), 2.0535(9), 2.0280(7), 1.9436(7), 1.9375(13), 1.8546(17), 1.8253(8), 1.8808(7), 1.7843(2), 1.7284(5), 1.7202(10), 1.7069(7), 1.6698(3), 1.6600(3), 1.5314(3), 1.5200(6), 1.4806(6), 1.4490(3), 1.4420(2), 1.4088(2), 1.3966(1), 1.3766(7).

This mineral has a nearly zeolitic structure with only apical hydroxyl ions of the Si_2O_7 groups contained within the structure, which interrupt the three-dimensional linkage of the Si–Al tetrahedra. Each four-member ring of one layer is connected to two rings, one from the preceding layer and one from the subsequent one, through four Al tetrahedra. This results in cages defined by two four- and four six-member rings. Each of the cages shares two planar four-member rings, which are rotated by 45°. Two identical cages are located above and below, forming columns parallel to c. Each of these columns is laterally linked through Si_2O_7 groups to four neighboring columns, thus forming an open three-dimensional network.

Intersecting channels have a free diameter of 4.2 A and run parallel to the c axis. Six- and ten-member rings are present; however, they are bisected by Si_2O_7 groups and obstructed by Ca cations. These Ca cations lie in the cavities where two tetrahedral framework is interrupted and are coordinated by two oxygen atoms and four hydroxyl ions.

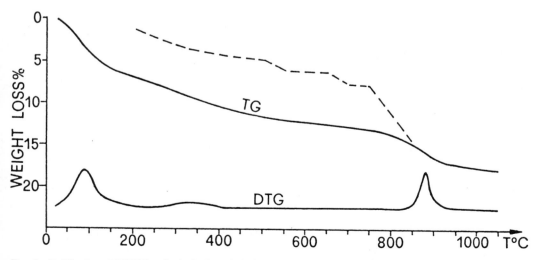

Roggianite Fig. 2:   TG/DTG and rehydration (dashed line) curves of roggianite (*Mineral. Mag.* 52:201(1988)).
(Reproduced with permission of the Mineralogical Society)

The distribution of Si and Al in the structure appears to be ordered in the three independent tetrahedral positions. The average interatomic distances are Si1–O = 1.62 Å, Si2–O = 1.63 Å, and Al–O = 1.715 Å.

Thermal Properties

The presence of the hydroxyls in this structure can be readily observed by the DTA curve and by the ignition loss. There is a small endothermic peak at 113°C and a weight loss of 4.49% at 100°C. A sharp endothermic peak is observed at 874°C with a weight loss of 17.24% at 1000°C (*Mineral. Mag.* 52:201(1988)).

Atomic coordinates for roggianite (*Proc. 5th IZC* 205(1980)):

	x	y	z
Ca	0.1771(2)	0.1771(2)	1/4
Si1	0.1047(5)	0.2809(5)	0
Si2	0.1161(5)	0.0411(4)	0
Al	0.2169(4)	0	1/4
O1	0.1339(11)	0.3661(11)	0
O2	0.1641(7)	0.0581(8)	0.1469(15)
O3	0.0586(8)	0.2671(8)	0.1446(17)
O4	0.0429(12)	0.0939(12)	0
OH1	0.1740(13)	0.2233(12)	0
OH2	0.2942(8)	0.2058(8)	0.1512(23)
H$_2$O1*	0.4258(30)	0.0742(30)	0.1950(76)
H$_2$O2*	0.3853(78)	0.0773(77)	0

*H$_2$O occupancy factors are: 69(7)5 for H$_2$O1 and 34(6)% for H$_2$O2, respectively.

S

S
(US3,054,657(1962))
Union Carbide Corporation

Related Material: gmelinite type

Structure

CHEMICAL COMPOSITION: 0.9 Na$_2$O : Al$_2$O$_3$: 4.6–5.9 SiO$_2$: 6–7 H$_2$O
DENSITY: 2.08 g/cc

X-Ray Powder Diffraction Data: (d(Å) (I/I_0))
11.8(77), 7.73(19), 7.16(100), 5.96(9), 5.03(72), 4.50(46), 4.12(79), 3.97(20), 3.44(62), 3.305(13), 3.236(23), 2.973(80), 2.858(47), 2.693(19), 2.603(39), 2.126(11), 2.089(39), 1.910(12), 1.809(40), 1.722(32).

Zeolite S is thought to be a gmelinite-type structure.

Synthesis

ORGANIC ADDITIVES
none

Zeolite S crystallizes from a reactive sodium aluminosilicate gel with a composition of: 3.2 Na$_2$O : 6 SiO$_2$: 80 H$_2$O. Crystallization occurs after 17 hours at 100°C.

Infrared Spectrum

Mid-infrared Vibrations (cm^{-1}): (SiO$_2$/Al$_2$O$_3$ = 2.5) 1140wsh, 1020s, 770vwsh, 722mw, 690vwsh, 631-595mb, 518mb, 448m, 424ms, 370vwsh (*ACS* 101:201(1971)).

Adsorption

Adsorptive properties of activated synthetic zeolite S (US3,054,657 (1962)):

Adsorbate	Pressure, mm Hg	Temp., °C	Wt% adsorbed
Water	0.012	25	10.4
	0.076	25	15.2
	4.5	25	21.5
	24	25	28.0
	2	100	2.8
	4.5	100	4.2
	14	100	11.2
	22	100	13.4

(continued)

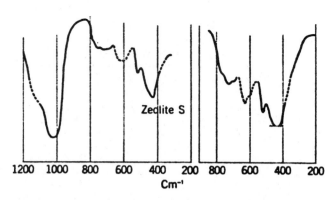

S Fig. 1: Mid-infrared spectrum of zeolite S (*Advances in Chemistry Series* 101, *Molecular Sieve Zeolites*, Flanigan, E. M, Sand, L. B., eds., 201(1971). (Reproduced with permission of the American Chemical Society)

Adsorptive properties of activated synthetic zeolite S (US3,054,657 (1962)):

Adsorbate	Pressure, mm Hg	Temp., °C	Wt% adsorbed
Agron	1.8	−196	4.7
	24	−196	7.8
	130	−196	9.0
CO_2	6	25	5.4
	24	25	8.7
	100	25	11.7
	700	25	13.9
Nitrogen	110	−196	4.4
	8.5	−196	4.7
	700	−196	6.2
	700	−78	2.5
Propane	700	25	2.9
Propylene	700	25	4.7
Benzene	65	25	<1
Ammonia	0.5	25	5.5
	46	25	7.7
	700	25	9.4
SO_2	0.02	25	7.5
	7	25	16.1
	100	25	18
	700	25	21.2
Oxygen	140	−196	4.9
Krypton	18	−183	5.5
n-pentane	404	25	1.3

Sacrofanite (natural)

Related Materials: cancrinite

Sacrofanite is a cancrinite-type phase with unit cell parameters of $a = 12.865$ and $c = 74.24$ Å, with an unknown layer stacking sequence (*Proc. NATO Study Institute of Feldspars and Feldspathoids*, Rennes, France, 1983).

SAPO
US4,440,871(1984))
Union Carbide Corporation

Chemical Composition: $mR : (Si_xAl_yP_z)O_2$. See SAPO Figs. 1 & 2.

SAPO-5 (AFI)
(US4,440,871(1984))
Union Carbide Corporation

Related Materials: $AlPO_4$-5

Structure

Chemical Composition: silicoaluminophosphate. See SAPO.

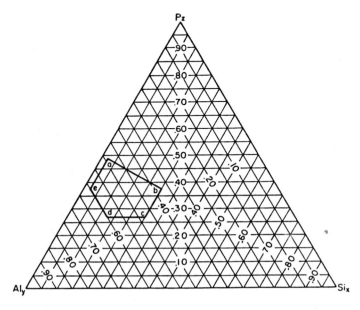

SAPO Fig. 1:   Composition range of SAPO molecular sieve crystals (US4,440,871(1984)).

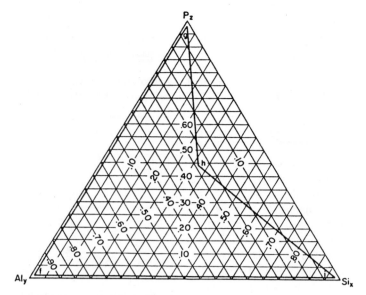

SAPO Fig. 2:   Composition range of the reaction mixture for SAPO molecular sieves (US4,440,871(1984)).

X-Ray Powder Diffraction Data: $(d(\text{Å}) \, (I/I_0))$ 11.8(100), 6.86(12), 5.91(26), 4.46(61), 4.21(53), 3.96(77), 3.59(5), 3.43(30), 3.07(17), 2.96(19), 2.66(5), 2.59(16); (after thermal treatment to 600°C) 11.79(100), 6.81(27), 5.91(11), 4.46(42), 4.17(62), 3.93(96), 3.56(4), 3.427(44), 3.058(23), 2.959(23), 2.652(6), 2.592(17).

SAPO-5 exhibits properties characteristic of a large-pore molecular sieve. For a description of the structure see $AlPO_4$-5.

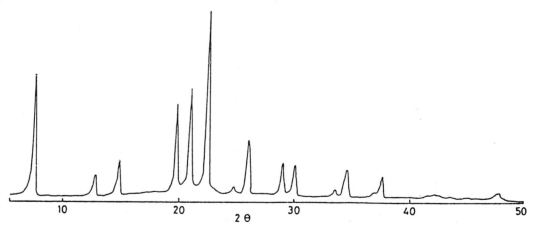

SAPO-5 Fig. 1:   X-ray powder diffraction pattern for SAPO-5. (Reproduced with permission of M. Chaudhary, Georgia Tech)

Synthesis

ORGANIC ADDITIVES
tetraethylammonium$^+$
tetrapropylammonium$^+$
tri-*n*-propylamine
tetrabutylammonium$^+$
tetrapropylammonium$^+$/tetramethylammonium$^+$
diethylethanolamine (US4,440,871(1984))

SAPO-5 is prepared from a reactive gel composition with the following composition range: R : 0.3–0.6 SiO_2 : Al_2O_3 : P_2O_5 : 40–60 H_2O. This structure does not appear to be strongly selective for the organic used in the crystallization mixture, as several different amines have been successfully used. Crystallization occurs between 150 and 225°C after 24 to 168 hours. Higher temperatures promote more rapid crystallization. Impurity phases generally encountered are of the quartz type, which are observed with extended crystallization times, especially at the higher temperatures. Entrapment of the organic additive generally is encountered.

Thermal Properties

This material appears thermally stable, as heating to 600°C results in little change in the X-ray powder diffraction pattern.

Absorption

Adsorption properties of SAPO-5 (US4,440,871 (1984)):

	Kinetic diameter, Å	Pressure, Torr	Temp., °C	Wt.% adsorbed
O_2	3.46	100	−183	14.5
O_2		750	−183	19.8
Cyclohexane	6.0	60	24	10.9
· Neopentane	6.2	743	24	7.6
H_2O	2.65	4.6	24	14.7
H_2O		20	24	31.3

SAPO-8 (AET)

SAPO-8 has been identified as the thermally treated form of MCM-9 (*Abstracts MRS Meeting,* Fall 1990). Conversion of MCM-9 to SAPO-8 occurs at temperatures above 100°C. It is topologically related to $AlPO_4$-8.

SAPO-11 (AEL)
(US4,440,871(1984))
Union Carbide Corporation

Related Materials: $AlPO_4$-11

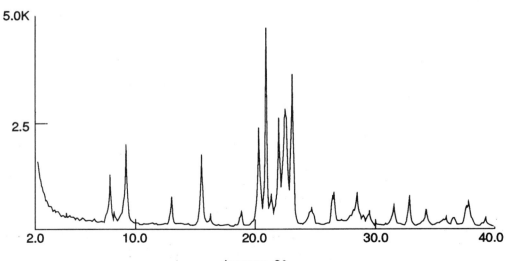

SAPO-11 Fig. 1:   X-ray powder diffraction pattern for SAPO-11. (Reproduced with permission of M. Chaudhary, Georgia Tech

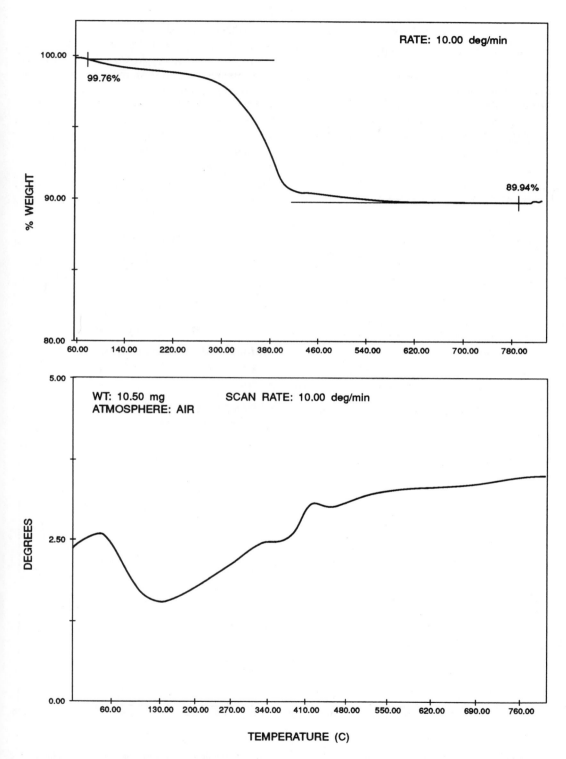

SAPO-11 Fig. 2:   TGA (top)/DTA (bottom) of SAPO-11 prepared in the presence of di-*n*-propylamine. (Reproduced with permission of M. Chaudhary, Georgia Tech)

Structure

Chemical Composition: silicoaluminophosphate. See SAPO.

X-Ray Powder Diffraction Data: $(d(\text{Å})\ (I/I_0))$
10.98(20), 9.41(36), 6.76(13), 5.66(23), 5.44(3), 4.68(5), 4.35(36), 4.23(100), 4.02(54), 3.95–3.92(56), 3.84(66), 3.63–3.60(8), 3.38(19), 3.28(1), 3.121(14), 3.079(3), 3.033(6), 2.840(8), 2.730(13), 2.629(8), 2.512(3), 2.475(3), 2.389–2.380(10), 2.292(3), 2.238(2), 2.113(6), 2.019(4), 1.941(1), 1.870(2), 1.807(3), 1.684(4).

SAPO-11 exhibits properties characteristic of a medium-pore molecular sieve. See AlPO$_4$-11 for a description of the framework topology.

Synthesis

ORGANIC ADDITIVES
di-*n*-propylamine
TBA$^+$/di-*n*-propylamine

SAPO-11 crystallizes from a reactive aluminophosphate gel with a batch composition of: 1.9 Pr$_2$NH : 0.1 SiO$_2$: Al$_2$O$_3$: P$_2$O$_5$: 42 H$_2$O. Crystallization requires an oven temperature of 150 to 200°C, with crystals resulting after 24 to 133 hours. At the lower temperatures, AlPO$_4$-H3 is observed as a common impurity phase. SAPO-11 also crystallizes as large crystals from mixtures containing HF. Depending on the HF/Al$_2$O$_3$ and the SiO$_2$/Al$_2$O$_3$, SAPO-31 and SAPO-41 also appear as either a minor or a major phase in this system.

Thermal Properties

SAPO-11 retains its framework structure upon thermal treatment to 550°C, which is necessary to remove the organic.

Infrared Spectrum

Mid-infrared Vibrations (cm^{-1}): 1125(s), 740(m), 710(m), 590(w), 550(m), 470(ms) (From author).

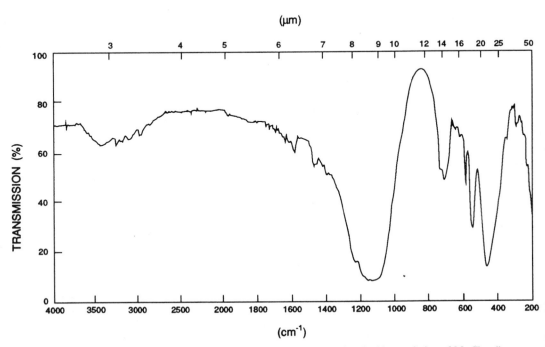

SAPO-11 Fig. 3:   Infrared spectrum of SAPO-11, as-synthesized. (Reproduced with permission of M. Chaudhary, Georgia Tech)

Adsorption

Adsorption properties of SAPO-11:

Adsorbate	Kinetic diameter, Å	Pressure, Torr	Temp., °C	Wt% adsorbed
O_2	3.46	102	−183	8.4
O_2		743	−183	10.9
Cyclohexane	6.00	52	24.6	4.5
Neopentane	6.2	300	24.8	0.8
H_2O	2.65	4.6	23.9	10.5
H_2O		20.2	23.2	15.4

SAPO-16 (AST)
(US4,440,871(1984))
Union Carbide Corporation

Related Materials: $AlPO_4$-16

Structure

Chemical Composition: silicoaluminophosphate. See SAPO.

X-Ray Powder Diffraction Data: $(d(\text{Å})\ (I/I_0))$
7.73(54), 5.11(4), 4.72(51), 4.03(100), 3.345(20), 3.058(6), 2.993(25), 2.739(3), 2.578(4), 2.365(8), 2.259(3), 2.040(2), 1.877(6), 1.859(1), 1.746(2), 1.675(2).

SAPO-16 exhibits properties characteristic of a small-pore molecular sieve. See $AlPO_4$-16 for a description of the framework topology.

Synthesis

ORGANIC ADDITIVES
quinuclidine

SAPO-16 crystallizes from a batch composition of: 1.0 Q : Al_2O_3 : P_2O_5 : 0.3 SiO_2 : 50 H_2O. Crystallization occurs at temperatures of 200°C after 48 hr to 338 hours. The source of aluminum is $Al(OC_3H_7)_3$, and that of phosphorus is H_3PO_4.

SAPO-17 (ERI)
(US4,440,871(1984))
Union Carbide Corporation

Related Materials: $AlPO_4$-17
erionite

Structure

Chemical Composition: silicoaluminophosphate. See SAPO.

X-Ray Powder Diffraction Data: $(d(\text{Å})\ (I/I_0))$
11.4*(100), 9.03(5), 6.61(60), 5.70*(65), 5.31(5), 4.93(25), 4.51(10), 4.31(100), 4.15(sh), 3.80(20), 3.507(15), 3.302(24), 3.255(5), 3.110(5), 2.921(sh), 2.853(10), 2.797(20), 2.683(5), 2.491(10), 2.465(10), 2.254–2.238(40), 1.977(5), 1.834(5), 1.794(15), 1.701(5), 1.656(5) (*may contain impurity).

SAPO-17 exhibits properties characteristic of a small-pore erionite molecular sieve. See $AlPO_4$-17 for a description of the framework topology.

Synthesis

ORGANIC ADDITIVES
quinuclidine
cyclohexylamine

SAPO-17 crystallizes from a batch composition of: 1.0 Q : 1.3 Al_2O_3 : P_2O_5 : 0.6 SiO_2 : 60 H_2O. Crystallization occurs at temperatures of 200°C after 50 hours (using cyclohexylamine as the organic additive) to 338 hours (with quinuclidine). The source of aluminum is $Al(OC_3H_7)_3$, and that of phosphorus is H_3PO_4.

Thermal Properties

SAPO-17 is stable to 550°C, the temperature necessary to remove the organic from the pores of the structure.

Adsorption

Adsorption capacities measured for SAPO-17 (US4,440,871 (1984)):

Adsorbate	Kinetic diameter, Å	Pressure, Torr	Temp., °C	Wt% adsorbed
O_2	3.46	98.5	−183	21.5
O_2		740	−183	29.4
n-Hexane	4.3	53.5	24	10.3
H_2O	2.65	4.6	23	25.2
H_2O		19.4	24	35.0
Isobutane	5.0	400	24	1.1

SAPO-20 (SOD)
(US4,440,871(1984))
Union Carbide Corporation

Related Materials: AlPO₄-20
sodalite

Structure

Chemical Composition: silicoaluminophosphate. See SAPO.

X-Ray Powder Diffraction Data: $(d(\text{Å})\,(I/I_0))$
6.28(40),   4.46(41),   4.00(5),   3.65(100),
3.164(13),   2.840(11),   2.585(14),   2.398(2),
2.243(4), 2.110(6), 1.908(5), 1.759(10).

SAPO-20 exhibits properties characteristic of the very small-pore system of sodalite.

Synthesis

ORGANIC ADDITIVES
TMA⁺

SAPO-20 can be crystallized from a batch composition of: 0.75 TMA_2O : Al_2O_3 : P_2O_5 : SiO_2 : 50 H_2O. Crystallization occurs at temperatures of 125°C after 68 hours. The source of aluminum is pseudoboehmite, and that of phosphorus is H_3PO_4.

Thermal Properties

SAPO-20 is stable to 550°C, the temperature necessary to remove the organic from the pores of the structure.

Infrared Spectrum

See SAPO-20 Figure 1.

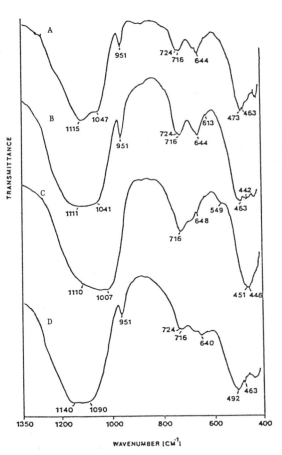

SAPO-20 Fig. 1:   Mid-infrared spectra of the sodalite SAPO molecular sieves: A, B, C, and D with Si/Al ratios of 0.1/0.55, 0.48/0.35, 0.41/0.44, and O (*JACS* 110,2127(1988)). (Reproduced with permission of the American Chemical Society)

Adsorption

Adsorption capacities measured for SAPO-20:

Adsorbate	Kinetic diameter, Å	Pressure, Torr	Temp., °C	Wt% adsorbed
O_2	3.46	100	− 183	3.2
O_2		761	− 183	12.8
H_2O	2.65	4.6	25.1	14.6
H_2O		20.0	25.0	25.1

SAPO-31 (ATO)
(US4,440,871(1984))
Union Carbide Corporation

Structure

Chemical Composition: silicoaluminophosphate. See SAPO.

X-Ray Powder Diffraction Data: $(d(Å)\ (I/I_0))$
11.5(sh), 10.4(100), 9.94(sh), 9.21(sh), 9.03(3), 6.89(1), 6.03(7), 5.50(3), 5.20(10), 4.81(2), 4.37(34), 4.15(sh), 4.03(37), 3.93(81), 3.81(3), 3.548(3), 3.466(4), 3.198(11), 3.008(8), 2.885(1), 2.823(18), 2.736(1), 2.557(7),

2.481(2), 2.417(2), 2.392(2), 2.350(2), 2.292(3), 2.276(1), 2.238(3), 2.094(1), 2.038(1), 2.014(2), 1.929(3), 1.910(2), 1.873(2), 1.852(1), 1.797(1), 1.771(4), 1.653(1).

SAPO-31 exhibits properties characteristic of a large-pore molecular sieve.

Synthesis

ORGANIC ADDITIVES
di-*n*-propylamine

SAPO-31 can be crystallized from a batch composition of: DPA : 0.9 Al_2O_3 : P_2O_5 : 0.1 SiO_2 : 49 H_2O. Crystallization occurs at temperatures of 150 to 200°C after 52 to 133 hours. The source of silica is aqueous silica sol or fumed silica, the source of aluminum is aluminum isopropoxide or pseudoboehmite, and the source of phosphorus is phosphoric acid.

Thermal Properties

SAPO-31 is stable to 550°C for 2 hours, the temperature necessary to remove the organic from the pores of the structure.

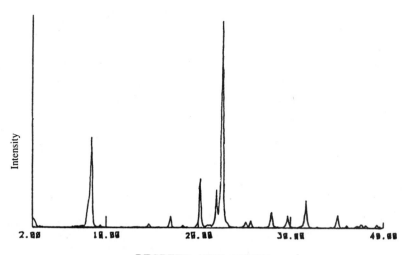

SAPO-31 Fig. 1:   X-ray powder diffraction pattern for SAPO-31. ((Reproduced with permission of M. Chaudhary, Georgia Tech)

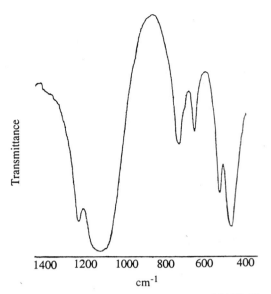

SAPO-31 Fig. 2: Mid-infrared spectrum of SAPO-31. (Reproduced with permission of M. Chaudhary, Georgia Tech)

Infrared Spectrum

Mid-infrared Vibrations (cm^{-1}): 1240(s), 1130(s), 740(m), 650(m), 530(m), 470(s), 390(w).

Adsorption

Adsorption capacities measured for SAPO-31:

Adsorbate	Kinetic diameter, Å	Pressure, Torr	Temp., °C	Wt% adsorbed
O_2	3.46	99	−183	8.8
O_2		740	−183	
Cyclohexane	6.0	49	25	7.2
Neopentane	6.2	400	24	5.9
H_2O	2.65	4.6	23	6.2
H_2O		19.4	24	21.1

SAPO-34 (CHA)
(US4,440,871(1984))
Union Carbide Corporation

Related Materials: AlPO$_4$-34
                      chabazite

Structure

Chemical Composition: silicoaluminophosphate. See SAPO.

X-Ray Powder Diffraction Data: (d(Å) (I/I_0)) 9.21(100), 6.81(17), 6.30(23), 5.50(33), 4.97(75), 4.67(2), 4.29(99), 4.03(4), 3.85(10), 3.57(76), 3.43(19), 3.220(3), 3.170(12), 3.038(4), 2.912(67), 2.880(28), 2.763(2), 2.683(6), 2.596(14), 2.495(11), 2.270(4), 2.085(3), 1.910(6), 1.866(7), 1.852(5), 1.802(8), 1.722(6), 1.691(4), 1.645(4).

SAPO-34 exhibits properties characteristic of a small-pore molecular sieve.

Synthesis

ORGANIC ADDITIVES
TEA$^+$
i-PrNH$_2$

SAPO-34 can be crystallized from a batch composition of: 0.5 TEA$_2$O : Al$_2$O$_3$: P$_2$O$_5$: 0.6 SiO$_2$: 52 H$_2$O. Crystallization occurs at temperatures of 200°C after 48 hours or 150°C after 336 hours. The sources of aluminum are pseudoboehmite or aluminum isopropoxide, that of phosphorus is H$_3$PO$_4$, and that of silica is aqueous silica sol or fumed silica.

Thermal Properties

SAPO-34 is stable to 550°C, the temperature necessary to remove the organic from the pores of the structure.

Adsorption

Adsorption capacities of SAPO-34:

Adsorbate	Kinetic diameter, Å	Pressure, Torr	Temp., °C	Wt% adsorbed
O_2	3.46	104	−183	25.1
O_2		746	−183	36.6
n-Hexane	4.3	46	23.4	11.0
H_2O	2.65	4.6	23.0	30.1
H_2O		19.5	22.8	42.3

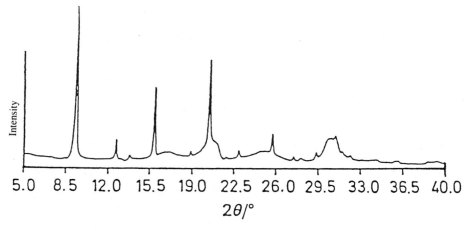

SAPO-34 Fig. 1:   X-ray powder diffraction pattern for SAPO-34 (*JCS Faraday Trans.* 86:425(1990)). (Reproduced with permission of the Chemical Society)

SAPO-35 (LEV)
(US4,440,871(1984))
Union Carbide Corporation

Related Materials: levyne

Structure

Chemical Composition: silicoaluminophosphate. See SAPO.

X-Ray Powder Diffraction Data: $(d(\text{Å})\,(I/I_0))$
13.0(2),   10.78(2),   10.46(14),   8.04(100),
7.76(17), 6.53(89), 5.50(4), 5.10(24), 4.75(3),
4.23(29), 4.00(63), 3.87(4), 3.77(15), 3.555(13),
3.427(9), 3.267(20), 3.121(42), 3.028(10),
2.921(2), 2.818(6), 2.763–2.747(32), 2.592(7),
2.536(4), 2.475(2), 1.899(2), 1.768(3).

SAPO-35 exhibits properties characteristic of a small-pore molecular sieve.

Synthesis

ORGANIC ADDITIVES
quinuclidine

SAPO-35 can be crystallized from a batch composition of: 1.0 quin : Al_2O_3 : P_2O_5: 0.6 SiO_2 : 60 H_2O. Crystallization occurs at temperatures of 150°C after 48 hours. The source

of aluminum is aluminum isopropoxide, that of phosphorus is H_3PO_4, and that of silica is aqueous silica sol.

Thermal Properties

SAPO-35 is stable to 600°C, the temperature necessary to remove the organic from the pores of the structure.

Adsorption

Adsorption capacities measured for SAPO-35:

Adsorbate	Kinetic diameter, Å	Pressure, Torr	Temp., °C	Wt% adsorbed
O_2	3.46	98	−183	15.3
O_2		746	−183	30.3
n-Hexane	4.3	48	25	0.7
Isobutane	5.0	101	25	10.2
H_2O	2.65	4.6	22	22.2
H_2O		19	24	47.7

SAPO-37 (FAU)
(US4,440,871(1984))
Union Carbide Corporation

Related Materials: faujasite

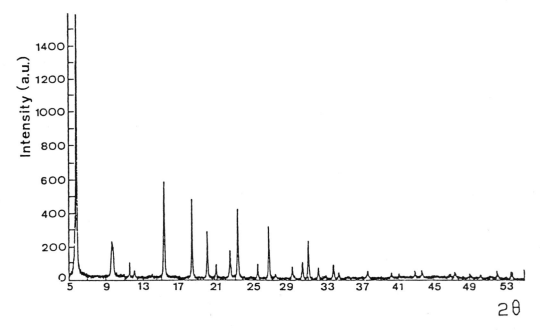

SAPO-37 Fig. 1: X-ray powder diffraction pattern of as-synthesized SAPO-37 (*SSSC* 49A:215(1989)). (Reproduced with permission of Elsevier Publishers)

Structure

Chemical Composition: silicoaluminophosphate. See SAPO.

X-Ray Powder Diffraction Data: $(d(\text{Å})\ (I/I_0))$
14.25(100), 8.74(22), 7.44(5), 5.68(42), 4.80(34), 4.40(16), 4.19(4), 3.92(11), 3.79(39), 3.59(1), 3.47(6), 3.314(27), 3.232(2), 3.038(7), 2.921(9), 2.867(18), 2.780(5), 2.714(2), 2.644(7), 2.607(3), 2.380(6), 2.233(2), 2.191(2), 2.099(1), 2.062(3).

SAPO-37 exhibits properties characteristic of the large-pore faujasite molecular sieve.

Synthesis

ORGANIC ADDITIVES
TPA^+/TMA^+

SAPO-37 can be crystallized from a batch composition of: 1.0 TPA_2O : 0.025 TMA_2O : Al_2O_3 : P_2O_5 : 0.4 SiO_2 : 50 H_2O. Crystallization occurs at temperatures of 200°C after 24 hours.

The source of aluminum is pseudoboehmite, that of phosphorus is H_3PO_4, and that of silica is fumed silica.

Thermal Properties

SAPO-37 is stable to 600°C, the temperature necessary to remove the organic from the pores of the structure.

Adsorption

Adsorption capacities measured for SAPO-37:

Adsorbate	Kinetic diameter, Å	Pressure, Torr	Temp., °C	Wt% adsorbed
O_2	3.46	100	-183	35.0
O_2		750	-183	42.9
Cyclohexane	6.0	60	24	23.2
Nepentane	6.2	743	24	14.8
H_2O	2.65	4.6	24	35.3

SAPO-40 (AFR)
(US4,440,871(1984))
Union Carbide Corporation

Structure

Chemical Composition: silicoaluminophosphate. See SAPO.

X-Ray Powder Diffraction Data: $(d(\text{Å})\ (I/I_0))$
11.63(18), 11.01(100), 7.12(18), 6.47(43), 6.32(12), 5.50(1), 5.11(7), 4.80(14), 4.50(6), 4.36(13), 4.15(10), 4.10(6), 3.88(4), 3.75(19), 3.68(5), 3.61(1), 3.264(22), 3.204(15), 3.176(4), 3.123(1), 2.946(3), 2.920(2), 2.878(3), 2.817(4), 2.769(3), 2.692(2), 2.654(2), 2.559(2), 2.507(3).

SAPO-40 exhibits properties characteristic of a large-pore molecular sieve.

Synthesis

ORGANIC ADDITIVES
TPA^+

SAPO-40 can be crystallized from a batch composition of: 1.0 TPA_2O : Al_2O_3 : P_2O_5 : 0.4 SiO_2 : 50 H_2O. Crystallization occurs at temperatures of 200°C after 96 hours. The source of aluminum is pseudoboehmite, that of phosphorus is H_3PO_4, and that of silica is fumed silica.

Thermal Properties

SAPO-40 is stable to 700°C, the temperature necessary to remove the organic from the pores of the structure.

Adsorption

Adsorption capacities measured for SAPO-40:

Adsorbate	Kinetic diameter, Å	Pressure, Torr	Temp., °C	Wt% adsorbed
O_2	3.46	100	−183	21.8
O_2		750	−183	24.4
Cyclohexane	6.0	60	24	8.0
Nepentane	6.2	743	24	5.1
H_2O	2.65	4.6	24	22.7
H_2O		20	24	31.5
Isobutane	5.0	697	24	7.0
SF_6	5.5	400	24	11.6

SAPO-41 (AFO)
(US4,440,871(1984))
Union Carbide Corporation

Structure

Chemical Composition: silicoaluminophosphate. See SAPO.

X-Ray Powder Diffraction Data: $(d(\text{Å})\ (I/I_0))$
13.19(24), 9.21(25), 6.51(28), 4.87(10), 4.33(10), 4.21(100), 4.02(82, 3.90(43), 3.85(30), 3.52(20), 3.47(28), 3.048(23), 2.848(10), 2.706(7), 2.392(15), 2.362(7), 2.276(5), 2.103(8), 1.855(8), 1.774(8).

SAPO-41 exhibits properties characteristic of a molecular sieve with pore openings between 6.00 and 6.2 Å.

Synthesis

ORGANIC ADDITIVES
TBA^+

SAPO-41 can be crystallized from a batch composition of: TBA_2O: Al_2O_3 : P_2O_5 : 0.4 SiO_2 : 98.7 H_2O. Crystallization occurs at temperatures of 200°C after 144 hours. The source of silica is fumed silica, that of aluminum is pseudoboehmite, and that of phosphorus is phosphoric acid.

Thermal Properties

SAPO-41 is stable to 700°C for 2 hours, the temperature necessary to remove the organic from the pores of the structure.

Adsorption

Adsorption capacities measured for SAPO-41:

Adsorbate	Kinetic diameter, Å	Pressure, Torr	Temp., °C	Wt% adsorbed
O_2	3.46	100	−183	9.3
O_2		750	−183	11.8
Cyclohexane	6.0	60	24	4.2
Nepentane	6.2	743	24	1.2
H_2O	2.65	4.6	24	10.4
H_2O		20.0	24	21.9

SAPO-42 (LTA)
(US4,440,871(1984))
Union Carbide Corporation

Related Material: Linde type A

Structure

Chemical Composition: silicoaluminophosphate. See SAPO.

X-Ray Powder Diffraction Data: $(d(Å) (I/I_0))$
11.9(71), 8.51(55), 6.97(61), 6.37(7), 5.42(31), 4.96(13), 4.13–4.06(68), 3.85(17), 3.67(100), 3.376(29), 3.255(83), 2.955(75), 2.876(15), 2.722(19), 2.660(9), 2.603(37), 2.491(19), 2.436(5), 2.347(5), 2.227(7), 2.158(11), 2.125(6), 2.096(3), 2.065(1), 2.036(9), 1.907(8), 1.884(4), 1.859(1), 1.841(6), 1.822(4), 1.746(3), 1.728(16), 1.710(2), 1.679(16), 1.664(2).

SAPO-42 exhibits properties characteristic of a small-pore type A molecular sieve.

Synthesis

ORGANIC ADDITIVES
TMA^+/Na^+
TEA^+

SAPO-42 can be crystallized from a batch composition of: 1.2 Na_2O : 1.1 TMA_2O : 1.66 Al_2O_3 : 0.66 P_2O_5 : 4.0 SiO_2 : 95 H_2O. Crystallization occurs at temperatures of 100°C after 480 hours. Aging and the lower temperatures appear to facilitate the crystallization of this molecular sieve. Without the lower temperatures and aging, SAPO-20 is crystallized under identical conditions. In the presence of TEA^+ and in the absence of Na^+, crystallization occurs at 200°C after 168 hours.

Thermal Properties

SAPO-42 is stable to 550°C, the temperature necessary to remove the organic from the pores of the structure.

Adsorption

Adsorption capacities measured for SAPO-42:

Adsorbate	Kinetic diameter, Å	Pressure, Torr	Temp., °C	Wt% adsorbed
O_2	3.46	100	−183	12.6
O_2		740	−183	17.0
n-Hexane	4.3	53.5	24	7.4
Isobutane	5.0	751	24	1.0
H_2O	2.65	4.6	23	15.5
H_2O		19.4	24	21.0

SAPO-44 (CHA)
(US4,440,871(1984))
Union Carbide Corporation

Related Material: chabazite

Structure

Chemical Composition: silicoaluminophosphate. See SAPO.

X-Ray Powder Diffraction Data: $(d(\text{Å}) (I/I_0))$
11.8(2), 9.31(100), 8.08(4), 6.81(31), 6.66(1),
6.44(3), 5.95(1), 5.49(51), 5.10(9), 4.67(6),
4.51(1), 4.26–4.21(98), 4.06(25), 3.95–3.92(7),
3.85(12), 3.626(55), 3.440–3.401(22), 3.314(1),
3.198(10), 3.132(2), 3.079(1), 3.008(4),
2.959(18), 2.894(80), 2.831(1), 2.784(2),
2.751(3), 2.714(5), 2.667(1), 2.678(3),
2.522(11), 2.338(1), 2.298(1), 2.259(2), 2.137–
2.122(4), 2.071(3), 2.040(1), 2.006(1), 1.965(1),
1.922(2), 1.888(6), 1.866(5), 1.807(9), 1.784(1),
1.752(2), 1.698(8).

SAPO-44 exhibits properties characteristic of a small-pore molecular sieve.

Synthesis

ORGANIC ADDITIVES
cyclohexylamine

SAPO-44 can be crystallized from a batch composition of: $C_6H_{11}NH_2$: Al_2O_3 : P_2O_5 : 0.6 SiO_2 : 50 H_2O. Crystallization occurs at temperatures of 200°C after 52 hours. The source of silica is aqueous silica sol, that of aluminum is aluminum isopropoxide, and that of phosphorus is phosphoric acid.

Thermal Properties

SAPO-44 is stable to 550°C for 2 hours, the temperature necessary to remove the organic from the pores of the structure.

Adsorption

Adsorption capacities measured for SAPO-44:

Adsorbate	Kinetic diameter, Å	Pressure, Torr	Temp., °C	Wt% adsorbed
O_2	3.46	98	−183	25.5
O_2		746	−183	32.3
n-Hexane	4.3	48	23.9	3.6
Isobutane	5.0	101	25.4	0.0

SAPO-47 (CHA)
(J. Phys. Chem. 93:6516(1989))
Union Carbide Corporation

Related Material: chabazite

Structure

CHEMICAL COMPOSITION: $Al_6(Si_{1.4}P_{4.6})O_{24}*$ $1.4C_5H_{12}NH_2*2.5H_2O$
SYMMETRY: rhombohedral
SPACE GROUP: $R\bar{3}$
UNIT CELL CONSTANTS (Å): $a = b = c = 9.3834$ Å
$\alpha = \beta = \gamma = 94.085°$
DENSITY: 1.82 g/cm³

Atomic positions of SAPO-47 model (*J. Phys. Chem.* 93:6516(1989)):

Atom	x	y	z
Al	0.09888(9)	0.33196(9)	0.87510(10)
P	0.33245(8)	0.10847(8)	0.87466(8)
O(1)	0.2580(3)	−0.2704(2)	−0.0127(3)
O(2)	0.1519(2)	−0.1445(3)	0.4911(2)
O(3)	0.2542(2)	0.2460(2)	0.8876(3)
O(4)	0.0291(3)	0.0104(3)	0.3178(3)
C(1)	−0.132(2)	0.524(3)	0.500(3)
N	0.009(2)	0.480(2)	0.551(2)
C(2)	0.112(2)	0.492(2)	0.440(3)
C(3)	0.189(2)	0.353(2)	0.424(3)
C(4)	0.334(2)	0.382(2)	0.359(2)
C(5)	0.449(2)	0.303(3)	0.439(3)

Synthesis

ORGANIC ADDITIVES
methylbutylamine

SAPO-47 crystallizes from a reactive silicoaluminophosphate gel using the procedures outlined in US4,440,871(1984).

Scapolites
(natural)

Scapolites are a group of feldspathoid-like aluminosilicates that form a solid-solution series ranging from compositions of $Na_4Al_3Si_9O_{24}Cl$, known as marialite, to $Ca_4Al_6Si_6O_{24}CO_3$, known as meionite (*Am. Mineral.* 73:119(1988)).

Schneiderite (LAU)

Obsolete name for laumontite.

Scolecite (NAT)
(natural)

Related Materials: natrolite

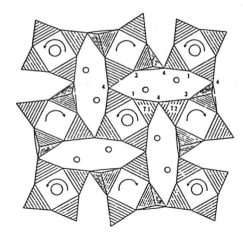

Scolecite Fig. 1S: Framework topology.

Structure

(*Zeo. Mol. Sieve.* 171(1974))

CHEMICAL COMPOSITION: $Ca_8(Al_{16}Si_{24}O_{80})*24$ H_2O
SYMMETRY: monoclinic
SPACE GROUP: Cc

UNIT CELL CONSTANTS (Å): $a_0 = 18.52$
$b_0 = 18.99$
$c_0 = 6.57$
$\beta_0 = 90°39'$

PORE STRUCTURE: eight-member rings 2.6 × 3.8 Å intersecting variable eight-member rings

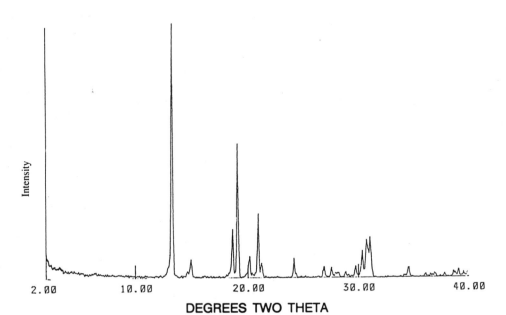

Scolecite Fig. 1:   X-ray powder diffraction pattern for scolecite from Pune, India. (Reproduced with permission of author)

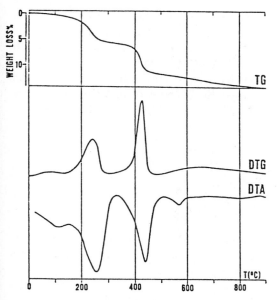

Scolecite Fig. 2:   Thermal curves of scolecite from Púna, India; in air, heating rate 20°C/min (*Nat. Zeo.* 49(1985)). (Reproduced with permission of Springer Verlag Publishers)

X-Ray Powder Diffraction Data: (*Nat. Zeo.* 318(1985)) ($d(\text{Å})$ (I/I_0)) 9.26(12), 6.55(62), 5.90(24), 5.37(5), 4.64(98), 4.38(34), 4.14(61), 3.953(6), 3.506(67), 3.277(18), 3.202(20), 3.182(46), 3.083(6), 2.948(38), 2.926(40), 2.859(100), 2.794(16), 2.677(78), 2.582(17), 2.564(10), 2.433(21), 2.326(4), 2.313(6), 2.284(6), 2.255(21), 2.190(24), 2.178(50), 2.124(9), 2.087(5), 2.067(14), 2.025(2), 1.976(3).

See natrolite for a description of the framework topology of scolecite.

Synthesis

ORGANIC ADDITIVE
none

Scolecite first was synthesized in 1960 from a reactive aluminosilicate gel with a batch composition of $CaO*Al_2O_3*3SiO_2$ with temperatures ranging between 230 and 285°C (*J. Geol.* 68:41(1960)). Scolecite can convert to analcime at 180°C after 24 to 160 hours (*Crys. Res. Technol.* 21:46(1986)).

Thermal Properties

There are two fundamental losses of water for scolecite, the first occurring at 240°C and the second appearing at 420°C. Above 420°C the structure degrades to form an amorphous phase. See Scolecite Figure 2.

Infrared Spectrum

See Scolecite Figure 3.

Ion Exchange

Scolecite exchanges with much difficulty, requiring fused salts at temperature between 180 and 270°C to obtain partial substitution. No exchange with NH_4^+ is observed (*Mineral. Mag.* 24:227(1936)).

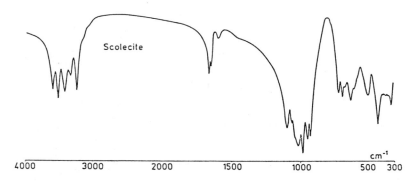

Scolecite Fig. 3:   Infrared spectrum of scolecite (*Nat. Zeo.* 348(1985)). (Reproduced with permission of Springer Verlag Publishers)

Seebachite (CHA)

Obsolete name for chabazite.

Sigma-1 (DDR)
(SSSC 37:57(1988))

Related Materials: deca-dodecasil 3R

Structure

Sigma-1 exhibits properties characteristic of a clathrasil. For a description of the framework topology, see deca-dodecasil 3R.

Synthesis

ORGANIC ADDITIVES
1-adamantaneamine

Sigma-1 crystallizes in a sodium oxide system at 180°C. This material crystallizes readily with a molar ratio of SiO_2/Al_2O_3 of around 60,

with low ratios of free OH/SiO_2 ($pH_{final} = 11.72$–12.10). At higher hydroxide concentrations, ZSM-5, cristobalite, and mordenite are observed. At different SiO_2/Al_2O_3 ratios, sigma-2 and Nu-3 are obtained instead. A typical batch composition in terms of molar oxide ratios is: $3\ Na_2O : 20\ AN : Al_2O_3 : 60\ SiO_2 : 2400\ H_2O$.

Thermal Properties

Weight loss is observed below 250°C due to water; combustion of the organic occurs between 250 and 800°C. The rate of loss is rapid at 420°C, with two stages being observed. Exotherms are noted at 472 and 610°C in the DTA. Between 250 and 800°C, 10% weight loss is observed.

Infrared Spectrum

Mid-infrared vibrations for sigma-1 (cm^{-1})
1070–1120s, 1220sh, 610w, 528, 517m, 462ms.

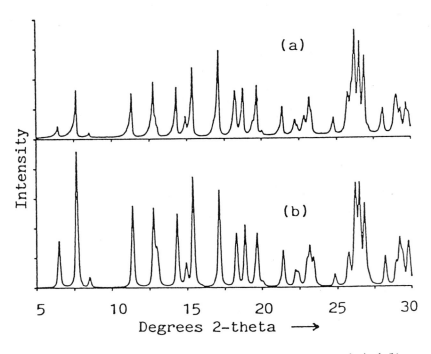

Sigma-1 Fig. 1:   X-ray powder diffraction pattern for sigma-1: (a) as synthesized; (b) simulated (*SSSC* 37:57(1988)). (Reproduced with permission of Elsevier Publishers)

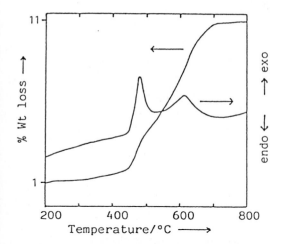

Sigma-1 Fig. 2: DTA/TGA of as-synthesized sigma-1 (*SSSC* 37:57(1988)). (Reproduced with permission of Elsevier Publishers)

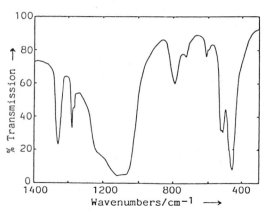

Sigma-1 Fig. 3: Infrared spectrum of H-sigma-1 (as a nujol mull) (*SSSC* 37:57(1988)). (Reproduced with permission of Elsevier Publishers)

Adsorption

Adsorption of various sorbates in H-sigma-1 (single-point method unless otherwise stated) (*SSSC* 37:57(1988)):

Sorbate	Kinetic diameter, Å	Temp., K	P/P_o	Wt% adsorbed
H_2O	2.65	293	0.5	10.5
N_2[a]	3.64	77	range	9.8
Methanol	3.80	295	0.3	9.2
Xenon[b]	3.96	293	range	20.9
n-Butane	4.30	273	0.5	0.1
i-Butane	5.00	193	0.5	0.3
Benzene	5.85	293	range	0.2

a. Alpha-plot method.
b. Extrapolation of isotherms of Langmuir type; density = 1.99 g/cm³, obtained by comparison of uptakes of Xe and N_2 on a range of zeolites.

Sigma-2 (SGT)
(*Zeo.* 9:140(1989))

Related Materials: ZSM-58

Structure

CHEMICAL COMPOSITION: $[Si_{64}O_{128}]*4R$
SYMMETRY: tetragonal

SPACE GROUP: I4₁/amd
UNIT CELL CONSTANTS (Å): $a = 10.2$
                          $c = 34.4$
PORE STRUCTURE: openings formed by six-member rings

X-Ray Powder Diffraction Data: (d(Å) (I/I_0))
9.81(40), 8.60(7), 7.64(2), 6.67(1), 5.71(1), 4.91(1), 4.54(99), 4.49(100), 4.43(2), 4.40(36), 4.30(11), 4.25(10), 3.82(2), 3.81(11), 3.62(4), 3.58(2), 3.40(18), 3.34(43), 3.29(24), 3.27(6), 3.18(2), 3.11(2), 3.06(10), 3.03(2), 2.99(15),

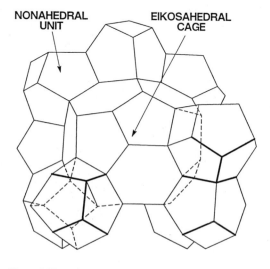

Sigma-2 Fig. 1S: Framework topology.

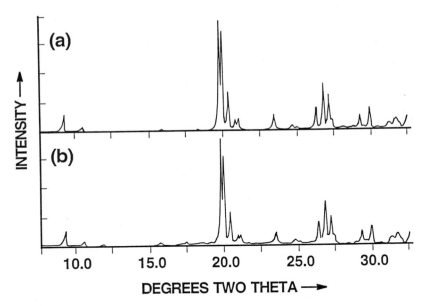

Sigma-2 Fig. 2:   X-ray powder diffraction pattern for "as-synthesized" sigma-2 (top) and calcined sigma-2 (bottom) (*Zeo.* 9:140(1989)). (Reproduced with permission of Butterworths Publishers)

2.93(1), 2.86(3), 2.85(2), 2.83(5), 2.82(7), 2.80(3), 2.76(8), 2.63(10), 2.59(5), 2.58(3), 2.56(4), 2.50(2), 2.39(2), 2.34(2), 2.29(1), 2.29(2), 2.27(1), 2.25(3).

The structure of sigma-2 is composed of non-asil ($4^3 5^6$) and large eikosasil ($5^{12} 6^8$) cages. The organic adamantanamine unit is located in the large cage and is disordered in the structure.

The borosilicate analog of sigma-2 also has been identified. With the same space group as the aluminosilicate, it has unit cell parameters of a = 10.2431 and c = 34.3184 Å (*PP 8th IZC* 20 (1989)).

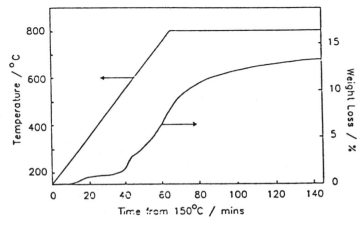

Sigma-2 Fig. 3:   Thermal analysis of as-synthesized sigma-2 (*Zeo.* 9:140(1989)). (Reproduced with permission of Butterworths Publishers)

Synthesis

ORGANIC ADDITIVES
1-adamantanamine

Sigma-2 crystallizes from a reactive aluminosilicate gel in the presence of either Na^+ or K^+ cations. In the presence of progressively lower SiO_2/Al_2O_3 ratios, sigma-1 and Nu-3 are observed to form. The batch compositions that result in the formation of sigma-2 are: 0–3 Na_2O : 0–3 K_2O : 0–1 Al_2O_3 : SiO_2 : 2400 H_2O. Crystallization occurs at temperatures between 140 and 200°C.

The crystallization of the boron analog of sigma-2 occurs at pH between 9 and 13 at 150°C for 10 to 30 days with either 1-adamantanamine or 2-adamantanamine. If aluminum or gallium is substituted for the boron under these conditions, sigma-1 (DDR) forms instead (*PP 8th IZC* 20(1989)).

Thermal Properties

The organic encapsulated within the large cages of sigma-2 is difficult to remove by calcination, resulting in black-colored materials even after long calcination times. The presence of aluminum within the framework enhances the ability of the organic to be removed from the cages. In the TGA, little loss in weight is observed below 450°C. The rapid loss in weight around 500°C is attributed to degradation of the adamantanamine. An exotherm is observed with a maximum at 560°C. The total weight loss is 13.2%, with 7.5% occurring up to 800°C. See Sigma-2 Figure 3.

Calcination studies on sigma-2 (*Zeo.* 9:140(1989)):

SiO_2/Al_2O_3	Calcination conditions (temp., °C, time, hr.)	Color	Wt% C
>1500	500, 48	black	8.2
870	500, 28; 700, 17	black	1.8
91	500, 48, 700, 24	dark grey	1.6
50	500, 48	yellow	4.2
	550, 16	buff	2.7
48	500, 48	yellow	4.6
	500, 48; 700, 24	white	0.3

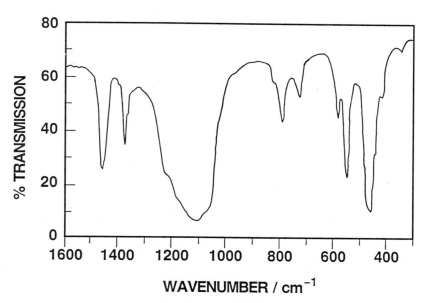

Sigma-2 Fig. 4:   Infrared spectrum of as-synthesized sigma-2 in nujol mull (*Zeo.* 9:140(1989)). (Reproduced with permission of Butterworths Publishers)

Infrared Spectrum

The infrared spectrum of as-synthesized sigma-2 with a SiO_2/Al_2O_3 ratio greater than 1000 has an intense stretching mode around 1100 cm^{-1}, with a side band at 1225 cm^{-1}, and bending modes at around 500 cm^{-1}; at 632(m), 550(s), 465(vs), 444(w), 420(w) (*Zeo.* 9:140(1989)). See Sigma-2 Figure 4.

Ion Exchange

No ion-exchange properties were observed for this structure. Even under reflux conditions with 1 M HCl solutions, the Na^+ and K^+ cations within the structure will not exchange (*Zeo.* 9:140(1989)).

Adsorption

Xenon, water, and *n*-hexane could not be adsorbed into the cavities of sigma-2.

Silicalite (MFI) (US4,061,724(1977)) Union Carbide Corporation

Related Materials: ZSM-5

Structure

(*Nature* 271:512(1978))

CHEMICAL COMPOSITION: $(TPA)_2O*48SiO_2*H_2O$
SYMMETRY: orthorhombic
SPACE GROUP: Pnma or $Pn2_1a$
UNIT CELL CONSTANTS (Å): $a = 20.06$
$b = 19.80$
$c = 13.36$

DENSITY: 1.76 g/cm^3
PORE STRUCTURE: intersecting 10-member rings 5.7 × 5.1 Å, and 5.4 Å

X-Ray Powder Diffraction Data: (d(Å) (I/I_0))
11.1(100),   10.02(64),   9.73(16),   8.99(1), 8.04(0.5), 7.42(1), 7.06(0.5), 6.68(5), 6.35(9), 5.98(14), 5.70(7), 5.57(8), 5.36(2), 5.11(2), 5.01(4), 4.98(5), 4.86(0.5), 4.60(3), 4.44(0.5), 4.35(5), 4.25(7), 4.08(3), 4.00(3), 3.85(59), 3.82(32),   3.74(24),   3.71(27),   3.64(12), 3.59(0.5), 3.48(3), 3.44(5), 3.34(11), 3.30(7), 3.25(3), 3.17(0.5), 3.13(0.5), 3.05(5), 2.98(10).

Silicalite is topologically related to ZSM-5, composed of 5–1 secondary building units and intersecting straight and zigzag channels. It has also been called silicalite-1.

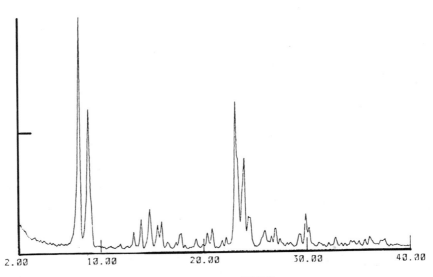

DEGREES TWO THETA

Silicalite Fig. 1:   X-ray powder diffraction pattern of silicalite. (Reproduced with permission of F. Testa, U. della Calabria)

Atom positions in silicalite (*Nature* 271:512(1978)):

Atom	x	y	z
Si(1)	0.4225(6)	0.5468(3)	0.3518(9)
Si(2)	0.3055(9)	0.5150(10)	0.1944(7)
Si(3)	0.1902(2)	0.5570(2)	0.3197(6)
Si(4)	0.0687(7)	0.5225(8)	0.1616(9)
Si(5)	0.2160(8)	0.4223(8)	0.4595(10)
Si(6)	0.3752(8)	0.4401(10)	0.4627(10)
Si(7)	0.0788(8)	0.3597(10)	0.1744(10)
Si(8)	0.1888(10)	0.3056(10)	0.3060(10)
Si(9)	0.3166(10)	0.3582(10)	0.1586(10)
Si(10)	0.4266(9)	0.3193(10)	0.3227(10)
Si(11)	0.1220(10)	0.3138(11)	0.9710(10)
Si(12)	0.0754(7)	0.1208(8)	0.1912(10)
Si(13)	0.1965(9)	0.1517(8)	0.3032(10)
Si(14)	0.3187(10)	0.1150(10)	0.1766(12)
Si(15)	0.4228(10)	0.1631(10)	0.3005(12)
Si(16)	0.1184(9)	0.1586(10)	0.9420(11)
Si(17)	0.2727(8)	0.3166(9)	0.9773(11)
Si(18)	0.2752(7)	0.1658(9)	0.9599(11)
Si(19)	0.2267(9)	0.0492(10)	0.4464(11)
Si(20)	0.3845(9)	0.0592(8)	0.4818(11)
Si(21)	0.4249(8)	0.9393(10)	0.3132(12)
Si(22)	0.3063(9)	0.9575(10)	0.1701(11)
Si(23)	0.1881(9)	0.9450(10)	0.3239(12)
Si(24)	0.0590(8)	0.9051(8)	0.1989(10)
O(1)	0.3834(16)	0.3370(17)	0.2262(21)
O(2)	0.3730(17)	0.5271(18)	0.2454(22)
O(3)	0.4837(18)	0.5483(17)	0.2972(23)
O(4)	0.2409(17)	0.5344(18)	0.2262(22)
O(5)	0.3047(18)	0.4414(19)	0.1849(23)
O(6)	0.1082(19)	0.5474(18)	0.2573(22)
O(7)	0.1234(18)	0.3419(19)	0.2806(23)
O(8)	0.2527(17)	0.3213(18)	0.2209(24)
O(9)	0.2995(18)	0.3313(17)	0.0822(23)
O(10)	0.0963(19)	0.3102(18)	0.0572(23)
O(11)	0.1306(17)	0.1508(19)	0.2555(24)
O(12)	0.4967(16)	0.1527(17)	0.2721(22)
O(13)	0.3214(17)	0.0614(18)	0.1352(23)
O(14)	0.3687(17)	0.1690(18)	0.2261(24)
O(15)	0.2481(18)	0.1473(16)	0.2212(23)
O(16)	0.3214(16)	0.1481(18)	0.0394(22)
O(17)	0.0836(16)	0.4454(19)	0.1702(24)
O(18)	0.9931(18)	0.3393(18)	0.1904(25)
O(19)	0.4274(17)	0.2462(19)	0.3529(20)
O(20)	0.3866(16)	0.4978(18)	0.4116(25)
O(21)	0.1827(17)	0.4747(18)	0.3853(22)
O(22)	0.4039(19)	0.3795(18)	0.4059(25)
O(23)	0.1907(18)	0.3629(18)	0.4021(24)
O(24)	0.2892(16)	0.4112(17)	0.4760(20)
O(25)	0.1802(16)	0.4479(18)	0.5685(20)
O(26)	0.4021(17)	0.4694(16)	0.5966(21)
O(27)	0.1974(18)	0.2503(18)	0.3459(20)
O(28)	0.2054(16)	0.3315(16)	0.9469(18)
O(29)	0.0696(15)	0.3590(16)	0.9334(20)
O(30)	0.1011(12)	0.2550(16)	0.9533(20)

Atom positions in silicalite (*Nature* 271:512(1978)):

Atom	x	y	z
O(31)	0.2768(16)	0.2320(18)	0.9137(19)
O(32)	0.0818(17)	0.1548(17)	0.0798(19)
O(33)	0.0766(18)	0.0362(18)	0.1714(22)
O(34)	0.2074(16)	0.1165(18)	0.4281(20)
O(35)	0.3970(16)	0.1289(16)	0.3892(21)
O(36)	0.1917(17)	0.1492(17)	0.9844(20)
O(37)	0.0959(17)	0.1096(16)	0.8882(22)
O(38)	0.2978(18)	0.3715(17)	0.8828(23)
O(39)	0.3030(19)	0.2375(16)	0.8621(20)
O(40)	0.1989(18)	0.0438(17)	0.5593(19)
O(41)	0.4144(18)	0.0564(17)	0.5818(22)
O(42)	0.3853(17)	0.9245(18)	0.2291(19)
O(43)	0.2588(16)	0.9544(18)	0.2675(19)
O(44)	0.1171(16)	0.9156(17)	0.2415(19)
O(45)	0.5020(16)	0.9461(17)	0.2986(22)
O(46)	0.3103(17)	0.0342(16)	0.4894(19)
O(47)	0.4198(16)	0.9896(17)	0.3833(22)
O(48)	0.1992(18)	0.9823(17)	0.4196(21)

Synthesis

ORGANIC ADDITIVE
TPA$^+$
hexane 1,6-diamine
Pip + TPA
Pip + TBA
Pip + TrBPe
Pip + TrBHp
Pip + TPeA
diethylamine

Silicalite first was prepared from a reactive gel containing TPA and silica at temperatures ranging between 100 and 200°C. The molar oxide composition of the synthesis gel is: $(TPA)_2O$: 6.2 Na_2O : 38.4 SiO_2 : 413 H_2O (US 4,061,724(1977)). In a gel composition of: 0.99 TPABr : 0.026 Na_2O : SiO_2 : 24.8 H_2O, large crystals of silicalite can be prepared. The crystal size shows a strong dependence on the temperature of crystallization, with the larger 30-micron-size crystals formed at 200°C and 150°C producing crystals that have a maximum length of 20 microns. Crystallization also is much more rapid at the higher temperatures (*Zeo.* 9:136(1989)).

Silicalite has been crystallized from the mixture: x HEXDM (1,6 hexanediamine) : 20 SiO_2 : w H_2O (x = 20, 50; w = 250, 1000). At lower temperatures (less than 150°C) and high HEDM/SiO_2 ratios, crystallization of silicalite is preferred. At higher temperatures, ZSM-48 forms. Stirring the crystallizing mixture also encourages the formation of silicalite. The addition of small amounts of TPA further improves the range over which this structure will crystallize. The pH of the solution, prior to and after crystallization is complete, is between 12.6 and 13.3. The crystal habit appears to be blocks and aggregates 7 × 3 × 3 microns in size. Thermal analysis of the material shows the occlusion of the organic within the channel system (*Zeo.* 9:495(1988)).

In the mixed organic system, Pip + (TPA, TBA, TrBPe, TrBHp, and TPeA) with the following composition: 10 Pip : 3 organic : 20 SiO_2 : 250 H_2O, silicalite will form as the only phase after crystallization times ranging from 7 days (Pip + TPA) to 63 days (Pip + TrBHpA). The pH of this system ranges from 11.4 initially, decreasing to a final pH value around 11 for Pip + TPA. The presence of the piperazine appears to increase the time needed for this zeolite structure to crystallize (*Zeo.* 8:501(1988)). The primary effect these mixed organic-containing mixtures have on silicalite products is modification of the morphology. The Pip + TPA silicalite has very large twinned crystals, greater than 25 microns in length. The Pip + TBA–containing system has long mixed dendrital crystals and longer 50-micron needles, with less twinning observed when TPeA is used.

The neutral diethylamine also will encourage the growth of silicalite but only if sodium also is present in the crystallization mixture: y DEA: x NaOH: 20 SiO_2 : w H_2O (w = 250, 1000; x = 0, 1, 3, 5; y = 20, 50, 100). The time required for crystals to form from this system is over 14 days (*Zeo.* 8:508(1988)). The other phase crystallizing is Nu-10 in the absence of Na^+.

The size of the crystals of silicalite can be varied by using the following batch compositions (*Zeo.* 8:416(1988)):

Oxide composition (170°C, unstirred)	Crystal size (macrons)		Aspect ratio
	Length	Width	
4Na$_2$O:8TPABr:100SiO$_2$:1000OH$_2$O	12	13	0.9
2Na$_2$O:8TPABr:100SiO$_2$:1000OH$_2$O	74	30	2.5
1Na$_2$O:8TPABr:100SiO$_2$:1000OH$_2$O	160	37	4.3
0.5Na$_2$O:8TPABr:100SiO$_2$:1000OH$_2$O	100	15	6.7
4TPA$_2$O:100SiO$_2$:1000OH$_2$O	24	20	1.2
2TPA$_2$O:4TPABr:100SiO$_2$:1000OH$_2$O	70	35	2
1TPA$_2$O:6TPABr:100SiO$_2$:1000OH$_2$O	105	30	3.5
0.5TPA$_2$O:7TPABr:100SiO$_2$:1000OH$_2$O	110	18	6.1

In this synthesis, made without the presence of sodium, nucleation and crystallization were ob-

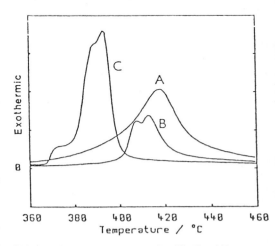

Silicalite Fig. 2:   DTA traces for silicalite: (A) HEXDM; (B) HEXDM,TPA; (C) TPA (*Zeo.* 8:495(1988)). (Reproduced with permission of Butterworths Publishers)

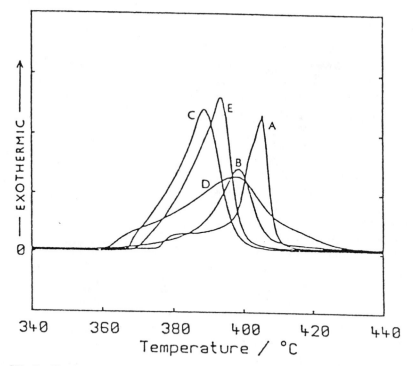

Silicalite Fig. 3: DTA traces of as-synthesized silicalite: (A) TPA-silicalite; (B) TBA-silicalite; (C) TPeA-silicalite; (D) TrBPe-silicalite; (E) TrBHp-silicalite (*Zeo.* 8:501(1988)). (Reproduced with permission of Butterworths Publishers)

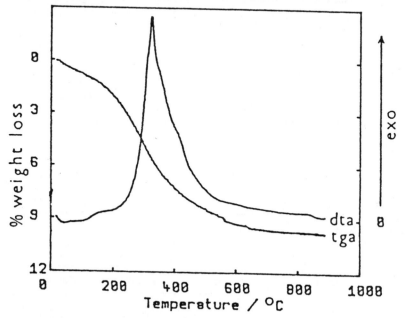

Silicalite Fig. 4: Thermal analysis of DEA-silicalite (*Zeo.* 8:508(1988)). (Reproduced with permission of Butterworths Publishers)

served to occur more rapidly than for the so-lutions using Na$^+$. In systems using Na$^+$, K$^+$, and TPA$^+$, a variation in the crystallinity was observed with varying K/Na + K in the reaction mixture. The presence of alkali salt in the crystallization mixture influences the kinetics of formation, the dissolution of the gel phase, and the morphology of the final crystals (*J. Catal.* 60:241(1979); *Proc. 5th IZC* 64(1980). In systems that contain NH$_4^+$, TPA$^+$, and an alkali cation, nucleation decreased with increasing alkali concentration. In the presence of Li$^+$ cations, very large uniform (140 × 40–micron) crystals were observed to form (*Zeo.* 3:219 (1983)). If sodium salt was the only source of alkali cations, large spherical crystal aggregates, 80 microns in diameter, could be produced. In systems that introduce the alkali as the hydroxide, it has been noted that the OH anions could be regulated to produce crystals of various sizes. The lower alkalinity produced larger crystals than those from systems with higher alkalinity (*JCS Faraday Trans.* 1:82:785(1986)). The length of the crystals has been shown to be inversely related to the pH (EP 93519(1983)).

"Gel particle" silicalite is prepared through the transformation of particles of silica gel. The gel composition consists of: 10 Pip : 2 TPABr : 20 SiO$_2$: 250–1000 H$_2$O, with crystallization occurring around 150°C within 24 hours. (*Zeo.* 7:135(1987)). The process involves the formation of a dense layer of small-micron-size crystals at the surface of the gel particle. The formation of this layer keeps the particles intact.

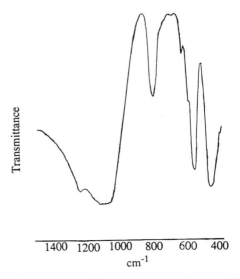

Silicalite Fig. 5:   Infrared spectrum of silicalite in the mid-infrared region. (Reproduced with permission of B. Duncan, Georgia Tech)

The interior of the gel particle contains 5-micron, loosely packed crystals.

Thermal Properties

See Silicalite Figures 2, 3 and 4.

Infrared Spectrum

See Silicalite Figure 5.

Thermal analysis of HEXDM silicalite and HEXDM,TPA silicalite:

Organic	DTA peak, °C	Wt losses from TGA (%)			
		25–180	180–600	600–900	25–900(°C)
HEXDM	417	0.25	13.2	1.8	15.25
HEXDM + Na	426,431	0.3	13.9	0.7	14.9
HEXDM + TPA	430	0.35	12.85	2.1	15.3
HEXDM + TPA′	407	0.0	12.1	1.4	13.5
HEXDM + TPA″	389	0.4	12.8	1.9	15.1

Thermal analysis results for "as-synthesized" silicalite
(*Zeo.* 8:501(1988)):

Organic	DTA peak temp., °C	TGA wt loss (%) 0–300	300–600	600–900 °C
TPA	399	0.01	11.77	0.79
TPA	393	0.14	11.75	0.79
TPA	405	0.02	11.38	0.60
TBA	400	0.16	11.80	0.79
TBA	382	0.26	11.93	1.56
TBA	392	0.25	12.16	0.71
TrBPeA	397	0.23	11.29	0.40
TrBHp	393	0.30	11.76	0.69
TPeA	388	0.20	12.02	0.56
TPeA	402	0.30	10.91	0.57
TPeA	408	0.36	11.38	0.90

Adsorption

For the adsorption of ethanol on silicalite it has been reported that this molecular sieve adsorbs 8 wt% ethanol at low ethanol concentrations (*J. Catal.* 59:114(1979)). In an examination of straight chain alcohols and their isomers, it has been reported that the saturated alcohol adsorption capacity increases with increasing chain length of the straight chain alcohols and decreases for the three isomers of butanol (*J. Chem. Tech. Biotechnol.* 31:732(1981)):

1-butanol > 2-butanol > i-butanol

In liquid equilibrium adsorption isotherms of cyclohexane, 1-heptene, benzene, cyclohexene,

Adsorption volumes in silicalite (*Nature* 271:512(1978)):

Adsorbate	T(°C)	Kinetic diameter, Å	V_p (cm/g)	V_t	Molecules adsorbed per unit cell
H_2O	RT	2.65	0.47	0.083	15.1
O_2	−183	3.46	0.185	0.326	38.0
CH_3OH	RT	3.8	0.193	0.340	27.6
n-Butane	RT	4.3	0.190	0.334	10.9
n-Hexane	RT	4.3	0.199	0.350	8.8
SF_6	RT	5.5	0.167*	—	10.9
C_6H_6	RT	5.85	0.134	0.236	8.7
Neopentane	RT	6.2	0.029	0.051	1.4

V_p = total micropore volume in activated silicalite from saturation capacity, calculated using normal liquid densities at the adsorption temperature. Void fraction: $V_t = V_p \times d$ (g/cm³), where d is the measured density, 1.76 g/cm³. All samples activated by calcination in air at 600°C followed by vacuum activation (10^{-3} Torr). Adsorption measurements all by gravimetric McBain-Bakr balance technique. RT = room temperature. * At 760 Torr.

Adsorption capacity and heat of adsorption of hydrocarbons on silicalite (T = 20°C) (*Zeo.* 3:118(1983)):

Adsorbate	Kinetic diameter, Å	Adsorption capacity (mmol/g)	(molecules/uc)	V_p	q_{st}
Benzene	5.85	1.42	8.2	0.126	80
Cyclohexane	6.0	—	—	—	86*
Hexane	4.3	1.42	8.2	0.185	87
p-Xylene	5.85#	1.06	6.1	0.13	41
Ethylbenzene	5.85#	1.04	6.0	0.14	44
m-Xylene	6.8	0.69	4.0	0.085	—
o-Xylene	6.8##	0.51	2.9	0.062	48

*Estimated by isotherms at 100 and 200°C; adsorption amount = 0.16 mmol/g.
#Taken as the same as benzene.
##Taken as the same as *m*-xylene.

and *n*-octane at 30°C in hexane solvent, the equilibrium adsorption capacities decrease in the order (*AICHE Sym. Ser.* 81:41):

1-heptene > cyclohexene > benzene > *n*-octane > cyclohexane

Silica-sodalite (SOD)

See sodalite.

Silicalite-2

Silica polymorph of ZSM-11.

Skolezit (NAT)
(natural)

See scolecite.

Sloanite (LAU)

Obsolete name for laumontite.

Small-Port Mordenite

A term used to describe mordenite samples that did not exhibit properties characteristic of the large-pore framework topology. See mordenite.

Sn-A
(*Solid State Ionics* 32/33:423(1989))

Structure

CHEMICAL COMPOSITION: $Na_8Sn_3Si_{12}O_{34} \cdot nH_2O$
SYMMETRY: orthorhombic
UNIT CELL CONSTANTS (Å): $a = 7.940$
$b = 10.338$
$c = 11.591$

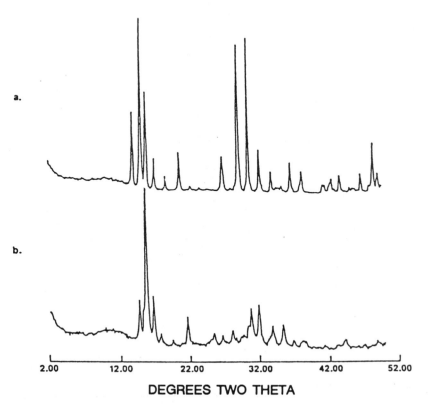

Sn-A Fig. 1:   X-ray powder diffraction pattern for Sn-A (a) hydrated and (b) dehydrated (*Solid State Ionics* 32/33:423(1989)). (Reproduced with permission of Elsevier Publishers)

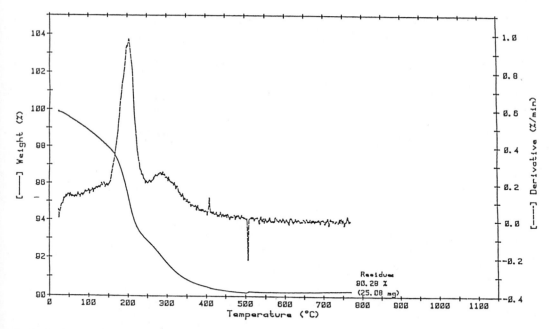

Sn-A Fig. 2:   TGA/DTA for Sn-A (*Solid State Ionics* 32/33:423(1989)).

X-Ray Powder Diffraction Data: $(d(\text{Å})\ (I/I_0))$
6.33(29), 5.83(100), 5.56(42), 5.19(12), 4.73(8),
4.28(18), 3.30(23), 3.05(73), 2.91(69), 2.77(27),
2.64(11), 2.45(15), 2.35(9).

Sn-A exhibits properties characteristic of a condensed felspathoid.

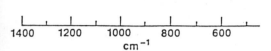

Sn-A Fig. 3:   Infrared spectrum of Sn-A (*Solid State Ionics* 32/33:423(1989)). (Reproduced with permission of Elsevier Publishers)

Synthesis

ORGANIC ADDITIVE
none

Sn-A crystallizes from a basic sodium stanosilicate gel with a molar oxide composition of: $2\ Na_2O : SnO_2 : 4\ SiO_2 : 80\ H_2O$. Crystallization occurs at 200°C after 21 days.

Thermal Properties

A 10% weight loss is observed for this structure upon thermal treatment.

Infrared Spectrum

The infrared spectrum exhibits a very sharp and complex band in the T–O bending region (400–500 cm^{-1}) and symmetric stretching vibrations between 650 and 800 cm^{-1}. OH stretches are observed around 1600 to 1800 cm^{-1}, as well as water vibrations at 3200 to 3600 cm^{-1}.

Sn-B
(*Solid State Ionics* 32/33:423(1989))

Structure

CHEMICAL COMPOSITION: $Na_8Sn_3Si_{12}O_{34}*nH_2O$
SYMMETRY: orthorhombic
UNIT CELL CONSTANTS (Å): a = 7.940
                            b = 10.338
                            c = 11.591

X-Ray Powder Diffraction Data: (d(Å) (I/I_0))
6.33(29), 5.83(100), 5.56(42), 5.19(12), 4.73(8),
4.28(18), 3.30(23), 3.05(73), 2.91(69), 2.77(27),
2.64(11), 2.45(15), 2.35(9).

Sn-A exhibits properties characteristic of a
condensed felspathoid and contains both tetrahedral silicon and octahedral tin.

Synthesis

ORGANIC ADDITIVE
none

Sn-B crystallizes from a basic sodium stanosilicate gel with a molar oxide composition of:
$Li_2O : Na_2O : SnO_2 : 4 SiO_2 : 80 H_2O$. Crystallization occurs at 200°C after 16 days. Fumed
silica is used as the source of silica, and tin
chloride as the source of tin.

Infrared Spectrum

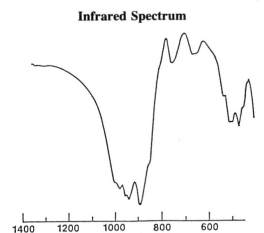

Sn-B Fig. 2:   Infrared spectrum of Sn-B (*Solid State
Ionics* 32/33:423(1989)). (Reproduced with permission of
Elsevier Publishers)

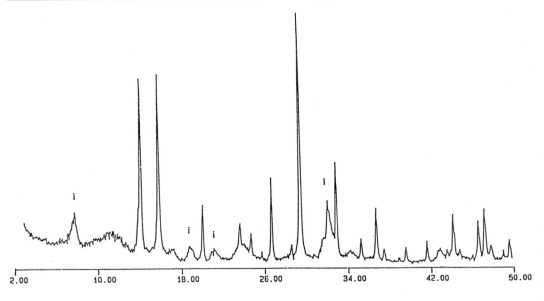

DEGREES TWO THETA

Sn-B Fig. 1:   X-ray powder diffraction pattern for Sn-B (*Solid State Ionics* 32/33:423(1989)). (Reproduced with
permission of Elsevier Publishers)

SnAPO-5 (AFI)
(*Zeo.* 10:680(1990))

Related Materials: AlPO$_4$-5

Structure

Chemical Composition: (Al$_{49.7}$ P$_{47}$ Sn$_4$)O$_{200}$.

The X-ray powder diffraction pattern is similar to that of AlPO$_4$-5.

Synthesis

ORGANIC ADDITIVE
triethylamine

SnAPO-5 is prepared by throughly mixing SnCl$_4$*2H$_2$O with hydrolyzed aluminum isopropoxide and phosphoric acid. Triethylamine is used as the organic additive. The resulting batch composition that crystallizes this structure is: 1.4 Et$_3$N : 0.2 SnO$_2$: 1.0 Al$_2$O$_3$: 1.0 P$_2$O$_5$: 40 H$_2$O. The gel is aged for 96 hours at 90°C prior to the addition of the organic. Crystallization of the final product occurs at 200°C after 24 hours.

Adsorption

Comparison of adsorption capacities for AlPO$_4$-5 and SnAPO-5 (*Zeo.* 10:680(1990)):

Sample	N$_2$ capacity, cm^3/100 g	H$_2$O capacity, cm^3/100 g
AlPO$_4$-5	12.5	21.7
SnAPO-5	4.7	19.5

N$_2$ T = 78K, P/P_o = 0.25.
H$_2$O T = 298K, P/P_o = 0.8.

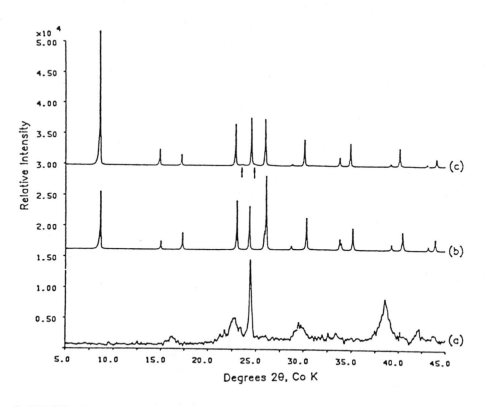

SnAPO-5 Fig. 1:   X-ray powder diffraction pattern of SnAPO-5 after (a) 90°C pretreatment, (b) 200°C crystallization, (c) 500°C calcination stages of preparation (*Zeo.* 10:680(1990)). (Reproduced with permission of Butterworths Publishers)

SnS-1
(*SSSC* 49A:375(1989))

Structure

Chemical Composition: tin sulfide.

X-Ray Powder Diffraction Data: (d(Å) (I/I_0))
11.43(9), 9.45(22), 8.47(100), 6.81(9), 5.06(10),
4.73(7), 4.28(5), 4.23(8), 3.12(4), 2.76(5),
2.37(3), 2.31(4).

Synthesis

ORGANIC ADDITIVE
tetramethylammonium$^+$

SnS-1 crystallizes from a reactive gel with
the following composition: 0.5 $RHCO_3$: SnS_2
: 14 H_2O. Crystallization occurs after 65 hours
at 150°C. The initial pH of the reactive slurry
is 7 to 9, with a pH of 0.5 to 3 upon crystal
formation.

Sodalite (SOD)
(natural)

Related Materials: $AlPO_4$-20
basic sodalite
bicchulite
danalite
G
genthelvite
hauyn
helvin
hydroxosodalite
lazurite
nosean
silica sodalite
TMA-sodalite
tugtupite

Sodalite and its variants represent a transition
between the nonzeolitic felspathoids and the
zeolites, depending on the composition and guest
species contained within the cages of this structure.

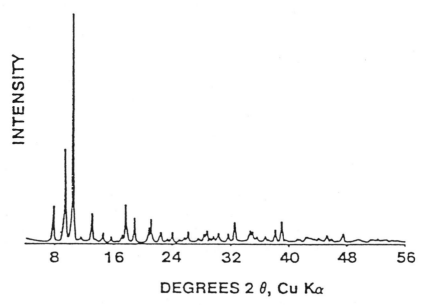

SnS-1 Fig. 1:   X-ray powder diffraction pattern for SnS-1 (*SSSC* 49A:375(1989)).
(Reproduced with permission of Elsevier Publishers)

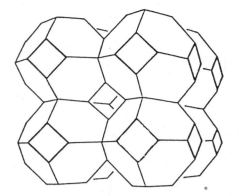

Sodalite Fig. 1S: Framework topology.

Structure

CHEMICAL COMPOSITION: $Na_6[Al_6Si_6O_{24}]*2NaCl$
SYMMETRY: cubic
SPACE GROUP: $P\bar{4}3n$
UNIT CELL CONSTANTS (Å): $a = 8.9$
PORE STRUCTURE: six-member-ring pore openings
  only

X-Ray Powder Diffraction Data: $(d(Å)\ (I/I_0))$
6.388(40), 4.823(12), 4.506(35), 4.028(13),
3.686(94), 3.294(10), 3.188(8), 2.927(12),
2.852(8), 2.819(14), 2.603(100), 2.476(8),
2.411(37), 2.252(12), 2.125(57), 2.094(5),
2.054(7), 2.006(7), 1.840(15), 1.822(6),
1.769(80), 1.646(20), 1.589(60), 1.546(58),
1.502(40), 1.463(48), 1.391(14), 1.360(45),
1.329(21), 1.302(35), 1.275(18), 1.227(55),
1.205(12), 1.184(10), 1.145(15).

The framework structure of sodalite consists of a cubic array of beta cages; it represents one of the more condensed structures of the zeolite frameworks. It can have a full range of aluminum concentrations, from pure aluminum to a pure silica form. In addition, a number of other ions have been shown to substitute into this framework lattice. This behavior has been attributed to the flexibility of the framework, where the T–O–T angle can range from 125° to 155° (*Hydrotherm. Chem. Zeo.* 310(1982)).

The space-filling character of any guest molecules or ions is a well-accepted model for sodalite: $M_8[AlSiO_4]_6X_2$, with M = Li, Na, K,

Rb, Ca, Sr, Ba, Cd, Cu, and others, and X = Cl, Br, I, OH, or $XO_4 = SO_4$, WO_4, MoO_4, ClO_4, etc. (*Mineral. Mag.* 38:593(1972); *Acta Crystallogr.* B40:185(1984); *Acta Crystallogr.* B40:6(1984)).

The pure silica end member also can be prepared by using a variety of organic additives. The trapping of that organic in the cages can notably change the intensities in the lines of the X-ray powder diffraction pattern (*SSSC* 49A:237(1989)).

Atomic positional parameters for sodalite (*Zeo.* 6:367(1986)):

Atom	Occupancy	x	y	z
(a) $Na_6[AlSiO_4]_6*8H_2O$ at 298K				
Si	1.0	1/4	0	1/2
Al	1.0	1/4	1/2	0
O(1)	1.0	0.1366(7)	0.4338(5)	0.1490(7)
O(2)	1.0	0.3753(5)	0.3753(5)	0.3753(5)
Na	0.75	0.1504(6)	0.1504(6)	0.1504(6)
(b) $Na_6[AlSiO_4]_6$ at 673K				
Si	1.0	1/4	0	1/2
Al	1.0	1/4	1/2	0
O	1.0	0.1450(9)	0.489(2)	0.1550(9)
Na	0.75	0.235(1)	0.235(1)	0.235(1)

The crystal structure of $Na_6[AlSiO_4]_6*8H_2O$ is strongly determined by hydrogen bonding between the nonframework hydrate water oxygens (donors) and the framework oxygens (acceptors). The hydrogen bonding scheme is destroyed during dehydration (*Zeo.* 6:367(1986)).

Synthesis

ORGANIC ADDITIVES
TMA^+
dioxane
trimethylamineH^+

Sodalite can be synthesized by various techniques, such as solid state reactions, hydrothermal growth, and transformation of zeolites. Hydroxysodalite is a synthetic analog of the mineral sodalite, whose typical formula is $Na_2O*Al_2O_3*2SiO_2*2.5H_2O$. This structure has been synthesized in the presence of Na, TMA,

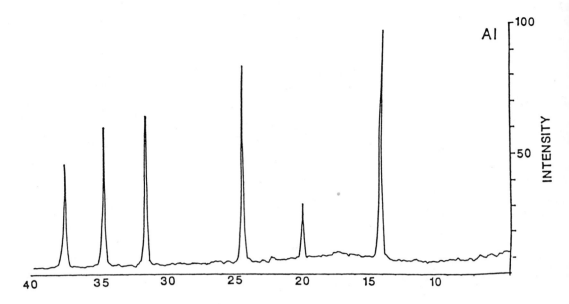

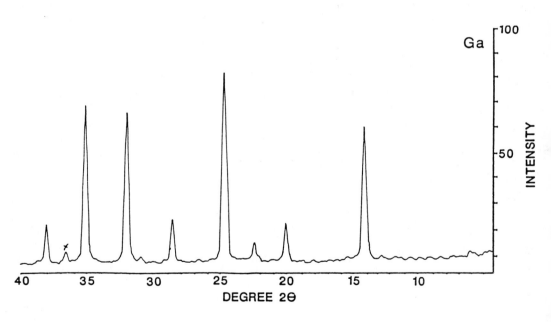

Sodalite Fig. 2:   X-ray powder diffraction pattern for sodalite (top) and gallosilicate sodalite (bottom). (Reproduced with permission of author

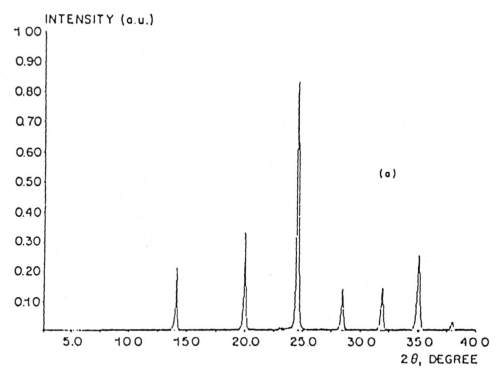

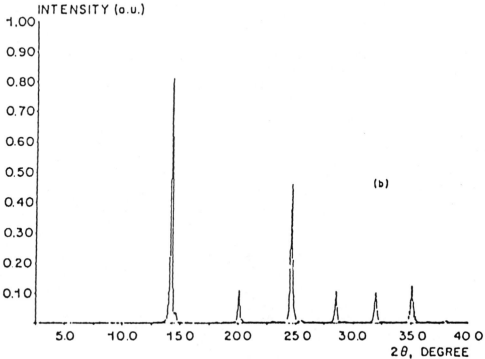

Sodalite Fig. 3:   X-ray powder diffraction pattern for trioxane-sodalite samples (a) as synthesizes and (b) after calcination (*SSSC* 49A:237(1989)). (Reproduced with permission of Elsevier Publishers)

Conversion of cubic P zeolite to TMA sodalite at 130°C (*Zeo.* 8:166(1988)):

	Initial state	Final state
Zeolite	Cubic P	TMA-sodalite
pH	12.39	12.83
Relative mass	1.00	2.00
SiO_2/Al_2O_3	6.00	13.00
Na_2O/Al_2O_3	1.03	0.56

trimethylamineH$^+$, (Na,TMA), (Na,K), (Na,Li), (Ca,TMA), and (Li,Cs,TMA), but the preferred cation appears to be sodium (*Zeo.* 1:130(1981)).

Hydrothermal crystallization in the system: $Na_2O : SiO_2 : Al_2O_3 : B_2O_5 : H_2O$ results in the dominant formation of tetrahydroborate sodalite, $Na_8[AlSiO_4]_6[B(OH)_4]_2$, in the temperature range 473 to 773K at pressures of 100 to 150 MPa and concentration of reactants in the ratio: $17\ Na_2O : 2\ SiO_2 : Al_2O_3 : x\ B_2O_5 : y\ H_2O$ ($x = 0.1$–10; $y = 260$–290). The $[B(OH)_4]^-$ anion is at the center of the sodalite cage (*Zeo.* 9:40(1989)). Elements such as germanium, gallium, beryllium, and iron have been shown to occupy framework sites in the sodalite system.

The pure silica end member can be prepared in the presence of TMA$^+$ cation, as well as the non-nitrogen-containing organic trioxane. The trioxane narrowly fits in the beta cages of this structure. Crystallization occurs from a batch composition of: $40\ SiO_2 : 0.4\ Al_2O_3 : 4.5\ Na_2O : 1000\ H_2O : 30$ trioxane (*SSSC* 49A:237(1989)).

Silica-sodalite can be prepared from non-aqueous systems in the presence of ethylene glycol, trapping the ethylene glycol within the cages of the structure (*Nature* 317:157(1985)).

The aluminate sodalite where the framework contains $[Al_{12}O_{24}]^{-12}$ has been prepared. The tetrahedral cage anions are XO_4, where X = S, Cr, Mo, and W. The aluminate sodalite is a prime example of the violation of Lowensteins's rule. This material was prepared by twice firing pellets of stoichiometric mixtures of the appropriate oxides at 1340 to 1400°C for 12 hours (*Phys. Chem. Minerals* 14:419(1988)) and exhibits a strong interaction between the framework and the XO$_4$ unit (*J. Incl. Phenom.* 5:279(1987)).

Thermal Properties

Upon dehydration of basic hydroxysodalites, partial collapse of the framework structure is observed upon release of the hydrated water (*J. Am. Chem. Soc.* (56:523(1973)). In the non-basic sodalite, the hydrogen bonding scheme in this structure is destroyed during a two-step dehydration at 400 to 500K. Thermal treatment of silica sodalite does not completely remove the ethylene glycol from the cages of the material, even upon heating to 1000°C (*Nature* 317: 357(1985)).

Salt-filled sodalites and their stability (*JCS* A1516 (1970)):

Stability in DTA	Salt
Stable above 850°C	Na_2SO_3
	Na_2SO_4
Stable at 850°C	NaCl
	NaBr
	NaI
Lattice breakdown at 850°C	NaOH
	CH_3COONa
	Na_2CO_3
	$NaHCO_3$
	$Na_2C_2O_4$
	Na_2WO_4
	Na_3PO_4
	$NaClO_4$
	$NaClO_3$
	Na_2S

Infrared Spectra

Infrared absorption bands (cm^{-1} for TMA-sodalite and Ga-TMA-sodalite (*JCS Dalton Trans.* 623 (1987)):

Assignment	TMA-sodalite	Ga-TMA-sodalite
H$_2$O	3430	3430
TMA+	—	3040
H$_2$O	1650	1650
TMA+	1490	1490
TMA+	1420	1420
	1080	1020
TMA+	960	—
	760	760
	720	670
	600	595
	460	450
	—	330

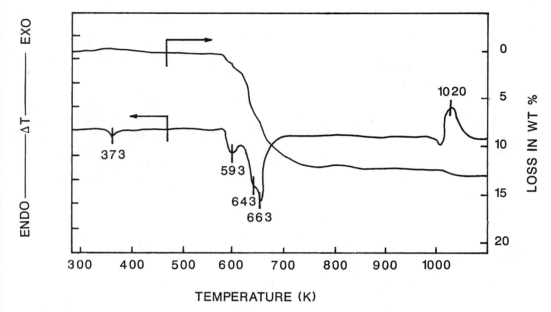

Sodalite Fig. 4: DTA/TG profiles of sodium gallosilicate sodalite (*Zeo*. 5:11(1985)). (Reproduced with permission of Butterworths Publishers)

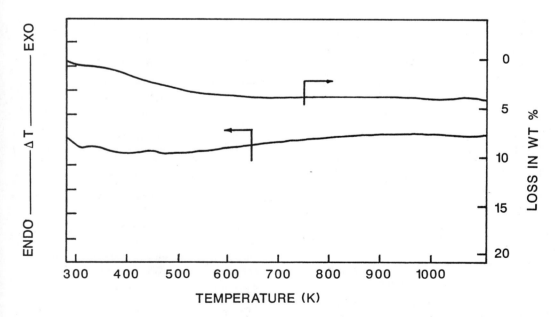

Sodalite Fig. 5: DTA and TG profiles of normal hydroxysodalite (*Zeo*. 5:11(1985)). (Reproduced with permission of Butterworths Publishers)

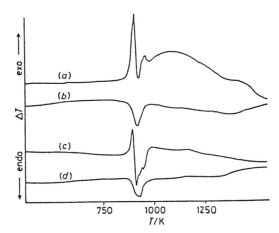

Sodalite Fig. 6:   DTA curves for TMA-sodalite (a) in air and (b) in a nitrogen atmosphere, and for gallium TMA-sodalite (c) in air and (d) in a nitrogen atmosphere (*JCS Dalton Trans.* 641(1987)). (Reproduced with permission of the Chemical Society)

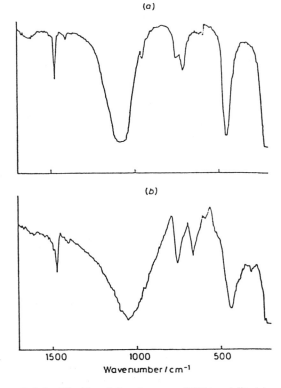

Sodalite Fig. 7:   Infrared spectra of TMA-sodalite (a) and gallium TMA-sodalite (b) (*JCS Dalton Trans.* 641(1987)). (Reproduced with permission of the Chemical Society)

Adsorption

This structure adsorbs 18 wt% water and ethanol in very small amounts (*Dokl. Akad. Nauk SSSR* 154:419(1964)). Sodalite also can intercalate sodium hydroxide as well as a variety of other salts, as described above. The thermal motion of water in the hydroxysodalite was examined as a function of temperature between 120 and 420K. Below 160K the water molecules are rigid, with a reorientation time greater than about 10 microseconds. Above 160K thermal motion becomes extremely fast, and transport between cages is observed (*Zeo.* 3:209(1983)).

Sodalite hydrate (SOD)

See sodalite.

Species P (KFI)
(*JCS* London A:2735(1970))

Related Materials: ZK-5

Species P is a name for a synthetic salt-bearing aluminosilicate compositionally related to the synthetic zeolite ZK-5 (*Z. Kristallogr.* 135:374 (1972)).

Structure

(*Z. Kristallogr.* 135:374(1972))

CHEMICAL COMPOSITION: $Ba_{15}[Al_{30}Si_{66}O_{192}]^*$
1.7$Ba(OH)_2$*12.5$BaCl_2$*35H_2O
SYMMETRY: cubic
UNIT CELL CONSTANT (Å): $a = 18.65$
PORE STRUCTURE: eight-member-ring openings 3.9 × 3.9 Å

X-Ray Powder Diffraction Data: (d(Å) (I/I_0))
13.29(vvs), 9.37(mw), 4.99(—), 4.66(m), 4.40(vw), 4.17(vs), 3.98(w), 3.81(s), 3.65(m), 3.30(vw), 3.19(s), 3.10(vw), 3.02(vvs), 2.94(vvw), 2.87(vw), 2.80(s), 2.74(vvw), 2.69(m), 2.58(vvw), 2.53(s), 2.49(w), 2.45(vvw), 2.36(vw), 2.32(w).

The framework structure of species P is related to that of ZK-5, built up of 26-hedral cavities, linked through pairs of six-member ring.

Atom parameters for species P (Z. *Kristallogr.* 135:374 (1972)):

Atom	x	y	z	Fractional Occupancy
Si,Al	0.0814(20)	0.1937(20)	0.3211(20)	
O(1)	0.1268(30)	0.1268	0.3156(50)	
O(2)	0.2574(40)	0.2574	0.4039(40)	
O(3)	0	0.1834(40)	0.3400(40)	
O(4)	1/4	0.1169(60)	0.3831	
Ba(1)	0	1/4	1/2	0.60(1)
Ba(2)	0	0.3877(20)	0	0.39(1)
Ba(3)	0.1194(30)	0.1194	0	0.18(1)
Ba(4)	0.1413(10)	0.1413	0.1413	0.34(1)
Cl(1)	0.4060(20)	0.4060	0	0.87(2)
Cl(2)	0	0.1540(150)	0	0.20(4)
Cl(3)	0	0.2543(150)	0	0.30(5)

These form hexagonal prisms and intermediate 18-hedral cavities.

The nonframework species were located in the 18-hedra as well as the 26-hedra of this structure. No definite positions were assigned to the adsorbed water molecules. Ba(1) is located at the center of the puckered eight-ring shared jointly by two 18-hedra, resulting in two sets of four barium–oxygen contacts. Four halogen atoms occupy the 18-hedra, with two approaching Ba(1) from each side. Ba(2) occupies a fourfold axis within the 18-hedra close to the planar eight-member rings. Ba(4) occupies sites that are close to three of the six oxygen atoms in the six-rings of the 26-hedra cavity.

Synthesis

ORGANIC ADDITIVES
none

Species P is prepared from the addition of $BaCl_2$ at 250°C after 4 days, starting with chabazite or zeolite Y as well as an aluminosilicate gel.

Species Q (KFI)
(*JCS* London A:2735(1970))

Related Materials: ZK-5

Species Q is a name for a synthetic salt-bearing aluminosilicate compositionally related to the synthetic zeolite ZK-5 (Z. *Kristallogr.* 135: 374(1972)).

Structure

(Z. *Kristallogr.* 135:374(1972))

CHEMICAL COMPOSITION: $Ba_{15}[Al_{30}Si_{66}O_{192}]$*
  $1.6Ba(OH)_2*11.9BaBr_2*3OH_2O$
SYMMETRY: cubic
UNIT CELL CONSTANT (Å): $a = 18.66$
PORE STRUCTURE: eight-member-ring openings 3.9
  × 3.9 Å

X-Ray Powder Diffraction Data: (d(Å) (I/I_0))
13.32(vvs), 9.49(mw), 5.92(vvs), 5.41(vvw),
5.01(vw), 4.41(m), 4.19(m), 3.82(s), 3.67(w),
3.31(vw), 3.20(vs), 3.12(vw), 3.03(vvs),
2.88(m), 2.82(vs), 2.79(vw), 2.64(s), 2.59(vw),
2.54(s), 2.50(vw), 2.45(vw), 2.37(vw), 2.33(m).

See species P for a description of the structure.

Synthesis

ORGANIC ADDITIVES
none

Species Q is prepared by the addition of $BaBr_2$ at 230°C after 2 to 4 days, starting with synthetic analcime, natural chabazite, or zeolite Y, as well as an aluminosilicate gel.

Atom parameters for species Q (Z. *Kristallogr*. 135:374 (1972)):

Atom	x	y	z	Fractional Occupancy
Si,Al	0.0795(20)	0.2031(20)	0.3175(20)	
O(1)	0.1206(30)	0.1206	0.3253(40)	
O(2)	0.2471(40)	0.2471	0.3840(40)	
O(3)	0	0.1705(40)	0.3497(40)	
O(4)	1/4	0.1178(60)	0.3822	
Ba(1)	0	1/4	1/2	0.48(1)
Ba(2)	0	0.3821(20)	0	0.34(1)
Ba(4)	0.1430(10)	0.1430	0.1430	0.20(1)
Br(1)	0.3998(10)	0.3998	0	0.57(2)
Br(2)	0	0.2065(80)	0	0.18(4)

Sr-D (FER)
(*JCS* 485(1964))

Related Materials: ferrierite-type

Structure

Chemical Composition: Si/Al ca. 3.5.

X-Ray Powder Diffraction Data: (d(Å) (I/I_0))
9.49(75), 7.07(20), 6.96(15), 6.61(55), 5.77(15), 5.43(5), 4.96(15), 4.76(15), 3.99(45), 3.94(35), 3.86(25), 3.78(50), 3.74(10), 3.67(30), 3.555(10), 3.536(90), 3.483(100), 3.3389(15), 3.313(20), 3.142(55), 3.058(45), 2.960(25), 2.938(25), 2.897(35).

Synthesis

ORGANIC ADDITIVES
none

Sr-D crystallizes from a reactive aluminosilicate gel using hydrated alumina and silica sol in the presence of the Sr^{+2} cation. The batch composition that results in the formation of this phase is: $SrO : Al_2O_3 : 9\ SiO_2 : 485\ H_2O$. Crystallization required temperatures of 340°C and 10 days, and lath-like crystals resulted under these conditions.

Sr-F (GME)
(*JCS* 485(1964))

Related Materials: gmelinite-type

Structure

Chemical Composition: aluminosilicate.

X-Ray Powder Diffraction Data: (d(Å) (I/I_0))
11.9(ms), 9.40(w), 7.65(vw), 6.93(vw), 5.95(vvw), 5.54(vw), 5.14(w), 4.99(s), 4.53(mw), 4.32(m), 4.12(s), 3.98(m), 3.89(vw), 3.54(mw), 3.46(ms), 3.32(w), 3.19(w), 2.99(s), 2.93(ms), 2.87(m), 2.675(ms), 2.608(s), 2.302(m), 2.091(ms), 2.050(vvw), 1.947(vvw), 1.912(vvw), 1.801(m), 1.727(s), 1.694(w).

Synthesis

ORGANIC ADDITIVES
none

Sr-F crystallizes from a reactive aluminosilicate gel using hydrated alumina and silica sol in the presence of the Sr^{+2} cation. The batch composition that results in the formation of this phase is: $SrO : Al_2O_3 : 4.4\ SiO_2 : 485\ H_2O$. Crystallization required temperatures of 205°C and 6 days. Hexagonal crystals resulted, but reproducibility of this synthesis was found to be difficult.

Sr-G (CHA)
(*JCS* 485(1964))

Related Materials: chabazite-type

Structure

Chemical Composition: aluminosilicate.

X-Ray Powder Diffraction Data: $(d(\text{Å}) (I/I_0))$
9.43(m), 6.80(m), 5.54(m), 5.12(m), 4.48(vvw), 4.37(vw), 4.30(ms), 3.95(m), 3.68(w), 3.21(w), 3.12(m), 2.95(vw), 2.91(s), 2.82(vvw), 2.78(vw), 2.74(vvw), 2.54(m), 2.28(w), 2.09(mw), 1.81(mw), 1.71(mw).

Synthesis

ORGANIC ADDITIVES
none

Sr-G crystallizes from a reactive aluminosilicate gel using hydrated alumina and silica sol in the presence of the Sr^{+2} cation. The batch composition that results in the formation of this phase is: $SrO : Al_2O_3 : 3 SiO_2 : 485 H_2O$. Crystallization required temperatures of 150°C and 35 days.

Sr-I (ANA)
(*JCS* 485(1964))

Related Materials: may be analcime-type

Structure

Chemical Composition: aluminosilicate.

X-Ray Powder Diffraction Data: $(d(\text{Å}) (I/I_0))$
5.60(s), 4.87(mw), 3.68(vw), 3.44(vs), 2.93(s), 2.806(vw), 2.692(m), 2.509(m), 2.423(m), 2.227(m), 1.906(ms), 1.757(ms), 1.719(vw), 1.695(vw), 1.596(mw), 1.415(ms), 1.360(ms).

Synthesis

ORGANIC ADDITIVES
none

Sr-I crystallizes from a reactive aluminosilicate gel using hydrated alumina and silica sol in the presence of the Sr^{+2} cation. The batch composition that results in the formation of this phase is: $SrO : Al_2O_3 : 3 SiO_2 : 485 H_2O$. Crystallization required temperatures of 380°C and 3 days. The resulting product was impure and of limited reproducibility.

Sr-M (MOR)
(*JCS* 485(1964))

Related Materials: mordenite-type

Structure

Chemical Composition: aluminosilicate.

X-Ray Powder Diffraction Data: $(d(\text{Å}) (I/I_0))$
13.56(mw), 9.10(m), 6.63(m), 6.43(w), 6.07(vw), 5.86(m), 4.76(mw), 4.55(m), 4.32(vw), 4.18(m), 4.00(ms), 3.84(vw), 3.76(w), 3.64(w), 3.48(s), 3.40(ms), 3.23(s), 3.04(mw), 2.92(s), 2.69(mw), 2.563(ms), 2.523(ms), 2.182(w), 2.090(w), 1.886(w), 1.821(w), 1.798(vw), 1.705(vw), 1.638(vw).

Synthesis

ORGANIC ADDITIVES
none

Sr-M crystallizes from a reactive aluminosilicate gel using hydrated alumina and silica sol in the presence of the Sr^{+2} cation. The batch composition that results in the formation of this phase is: $SrO : Al_2O_3 : 7 SiO_2 : 485 H_2O$. Crystallization required temperatures of 300°C and 5 days.

Sr-Q (YUG)
(*JCS* 485(1964))

Related Materials: yugawaralite-type

Structure

Chemical Composition: Si/Al = 3.

X-Ray Powder Diffraction Data: (d(Å) (I/I_0))
7.80(5), 6.91(25), 6.28(15), 5.85(95), 4.74(80), 4.65(40), 4.30(45), 4.16(25), 3.93(10), 3.76(30), 3.48(5), 3.30(20), 3.26(70), 3.231(5), 3.137(10), 3.105(35), 3.030(100), 2.928(60), 2.76(20), 2.735(20), 2.650(25).

Synthesis

ORGANIC ADDITIVES
none

Sr-Q crystallizes from a reactive aluminosilicate gel using hydrated alumina and silica sol in the presence of the Sr^{+2} cation. The batch composition that results in the formation of this phase is: SrO : Al_2O_3 : 8 SiO_2 : 485 H_2O. Crystallization required temperatures of 340°C and 5 to 7 days.

Sr-R (HEU)
(*JCS* 485(1964))

Related Materials: heulandite-type
(low purity)

Structure

Chemical Composition: aluminosilicate.

X-Ray Powder Diffraction Data: (d(Å) (I/I_0))
9.04(m), 7.99(m), 5.28(mw), 5.23(ms), 4.66(w), 4.37(w), 3.98(vs), 3.92(w), 3.84(vw), 3.73(vw), 3.58(mw), 3.44(w), 3.41(w), 3.33(w), 3.18(mw), 3.13(mw), 3.08(vw), 2.97(ms), 2.80(m), 2.74(m), 2.67(vw), 2.28(vw), 1.96(vw), 1.83(vw).

Synthesis

ORGANIC ADDITIVES
none

Sr-R crystallizes from a reactive aluminosilicate gel using hydrated alumina and silica sol

in the presence of the Sr^{+2} cation. The batch composition that results in the formation of this phase is: SrO : Al_2O_3 : 9 SiO_2 : 485 H_2O. Crystallization required temperatures of 250°C and 23 days.

SSZ-13 (CHA)
(US4,544,538(1985))
Chevron Research Company

Related Materials: chabazite

Structure

Chemical Composition: Si/Al = 6.

X-Ray Powder Diffraction Data: (d(Å) (I/I_0))
9.120(39), 6.790(12), 6.280(12), 5.420(54), 4.960(26), 4.620(2), 4.210(100), 4.917(5), 3.926(6), 3.826(6), 3.552(36), 3.401(22), 3.187(5), 2.877(59), 2.858(19), 2.814(2), 2.741(2), 2.576(5), 2.483(6).

SSZ-13 exhibits properties characteristic of a small-pore, eight-member-ring molecular sieve.

Comparison of rhombohedral unit cell parameters for SSZ-13 and Ca-chabazite (*Zeo.* 8:166 (1988):

Zeolite	a (Å)	α	Vol/cell (A^3)
SSZ-13 as prepared	9.291	93.92	796
SSZ-13 calcined (550°C)	9.269	94.33	789
Ca-chabazite	9.420	94.47	828

Synthesis

ORGANIC ADDITIVE
N,N,N-trimethylaminoadamantane$^+$

SSZ-13 crystallizes from a reactive aluminosiliciate gel through initial conversion to cubic zeolite P at pH around 12.35 and SiO_2/Al_2O_3 starting at 6. Crystallization occurs at 130°C with the organic additive being incorporated into the final product (*Zeo.* 8:408(1988)).

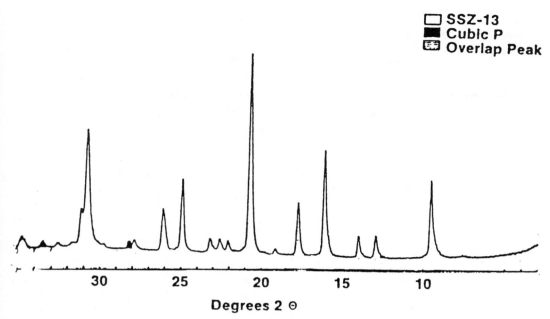

☐ SSZ-13
■ Cubic P
⊞ Overlap Peak

Degrees 2 Θ

SSZ-13 Fig. 1:   X-ray powder diffraction pattern of SSZ-13 (*Zeo.* 8:166(1988)). (Reproduced with permission of Butterworths Publishers)

SSZ-15
(US4,610,854(1986))
Chevron Research Company

Structure

Chemical Composition: 0.5–1 R_2O : 0–0.5 M_2O : Al_2O_3 : > 5 SiO_2.

X-Ray Powder Diffraction Data: ($d(\text{Å})$ (I/I_0))
11.16(37), 9.26(63), 7.53(5), 6.83(8), 5.96(8), 5.81(8), 4.62(74), 4.47(26), 4.31(100), 4.07(11), 3.99(40).

Synthesis

ORGANIC ADDITIVE
cyclopentyl trimethylammonium$^+$

SSZ-15 crystallizes from a reactive alumi-nosilicate gel in the presence of cyclopentyltri-

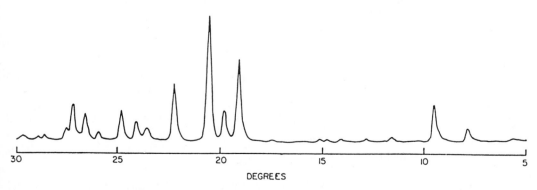

DEGREES

SSZ-15 Fig. 1:   X-ray powder diffraction pattern for SSZ-15 (US4,610,854(1986)).

methylammonium$^+$ cation in a ratio of R/Al_2O_3 between 5 and 20. The SiO_2/Al_2O_3 ratio of the crystallizing gel is high (greater than 80), thus producing a very high-silica molecular sieve. The OH/SiO_2 is greater than 0.40, with crystallization requiring 3 to 6 days at 150°C. Quartz and magadiite are two reported impurity phases observed in this synthesis (US4,620,854(1986)). Calcination to 704°C does not alter the X-ray diffraction pattern significantly; however, the organic, which is trapped within the pores of the material, is not completely removed by calcination.

SSZ-16
(US4,508,837(1985))
Chevron Research Company

Structure

Chemical Composition: 0.5–1.0 R_2O : 0–0.5 M_2O : Al_2O_3 : > 5 SiO_2.

X-Ray Powder Diffraction Data: ($d(Å)$ (I/I_0)) 11.74(32), 10.13(81), 7.63(46), 6.77(14), 5.87(22), 5.64(62), 5.60(49), 5.07(68), 5.02(54), 4.84(5), 4.62(5), 4.44(19), 4.34(100), 4.07(95), 3.917(22), 3.800(32), 3.418(59), 3.394(24), 3.317(5), 3.222(68), 3.100(11), 2.979(16), 2.912(73).

From constraint index data, SSZ-16 exhibits properties characteristic of a small-pore material.

Synthesis

ORGANIC ADDITIVE
1,4-di(1-azoniabicyclo[2.2.2]octane)butyl dibromide

SSZ-16 crystallizes from reactive aluminosiliciate gels with a SiO_2/Al_2O_3 ratio between 15 and 200. Crystallization occurs readily at 140°C after 3 to 6 days. Generally the cubic P zeolite is observed to form first, prior to the formation of SSZ-16 after 3 days. Other impurity phases observed include mordenite. Both stirred and unstirred autoclaves have been utilized in this synthesis with little observable difference. The pH range of crystallization is between 12.2 and 12.8. The organic additive is incorporated into the final crystalline product (*Zeo.* 8:409(1988)).

SSZ-17 (LEV)
(*Zeo.* 8:406(1988))
Chevron Research Company

Related Materials: levyne

Structure

Chemical Composition: aluminosilicate.

X-Ray Powder Diffraction Pattern: ($d(Å)$) (I/I_0)) 10.11(8), 8.01(33), 7.56(1), 6.56(19), 5.50(10), 5.07(79), 4.94(14), 4.69(6), 4.62(2), 4.39(3.5), 4.21(56), 4.01(100), 3.78(35), 3.54(6), 3.42(3), 3.27(18), 3.18(2), 3.12(48), 3.03(9).

Synthesis

ORGANIC ADDITIVE
N-methylquinuclidine$^+$

SSZ-17 crystallizes at 135°C after 4 to 5 days in the presence of the substituted quinuclidine, used to encourage crystallization. The pH range for this system is between 12.2 and 12.8. The organic is observed to be present, incorporated into the final crystalline product (*Zeo.* 8:406(1988)).

SSZ-19
(US4,510,256(1985))
Chevron Research Company

Structure

Chemical Composition: 0.5–1.0 R_2O : 0–0.5 M_2O : Al_2O_3 : > 6 SiO_2.

X-Ray Powder Diffraction Data: ($d(Å)$ (I/I_0)) 11.22(22), 7.55(50), 6.49(40), 5.59(13), 4.93(70), 4.62(40), 4.25(12), 3.92(100), 3.75(56), 3.25(70), 2.98(16), 2.91(22).

Characterization of this material indicates only the presence of a small-pore system.

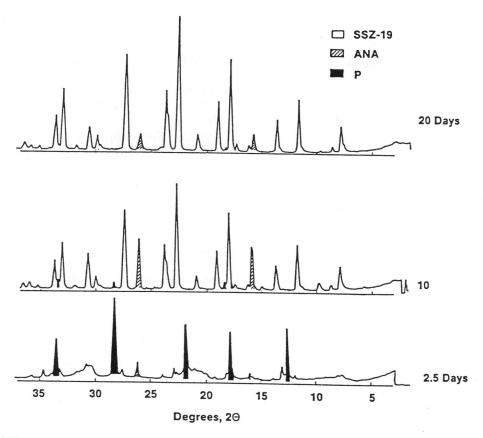

SSZ-19 Fig. 1:   X-ray powder diffraction pattern for SSZ-19 with some analcime and P impurities (*Zeo.* 8:409(1988)). (Reproduced with permission of Butterworths Publishers)

Synthesis

ORGANIC ADDITIVES

N,N,N-trimethylcyclopentylammonium iodide
N-ethylquinuclidinium iodide
N,N,N-trimethylneopentylammonium iodide
1-azoniaspiro [4,4] nonylbromide
4-adamantane-ammonium halide

SSZ-19 crystallizes from high-silica aluminosilicate gels with a SiO_2/Al_2O_3 greater than 15 and an inorganic oxide/Al_2O_3 between 10 and 30. The organic additives are complex quaternary amines, with 1-azoniaspiro [4,4] nonyl bromide producing the highest-purity products. The OH/SiO_2 falls between 0.95 and 1.1. A common impurity found in the product is analcime. Crystallization occurs readily after 6 days at 150°C.

Using the adamantane-ammonium cation as the organic, SSZ-19 crystallizes after 3 to 4 days at 145°C. It has been observed in this system that mixtures of analcime and zeolite P crystallize prior to the onset of SSZ-19 crystallization (*Zeo.* 8:409(1988)). When N-ethylquinuclidine is used, the initial formation of a transient cubic P phase is observed after several days. After 20 days the final crystals of SSZ-19 are formed (*Zeo.* 8:409(1988)).

Adsorption

SSZ-19 adsorbs 11.5 wt% water after 24 hours and adsorbs no *n*-hexane (US4,510,256(1985)).

SSZ-23
(EP 231,018(1987))
Chevron Research Company

Structure

Chemical Composition: 0.1–3 R_2O : 0.1–2 M_2O : Al_2O_3 : > 50 SiO_2.

X-Ray Powder Diffraction Data   (d(Å) (I/I_0))
10.85(100), 10.31(45), 9.31(55), 8.39(40), 5.04(45), 4.79(80), 4.52(65), 4.43(65), 4.13(100), 4.011(50), 3.914(70), 3.580(45).

SSZ-24 (AFI)
(EP 231,019(1987))
Chevron

Related Materials: AlPO$_4$-5

Structure

Chemical Composition: 0.1–10 R_2O : 0.1–5.0 M_2O : Al_2O_3 : > 100 SiO_2.

X-Ray Powder Diffraction Data: (d(Å) (I/I_0))
11.79(98), 6.81(10), 5.906(46), 4.459(100), 4.184(36), 3.927(87), 3.544(4), 3.406(44), 3.041(13), 2.949(30), 2.631(5), 2.574(20), 2.404(4), 2.351(7), 2.152(4), 2.118(4).

Lattice parameters Å for the hexagonal unit cells of AlPO$_4$-5 and SSZ-24 (template for AlPO$_4$-5 is tetrapropylammonium hydroxide) (*ACS* 368:236 (1988):

	a	b
AlPO$_4$-5		
with template	13.726	8.484
calcined	13.77	8.38
SSZ-24		
with template	13.62	8.296
calcined	13.62	8.324

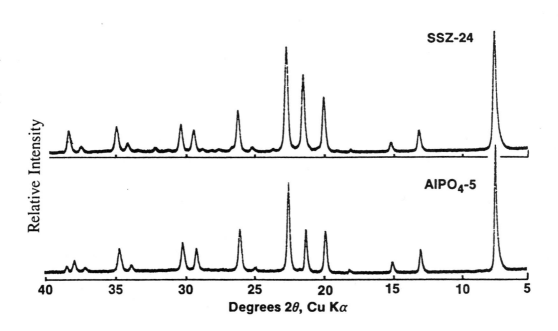

SSZ-24 Fig. 1:   X-ray powder diffraction pattern for SSZ-24 and AlPO$_4$-5 (*ACS Symposium Series* 368:236(1988), *Perspectives in Molecular Sieve Science*, Flank, W. H., Whyte, T. E., Jr., eds.). (Reproduced with permission of the American Chemical Society)

SSZ-24 is isostructural with AlPO$_4$-5, containing a unidimensional 12-member-ring channel system.

Synthesis

ORGANIC ADDITIVE
N,N,N-trimethyl-1-adamantammonium hydroxide

SSZ-24 crystallizes from a silicate gel system where the SiO$_2$/Al$_2$O$_3$ ratio is greater than 50. The OH/SiO$_2$ ratios range between 0.2 and 0.3, and the organic-to-SiO$_2$ ratio is between 0.1 and 0.2. Sodium is the alkali cation employed in this crystallization. The crystallization occurs readily in a stirred autoclave at 160°C after 6 days.

Infrared Spectrum

Mid-infrared Vibrations (cm^{-1}): 1213sh, 1104(s), 1060(s), 811(m), 712(w), 692(w), 621(mw), 567(w), 518(w), 460(ms).

Adsorption

Adsorption properties at 127°C in units of mg/g (*ACS* 368:236 (1988)):

	AlPO$_4$-5	SAPO-5	SSZ-24
2,2-DMB[1]	27	21	26
3MP[2]	13	12	15
n-C6[3]	6	9	12

1. 2,2-dimethylbutane; 2. 3-methylpentane; 3. n-hexane.

SSZ-25
(EP 231,019(1987))
Chevron Research Company

Related Materials: MCM-22
                        PSH-3

Structure

Chemical Composition: 0.1–2 R$_2$O : 0.1–2.0 M$_2$O : Al$_2$O$_3$: > 20 SiO$_2$.

X-Ray Powder Diffraction Data: (as-synthesized) (d(Å) (I/I$_0$)) 29.0(20), 13.77(100), 12.31(100), 11.22(47), 9.19(53), 5.63(27),

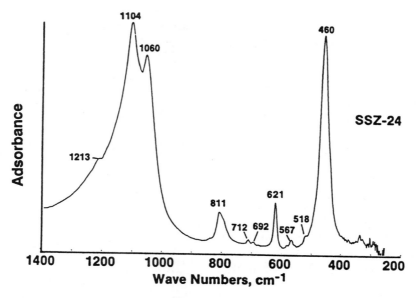

SSZ-24 Fig. 2:   Mid-infrared vibrations for SSZ-24 (*ACS Symposium Series* 368:236(1988), *Perspectives in Molecular Sieve Science*, Flank, W. H., Whyte, T. E., Jr., eds.). (Reproduced with permission of the American Chemical Society)

4.58(47), 3.94(50), 3.86(30), 3.42(73), 3.32(33); (calcined) 25.5(17), 12.30(100), 11.00(55), 8.78(63), 6.17(40), 5.51(17), 3.90(38), 3.47(20), 3.417(65).

Synthesis

ORGANIC ADDITIVE
N,N,N-trimethyl-1-adamantammonium hydroxide

SSZ-25 crystallizes from an aluminosilicate gel system where the SiO_2/Al_2O_3 ratio is preferably between 30 and 100. The OH/SiO_2 ratios range between 0.2 and 0.4, and the organic-to-SiO_2 ratio is between 0.15 and 0.3. Sodium is the alkali cation employed in this crystallization. The crystallization occurs readily in a stirred autoclave at 175°C after a week (US 4,826,667(1989)).

SSZ-26
(PCT WO 89/09185(1989))
Chevron Research Company

Structure

Chemical Composition: 0.1–2.0 R_2O : 0.1–2.0 M_2O : Al_2O_3 : 10–200 SiO_2.

X-Ray Powder Diffraction Data: (as-synthesized) (d(Å) (I/I_0)) 11.36(100), 4.389(63), 4.158(25), 4.042(53), 3.890(48sh), 3.867(64), 3.365(33); (calcined 540°C) (d(Å) (I/I_0)) 11.36(100), 4.392(18), 4.164(5), 4.044(15), 3.878(13sh), 3.853(19), 3.366(12).

Synthesis

ORGANIC ADDITIVES:
hexamethyl [4.3.3.0] propellane-8,11-diammonium^{+2}

SSZ-26 crystallizes from a reactive aluminosilicate gel in a ratio (SiO_2/Al_2O_3) between 20 and 100. Sodium aluminate and fumed silica are preferred sources of aluminum and silicon. Ranges over which this structure is observed to form include:

OH/SiO_2: 0.20 to 0.50
Organic/SiO_2: 0.05 to 0.20
Na^+/SiO_2: 0.15 to 0.30
Water/SiO_2: 25 to 60
Organic/organic + Na^+: 0.30 to 0.67

Crystallization occurs between 150 and 170°C after 5 to 10 days in a slowly stirred autoclave. Rapid stirring results in co-crystallization with another zeolite phase. Impurity phases include Y zeolite, analcime, and quartz.

Thermal Properties

No structural degradation was reported to 820°C.

Adsorption

N_2 BET adsorption gives a surface area of 560 m^2/g for SSZ-26 and a micropore volume of 0.19 cc/g.

SSZ-31
(WP WO 90/04567(1990))
Chevron Research Company

X-Ray Powder Diffraction Data: as synthesized		Calcined SSZ-31	
(d/n)	I/I_0	d(Å)	I/I_0
20.7	5	17.5	2
14.49	6	14.49	27
12.01	30	11.96	96
10.81	11	10.80	43
8.25	1	8.55	1
7.36	1	8.18	6
6.18	1	7.25	2
6.02	1	6.13	14
5.57	2	5.97	9
5.08	7	5.54	1
4.811	9	5.06	5
4.374	15	4.806	6
4.206	69	4.360	13
4.156	9	4.210	64
3.997	100	4.127	4
3.921	7	3.969	100
3.600	23	3.742	1
3.535	11	3.583	14

(d/n)	I/I_0	d(Å)	I/I_0
3.466	5	3.534	4
3.339	9	3.401	14
3.278	5	3.326	6
3.220	5	3.220	2
3.167	2	3.164	1
3.103	4	3.084	3
3.079	3	3.060	3
3.028	2	2.995	3
2.996	5	2.979	3
2.925	2	2.885	7
2.894	11	2.770	3
2.783	5	2.726	3
2.734	6		

Synthesis

SSZ-31 is made using, N,N,N-trimethyl-8-ammonium tricyclo[5.2.1.0^{2,6}]decane under SSZ-24 reaction conditions; $SiO_2/Al_2O_3 \geq 100$.

SSZ-33
(US4,963,337(1990))
Chevron Research Company

Structure

Chemical Composition: 1.0–5 R_2O : 0.1–1.0 M_2O : W_2O_3 : > 20 SiO_2, where W is boron or mixtures with aluminum, gallium, and iron.

X-Ray Powder Diffraction Data: (d(Å) (I/I_0))
11.25(90), 10.58(2), 6.23(13), 5.62(7), 5.29(10), 4.336(100), 4.139(40), 4.035(90), 3.837(64), 3.526(13), 3.323(40), 3.116(12), 3.060(10), 2.920(8).

Synthesis

ORGANIC ADDITIVE
N,N,N-trimethyl-8-ammonium tricyclo[5.2.1.0^{2,6}]decane

The borosilicate SSZ-33 crystallizes from a reactive borosilicate gel with a preferred composition in which SiO_2/W_2O_3 is between 30 and 60, OH/SiO_2 is between 0.2 and 0.3, R/SiO_2 is

between 0.1 and 0.25, and H_2O/SiO_2 is between 25 and 60. Crystallization occurs at 160°C after 10 days.

Stellerite (STI)
(natural)

Related Materials: stilbite

Stellerite identifies the calcium-rich phase of the natural mineral stilbite.

Structure

(*Nat. Zeo.* 284(1985))

CHEMICAL COMPOSITION: $Ca_4(Al_8Si_{28}O_{72})*28H_2O$
SYMMETRY: monoclinic
SPACE GROUP: Fmmm
UNIT CELL CONSTANTS (Å): $a = 13.60$
                        $b = 18.22$
                        $c = 17.84$
PORE STRUCTURE: 8-member rings (2.7 × 5.6 Å)
  intersecting 10-member rings (4.9 × 6.1 Å)

X-Ray Powder Diffraction Pattern: (*Nat. Zeo.* 343(1985)) (d(Å) (I/I_0)) 9.03(100), 6.37(<1), 5.44(2), 5.41(3), 5.29(4), 4.65(15), 4.56(4), 4.47(2), 4.28(6), 4.06(45), 4.01(6), 3.784(1), 3.734(5), 3.482(3), 3.397(7), 3.181(7), 3.100(3),

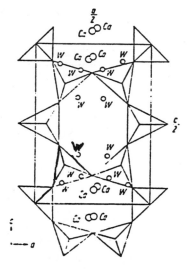

Stellerite Fig. 1S: Framework topology.

3.028(23), 3.003(10), 2.976(1), 2.875(2), 2.827(1), 2.804(1), 2.771(8), 2.703(2), 2.608(2), 2.562(4), 2.546(1), 2.532(<1), 2.512(1), 2.508(1), 2.494(2), 2.485(2), 2.452(1), 2.351(2), 2.318(1), 2.267(1), 2.239(1), 2.233(1), 2.223(1), 2.206(1), 2.201(1), 2.168(<1), 2.161(<1), 2.124(1), 2.120(1), 2.097(1), 2.078(<1), 2.066(2), 2.037(2), 2.031(2), 1.978(<1).

Atom coordinates ($\times$ 10^4) for natural stellerite (*Cryst. Res. Technol.* 21:1029 (1986)):

Atom	x	y	z	Occupation
Ca	5000	0	2908(1)	0.25
Si(1)	3858	3074(1)	3768(1)	0.69
Al(1)	3858(1)	3074(1)	3768(1)	0.31
Si(2)	3016(1)	4114(1)	5000	0.50
Si(3)	3883(1)	1838(1)	50000	0.38
Al(3)	3883(1)	1838(1)	50000	0.12
Si(4)	2500	2500	2500	0.16
Al(4)	2500	2500	2500	0.09
O(1)	1435(2)	−1198(2)	5763(2)	1.00
O(2))	1275(2)	−2680(2)	5748(2)	1.00
O(3)	0	−1848(2)	6514(2)	0.50
O(4)	1817(2)	−1966(2)	6984(2)	1.00
O(5)	3150(3)	−1135(2)	5000	0.50
O(6)	1948(4)	0	5000	0.25
O(7)	5000	−1502(3)	5000	0.25
W(1)	0	1303(7)	1982(7)	0.20
W(2)	0	0	808(12)	0.05
W(3)	1160(14)	0	3201(10)	0.24
W(4)	4539(12)	0	1599(10)	0.20
W(5)	4394(10)	0	4202(8)	0.21
W(6)	4483(11)	1201(7)	2985(7)	0.32
W(7)	3395(12)	743(9)	3138(11)	0.63
W(8)	3659(17)	0	3774(15)	0.17
W(9)	3489(27)	0	2082(20)	0.19

See stilbite for a description of the framework topology of stellerite. No Si/Al ordering is observed in this mineral.

Thermal Properties

The thermal behavior of stellerite is similar to that of stilbite with losses due to water occurring at 70, 175, 250, and 680°C.

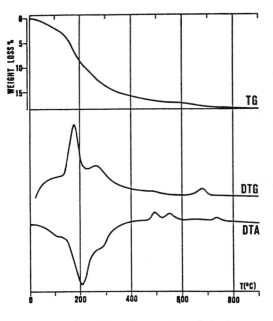

Stellerite Fig. 2:   Thermal curves for stellerite from Villanova, Monteleone, Sardinia, Italy; in air, heating rate 20°C/min (*Nat. Zeo.* 295(1985)). (Reproduced with permission of Springer Verlag Publishers)

Adsorption

Adsorption properties of natural stellerite (*Zeo. Mol. Sieve.* (1974)):

Adsorbate	Activation, temp., °C	Adsorption temp. °C	Pressure, Torr	Wt % adsorbed
H$_2$O	250	298	25	17
CO$_2$	250	298	707	9.7
O$_2$	250	77	NA	

Stilbite (STI) (natural)

Related Materials: stellerite
barrerite

Stilbite was named in 1801 by Hauy, from the Greek work for luster (*Nat. Zeo.* 285(1985)). Though the name stilbite was used to describe a number of orthorhombic lamellar zeolites in-

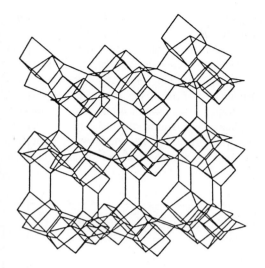

Stilbite Fig. 1S: Framework topology.

cluding heulandite, it was not until 1822 that it was used to designate a distinct structure.

Structure

(Nat. Zeo. 284(1985))

CHEMICAL COMPOSITION: $NaCa_4(Al_9Si_{27}O_{72})*30\ H_2O$
SYMMETRY: monoclinic
SPACE GROUP: C2/m
UNIT CELL CONSTANTS (Å): $a = 13.61$
$b = 18.24$
$c = 11.27$
$\beta = 127°51'$
(pseudo-orthorhombic cell)
SPACE GROUP: F2/m
UNIT CELL CONSTANTS (Å): $a_0 = 13.61$
$b_0 = 18.24$
$c_0 = 17.80$
$\beta_0 = 90°45'$
PORE STRUCTURE: 10-member rings (4.9 × 6.1 Å) and 8-member rings (2.7 × 5.6 Å)

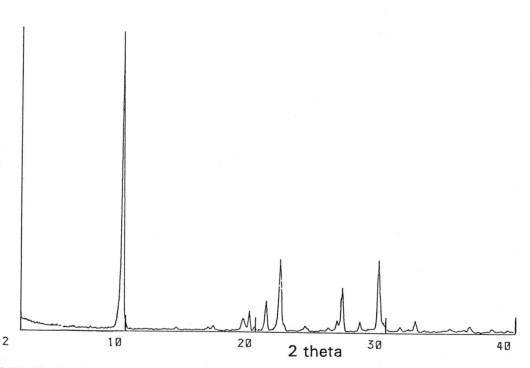

Stilbite Fig. 2:   X-ray powder diffraction pattern of stilbite from India, large crystals, some preferred orientation observed. (Reproduced with permission of author)

X-Ray Powder Diffraction Data: (*Nat. Zeo.* 342(1985)) ($d(\text{Å})$ (I/I_0)) 9.12(100), 8.88(9), 6.83(2), 6.37(2), 5.46(1), 5.30(8), 5.23(1), 4.68(16), 4.63(15), 4.56(3), 4.44(2), 4.30(9), 4.27(7), 4.06(58), 4.00(10), 3.788(2),, 3.756(6), 3.699(6), 3.504(3), 3.474(5), 3.456(2), 3.415(10), 3.396(8), 3.368(5), 3.199(13), 3.172(7), 3.122(5), 3.088(2), 3.028(36), 2.990(10), 2.965(4), 2.877(3), 2.811(4), 2.780(21), 2.751(1), 2.730(8), 2.687(1), 2.591(4), 2.577(6), 2.557(5), 2.537(3), 2.530(2), 2.510(4), 2.484(3), 2.474(3), 2.468(4), 2.444(3), 2.352(4), 2.311(2br), 2.276(1), 2.255(1), 2.239(2br), 2.216(2), 2.160(2), 2.122(3), 2.106(5), 2.101(4), 2.066(3), 2.056(4), 2.028(4), 1.995(1).

The structures of stilbite, stellerite, and barrerite are examples of the reduction of topological framework symmetry caused by repulsions

Atomic coordinates for natural stilbite (*Zeo.* 7:163 (1987)):

Atom	x/a	y/n	z/c
T(1)	0.3625(1)	0.3046(1)	0.1214(1)
T(2)	0.1347(1)	0.3097(1)	0.1304(1)
T(3)	−0.0529(1)	0.0893(1)	0.2426(1)
T(4)	−0.1384(1)	0.3169(1)	0.2426(1)
T(5)	0.000	0.2615(1)	0.000
O(1)	0.4226(2)	0.2916(2)	0.0432(2)
O(2)	0.0579(3)	0.3157(2)	0.0589(2)
O(3)	−0.1253(3)	0.2667(2)	0.1754(2)
O(4)	−0.0989(3)	0.1196(2)	0.1611(2)
O(5)	0.1211(3)	0.2302(2)	0.1719(2)
O(6)	0.1158(3)	0.3787(2)	0.1858(2)
O(7)	0.2477(2)	0.3156(2)	0.0953(2)
O(8)	0.0646(2)	0.1132(2)	0.2505(2)
O(9)	−0.0567(4)	0.000	0.2415(4)
O(10)	0.7500	0.3524(3)	0.2500
Ca	0.2340(2)	0.000	0.0471(1)
Na(1)	0.4818(9)	0.0665(6)	0.0188(7)
W(1)	0.0725(14)	0.0597(9)	0.0436(11)
W(2)	0.2732(5)	0.1266(5)	0.0523(4)
W(3)	0.1923(7)	0.000	0.1772(6)
W(4)	0.3873(9)	0.000	0.1146(8)
W(5)	0.3692(20)	0.500	0.0638(14)
W(6)	0.1236(8)	0.500	0.0534(8)
W(7)	0.1894(9)	0.000	−0.0802(7)
W(8)	0.0854(11)	0.0827(9)	0.38(3)

of the extra-framework cations (*Tschermaks. Mineral. Petrogr. Mitt.* 26:39(1976)). The sodium ions are responsible for the real symmetry of these structures. In stellerite, because of the absence of sodium, only one extra-framework site is present, and the real symmetry is the topological one, Fmmm. In barrerite, which has a space group Amma, the orthorhombic framework is a result of the electrostatic repulsion between the monovalent cations present in the

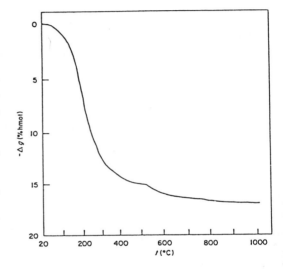

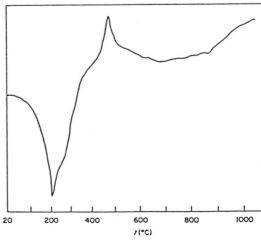

Stilbite Fig. 3: Thermal curves for stilbite: TGA (top), DTA (*Zeo.* 5:389(1985)). (Reproduced with permission of Butterworths Publishers)

site that is equivalent to the Na site of stilbite, and also in a new site.

The stilbite framework is composed of 4–4–1 building units arranged to form an open framework structure with intersecting 8- and 10-member rings. The Ca^{+2} cation is coordinated with eight molecules of water and Na$^+$ with four water molecules and two oxygen atoms of the framework. There are five unique T atom sites within the framework.

Thermal Properties

Stilbite exhibits three peaks in the range between room temperature and 500°C; the first appears at 70°C, followed by a large water loss at 175°C with a further water loss at 250°C. Dehydroxylation occurs at 500°C. Total loss in weight is 16.8%, with a major loss between 20 and 200°C of 14.8% (*Zeo.* 5:389(1985)). See Stilbite Figure 3.

Infrared Spectrum

Infrared absorption spectra of stilbite (*Zeo.* 5:389 (1985)):

	cm^{-1}
Antisym. stretching OH	3660–3360 (s)
Sym. stretching OH	3270–3150 (s)
Bending H$_2$O	1650 (s)
External TO$_4$	1300 (w)
	1150 (w)
Antisym. stretching T–O	1070–970 (vw)
Sym. stretching T—O	800 infl
	770 infl
	670 (vw)
Liberation H$_2$O	600–530 (m)

vs = very strong, s = strong, m = medium, w = weak, vw = very weak, infl = inflexion.

Ion Exchange

Stilbite has a high selectivity for cesium and potassium. Increasing the temperature from 23°C

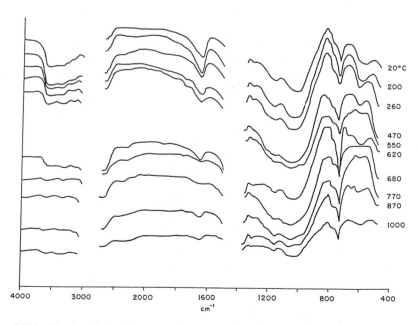

Stilbite Fig. 4:   Infrared spectrum changes as a function of temperature between 20 and 1000°C (*Zeo.* 5:389(1985)). (Reproduced with permission of Butterworths Publishers)

to 85°C produces a large drop in selectivity for cesium over sodium (*Zeo. Mol. Sieve.* 564(1974)).

Adsorption

Adsorption properties of natural stilbite (*Zeo. Mol. Sieve.* 626 (1974)):

Adsorbate	Activation °C	Temperature K	Pressure	Wt% adsorbed
CO_2	250	298	700 Torr	7.7
O_2	250	77	none adsorbed	
Ar	180	453	800 atm	31 cc/g
Kr	200	473	300 atm	30 cc/g

SUZ-2
(EP 353,914(1988))
British Petroleum

Structure

Chemical Composition: $0.5–1.5 M_2O : Al_2O_3 : > 5 SiO_2$.

X-Ray Powder Diffraction Data: (calcined, H^+ form) ($d(\text{Å})$ (I/I_0)) 11.4(s), 9.5(s), 7.1(w/m), 7.0(s), 5.8(m), 5.4(s), 4.8(m), 4.0(m/s), 3.8(vs), 3.6(s), 3.5(m), 3.1(s), 3.05(w).

Synthesis

ORGANIC ADDITIVES
quinuclidine/TEA$^+$

SUZ-2 crystallizes from reactive hydrogels with molar compositions such as: 1.23 Na_2O : 1.37 K_2O : 4.6 Q : 4.2 TEAOH : Al_2O_3 : 33.9 SiO_2 : 366 H_2O. Ranges of SiO_2/Al_2O_3 with a minimum of 10 result in the formation of this material. Crystallization occurs at 180°C under agitation after 6 days. Mordenite and quartz regularly are found as impurity phases in this system.

SUZ-4
(EP 353,915(1990))
British Petroleum

Structure

Chemical Composition: m $(M_{2/a}O)$: Al_2O_3 : at least 5 SiO_2.

X-Ray Powder Diffraction Data: ($d(\text{Å})$ (I/I_0)) 11.5(vs), 7.50(m), 7.20(w), 5.88(s), 5.73(m), 4.75(m), 4.58(ms), 4.00(w), 3.95(m), 3.81(m), 3.75(w), 3.67(w), 3.58(s), 3.55(s), 3.49(s), 3.48(m), 3.14(m), 2.97(w), 2.93(m), 2.91(w).

Synthesis

ORGANIC ADDITIVE
quinuclidine/TEA$^+$

SUZ-4 crystallizes from a reactive hydrogen with a molar composition of: 1.23 Na_2O : 2.84 K_2O : 2.9 Q : 5.2 TEAOH : 1.0 Al_2O_3 : 21.2 SiO_2 : 528 H_2O : 40 CH_3OH. Crystallization occurs at 180°C after 48 hours. The resulting crystals are stable to the calcination temperatures of 550°C used to remove the organic trapped within the structure. Mordenite is observed as a minor impurity phase in this system.

Svetlozarite

Svetlozarite is the name of a dachiardite mineral containing a stacking disorder.

T

T (ERI/OFF)
(US2,950,952(1960))
Union Carbide Corporation

Related Materials: offretite/erionite
OE

See also Linde T.

Structure

(Zeo. Mol. Sieve. 173 (1974))

CHEMICAL COMPOSITION: $(NaK_{2.8})[(AlO_2)_4$
$(SiO_2)_{14}]* 14H_2O$
SYMMETRY: orthorhombia
UNIT CELL DIMENSIONS (Å): $a = 6.62$
$b = 11.5$
$c = 15.1$
FRAMEWORK DENSITY: 1.50 g/cc
VOID FRACTION (determined from water content): 0.40
PORE STRUCTURE: three-dimensional, 3.6 × 4.8 Å

X-Ray Powder Diffraction Data: $(d(Å)(I/I_0))$
11.45(100), 9.18(4), 7.54(13), 6.63(54), 6.01(2),
5.74(6), 4.99(2), 4.57(8), 4.34(45), 4.16(3),
4.08(2), 3.82(16), 3.76(56), 3.67(1), 3.59(30),
3.42(2), 3.31(16), 3.18(12), 3.15(18), 2.93(11),
2.87(38), 2.85(45), 2.68(11), 2.61(2), 2.51(8),
2.49(13), 2.30(2), 2.21(6), 2.12(5), 2.09(3),
1.99(2), 1.96(2), 1.89(8), 1.87(2), 1.84(3),
1.78(8), 1.77(5), 1.75(2), 1.71(3), 1.66(9),
1.59(5), 1.52(1), 1.51(2), 1.47(3), 1.41(1),
1.39(3).

Zeolite T is related to both offretite and erionite, as it is a disordered intergrowth of the two types of structures. A larger portion of the material consists of the offretite structure. X-ray powder diffraction patterns show the 1 odd reflections of erionite to be very weak in this zeolite. The positions of the cations are in the D6R units, the gmelenite cages, and the single six-rings within the main cavities (*Izv. Akad. Nauk. SSSR Ser. Khim.* 6:116(1965)).

Synthesis

ORGANIC ADDITIVES
none

Zeolite T crystallization is very sensitive to changes in the ratio of OH/SiO_2 in the reaction mixture. The batch composition that results in the formation of this phase was a $Na_2O/Na_2O + K_2O$ ratio between 0.7 and 0.8, a M_2O/SiO_2 from 0.4 to 0.5, a SiO_2/Al_2O_3 ratio ranging between 20 and 28, and a H_2O/M_2O ratio from 20 to 51. Crystallization occurs at 100°C after 166 hours.

Infrared Spectrum

Mid-infrared Vibrations (cm^{-1}): (SiO_2/Al_2O_3 = 7.0) 1156(wsh), 1059(s), 1010(s), 771(w), 718(w), 623(mw), 575(w)m 467(mw), 433(ms), 410(vwsh), 366(wsh)(*ACS* 101:201(1971).

Adsorption

Adsorption properties of zeolite T (US2,950,952(1960)):

Adsorbate	Temp., °C	Pressure, Torr	Wt% adsorbed
H_2O	25	0.1	7.5
		4.5	16.2
		20	18.2
O_2	−196	0.1	8.4
		10	13.5
		100	15.8
Argon	−196	0.1	8.0
		10	13.5
		120	16.4
Propane	25	10	1.8
		100	2.5
		700	2.7
Propylene	25	10	3.7
		100	4.7
		700	5.5

(continued)

Adsorbate	Temp., °C	Pressure, Torr	Wt% adsorbed
n-Pentane	25	10	6.6
		100	8.1
		400	11.2
Cyclopropane	25	10	0.1
		100	0.8
		700	1.6
Isobutane	25	700	0.5
Thiophene	25	20	0.9
		70	4.3
Benzene	50	70	1.4
	25	27	0.9
		75	2.5
CO$_2$	25	1	3.2
		100	7.3
		700	10.0
Butene-1	25	10	1.7
		100	2.7
		700	3.3
NH$_3$	25	1	2.4
		100	6.6
		700	7.8
Krypton	−183	1	17.0
		10	23.1
		18	23.5
Cyclohexane	25	68	0.8

TEA-MOR (MOR)
(US4,052,472(1977))
Mobil Oil Corporation

TEA-MOR, designating a silica-enriched form of mordenite, was first reported by Givens, Plank, and Rosinski of Mobil in 1977 (US 4,052,472 (1977)).

TEA-Silicate (MTW)
(US4,104,294(1978))
Union Carbide Corporation

Related Materials: ZSM-12

Structure

Chemical Composition: SiO$_2$.

X-Ray Powder Diffraction Data: (d(Å) (I/I_0)) 11.9(60), 10.2(26), 4.98(5), 4.77(18),

4.29(100), 3.88(84), 3.66(16), 3.49(24), 3.39(32), 3.21(10), 3.06(8), 2.89(5), 2.65(5).

See ZSM-12 for a description of the framework topology.

Synthesis

ORGANIC ADDITIVES
TEA$^+$

TEA-silicate crystallizes from a reactive basic silica gel with a molar oxide ratio within the range: R$_2$O : 0–8 M$_2$O : 12–40 SiO$_2$: 100–500 H$_2$O, where R represents the tetraethylammonium cation. Crystal formation requires temperatures between 125 and 150°C and from 70 to 250 hours.

Adsorption

TEA-silicate adsorbs at least 2 wt% neopentane, characteristic of its large-channel system.

Tetragonal Edingtonite (EDI)
(*N. Jb. Miner. Mh.* 373(1984))

See edingtonite.

Tetragonal Natrolite (NAT)

See tetranatrolite.

Tetranatrolite (NAT)
(natural)

Related Materials: natrolite

Tetranatrolite or tetragonal natrolite has been used to describe samples found at Mt. St. Hilaire, Quebec in 1980 and also in Greenland in 1969. (Nat. Zeo. 35(1985)).

Structure

CHEMICAL COMPOSITION: Na$_{15}$(Al$_{16}$Si$_{24}$O$_{80}$)*
16H$_2$O
SYMMETRY: tetragonal
SPACE GROUP: I$\bar{4}$2d

UNIT CELL CONSTANTS (Å):   a = 13.10
c = 6.63
PORE STRUCTURE: small pore eight member rings

X-Ray Powder Diffraction Data: $(d(\text{Å})(I/I_0))$
6.53(50),   5.90(100),   4.61(25),   4.38(50),
4.12(25),   3.171(50),   3.114(25),   2.949(25),
2.914(5),   2.851(100).

Theta-1 (TON)
(EP 57,049(1982))
(British Petroleum)

Related Materials: ISI-1
KZ-2
Nu-10
ZSM-22

Structure

(*Nature* 312:533(1984))

CHEMICAL COMPOSITION: $0.9 \pm 0.2 \, M_{2/n}NO$:
$Al_2O_3 : X \, SiO_2 : y \, H_2O$ n = cation valence X < 10;
Y/x between 0 & 25:
SYMMETRY: orthorhombic
SPACE GROUP: Cmc21
UNIT CELL CONSTANTS (Å): $a = 13.836$
$b = 17.415$
$c = 5.042$

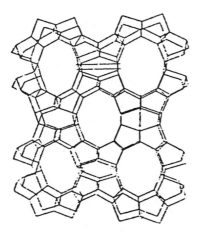

Theta-1 Fig. 1S: Framework topology.

FRAMEWORK DENSITY: 20.5 T/1000 Å³
DENSITY: 2.05 g/cm³
VOID VOLUME: 0.21
PORE STRUCTURE: unidimensional 10-member rings
4.4 × 5.5 Å

X-Ray Powder Diffraction Data: $(d(\text{Å})(I/I_0))$
10.85(100),   8.71(16),   6.93(16),   5.42(10),
4.57(10), 4.36(77), 3.70(74), 3.61(49), 3.46(23),
2.52(22).

The theta-1 structure can be constructed from complex 5–1 building units, making it part of the mordenite group of zeolite structures. The 10-member-ring elliptical channels are unidimensional in this structure. There are four crystallographically unique T atoms with relative occupancy of 4:4:8:8 (*Nature* 312:533(1984)).

Fractional atomic coordinates for theta-1 (*Acta Crystallogr.* C41:1391 (1985)):

Atom	x	y	z
Si(1)	0.2838(9)	0.0503(7)	0.250
Si(2)	0.201(1)	0.2157(8)	0.312(4)
Si(3)	0.500	0.218(1)	0.824(6)
Si(4)	0.500	0.118(1)	0.320(7)
O(11)	0.264(2)	−0.010(2)	0.493(8)
O(12)	0.214(2)	0.125(2)	0.309(8)
O(14)	0.404(2)	0.065(1)	0.280(9)
O(22)	0.257(3)	0.257(2)	0.534(7)
O(23)	0.407(2)	0.269(2)	0.795(10)
O(34)	0.500	0.152(3)	0.628(10)
O(43)	0.500	0.187(2)	0.132(9)

Synthesis

ORGANIC ADDITIVE:
diethanolamine
triethylenetetramine
methanol

Theta-1 crystallizes preferrably from a reactive gel with batch compositions containing a SiO_2/Al_2O_3 ratio greater than 40:1 and MOH/H_2O ratio greater than 2×10^{-3}:1. Crystallization can occur within 24 hours depending on the initial composition of the mixture. Crystallization takes place at 170°C. ZSM-5 and crystobalite have been observed as impurity phases in this synthesis.

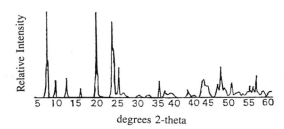

Theta-1 Fig. 2:   X-ray powder diffraction pattern for theta-1 (*Acta Crystallogr.* C41:1391(1985)). (Reproduced with permission of the International Union of Crystallography)

Infrared Spectrum

The presence of five-rings is supported by a band present in this material at 1230 to 1210 cm^{-1} (*Nature* 312:533(1984)).

Adsorption

Adsorption properties of theta-1 (*Nature* 312:533 (1981)):

Adsorption	Kinetic diameter, Å	Vg cm/g	Vf
N$_2$	3.64	0.1	0.21
H$_2$O	2.65	0.058	0.12
n-Hexane	4.3	0.089	0.18
p-Xylene	5.85	0.062	0.13
Cyclohexane	6.0	0.021	0.04
m-Xylene	6.0	0.031	0.06

Void volume, v_p (cm^3/g), is the total micropore volume in hydrogen form of theta-1, calculated by using normal liquid density at the adsorption temperature. The void fraction is $v_t = v_p \times d_c$, where d_c (2.05g/cm^3) is the measured crystal density. All samples were activated by calcination at 550^{0}C for 1 hour. Adsorption measurements were made by microbalance techniques at room temperature with relative pressure $P/P_0 = 0.5$. N$_2$ adsorption was measured by a single-point-Brunauer-Emmett-Teller test, $- 183$°C, $P/P_0 = 0.9$.

Theta-3 (MTW)
(EP 162,719(1985))
British Petroleum Company

Related Materials: ZSM-12

Structure

Chemical Composition: 0.9 ± 0.2 M$_{2/n}$O : Al$_2$O$_3$: at least 20 SiO$_2$: y H$_2$O.

X-Ray Powder Diffraction Data: (d(Å)(I/I_0)) 11.9(vs), 11.7(vs), 10.1(m), 4.76(m), 4.69(m), 4.24(vs), 3.98(vs), 3.88(vs), 3.82(vs), 3.45(w), 3.39(m), 3.32(w), 3.04(w), 2.89(w), 2.51(w).

Synthesis

ORGANIC ADDITIVES
BzNR$_3$$^+$ (Bz = benzyl or substituted benzyl; R = alkyl)

Theta-3 is prepared from a reactive gel within a composition range of: 0.006–0.025 Al$_2$O$_3$: 0.2–0.5 Na$^+$: 0.02–0.2 OH$^-$: 20–150 H$_2$O. Crystallization occurs between 24 and 80 hours at temperatures between 120 and 190°C. Colloidal silica (Ludox® As40) is used as the source of silica, and sodium aluminate (40% Al$_2$O$_3$) is used as the source of aluminum.

Thomsonite (THO)
(natural)

Related Materials: Ca-I

Thompsonite was the name given by Brooke in 1820 to a mineral found in Dunbartonshire, Scotland (*Nat. Zeo.* (1985)).

Structure

(*Zeo. Mol. Sieve.* 174 (1974))

CHEMICAL COMPOSITION: Na$_4$Ca$_8$(Al$_{20}$Si$_{20}$O$_{80}$)* 24H$_2$O
SYMMETRY   orthorhombic
SPACE GROUP: Pcnn
UNIT CELL CONSTANTS (Å): a = 13.05
                           b = 13.09
                           c = 13.22
FRAMEWORK DENSITY: 1.67 g/cc
VOID FRACTION (determined from water content): 0.32
PORE STRUCTURE: two-dimensional, 2.6 × 3.9 Å, and 2.2 × 4.0 A, eight-member rings

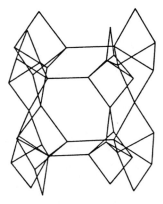

Thomsonite Fig. 1S: Framework topology.

Positional parameters for thomosonite (*Zeo.* 5:74 (1985)):

Atom	x	y	z($\times$ 10^5)
NaCa	5820(7)	50360(11)	36141(7)
Ca	49939(28)	47837(11)	49952(27)
Si(1)	25000	25000	68850(11)
Al(1)	25000	75000	69049(13)
Si(2)	11291(6)	69558(7)	50065(7)
Al(2)	11939(8)	30543(9)	49655(8)
Si(3)	30820(7)	38501(7)	37805(7)
Al(3)	30947(8)	62405(8)	38045(8)
O(1)	16863(6)	31051(6)	61839(5)
O(2)	15818(5)	69179(6)	61444(5)
O(3)	31248(6)	33136(5)	75655(5)
O(4)	31152(6)	65725(5)	76318(5)
O(5)	216(6)	63793(4)	50143(7)
O(6)	18380(5)	62980(6)	42414(6)
O(7)	19038(5)	38476(6)	41668(6)
O(8)	10438(5)	81313(6)	46106(6)
O(9)	11829(6)	18031(6)	45256(6)
O(10)	35586(4)	49928(6)	38346(4)
OW(1)	12597(6)	50192(9)	18859(7)
OW(2)	39173(7)	49760(10)	63937(6)
OW(3)	0	64972(12)	75000
OW(4)	0	34436(11)	75000

X-Ray Powder Diffraction Data: (*Nat. Zeo.* 319 (1985)) (d(Å)(I/I_0)) 6.63(100), 5.87(59), 4.74(74), 4.62(49), 4.40(35), 4.38(37), 4.34(8), 4.22(31), 4.16(20), 3.643(10), 3.308(4), 3.222(14), 3.187(10), 3.158(14), 3.083(7), 3.065(4), 2.991(12), 2.934(31), 2.902(40), 2.889(54), 2.882(48), 2.858(42), 2.607(3), 2.584(9), 2.579(7), 2.479(8), 2.448(2), 2.440(3), 2.421(8), 2.371(1), 2.336(1), 2.322(3), 2.3164, 2.296(4), 2.293(4), 2.272(4), 2.254(3), 2.248(3), 2.208(9), 2.170(3), 2.143(1), 2.111(3), 2.078(5), 2.063(1), 2.032(4), 1.993(3).

Thompsonite consists of an aluminosilicate framework constructed of crosslinked 4–1 type chains. The topologic cell of highest symmetry is replaced by a topochemical supercell when alternate tetrahedral nodes are occupied by Si and Al. Rotation of the 4–1 chains together with some distortion from the ideal shape allows bonding to the extra-framework ions and water molecules. Both sites for extra-framework cations are associated with elliptical eight-rings. The Na,Ca site contains 0.5 Na and 0.5 Ca and is surrounded by a square antiprism containing four water-oxygens in one square. Each square of water-oxygens is shared with an adjacent antiprism, and one edge between framework-oxygens is shared with a second antiprism to give an infinite chain parallel to c. There is no evidence for long-range alternations of Na and Ca atoms.

Thermal Properties

The DTG shows four peaks of weight loss under 450°C, plus one shoulder. This is in agreement with the presence of four water sites in the structure; a broad peak occuring between 500 and 700°C is attributed to dehydroxylation (*Nat. Zeo.* 61(1985)).

The vacuum treatment of thomsonite up to 100°C results in complete disappearance of the absorption at 1620 cm^{-1} from the infrared spectrum. Increasing the temperature to 200°C results in the disappearance of the band at 1680

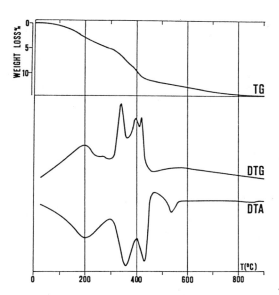

Thomsonite Fig. 2: Thermal curves of thomsonite; in air, heating rate 20°C/min. (*Nat. Zeo.* 60(1985)). (Reproduced with permission of Springer Verlag Publishers)

Infrared Spectrum

See Thomsonite Figure 3.

Tiptopite (CAN) (natural)

Related Materials: cancrinite

Tiptopite is a natural berrylophosphate mineral that was discovered in the Black Hills of South Dakota in 1985 (*Canad. Mineral.* 23:43(1985)).

Structure

CHEMICAL COMPOSITION: Li,K,Na,Ca,___, $Be_6(PO_4)_6(OH)_4$
SYMMETRY: hexagonal
SPACE GROUP: $P6_3$
UNIT CELL CONSTANTS (Å): $a = 11.655(5)$
                           $c = 4.692(2)$
PORE STRUCTURE: typical of cancrinite, 12-member rings 5.9 Å in diameter

cm^{-1}, attributed to the second step in dehydration. With vacuum treatment at 150°C, a broad band at 1590 cm^{-1} appears, which remains until 300°C. The broad vibration bands of the OH groups disappear upon heating between 450 and 700°C.

Tiptopite is isotypic with basic (hydroxyl) cancrinite. The framework is composed of BeO_4 and PO_4 units with complete ordering of Be and P. In the channels are the large cations and two nonframework oxygens.

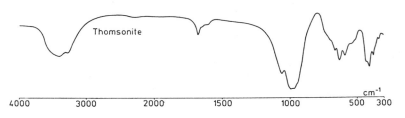

Thomsonite Fig. 3: Infrared spectrum of thomsonite (*Nat. Zeo.* 348(1985)). (Reproduced with permission of Springer Verlag Publishers)

Positional coordinates in tiptopite (*Am. Mineral.* 72:816 (1987)):

	x	y	z
P	0.0851(2)	0.4167(2)	3/4
Be	0.338(1)	0.418(1)	0.746(5)
O1	0.1902(6)	0.3843(6)	0.6518(15)
O2	0.1378(6)	0.5647(5)	0.7027(17)
O3	0.0547(6)	0.3801(5)	0.0688(15)
O4	0.3302(6)	0.3741(6)	0.0862(17)
M1	1/3	2/3	0.253(6)
M2	0.2207(8)	0.1137(7)	0.7486(30)
O5	0	0	0.253(6)
O6	0.064(3)	0.130(4)	0.378(9)
O7	0.066(4)	0.130(4)	0.097(10)
O8	0.168(4)	0.322(4)	0.236(20)

TiZnAPO-31
(EP 158,977(1985))
Union Carbide Corporation

Related Materials: AlPO₄-31

Structure

Chemical Composition: See XAPO.

X-Ray Powder Diffraction Data: $(d(\text{Å})(I/I_0))$ 10.40–10.28(m–s), 4.40–4.37(m), 4.06–4.02 (w–m), 3.93–3.92(vs), 2.823–2.814(w–m).

TiZnAPO-31 exhibits properties characteristic of a medium-pore molecular sieve. See AlPO₄-31 for a description of the framework topology.

Synthesis

ORGANIC ADDITIVE
di-*n*-propylamine

TiZnAPO-31 crystallizes from a reactive gel with a batch composition of: 1.0–2.0 R : 0.05–0.2 M_2O_q : 0.5–1.0 Al₂O₃ : 0.5–1.0 P₂O₅ : 40–100 H₂O (R represents the organic additive; *q* denotes the oxidation state of the titanium and zinc (M)).

Adsorption

Adsorption properties of TiZnAPO-31:

Adsorbate	Kinetic diameter, Å	Pressure Torr	Temperature °C	Wt% adsorbed*
O₂	3.46	100	− 183	4
O₂		750	− 183	6
Cyclohexane	6.2	90	24	3
Neopentane	6.2	700	24	3
H₂O	2.65	4.3	24	3
H₂O		20	24	10

*Typical amount adsorbed.

Tl-B (ABW)
(*JCS* 1949:1253(1949))

See A(BW).

Tl-C (ABW)
(*JCS Dalton Trans.* 934(1974))

See A(BW).

TMA-E (EAB)
(*JCS* A 1970:1470(1970))

TMA-E first was identified to be an erionite-type structure, but later was shown to be unique.

Structure

(*J. Solid State Chem.* 37:204(1981))

CHEMICAL COMPOSITION: TMA₂Na₇[Al₉Si₂₇O₇₂]* 26H₂O
SYMMETRY: hexagonal
SPACE GROUP: P6₃/mmc
    UNIT CELL CONSTANTS (Å): $a = 13.3$
                               $c = 15.2$

PORE STRUCTURE: eight-member rings 3.7 × 5.1 Å

X-Ray Powder Diffraction Data: (*JCS* A 1970:1470(1970)) $(d(\text{Å})(I/I_0))$ 11.58(2), 9.20(78), 7.63(17), 6.65(56), 6.30(18), 5.77(1), 5.46(18),

TMA-E Fig. 1S: Framework topology.

5.018(2),   4.650(2),   4.588(38),   4.341(5),
4.179(59), 3.842(35), 3.792(3), 3.776(100),
3.716(1),   3.616(64),   3.425(2),   3.332(20),
3.190(2),   3.181(13),   3.120(5),   3.072(3),
3.053(2),   2.942(23),   2.876(9),   2.863(31),
2.825(58), 2.704(1), 2.691(20), 2.642(15),
2.506(19),   2.492(6),   2.475(5),   2.438(2),
(2.385)(1), 2.342(3), 2.253(4), 2.214(10),
2.121(10),   1.994(4),   1.969(3),   1.901(1),
(1.915)(3), 1.901(6), 1.842(12), 1.785(18),
1.743(1),   1.709(3),   1.672(10,   1.660(20),
1.593(9), (1.575)(1), 1.561(3), 1.529(4).

The TMA-E framework consists of parallel six-rings in an ABBACC sequence. This differs somewhat from the stacking sequence found in erionite, which is AABAAC. The smaller cages in this structure are gmelinite-type. The aluminum atoms are thought to be located in the double six-rings, whereas the single x-rings may contain silicon only.

Refined atomic coordinates of the framework atoms of (Na,TMA)-E at room temperature, 200°C, and 350°C[a] (*J. Solid State Chem.* 37:204 (1981)):

Atom	x	y	z
T1	0.233(2)	0	0
	0.249(4)	0	0
	0.249(5)	0	0
T2	0.425(2)	0.093(2)	0.146(1)
	0.430(4)	0.088(7)	0.150(5)
	0.086(4)	0.086(4)	0.153(4)
O1	0.309(2)	−0.003(2)	0.088(2)
	0.305(5)	−0.008(6)	0.087(5)
	0.325(7)	0.005(9)	0.082(5)
O2	0.212(2)	0.106(1)	0.006(4)
	0.202(8)	0.101(4)	−0.026(9)
	0.214(6)	0.107(3)	0.011(8)
O3	0.478(2)	0.239(1)	0.117(3)
	0.406(3)	0.203(2)	0.150(5)
	0.428(6)	0.214(3)	0.162(6)

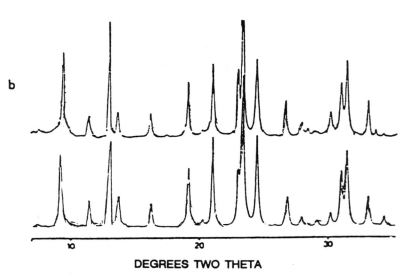

DEGREES TWO THETA

TMA-E Fig. 2:   Observed top and calculated bottom. X-ray powder diffraction pattern for TMA-E (*J. Solid State Chem.* 37:204(1981)). (Reproduced with permission of Academic Press)

Atom	x	y	z
O4	0.399(3)	0.081(4)	1/4
	0.375(6)	0.014(5)	1/4
	0.373(8)	0.009(4)	1/4
O5	0.536(2)	0.072(4)	0.106(3)
	0.543(2)	0.090(6)	0.109(5)
	0.576(4)	0.152(8)	0.127(7)

[a]Based on space group P6₃mmc (with estimated standard deviations in parenthesis).

Synthesis

Hexagonal unit cell constants (± 0.01 Å) of (Na,TMA)-E at different temperatures (*J. Solid State Chem.* 37:204 (1981)):

Temperature (°C)	a (Å)	c (Å)
20	13.28	15.21
220	13.00	15.47
350	12.86	15.51
490	(12.65)	(15.50)[a]

[a]Values calculated from a = 8.948 Å of the cubic sodalite-type product.

ORGANIC ADDITIVE
tetramethylammonium⁺

TMA-E crystallizes from a reactive gel with a molar oxide ratio of: 3.51 Na_2O : 3.63 TMA_2O : Al_2O_3 : 13.60 SiO_2 : 372 H_2O. Crystallization occurs at 80°C after 4 to 5 days. Crystals exhibit a platelike morphology, invariably twinned. Thermal treatment of this structure results in a solid state transformation to sodalite (*JCS* A 1970:1470(1970)).

Thermal Properties

An irreversible transformation of TMA-E to a sodalite-type structure is observed in this system

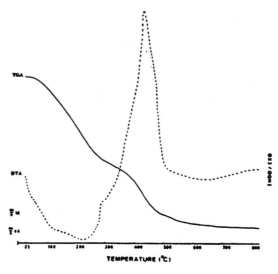

TMA-E Fig. 3:   TGA/DTA of TMA-E (*Zeo.* 4:226(1984)). (Reproduced with permission of Butterworths Publishers)

when it is heated in air to 350°C. This transformation appears to be more pronounced at higher aluminum concentrations or when the inorganic cation in the structure is Na^+. In a sample TMA-E made with K^+, the material is stable to temperatures of over 500°C. When it is heated under dry N_2, the temperature range over which this transformation is observed is shifted (*Zeo.* 4:226(1984))

TMA-Gismondine (GIS)

Gismondine structure synthesized in the presence of the tetra methyl ammonium organic amine cation.

TMA-O (OFF)

See offretite.

TPZ-3 (EUO)
(EP 51,318(1982))
Teijin Petrochemical Industries

Related Materials: EU-1

Structure

Chemical Composition: 0.5–4 $M_{2/n}O$: Al_2O_3 : at least 10 SiO_2.

X-Ray Powder Diffraction Data: $(d(\text{Å})(I/I_0))$
11.29(42,1), 10.28(23.5), 9.94(9.4), 6.943(3.1), 5.829(4.4), 5.644(4.0), 4.901(1.9), 4.671(37.5), 4.339(100), 4.022(60), 3.834(32.9), 3.722(23.5), 3.440(14.4), 3.376(15.4), 3.278(34), 3.164(3.9), 3.100(4.4), 2.959(5.6), 2.710(3.9), 2.546(7.9), 2.481(2.5), 2.414(4.6), 2.327(2.9), 2.301(2.9).

Synthesis

ORGANIC ADDITIVES
N,N,N,N′,N′, N′-hexamethyl-1,6-hexanediamine

Crystallization of TPZ-3 occurs after 6 to 8 days with temperatures ranging around 160°C. Ratios of SiO_2/Al_2O_3 between 50 and 200 have been investigated, with the higher ratios resulting in the crystallization of mixtures of TPZ-3 and quartz. The OH/SiO_2 is between 0.008 and 0.5. At high aluminum contents, analcite and mordenite predominate. In the absence of the organic additive, only quartz and mordenite crystallize.

Triclinic Bikitaite (BIK)
(Jb. Miner. Mh. 241(1986))

See bikitaite.

TS-1 (MFI)
(US4,410,501(1983))
Snamprogetti S.P.A.

Related Materials: ZSM-5

Structure

Chemical Composition: $x\,TiO_2 : (1-x)\,SiO_2$ $(x = 0.0005-0.04)$.

X-Ray Powder Diffraction Data: $(d(\text{Å})(I/I_0))$
11.14(vs), 9.99(s), 9.74(m), 6.702(w),

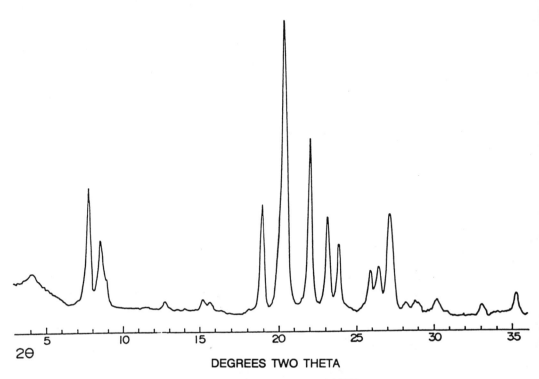

TPZ-3 Fig. 1:   X-ray powder diffraction pattern for TPZ-3 (EP 51,318(1982)).

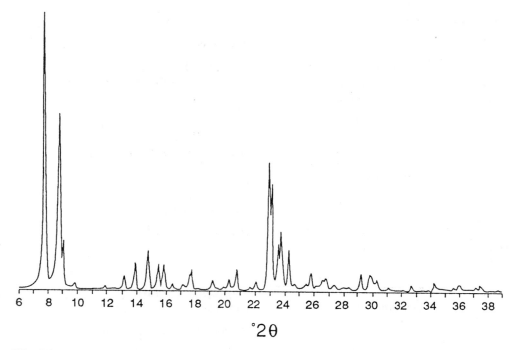

TS-1 Fig. 1:   X-ray powder diffraction pattern for TS-1 (US4,410,501 (1983)).

6.362(mw), 5.993(mw), 5,698(w), 5.574(w), 5.025(w), 4.980(w), 4.360(w), 4.260(mw), 3.855(s), 3.819(s), 3.751(s), 3.720(s), 3.646(m), 3.444(w), 3.318(w), 3.051(mw), 2.988(mw), 2.946(w), 2.014(mw), 1.994(mw).

See ZSM-5 for a description of the framework topology.

Synthesis

ORGANIC ADDITIVE
tetrapropylammonium$^+$

TS-1 crystallizes from a titanosilicate gel with a TiO$_2$/SiO$_2$ molar oxide ratio between 35 and 65, an OH/SiO$_2$ between 0.3 and 0.6, and a R/SiO$_2$ ratio between 0.4 and 1.0. Crystallization occurs after 10 days at 175°C.

Infrared Spectrum

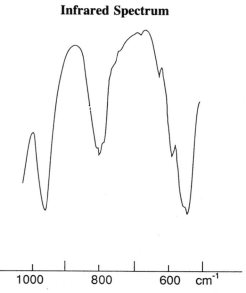

TS-1 Fig. 2:   Infrared spectrum showing the extra vibration occurring at 950 cm^{-1} attributed to the presence of framework titanium (US4,410,501(1983)).

Tschernichite (BEA*)
(natural)

Related Structures: beta

Tschernichite is found near Goble, Oregon and first was reported with boggsite in 1990 (*Am. Mineral.* 75:1200(1990)).

Structure

(*JCS Chem. Commun.* 363(1991))

CHEMICAL COMPOSITION: $CaSi_6Al_2O_{16}*8H_2O$
SYMMETRY: tetragonal
UNIT CELL CONSTANTS (Å): a = 12.88
                          c = 25.02

X-Ray Powder Diffraction Data: ($d(Å)(I/I_0)$)
25.98(2), 12.52(10), 11.36(32), 6.24(7), 5.63(2), 4.97(2), 4.74(2), 4.45(2), 4.22(14), 4.12(1), 4.02(100), 3.971(4), 3.708(6), 3.651(2), 3.569(13), 3.500(7), 3.333(5), 3.156(16), 3.062(15), 2.990(7), 2.950(1), 2.818(2), 2.730(10), 2.617(2), 2.527(5), 2.439(3), 2.300(1), 2.220(2), 2.191(2), 2.164(2), 2.114(16).

Thermal Properties

See Tschernichite Figure 2.

Infrared Spectrum

See Tschernichite Figure 3.

TSZ (MFI)
(EP 101,232(1984))

See ZSM-5.

TSZ-III (MFI)
(EP 170,751(1986))

See ZSM-5.

Tugtupite (SOD)
(natural)

Related Materials: sodalite

Tugtupite originally was referred to as beryllium sodalite or beryllo-sodalite; however, tugtupite was the name suggested based on the origin of the mineral samples (Tugtup Agtakorfia, South Greenland) (*Acta Crystallogr.* 20:812(1966)).

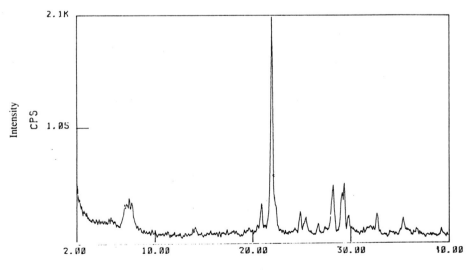

DEGREES TWO THETA

Tschernichite Fig. 1:   X-ray powder diffraction pattern of Tschernichite. (Reproduced with permission of A. Long, Georgia Tech)

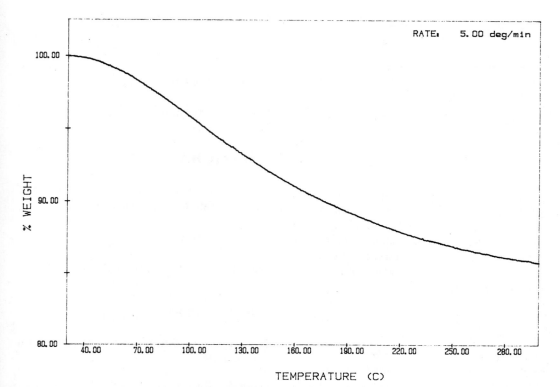

RATE: 5.00 deg/min

Tschernichite Fig. 2: TGA of Tschernichite. (Reproduced with permission of F. Testa, U. della Calabria)

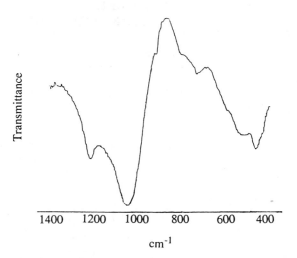

Tschernichite Fig. 3: Mid-infrared spectrum of Tschernichite.
(Reproduced with permission of B. Duncan, Georgia Tech)

473

Structure

(Acta Crystallogr. 20:812(1966)).

CHEMICAL COMPOSITION: $Na_8Al_2Be_2Si_8O_{24}(Cl,S)_2$
SYMMETRY: tetragonal
SPACE GROUP: $I\bar{4}$
        UNIT CELL CONSTANTS (Å): $a = 8.583$
                                    $c = 8.817$

In the chemical formula of this mineral the sulfur content accounts for only 5% of the (Cl,S) content. The structure of tugtupite consists of MO_4 tetrahedra linked, where M is Be, Al, and Si. One third of the rings contain only Si; the rest have Be, Si, and Al alternating. The Be- and Al-containing tetrahedra are not distorted, but the SiO_4 units are distorted. Of the four O atoms in this unit, two are bound to other Si atoms with one coordinated to a Be and the other to an Al. The sodium and chloride ions are contained within the beta cages.

Atomic coordinates for tugtupite (*Acta Crystallogr.* 20:812 (1966)):

Atom	x/a	y/b	z/c
O(1)	0.1471(8)	0.1332(8)	0.4431(13)
O(2)	0.3467(9)	0.0362(9)	0.6512(15)
O(3)	0.4261(8)	0.1506(8)	0.1347(14)
Na	0.1575(5)	0.1970(5)	0.1815(8)
Si	0.0134(3)	0.2535(3)	0.4956(5)
Be	0.000	0.5000	0.2500
Al	0.0000	0.5000	0.7500
Cl	0.0000	0.0000	0.0000

Type A (LTA)

See Linde type A.

Type B (GIS)

A disused term for synthetic gismondine.

Type C (ANA)

A term used by Union Carbide workers for synthetic analcime.

Type G (CHA)

Synthetic chabazite prepared in the presence of potassium cations (see chabazite).

Type R (CHA)

Synthetic chabazite prepared in the presence of sodium cations (see chabazite).

Type W (MER)

Type W was used to identify a synthetic form of zeolite W.

Type Y (FAU)

See Linde type Y.

Type Y (FAU)

See Linde type Y.

TZ-01 (MFI)
(EP 57,016(1982))

See ZSM-5.

U

Ultramarine blue (SOD)

Related Materials: sodalite

Ultramarine blue or "blue from beyond the sea" has a brilliant blue color and belongs in the ultramarine pigment family. The word "ultramarine" is derived from Latin, where *ultra* means over or from beyond and *mare* means ocean or sea. The natural source of the ultramarine blue is the mineral lapis lazuli, which is composed of three different sodalites—hayne, sodalite, and lazurite; the distinguishing chemicals for these sodalites are sodium sulfate, sodium chloride, and sodium sulfide, respectively.

Structure

Ultramarine blue is a sodium-alumino-sulfo-silicate that is closely related to the sodalite crystal structure (see sodalite). The typical composition of ultramarine blue is: 37–50% SiO_2 : 23–29% Al_2O_3 : 19–23% Na_2O : 8–14% S.

Thermal Properties

Ultramarine blue is heat-resistant up to temperatures greater than 350°C.

Ultrasil (MFI)
(*J. Phys. Chem.* 55:1175(1981))

See ZSM-5.

Ultrazet (MFI)
(*Catal. Lett.* 20:431(1982))

See ZSM-5.

USC-4 (MFI)
(US4,325,929(1982))
Union Oil

USC-4 is a silica polymorph of the ZSM-5 type structure.

USI-108 (MFI)
(US4,423,020(1983))

See ZSM-5.

V

VAPO-5 (AFI)
(EP 158,976(1985))
Union Carbide Corporation

Related Materials: AlPO$_4$-5

Structure

Chemical Composition: See FCAPO.

X-Ray Powder Diffraction Data: $(d(\text{Å})\,(I/I_0))$
12.1–11.56(m–vs), 4.55–4.46(m–s), 4.25–
4.17(m–vs), 4.00–3.93(w–vs), 3.47–3.40
(w–m).

VAPO-5 exhibits properties characteristic of a large-pore molecular sieve. See AlPO$_4$-5 for a description of the framework topology.

Synthesis

ORGANIC ADDITIVE
tripropylamine

VAPO-5 crystallizes from a reactive gel with a batch composition of: 1.0–2.0 R : 0.05–0.2 V$_2$O$_5$: 0.5–1.0 Al$_2$O$_3$: 0.5–1.0 P$_2$O$_5$: 40–100 H$_2$O (R represents the organic additive).

Adsorption

Adsorption properties of VAPO-5:

Adsorbate	Kinetic diameter, Å	Pressure Torr	Temperature °C	Wt% adsorbed*
O$_2$	3.46	100	−183	7
O$_2$		750	−183	10
Neopentane	6.2	700	24	4
H$_2$O	2.65	4.3	24	4
H$_2$O		20	24	12

*Typical amount adsorbed.

VAPO-11 (AEL)
(EP 158,976(1985))
Union Carbide Corporation

Related Materials: AlPO$_4$-11

Structure

Chemical Composition: See FCAPO.

X-Ray Powder Diffraction Data: $(d(\text{Å})\,(I/I_0))$
9.51–9.17(m–s), 4.40–4.31(m–s), 4.25–4.17(s–vs), 4.04–3.95(m–s), 3.95–3.92(m–s), 3.87–3.80(m–vs).

VAPO-11 exhibits properties characteristic of a medium-pore molecular sieve. See AlPO$_4$-11 for a description of the framework topology.

Synthesis

ORGANIC ADDITIVE
di-*n*-propylamine

VAPO-11 crystallizes from a reactive gel with a batch composition of: 1.0–2.0 R : 0.05–0.2 V$_2$O$_5$: 0.5–1.0 Al$_2$O$_3$: 0.5–1.0 P$_2$O$_5$: 40–100 H$_2$O (R refers to the organic additive).

Adsorption

Adsorption properties of VAPO-11:

Adsorbate	Kinetic diameter, Å	Pressure Torr	Temperature °C	Wt% adsorbed*
O$_2$	3.46	100	−183	5
O$_2$		750	−183	6
Cyclohexane	6.0	90	24	4
H$_2$O	2.65	4.3	24	6
H$_2$O		20	24	8

*Typical amount adsorbed.

Variscite
(natural)

Variscite first was described as peganite in 1830, and then as variscite in 1837 (*Encyclopedia of Minerals* 648(1974)). A natural mineral, variscite is found in deposits of phosphatic meteoric water on alumina-rich rocks. It also is found associated with crandallite and other phosphate minerals.

Structure

(*Acta Crystallogr*. B33:263(1977)

CHEMICAL COMPOSITION: $AlPO_4*2H_2O$
SYMMETRY: orthorhombic
SPACE GROUP: Pbca
UNIT CELL CONSTANTS (Å): $a = 9.87$
                                        $b = 9.57$
                                        $c = 8.52$
DENSITY: 2.57 g/cc
PORE STRUCTURE: very small-pore, condensed
  material

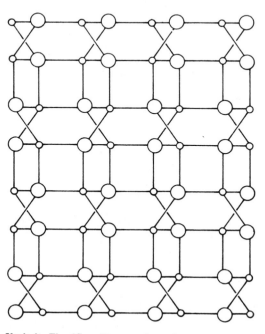

Variscite Fig. 1S: Framework topology.

X-Ray Powder Diffraction Data: (*Bull. Chem. Soc. Fr.* 1762(1961)) (synthetic variscite) (d(Å) (I/I_0))   5.38(vs), 4.83(s), 4.28(vs), 3.90(s), 3.64(m), 3.21(m), 3.05(vs), 2.92(s), 2.87(s), 2.64(m), 2.58(m), 2.49(m), 2.45(m), 2.40(m), 2.34(m), 2.29(w), 2.20(w), 2.14(mw), 2.085(m), 2.050(w), 2.018(w), 1.960(m), 1.945(m), 1.918(w), 1.860(m), 1.843(w), 1.780(w), 1.753(m), 1.718(w), 1.605(m), 1.592(m), 1.576(m), 1.559(m), 1.520(ms), 1.500(m).

The PO_4 tetrahedra share vertices with four $AlO_4(OH_2)_2$ octahedra in the variscite structure, resulting in a three-dimensional network of six-member rings. The Al–O distances are 1.963(4) and 1.909(4) Å for the two water molecules, which coordinate to the aluminum in the structure in a *cis* configuration. Hydrogen bonding is observed with the phsophate oxygen atoms. Only three of the four water hydrogens participate in hydrogen bonding in this structure.

Atomic parameters for variscite (*Acta Crystallogr*. B33:263 (1977)):

	x	y	z
Al	13389(7)	15500(8)	16841(6)
P	14779(6)	46844(6)	35284(6)
O1	11180(16)	29870(19)	31525(17)
O2	4030(17)	58186(21)	29453(17)
O3	28545(16)	51247(20)	29006(16)
O4	14997(16)	47916(19)	51224(16)
OW1	6041(19)	32564(23)	5469(19)
OW2	30726(18)	23597(21)	11499(19)
H11	63(5)	310(5)	−28(4)
H12	−12(5)	371(5)	68(4)

Synthesis

ORGANIC ADDITIVES
none

Synthetic variscite is prepared under low-temperature (ca. 80–100°C) hydrothermal conditions with phosphorous oxide to alumina ratios of 2.73. A very acidic pH in the reactive gel (ca. 1) is observed, and crystallization is observed between 8 and 24 hours (*Bull. Chem.*

Soc. Fr. 1762(1961)). Other phases present in the crystallization mixture include metavariscite, AlPO$_4$-H1, -H2, and -H3. The amounts of these other phases formed depend on the time of crystallization and the degree of dilution. In crystallization mixtures employing organic additives, variscite will crystallize between 55 and 125°C and usually in combination with H3 and/or metavariscite (*6th IZC* 97(1983)). In these systems the organic does not appear to play a role in the crystallization of the structure.

Thermal Properties

Upon heating to between 100 and 200°C, variscite is transformed to AlPO$_4$-B. TGA studies show that the loss of both associated waters occurs between these temperatures. AlPO$_4$-B will rehydrate to variscite. In addition, partial rehydration can occur to produce AlPO$_4$-H5

(H$_2$O/AlPO$_4$ = 1). When AlPO$_4$-B is heated to 400°C, further condensation occurs with the formation of tridimite and small quantities of cristobalite.

Viseite (ANA)
(natural)

(See analcime.) Viseite is a natural silicaluminophosphate related to analcime, with the composition Na$_2$Ca$_{10}$(Al$_{20}$Si$_6$P$_{10}$O$_{60}$(OH)$_{36}$)*16H$_2$O. It has cubic symmetry with a = 13.65 A (*Ann. Soc. Geol. Belg. Bull.* 66:b53(1943)).

VPI-5 (VFI)
(*Nature* 331:698(1988)

Related Materials: AlPO$_4$-Hl
                      MCM-9
                      AlPO$_4$-54
                      AlPO$_4$-Hl(GTRI)

Structure

(*Zeo.* 11:308(1991))

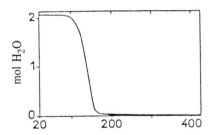

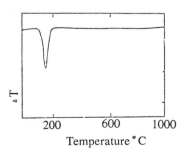

Variscite Fig. 2:   (bottom) DTA of synthetic variscite and (top) TGA of synthetic variscite (*J. Appl. Chem.* 13:17(1963)); *Bull. Chem. Soc. Fr.* 1762(1961)). (Reproduced with permission of Société Française de Chémie)

CHEMICAL COMPOSITION: AlPO$_4$
SYMMETRY: hexagonal
SPACE GROUP: P6$_3$
UNIT CELL DIMENSIONS (Å): a = 18.9752
                           c = 8.1044
PORE STRUCTURE: unidimensional 18-member ring, 12–13 Å pore opening

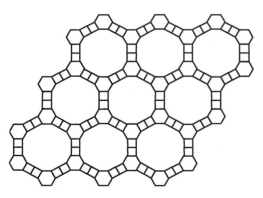

VPI-5 Fig. 1S: Framework topology.

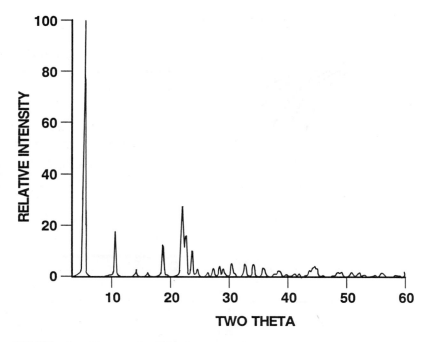

VPI-5 Fig. 2:   X-ray powder diffraction pattern of VPI-5 (*Zeo.* 8:362(1988)). (Reproduced with permission of Butterworths Publishers)

X-Ray Powder Diffraction Data: (*Zeo.* 8:362(1989)) ($d(\text{Å})$ (I/I_0)) 16.43(100), 9.49(2), 8.23(14), 6.21(6), 5.48(2), 4.75(6), 4.08(20), 4.05(22), 3.97(14), 3.94(15), 3.77(10), 3.64(4), 3.41(2), 3.28(16), 3.17(5), 3.08(7), 3.03(4), 2.95(8), 2.90(5), 2.74(7), 2.63(2), 2.50(3), 2.35(3).

The VPI-structure consists of unidimensional 18-member-ring channels. The channels are formed from six-rings alternating with two four-member rings. The aluminum and phosphorus in this type of structure are found to occupy tetrahedral oxide framework positions. The structure is similar to the theoretical net 81(1) of Smith and Dytrych (*Nature* 309:607(1984)). The 18-member-ring pore opening is circular with a diameter around 12 to 13 Å.

In the hydrated state the aluminum ions present in the double four member ring occupy octahedral sites. Two of the six oxygen atoms completing the coordination are from water molecules.

Positional parameters for VPI-5 (*Zeo.* 11:308 (1991)):

Atom	x	y	z
Al(1)	0.385(1)	−0.001(2)	0.239(3)
Al(2)	0.474(1)	−0.173(1)	0.204(4)
Al(3)	0.654(1)	0.169(1)	0.198(3)
P(1)	0.550(1)	−0.001(1)	0.358(3)
P(2)	0.311(1)	−0.185(1)	0.320(3)
P(3)	0.507(1)	0.185(1)	0.309(4)
O(1)	0.470(2)	0.001(3)	0.339(4)
O(2)	0.431(2)	0.009(2)	0.037(4)
O(3)	0.322(2)	−0.112(2)	0.223(4)
O(4)	0.430(2)	0.111(2)	0.255(5)
O(5)	0.540(2)	−0.074(2)	0.264(5)
O(6)	0.619(2)	0.072(2)	0.278(4)
O(7)	0.512(2)	−0.231(2)	0.280(5)
O(8)	0.378(1)	−0.206(2)	0.280(4)
O(9)	0.474(2)	−0.175(2)	−0.015(4)
O(10)	0.311(1)	−0.168(2)	0.500*
O(11)	0.575(1)	0.195(2)	0.198(4)
O(12)	0.739(1)	−0.236(1)	0.315(5)

*Held fixed in least-squares refinement.

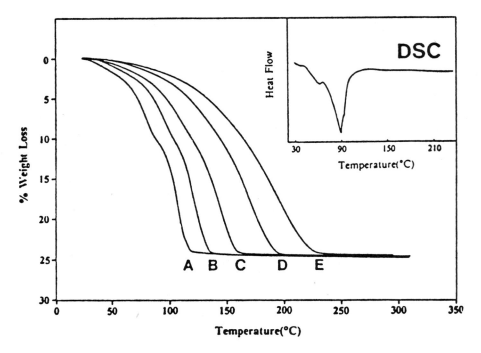

VPI-5 Fig. 3:   Thermogravimetric analysis of VPI-5. Heating rate: (A) 1°C/min (B) 2°C/min, (C) 8°C/min, (D) 25°C/min, (E) 30°C/min. Insert: differential scanning calorimetry of VPI-5; heating rate 1°C/min (*JACS* 111:3919(1989)). (Reproduced with permission of the American Chemical Society)

Synthesis

ORGANIC ADDITIVES
n-dipropylamine
tetrabutylammonium $^+$
n-dibutylamine

This aluminophophate was prepared from a gel composition of: DPA: Al_2O_3: P_2O_5: 40 H_2O, or TBA(OH): Al_2O_3: P_2O_5: 50 H_2O, using pseudoboehmite as the source of aluminum with temperatures ranging between 140 and 150°C for 24 hours. The starting pH is dependent on the nature of the organic additive. The initial pH of the reaction mixture begins around 4 to 6 and ends around 7 (*ACS Sym. Ser.* 398:291(1989)). When *n*-dibutylamine is used to encourage crystallization of this material, aluminum isopropoxide is used as the source of aluminum from a similar gel composition: DBA: Al_2O_3: P_2O_5: 40 H_2O. Crystallization occurs after 8 hours under these conditions (*Appl. Catal.* 56:L21 (1989)).

Thermal Properties

See VPI-5 Figure 3.

Infrared Spectrum

Mid-infrared Vibrations (cm^{-1}):   1265 m, 1160s, 1055s, 750mw, 610mw, 515m, 465m.

Adsorption

Adsorption capacity of VPI-5 at $P/P_0 = 0.4$ (JACS 111:3919 (1989)):

Adsorbate	Kinetic diameter, Å	Capacity, cm 3/g
HO_2	2.65	0.351
O_2	3.46	0.228
n-Hexane	4.30	0.198
Cyclohexane	6.0	0.156
Neopentane	6.2	0.148
Triisopropylbenzene	8.5	0.117

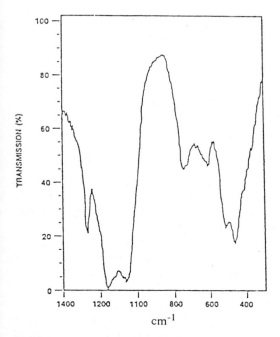

TRANSMISSION (%)

cm⁻¹

VPI-5 Fig. 4:   Infrared spectrum of VPI-5.
(Reproduced with permission of K. Vinje, U. of Oslo)

Water adsorption properties of VPI-5 and (b) VPI-5
calcined to $AlPO_4$-8; kinetic diameter of H_2O = 2.65 Å
(from author):

Pressure Torr	% Wt adsorbed (20°C)	% Wt adsorbed (35°C)	% Wt adsorbed (50°C)
(a) H_2O adsorption data for VPI-5:			
0.5	2.41	0.92	0.0
1.0	10.02	2.19	0.42
2.0	17.78	10.86	1.62
5.0	20.75	16.82	9.55
10.0	22.42	19.15	13.70
17.0	24.15	20.20	15.10
(b) H_2O adsorption data for VPI-5 calcined ($AlPO_4$-8):			
0.5	1.76	0.96	0.59
1.0	3.35	1.69	0.98
2.0	5.97	3.09	1.56
5.0	15.37	7.91	3.13
10.0	17.91	14.81	5.91
17.0	19.45	16.93	11.52

W

W

See Linde W.

Wairakite (ANA)
(natural)

Related Materials: analcime

Wairakite is a calcium ion–containing zeolite with the analcime structure. It first was discovered by Steiner in 1955, in Wairakei, New Zealand (*Nat. Zeo.* 76 (1985)).

Structure

(*Nat. Zeo.* 76(1985))

CHEMICAL COMPOSITION: $Ca_8(Al_{16}Si_{32}O_{96})*16H_2O$
SYMMETRY: monoclinic-pseudocubic
SPACE GROUP: Ia or Ia/2
UNIT CELL CONSTANTS (Å): $a = 13.69$
$b = 13.56$
$c = 13.56$
$\beta = 90.5$
PORE STRUCTURE: highly distorted eight-member rings

See analcime for a description of the structure.

X-Ray Powder Diffraction Data: (*Zeo. Mol. Sieve.* 240(1974)) ($d(Å)$ (I/I_0)) 6.85(40), 5.57(80), 4.84(40), 3.64(30), 3.42(60), 3.39(100), 3.21(<10br), 3.04–3.06(10br), 2.909(50), 2.897(30), 2.783(10), 2.770(10), 2.680(40), 2.67(10), 2.50(<10), 2.489(40), 2.418(30), 2.35(<10br), 2.26–2.28(10br), 2.215(40), 2.17(<10), 2.147(10), 2.115(10), 2.095(<10), 1.996(20), 1.93(<10br), 1.886–1.895 (30br), 1.867 (10), 1.844 (10), 1.822 (<10br), 1.722–1.732 (40br), 1.708 (<10), 1.696 (<10), 1.680 (20br), 1.66 (<10br), 1.612 (10br), 1.595 (<10), 1.986 (20).

Synthesis

ORGANIC ADDITIVES
none

Several attempts have been made to synthesize wairakite from calcium-containing gels. A crystalline phase thought to be wairakite was formed from $Ca(OH)_2$, hydrated alumina, and silica gel at 300°C (*Chemie der Erde* 10:129 (1936)). Preparation through ion exchange of analcime was only partially successful, as only limited replacement takes place (*JCS* 2342 (1950)). Decomposition of mordenite at 325 to 400°C has resulted in the generation of wairakite (*Am. Mineral.* 43:476 (1958)). Reactive mixtures that will result in this phase include: $CaO:Al_2O_3:4.7SiO_2$. Temperatures range between 200 and 400°C, requiring 4 weeks of crystallization (*Cryst. Res. Technol.* 20:K90 (1985)).

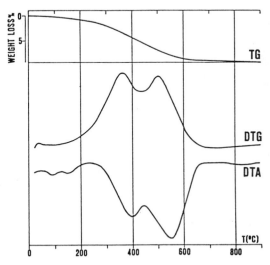

Wairakite Fig. 1: Thermal curves of wairakite from Wairakei, New Zealand, in air, heating rate 20°C/min (*Nat. Zeo.* 90(1985)). (Reproduced with permission of Springer Verlag Publishers)

Higher temperatures appear to produce better-quality cubic crystals (*JCS* 983(1961)). Laumontite, which has the same chemical formula as wairakite but more zeolitic water, also can be formed in this system at high temperatures. Anorthite and quartz also are high-temperature-phase products.

Thermal Properties

Water desorption occurs in two steps in wairakite, at 360 and at 500°C.

Wellsite (PHI)
(natural)

Related Materials: phillipsite

Wellsite is an intermediate member of the phillipsite-harmotome group of mineral zeolites, composed of a mixed population of Ba and K with very low Na/Ca ratios. It sometimes occurs with chabazite, phillipsite, analcime, leonhardite, and calcite (*Nat. Zeo.* 143(1985)).

Structure

X-Ray Powder Diffraction Data: (*Nat. Zeo.* 328(1985)) ($d(\text{Å})$ (I/I_0)) 8.15(19), 7.13(100), 6.41(19), 5.36(24), 5.05(23), 4.96(19), 4.30(16), 4.11(65), 4.07(26), 3.962(5), 3.922(5), 3.684(3), 3.474(5), 3.263(38), 3.220(31), 3.189(100), 3.138(58), 3.093(7), 2.934(23), 2.752(30), 2.702(27), 2.690(47), 2.669(20), 2.569(7), 2.537(6), 2.528(9), 2.479(3), 2.388(5), 2.337(7), 2.311(4), 2.250(7), 2.228(4), 2.159(5), 2.072(5), 2.058(5), 2.003(3), 1.963(7).

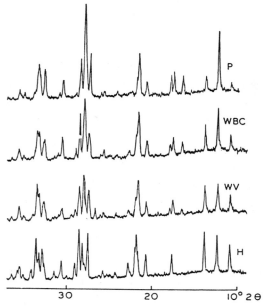

Wellsite Fig. 1:   X-ray powder diffraction pattern of phillipsite from Monte Somma (P), wellsite from Buck Creek (WBC), wellsite from Vezna (WV), and harmotome from Korsnas (H). Note the prominent changes in relative intensities of several reflections, particularly in the groups located at 10–14°, 16–18°, 20–24°, and 27–29° (*N. Jb. Miner. Abh.* 128:312(1977)). (Reproduced with permission of *Neues Jahrbuch für Mineralogie*)

Thermal Properties

The thermal curves of wellsite appear very similar to those recorded for harmotome. The significant difference occurs with the last water loss peak, which shifts to a higher (ca. 350°C) temperature (*Nat. Zeo.* 144(1985)). See Wellsite Figure 2.

Unit cell dimensions of wellsite samples (*N. Jb. Miner. Abh.* 128:312 (1977)):

	a_m (Å)	b_m (Å)	c_m (Å)	β	V_m (Å^3)	c_o (Å)	V_o (Å^3)
Vezna	9.882	14.222	8.704	124.48	1004.5	14.294	2008.9
M. Calvarina	9.909	14.246	8.710	124.41	1011.1	14.327	2022.5
Bucks Creek	9.878	14.246	8.696	124.54	10,003.5	14.263	2007.1
Kurtsy	9.869	14.214	8.679	124.51	999.1	14.243	1998.0

m = monoclinic; o = orthorhombic.

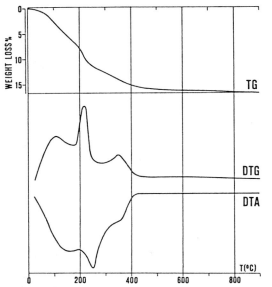

Wellsite Fig. 2:   Thermal curves of wellsite from Monte Calvarina, Italy; in air, heating rate 20°C/min (*Nat. Zeo.* 145(1985)). (Reproduced with permission of Springer Verlag Publishers)

Willhendersonite (CHA) (natural)

Related Materials: chabazite

Willhendersonite is the name given in 1984 to the potassium calcium aluminosilicate with the chabazite structure after William A. Henderson, who determined the material to be compositionally unique (*Nat. Zeo.* 176(1985)).

Structure

(*Nat. Zeo.* 175(1985))

CHEMICAL COMPOSITION: $K_2Ca_2(Al_6Si_6O_{24})*10H_2O$
SYMMETRY: triclinic
SPACE GROUP: P$\bar{1}$
UNIT CELL CONSTANTS (Å): $a = 9.20$
$b = 9.18$
$c = 9.49$
$\alpha = 92°36'$
$\beta = 92°26'$
$\gamma = 90°3'$
PORE STRUCTURE: intersecting eight-member rings

X-Ray Powder Diffraction Data: (*Nat. Zeo.* 334(1985)) (d(Å) (I/I_0)) 9.16(100), 5.18(30), 4.71(5), 4.57(5), 4.27(2), 4.09(40), 3.93(20), 3.82(20), 3.71(30), 3.06(10), 3.01(10), 2.907(60), 2.804(50), 2.746(2), 2.674(1), 2.538(20), 2.508(20), 2.429(10), 2.264(15), 2.209(1), 2.163(15), 2.078(10), 2.042(10), 2.004(5), 1.979(5).

X

X

See Linde Type X

XAPO
(EP 158,977(1985))
Union Carbide Corporation

"XAPO" is an acronym that designates metal-substituted aluminophosphate molecular sieves where X can be iron or titanium in combination with cobalt, magnesium, manganese, or zinc. The numerical designation after the acronym, as in XAPO-5, represents a topological code. See XAPO Figure 1.

XAPO-5 (AFI)
(EP 158,977(1985))

Related Materials: AlPO$_4$-5

X-Ray Powder Diffraction Data: $(d(Å)\ (I/I_0))$
12.1–11.56(m–vs), 4.55–4.46(m–s), 4.25–

4.17(m–vs), 4.00–3.93(w–vs), 3.47–3.40 (w–m).

XAPO-5 materials exhibit properties characteristic of large-pore molecular sieves.

XAPO-11 (AEL)
(EP 158,977(1985))

Related Materials: AlPO$_4$-11

X-Ray Powder Diffraction Data: $(d(Å)\ (I/I_0))$
9.51–9.17(m–s), 4.40–4.31(m–s), 4.25–4.17(s–vs), 4.04–3.95(m–s), 3.95–3.92(m–s), 3.87–3.80(m–vs).

XAPO-11 materials exhibit properties characteristic of medium-pore molecular sieves.

XAPO-14
(EP 158,977(1985))

Related Materials: AlPO$_4$-14

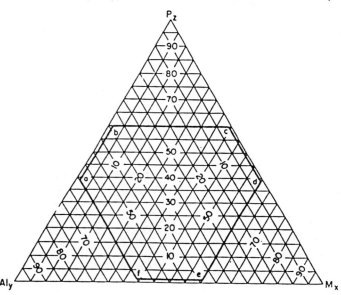

XAPO Fig. 1: Product compositional range (broad) for XAPO molecular sieves (EP 158,977(1985)).

X-Ray Powder Diffraction Data: $(d(\text{Å})\ (I/I_0))$
10.3–9.93(vs), 6.81(w), 4.06–4.00(w), 3.51(w), 3.24(w), 3.01(w).

XAPO-14 exhibits properties of small pore molecular sieves.

XAPO-16 (AST)
(EP 158,977(1985))

Related Materials: AlPO$_4$-16

X-Ray Powder Diffraction Data: $(d(\text{Å})\ (I/I_0))$
7.83–7.63(m–vs),   4.75–4.70(w–s),   4.06–3.99(m–vs),   3.363–3.302(w–m),   3.008–2.974(w–m).

XAPO-16 materials exhibit properties of very small-pore molecular sieves.

XAPO-17 (ERI)
(EP 158,977(1985))

Related Materials: AlPO$_4$-17
                      erionite

X-Ray Powder Diffraction Data: $(d(\text{Å})\ (I/I_0))$
11.5–11.4(vs), 6.61(s–vs), 5.72–5.70(s), 4.52–4.51(w–s), 4.33–4.31(vs), 2.812–2.797(w–s).

XAPO-17 materials exhibit properties of small-pore molecular sieves.

XAPO-18 (AEI)
(EP 158,977(1985))

Related Materials: AlPO$_4$-18

X-Ray Powder Diffraction Data: $(d(\text{Å})\ (I/I_0))$
9.21–9.16(vs), 5.72–5.70(m), 5.25–5.19(m), 4.41–4.39(m), 4.24–4.22(m), 2.814–2.755(m).

XAPO-18 materials exhibit properties characteristic of small-pore molecular sieves.

XAPO-20 (SOD)
(EP 158,977(1985))

Related Materials: AlPO$_4$-20
                      sodalite

X-Ray Powder Diffraction Data: $(d(\text{Å})\ (I/I_0))$
6.46–6.22(m–vs),   4.54–4.44(w–s),   3.70–3.63(m–vs),   2.614–2.546(vw–w),   2.127–2.103(vw–w).

XAPO-20 materials exhibit properties characteristic of very small-pore molecular sieves.

XAPO-31 (ATO)
(EP 158,977(1985))

Related Materials: AlPO$_4$-31

X-Ray Powder Diffraction Data: $(d(\text{Å})\ (I/I_0))$
10.40–10.28(m–s), 4.40–4.37(m), 4.06–4.02 (w–m), 3.93–3.92(vs), 2.823–2.814(w–m).

XAPO-31 materials exhibit properties characteristic of medium-pore molecular sieves.

XAPO-33 (ATT)
(EP 158,977(1985))

Related Materials: AlPO$_4$-33

X-Ray Powder Diffraction Data: (as-synthesized form) $(d(\text{Å})\ (I/I_0))$ 9.56–9.26(w–m), 7.08–6.86(vs), 5.25–5.13(w–m), 4.34–4.25(w–m), 3.73–3.67(w–m),   3.42–3.38(w–m),   3.27–3.23(vs); (calcined form) $(d(\text{Å})\ (I/I_0))$ 6.73–6.61(vs), 4.91–4.83(m), 4.82–4.77(m), 3.36–3.34(m), 2.80–2.79(m).

XAPO-33 materials exhibit properties characteristic of small-pore molecular sieves.

XAPO-34 (CHA)
(EP 158,977(1985))

Related Materials: AlPO$_4$-34
                     chabazite

X-Ray Powder Diffraction Data: $(d(\text{Å})\ (I/I_0))$
9.41–9.17(s–vs),   5.57–5.47(vw–m),   4.97–4.82(w–s), 4.37–4.25(m–vs), 3.57–3.51(vw–s), 2.95–2.90(w–s).

XAPO-34 materials exhibit properties characteristic of small-pore molecular sieves.

XAPO-35 (LEV)
(EP 158,977(1985))

Related Materials: levyne

X-Ray Powder Diffraction Data: $(d(\text{Å})\ (I/I_0))$
8.19–7.97(m),   5.16–5.10(s–vs),   4.23–4.18 (m–s), 4.08–4.04(vs), 2.814–2.788(m).

XAPO-35 materials exhibit properties characteristic of small-pore molecular sieves.

XAPO-36 (ATS)
(EP 158,977(1985))

Related Materials: MAPO-36

X-Ray Powder Diffraction Data: (d(Å) (I/I_0))
11.5–11.2(vs), 5.47–5.34(w–m), 4.70–4.60 (m–s), 4.31–4.27(w–s), 4.08–4.04(m), 4.00–3.95(w–m).

XAPO-36 materials exhibit properties characteristic of large-pore molecular sieves.

XAPO-37 (FAU)
(EP 158,977(1985))

Related Materials: SAPO-37
faujasite

X-Ray Powder Diffraction Data: (d(Å) (I/I_0))
14.49–14.03(vs), 5.72–5.64(w–m), 4.80–4.72(w–m), 3.79–3.75(w–m), 3.31–3.29(w–m).

XAPO-37 materials exhibit properties characteristic of large-pore molecular sieves.

XAPO-39 (ATN)
(EP 158,977(1985))

Related Materials: AlPO$_4$-39

X-Ray Powder Diffraction Data: (d(Å) (I/I_0))
9.41–9.21(w–m), 6.66–6.51(m–vs), 4.93–4.82(m), 4.19–4.13(m–s), 3.95–3.87(s–vs), 2.96–2.93(w–m).

XAPO-39 molecular sieves exhibit adsorption properties of small-pore molecular sieves.

XAPO-40 (AFR)
(EP 158,977(1985))

Related Materials: SAPO-40

X-Ray Powder Diffraction Data: (d(Å) (I/I_0))
11.79–11.48(vw–m), 11.05–10.94(s–vs), 7.14–7.08(w–vs), 6.51–6.42(m–s), 6.33–6.28(w–m), 3.209–3.187(w–m).

XAPO-40 materials exhibit properties characteristic of large-pore molecular sieves.

XAPO-41 (AFO)
(EP 158,977(1985))

Related Materials: SAPO-41

X-Ray Powder Diffraction Data: (d(Å) (I/I_0))
6.51–6.42(w–m), 4.33–4.31(w–m), 4.21–4.17(vs), 4.02–3.99(m–s), 3.90–3.86(m), 3.82–3.80(w–m), 3.493–3.440(w–m).

XAPO-41 materials exhibit properties characteristic of medium-pore molecular sieves.

XAPO-42 (LTA)
(EP 158,977(1985))

Related Materials: SAPO-42
Linde type A

X-Ray Powder Diffraction Data: (d(Å) (I/I_0))
12.36–11.95(m–vs), 7.08–6.97(m–s), 4.09–4.06(m–s), 3.69–3.67(vs), 3.273–3.255(s), 2.974–2.955(m–s).

XAPO-42 materials exhibit properties characteristic of small-pore molecular sieves.

XAPO-44 (CHA)
(EP 158,977(1985))

Related Materials: chabazite
SAPO-44

X-Ray Powder Diffraction Data: (d(Å) (I/I_0))
9.41–9.26(vs), 6.81–6.76(w–m), 5.54–5.47(w–m), 4.31–4.26(s–vs), 3.66–3.65(w–vs), 2.912–2.889(w–s).

XAPO-44 materials exhibit properties characteristic of small-pore molecular sieves.

XAPO-46 (AFS)
(EP 158,977(1985))

Related Materials: MgAPSO-46

X-Ray Powder Diffraction Data: (d(Å) (I/I_0))
12.3–10.9(vs), 4.19–4.08(w–m), 3.95–3.87(vw–m), 3.351–3.278(vw–w), 3.132–3.079(vw–w).

XAPO-46 materials exhibit properties characteristic of large-pore molecular sieves.

XAPO-47
(EP 158,977(1985))

Related Materials: MgAPO-47

X-Ray Powder Diffraction Data: $(d(\text{Å}) (I/I_0))$
9.41(vs), 5.57–5.54(w–m), 4.33–4.31(s), 3.63–
3.60(w), 3.45–3.44(w), 2.940–2.931(w).

XAPO-47 materials exhibit properties characteristic of small-pore molecular sieves.

Y

See Linde type Y.

Yugawaralite (YUG) (natural)

Related Materials: Sr-Q

The name yugawaralite was proposed by the Japanese workers Sakurai and Hayashi in 1952 for the locality in which this zeolite first was found (Yugaware Hot Springs in Japan) (*Nat. Zeo.* 110 (1985)). Since its initial discovery, crystals of this rare zeolite have been found in Italy, Alaska, India, and Iceland.

Structure

(*Nat. Zeo.* 110(1985))

CHEMICAL COMPOSITION: $Ca_2[(AlO_2)_4(SiO_2)_{12}]*$ $18H_2O$
SYMMETRY:
SPACE GROUP: Pc
UNIT CELL CONSTANTS (Å): $a = 6.700(1)$
$\qquad\qquad\qquad\qquad b = 13.972$
$\qquad\qquad\qquad\qquad c = 10.039$
$\qquad\qquad\qquad\qquad \beta = 111.07$
DENSITY: 1.81 g/cc
PORE STRUCTURE: two-dimensional eight-rings 3.6 × 2.8 Å

X-Ray Powder Diffraction Data: (*Nat. Zeo.* 323(1985)) (d(Å) (I/I_0)) 7.79(20), 6.99(60), 6.26(25), 5.82(90), 5.62(5), 4.68(85), 4.65(85), 4.45(10), 4.41(30), 4.30(65), 4.18(30), 3.89(15), 3.87(5), 3.78(25), 3.75(5), 3.30(5), 3.27(5), 3.235(55), 3.198(10), 3.135(20), 3.056(100), 2.997(10), 2.937(30), 2.907(60), 2.864(10), 2.763(30), 2.720(35), 2.706(5), 2.680(25), 2.650(20), 2.638(15), 2.603(5), 2.587(15), 2.562(5), 2.513(15).

The framework structure of yugawaralite consists of four-, five-, and eight-member rings. The two-dimensional channel system lies in the *ac* plane and is a small-pore zeolite possessing eight-ring openings of about 3.5 Å (*Z. Kristallogr.* 125:220(1967); *Am. Mineral.* 50:484(1965); *Acta Crystallogr.* B 25:1183(1069); *Z. Kristallogr.* 130:88 (1969)). Complete ordering of the Si and Al is observed. The calcium ions lie in the channel intersection; they are located on the side of the cavity and are asymmetrically bonded to four framework oxygens and four water oxygens. There are seven different water oxygens in the strucutre, with only two fully occupied oxygen sites and a multitude of alternative hydrogen sites.

Atomic coordinates ($\times 10^5$) for Yugawaralite:

	x	y	z
Si1	34121(25)	14798(8)	98083(19)
Si2	71051(24)	3650(8)	19124(19)
Si3	40911(25)	12450(8)	69431(19)
Si4	2736(24)	47598(8)	43748(19)
Si5	36039(24)	37327(8)	96046(19)
Si6	74211(23)	49757(8)	62097(18)
Al1	0*	710(9)	0*
Al2	39614(28)	35598(9)	65361(20)

(*continued*)

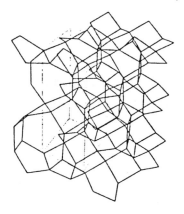

Yugawaralite Fig. 1S: Framework topology.

	x	y	z
O1	10689(22)	10638(6)	94841(17)
O2	85591(23)	4834(6)	9817(17)
O3	19153(22)	7642(6)	59056(17)
O4	50365(22)	10778(6)	13425(16)
O5	43489(23)	11736(7)	86004(17)
O6	61915(23)	7317(6)	67971(17)
O7	84451(22)	6416(6)	35574(17)
O8	33904(23)	26252(6)	−394(18)
O9	39913(23)	23277(5)	63939(17)
O10	16380(23)	42664(6)	98834(17)
O11	83032(22)	48293(6)	49303(17)
O12	17154(22)	38236(6)	49904(17)
O13	57911(22)	41154(6)	8345(17)
O14	36053(23)	39320(6)	80583(17)
O15	62998(22)	40026(6)	64000(17)
O16	93634(22)	47014(6)	26478(16)
Ca	5134(24)	21654(7)	42364(19)
OW1	98759(34)	24929(13)	17041(25)
OW1A	4165(297)	27602(96)	20621(141)
H11	95821(60)	19582(22)	10667(37)
H12	3505(65)	30007(25)	12817(42)
H13	5913(412)	34986(134)	21315(240)
OW2	90310(24)	23331(7)	61844(18)
H21	87938(55)	18305(18)	67456(38)
H22	81676(40)	28638(15)	62799(32)
OW3	70811(23)	29016(7)	32835(18)
H31	59175(71)	27280(50)	34773(68)
H32	65286(127)	33338(60)	39098(76)
H33	66062(41)	32229(17)	23748(28)
OW4	36902(36)	15938(14)	36919(24)
OW4A	33283(237)	12668(97)	35814(155)
H41	39007(48)	14373(22)	28329(32)
H42	49359(83)	13517(40)	44553(52)
H43	45146(288)	15376(424)	40785(405)
OW5	81646(142)	25961(65)	92491(117)
H51	90889(282)	20607(123)	91977(216)
H52	87863(581)	31721(154)	90312(424)

*Coordinate fixed to define origin.

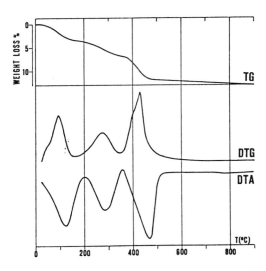

Yugawaralite Fig. 2:   Thermal curves of yugawaralite from Sardinia, Italy; in air, heating rate 20°C/min (*Nat. Zeo.* 116(1985)). (Reproduced with permission of Springer Verlag Publishers)

Thermal Properties

Water is lost from yugawaralite in three steps: between 70 and 100°C, between 200 and 300°C, and at 450°C. These steps correspond to the loss of two, two, and four water molecules per unit cell. At high temperatures the crystals melt to form a glass (970°C). With further increase of the temperature to 1050°C, recrystallization to anorthite and cristobalite is observed.

Ion Exchange

The ion exchange capacity for yugawaralite was found to be 0.045 meq/g, which represents 1% of the maximum value (3.34 meq/g). Higher capacities were observed in the synthetic material (30% of theoretical), with 0.14 to 1.0 meq reported. A strong selectivity for Sr is observed (*Mat. Res. Bull.* 2:951(1967)).

Z

Z
(US2,972,516(1961))
Union Carbide Corporation

Related Materials: K-F

Structure

(*Zeo. Mol. Sieve.* 370(1974))

CHEMICAL CONTENTS: K₂Al₂SiO₈*3H₂O
VOID VOLUME (cc/g):   0.14
PORE STRUCTURE (based on adsorption):   2.6 Å

X-Ray Powder Diffraction Data: (d(Å) (I/I₀))
7.45(vvs), 4.78(vw), 3.98(m), 3.47(m), 3.29(m),
3.09(vs), 2.97(s), 2.82(vs), 2.35(mw), 2.20(w),
2.11(m), 1.85(m), 1.74(m), 1.59(m), 1.56(m).

Synthesis

ORGANIC ADDITIVES
none

Zeolite Z crystallizes from reactive aluminosilicate mixtures with the composition in mo-
lar ratios of: $K_2O/SiO_2 = 0.58$, $SiO_2/Al_2O_3 = 1$ to 7, and $H_2O/K_2O = 316$. Crystallization
occurs readily after 3 to 4 days at 120°C
(US2,972,516(1961); *Dokl. Akad. Nauk SSSR*
157:913(1964)). Crystallization also can occur
at temperatures as low as 80°C (*Rend. Accad.
Sci. Fis. Mat.* Naples 35:165(1968)).

Thermal Properties

Zeolite Z maintains its structure during hydration at 300°C but decomposes at 600°C after 3
days.

Adsorption

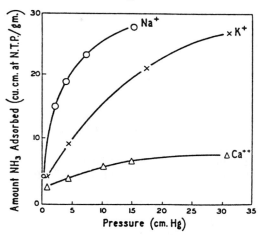

Z Fig. 1:   Adsorptive capacities of ion-exchanged Z
(US2,972,516(1961)).

Z-21 (LTN)
(Ger. 1,935,861(1970))
W. R. Grace and Co.

Related Materials: Linde type N

Structure

(*Zeo. Mol. Sieve.* 272(1974))

CHEMICAL COMPOSITION: Na₂O*Al₂O₃*1.7–
2.1SiO₂*9H₂O
SYMMETRY: cubic
UNIT CELL CONSTANT (Å): $a = 36.7$
VOID VOLUME (cc/g):   0.14
PORE STRUCTURE: based on adsorption
measurements, 2.6 Å

X-Ray Powder Diffraction Data: (Ger.
1,935,861(1970))   (d(Å)   (I/I₀))   21.12(50),
12.98(75),   11.07(70),   9.18(35),   8.42(10),
7.50(3), 7.06(10), 6.51(100), 6.23(50), 5.82(20),
5.60(10), 5.30(5), 5.15(15), 4.79(15), 4.50(10),

4.34(20), 4.24(3), 4.12(15), 4.04(70), 3.92(10), 3.76(45), 3.70(75), 3.61(5), 3.55(10), 3.43(10), 3.369(15), 3.31(1), 3.21(15), 3.15(2), 3.12(30), 3.03(15), 2.9(45), 2.95(0), 2.90(2), 2.88(5), 2.82(15), 2.77(10), 2.75(45).

Synthesis

ORGANIC ADDITIVES
none

Zeolite Z-21 crystallizes from a gel composition with the following molar ratios: 0.3–6.0 $SiO_2 : Al_2O_3 : 7–43 Na_2O : 117–720 H_2O$. Crystallization occurs readily at 90°C after 1 hour. Sodium silicate and aluminum trihydrate are effective sources of silica and alumina in this synthesis.

ZAPO-5 (AFI)
(US4,567,029(1986))
Union Carbide Corporation

Related Materials: AlPO$_4$-5

Structure

Chemical Composition: zinc aluminophosphate. See XAPO.

X-Ray Powder Diffraction Data: (d(Å) (I/I_0)) 11.95(100), 6.86(shared peak), 5.91(19), 4.48(46), 4.27(43), 3.97(75), 3.65(shared peak), 3.41(shared peak), 3.08(12), 2.959(21), 2.660(4), 2.578(14), 2.423(4), 2.392(9) (a small impurity of ZAPO-11 appears in this example, with some peaks shared with this impurity).

ZAPO-5 exhibits properties characteristic of a large-pore molecular sieve. For a detailed description of the framework topology of ZAPO-5, see AlPO$_4$-5.

Synthesis

ORGANIC ADDITIVES
diisopropylamine
diethylethanolamine
cycloyhexylamine

ZAPO-5 crystallizes from a reactive gel composition with a molar oxide ratio of: 1.0 R : 0.4 ZnO : 0.8 Al_2O_3 : P_2O_5 : 0.8 CH_3COOH : 50 H_2O (R represents the organic additive). Zinc acetate is an effective source of zinc, and hydrated alumina is the source of aluminum. Crystallization occurs readily after 24 hours at 150°C.

ZAPO-11 (AEL)
(US4,567,029(1986))
Union Carbide Corporation

Related Materials: AlPO$_4$-11

Structure

Chemical Composition: zinc-aluminophosphate. See XAPO.

X-Ray Powder Diffraction Data: (d(Å) (I/I_0)) 10.92(37), 9.41(67), 6.81(16), 5.72(46), 5.51(5), 4.70(9), 4.40(52), 4.23(73), 4.04(69), 3.99(57), 3.93(69), 3.87(100), 3.66(12), 3.62(15), 3.39(37), 3.18(12), 3.12(28), 3.038(11), 2.849(13), 2.747(32), 2.629(14), 2.481(9), 2.392(14), 2.298(7), 2.036(8), 1.807(7).

ZAPO-11 exhibits properties characteristic of a medium-pore molecular sieve. For a detailed description of the framework topology of ZAPO-11, see AlPO$_4$-11.

Synthesis

ORGANIC ADDITIVES
diisopropylamine
dipropylamine

ZAPO-11 crystallizes from a reactive gel composition with a molar oxide ratio of: 1.0 R : 0.4 ZnO : 0.8 Al_2O_3 : P_2O_5 : 0.8 CH_3COOH : 50 H_2O (R represents the organic additive). Zinc acetate is an effective source of zinc, and hydrated alumina is the source of aluminum. Crystallization occurs readily after 24 hours at 200°C.

ZAPO-14
(US4,567,029(1986))
Union Carbide Corporation

Related Materials: $AlPO_4$-14

Structure

Chemical Composition: zinc-aluminophosphate. See XAPO.

X-Ray Powder Diffraction Data: $(d(Å) (I/I_0))$
9.94(100), 7.97(22), 6.81(27), 5.61(22), 4.96(12), 4.27(10), 4.08(17), 4.00(22), 3.93(32), 3.41(11), 3.30(4), 3.03(16), 2.950(14), 2.667(6).

ZAPO-14 exhibits properties characteristic of a small-pore molecular sieve. For a detailed description of the framework topology of ZAPO-14 see $GaPO_4$-14.

Synthesis

ORGANIC ADDITIVES
isopropylamine

ZAPO-14 crystallizes from a reactive gel composition with a molar oxide ratio of: 1.0 R : 0.167 ZnO : 0.917 Al_2O_3 : P_2O_5 : 0.33 CH_3COOH : 50 H_2O (R represents the organic additive). Zinc acetate is an effective source of zinc, and hydrated alumina is the source of aluminum. Crystallization occurs readily after 168 hours at 150°C.

ZAPO-34 (CHA)
(US4,567,029(1986))
Union Carbide Corporation

Related Materials: chabazite
                    $AlPO_4$-34

Structure

Chemical Composition: zinc-aluminophosphate. See XAPO.

X-Ray Powder Diffraction Data: $(d(Å) (I/I_0))$
9.31(99), 6.86(21), 6.33(15), 5.54(45), 4.98(24), 4.33(100), 4.00(7), 3.88(7), 3.56(36), 3.47(26), 3.24(12), 3.16(13), 3.03(12), 2.93(55),

2.876(36), 2.614(14), 2.488(10), 2.281(5), 2.443(5), 1.918(6), 1.866(8), 1.804(6), 1.728(5).

ZAPO-34 exhibits properties characteristic of a small-pore molecular sieve. For a description of the framework topology of ZAPO-34, see $AlPO_4$-34.

Synthesis

ORGANIC ADDITIVES
tetraethylammonium$^+$

ZAPO-34 crystallizes from a reactive gel composition with a molar oxide ratio of: 1.0 R : 0.4 ZnO : 0.8 Al_2O_3 : P_2O_5 : 0.8 CH_3COOH : 50 H_2O (R represents the organic additive). Zinc acetate is an effective source of zinc, and hydrated alumina is the source of aluminum. Crystallization occurs readily after 68 hours at 100°C. ZAPO-14 has been noted as an impurity in some crystallizations.

Adsorption

Adsorption properties of ZAPO-34:

Adsorbate	Kinetic diameter, Å	Pressure, Torr	Temp., °C	Wt% adsorbed
O_2	3.46	12	−183	18
O_2		704	−183	22.2
Butane	4.3	692	24	8.3
Xenon	4.0	754	23	17.5
H_2O	2.65	20	23	27.8

ZAPO-35 (LEV)
(US4,567,029(1986))
Union Carbide Corporation

Related Materials: SAPO-35

Structure

Chemical Composition: zinc-aluminophosphate. See XAPO.

X-Ray Powder Diffraction Data: (d(Å) (I/I_0))
10.40(14),   8.19(42),   6.66(35),   5.57(8),
5.01(10sh),   4.19(49),   4.08(100),   3.56(6),
3.35(23),   3.12(34),   2.814(48),   2.592(12),
2.529(7), 2.151(2), 1.877(9), 1.852(7), 1.781(7),
1.667(7).

Synthesis

ORGANIC ADDITIVES
quinuclidine

ZAPO-35 crystallizes from a reactive gel composition with a molar oxide ratio of: 1.0 R : 0.5 ZnO : 0.65 Al_2O_3 : P_2O_5 : 1.0 CH_3COOH : 3.9 C_3H_7OH : 55 H_2O (R represents the organic additive). Zinc acetate is an effective source of zinc, and aluminum isopropoxide is the source of aluminum. Crystallization occurs readily after 72 hours at 150°C.

ZAPO-36 (ATS)
(US4,567,029(1986))
Union Carbide Corporation

Related Materials: related metal-substituted $AlPO_4$'s

Structure

Chemical Composition: zinc-aluminophosphate. See XAPO.

X-Ray Powder Diffraction Data: (d(Å) (I/I_0))
11.3(100) 10.9(sh), 6.56(4), 5.61(sh), 5.40(32),
4.67(49), 4.29(36), 4.11(sh), 4.06(44), 3.99(40),
3.90(sh), 3.75(11), 3.29(16), 3.16(13), 3.09(11),
2.96(7), 2.814(10), 1.585(20), 2.529(4).

ZAPO-36 exhibits properties characteristic of a large-pore molecular sieve.

Synthesis

ORGANIC ADDITIVES
tripropylamine

ZAPO-36 crystallizes from a reactive gel composition with a molar oxide ratio of: 1.0

R : 0.167 ZnO : 0.917 Al_2O_3 : P_2O_5 : 0.33 CH_3COOH : 39.8 H_2O (R represents the organic additive). Zinc acetate is an effective source of zinc, and hydrated alumina is the source of aluminum. Crystallization occurs readily after 24 hours at 150°C.

ZAPO-44 (CHA)
(US4,567,029(1986))
Union Carbide Corporation

Related Materials: chabazite
                        SAPO-44

Structure

Chemical Composition: zinc-aluminophosphate. See XAPO.

X-Ray Powder Diffraction Data: (d(Å) (I/I_0))
9.41(97), 6.81(28), 6.46(5), 5.51(46), 5.13(3),
4.70(6), 4.34(90), 4.10(48), 3.93(13), 3.87(15),
3.66(100),   3.41(33),   3.21(14),   3.018(10),
2.979(27), 2.894(60), 2.755(8), 2.722(10),
2.578(6),   2.529(23),   2.344(8),   2.132(8),
1.895(10),   1.875(6),   1.817(13),   1.765(5),
1.707(10).

ZAPO-44 exhibits properties characteristic of a small-pore molecular sieve.

Synthesis

ORGANIC ADDITIVES
cyclohexylamine

ZAPO-44 crystallizes from a reactive gel composition with a molar oxide ratio of: 1.0 R : 0.4 ZnO : 0.8 Al_2O_3 : P_2O_5 : 0.8 CH_3COOH : 50 H_2O (R represents the organic additive). Zinc acetate is an effective source of zinc, and hydrated alumina is the source of aluminum. Crystallization occurs readily after 24 hours at 200°C.

ZAPO-47 (CHA)
(US4,567,029(1986))
Union Carbide Corporation

Related Materials: chabazite
                        SAPO-47

Structure

Chemical Composition: zinc-aluminophosphate. See XAPO.

X-Ray Powder Diffraction Data: $(d(\text{Å}) (I/I_0))$
9.41(100), 6.86(15), 6.42(6), 5.57(34), 5.10(8), 4.70(4), 4.33(85), 4.10(8), 3.88(10), 3.63(22), 3.45(21), 3.24(12), 3.038(7), 2.959(14), 2.931(49), 2.849(5), 2.698(2), 2.600(9), 2.536(5), 2.356(4), 2.281(4), 2.137(2), 1.910(4), 1.887(6), 1.824(6), 1.725(6), 1.704(2).

ZAPO-47 exhibits properties characteristic of a small-pore molecular sieve.

Synthesis

ORGANIC ADDITIVES
diethylethanolamine

ZAPO-47 crystallizes from a reactive gel composition with a molar oxide ratio of: 1.0 R : 0.167 ZnO : 0.917 Al_2O_3 : P_2O_5 : 0.33 CH_3COOH : 5.5 C_3H_7 OH : 40 H_2O (R represents the organic additive). Zinc acetate is an effective source of zinc, and aluminum isopropoxide is the source of aluminum. Crystallization occurs readily after 144 hours at 150°C.

ZAPO-50 (AFY)
(US4,853,197(1989))
UOP

Structure

Chemical Composition: zinc aluminophosphate. See MeAPO.

X-Ray Powder Diffraction Data: $(d(\text{Å}) (I/I_0))$
11.00(100), 9.00(30), 6.98(4), 6.38(12), 5.21(10), 3.79(42), 3.68(44), 3.49(6), 3.41(11), 3.07(10), 3.01(13), 2.91(6), 2.86(3), 2.72(5), 2.64(3), 2.61(2), 2.44(3), 2.41(4), 2.19(2), 1.84(3), 1.74(3).

See CoAPO-50 for a description of the framework topology.

Synthesis

ORGANIC ADDITIVES
di-*n*-propylamine

ZAPO-50 crystallizes from a reactive mixture with a molar oxide ratio of: 2.5 R : 0.3 ZnO : 0.85 Al_2O_3 : 1.0 P_2O_5 : 0.6 CH_3COOH : 50 H_2O. Crystallization occurs at 150°C after 144 hours.

ZBH (MFI)
(EP 77,946(1983))
BASF

Related Materials: ZSM-5
AMS-1B

Structure

Chemical Composition: $M_{2/n}O$: B_2O_3 : $\geqslant 10$ SiO_2 : 0–80 H_2O (M = alkali or alkaline earth metal cation; n = charge on the cation).

X-Ray Powder Diffraction Data: $(d(\text{Å}) (I/I_0))$
11.0746(100), 9.9725(68), 9.7271(19), 6.6567(10), 6.3220(17), 5.9489(22), 5.6740(11), 5.5405(18), 5.0022(10), 4.9479(10), 4.5883(8), 4.3352(10), 4.2392(11), 3.9747(5), 3.8322(96), 3.7980(70), 3.7282(30), 3.6940(46), 3.6254(25), 3.4652(5), 3.4174(9), 3.3654(6), 3.3371(10), 3.2907(10), 3.2429(3), 3.1188(4), 3.0303(12), 2.9691(13), 2.9259(7), 2.8543(4), 2.7151(5), 2.5928(5), 2.4984(3), 2.4746(5), 2.4013(4), 2.3831(5), 2.3106(3), 1.9999(10), 1.9825(9), 1.9418(3), 1.9062(3), 1.8621(5), 1.8289(3), 1.7567(3), 1.6615(3), 1.6544(3), 1.6501(3), 1.4807(3), 1.4539(4), 1.4377(4), 1.4149(4), 1.4046(3), 1.3864(4), 1.3738(4), 1.3556(3).

Synthesis

ORGANIC ADDITIVES
ethylene glycol dimethyl ether
diethylene glycol dimethyl ether
triethylene glycol dimethyl ether
tetraethylene glycol dimethyl ether
CH_3O-$(CH_2CH_2O)_5$-C_3H_7
CH_3O-$(CH_2CH_2O)_2$-C_3H_7

ZBH crystallizes from a reactive borosilicate gel with a SiO_2/B_2O_3 ratio of 11 to 40 in the presence of sodium cations with ether as the organic additive. Crystallization occurs over a temperature range of 80 to 140°C.

ZBM-30 (MFI)
(EP 46,504(1982))
BASF

Related Materials: ZSM-5
AMS-1B

Structure

Chemical Composition: $2/n$ R : M_2O_3 : m SiO_2 : 0–160 H_2O (R = neutral amine; n = number of amine groups in R; m = 96–$[40n/L_R$ + K/ $5n/K_R$ + K]; L_R = size of the amine in A).

X-Ray Powder Diffraction Data: $(d(\text{Å})$ $(I/I_0))$ 11.60(36), 10.02(15), 7.08(5), 6.05(6), 5.80(11), 4.18(100), 3.87(79), 3.60(4), 3.56(6), 3.40(5), 3.35(7), 3.06(3), 2.84(16), 2.09(8), 2.07(4).

Synthesis

ORGANIC ADDITIVES
dipropylenetriamine
hexamethylenediamine
dihexamethylenetriamine
triethylenetetramine
propylenediamine

Crystallization of ZBM-30 occurs at elevated temperatures (150°C) using neutral organic amine additives. Both the aluminosilicate and the borosilicate composition have been prepared.

Zeagonite (GIS)

Earliest name for the mineral gismondine.

Zeolon (MOR)

An industrially used name referring to mordenite prepared by the Norton Company. See mordenite.

Zh (SOD)
(*Dokl. Akad. Nauk. SSSR* 147:1118(1962))

Sodalite hydrate. See sodalite.

ZK-4 (LTA)
(US3,314,752(1967))
Mobil Oil Corporation

Related Materials: Linde type A

Structure

(*Zeo. Mol. Sieve.* 179(1974))

CHEMICAL COMPOSITION: $Na_8TMA[(AlO_2)_9$ $(SiO_2)_{15}]*28H_2O$
SYMMETRY: cubic
SPACE GROUP: Pm3m
UNIT CELL CONSTANTS (Å): a = 12.191
FRAMEWORK DENSITY: 1.3 g/cc
VOID FRACTION (determined from water content): 0.47
PORE STRUCTURE: three-dimensional eight-member rings 4.2 Å

X-Ray Powder Diffraction Data: $(d(\text{Å})$ $(I/I_0))$ 12.08(100), 9.12(29), 8.578(73), 7.035(52), 6.358(15), 5.426(23), 4.262(11), 4.062(49), 3.662(65), 3.391(30), 3.254(41), 2.950(54), 2.725(10), 2.663(7), 2.593(15), 2.481(2),

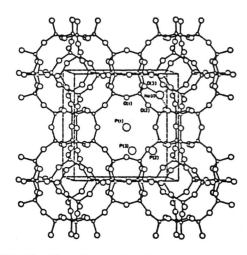

ZK-4 Fig. 2S: Framework topology.

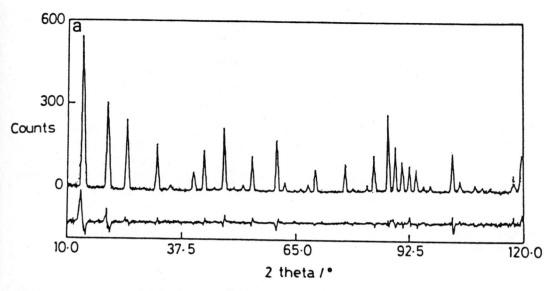

ZK-4 Fig. 2:   X-ray powder diffraction pattern for ZK-4 (*Zeo.* 6:449(1986)). (Reproduced with permission of Butterworths Publishers)

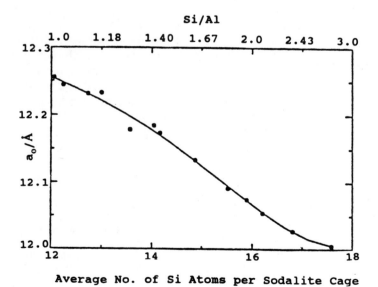

ZK-4 Fig. 3:   Cubic lattice parameter vs. composition for the hydrated sodium form of ZK-4 (*ACS Symposium Series* 218:267(1983), *Intrazeolite Chemistry*, Stuckey, C., Dwyer, J., eds.). (Reproduced with permission of the American Chemical Society)

2.435(1), 2.341(2), 2.225(2), 2.159(4), 2.121(5), 2.085(2), 2.061(2), 2.033(5), 1.90(2), 1.880(2), 1.828(1), 1.813(1), 1.795(1), 1.735(1), 1.720(5), 1.703(1), 1.669(2), 1.610(1), 1.581(2), 1.559(1).

ZK-4 is a higher-silica composition structurally related to zeolite A. It also is similar to N-A. See Linde type A for a description of the overall framework topology.

The difference between the X-ray powder diffraction pattern of ZK-4 and Linde type A is the absence of superlattice reflections in the former, which arise from the ording of Si and Al in type A. No vacancies in the framework are detected. The sodium cations are found in the six-ring windows, coordinated to three oxygen atoms of the framework at 2.311 Å. No cations occupy the eight-ring pores; their absence accounts for the ability of this zeolite to adsorb linear hydrocarbons. Neutron diffraction studies of ZK-4 reveal the presence of detrital material, one at the center of the eight-ring window and two in the alpha cage, thought to be residual TMAOH, water, or an aluminum- or silicon-containing species. The presence of this detrital material is consistent with the hydrocarbon adsorption properties because only 30% of the windows appear to be occupied in this structure (*Zeo.* 6:449 (1986)).

Atom position parameters using the asymmetric Voight refinements (*Zeo.* 6:449(1986)):

Atom	x	y	z
T	0.0	0.1844(8)	0.3720(7)
O(1)	0.0	0.2208(7)	0.5
O(1)	0.0	0.2924(4)	0.2924(4)
O(3)	0.1118(5)	0.1118(5)	0.3386(5)
Na(1)	0.2114(9)	0.2114(9)	0.2114(9)
P(1)	0.0	0.5	0.5
P(2)	0.300(2)	0.300(2)	0.437(3)
P(3)	0.303(9)	0.5	0.5

Synthesis

ORGANIC ADDITIVE
TMA^+

Zeolite ZK-4 crystallizes from the following molar oxide ratios:

$$SiO_2/Al_2O_3 = 4\text{--}11$$
$$Na_2O + TMA_2O/Al_2O_3 = 9\text{--}30$$
$$Na_2O/Al_2O_3 = 1\text{--}1.5$$
$$H_2O/Al_2O_3 = 100\text{--}350$$

Crystallization occurs between 24 and 72 hours with times as short as 13 hours in the presence of seed crystals. Temperatures are in the range of 100°C. Sodalite can be a major impurity phase at long crystallization times and/or varying hydroxide content.

Thermal Properties

ZK-4 is thermally stable to 600°C, the temperature necessary to remove the organic from the structure.

Ion Exchange

ZK-4 can be prepared in the potassium and sodium form by direct treatment of the calcined material with aqueous NaOH or KOH. The lower cation content in ZK-4 increases the effective size of the apertures for adsorption; in the dehydrated form adsorption is equivalent to that of the Ca-form of zeolite A (*Zeo. Mol. Sieve.* 576(1974)).

Adsorption

Comparison of the adsorption properties of Linde type A and ZK-4 (US3,314,752(1967)):

	Sorption (g/100g zeolite)			
	H_2O	n-C_6H_{14}	Cyclohexane	3-Methylpentane
4A	24	<1	<1	<1
ZK-4	20–25	8–13	<1	<1

ZK-5 (KFI)
(US3,247,195(1966))
Mobil Oil Corporation

Related Materials: species P
species Q
P-[Cl]
Q-[Br]

Structure

(*Zeo. Mol. Sieves.* 179(1974))

CHEMICAL COMPOSITION: $Na_{30}[Al_{30}Si_{66}O_{192}]*98H_2O$
SYMMETRY: cubic
SPACE GROUP: Im3m
UNIT CELL CONSTANT (Å): $a = 18.7$
PORE STRUCTURE: three-dimensional eight-member-ring nonintersecting dual channel system 3.9 × 3.9 Å

X-Ray Powder Diffraction Data: (d(Å) (I/I_0))
13.3(18), 9.41(100), 6.62(6), 5.93(41), 5.41(48), 5.03(2), 4.69(6), 4.41(50), 4.19(34), 3.98(22), 3.81(18), 3.66(6), 3.41(13), 3.21(35), 3.02(28), 2.94(21), 2.88(2), 2.81(26), 2.75(9), 2.64(11), 2.59(2), 2.54(9), 2.45(3), 2.37(1), 2.30(2), 2.20(3), 2.17(2), 2.14(1), 2.06(1), 2.04(2), 2.02(3), 1.97(0.5), 1.93(2), 1.89(2), 1.87(5).

The ZK-5 framework structure can be described as joining truncated cuboctahedra, or alpha cages, through double six-member rings, producing a body-centered array. There are two truncated cuboctahedra per unit cell. The main cavities are similar to those found in zeolite A,

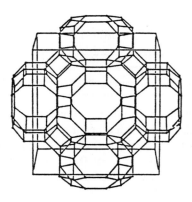

ZK-5 Fig. 1S: Framework topology.

and thus both zeolites produce similar adsorption properties. The cation positions in the structure are sensitive to temperature and water content. At room temperature 16 of the 30 sodium ions in the unit cell of the hydrated zeolite are located in the alpha cage near the center of the

Atomic parameters of ZK-5 at room temperature with corresponding values at 150°C in parentheses (*Z. Kristallogr.* 121:211(1965)):

Atom	x	y	z
Si,Al	0.084(0.085)	0.2015(0.204)	0.322(0.323)
O(1)	0.1245(0.1265)	0.1245(0.1265)	0.321(0.325)
O(2)	0.2525(0.251)	0.2525(0.251)	0.4045(0.3985)
O(3)	0(0)	0.183(0.187)	0.327(0.328)
O(4)	1/4(1/4)	0.1125(0.1095)	0.3875(0.3905)
Na	0.1315	0.1315	0.1315

Positional parameters for (1) ZK5RT, (2) ZK5220, (3) DZK5, and (4) CsZK5 refinements (*Z. Kristallogr.* 165:175(1983)):

	(1)	(2)	(3)	(4)
T				
x	0.0825(3)	0.0828(4)	0.0843(4)	0.0850(3)
y	0.2023(3)	0.2029(5)	0.2029(4)	0.2026(5)
z	0.3211(3)	0.3219(5)	0.3188(5)	0.3168(5)
O(1) $x = y$				
y	0.1280(2)	0.1277(3)	0.1258(3)	0.1251(3)
z	0.3145(3)	0.3140(5)	0.3163(4)	0.3153(5)
O(2) $x = y$				
y	0.2522(2)	0.2522(3)	0.2551(3)	0.2551(4)
z	0.4081(3)	0.4082(4)	0.4004(4)	0.4020(4)
O(3) $x = 0$				
y	0.1783(3)	0.1785(4)	0.1862(4)	0.1852(5)
z	0.2269(3)	0.3390(4)	0.3322(4)	0.3327(5)
O(4) $x = 1/4; y = 1/2 - z$				
z	0.3915(2)	0.3909(3)	0.3867(3)	0.3860(3)
Cs $x = y = 0$				
z	0.314(1)	0.314(1)	0.336(2)	0.327(2)
population[a]	9.7(3)	9.0(4)	3.9(3)	6.0(2)
K(1) $x = 0; y = 1/4; z = 1/2$				
population[a]	9.4(4)	9.9(6)	0.8(3)	2.7(3)
K(2) $x = y = z$				
z	0.150(3)	0.152(9)	—	—
population[a]	3.6(5)	1.9(5)	—	—

[a]Atoms/unit cell.

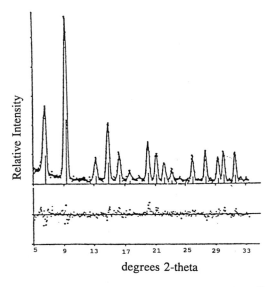

degrees 2-theta

ZK-5 Fig. 2:   Diffraction pattern for ZK-5 (*Z. Kristallogr.* 165:175(1983)). (Reproduced with permission of Oldenbourg Verlag Publishers)

six-rings on the threefold axis. Heat treatment to 150°C produces a shift in the sodium ions, as the ones near the six-ring sites move to another (undetermined) site (*Zeo. Mol. Sieve.* 90(1974)). With Cs,K-exchanged ZK-5, the Cs occupies the flat eight-ring sites, and the K prefers the puckered eight-ring sites. Distortion of the pore openings is observed upon extraction of the potassium cations. The Cs atom either moves toward the alpha cage if K is present or, to relax, toward the gamma cage if K is extracted (*Z. Kristallogr.* 165:175(1983)). The site preferences of the two cations, Cs and K, suggest the possibility of producing two distinct adsorbents in this zeolite.

Distances between atoms in the framework of the dehydrated materials change little upon heating. This behavior is not observed in the hydrated material, as large atomic shifts are observed upon dehydration. Only for exchange with the small proton (deuterium) is the change in the framework significant.

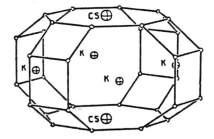

ZK-5 Fig. 3:   The Cs and K positions in the gamma cage of ZK-5. For clarity, oxygen is omitted from the drawing of the framework (*Z. Kristallogr.* 165:175(1983)). (Reproduced with permission of Oldenbourg Verlag Publishers)

Synthesis

ORGANIC ADDITIVE
1,4-dimethyl-1,4-diazoniabicyclo[2.2.2]octane (DABCO)

ZK-5 can be prepared from a reactive aluminosilicate gel with a batch composition having a SiO_2/Al_2O_3 ratio between 2.5 and 15; Na_2O/Na_2O + DABCO from 0.01 to 0.25, H_2O/Na_2O + DABCO falling between 25 and 50, and Na_2O + DABCO/SiO_2 between 1 and 2. This gives a batch composition with a molar ratio of: 5.9 Na_2O : 45 DABCO : 4.0 Al_2O_3 : 45.1 SiO_2.

Crystallization occurs after 8 days at temperatures between 95 and 100°C.

This zeolite also has been prepared in the presence of Cs and K and the absence of organics (US3,720,753(1973)). ZK-5 also can be prepared in the presence of Li cations in addition to Na cations at a SiO_2/Al_2O_3 ratio of 5 (USSR 1,115,564(1985)). It has been synthesized in the presence of Ba,(Ba,Na),(K,Ba), (Li,Cs,TMA)–containing reaction mixtures and shows a preference for the presence of barium (*Zeo.* 1:130(1981)). See species P and species Q for a description of these synthesis.

Thermal Properties

ZK-5 is thermally stable to at least 500°C, the temperature needed to remove the organic trapped within the pores of the structure. In the NH_4^+ form it can be deep-bed-calcined to produce a more dealuminated form containing trapped Al–O–OH clusters (*Zeo.* 6:378(1986)).

Adsorption

Adsorption properties of ZK-5 (US3,247,195(1966)):

Adsorbate	Wt% adsorbed
H_2O	20.6
Cyclohexane	0.9
n-Hexane	12.7

ZK-14 (CHA)
(*Molec. Sieves Soc. Chem. Ind.* 85(1968))

Related Materials: chabazite

Structure

(*Zeo.* 4:218(1984))

CHEMICAL COMPOSITION:
$Na_{10.6}K_{0.5}[Si_{24.9}Al_{11.1}O_{72}]*32H_2O$
SPACE GROUP: R3m (room temperature)

UNIT CELL CONSTANTS (Å): $a = 13.822$
$c = 15.155$
(room temperature)
at 220°C C2/m $a = 12.87$
$b = 13.51$
$c = 15.45$
at 350°C C2/m $a = 13.12$
$b = 13.57$
$c = 15.81$
PORE STRUCTURE: eight-member rings

X-Ray Powder Diffraction Data: (d(Å) (I/I_0))
11.78(2), 9.41(7), 6.89(3), 5.55(4), 5.031(8), 4.669(3), 4.332(27), 3.979(3), 3.880(21), 3.602(20), 3.449(8), 3.236(4), 3.188(14), 3.130(4), 2.930(100), 2.894(53), 2.857(2), 2.775(5), 2.696(6), 2.607(22), 2.576(10), 2.517(10), 2.364(1), 2.316(6), 2.299(13), 2.168(8), 2.128(2), 2.092(18), 2.066(8), 1.914(5), 1.875(20), 1.811(41), 1.772(14), 1.725(41), 1.702(8), 1.680(18), 1.651(26), 1.523(10), 1.518(12).

The structure of ZK-14 is based on a chabazite-type framework generated by stacking parallel six-rings of tetrahedra in an AABBCC stacking sequence. A minor structural change is observed in this material upon thermal treatment between 200 and 800°C. In the potassium-exchanged ZK-14, potassium cations are positioned at the center of one of two symmetrically inequivalent double six-rings coordinated by six oxygens. The other two sites where potassium is located are in two symmetrically inequivalent eight-rings. The population parameter of one eight-ring site is only 0.6 and is proposed to be a result of alternate occupancy of that site.

The sodium form of ZK-14 transforms topotactially to sodalite above 600°C, a transformation dependent on the aluminum content, number of protons, and amount of water present. The potassium form of this material does not transform. This lack of transformation is attributed to the presence of the K^+ cations in the eight-member-ring sites. The transformation can be represented as: AABBCC → ABCABC. Natural chabazite in the Na-exchanged form also transforms to sodalite phases at 360°C in wet N_2. Stacking faults are observed in the ZK-14 structure based on the presence of both broad

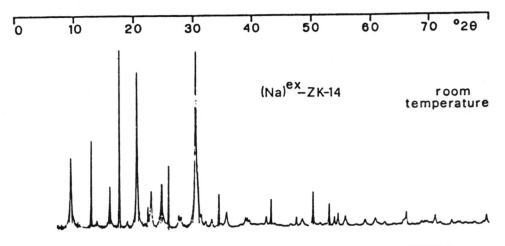

ZK-14 Fig. 1:   X-ray powder diffraction pattern for sodium-exchanged ZK-14 (*Zeo.* 4:218(1984)). (Reproduced with permission of Butterworths Publishers)

and narrow reflections, which are observed in the powder diffraction pattern. The hexagonal cell is observed only at room temperature; the transformation is to monoclinic symmetry at higher temperatures.

Positional parameters of Na-ZK-14 at room temperature, 220°C, and 350°C (*Zeo.* 4:218(1984)):

Atom	x	y	z	Population*
room temperature, R3m (hexagonal axis)				
T(1)	−0.0013(5)	0.2245(4)	0.1053(2)	
O(1)	0	0.2697(7)	0	
O(2)	−0.0238(9)	0.3095	1/6	
O(3)	0.1248(5)	0.2496	0.1282(9)	
O(4)	−0.0932(5)	0.0932	0.139(1)	
Na(1)	0.440(1)	0	0	0.70(1)
Na(2)		0	0.124(4)	0.30(2)
W(1)	0.1032(7)	0.2063	0.470(1)	1
W(2)	0.7293(5)	0.4587	0.016(1)	0.84(2)
W(3)	0	0	0.456(4)	0.60(2)
W(4)	0	0	0.768(3)	0.36(4)
220°C, C2/m				
T(11)	0.001(3)	0.237(1)	0.101(3)	
T(12)	0.363(3)	0.124(1)	0.340(2)	
T(13)	0.340(3)	0.115(1)	0.131(2)	
O(11)	0	0.282(3)	0	
O(12)	0.392(4)	0.157(2)	0.260(3)	
O(21)	−0.021(4)	0.326(3)	0.157(4)	

Positional parameters of Na-ZK-14 at room temperature, 220°C, and 350°C (*Zeo.* 4:218(1984)):

Atom	x	y	z	Population*
220°C, C2/m (*cont.*)				
O(22)	1/2	0.160(4)	1/2	
O(31)	0.197(4)	0.199(2)	0.250(3)	
O(32)	0.339(5)	0	0.111(5)	
O(41)	−0.155(4)	0.153(2)	0.035(3)	
O(42)	0.293(5)	0	0.306(5)	
Na(11)	0.735(5)	0.266(2)	0.389(3)	1.26(1)
Na(2)	0.007(11)	0	0.143(8)	0.68(1)
W(12)	−0.212(5)	0	0.274(4)	1.20(1)
W(22)	0.160(12)	0	0.143(8)	0.51(1)
350°C, C2/m				
T(11)	0.0042(7)	0.2561(4)	0.0965(5)	
T(12)	0.3442(6)	0.1125(3)	0.3278(4)	
T(13)	0.3497(6)	0.1193(3)	0.1355(5)	
O(11)	0	0.3203(6)	0	
O(12)	0.4292(9)	0.1372(5)	0.2924(7)	
O(21)	−0.0283(7)	0.3315(4)	0.1491(6)	
O(22)	1/2	0.1347(7)	1/2	
O(31)	0.1993(7)	0.2072(4)	0.2422(6)	
O(32)	0.308(1)	0	0.0850(9)	
O(41)	−0.1454(8)	0.1619(5)	0.0205(6)	
O(42)	0.256(1)	0	0.289(1)	
Na(11)	0.7235(9)	0.2619(4)	0.3904(7)	1.19(1)
W(21)	0.663(1)	0.5645(6)	0.7249(8)	0.47(1)
W(5)	0.304(2)	0	0.686(2)	0.53(1)

*Population parameters of all nonframework atoms based on O^{2-} (10 electrons).

Synthesis

ORGANIC ADDITIVES
tetramethylammonium$^+$ (TMA)

Zeolite ZK-14 crystallizes from a reactive aluminosilicate mixture with a composition of: 4.61 K_2O : 15.91 TMA_2O : SiO_2 : 2342 H_2O. A temperature of 70°C has been used, with crystallization at this temperature complete after 11 to 12 days. Precipitated silicic acid was used as the silica source (*Zeo.* 4:218(1984)).

ZK-19 (PHI)
(*Am. Mineral.* 54:1607(1969))

Related Materials: phillipsite

Structure

SYMMETRY: orthorhombic
UNIT CELL CONSTANTS (Å): $a = 9.96(5)$
$b = 14.25(2)$
$c = 14.25(2)$

X-Ray Powder Diffraction Data: (*Zeo. Mol. Sieve.* 370(1974)) (d(Å) (I/I_0)) 8.17(19), 7.13(88), 5.37(19), 5.03(11sh), 4.98(23), 4.29(12), 4.13(7sh), 4.08(26), 3.68(3), 3.26(15sh), 3.23(13sh), 3.18(100), 3.15(16sh), 2.94(30), 2.89(5), 2.74(23), 2.685(31), 2.56(6), 2.54(4), 2.51(6), 2.39(5), 2.34(7), 2.24(5), 2.05(6).

The X-ray powder diffraction pattern of ZK-19 is very similar to that of natural sodium/potassium phillipsite. ZK-19 represents a more potassium-rich phillipsite (*Am. Mineral.* 54:1607(1969)).

Synthesis

ORGANIC ADDITIVE
none

ZK-19 crystallizes from a reactive aluminosilicate gel with a SiO_2/Al_2O_3 ratio between 8 and 10, resulting in a structure with a ratio of 4.8. Tripotassium phosphate, sodium aluminate, and water glass are used as starting materials. The ratio of $Na_2O/(Na_2O + K_2O)$ ranges between 0.3 and 0.85. If trisodium phosphate is substituted for the potassium phosphate, a phillipsite with a SiO_2/Al_2O_3 ratio between 4.1 and 4.8 crystallizes instead. In more aluminum-rich gels the resulting ZK-19 crystals will have a SiO_2/Al_2O_3 between 3 and 4.5.

Thermal Properties

The structure of ZK-19 is destroyed upon calcination to 350°C for either the sodium or the calcium form. At higher SiO_2/Al_2O_3 ratios, in the potassium form, the structure is more thermally stable.

Adsorption capacities for zeolite ZK-19 (*Am. Mineral.* 54:1607(1969)):

SiO_2/Al_2O_3	Na/Na + K	Adsorption, g/100g^a	
		n-C_6H_{14}	H_2O
3.26	0.32	0.94	3.2
3.35	0.39	0.31	1.0
3.90	0.27	0.25	1.5
4.19	0.43	0.51	5.7
4.26	0.35	0.59	11.4
4.30	0.51	0.62	7.0
4.33	0.24	0.66	13.0
4.74	0.46	0.69	5.6
4.79	0.14	0.49	12.3
4.90	0.38	0.25	9.6
5.08	0.37	0.20	11.9
5.10	0.26	0.34	13.7
5.16	0.28	0.10	12.6
5.39	0.16	0.34	13.4
6.00	0.26	0.10	12.9
6.22	0.18	0.11	12.8
Potassium-exchanged form:			
3.26	0.002	0.31	11.9
3.35	0.006	1.58(?)	11.8
3.90	0.002	0.21	12.2
4.74	0.009	0.74	12.0
Sodium-exchanged form:			
4.79	0.87	0.35	1.42
5.16	0.92	0.29	0.24
6.00	0.82	0.30	0.43

^{a}The sorption capacities were measured at 25°C and 20 mm Hg (n-C_6H_{14}) or 12 mm Hg (H_2O), respectively.

Adsorption

Calcium-exchanged ZK-19 (*Am. Mineral.* 54:1607(1969)):

SiO$_2$/ Al$_2$O$_3$	2Ca/Na + K + 2Ca	K/Na + K + 2Ca	Adsorption, g/100g^a	
			n-C$_6$H$_{14}$	H$_2$O
4.19	0.78	0.22	0.33	0.82
4.34	0.66	0.26	0.58	0.45
4.79	0.49	0.49	0.45	6.6
5.08	0.58	0.38	0.51	2.48
5.16	0.71	0.28	0.26	1.70
5.39	0.46	0.50	0.43	4.80
6.22	0.57	0.40	0.14	8.65

^{a}The sorption capacities were measured at 25°C and 20 mm Hg (*n*-C$_6$H$_{14}$) or 12 mm Hg (H$_2$O), respectively.

ZK-20 (LEV)
(US3,459,676(1969))
Mobil Oil Corporation

Related Materials: levyne
SAPO-35

Structure

Chemical Composition: 0.1–0.2 R$_2$O : 0.8–0.9 Na$_2$O : Al$_2$O$_3$: 4–5 SiO$_2$: 1–5 H$_2$O.

X-Ray Powder Diffraction Data: (d(Å) (I/I_0))
14.2(vw), 10.4(m), 9.5(w), 8.2(S), 7.7(w), 6.7(m), 5.2(s), 4.3(s), 4.1(vs), 3.86(m–s), 3.62(w), 3.48(w), 3.34(w), 3.18(s), 3.10(m), 2.87(m), 2.81(s–vs), 2.64(m), 2.59(vw), 2.52(w), 2.41(m), 2.33(w–m), 2.18(vw), 2.14(m), 2.07(vw), 2.04(w), 1.96(w), 1.93(w), 1.90(w), 1.88(vw), 1.86(w), 1.80(m), 1.69(w), 1.68(m), 1.60(m), 1.555(m), 1.545(m), 1.435(m), 1.400(m).

ZK-20 appears to be isostructural with the natural zeolite levyne.

Synthesis

ORGANIC ADDITIVES
1-methyl-1-azonia-4-azabicyclo[2.2.2]octane$^+$

ZK-20 crystallizes from a reactive aluminosilicate gel in the presence of the organic cation 1-methyl-1-azonia-4-azabicyclo [2.2.2] octane at a SiO$_2$/Al$_2$O$_3$ ratio between 4 and 11, a Na$_2$O/Na$_2$O + [C$_7$H$_{15}$N$_2$]$_2$O between 0.05 and 0.15, and a H$_2$O/Na$_2$O + [C$_7$H$_{15}$N$_2$]$_2$O between 20 and 50. Crystallization occurs at temperatures around 100°C. Silica gel and sodium aluminate are used as the sources of silica and alumina, respectively. Crystallization requires 5 days for completion.

ZK-21 (LTA)
(US3,355,246(1967))
Mobil Oil Corporation

Related Materials: Linde type A

Structure

Chemical Composition: 1.0 ± M$_{2/n}$O : Al$_2$O$_3$: y SiO$_2$: 0.01(y + 2)/48 P$_2$O$_5$ (y = 1.9–4.5).

X-Ray Powder Diffraction Data: (d(Å) (I/I_0))
12.16(100), 8.65(68), 7.07(63), 5.48(29), 5.02(4), 4.33(15), 4.07(63), 3.86(3), 3.675(83), 3.389(30), 3.264(59), 2.952(77), 2.878(15), 2.732(10), 2.664(5), 2.601(32), 2.491(9), 2.441(4), 2.395(2), 2.348(6), 2.227(3), 2.154(10), 2.122(4), 2.093(3), 2.061(3), 2.033(9), 1.904(7), 1.881(5), 1.839(2), 1.818(2), 1.740(2), 1.720(14), 1.706(2), 1.674(8), 1.659(2), 1.615(4), 1.586(5), 1.560(5).

ZK-21 appears to be isostructural with the Linde zeolite type A.

Synthesis

ORGANIC ADDITIVES
tetramethylammonium$^+$ (TMA)

ZK-21 crystallizes from a reactive aluminosiliciate gel in the presence of phosphoric acid and TMA$^+$. The reactant composition that produces ZK-21 after 96°C and 5 days is: 5.72 Na$_2$O : 8.28 TMA$_2$O : 4.0 SiO$_2$: Al$_2$O$_3$: 6.58 H$_2$O. The crystals formed contain a SiO$_2$/Al$_2$O$_3$ ratio of 3.26. Sodium aluminate was used as

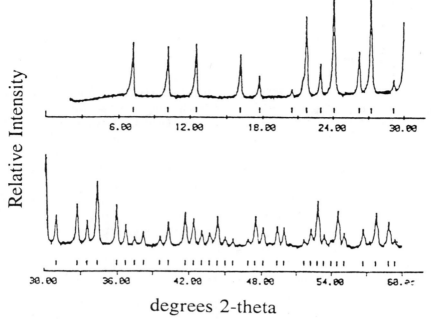

ZK-21 Fig. 1:   X-ray powder diffraction pattern of ZK-21 (*Zeo.* 10:2(1990)). (Reproduced with permission of Butterworths Publishers)

the source of aluminum; sodium metasilicate was the source of silica as well as water glass and colloidal silica. Aluminum phosphate dihydrate and phosphoric acid were the sources of additional aluminum as well as phosphorus.

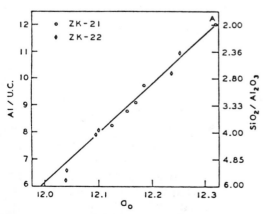

ZK-21 Fig. 2:   Lattice parameters of zeolites ZK-21 and ZK-22 (*Inorg. Chem.* 10:2488(1971)). (Reproduced with permission of the American Chemical Society)

Zeolite Y and sodalite are observed as impurity phases in this system.

Adsorption

ZK-21 adsorbs between 0.9 and 1.54 g cyclohexane, 11 to 13.2 g *n*-hexane, and 21 to 24.6 g of water per 100 g zeolite (*Inorg. Chem.* 10:2488(1971)).

ZK-22 (LTA)
US3,791,964(1974))

Related Materials: Linde type A
ZK-21

Structure

ZK-22 is differentiated from ZK-21 in that ZK-22 contains TMA as a cation to balance part of the framework charge and has a higher SiO_2/Al_2O_3 ratio than ZK-21. ZK-22 is also isostructural with the Linde zeolite type A.

Synthesis

ORGANIC ADDITIVES
tetramethylammonium$^+$ (TMA)

ZK-21 crystallizes from a reactive alumi-nosilciate gel in the presence of phosphoric acid and TMA$^+$. The reactant composition that produces ZK-22 after 90°C and 74 days is: 6.7 Na_2O : 2.7 TMA_2O : 6.0 SiO_2 : Al_2O_3 : 404 H_2O. The crystals formed contain a SiO_2/Al_2O_3 ratio of 5.34. Sodium aluminate was used as the source of aluminum; sodium metasilicate, water glass and colloidal silica were the sources of silica. Aluminum phosphate dihydrate and phosphoric acid were the sources of additional aluminum as well as phosphorus. Zeolite Y and sodalite are observed as impurity phases in this system.

Adsorption

ZK-22 adsorbs between 0.9 and 1.54 g cyclo-hexane, 11 to 14.8 g n-hexane, and 21 to 24.6 g of water per 100 g zeolite (*Inorg. Chem.* 10:2488(1971)).

ZKQ-1B (MFI)
(EP 148,038(1984))

See ZSM-5.

ZKU-2 (ERI/OFF)
(US4,440,328(1983))

Related Materials: ZSM-34

Structure

Chemical Composition: $M_{2/n}O$: Al_2O_3 : 8–70 SiO_2.

X-Ray Powder Diffraction Data: ($d(Å)$ (I/I_0))
11.50(100),   7.59(19),   6.67(70),   6.36(33),
5.78(18), 4.59(23), 4.49(7), 4.29(82), 4.17(8),
3.84(29), 3.77(119), 3.66(20), 3.59(74), 3.43(3),
3.33(25), 3.19(16), 3.16(38), 2.95(15), 2.87(76),
2.95(91), 2.69(30), 2.51(16), 2.49(20), 2.20(10),
1.89(14), 1.78(16), 1.66(21), 1.59(81).

Synthesis

ORGANIC ADDITIVE
tetraalkyl ammonium, with alkyl either methyl or ethyl

ZKU-2 crystallizes from a reactive gel with a batch composition of: SiO_2/Al_2O_3 between 8 and 70; OH/SiO_2 between 0.2 and 1.6; H_2O/SiO_2 between 10 and 30; M/SiO_2 between 0.2 and 1.7; K_2O/M_2O between 0.05 and 0.3; R/SiO_2 between 0.0086 and 0.20 (M = K + Na). Crystallization occurs rapidly in the presence of seed crystals at temperatures between 150 and 270°C after 0.5 to 20 hours. The crystals have an orbicular regular crystal form.

ZKU-4 (ERI/OFF)
US4,440,328(1983))

Related Material: ZSM-34

Structure

Chemical Composition: $M_{2/n}O$: Al_2O_3 : 8–70 SiO_2.

X-Ray Powder Diffraction Data: ($d(Å)$ (I/I_0))
11.62(100),   7.61(24),   6.67(59),   6.35(47),
5.77(11), 4.58(16), 4.49(22), 4.36(60), 4.15(5),
3.84(26), 3.77(97), 3.66(56), 3.60(60), 3.43(4),
3.33(16), 3.30(3), 3.19(12), 3.17(35), 2.94(12),
2.88(53), 2.69(85), 2.59(8), 2.51(14), 2.50(16),
2.21(8), 1.91(11), 1.78(13), 1.76(6), 1.66(13),
1.59(13).

Synthesis

ORGANIC ADDITIVE
tetraalkyl ammonium, with alkyl either methyl or ethyl

ZKU-4 crystallizes from a reactive gel with a batch composition of: SiO_2/Al_2O_3 between 8 and 70; OH/SiO_2 between 0.2 and 1.6; H_2O/SiO_2 between 10 and 30; M/SiO_2 between 0.2 and 1.7; K_2O/M_2O between 0.05 and 0.3; R/SiO_2 between 0.0086 and 0.20 (M = K + Na). Crystallization occurs rapidly in the presence of seed crystals at temperatures between 150 and

270°C after 0.5 to 20 hours. The crystals have a rice-grain morphology.

ZMQ-TB (MFI)
(EP 104,107(1983))

See ZSM-5.

ZnAPSO
(EP 158,975(1985))
Union Carbide Corporation

ZnAPSO represents a family of zinc silicoaluminophosphates with the batch composition: m R : $(Zn_w Al_x P_y Si_z)O_2$. See ZnAPSO Figures 1, 2, and 3.

Synthesis

See ZnAPSO Figures 1, 2 and 3.

ZnAPSO-5 (AFI)
(EP 158,975(1985))
Union Carbide Corporation

Related Materials: AlPO$_4$-5

Structure

Chemical Composition: Zinc silicoaluminophosphate. See ZnAPSO.

X-Ray Powder Diffraction Data: ($d(Å)$ (I/I_0))
12.28–11.91(vs), 4.58–5.58(m), 4.23–4.19(m), 3.971–3.952(m–s), 3.466–3.427(w–m).

Zn-APSO-5 exhibits properties characteristic of a large-pore molecular sieve. See AlPO$_4$-5 for a description of the framework topology.

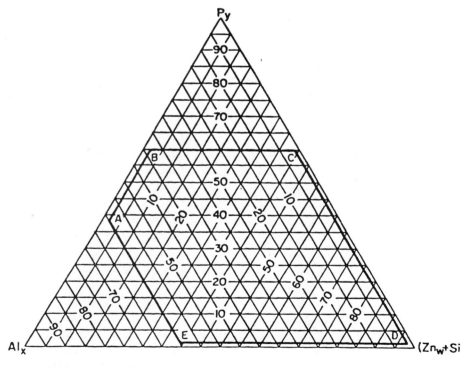

ZnAPSO Fig. 1:   Ternary diagram indicating the compositional parameters for the ZnAPSO molecular sieves (EP 158,975(1985)).

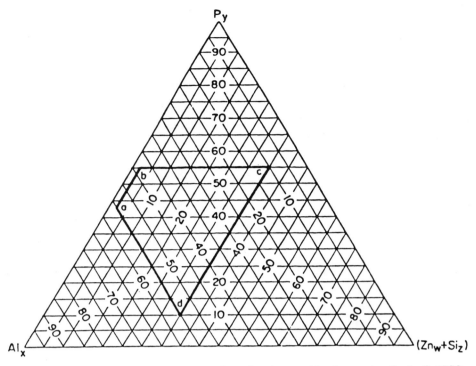

ZnAPSO Fig. 2: Ternary diagram indicating the preferred compositional parameters for the ZnAPSO molecular sieves (EP 158,975(1985)).

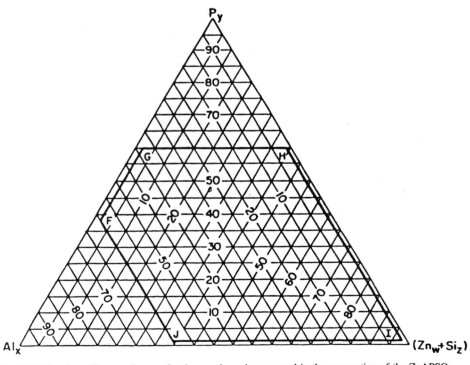

ZnAPSO Fig. 3: Ternary diagram for the reaction mixtures used in the preparation of the ZnAPSO molecular sieves (EP 158,975(1985)).

Synthesis

ORGANIC ADDITIVES
tripropylamine
tetrapropylammonium$^+$
tetraethylammonium$^+$
diethylethanolamine

ZnAPSO-5 crystallizes from a reactive gel with a molar oxide ratio of: R : 0.1–0.2 ZnO : Al_2O_3 : P_2O_5 : 0.6 SiO_2 : 50 H_2O (R represents the organic additive). Crystallization occurs after 42 hours at 150 to 200°C. Co-crystallizing phases observed in this system include ZnAPSO-34, -36, -44, and -47.

Adsorption

Adsorption properties of ZnAPSO-5:

Adsorbate	Kinetic diameter, Å	Pressure, Torr	Temperature, °C	Wt% adsorbed*
O_2	3.46	100	−183	7
O_2		750	−183	10
Neopentane	6.2	700	24	4
H_2O	2.65	4.3	24	4
H_2O		20	24	12

*Typical amount adsorbed.

ZnAPSO-11 (AEL)
(EP 158,975(1985))
Union Carbide Corporation

Related Materials: $AlPO_4$-11

Structure

Chemical Composition: zinc-silicoaluminophosphate. See ZnAPSO.

X-Ray Powder Diffraction Data: (d(Å) (I/I_0)) 9.51–9.17(m–s), 4.40–4.31(m–s), 4.25–4.17(s–vs), 4.04–3.95(m–s), 3.95–3.92(m–s), 3.87–3.80(m–vs).

ZnAPSO-11 exhibits properties characteristic of a medium-pore molecular sieve. See $AlPO_4$-11 for a description of the framework topology.

Synthesis

ORGANIC ADDITIVE
di-n-propylamine

ZnAPSO-11 crystallizes from a reactive gel with a molar oxide ratio of: R : 0.1 ZnO : Al_2O_3 : P_2O_5 : 0.6 SiO_2 : 50 H_2O (R represents the organic additive). (See ZnAPSO.) Crystallization occurs after 42 hours at 150 to 200°C. Co-crystallizing phases observed in this system include ZnAPSO-46 and -39.

Adsorption

Adsorption properties of ZnAPSO-11:

Adsorbate	Kinetic diameter, Å	Pressure, Torr	Temperature, °C	Wt% adsorbed*
O_2	3.46	100	−183	5
O_2		750	−183	6
Cyclohexane	6.0	90	24	4
H_2O	2.65	4.3	24	6
H_2O		20	24	8

*Typical amount adsorbed.

ZnAPSO-20 (SOD)
(EP 158,975(1985))
Union Carbide Corporation

Related Materials: $AlPO_4$-20
sodalite

Structure

Chemical Composition: Zinc silicoaluminophosphate. See ZnAPSO.

X-Ray Powder Diffraction Data: (d(Å) (I/I_0)) 6.39–6.33(m), 4.52–4.48(m), 3.685–3.663(vs), 3.187–3.170(w), 2.853–2.840(w), 2.600–2.589(m–w).

ZnAPSO-20 exhibits properties characteristic of a very small-pore material.

Synthesis

ORGANIC ADDITIVE
tetramethylammonium$^+$

Zn-APSO-20 crystallizes from a reactive gel with a molar oxide ratio of: R : 0.2 ZnO : Al$_2$O$_3$: P$_2$O$_5$: 0.6 SiO$_2$: 50 H$_2$O (R represents the organic additive). (See ZnAPSO.) Crystallization occurs between 46 and 166 hours at temperatures ranging between 150 and 200°C. Co-crystallizing phases observed in this system include ZnAPSO-43.

ZnAPSO-31 (ATO)
(EP 158,975(1985))
Union Carbide Corporation

Related Materials: AlPO$_4$-31

Structure

Chemical Composition: Zinc silicoaluminophosphate. See ZnAPSO.

X-Ray Powder Diffraction Data: (d(Å) (I/I_0))
10.53–10.40(m), 4.40–4.37(m), 4.171(w), 4.036(m), 3.952–3.934(vs), 2.831–2.820(w–m).

Zn-APSO-31 exhibits properties characteristic of a medium-pore molecular sieve. See AlPO$_4$-31 for a description of the framework topology.

Synthesis

ORGANIC ADDITIVE
di-n-propylamine

ZnAPSO-31 crystallizes from a reactive gel with a molar oxide ratio of: R : 0.2 ZnO : Al$_2$O$_3$: P$_2$O$_5$: 0.6 SiO$_2$: 50 H$_2$O (R represents the organic additive). (See ZnAPSO.) Crystallization occurs after 42 hours at temperatures between 150 and 200°C. Co-crystallizing phases observed in this system include ZnAPSO-46.

ZnAPSO-34 (CHA)
(EP 158,975(1985))
Union Carbide Corporation

Related Materials: chabazite
                        AlPO$_4$-34

Structure

Chemical Composition: Zinc silicoaluminophosphate. See ZnAPSO.

X-Ray Powder Diffraction Data: (d(Å) (I/I_0))
(as-synthesized) 9.19(100), 6.84(16), 6.25(14), 5.50(42), 4.90(22), 4.30(91), 3.978(5), 3.842(5), 3.521(25), 3.437(18), 3.281(5), 3.135(6), 3.015(5), 2.920(33), 2.856(23), 2.755(2), 2.602(7), 2.468(5), 2.320(4), 2.267(5), 2.097(4), 2.077(4)*, 1.908(5), 1.856(8), 1.827(4), 1.791(4), 1.723(3), 1.697(3), 1.645(3), (* represents an impurity peak); (after calcination in air to 500°C) 9.27(100), 6.85(24), 5.49(13), 4.94(10), 4.28(30), 4.004(2), 3.828(5), 3.533(9), 3.411(12), 3.138(4), 2.896(16), 2.852(9).

ZnAPSO-34 exhibits properties characteristic of a small-pore molecular sieve.

Synthesis

ORGANIC ADDITIVE
tetraethylammonium$^+$

ZnAPSO-34 from a reactive gel with a molar oxide ratio of: R : 0.1 ZnO : 0.95–1.0 Al$_2$O$_3$: 0.7–1.0 P$_2$O$_5$: 0.6 SiO$_2$: 50 H$_2$O (R represents the organic additive). (See ZnAPSO.) Crystallization occurs after 17 hours at temperatures between 100 and 200°C. Co-crystallizing phases observed in this system include ZnAPSO-5.

Adsorption

Adsorption properties of ZnAPSO-34:

Adsorbate	Kinetic diameter, Å	Pressure, Torr	Temperature, °C	Wt% adsorbed
O$_2$	3.46	99	−183	14.5
O$_2$		725	−183	25.8
Isobutane	5.0	100	22.8	0.8
n-Hexane	4.3	98	23.3	13.3
H$_2$O	2.65	4.6	23.2	19.9
H$_2$O		17.8	23.5	30.1

ZnAPSO-35 (LEV)
(EP 158,975(1985))
Union Carbide Corporation

Related Materials: levyne
                        SAPO-35

Structure

Chemical Composition: Zinc silicoaluminophosphate. See ZnAPSO.

X-Ray Powder Diffraction Data: (d(Å) (I/I_0))
10.27(20), 8.44sh*, 8.08(47), 7.80(4), 6.66(39), 5.57(10), 5.13(72), 4.98(sh), 4.20(48), 4.06(100), 3.841(19), 3.762(3), 3.552(4), 3.325(22), 3.107(30<), 3.069(sh), 2.788(43), 2.582(9), 2.530(3), 2.507(5), 2.382(5), 2.889(4), 2.134(6), 2.096(4), 1.873(11), 1.845(8), 1.773(6), 1.661(6) (*represents an impurity peak).

ZnAPSO-35 exhibits properties characteristic of a small-pore molecular sieve.

Synthesis

ORGANIC ADDITIVE
quinuclidine

ZnAPSO-35 crystallizes from a reactive gel with a molar oxide ratio of: R : 0.2 ZnO : 0.9

Al$_2$O$_3$: 0.7 P$_2$O$_5$: 0.6 SiO$_2$: 50 H$_2$O (R represents the organic additive). (See ZnAPSO.) Crystallization occurs after 46 hours at temperatures between 150 and 200°C.

Adsorption

Adsorption properties of ZnAPSO-35*:

Adsorbate	Kinetic diameter, Å	Pressure, Torr	Temperature, °C	Wt% adsorbed
O$_2$	3.46	99	−183	10.2
O$_2$		725	−183	19.1
Isobutane	5.0	100	22.8	0.8
n-Hexane	4.3	98	23.3	8.6
H$_2$O	2.65	4.6	23.2	17.2
H$_2$O		17.8	23.5	26.3

*Calcined in air at 500°C for 1.75 hours prior to activation.

ZnAPSO-36 (ATS)
(EP 158,975(1985))
Union Carbide Corporation

Related Materials: related metal-substituted
                     aluminophosphate
                     molecular sieves

Structure

Chemical Composition: Zinc silicoaluminophosphate. See ZnAPSO.

X-Ray Powder Diffraction Data: (d(Å) (I/I_0))
11.14–11.04(100), 10.76–10.68(0–sh), 6.53–6.50(3–4), 5.60–5.56(10–12), 5.38–5.36 (18–31), 4.65–4.62(19–22), 4.28–4.25(17–39), 4.09–4.08(10–17), 4.027–4.017(14–17), 3.865–3.859(3–4), 3.728–3.707(3–6), 3.273–3.260(9–15), 3.152–3.142(6–9), 2.970–2.940(4–6),

2.803–2.788(6–11), 2.698–2.665(1–2), 2.580–2.572(7–10), 2.504–2.497(2–6), 2.384–2.380(2), 2.246–2.232(1–3), 2.180–2.176(1–2), 2.142–2.137(0–2), 1.779–1.776(2), 1.697(0–1), 1.658–1.648(1–2).

ZnAPSO-36 exhibits properties characteristic of a large-pore molecular sieve.

Synthesis

ORGANIC ADDITIVE
tripropylamine

ZnAPSO-36 crystallizes from a reactive gel with a molar oxide ratio of: R : 0.1 ZnO : 1.0 Al$_2$O$_3$: 1.0 P$_2$O$_5$: 0.6 SiO$_2$: 50 H$_2$O (R represents the organic additive). (See ZnAPSO.) Crystallization occurs between 42 and 183 hours at temperatures between 150 and 200°C. Zn-APSO-5 appears as a common impurity phase in this system.

ZnAPSO-39 (ATN)
(EP 158,975(1985))
Union Carbide Corporation

Related Materials: AlPO$_4$-39

Structure

Chemical Composition: Zinc silicoaluminophosphate. See ZnAPSO.

X-Ray Powder Diffraction Data: (d(Å) (I/I_0)) 9.46–9.36(60–72), 6.73–7.73(22–40), 4.85–4.82(16–40), 4.21–4.19(83–100), 3.909–3.892(85–100), 3.389–3.314(12–40), 3.164–3.153(5–8), 3.110–3.100(19–20), 3.008–2.998(11–32), 2.979–2.959(11–25), 2.730–2.718(8–12), 2.600–2.589(5–6), 2.465–2.462(4–12), 2.377–2.362(3–10), 2.222–2.204(0–4).

ZnAPSO-39 exhibits properties characteristic of a small-pore molecular sieve. See MAPO-39 for a description of the framework topology.

Synthesis

ORGANIC ADDITIVE
di-*n*-propylamine

ZnAPSO-39 crystallizes from a reactive gel with a molar oxide ratio of: R : 0.1 ZnO : 1.0 Al$_2$O$_3$: 1.0 P$_2$O$_5$: 0.6 SiO$_2$: 50 H$_2$O (R represents the organic additive). (See ZnAPSO.) Crystallization occurs between 42 and 183 hours at temperatures between 150 and 200°C. Zn-APSO-11 and -46 appear as common impurity phases in this system.

ZnAPSO-43 (GIS)
(EP 158,975(1985))
Union Carbide Corporation

Related Materials: gismondine

Structure

Chemical Composition: Zinc silicoaluminophosphate. See ZnAPSO.

X-Ray Powder Diffraction Data: (d(Å) (I/I_0)) 7.20–7.11(66–100), 5.28–5.23(0–10), 4.27–4.24(10–13), 4.095–4.068(0–48), 3.308–3.290(82–100), 3.116–3.105(11–23), 2.763–2.751(18–20), 2.191(0–4), 2.017–2.010(8–15), 1.792–1.786(0–7), 1.710–1.707(0–8).

ZnAPSO-43 exhibits properties characteristic of a small-pore molecular sieve.

Synthesis

ORGANIC ADDITIVE
tetramethylammonium$^+$

ZnAPSO-43 crystallizes from a reactive gel with a molar oxide ratio of: R : 0.2 ZnO : 0.9 Al$_2$O$_3$: 0.7 P$_2$O$_5$: 0.6 SiO$_2$: 50 H$_2$O (R represents the organic additive). (See ZnAPSO.) Crystallization occurs between 46 and 165 hours at temperatures between 150 and 200°C. Zn-APSO-20 appears as a common impurity phase in this system.

ZnAPSO-44 (CHA)
(EP 158,975(1985))
Union Carbide Corporation

Related Materials: chabazite
                        SAPO-44

Structure

Chemical Composition: Zinc silicoalumino-phosphate. See ZnAPSO.

X-Ray Powder Diffraction Data: $(d(\text{Å}) \, (I/I_0))$
9.41–9.25(100), 6.86–6.78(8–34), 6.51–6.30 (3–5),   5.54–5.47(14–21),   5.14–4.94(0–6), 4.68–4.66(0–5),   4.30–4.27(35–52),   4.15–4.08(18–32),   3.943–3.926(4),   3.842–3.826 (5–10), 3.663–3.541(22–47), 3.414–3.395(8–14), 3.220–3.143(7–9), 2.998(0–3), 2.974(0–13), 2.908–2.889(23–31), 2.743–2.730(0–3), 2.714 (0–6), 2.571(0–2), 2.553(0–2), 2.543–2.522 (9–10), 2.327(0–2), 2.292–2.243(0–2), 2.249 (0–1), 2.146–2.137(0–3), 2.127(0–2), 2.071 (0–1), 1.888(0–3), 1.872–1.866(0–5), 1.817–1.811(0–5), 1.759(0–1), 1.704–1.698(0–7).

ZnAPSO-44 exhibits properties characteristic of a small-pore molecular sieve.

Synthesis

ORGANIC ADDITIVE
cyclohexylamine

ZnAPSO-44 crystallizes from a reactive gel with a molar oxide ratio of: R : 0.2 ZnO : 0.9 Al_2O_3 : 0.7 P_2O_5 : 0.6 SiO_2 : 50 H_2O (R represents the organic additive). (See ZnAPSO.) Crystallization occurs between 40 and 158 hours at temperatures between 150 and 200°C. Zn-APSO-5 appears as a common impurity phase in this system.

Adsorption

Adsorption properties of ZnAPSO-44*:

Adsorbate	Kinetic diameter, Å	Pressure, Torr	Temperature, °C	Wt% adsorbed
O_2	3.46	99	−183	10.3
O_2		745	−183	19.8
n-Hexane	4.3	98	23.3	9.7
Isobutane	5.0	100	22.8	0.8
H_2O	2.65	4.6	23.1	14.0
H_2O		17.8	23.1	24.0

*Calcined in air at 500°C for 1.75 hours prior to activation.

ZnAPSO-46 (AFS)
(EP 158,975(1985))
Union Carbide Corporation

Related Materials: MgAPSO-46

Structure

Chemical Composition: Zinc silicoalumino-phosphate. See ZnAPSO.

X-Ray Powder Diffraction Data: $(d(\text{Å}) \, (I/I_0))$
13.60–13.19(7–10), 11.63–11.42(100), 8.67(0–1), 6.76–6.63(10–20), 6.46–6.41(4–5), 5.95–9.51(4–5), 5.83–5.77(5–7), 5.34–5.28(7), 5.11–5.07(0–1), 4.51–4.44(2–3), 4.37–4.33(6–11), 3.934–3.896(32–58), 3.723–3.700(2–3), 3.548–3.520(0–1),   3.333–3.302(10–12),   3.220–3.187(3–4), 3.175–3.152(2–3), 3.121–3.089(8–11),   3.008–2.988(6–9),   2.885–2.870(3–4), 2.831–2.813(0–1), 2.730–2.706(3–4), 2.626–2.607(2–4).

ZnAPSO-46 exhibits properties characteristic of a large-pore molecular sieve.

Synthesis

ORGANIC ADDITIVE
di-n-propylamine

ZnAPSO-46 crystallizes from a reactive gel with a molar oxide ratio of: R : 0.1 ZnO : 1.0

Al_2O_3 : 1.0 P_2O_5 : 0.6 SiO_2 : 50 H_2O (R represents the organic additive). (See ZnAPSO.) Crystallization occurs between 42 and 183 hours at temperatures between 150 and 200°C. ZnAPSO-11 and -39 appear as common impurity phases in this system.

ZnAPSO-47 (CHA)
(EP 158,975(1985))
Union Carbide Corporation

Related Materials: chabazite
                        SAPO-47

Structure

Chemical Composition: Zinc silicoaluminophosphate. See ZnAPSO.

X-Ray Powder Diffraction Data: $(d(\text{Å}) (I/I_0))$ (as synthesized) 11.88(2*), 9.35(93), 6.87(17), 6.38(7), 5.54(42), 5.03(11), 4.67(3), 4.31(200), 4.07(7), 3.97(6), 3.867(11), 3.600(21), 3.439(23), 3.228(10), 3.188(3), 3.029(5), 2.922(49), 2.894(sh), 2.839(3), 2.772(2), 2.689(3), 2.600(10), 2.573(2), 2.516(4), 2.344(3), 2.273(4), 2.126(3), 2.089(2), 2.019(2), 1.909(4), 1.873(5), 1.807(5), 1.721(5), 1.684(2), 1.642(5); (after calcination at 500°C in air) 11.78(11*), 9.17(100), 6.78(25), 6.26(3), 5.46(10), 4.93(8), 4.61(3), 4.49(2), 4.26(27), 4.18(sh), 3.950(8), 3.816(4), 3.533(8), 3.399(10), 3.187(2), 3.126(3), 2.998(2), 2.885(18), 2.849(sh), 2.571(2) (*indicates an impurity peak).

ZnAPSO-47 exhibits properties characteristic of a small-pore molecular sieve.

Synthesis

ORGANIC ADDITIVE
diethylethanolamine

ZnAPSO-47 crystallizes from a reactive gel with a molar oxide ratio of: R : 0.2 ZnO : 0.9 Al_2O_3 : 0.7 P_2O_5 : 0.6 SiO_2 : 50 H_2O (R represents the organic additive). (See ZnAPSO.) Crystallization occurs between 40 and 158 hours at temperatures between 150 and 200°C. Zn-

APSO-5 appears as a common impurity phase in this system.

Adsorption

Adsorption properties of ZnAPSO-47*:

Adsorbate	Kinetic diameter, Å	Pressure, Torr	Temperature, °C	Wt% adsorbed
O_2	3.46	99	−183	13.9
O_2		725	−183	23.0
n-Hexane	4.3	98	23.3	7.8
Isobutane	5.0	100	22.8	0.7
H_2O	2.65	4.6	23.1	18.8
H_2O		17.8	23.1	27.0

*Calcined in air at 500°C for 1.75 hours prior to activation.

ZSM-2 (FAU)
(US3,411,874(1968))
Mobil Oil Corporation

Structure

(*Zeo. Mol. Sieve.* 259 (1974)

Chemical Composition: $M_{2n}O*Al_2O_3*3.3-4.0SiO_2*ZH_2O$

X-Ray Powder Diffraction Data: (US3,411,874(1968)) $(d(\text{Å}) (I/I_0))$ 14.0(83), 13.8(76), 12.2(42), 8.70(47), 7.34(37), 7.07(41), 6.83(8), 5.63(85), 5.48(10), 5.37(16), 4.86(5), 4.70(16), 4.56(7), 4.34(33), 4.24(5), 4.14(35), 4.04(43), 3.97(7), 3.71(11), 3.66(39), 3.60(10), 3.51(4), 3.40(11), 3.38(16), 3.36(8), 3.28(27), 3.19(8), 3.15(41), 3.01(100), 2.97(16), 2.94(18), 2.90(22), 2.88(3), 2.84(5), 2.79(9), 2.75(10), 2.69(23), 2.61(6), 2.58(5), 2.54(3), 2.47(12), 2.43(3), 2.34(10).

Synthesis

ORGANIC ADDITIVES
none

ZSM-2 is prepared from lithium aluminosilicate glass with an oxide ratio of: 4 Li_2O : 1 Al_2O_3 : 9 SiO_2. Initially it stood in aqueous solution for 48 hours; crystallization occurred after 30 days of heating at 60°C. With changes in the Li and the SiO_2 content of the initial reaction mixture, ZSM-2 also could be obtained after lengthier crystallization times. Impurity phases include both the unreacted glass and lithium silicate.

Adsorption

Adsorption properties of ZSM-2 (US3,411,874(1968)):

Wt% cyclohexane	Wt% water
15	22

ZSM-3 (FAU/EMT)
(US3,415,736(1968))
Mobil Oil Corporation

Structure

Chemical Composition: (0.05–0.8)Li_2O*(0.2–0.95)Na_2O*Al_2O_3 *(2–6)SiO_2*(0–9)H_2O.

X-Ray Powder Diffraction Data: $(d(\text{Å})\ (I))$
15.26(62), 14.16(124), 13.19(26), 11.86(4), 9.21(2), 8.76(44), 8.00(3), 7.61(9), 7.41(32), 7.22(7), 7.03(15), 6.89(2), 5.94(4), 5.72(37), 5.62(25), 5.48(6), 5.15(4), 5.02(7), 4.90(3), 4.82(2), 4.78(2), 4.75(18), 4.70(1), 4.62(3), 4.58(1), 4.49(5), 4.38(51), 4.19(14), 4.16(6), 4.11(13), 3.99(1), 3.96(2), 3.93(7), 3.85(14), 3.78(19), 3.72(23), 3.52(1), 3.47(10), 3.40(10), 3.31(42), 3.22(16), 3.18(17), 3.02(60), 2.98(11), 2.92(31), 2.87(10), 2.85(8), 2.80(9), 2.73(11), 2.70(11).

Synthesis

ORGANIC ADDITIVES
none

ZSM-3 is prepared from reactive gels containing sodium and lithium cations. A precursor aluminosilicate solution is prepared initially. The precipitate is filtered, dried, and mixed with a lithium hydroxide solution, with heating to 100°C following. Crystallization occurs after 16 hours at this temperature. At a lower temperature (60°C), crystallization times are significantly increased, requiring 5 days. Without the precursor step, zeolite P and sodalite are formed instead

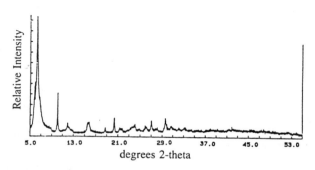

ZSM-3 Fig. 1:   X-ray powder diffraction pattern for ZSM-3 *SSSC* 37:123(1988)). (Reproduced with permission of Elsevier Publishers)

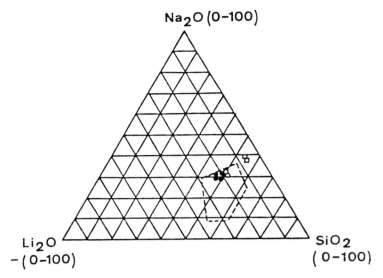

ZSM-3 Fig. 2:   Molar ratios of Na_2O, Li_2O, and SiO_2 in the mixtures for the synthesis of ZSM-3. The SiO_2/Al_2O_3 ratio varies from 12.9 to 16.7. The open symbols refer to pure phase ZSM-3, the closed symbols to ZSM-3 contaminated with lithium silicate; the area circumscribed by the dashed line corresponds to the preferential compositions according to the patent. (*SSSC* 37:123(1988)). (Reproduced with permission of Elsevier Publishers)

(*ACS* 101:109(1971)). In the presence of excess lithium cations (Li/Li + Na > 0.5), lithium silicate is the major product (*SSSC* 37:123(1988)).

Thermal Properties

The as-synthesized ZSM-3 does not lose crystallinity on heating to 500°C. Exotherms are observed without accompanying weight loss at 780°C. When it is heated in the NH_4^+ form, crystallinity is lost at 400°C (*SSSC* 37:123(1988)).

Infrared Spectrum

The hydroxyl stretching region of ZSM-3 in the H^+-exchanged form contains a high frequency band at 3650 cm^{-1} and a low frequency band at 3545 cm^{-1}. These bands are similar to those observed for the faujasite zeolites. A third band at 3745 cm^{-1} also is observed and is attributed to the crystal-terminating silanol groups (*SSSC* 37:123(1988)).

Adsorption

Adsorption properties of ZSM-3 (*ACS* 101:109(1971)):

SiO_2/Al_2O_3	% cyclohexane	% water
3.05	17.6	30.2
2.99	18.0	29.5
3.6	18.9	29.1

Ion Exchange

See ZSM-3 Figures 3 and 4.

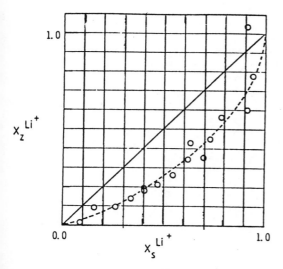

ZSM-3 Fig. 3: Ion-exchange isotherms for ZSM-3: Li^+–Na^+; total normality = 25°C; SiO_2/Al_2O_3 = 3.25; contact time = 48 hours; X_z = equivalents of lithium in zeolite/gram atom Al; X_s = equivalents of lithium in solution/total equivalents in solution (*SSSC* 37:123(1988)). (Reproduced with permission of Elsevier Publishers)

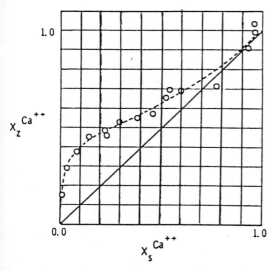

ZSM-3 Fig. 4: Ion-exchange isotherms for ZSM-3: Ca^{+2}–Na; total normality = 0.1; temperature = 25°C; SiO_2/Al_2O_3 = 3.25; contact time = 48 hours; X_z = equivalents of calcium in zeolite/gram atoms Al; X_s = equivalents of calcium in solution/total equivalents in solution (*SSSC* 37:123(1988)). (Reproduced with permission of Elsevier Publishers)

ZSM-4 (MAZ)
(BP 1,117,568(1968))
Mobil Oil Corporation

Related Materials: mazzite

Structure

CHEMICAL COMPOSITION: $M_{2/n}O : Al_3O_3 : 6–15$ $SiO_2 : 0–5 H_2O$
PORE STRUCTURE: 12-member rings 7.4 Å in diameter

X-Ray Powder Diffraction Data: $(d(\text{Å}) (I/I_0))$
16.7(w), 9.12(s), 7.90(m), 6.92(m), 5.99(ms), 5.37(vs), 5.28(w), 4.70(ms), 4.39(vw), 3.93(w), 3.79(s), 3.71(w), 3.62(w), 3.52(ms), 3.44(w), 3.15(m), 3.08(w), 3.04(w), 2.92(m), 2.65(w), 2.62(w), 2.52(w), 2.37(w), 2.28(w), 3.24(vw), 2.10(vw), 2.08(vw), 2.03(vw), 1.98(w).

For a description of the structure see mazzite.

Synthesis

ORGANIC ADDITIVES
tetramethylammonium$^+$
pyrrolidine
choline chloride
diazobicyclooctane

ZSM-4 can be prepared from reaction mixtures that have been heated at 95°C for 20 to 186 days with R_2O/R_2O + Na_2O in the range of 0.05 to 0.7, R_2O + Na_2O/SiO_2 between 0.15 and 0.50, SiO_2/Al_2O_3 between 6 and 20, and H_2O/R_2O + Na_2O between 30 and 150. Sodium silicate is used as the source of silica, and either sodium aluminate or aluminum chloride is used as the source of aluminum.

Thermal Properties

Thermal treatment of this material above 500°C showed an apparent loss in crystallinity based on a loss in cracking activity with the time of thermal treatment at this temperature (5 min to 25 min).

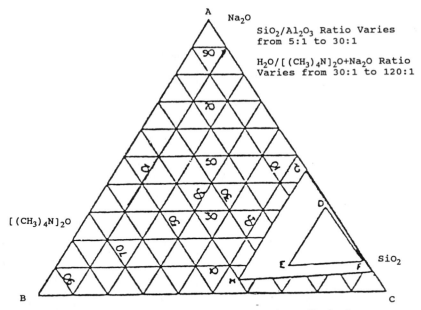

ZSM-4 Fig. 1:   Ternary phase diagram of the region of crystallization for ZSM-4 using TMA as the organic additive (BP 1,117,568(1968)).

Adsorption

Adsorption capacities for ZSM-4 sample crystallized after 7 days (BP 1,117,568(1968)):

Wt% cyclohexane 20 mm Hg	Wt% water 12 mm Hg
4.6	13.6

ZSM-5 (MFI)
(VP 22.516(1968))
Mobil Oil Corporation

Related Materials: AMS-1B
AZ-1
Bor-C
Encilite
FZ-1
LZ-105
Nu-4
Nu-5
TS-1
TSZ
TSZ-III
TZ-01
USC-4
USI-108
ZBH
ZKQ-1B
ZMQ-TB
silicalite
silicalite-1

Structure

(*J. Phys. Chem.* 85:2238(1981))

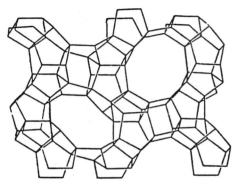

ZSM-5 Fig. 1S: Framework topology.

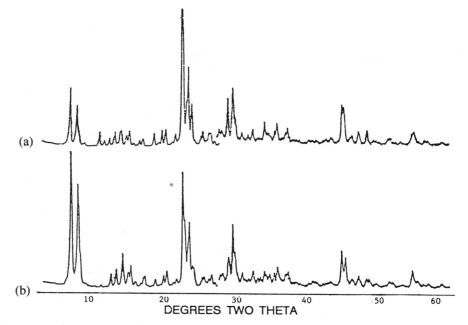

ZSM-5 Fig. 2:   X-ray powder diffraction pattern for ZSM-5: (a) as-synthesized form; (b) ammonium-exchanged form (scale factor changes just below 28° 2 theta) (*J. Phys. Chem.* 83:2777(1979)). (Reproduced with permission of the American Chemical Society)

CHEMICAL COMPOSITION: (TPA,Na)$_2$O*Al$_2$O$_3$*5–100SiO$_2$*4H$_2$O (US3,702,886(1972))

SYMMETRY: orthorhombic

SPACE GROUP: Pnma

UNIT CELL CONSTANTS (Å): $a = 20.1$
$b = 19.9$
$c = 13.4$

VOID VOLUME: 0.10 g/cc

PORE STRUCTURE: three-dimensional intersecting 10-member rings 5.3 × 5.6 A and 5.1 × 5.5 Å

X-Ray Powder Diffraction Data: $(d(Å)\,(I/I_0))$
11.1(s),   10.00(s),   7.4(w),   7.1(w),   6.3(w), 6.04(w),   5.97(w),   5.56(w),   5.01(w),   4.60(w), 4.25(w),   3.85(vs),   3.71(s),   3.04(w),   2.99(w), 2.94(w).

Zeolite ZSM-5 can be constructed from five-member-ring building units. These units link together to form chains and the interconnection of these chains leads to the formation of the channel system in the structure. The combination of these building units results in a framework containing two intersecting channel systems, one sinusoidal and the other straight. The

pore openings are elliptical 10-member rings (*J. Phys. Chem.* 85:2238(1981)). There are 12 unique T atom sites in ZSM-5.

This structure is successfully described in the orthorhombic space group Pnma; however, it shows a reversible phase transition at about 330K (though the temperature is dependent on the

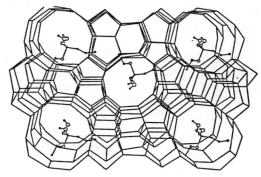

ZSM-5 Fig. 3:   Packing diagram with Si atoms as framework viewed along <010> axis horizontal (*Zeo.* 6:35(1986)). (Reproduced with permission of Butterworths Publishers)

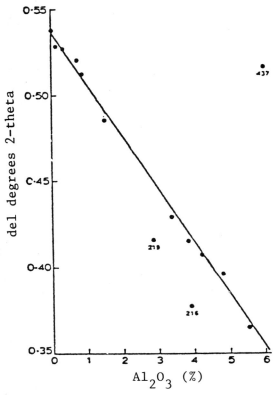

ZSM-5 Fig. 4: Peak spacing delta (deg. 2 theta CuK alpha radiation) vs. the bulk Al_2O_3 content (%) of HZSM-5 samples equilibrated at 75% relative humidity. The relationship follows: percentage Al_2O_3 = 16.5 − 30.8 delta (*J. Catal.* 72:373(1981)). (Reproduced with permission of Academic Press)

Positional parameters for ZSM-5 (*J. Phys. Chem.* 85:2228(1981)):

Atom	x	y	z
T(1)	0.4232(5)	0.0605(7)	−0.3306(9)
T(2)	0.3090(9)	0.0281(5)	−0.1849(11)
T(3)	0.2788(5)	0.0602(8)	0.0312(10)
T(4)	0.1220(6)	0.0641(7)	0.0326(9)
T(5)	0.0728(7)	0.0278(6)	−0.1822(13)
T(6)	0.1884(8)	0.0573(7)	−0.3249(9)
T(7)	0.4232(7)	−0.1695(7)	−0.3258(13)
T(8)	0.3076(9)	−0.1286(7)	−0.1848(11)
T(9)	0.2748(6)	−0.1722(6)	0.0366(10)
T(10)	0.1199(6)	−0.1729(6)	0.0269(10)
T(11)	0.0702(6)	−0.1304(6)	−0.1826(10)
T(12)	0.1894(8)	−0.1742(5)	−0.3172(9)
O(1)	0.3728(14)	0.0665(15)	−0.2397(19)
O(2)	0.3095(13)	0.0621(12)	−0.0761(13)
O(3)	0.1993(18)	0.0605(16)	0.0263(18)
O(4)	0.1000(15)	0.0571(26)	−0.0817(23)
O(5)	0.1183(10)	0.0592(15)	−0.2697(13)
O(6)	0.2407(14)	0.0440(16)	−0.2418(21)
O(7)	0.3750(15)	−0.1509(16)	−0.2361(21)
O(8)	0.3079(18)	−0.1514(11)	−0.0681(17)
O(9)	0.1981(17)	−0.1541(10)	0.0324(17)
O(10)	0.0909(16)	−0.1652(18)	−0.0751(25)
O(11)	0.1230(18)	−0.1427(22)	−0.2632(26)
O(12)	0.2529(14)	−0.1660(14)	−0.2393(21)
O(13)	0.3112(17)	−0.0519(21)	−0.1923(20)
O(14)	0.0790(12)	−0.0526(22)	−0.1743(20)
O(15)	0.4091(13)	0.1258(15)	−0.3870(21)
O(16)	0.4160(15)	0.0011(18)	−0.4177(24)
O(17)	0.3976(15)	−0.1300(16)	−0.4191(22)
O(18)	0.1847(11)	0.1295(12)	−0.3824(16)
O(19)	0.2037(16)	0.0030(15)	−0.4020(22)
O(20)	0.1910(20)	−0.1259(17)	−0.4147(22)
O(21)	−0.0016(12)	0.0425(13)	−0.2056(19)
O(22)	−0.0016(14)	−0.1565(13)	−0.2078(20)
O(23)	0.4277(19)	−0.25	−0.3503(28)
O(24)	0.2045(17)	−0.25	−0.3476(23)
O(25)	0.2861(14)	−0.25	0.0590(23)
O(26)	0.1089(23)	−0.25	0.0580(34)
Ox1	0.1746(33)	0.75	0.3515(50)

aluminum content to monoclinic symmetry. Below this temperature, the symmetry is monoclinic, and it is orthorhombic above the transition temperature (*JCS Chem. Commun.* 1433(1984); *J. Phys. Chem.* 89:1070(1985)). At room temperature this change in symmetry is observed upon adsorption of molecules into the pores (*J. Phys. Chem.* 83:2777(1979); *JCS Chem. Commun.* 541(1984); *Pure Appl. Chem.* 58:1367(1986)).

The TPA cations of this structure are located at the channel intersections with a conformation that is different from that of free TPABr. The C3 groups have larger thermal parameters, indicating some freedom of motion, and each extends into the 10-member ring window. (*JACS* 104:5971(1982); *Zeo.* 6:35(1986)).

Synthesis

The ZSM-5 structure can readily be prepared under a wide range of conditions. It crystallizes readily, with the SiO_2/Al_2O_3 ratio ranging from 20 to that of the pure silica polymorph silicalite. Generally the higher the SiO_2/Al_2O_3 ratio, the more facile the crystallization. (See silicalite for a discussion of the pure silica end member of this series of materials.) Numerous synthesis procedures have been described, and a wide range of organic additives have been employed in encouraging the formation of this structure. In the presence of the TA cation this structure is very readily formed over the largest SiO_2/Al_2O_3 ratios. Some of the literature concerning the role of the organic in encouraging crystal growth is confusing, as the aluminosilicate with the ZSM-5 structure also can crystallize in the absence of any organic additive when the SiO_2/Al_2O_3 ratio of the gel is between 70 and 90. The role of some of the organic additives is thought to be that of hydrophobic void filler, such as hexane-1, 6-diol and piperazine (*Zeo.* 6:111(1986)). Under certain conditions it has been observed

---

Use of various organics in the synthesis of ZSM-5 structure:

Organic	Citation
TPA	US3,702,886(1972)
TEA	*PP IZC* 10(1989)
DEA	*Zeo.* 8:508(1988)
tripropylamine	*PP IZC* 10(1989)
diethylamine	*PP IZC* 10(1989)
DMEPA	US4,565,681(1986)
HEXDM	*Zeo.* 8:495(1988)
1,6-hexanediol	EP 42,225(1981)
no alkalki	Ger. 3,021,580(1981)
propylamine + propylbromide	JP 85,71,517(1985)
tripropylamine + methyliodide	JP 85,71,517(1985)
tributylamine + methyliodide	JP 85,71,517(1985)
triethyl-n-propylamminium$^+$	*Zeo.* 2:313(1982)
diquat phosphonium^{+2}	Ger. 3,333,124(1985)
alkyltropinium	US4,592,902(1986)
fluorocarbon-carboxylic acid	JP 84,73,425(1984)
fluorocarbon-sulfonic acid	JP 84,73,425(1984)
seeds	JP 85,71,519(1985)
pyrrolidine	*Zeo.* 3:282(1983)
piperazine	*Zeo.* 8:501(1988)
Pip + TPA	*Zeo.* 8:501(1988)
Pip + TBA	*Zeo.* 8:501(1988)
Pip + TrBPe	*Zeo.* 8:501(1988)
monoethylene glycol dimethylether	EP 51,741(1982)
cyanoalkanes	JP 86,68,318(1986)
cyanoalkenes	JP 86,68,318(1986)
N-2-hydroxyethyl-piperazine-N-2-ethanesulfonic acid	US4,639,360(1987)

in the TPA system that silicalite may be the initial crystallizing species, dissolving and recrystallization to form the more aluminum-rich species (*Zeo.* 5:295(1984)).

Starting with a batch composition of: 12 Na_2O : 4.5 TPA_2O : Al_2O_3 : 90 SiO_2 : 2000 H_2O, which will produce spherulitic aggregates on the order of 1 micron, additives to this system can promote the formation of larger crystals of this structure. Substitution of sodium salts for sodium hydroxide will yield 80-micron crystals. The addition of organic additives also will modify the crystal habit (*Zeo.* 3:219(1983)). The largest crystals produced to date were from gels containing NH_4HF_2 and NH_4F, in addition to TPA (*SSSC* 28:121(1986)). Under conditions of high gravity, very large 800-micron-size crystals also have been prepared (*PP IZC* 13(1989)). In the presence of 4-picoline N-oxide or tripropylamine N-oxide, 100% crystallinity is not obtained; however, the presence of a crystalline second phase that is not thermally stable has been used to prepare, in situ, bindered material (US4,430,314(1984)). Synthesis from type A zeolite and excess silicate has been claimed (JP 86,155,214(1986)). A common impurity found in the ZSM-5 synthesis is quartz-type phases, and in aluminum-rich systems, mordenite.

Crystallization in general is more rapid in the presence of seeds when the reactive gel does not contain an organic amine (*PP IZC* 7(1989)). Sucrose has been found to inhibit the crystallization of this structure (JP 85,71,518(1985)).

Thermal Properties

The peak at 360°C for TPA-ZSM-5 thermal decomposition is attributed to external TPA species on the surface of the crystal associated with surface aluminum species (*Zeo.* 8:209(1988)). By following the thermal decomposition of TPA ions with different techniques, at least two types of TPA framework species could be evidenced, a TPA counterion to the framework negative charge located near the Al atom and a less strongly bound species released at lower temperature (*Calorim Anal. Therm.* 14:20(1984)). Desorption peaks between 400 and 470°C are associated with the loss of TPA from the crystal (*Zeo.*

Chemical composition of sonicated ZSM-5 samples obtained from the system: 4.5(Li,Na,K)2O : Al₂O₃ : 90 SiO₂ :
30,000 H₂O : 2 TPABr (*Zeo.* 7:549 (1987)):

Alkali	TPA	Li	Na	K	(TPA + M)	Al	H₂O	Si/Al (per unit cell)
Li	2.7	0.9	—	—	3.6	1.9	4.1	49.5
Na	2.7	—	2.4	—	5.1	2.2	3.5	42.6
K	3.0	—	—	0.9	3.9	2.1	2.6	44.7
Li–Na	2.8	0.4	1.0	—	4.2	2.3	3.8	40.7
Li–K	3.0	0.5	—	0.6	4.1	2.3	2.9	40.7
Na-K	3.0	—	1.8	1.2	6.0	2.2	2.6	42.6

TPA results obtained from TG measurements, inorganic cations from atomic absorption.

4:2(1984)). The decomposition range for NH₄-
ZSM-5 is between 553 and 893K (*Zeo.*
3:249(1983)).

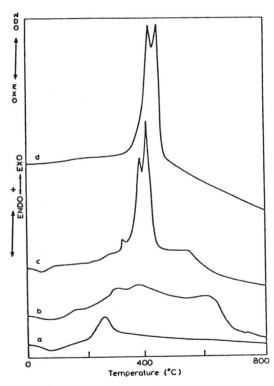

ZSM-5 Fig. 5:   DTA curves for as-synthesized ZSM-5:
(a) NaZSM-5; (b) (Na<Pipz)ZSM-5; (c)
(Na,HXDM)ZSM-5; (d) (Na,TPA)ZSM-5 (*Zeo.*
6:111(1986)). (Reproduced with permission of
Butterworths Publishers)

Ion Exchange

All univalent cations achieved 100% exchange
in ZSM-5. The thermodynamic affinity for exchange is:

$$Cs > Rb = NH_4 = H_3O > K > Na > Li$$

for ZSM-5 with a SiO₂/Al₂O₃ ratio of 78. Mul-
tivalent cations are unable to achieve 100% exchange in this structure. The maximum exchange increases through the series Ca < Sr <
Ba and is better achieved at higher temperatures
than at room temperature. The maximum exchange for La is proposed to be similar to that
for Ca (*Chem. Age India* 37:353(1986)).

Organic ions have exceptionally high selectivities, with full capacity exchange readily at-
tained for all organic ions except for those that
are larger than the 10-member-ring openings in
this structure (*Zeo.* 8:423(1988)). The selectivity sequence for organic amines correlates with
ionic size:

$$Na \ll MA < WA < PA < BA$$
$$Na \ll M_2A < E_2A < P_2A < B_2A$$
$$MA < M_2A < TMA$$

Copper ions can be exchanged into this zeolite
from either CuO, CuF₂, or Cu₃(PO₄)₂ by heating
the solids together in the absence of water at
temperatures between 550 and 800°C (*Zeo.*
6:175(1986)).

Thermogravimetric analysis of bases sorbed on HZSM-5 (containing 2.7 Al/U.C.) (*Zeo.* 5: 384 (1985)):

	Peak temperature (°C)					Weight loss (%)[d]		Molecules/uc[e]					
	1a	1b	2a	2b	2c	1	2 total	1	2a	2b	2c	2[f] total	2[g] total corr.
Ammonia[a]	150	—	360	—	—	2.03	1.25	6.9	4.2	—	—	4.2	4.2
Ammonia[b]	—	200	360	—	—	0.23	0.91	0.8	3.1	—	—	3.1	3.1
Ammonia[c]	—	—	360	—	—	—	0.80	—	2.7	—	—	2.7	2.7
Methylamine	140	—	400	470	520	3.57	2.09	6.6	1.5	1.2	0.3	3.9	3.0
Ethylamine	100	180	400	490	—	5.94	2.19	7.6	2.8	—	—	2.8	2.8
n-Propylamine	80	210	370	430	—	8.15	2.38	8.0	2.3	—	—	2.3	2.3
n-Butylamine	110	220	370	430	—	7.88	3.64	5.8	2.0	0.5	—	2.9	2.5
Pyridine	100	—	600	—	—	8.11	3.00	5.9	2.2	—	—	2.2	2.2

[a]Anhydrous ammonia sorbed on H-ZSM-5
[b]Sample from (a), after heating at 190°C for ~ 16 h
[c]Ammonium ion exchanged H-ZSM-5
[d]The weight loss is given as a percentage of the zeolite weight at 600°C
[e]The number of base molecules per unit cell calculated from the % weight loss
[f]Calculated assuming no amine reactions have occurred
[g]Calculated by estimating the proportion of monoamine (2a), diamine (2b) and triamine (2c) from t.d./m.s. data

Infrared Spectrum

Vibrational data of silicalite-1 and ZSM-5 (Si/Al = 20.7).l.r. is used to indicate infrared spectroscopy, i.r. FS for infrared Fourier spectroscopy, and RS for Raman spectroscopy (*Zeo.* **7:249 (1987)**)

Silicalite-1					ZSM-5			
293 K (i.r.)	77 K (i.r.)	77 K (i.r. FS)	15 K (i.r. FS)	293 K (RS)	293 K (i.r.)	77 K (i.r.)	293 K (RS)	Assignment
1230m	1230m	1232m						Asymmetric stretch
		1222sh		1220vw	1225w	1225w	1220w	ν_{as}(T–O–T)
1180sh	1181sh	1170sh		1180vw	1180sh	1180sh	1185vw	
1140sh	1140sh	1140		1155vw				
1115vs		1113vs		1125w			1137vw	
	1105vs	1105sh			1105vs	1105vs		
		1095sh		1095vw			1094vw	
1085sh	1085sh	1086sh						
		1073sh					1074w	
		1067sh		1045w	1065sh	1065sh	1033vw	
		974wsh		975w			946w	

(continued)

Vibrational data of silicalite-1 and ZSM-5 (Si/Al = 20.7).l.r. is used to indicate infrared spectroscopy, i.r. FS for infrared Fourier spectroscopy, and RS for Raman spectroscopy (*Zeo.* **7:249 (1987)**)

Silicalite-1					ZSM-5			
293 K (i.r.)	77 K (i.r.)	77 K (i.r. FS)	15 K (i.r. FS)	293 K (RS)	293 K (i.r.)	77 K (i.r.)	293 K (RS)	Assignment
	820vw		826m	825w			825w	Symmetric stretch ν_s(T–O–T)
815w	812w		814m			812w		
800w	798w		801m	800w		800m	793w	
785vw	795m		786s		795m	786m		
			784sh					
			758w	760vw			759vw	
			752w					
			745vw					
			739sh				740vw	
			736m	735vw				
			734sh					
			730vw					
			727vw					
			721vw				714w	
			704vw	700vw				
			697w					Bending vibrations δ(O–T–O) and δ(T–O–T)
			687m					
			667vw					
	645w		649w	655w			648vw	
630	635w		635s		630sh	630w	630w	
	625vw			625w				
	615w		615m					
598vw	602w		602s					
590vw	590w		588s					
			580sh					
			572s					
	565sh		566s					
			564sh					
	557s		559s	560m				
			553s				555w	
550s			552sh			550s		
	542sh		542s		545s	545sh		
530sh	535sh		534sh					
	525sh		525m	525m		528sh	525m	
490vw	490w		496s	485w			480w	
			473s	470vw			475w	
	460sh		466s			460sh	469vw	
			457sh					
			452s	450vw			450vw	
			446s		448s	448s		
440s	440s		440sh					
			433s				435vw	
			431sh	430vw				
			422m					
			420sh					

Vibrational data of silicalite-1 and ZSM-5 (Si/Al = 20.7).l.r. is used to indicate infrared spectroscopy, i.r. FS for infrared Fourier spectroscopy, and RS for Raman spectroscopy (*Zeo.* **7:249 (1987)**)

Silicalite-1					ZSM-5			Assignment
293 K (i.r.)	77 K (i.r.)	77 K (i.r. FS)	15 K (i.r. FS)	293 K (RS)	293 K (i.r.)	77 K (i.r.)	293 K (RS)	
415sh	414w		414m		413sh	415sh		
			411sh					
			408sh					
398sh			400w			400sh		
			393sh					
			389m		390sh	386vw	385m	
380vw			379w	380vs				
			375sh					
360vw			362m	365w				
			358m			355sh	355m	
			350m	354s				
			341w					
			335m	338w			340vw	
				290vw			290w	Bending vibrations
				285w			285w	(O–T–O) and
				265w			264vw	(T–O–T)
				255m			255w	
				240vw			245w	
				204vw			218w	
				185vw			192vw	Translation of Si/Si
				180w			173w	and Al/Al lattice
				165w			163w	modes, rotation of
				160vw			160w	the fragment of
				140w			138w	zeolite chain
				135w				
				104s			104s	
							65vw	
				50w			50m	
				47vw			43vw	
				38vs			38vs	
				30vw				
				28.6m			28.6m	

vs = very strong, s = strong, m = medium, w = weak, vw = very weak, sh = shoulder

Adsorption

In the adsorption of paraffinic hydrocarbons on HZSM-5, linear and branched hydrocarbons behave as condensed liquids, filling the entire pore structure. This is also true for the cyclic paraffins, but less of the pore structure is accessible (*Zeo.* 7:438(1987)). Crystal morphology can influence the sorption and the catalytic properties of this zeolite (*Zeo.* 4:337(1984)).

The heats of immersion for alkanes, linear alkenes, and alcohols are strongly dependent on the chain length. The heats of adsorption can be evaluated from the heats of immersion and condensation by the following relationship (*Chem. Technik.* 35:144 (1983)):

$$(-) \Delta AH = (-) \Delta IH + (-) \Delta LH$$

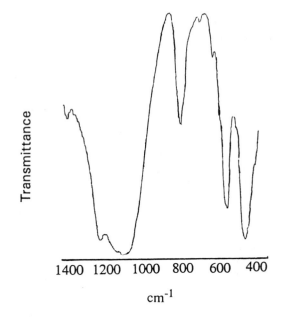

ZSM-5 Fig. 6: Mid-infrared spectrum of ZSM-5. (Reproduced with permission of author)

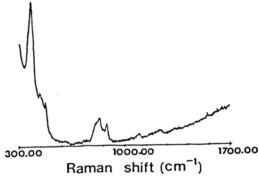

ZSM-5 Fig. 7: Raman spectrum of calcined ZSM-5 crystals (*J. Phys. Chem.* 91:4329(1987)). (Reproduced with permission of the American Chemical Society)

Adsorption of normal paraffins at 2 kPa on ZSM-5 (*Zeo.* 7:438 (1987)):

n-Alkane	Temp. (K)	Wt% ads.	Liq. density (g/cm³)
Methane	79	7.40	0.460
	158	4.10	0.340
	186	2.35	0.245
	253	0.75	0.000
	273	0.28	0.000
	295	0.04	0.000
Ethane	158	9.70	0.577
	186	7.80	0.550
	253	4.15	0.460
	273	3.90	0.420
	295	1.35	0.355

Adsorption of normal paraffins at 2 kPa on ZSM-5 (*Zeo.* 7:438 (1987)):

n-Alkane	Temp. (K)	Wt% ads.	Liq. density (g/cm³)
Propane	186	9.40	0.635
	253	8.10	0.555
	273	7.50	0.530
	294	6.10	0.500
	323	3.50	0.450
	373	1.15	0.000
Butane	253	9.60	0.620
	272	9.00	0.605
	295	8.60	0.580
	325	7.60	0.535
	373	4.50	0.470
	414	2.15	0.355
Pentane	253	10.40	0.664
	293	10.20	0.624
	295	10.00	0.624
	323	9.40	0.586
	373	7.00	0.540
Hexane	295	11.60	0.656
	323	11.00	0.632
	373	9.00	0.580
	433	5.70	0.506
Heptane	295	12.80	0.684
	323	11.70	0.660
	346	9.50	0.640
	373	8.00	0.614
	478	5.50	0.468
Octane	295	12.15	0.700
	373	7.80	0.636
	458	6.10	0.550
	491	5.10	0.510

Branched paraffin adsorption at 2kPa:

	Temp. (K)	Wt% ads.	Liq. density (g/cm³)
2-Methylpropane	273	5.6	0.583
	294	4.2	0.555
	323	3.6	0.520
	373	2.2	0.433
2-Methylbutane	272	8.8	0.638
	296	7.0	0.615
	323	5.4	0.587
	373	4.2	0.528
2-Methylpentane	273	9.4	0.675
	295	7.5	0.655
	323	6.6	0.628
	373	6.0	0.575
3-Methylpentane	273	10.1	0.668
	295	8.4	0.662
	323	6.9	0.635
	373	5.9	0.581

Adsorption of cycloparaffins at 1 kPa on HZSM-5:

n-Alkane	Temp. (K)	Wt% ads.	Liq. density (g/cm³)
Cyclopropane	273	7.4	0.633
	293	6.1	0.605
	323	chemical reaction	
Cyclopentane	273	10.0	0.760
	296	7.2	0.737
	323	5.2	0.710
	373	3.6	0.655
Cyclohexane	295	4.8	0.773
Methylcyclopentane	295	8.6	0.748
Methylcyclohexane	296	6.7	0.769

Intracrystalline sorption capacity of the extruded HZSM-5 zeolite for different liquid sorbates at 303K (*Zeo.* 6:206 (1986)):

Sorption	Intracrystalline sorption* (Q)		
	Period allowed for sorption (d)	mmol/g	Number of molecules per unit cell
Methanol	1.0	4.40	25.5
n-Hexane	1.0	1.36	7.9
Benzene	1.0	1.35	7.8
p-Xylene	2.0	1.20	7.0
o-Xylene	60.0	0.88	5.1
m-Xylene	60.0	0.89	5.2
Mesitylene	112.0	0.70	4.0

*Corrected for the presence of binders (20% kaolinite and bentonite) in the extruded zeolite.

Adsorption of amines at 150°C and 25°C (*PP 7th IZC* 2c-2 (1986)):

Molecule	Length (nm)	Strong sorption at 150°C (molecules/ u.c.)	Sorption capacity at 25°C (molecules/ u.c.)
$(CH_3)_3N$	0.49	2.4	9.0
$(C_2H_5)_2NH$	0.74	2.3	9.1
$(n\text{-}C_3H_7)_2NH$	0.99	2.3	7.4
$(n\text{-}C_4H_9)_2NH$	1.23	2.4	5.7
$(CH_3)_2NCH_2NHC_2H_5$	0.98	1.1	

ZSM-5/ZSM-11
(US4,289,607(1981))
Mobil Oil Corporation

Structure

Chemical Composition: $0.9 \pm 0.3\ M_2O/n$: Al_2O_3 : at least $5\ SiO_2$.

The ZSM-5 and ZSM-11 can be found as intergrowths under certain crystallization conditions. The difference between ZSM-5 and ZSM-11 can be seen in the region around degrees two theta of 45.

ZSM-6
(US4,187,283(1980))
Mobil Oil Corporation

Related Materials: ZSM-47

Structure

Chemical Composition: SiO_2/Al_2O_3 200–1000.

X-Ray Powder Diffraction Data: $(d(\text{Å})\ (I/I_0))$
11.6(20), 9.0(30), 8.33(70), 6.64(40), 6.55(20), 6.27(70), 5.77(20), 5.41(20), 4.65(20), 4.48(70), 4.33(70), 4.24(20), 4.17(20), 4.11(40),

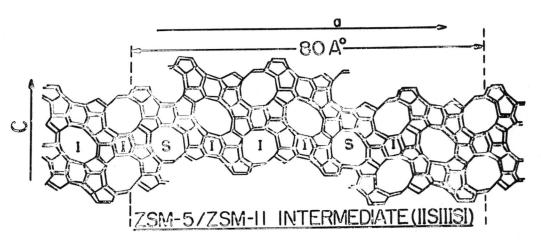

ZSM-5/ZSM-11 Fig. 1S: Intergrowth structure.

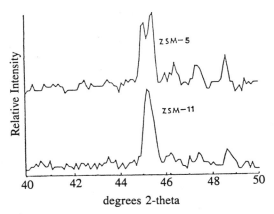

ZSM-5/ZSM-11 Fig. 2:   Slow X-ray scan at 45° 2 theta to identify (a) ZSM-11 and (b) ZSM-5 (*SSSC* 28: 890)1986)). (Reproduced with permission of Elsevier Publishers)

4.05(100), 4.01(40), 3.97(70), 3.91(100), 3.85(40), 3.56(40).

ZSM-10
(US3,692,470(1972))
Mobil Oil Corporation

Synthesis

ORGANIC ADDITIVE
TMA-Br

ZSM-6 crystallizes from a batch composition with Na_2O/Al_2O_3 around 21.6, R/Al_2O_3 around 18, and H_2O/Al_2O_3 of 1000. Crystallization occurs at 200° after 24 hours.

ZSM-8
(BP 1,334,243(1973))
Mobil Oil Corporation

Related Materials: ZSM-5/ZSM-11

Structure

X-Ray Powder Diffraction Data: (d(Å) (I/I_0))
11.10(46), 10.0(42), 9.7(10), 7.42(10), 6.35(12), 5.97(12), 5.69(9), 5.56(13), 4.25(18), 4.07(20), 4.00(10), 3.85(100), 3.82(57), 3.75(25), 3.71(30), 3.64(26), 3.43(9), 3.34(18), 3.31(8), 3.04(10).

ZSM-8 is thought to be an intergrowth of ZSM-5 with ZSM-11 (*Appl. Catal.* 5:109 (1983)).

Synthesis

ORGANIC ADDITIVE
tetraethylammonium cation

ZSM-8 crystallizes from a glass-lined autoclave after 7 days at 180°C. Ludox is used as a source of silica and sodium aluminate as the source of aluminum. Na_2O/Al_2O_3 is near 1, and TEA_2O/Al_2O_3 is around 3.

Structure

CHEMICAL COMPOSITION: $(R,K_2)O*Al_2O_3*5–7SiO_2*9H_2O$
VOID VOLUME: 0.20 cc/g
PORE STRUCTURE: (based on adsorption studies) 6+ Å

X-Ray Powder Diffraction Data: (d(Å) (I/I_0))
15.85(58), 13.92(42), 10.22(13), 7.87(22), 7.55(56), 7.04(13), 6.29(35), 5.96(22), 5.46(31), 5.25(15), 5.06(25), 4.50(76), 4.41(67), 4.32(27), 3.87(91), 3.64(100), 3.54(56), 3.47(25), 3.42(27), 3.32(13), 3.22(16), 3.16(31), 3.10(67), 3.04(73), 2.89(89), 2.73(48), 2.69(15), 2.57(15).

Synthesis

ORGANIC ADDITIVE
1,4-dimethyl-1,4-diazoniabicyclo(2,2,2)octane^{+2}

ZSM-10 is prepared from reactive mixtures with the following composition range: SiO_2/Al_2O_3 = 13–17; H_2O/K_2O + RO = 60–100, K_2O/SiO_2 = 0.27–0.35; and RO/SiO_2 = 0.1 to 0.15 (R = the organic additive). Crystallization occurs at the 100° between 2 and 250 hours. This structure could not be crystallized in reactive gels containing sodium in place of potassium.

Adsorption

Adsorption properties of ZSM-10 (US3,692,470 (1972)):

H_2O	Cyclohexane	*n*-Hexane
13.7	7.4	9.3

ZSM-11 (MEL)
(US3,709,979(1973))
Mobil Oil Corporation

Related Materials: silicalite-2

Structure

(*Nature* 275:119(1978)

CHEMICAL COMPOSITION: $0.9 \pm 0.3\ M_2O : Al_2O_3$
: $20–90\ SiO_2 : 6–12\ H_2O$ (US3,709,979(1973))
SYMMETRY: tetragonal
SPACE GROUP: I̅4m2
UNIT CELL CONSTANTS: (Å) $a = 20.1$
$c = 13.4$
PORE STRUCTURE: 10-member rings 5.3×5.4 Å

X-Ray Powder Diffraction Data: (d(Å) (I/I_0))
11.19(27), 10.07(23), 7.50(2), 7.25(1), 7.11(1),
6.73(3), 6.42(2), 6.33(2), 6.09(3), 6.03(5),
5.75(2), 5.61(5), 5.16(1), 5.03(3), 4.62(4),
4.48(1), 4.37(9), 4.28(2), 4.08(2), 4.00(4),
3.86(100), 3.73(39), 3.68(5), 3.49(6), 3.41(3),
3.35(5), 3.27(1), 3.19(<1), 3.14(<1), 3.07(6),
3.00(10), 2.87(2), 2.80(<1), 2.62(3), 2.56(1),
2.51(2), 2.50(4), 2.46(1), 2.42(2), 2.40(2),
2.35(2), 2.28(<1), 2.24(<1), 2.21(<1),
2.18(<1), 2.14(<1), 2.12(2), 2.09(2), 2.08(2),
2.01(21), 1.99(2), 1.97(4), 1.93(5), 1.88(7),
1.85(1), 1.82(1), 1.78(3), 1.76(2), 1.74(1),
1.72(1), 1.71(1), 1.68(7), 1.63(1), 1.62(2),
1.60(<1), 1.57(1), 1.65(1).

Positional parameters for ZSM-11 framework (*Nature* 275:119 (1978)):

Atom	Position	x	y	z
T1	8g	0.0774	0.0774	0.0
T2	16j	0.1263	0.1930	0.1263
T3	16j	0.0784	0.2245	0.3477
T4	16j	0.2833	0.1868	0.1321
T5	16j	0.3118	0.0742	0.0089
T6	8g	0.1796	0.1796	0.5000
T7	16j	0.0770	0.3811	0.3610
O1	8i	0.0924	0.0	0.0208
O2	16j	0.0989	0.1207	0.0954
O3	16j	0.1115	0.2063	0.2422
O4	16j	0.2051	0.1965	0.1065
O5	16j	0.3157	0.1273	0.0981
O6	8i	0.3068	0.0	0.0533
O7	8h	0.1922	0.1067	0.2500
O8	16j	0.1126	0.1810	0.4342
O9	16j	0.2431	0.1814	0.4271
O10	16j	0.0895	0.3023	0.3707
O11	8i	0.0	0.2087	0.3441
O12	8i	0.0	0.3970	0.3836
O13	16j	0.0892	0.2480	0.0596
O14	16j	0.1230	0.4198	0.4400
O15	8h	0.0951	0.5059	0.2500

The ZSM-11 framework contains two intersecting channel systems defined by 10-member-ring openings. Unlike the ZSM-5 structure, the intersecting channels are straight and have the same size elliptical pore mouth. The structure is composed of four-, five-, and six-member rings with the linkage of the chains occurring only through four- and six-member rings.

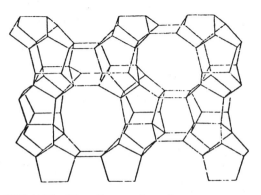

ZSM-11 Fig. 1S: Framework topology.

Synthesis

ORGANIC ADDITIVES
tetraalkylammonium$^+$ with alkyl groups between one and
seven carbon atoms
tetrabutylammonium$^+$
diamines (US4,108,881(1987))

ZSM-11 crystallizes from a preferred range of batch compositions: SiO_2/Al_2O_3 of 20 to 90; Na_2O/SiO_2 between 0.02 and 0.40; R/SiO_2 between 0.02 and 0.15; H_2O/Na_2O between 100

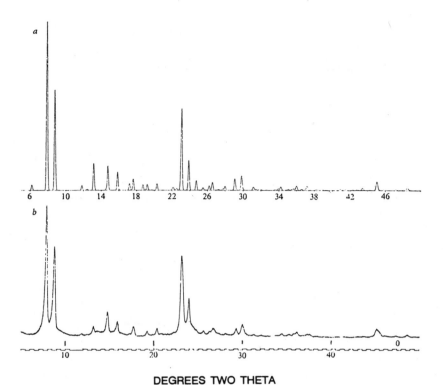

ZSM-11 Fig. 2:   X-ray powder diffraction patterns: (a) computed and (b) of calcined ZSM-11 (*Nature* 275:119(1978)). (Reproduced with permission of MacMillan Magazines Ltd.)

and 600. Crystallization occurs readily after several days at temperatures between 100 and 200°C, but the lower temperatures are preferred. With diamines used as the organic additive, the C_8 diamine ($NH_2(CH_2)_8NH_2$) produces the higher crystalline product. Impurities observed include ZSM-5 and alpha quartz.

Thermal Stability

The ZSM-11 structure has a high thermal stability, maintaining its structure beyond the calcination temperatures of 600°C.

Adsorption

Adsorption properties of ZSM-11 (*Zeo*. 4:337 (1984)):

Adsorbate	Kinetic diameter, Å	Sorbate uptake, mol/kg
n-Hexane	4.3	1.41
3-Methylpentane	5.58	0.87
2,2-Dimethylbutane	6.2	not adsorbed
Cyclohexane	6.0	0.15 (slow adsorption)
Benzene	5.85	1.35
m-Xylene	6.8	0.36

ZSM-12 (MTW)
(US3,832,449(1974))
Mobil Oil Corporation

Related Materials: TEA-silicate
CHZ-5
Nu-13
theta-3
TPZ-12

Structure

(*Zeo.* 5:347(1985))

CHEMICAL COMPOSITION: $1.0 \pm M_{2/n}O : Al_2O_3$:
$20-100 \ SiO_2 : 0-60 \ H_2O$
SPACE GROUP: C2/m
UNIT CELL PARAMETERS: (Å) $a = 24.88$
$b = 5.02$
$c = 12.15$
$\beta = 107.7°$
DENSITY: $1.93 \ g/cm^3$
PORE STRUCTURE: 12-member-ring channels 5.7 ×
6.1 Å

X-Ray Powder Diffraction Data: (d(Å) (I/I_0))
11.9(27), 11.6(10), 11.15(10), 10.02(35),
9.72(5), 8.93(2), 8.55(2), 7.43(3), 7.08(3),
6.66(1), 6.37(2), 6.02(5), 5.86(3), 5.69(4),
5.57(5), 5.00(2), 4.96(5), 4.82(3), 4.75(14),
4.70(11), 4.45(6), 4.28(100), 4.10(8), 3.98(14),
3.85(67), 3.75(5), 3.71(9), 3.65(7), 3.545(1),
3.485(16), 3.39(20), 3.36(11), 3.20(10), 3.16(3),
3.135(2), 3.08(1), 3.06(3).

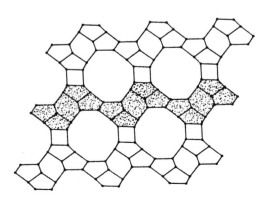

ZSM-12 Fig. 1S: Framework topology.

DLS-atomic coordinates of ZSM-12 ($\times 10^4$) (*Zeo.* 5:346
(1985)):

Atom	x	y	z
T1	0714	0000	1611
T2	0613	0000	9116
T3	3694	0000	7085
T4	3797	0000	1186
T5	2942	0000	8683
T6	2788	0000	2127
T7	2826	0000	4751
O1	0947	0000	0494
O2	0044	0000	1077
O3	0913	2495	2439
O4	3164	0000	7575
O5	0782	2503	−1513
O6	3453	0000	5684
O7	2500	2500	5000
O8	2465	2494	1403
O9	3420	0000	2038
O10	3446	0000	−0140
O11	2831	0000	3454

The ZSM-12 structure contains features of
the ferrierite framework and the mordenite
framework. Structurally it belongs to the mordenite group. It contains both four-member rings
and five-member rings within its structure. The
opening in this structure is a 12-member ring,
which is highly nonplanar.

Synthesis

ORGANIC ADDITIVES (*Mol. Sieve Princ.* (1989))
TEA
N-containing polymers
TEMA
$(PhCH_2)Me_3N$
$(PhCH_2)_2Me_2N$
Et_2Me_2N

ZSM-12 structure is claimed to crystallize in
a preferred range of: SiO_2/Al_2O_3 between 85 and
125; OH/SiO_2 of 0.15 to 0.25; R/R + Na of
0.28 to 0.90; H_2O/SiO_2 between 5 and 100.
Crystallization occurs readily between 150 and
170°C after 5 to 12 days. Crystobalite is a common impurity phase.

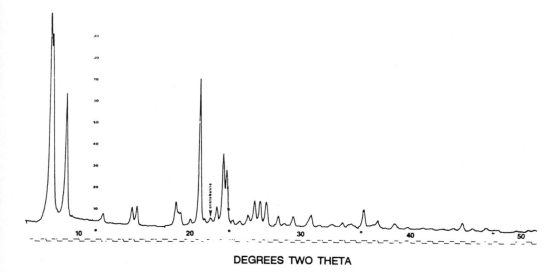

DEGREES TWO THETA

ZSM-12 Fig. 2:   X-ray powder diffraction pattern of ZSM-12 (*Zeo.* 5:347(1985)). (Reproduced with permission of Butterworths Publishers)

Thermal Properties

Because of the high silica content of this material, it is thermally stable to temperatures above the calcination temperature needed to burn the organic trapped in the pore system.

Adsorption

Adsorption properties of ZSM-12:

Cyclohexane	11.4 wt%
Hexane	9.3 wt%
H_2O	7.8 wt%

ZSM-14
(US4,610,855(1986))
Mobil Oil Corporation

Structure

Chemical Composition: 0.005–5 R_2O : 0.01– 5 $M_{2/n}O$: 0–0.2 Al_2O_3 : 100 SiO_2.

X-Ray Powder Diffraction Data: ($d(\text{Å})$ (I/I_0)) 15.49(100), 7.72(10), 7.31(8), 6.92(3), 5.68(2), 5.16(15), 5.01(8), 4.85(5), 4.55(5), 4.29(3), 4.10(2), 3.85(2), 3.66(25), 3.57(36), 3.46(87), 3.32(56), 3.26(20), 3.16(66), 2.988(3), 2.865(6), 2.825(8), 2.641(4), 2.578(3), 2.430(2), 2.348(4), 2.292(2), 2.069(2), 1.834(16), 1.789(3).

Synthesis

ORGANIC ADDITIVES
anionic organic sulfonate or carbonate

ZSM-14 crystallizes from a preferred range of compositions: SiO_2/Al_2O_3 of 1000 to infinity; H_2O/SiO_2 of 10 to 200; OH/SiO_2 between 0.05 and 2; M/SiO_2 of 0.05 to 1; R/SiO_2 of 0.02 to 2. Crystallization occurs readily at temperatures around 160°C after 72 hours.

ZSM-18
(US3,950,496(1976))
Mobil Oil Corporation

Structure

CHEMICAL COMPOSITION: 0.9 ± 0.3 $M_{2/n}O$:
  Al_2O_3 : 10–30 SiO_2
SPACE GROUP: $P\bar{6}_3/m$

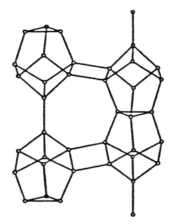

ZSM-18 Fig. 1S: Framework topology.

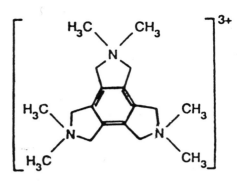

ZSM-18 Fig. 3:   Organic cation structure.

UNIT CELL CONSTANTS: (Å) a = 13.175
                            c = 15.848
FRAMEWORK DENSITY: 14.3
PORE STRUCTURE: unidimensional 12-member rings

X-Ray Powder Diffraction Data: $(d(\text{Å})\ (I/I_0))$
11.5(vs), 9.3(w), 7.9(m–s), 6.6(m), 5.1(m),
4.8(w), 4.64(w), 4.34(m), 4.18(vs), 3.97(s),
3.88(w), 3.80(s), 3.31(m), 3.24(m), 3.13(w),
3.06(m), 2.87(m), 2.50(w).

ZSM-18 contains three-, four-, six-, seven-, and twelve-member rings. The large channels in the system are composed of twelve-member rings and are unidimensional with seven-member-ring side pockets with dimensions of 2.8 ×

3.5 Å. The triquaternary organic cation occupies specific sites within the channel (*Science* 247:1319(1990)).

Synthesis

ORGANIC ADDITIVE
2,3,4,5,6,7,8,9-octahydro-2,2,5,5,8,8-hexamethyl-1 1H-benzo[1,2-c:3,4-c′:5,6-c″]tripyrrolium⁺ (See ZSM-18 Figure 3.)

ZSM-18 crystallizes from a range of: SiO_2/Al_2O_3 between 10 and 30; Na_2O/Al_2O_3 between 0.2 and 5; H_2O/Al_2O_3 between 200 and 1500; organic/Al_2O_3 between 1 and 10. Crystallization occurs at temperatures around 125 and 130°C after 5 days to 2 weeks. Analcime appears as an impurity phase in this system.

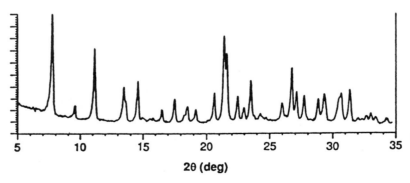

ZSM-18 Fig. 2:   X-ray powder diffraction pattern of ZSM-18 (*Science* 247:1319(1990)). (Reproduced with permission of the American Association for the Advancement of Science)

Thermal Properties

ZSM-18 is thermally stable to 500°C after 3 hours.

ZSM-20 (FAU/EMT)
(US3,972,983(1976))
Mobil Oil Corporation

Related Materials: faujasite/EMC-2

Structure

Chemical Composition: (0.3–0.6) R_2O : (0.4–0.7) M_2O : Al_2O_3 : (7–10) SiO_2.

X-Ray Powder Diffraction Data: ($d(\text{Å})$ (I/I_0))
14.85(88), 14.23(100), 11.87(46), 10.92(29), 8.66(21), 8.19(20), 7.66(7), 7.42(34), 5.85(6), 5.64(71), 5.31(26), 5.18(9), 4.97(10), 4.85(21), 4.72(24), 4.43(5), 4.34(35), 4.15(16), 3.97(63), 3.82(19), 3.75(44), 3.63(5), 3.58(5), 3.53(5), 3.44(9), 3.32(10), 3.28(39), 3.19(5), 3.11(5), 3.03(8), 3.00(5), 2.94(3), 2.89(27), 2.86(10), 2.84(35), 2.78(11), 2.75(5), 2.69(24), 2.61(24), 2.41(6), 2.37(15), 2.22(1), 2.20(1), 2.17(7), 2.14(6), 2.08(11), 2.05(8), 1.92(4), 1.89(4), 1.88(1), 1.84(5), 1.797(4), 1.759(6), 1.737(10), 1.691(10), 1.639(3), 1.600(3), 1.575(11), 1.549(2).

The broad peaks in the X-ray powder diffraction pattern are consistent with the presence of growth faults. These faults constitute the faujasite type structure. About half of the faults are in the EMT stacking sequence (*JCS Chem. Commun.* 493 (1989)).

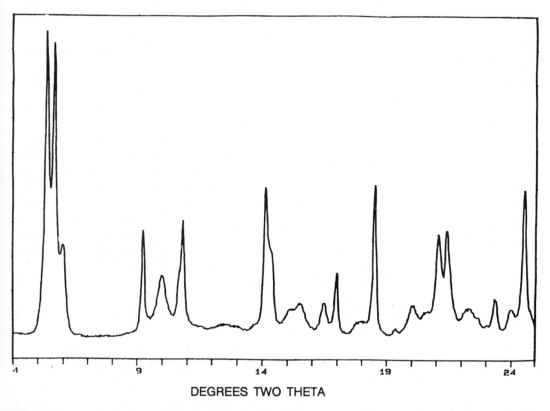

ZSM-20 Fig. 1:   X-ray powder diffraction pattern of ZSM-20. (Reproduced with permission of J. B. Higgins, Mobil)

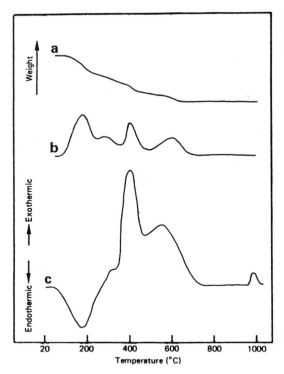

ZSM-20 Fig. 2:   Results of the combined TGA/DTG/DTA analysis of assynthesized ZSM-20 in an air flow: (a) TGA; (b) DTG; (c) DTA (*Zeo.* 7:180(1987)). (Reproduced with permission of Butterworths Publishers)

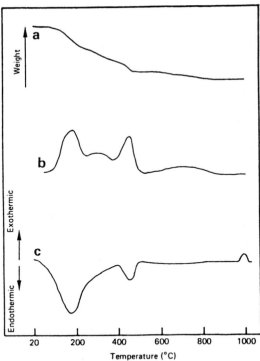

ZSM-20 Fig. 3:   Results of the combined TGA/DTG/DTA analysis of as-synthesized ZSM-20 in a nitrogen flow: (a) TGA; (b) DTG: (c) DTA (*Zeo.* 7:180(1987)). (Reproduced with permission of Butterworths Publishers)

Synthesis

ORGANIC ADDITIVE
tetraethylammonium cations

ZSM-20 crystallizes in a range of SiO_2/Al_2O_3 around 30. The H_2O/SiO_2 ratio is between 10 and 20, and the H_2O/OH between 15 and 30. Crystallization appears slow, requiring times in the range of 1 to 2 weeks before observable crystals appear. Crystallization temperatures are between 80 and 120°C. A co-crystallizing phase appears to be that of zeolite beta (*SSSC* 37:29(1988)).

Thermal Properties

Between room temperature and 220°C adsorbed water is lost (11.4%). The weight loss between

220 and 850°C is due to loss of the organic. Losses between 250 and 360°C are attributed to TEA on the outer surface of the crystals, the occluded organic in the pores decomposes between 360 and 500°C, and the TEA cations balancing some of the aluminum charge occur at > 500°C and correspond to 6%, 6.7%, and 3.5% of the initial weight. The framework is stable to temperatures of 1000°C (*Zeo.* 7:180 (1987)).

Infrared Spectrum

The mid-infrared region of the spectrum appears to be similar to that of the faujasite zeolite type Y.

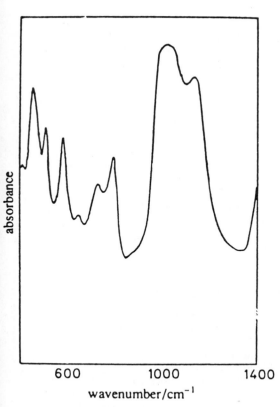

ZSM-20 Fig. 4:   Mid-infrared spectrum of Na$_{0.22}$(NH$_4$)ZSM-20 (*JCS Faraday Trans*. I, 85:2127(1989)). (Reproduced with permission of the Chemical Society)

Adsorption

Adsorption properties of ZSM-20 (US3,972,983 (1976)):

Adsorption	Wt%
Cyclohexane	19.7
n-Hexane	18.2
Water	31.6

ZSM-21 (FER)
(US4,046,859(1977))
Mobil Oil Corporation

Related Materials: ferrierite

Structure

Chemical Composition: 0.3–2.5 R$_2$O : 0–0.8 M$_2$O : Al$_2$O$_3$: > 8 SiO$_2$.

X-Ray Powder Diffraction Data: (d(Å) (I/I_0)) 9.5(vs), 7.0(m), 6.6(m), 5.8(w), 4.95(w), 3.98(s), 3.80(s), 3.53(vs), 3.47(vs), 3.13(w), 2.92(w).

The ZSM-21 family is distinguished from natural ferrierite by the X-ray powder diffraction lines. The line at 11.33 A is present in natural ferrierites but is of weak intensity in the ZSM-21 family. This has been attributed to differences in cation content (*Zeo*. 5:349(1985)).

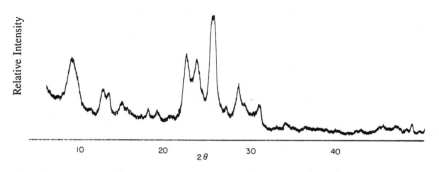

ZSM-21 Fig. 1:   X-ray powder diffraction pattern of ZSM-21 (US4,046,859(1977)).

Synthesis

ORGANIC ADDITIVE
$H_2N(CH_2)_2NH_2$

An example of the synthesis of a ZSM-21 family member has been provided in US4,046,859(1975)). For this example the SiO_2/Al_2O_3 ratio was 16.7 with an OH/SiO_2 of 0.22 and an H_2O/OH of 152. The ratio of the organic to the total inorganic cation + organic is 0.82. Crystallization was completed after 62 days at 100°C.

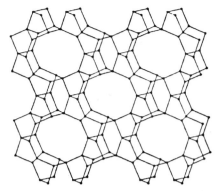

ZSM-22 Fig. 1S: Framework topology.

Adsorption

Adsorption properties of ZSM-21:

Adsorbate	Wt%
Cyclohexane	1.0
n-Hexane	5.4
Water	9.0

ZSM-22 (TON)
(US4,481,177(1984))
Mobil Oil Corporation

Related Materials: theta-1

Structure

(*Zeo*. 7:393(1987))

COMPOSITION: 24 $SiO_2((C_2H_5)_2NH$ (silica end member)
SYMMETRY: orthorhombic
SPACE GROUP: Cmcm
UNIT CELL PARAMETERS (Å): $a = 13.86$
$b = 17.41$
$c = 5.04$
PORE STRUCTURE: 5.5 × 4.5 Å, unidimensional, 10-member-ring channel system
DENSITY: 1.97g/cm³

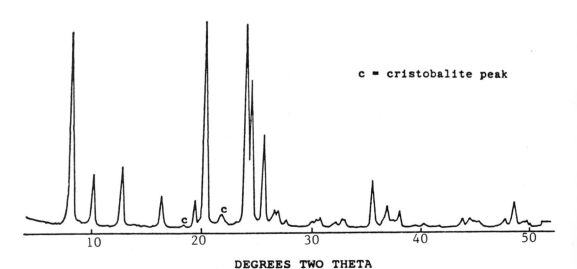

c = cristobalite peak

DEGREES TWO THETA

ZSM-22 Fig. 2:   X-ray powder diffraction pattern for ZSM-22 (*Zeo*. 5:349(1985)). (Reproduced with permission of Butterworths Publishers)

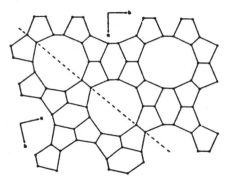

ZSM-22 Fig. 3:   Twinning of silica-ZSM-22 with twin plane (110); projection along (001) (*Zeo.* 7:393(1987)). (Reproduced with permission of Butterworths Publishers)

X-Ray Powder Diffraction Data: $(d(\text{Å}) (I/I_0))$
10.90(40), 8.70(10), 6.94(13), 5.40(8), 4.58(10), 4.36(100), 3.68(97), 3.62(65), 3.47(46), 3.30(5), 2.74(3), 2.52(19).

ZSM-22 is a 10-member-ring zeolite composed of five- and six-member rings. The structure contains ferrierite sheets similar to those of ZSM-5, ZSM-11, and ZSM-35. The six-ring sheets compare to those of bikitaite. The 10-member rings are unidimensional with an elliptical cross-section. There are four unique T sites in the unit cell (*Zeo.* 5:349(1985)). The

Atomic coordinates of silica-ZSM-22 (*Zeo.* 7:393 (1987)):

Atom	No. of pos.	x	y	z
Si1	4	0.0	0.2737(2)	0.25
Si2	4	0.0	0.3768(2)	0.7465(15)
Si3	8	0.2951(2)	0.0482(1)	0.2265(12)
Si4	8	0.2055(2)	0.2112(1)	0.1445(12)
O1	8	0.0923(4)	0.4268(4)	0.6977(20)
O2	8	0.0942(4)	0.2224(4)	0.2068(20)
O3	8	0.2712(4)	0.3789(4)	0.6639(26)
O4	8	0.2288(4)	0.4800(4)	0.0234(20)
O5	8	0.2705(5)	0.2584(4)	0.3535(19)
O6	4	0.0	0.3062(5)	0.5441(26)
O7	4	0.0	0.3450(5)	0.0493(24)
C1	2.3(2)	0.5	0.5195	0.1363
C2	2.8(2)	0.5	0.5124	0.3458

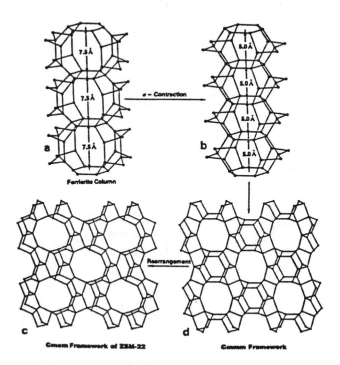

ZSM-22 Fig. 4:   Derivation of the ZSM-22 structure from the ferrierite structure. The diagram depicts the stages in the solution of the structure (*Zeo.* 5:349(1985)). (Reproduced with permission of Butterworths Publishers)

framework topology is closely related to those of ZSM-23 and ZSM-48.

Twinning of the high-silica end member has been observed with (110) as the twin plane. Such twinning is assumed to cause only slight distortions of the Si–O bonds and the Si–O–Si angles.

Synthesis

ORGANIC ADDITIVES (*Princ. Mol. Sieves* (1987))
1-aminobutane
diethylamine
ethylene diamine
1,3-diaminopropane
1,6-diaminohexane
1,4,8,11-tetra-aza-undecane
1,5,8,12-tetra-aza-undecane
1,5,9,13-tetra-aza-undecane
N-ethylpyridinium

ZSM-22 can be formed from an aluminosilicate gel with a SiO_2/Al_2O_3 ranging between 30 and 10,000. Using N-ethylpyridinium as the organic additive, the ZSM-22 structure crystallizes when R/SiO_2 is between 0.05 and 1.0 (US4,481,177(1984)). The preferred inorganic cation in this system is a larger monovalent cation such as potassium (or cesium). Crystallization occurs readily at 160°C (EP 102,716 (1984)).

Infrared Spectrum

Mid-infrared Vibrations (cm^{-1}): 1220sh, 1120sh, 1090vs, 810m, 785m, 640m, 555m, 460s.

ZSM-23 (MTT)
(US4,076,842(1978))
Mobil Oil Corporation

Related Materials: ISI-4
KZ-1

Structure

(*Zeo.* 5:352(1985))

CHEMICAL COMPOSITION: $Na_n[Al_nSi_{24-n}O_{48}]*~4H_2O$
SYMMETRY: orthorhombic
SPACE GROUP: Pmn2$_1$
UNIT CELL CONSTANTS: (Å) $a = 21.5$
                              $b = 11.1$
                              $c = 5.0$
PORE STRUCTURE: unidimensional 10-member ring
4.5 × 5.2 Å

X-Ray Powder Diffraction Data: (d(Å) (I/I_0))
11.91(47), 10.07(24), 9.51(1), 7.86(15), 7.50(4), 6.40(4), 6.15(4), 6.03(6), 5.71(5), 5.58(6), 5.44(6), 5.16(2), 5.00(5), 4.91(10), 4.54(54), 4.45(15), 4.37(15), 4.27(73), 4.17(21), 4.12(23),

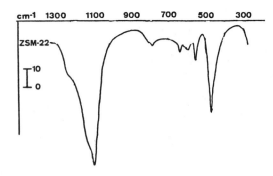

ZSM-22 Fig. 5:   Mid-infrared vibrations of ZSM-22 (*SSSC* 33:41(1989)). (Reproduced with permission of Elsevier Publishers)

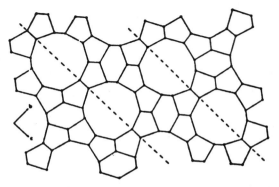

ZSM-23 Fig. 1S: Framework topology.

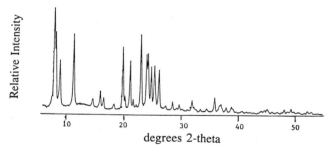

ZSM-23 Fig. 2:   X-ray powder diffraction pattern for calcined ZSM-23 (*Zeo*. 5:352(1985)). (Reproduced with permission of Butterworths Publishers)

4.07(50),   3.90(100),   3.83(31),   3.73(79), 3.63(58), 3.54(33), 3.45(40), 3.37(6), 3.32(6), 3.17(9), 3.06(7), 2.996(7), 2.956(2), 2.851(12), 2.787(2),   2.732(1),   2.640(4),   2.609(3), 2.537(29),   2.495(13),   2.472(12),   2.445(6), 2.401(6), 2.344(8), 2.316(2), 2.242(2), 2.201(1), 2.173(1), 2.138(1), 2.119(1), 2.095(2), 2.078(4), 2.060(3), 2.046(4), 2.029(7), 2.014(4), 2.003(3), 1.992(4), 1.973(1), 1.960(4), 1.944(2), 1.931(4), 1.912(4), 1.895(3), 1.881(6), 1.864(7), 1.850(3), 1.831(2).

ZSM-23 is a high-silica zeolite containing five-, six-, and ten-ring subunits that generate unidimensional ten-member-ring channels which are parallel to the short 5 A axis. The pore opening is a teardrop-shaped ten-member ring. This structure is thought to be similar to that of ZSM-22.

Synthesis

ORGANIC ADDITIVES
pyrrolidine
Diquat-7 ($(CH_3)_3N^+$-R_1-$N^+(CH_3)_3$; R = 7, saturated or unsaturated)

ZSM-23 was claimed to crystallize from a reactive aluminosilicate gel with a SiO_2/Al_2O_3

DLS-atomic coordinates ($\times$ 10^4) of ZSM-23 origin at mmn (*Zeo*. 5:352 (1985)):

Atom	x	y	z
T1	0000	2076	−0060
T2	0000	1700	−2696
T3	0000	1353	2379
T4	0000	0000	2936
T5	5000	0000	4642
T6	5000	1921	3638
T7	5000	1273	6030
O1	2500	2500	0000
O2	0000	1671	−1261
O3	0000	1647	1085
O4	0000	2375	−3243
O5	2498	1296	−3101
O6	0000	0611	2125
O7	2507	1540	3151
O8	2502	0000	3783
O9	5000	0612	5418
O10	5000	1838	5085

ratio between 55 and 70 with pyrrolidine as the organic additive and an OH/SiO_2 between 0.01 and 0.049. In the presence of Diquat-7 the range of SiO_2/Al_2O_3 was extended to 5000. The ratio of the organic to the total cation content in the gel ranges between 0.85 and 0.95. Crystallization takes place after 5 days at 160°C, producing pure ZSM-23. Common impurity phases,

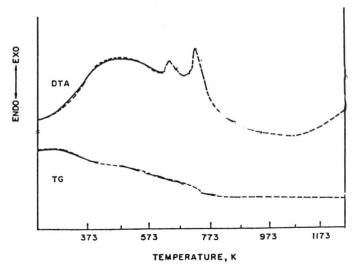

ZSM-23 Fig. 3:   DTA/TGA of ZSM-23 (*J. Catal.* 121:89(1990)). (Reproduced with permission of Academic Press)

which are dependent on the amount of hydroxide or the SiO_2/Al_2O_3 ratio, include cristobalite and ZSM-5 (*ACS Sym. Ser.* 398:560(1989)).

Crystallization of this structure also occurs under more acidic conditions in the presence of fluoride ion, with an F/SiO_2 ratio between 0.1 and 6 (EP 347,273(1989)).

The iron-containing analog of this material also has been prepared (*J. Catal.* 121:89(1980)).

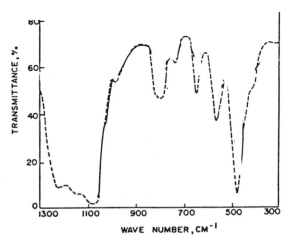

ZSM-23 Fig. 4:   Mid-infrared framework vibrations for ZSM-23 (*J. Catal.* 121:89(1990)). (Reproduced with permission of Academic Press)

Thermal Properties

The organic remaining trapped within the pores of the structure is removed through thermal treatment. Heating to 550°C does not result in a decrease in the X-ray crystallinity of this material.

Infrared Spectrum

See ZSM-23 Figure 4.

Adsorption

Adsorption properties of ZSM-23 (US4,076,842 (1978)):

	Cyclohexane	*n*-Hexane	Water
Wt%	1.4	5.3	5.5

ZSM-25
(US4,247,416(1981))
Mobil Oil Corporation

Structure

Chemical Composition: 0.01–0.4 R_2O : 0.9 ± 0.2 M_2O : Al_2O_3 : 6–10 SiO_2.

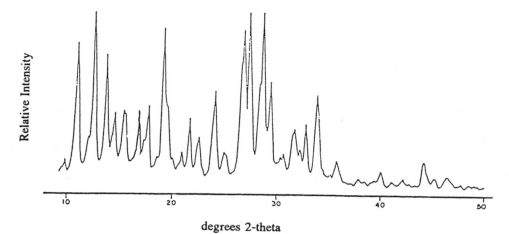

ZSM-25 Fig. 1:   X-ray powder diffraction pattern for ZSM-25 (US4,247,416(1978)).

X-Ray Powder Diffraction Data: $(d(\text{Å}) (I/I_0))$
16.06(2), 12.28(2), 11.63(2), 10.74(2), 9.25(4), 8.93(7), 8.08(67), 7.37(13), 7.04(86), 6.44(56), 6.22(5), 6.08(25), 5.76(20), 5.68(20), 5.49(2), 5.28(30), 5.13(14), 5.01(34), 4.79(6), 4.62(72), 4.54(26), 4.44(6), 4.32(3), 4.24(12), 4.10(29), 3.98(8), 3.94(15), 3.74(11), 3.70(44), 3.53(8), 3.54(9), 3.36(22), 3.31(67), 3.25(100). 3.13(34), 3.11(80), 3.04(50), 2.981(3), 2.945(8), 2.909(12), 2.838(11), 2.814(22), 2.772(13), 2.725(29), 2.666(9), 2.638(40), 2.603(4), 2.557(4), 2.507(10), 2.486(3), 2.395(2), 2.374(4), 2.348(3), 2.328(2), 2.307(2), 2.283(3), 2.252(7), 2.208(2), 2.193(2), 2.173(1), 2.156(2), 2.140(4), 2.118(1), 2.085(1), 2.075(1), 2.0505(12), 2.033(3), 2.011(2), 2.001(5), 1.965(3), 1.954(4), 1.946(4), 1.931(2), 1.902(3), 1.872(1), 1.853(2), 1.803(1), 1.774(35), 1.756(3), 1.737(7), 1.694(5), 1.687(6), 1.674(6), 1.650(2), 1.627(5), 1.613(3), 1.588(5), 1.566(3).

Synthesis

ORGANIC ADDITIVES
tetraethylammonium$^+$

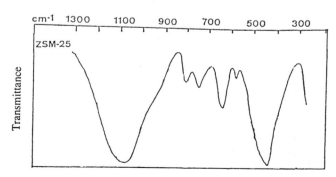

ZSM-25 Fig. 2:   Mid-infrared spectrum of ZSM-25 (*SSSC* 33:41(1989)).
(Reproduced with permission of Elsevier Publishers)

Zeolite ZSM-25 crystallizes from a range of SiO_2/Al_2O_3 between 5 and 15 in the presence of the TEA^+ cation. Crystallization occurs over a range of OH/SiO_2 of 0.3 and 0.7 with the R/R + M cation ratio between 0.4 and 1.0. Crystallization occurs readily, depending upon temperature, which ranges between 80 and 180°C, in from 6 hours to 150 days. The optimum temperature is between 121 and 139°C, which results in crystal formation after 4 days. Only small amounts of organic are retained in this structure.

Thermal Properties

ZSM-25 can be calcined to 500°C with minimal loss in crystallinity, with some broadening of the lines observed.

Infrared Spectrum

Mid-infrared Vibrations (cm⁻¹):   1075vs, 780s, 730s, 630s, 430s (*SSSC* 33:26(1989)).

Adsorption

Adsorption properties of ZSM-25 (US4,247,416 (1978)):

	n-Hexane	Cyclohexane	Water
Wt%	10.87	3.60	9.15

ZSM-34 (ERI/OFF)
(US4,086,186(1978))
Mobil Oil Corporation

Related Materials: erionite/offretite

Structure

Chemical Composition: 0.5–1.3 R_2O : 0–0.15 Na_2O : 0.10–0.5 K_2O : Al_2O_3 : 8–50 SiO_2.

X-Ray Powder Diffraction Data: (d(Å) (I/I_0)) 11.56(100), 9.21(3), 7.60(25), 6.62(52), 6.32(10), 5.74(31), 5.33(4), 4.98(10), 4.58(64), 4.33(61), 4.16(7), 3.82(55), 3.76(86), 3.59(86), 3.30(34), 3.16(40), 2.926(9), 2.853(84),

2.804(11), 2.679(16), 2.515(4), 2.488(21), 2.286(4), 2.200(7), 2.108(6), 2.080(4), 1.983(4), 1.956(3), 1.890(19), 1.865(5), 1.830(6).

ZSM-34 is a synthetic zeolite consisting of an intergrowth of offretite and erionite. The sorptive and catalytic properties are controlled by the nature and the distribution of the stacking faults and intergrowths in the material.

Offretite with erionite intergrowths (*Zeo.* 6:474 (1986)):

Zeolite	Erionite (%)
T-type	30
Si-rich offretite	10
ZSM-34	<10 (under detection limit)

Synthesis

ORGANIC ADDITIVE
choline

ZSM-34 crystallizes from a reactive aluminosilicate gel in the presence of the organic amine choline and a mixture of inorganic cations Na^+ and K^+. This structure will form over a gel SiO_2/Al_2O_3 range of 10 to 55 with an OH/SiO_2 of 0.3 to 0.8 and a $K_2O/(Na + K)_2O$ of 0.1 and 1.0. Crystallization occurs readily, depending on temperature (80 to 175°C), with crystals forming after 12 hours to 200 days. The SiO_2/Al_2O_3 ratio of the resulting crystal falls between 7.6 and 8 (*Zeo.* 6:474(1986)).

Thermal Properties

ZSM-34 retains its crystallinity after treatment to 500°C to remove the organic amine trapped within the pores.

Infrared Spectrum

Mid-infrared Vibrations (cm⁻¹):   1174(bwsh), 1065(s), 952(sh), 791(m), 729(bvw), 683(w), 640(mw), 556(w), 471(ms), 447(wsh), 420(wsh) (*Zeo.* 9:104(1989)).

Adsorption

Adsorption properties of ZSM-34 (US4,086,186 (1978)):

	Cyclohexane	n-Hexane	Water
Wt%	4.4	11.5	22.2

ZSM-35 (FER)
(US4,016,245(1977))
Mobil Oil Corporation

Related Materials: ferrierite

Structure

Chemical Composition: 0.3–2.5 R_2O : 0–0.8 M_2O : Al_2O_3 : > 8 SiO_2.

X-Ray Powder Diffraction Data: $(d(\text{Å})\,(I/I_0))$
11.48(5), 9.53(27), 7.07(25), 6.93(25), 6.61(27),
5.76(12), 5.67(4), 5.25(4), 4.95(10), 4.73(5),
4.58(4), 3.98(63), 3.93(39), 3.85(20), 3.77(51),
3.73(12), 3.66(33), 3.53(100), 3.47(80),
3.39(28), 3.31(24), 3.13(18), 3.04(8), 2.954(6),
2.917(2), 2.885(4), 2.848(3), 2.702(3), 2.638(7),
2.607(3), 2.573(4), 2.539(2), 2.473(4), 2.444(2),
2.401(4), 2.348(5), 2.309(2), 2.283(2), 2.137(2),
2.108(2), 2.025(5), 1.9991(8), 1.965(3),
1.945(4), 1.922(9), 1.862(6).

Synthesis

ORGANIC ADDITIVES
ethylenediamine
pyrrolidine

ZSM-35 crystallizes from a reactive aluminosilicate gel with a SiO_2/Al_2O_3 ranging between 12 and 60 in the presence of either ethylenediamine or pyrrolidine. Crystallization occurs with the OH/SiO_2 range between 0.07 and 0.49 after 10 days at 177°C. The only significant difference between the synthesis of ZSM-35 and that of ZSM-38 appears to be in the type of organic amine used.

Infrared Spectrum

Mid-infrared Framework Vibrations (cm^{-1}):
1225s, 1085vs, 810m, 580m.

Adsorption

Adsorption properties of ZSM-35 (US4,016,245 (1977)):

	Cyclohexane	n-Hexane	Water
Wt%	1.0	5.4	9.0

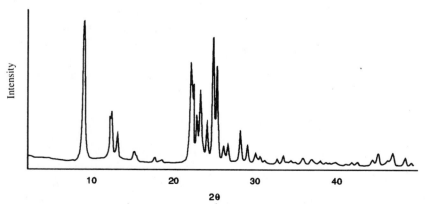

ZSM-35 Fig. 1:   X-ray powder diffraction pattern for ZSM-35 (US4,016,245(1977)).

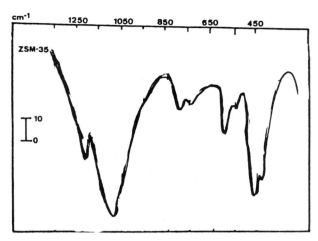

ZSM-35 Fig. 2: Infrared vibrations in the mid-infrared range (KBr pellets) for ZSM-35 (*SSSC* 33:43(1989)). (Reproduced with permission of Elsevier Publishers)

The adsorption properties for ZSM-35 suggest that pore blockage, possibly due to stacking faults within the structure, inhibits the adsorption of the larger organic hydrocarbons such as cyclohexane.

ZSM-38
(US4,046,859(1977))
Mobil Oil Corporation

Related Materials: ferrierite-type

Structure

Chemical Composition: 0.3–2.5 R_2O : 0–0.8 M_2O : Al_2O_3 : > 8 SiO_2.

X-Ray Powder Diffraction Data: ($d(Å)$ (I/I_0))
9.80(45), 9.12(29), 7.97(2), 7.08(34), 6.68(34), 6.03(15), 5.83(11), 5.00(13), 4.73(11), 4.37(7), 4.23(5), 4.01(82), 3.81(68), 3.69(23), 3.57(100), 3.51(100), 3.34(26), 3.17(48), 3.08(24), 3.00(17), 2.921(29), 2.734(5), 2.657(11), 2.596(6), 2.495(7), 2.436(3), 2.376(4), 2.332(4), 2.270(2), 2.127(4), 2.067(2), 2.036(7).

This member of the ZSM-21 family of zeolites can be distinguished from the natural ferrierite material by the lack of intensity in the line at 11.33 Å. Changes in intensities of these lines generally are related to the cation content of the material (*Zeo.* 5:349(1985)).

Synthesis

ORGANIC ADDITIVE
2-hydroxyethyltrimethylammonium cation

ZSM-38 crystallizes from a reactive aluminosilicate gel in the presence of 2-hydroxyethyltrimethylammonium and sodium cations. Crystallization occurs after 70 days as 100°C. The SiO_2/Al_2O_3 ratio of the starting gel is between 9 and 200 with an OH/SiO_2 between 0.05 and 0.5. The organic/organic + inorganic cation ratio falls between 0.2 and 1.0. The significant difference between ZSM-35 and ZSM-38 appears to be in the use of the organic amine in the synthesis of the structure (*SSSC* 33: 223 (1989)).

Adsorption

The adsorption properties of this material appear dependent on the synthesis. Adsorption capacities have been reported in the range for cyclohexane between 4.0 and 7.1 wt%, whereas *n*-hexane was adsorbed at 7.1 to 7.2 wt% by

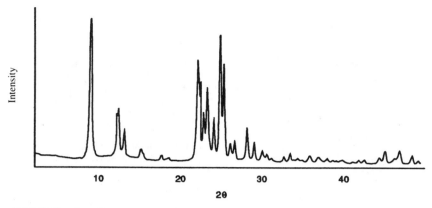

ZSM-38 Fig. 1:   X-ray powder diffraction pattern for ZSM-38 (US4,046,859(1977)).

these samples. Water adsorption ranged be-tween 11.2 and 14.7 wt%.

ZSM-39 (MTN)
(US4,287,166(1981))
Mobil Oil Corporation

Related Materials: Dodecasil-3C
                        Holdstite
                        CF-3

Structure

(*Nature* 294:341(1981))

CHEMICAL COMPOSITION: $S_{136}O_{272}*qR$ [R = TEAOH]
SYMMETRY: cubic
SPACE GROUP: Fd3m
UNIT CELL CONSTANT (Å): 19.4

ZSM-39 Fig. 1S: Framework topology.

X-Ray Powder Diffraction Data: $(d(Å)\,(I/I_0))$
11.5(5), 6.84(23), 6.45(2), 6.17(1), 5.83(93), 5.58(69), 5.35(1), 4.83(47), 4.43(36), 4.19(2), 4.12(1), 3.95(48), 3.89(2), 3.85(2), 3.72(100), 3.63(1), 3.42(42), 3.27(84), 3.23(10), 3.06(12), 2.954(8),   2.710(2),   2.520(9),   2.420(1), 2.365(10),   2.281(17),   2.236(2),   2.220(1), 2.166(3), 2.126(1), 1.975(3), 1.946(4), 1.899(2).

The ZSM-39 structure can be described as a framework of space-filling pentagonal dodeca-hedra (12-hedra) and hexakaidecahedra (16-hedra), making it the silicate structural analog of the 17 A gas hydrate (*Nature* 294:341(1981); *J. Chem. Phys.* 19:112(1951)). The arrange-ment of 16-hedra is the same as that of the carbon atoms in a diamond. The ZSM-39 frame-work consists entirely of five- and six-member rings, thus limiting its adsorption capacity. There are only three tetrahedral sites (T(1), T(2), and T(3)) in the unit cell. Structural defects such as

---

Atomic coordinates of ZSM-39 Fd3m with origin at $\overline{3}$m (*Nature* 294:341 (1981)):

	x	y	z
T1	0.125	0.125	0.125
T2	0.2195	0.2195	0.2195
T3	0.1823	0.1823	0.3728
O1	0.1722	0.1722	0.1722
O2	0.2024	0.2024	0.2967
O3	0.125	0.125	0.3743
O4	0.25	0.1573	0.4073

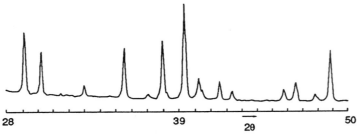

ZSM-39 Fig. 2:   X-ray powder diffraction pattern for ZSM-39 (*J. Incl. Phenom.* 5:355(1987)). (Reproduced with permission of Kluwer Academic Publishers)

twinning and stacking faults have not been observed (*Zeo.* 4:112(1984)).

Synthesis

ORGANIC ADDITIVES
TMA/TEA
TMA/piperazine
trimethylbenzylamine/piperazine
triethylbenzylamine/piperazine
2-aminopropane
2-amino-2-methylpropane
piperazine
pyrrolidine

amine/alcohol
glycerine
ethylene glycol
1,3-propanediol
morpholine
dioxane

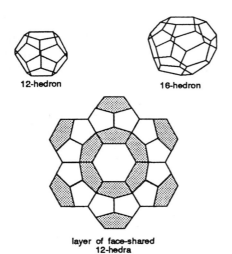

12-hedron

16-hedron

layer of face-shared
12-hedra

ZSM-39 Fig. 3:   Structural elements of ZSM-39; the 12-hedron, the association of the 12-hedra into layers of face-shared dodecahedra, and the 16-hedron formed via an ABC stacking of these layers (*SSSC* 33:331(1989)). (Reproduced with permission of Elsevier Publishers)

ZSM-39 crystallizes in a system containing TMA and TEA hydroxide, silica, and sodium aluminate. The temperature preferred for crystallization is 155°C after 330 hours (US 4,287,166(1981)). Large 450-micron crystals were prepared from gel systems containing amine–alcohol–SiO_2 and water. The alcohols used include glycerine, ethylene glycol, and 1,3-propanediol (*J. Inc. Phenom.* 5:533(1987)). The amount of aluminum present in the framework appears to be limited to about 0.2% (*Zeo.* 13:11(1983)). At higher pH ranges the crystallization results in the formation of quartz (*Zeo.* 7:133(1987)). The use of mixed systems such as amine/piperazine does not change the course of crystallization, but it does lower the temperature at which the structure will form in the presence of the piperazine, from 180 to 160°C (*Zeo.* 8:495(1988)). In the presence of TPA cations, piperazine-ZSM-39 recrystallizes to TPA-silicalite-1 (*JSC Chem. Commun.* 804(1987)).

Thermal Properties

ZSM-39 exhibits significant thermal stability based on its pure silicate framework composition. TGA analysis of the "as-synthesized" material shows a continuous weight loss from about 100°C to 1400°C. Such loss ranged from 7 to

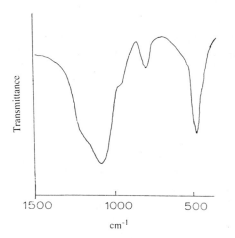

ZSM-39 Fig. 4:   Infrared spectrum (KBr pellets) of ZSM-39 (*Zeo.* 369(1984)). (Reproduced with permission of Butterworths Publishers)

up to 100 hours ZSM-39 sorbed variable amounts of water, between 1 and 5.5%.

ZSM-43
(US4,247,728(1981))
Mobil Oil Corporation

Structure

Chemical Composition: 0.6–$2.1\ M_{2/n}O : Al_2O_3$: at least $5\ SiO_2$.

X-Ray Powder Diffraction Data: ($d(\text{Å})$ (I/I_0)) 14.74(22), 9.94(6), 7.56(65), 7.20(8), 6.86(31), 6.55(7), 5.97(2), 5.06(6), 4.75(100), 4.45(5), 4.11(5), 3.80(62), 3.65(26), 3.53(30), 3.42(52), 3.31(59),   3.22(92),   3.07(49),   2.912(22), 2.847(28), 2.580(11), 2.529(24).

From the adsorption data, it is not possible conclusively to distinguish this material as a 10- or a 12-member ring molecular sieve.

9% in the aminopropane system. Little change in the X-ray diffraction pattern is observed, even after 100 hours at 950°C. The samples of ZSM-39 appear black after calcination, indicating the presence of some residual carbon trapped within the cavities. DTA in air of the as-synthesized material shows a small sharp endotherm followed by a very broad exotherm up to 900°C. The temperature of the endotherm varies with the amine used in the synthesis, as does the thermal stability (*Zeo.* 3:11(1983)).

Unlike melanophlogite—which is unstable to grinding, converting to quartz after 40 min in a mechanical mortar and pestle—ZSM-39 appears to be stable, with only slight broadening of peaks after 24 hours of grinding (*Zeo.* 3:12(1983)).

Synthesis

ORGANIC ADDITIVE
choline

ZSM-43 crystallizes from a reactive aluminosilicate gel using choline as the organic. Crystallization requires the presence of both Na^+ and Cs^+ in the reaction mixture. This material crystallizes over a very narrow SiO_2/Al_2O_3 range of 10 to 20. The OH/SiO_2 ratio is between 0.3 and 0.6, with the $R/R + M^+$ in the range of 0.4 to 0.6. Crystallization requires 20 to 170 days, with temperatures ranging between 100 and 150°C.

Infrared Spectrum

Mid-infrared Vibrations (cm⁻¹):   1090vs, 780w, 460s (*SSSC* 33:26(1989)).

Adsorption

No measurable uptake of He (less than 0.2%) was observed. Exposing ZSM-39 to water at 100% relative humidity for 24 hours also showed less than 0.2% adsorption. At elevated temperatures, between 100 and 160°C, for periods of

Adsorption

Static adsorption properties of ZSM-43 (US4,247,728 (1981)):

	n-Hexane	Cyclohexane	Water
Wt%	10–20	5	6.2

ZSM-45 (LEV)
(EP 107,370(1983))
Mobil Oil Corporation

Related Materials: levyne

Structure

Chemical Composition: 0.8–1.8 R_2O : 0–0.3 Na_2O : 0–0.5 K_2O : Al_2O_3 : greater than 8 SiO_2.

X-Ray Powder Diffraction Data: ($d(\text{Å})$ (I/I_0))
10.20(12), 7.97(41), 6.51(21), 5.47(11), 5.07(84), 4.93(15), 4.19(58), 4.00(100), 3.75(32), 3.52(7), 3.40(4), 3.26(18), 3.11(50), 3.02(9), 2.81(6), 2.75(39), 2.58(9), 2.09(7), 1.85(5), 1.76(9).

Adsorption and catalytic data identify this material as a small-pore molecular sieve. The material is structurally related to levyne, considered to be a siliceous member of this structure type.

Synthesis

ORGANIC ADDITIVES
dimethyldiethylammonium
cobaltinium
choline

ZSM-45 crystallizes from a preferred reactive aluminosilicate gel with: SiO_2/Al_2O_3 ranging between 10 and 80; OH/SiO_2 of 0.3 to 0.8; H_2O/OH of 20 to 80; R/R + M of 0.2 to 0.7; K/K + Na of 0.05 to 0.3.

Adsorption

Adsorption properties of ZSM-45:

Adsorbate	Pressure (Torr)	Adsorption capacity (wt%)
Cyclohexane	20	3.4
n-Hexane	20	12.9
Water	12	21.1

ZSM-47
(US4,187,283(1980))
Mobil Oil Corporation

Structure

Chemical Composition: aluminosilicate.

X-Ray Powder Diffraction Data: ($d(\text{Å})$ (I/I_0))
11.60(w), 9.00(w), 8.41(ms), 6.60(m), 6.27(ms), 5.71(w), 5.42(w), 4.65(w), 4.48(ms), 4.34(ms), 4.12(ms), 4.06(vs), 4.00(vs), 3.89(vs), 3.73(w), 3.55(m), 3.29(m), 3.15(w), 3.13(w), 3.03(w).

Adsorption and catalytic data identify this material as a small-pore molecular sieve.

Synthesis

ORGANIC ADDITIVES
TMA/choline

ZSM-47 crystallizes from a reactive aluminosilicate gel with SiO_2/Al_2O_3 ranging between 10 and 80. With the mixed organic amine system, TMA/choline crystallization occurs around 150°C after 7 days. The inorganic cation used in the synthesis is sodium preferably, with potassium also possible. The OH/SiO_2 of the gel ranges between 0.1 and 0.8.

Infrared Spectrum

Mid-infrared Vibrations (cm^{-1}): 1200sh, 1090vs, 775s, 530sh,w, 450vs.

Adsorption

Adsorption properties for ZSM-47 (US4,187,283 (1980)):

	Water	Cyclohexane	*n*-Hexane
Wt%	11.5	2.7	2.4

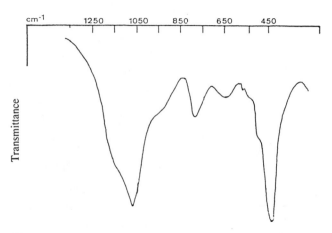

ZSM-47 Fig. 1: Mid-infrared spectrum for ZSM-48 (*SSSC* 33:43(1987)). (Reproduced with permission of Elsevier Publishers)

ZSM-48
(US4,397,827(1981))
Mobil Oil Corporation

Related Materials: EU-2

Structure

(*Zeo.* 5:355(1985))

CHEMICAL COMPOSITION: (0.05–5) Na$_2$O : (0.1–10) M$_{21/n}$ O : (0–4) Al$_2$O$_3$: (100) SiO$_2$
SYMMETRY: orthorhombic
UNIT CELL CONSTANTS (Å): a = 14.24
b = 20.14
c = 8.40

DENSITY: 1.98 g/cm^3
PORE STRUCTURE: medium-pore, unidimensional, 5.3 × 5.6 Å

Chemical Composition: (0.05–5) R$_2$O : (0.1–10) M$_{2/n}$O : (0–4) Al$_2$O$_3$: 100 SiO$_2$ [M is an alkali or alkaline earth cation, n the charge on the cation M].

X-Ray Powder Diffraction Data: (d(Å) (I/I_0))
11.81(74), 10.19(29), 7.19(7), 6.89(3), 6.10(7), 5.86(20), 5.61(4), 4.21(82), 4.08(9), 3.99(8), 3.89(100), 3.74(3), 3.62(3), 3.59(4), 3.37(4), 3.27(4), 3.07(4), 2.85(14), 2.46(4), 2.85(14), 2.46(4), 2.37(4).

ZSM-48 consists of a disordered structure containing ferrierite sheets linked via bridging oxygens. This framework has eight independent T atom sites. Two possible structures were proposed (*Zeo.* 5:355(1985)). These two structures were random intergrowths can be fitted to the X-ray diffraction pattern for ZSM-48. Both structures contain sheets of six-member rings normal to the b axis of the crystal. The walls of the 10-ring parallel channels that permeate the zeolite are composed of distorted six-member rings.

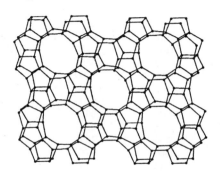

ZSM-48 Fig. 1S: Framework topology.

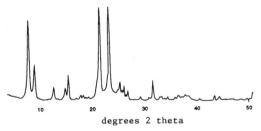

ZSM-48 Fig. 2:   X-ray powder diffraction pattern for ZSM-48 (*Zeo.* 5:355(1985)). (Reproduced with permission of Butterworths Publishers)

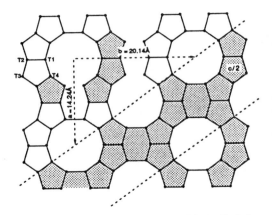

ZSM-48 Fig. 3:   Ferrierite sheet of ZSM-48. Shaded region indicates the location of the possible stacking fault in this structure (*Zeo.* 5:355(1985)). (Reproduced with permission of Butterworths Publishers)

UUDD-Cmcm atom parameters ($\times$ 10⁴) for one proposed topology of ZSM-48 (a = 14.24 Å; b = 20.14 Å; c = 8.40 Å) (*Zeo.* 5:355(1985)):

Atom	x	y	z
T1	0000	2251	5615
T2	0000	3752	5620
T3	1887	4306	4338
T4	1975	1868	4361
O1	0000	2226	7500
O2	0000	3707	7500
O3	1671	4437	2500
O4	2130	1736	2500
OI	0000	3004	4934
OII	0916	4150	5152
OIII	2310	5000	5000
OIV	2642	3747	4590
OV	2500	2500	5000
OVI	0896	1898	4833

UDUD-Imma atom parameters ($\times$ 10⁴) for second proposed topology of ZSM-48 (a = 8.40 Å, b = 14.24 Å; c = 20.14 Å) (*Zeo.* 5:355(1985)):

Atom	x	y	z
T1	3187	-2500	-0098
T2	1838	-2500	1316
T3	3143	-0527	1774
T4	1830	-0664	-0601
O1	5000	-2500	0172
O2	0000	-2500	1536
O3	5000	-0300	1704
O4	0000	-0965	-0518
OI	1965	-2500	0516
OII	2750	-1591	1595
OIII	2500	-0285	2500
OIV	2118	0143	1289
OV	2367	0000	0000

Synthesis

ORGANIC ADDITIVES
diquat-6
bis(N-methylpyridyl)ethylinium
diethylenetriamine
triethylenetetramine
tetraethylenepentamine
1,4,8,11-tetra-aza-undecane
1,5,9,13-tetra-aza-undecane
1,5,8,12-tetra-aza-undecane
1,3-diaminopropane
n-propylamine/TMA$^+$
hexane-diamine
triethylamine

Zeolite ZSM-48 crystallizes generally at high silica/alumina ratios in the presence of a wide variety of amines. Many of these amines have also been used to prepare other high-silica zeolites. Generally the structure forms at SiO_2/Al_2O_3 greater than 50, with the high-silica polymorph easily crystallized. The OH/SiO_2 falls between 0.05 and 0.4, with Na/SiO_2 and R/SiO_2 ratios of 0.1 to 1 and 0.05 to 1, respectively. Crystallization occurs readily at temperatures above 150°C. Depending on the temperature of crystallization and the nature of the organic additive, ZSM-48 or ZSM-22 will crystallize, with ZSM-48 crystallization preferred at the higher temperatures (*Zeo.* 8: 127(1988)). ZSM-5 also is observed to form under high hydroxide ratios.

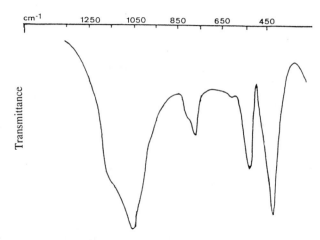

ZSM-48 Fig. 4:   Mid-infrared spectrum of ZSM-48 (*SSSC* 33:43(1987)).
(Reproduced with permission of Elsevier Publishers)

ZSM-48 crystallization is suppressed in the presence of boric acid and a neutral amine, with ferrierite and ZSM-5 crystallizing instead (*Zeo*. 8:127(1988)). The pH of the crystallization mixture is around 12.6 to 12.8. The crystal habbit appears to be bundles of needles under these conditions, between 20 and 45 microns in size (*Zeo*. 8:495(1988)).

ZSM-48 also has been prepared in a nonaqueous system using a dried aluminosilicate gel and hexanediamine or triethylamine as the organic/solvent. Crystallization began after 10 hours at 200°C.

Thermal Properties

ZSM-48 exhibits the high thermal stability characteristic of the high-silica zeolite molecular sieves.

Infrared Spectrum

Mid-infrared Vibrations for ZSM-48 (cm⁻¹):
1220sh, 1110vs, 790m, 555s, 475s, (*SSSC* 33:26(1987)).

Adsorption

Static adsorption results at room temperature for ZSM-48 (*SSSC* 28:547 (1986)):

	n-Hexane	Benzene	Cyclohexane	Mesitylene
Wt%	5.8	4.1	3.2	<0.1

ZSM-50 (EUO)
(EP 127,399(1984))

Related Materials: EU-1

Structure

See EU-1 for a description of the framework topology.

X-Ray Powder Diffraction Data: (d(Å) (I/I₀))
20.10(25), 11.10(75), 10.10(50), 9.70(25), 5.77(25), 5.61(25), 4.64(50), 4.35(50), 4.30(100), 4.00(75), 3.85(75), 3.70(50),

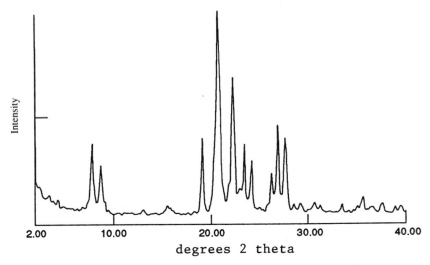

ZSM-50 Fig. 1:   X-ray powder diffraction pattern for as-synthesized ZSM-50. (Reproduced with permission of author)

3.42(25), 3.35(25), 3.27(50), 3.24(25), 2.94(25), 2.53(25).

3.77(w), 3.70(w–m), 3.57(w–m), 3.34(w–m), 3.27(m–s), 3.23(w), 2.69(w), 2.51(w), 2.31(w).

Synthesis

ORGANIC ADDITIVE
diquat-6 [Me$_3$N(CH$_2$)$_6$NMe$_3$]$^{+2}$

ZSM-50 crystallizes from a reactive gel in the presence of the diquanterary amine, diquat-6, and sodium. This structure crystallizes over a SiO$_2$/Al$_2$O$_3$ of 30 to 90, with the OH/SiO$_2$ ratio between 0.1 and 0.3.

ZSM-51 (NON)
(US4,568,654(1986))
Mobil Oil Corporation

Related Materials: nonasil

Structure

Chemical Composition: x RO$_2$: y M$_2$O : 0–7 Al$_2$O$_3$: 100 SiO$_2$.

X-Ray Powder Diffraction Data: (d(Å) (I/I_0))
11.12(w–m), 9.21(w–vs), 6.85(w), 5.84(w), 4.61(s–vs), 4.47(w–m), 4.31(vs), 3.98(m–s),

Synthesis

ORGANIC ADDITIVES (US4,568,654(1986); EP 226,674(1987))
cobalticinium
dimethylpiperidinium
trimethylene bis(trimethylammonium)
tetramethylpiperazinium
pyrrolidine (*RR 8th IZC* 14(1989))

ZSM-51 is prepared from a preferred range of batch compositions: SiO$_2$/Al$_2$O$_3$ = 15–25,000; OH$^-$/SiO$_2$ = 0.2–0.6; M$^+$/SiO$_2$ = 0.1–1.5; H$_2$O/SiO$_2$ = 20–100; R/SiO$_2$ = 0.05–1.5. R is the organic additive, and M$^+$ is generally sodium. Silica sol (30% silica) is used as the source of silica and aluminum sulfate as the source of aluminum. Colloidal silica also has been successfully used as a source of silica. Crystallization generally takes place in between 2 and 10 days, with cristobalite being an impurity phase observed in some samples. Temperatures used to crystallize this structure were generally around 160 to 180°C. Crystals are rod-like in shape and about 1 micron in length.

Thermal Properties

Three exothermal weight losses are observed for ZSM-51, occurring at 250, 510, and 910°C. The total weight lost is about 5%. Heating to 1000°C does not destroy the structure (*RR 8th IZC* 14(1989)).

ZSM-57 (MFS)
(EP 174,121(1986))
Mobil Oil Corporation

Structure

(*RR 8th IZC,* 132(1989))

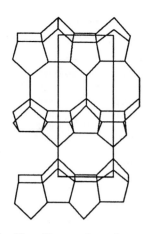

ZSM-57 Fig. 1S: Framework topology.

CHEMICAL COMPOSITION: 0–15 RO : 0–5 M_2O : 100 SiO_2 : 0.5–25 Al_2O_3
SPACE GROUP: Imm2
UNIT CELL CONSTANTS (Å): a = 7.45
b = 14.17
c = 18.77
PORE STRUCTURE: intersecting 8- and 10-member rings similar to those found in ferrierite

---

X-Ray Powder Diffraction Data: (d(Å) (I/I_0))
11.36(m–vs), 9.41(m–vs), 7.12sh(m–s), 6.95 (m–s), 5.74(m), 5.68sh(w–m), 5.42(m–s), 4.81 (w–m), 3.98(vw–m), 3.84sh(m–s), 3.79(vs), 3.64(w), 3.55(s), 3.48(s–vs), 3.36(w), 3.14(m–s), 3.06(w), 2.949(vw), 2.316(vw), 1.935(w).

The structure of ZSM-57 has been proposed, from synchrotron X-ray diffraction data and model building, to be related to ferrierite. It contains intersecting 8- and 10-member-ring channels that are linear. The difference between ZSM-57 and ferrierite lies in the four-rings in the ZSM-57 structure.

Synthesis

ORGANIC ADDITIVES
hexaethyl-diquat-5^{+2}
$[(C_2H_5)_3N(CH_2)_5N(C_2H_5)_3]^{+2}$

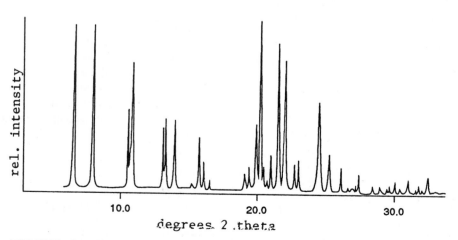

ZSM-57 Fig. 2:   X-ray powder diffraction pattern of ZSM-57 (*RR 8th IZC* 132(1989)).

DLS atomic coordinates of ZSM-57 ($\times$ 10^4) (*RR 8th IZC* 132 (1989)):

	x	y	z
T1	0000	0000	0000
T2	0000	2043	9486
T3	0000	6998	7986
T4	2941	0000	1176
T5	2057	1963	1848
T6	2939	7017	0427
T7	0000	5000	8668
T8	2945	5000	9848
O1	0000	0915	9503
O2	1754	0000	0473
O3	0000	2369	8678
O4	1755	2405	9876
O5	8255	7253	7524
O6	0000	5912	8173
O7	5000	0000	0946
O8	2531	0916	1632
O9	0000	2031	2079
O10	2441	2688	1221
O11	5000	7234	0276
O12	5000	5000	9611
O13	1748	5000	9149
O14	2570	5917	0319

ZSM-57 is prepared from reactive gels within the preferred range: SiO_2/Al_2O_3 = 40–100; H_2O/SiO_2 = 20–50; OH^-/SiO_2 = 0.1–0.5; M/SiO_2 = 0.1–2; R/SiO_2 = 0.1–1. R is the organic additive, and M is generally sodium cations. Crystallization takes place in between 4 and 7 days at 160°C, using silica sol (30% silica) as a source of silica and aluminum sulfate as the source of aluminum. The crystals are formed with a platelet morphology. Calcination to remove the template at 500°C did not result in degradation of the crystalline structure.

Adsorption

Adsorption properties of ZSM-57 (EP 174,121 (1986)):

	n-Hexane	Cyclohexane	Water
Wt%	7.1	4.7	6.7

Hydrocarbons, 20 Torr, water 12 Torr 25°C

ZSM-58 (SGT)
(EP 193, 282(1986))
Mobil Oil Corporation

Related Material: sigma-2

Structure

Chemical Composition: 0.1–2.0 R_2O : 0.02–1.0 $M_{2/n}O$: 0.1–2 Al_2O_3 : 100 SiO_2.

X-Ray Powder Diffraction Data: ($d(\text{Å})$ (I/I_0)) 13.7(w), 11.53(w–vs), 10.38(w), 7.82(w–vs), 6.93–6.79(w–vs), 6.19(w–vs), 5.94(w–m), 5.77(vs), 5.22(w), 5.18(vs), 4.86(m–s), 4.72(s), 4.57(w), 4.51(s), 4.43(w), 4.19(w), 4.15(m), 4.00(w), 3.97(w), 3.89(w), 3.84(m), 3.81(w–m), 3.59(w), 3.46(w–m), 3.41(s–vs), 3.36(s–vs), 3.32(m–s), 3.29(w), 3.17(w–m), 3.07(w–m), 3.05(w–m), 3.01(w–m), 2.88(w), 2.85(w), 2.75(w), 2.67(w), 2.60(w).

Synthesis

ORGANIC ADDITIVES
methyltropinium

ZSM-58 is prepared from a batch composition with the following preferred ratios: SiO_2/Al_2O_3 = 70–500; H_2O/SiO_2 = 10–100; OH^-/SiO_2 = 0.10–1.0; M/SiO_2 = 0.10–1.0; R/SiO_2 = 0.10–0.50. R is the methyltropinium ion, and the M is generally sodium. Crystallization occurs at 160°C after 4 days. The zeolite is isolated and dried at 110°C prior to X-ray analysis. Traces of an unidentified component in samples of ZSM-58 have been reported. The as-synthesized material contains between 3.8 and 5.0 methyltropinium cations per 100 TO_2 units. This material is stable to calcination at 538°C, the temperature used to decompose the organic.

ZYT-6 (CHA)
(*Acta Crystallogr.* C41:1698(1985))

Related Materials: SAPO-34
chabazite

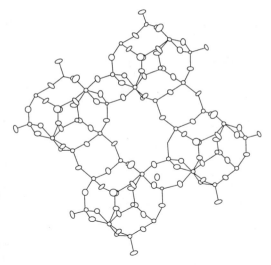

ZYT-6 Fig. 1S: Framework topology.

atoms of the hexagonal ring range from 2.75 to 3.10 A.

Positional parameters ($\times$ 10⁴) (*Acta Crystallogr.* C41:1698 (1985)):

	x	y	z
P(Si)1	38(3)	2261(3)	1082(3)
P(Si)2	9973(4)	7703(4)	8921(3)
Al1	2287(4)	2266(4)	1029(3)
Al2	7709(4)	7745(4)	8957(3)
O1	25(10)	2638(9)	92(6)
O2	1139(9)	2333(11)	1318(8)
O3	2005(10)	908(9)	1291(9)
O4	3247(9)	147(10)	1706(8)
O5	9950(11)	7474(13)	9904(8)
O6	8814(9)	7556(10)	8617(8)
O7	8095(10)	9112(9)	8779(9)
O8	6757(8)	9850(9)	8358(7)

Structure

CHEMICAL COMPOSITION: $H_3O^+ Al_2SiP_3O_{16}^{-}*nH_2O$
SYMMETRY: trigonal
SPACE GROUP: R3
UNIT CELL CONSTANTS (Å) a = 13.781(1)
c = 14.846(2)
DENSITY: 1.92
PORE STRUCTURE: eight-member-ring pore openings, 6.31 Å; cavity size (long and short axes) 11.69 Å and 10.63 Å

This structure is similar to that of the aluminosilicate chabazite. It is formed from four-, six-, and eight-member rings of AlO₄ and (P,Si)O₄ tetrahedra. The cavities are ellipsoidal and are linked by two six-member, six eight-member, and 12 four-member rings.

The double hexagonal rings in the structure contain two P atoms, two Al atoms, and eight 0 atoms, which are all crystallographically independent. Si substitutes for the P sites, resulting in an excess charge of -1 e per framework Si atom, which may be compensated by protonated water. The water O atoms and the O

Synthesis

ORGANIC ADDITIVE
morpholine

ZYT-6 was prepared hydrothermally from a synthesis gel composed of: P_2O_5 : Al_2O_3 : 2 morpholine : SiO_2.

3A (LTA)

The potassium form of Linde type A.

4A (LTA)

The sodium form of Linde type A.

5A (LTA)

The calcium form of Linde type A.

13X (FAU)

The sodium form of Linde type X.

Subject Index

Structure Codes Index

Organic Additives Index

TRIAMINES

TETRAALKYLAMMONIUM$^+$

OXYGEN CONTAINING ORGANICS

AMINO ALCOHOLS

OTHER NITROGEN CONTAINING ORGANICS

MIXED ORGANICS

RETURN TO: CHEMISTRY LIBRARY

100 Hildebrand Hall • 510-642-3753

LOAN PERIOD	1	2 1-MONTH USE 3
4	5	6

ALL BOOKS MAY BE RECALLED AFTER 7 DAYS.

Renewals may be requested by phone or, using GLADIS, type **inv** followed by your patron ID number.

DUE AS STAMPED BELOW.

OCT 28		

FORM NO. DD 10
3M 7-08

UNIVERSITY OF CALIFORNIA, BERKELEY
Berkeley, California 94720–6000